ADVANCES IN ECHO IMAGING USING CONTRAST ENHANCEMENT

Second Edition

Previous edition:
Advances in Echo Imaging using Contrast Enhancement, N.C. Nanda and
R. Schlief (Eds).
Kluwer Academic Publishers: 1993.

Advances in Echo Imaging Using Contrast Enhancement

Second Edition

edited by

NAVIN C. NANDA, MD

Professor of Medicine and Director, Heart Station/Echocardiography Laboratories, University of Alabama at Birmingham, Birmingham, Alabama, U.S.A.

Director, Echocardiography Laboratories, The Kirklin Clinic, University of Alabama, Health Services Foundation, Birmingham, Alabama, U.S.A.

REINHARD SCHLIEF, Dr Med Dipl Phys

Head of Magnetic Resonance and Ultrasound Contrast Media, Clinical Development Diagnostics of Schering AG, Berlin, Germany

and

BARRY B. GOLDBERG, MD

Professor of Radiology and Director, Department of Radiology, Thomas Jefferson University, Philadelphia, U.S.A.

KLUWER ACADEMIC PUBLISHERS
DORDRECHT / BOSTON / LONDON

Library of Congress Cataloging-in-Publication Data

ISBN 0-7923-4355-7

Published by Kluwer Academic Publishers,
P.O. Box 17, 3300 AA Dordrecht, The Netherlands.

Kluwer Academic Publishers incorporates the publishing programmes of
D. Reidel, Martinus Nijhoff, Dr W. Junk and MTP Press.

Sold and distributed in the U.S.A. and Canada
by Kluwer Academic Publishers,
101 Philip Drive, Norwell, MA 02061, U.S.A.

In all other countries, sold and distributed
by Kluwer Academic Publishers Group,
P.O. Box 322, 3300 AH Dordrecht, The Netherlands.

Printed on acid-free paper

Contents

Contents

List of contributors

Gary A. ABRAMS
UAB Liver Center, University of Alabama in Birmingham, 401 ZRB, 703 South 19th Street, Birmingham, AL 35294-0007, U.S.A.

Co-authors: Michael B. Fallon, Camilo R. Gomez and Navin C. Nanda

Aric A. AIAZIAN
Thoraxcenter, Bd 408, Erasmus University Rotterdam, P.O. Box 1738, 3000 DR Rotterdam, The Netherlands

Co-authors Chapter 4: Meine A. Taams, Folkert J. ten Cate and Jos R.T.C. Roelandt
Co-authors Chapter 31: Folkert J. ten Cate and Jos R.T.C. Roelandt

Albrecht BAUER
Clinical Development Diagnostics, Schering AG, P.O. Box 650311, D-13353 Berlin, Germany

Co-authors Chapter 43: Reinhard Schlief, Michael Zomack, Albrecht Urbank and Hans-Peter Niendorf
Co-authors Chapter 44: Marianne Mahler, Albrecht Urbank, Michael Zomack, Reinhard Schlief and Hans-Peter Niendorf

Shintaro BEPPU
School of Allied Health Sciences, Osaka University Faculty of Medicine, 2-2 Yamadaoka, Suita, Osaka 565, Japan

Ulrich BOGDAHN
Department of Neurology, University of Regensburg, Universitätsstrasse 84, D-93053 Regensburg, Germany

Co-authors: Georg Becker and Albrecht Bauer

Rodolfo CAMPANI
Institute of Radiology, University of Pavia, IRCCS Policlinico San Matteo, P. le Golgi 2, I-27100 Pavia (PV), Italy

Co-authors: Fabrizio Calliada, Olivia Bottinelli, Giulia Maresca, Enzo Angeli, Rizzardo Anguissola, Anna Bozzini and Alfreda La Fianza

xi

David O. COSGROVE
Consultant Radiologist, Department of Diagnostic Radiology, Hammersmith Hospital, Du Cane Road, London W12 0HS, U.K.

Co-authors: Barry B. Goldberg, Ji-Bin Liu and Martin J.K. Blomley

Nico DE JONG
Experimental Echocardiography, Room Ee 2302, Erasmus University Rotterdam, P.O. Box 1738, 3000 DR Rotterdam, The Netherlands

Anthony N. DeMARIA
Division of Cardiology, Room MN670, University of California at San Diego, 200 W. Arbor Drive, San Diego, CA 92103-8411, U.S.A.

Co-authors: Bruno Cotter and Koji Ohmori

Raimund ERBEL
Director, Department of Cardiology, Medical Clinic and Policlinic, University Clinic -GHS- Essen, Hufelandstrasse 55, D-45122 Essen 1, Germany

Co-authors: Roman Leischik, Christian Bruch, Rainer Zotz, Susanne Mohr-Kahaly, Frank Schön, Eckard Steinmetz and Rüdiger Brennecke

Steven B. FEINSTEIN
Section of Cardiology, (M/D) 787, University of Illinois at Chicago, 840 S. Wood Street, Box 698, Chicago, IL 60612-7323, U.S.A.

Co-authors: Jacqueline Wiewall Winkelmann and Romy Jill Block

Günter FÜRST
Institute of Diagnostic Radiology, Heinrich-Heine-University, Moorenstrasse 5, D-40225 Dusseldorf, Germany

Co-author: Matthias Sitzer

George GIRAUD
Division of Pediatric Cardiology, Oregon Health Sciences University, 3181 S.W. Sam Jackson Park Road, UHN60, Portland, OR 97201-3098, U.S.A.

Co-author: David J. Sahn

Barry B. GOLDBERG
Department of Radiology, Thomas Jefferson University, 132 South 10 Street, Philadelphia, PA 19107-5244, U.S.A.

Dirk HAUSMANN
Division of Cardiology, Hannover Medical School, Konstanty-Gutschow-Str 8, D-30625 Hannover, Germany

Co-authors: Andreas Mügge, Henning Kühn and Werner G. Daniel

Toshiko HIRAI
Departments of Oncoradiology and Radiology, Nara Medical University, 840, Shijo-cho, Kashiwara City, Nara Prefecture 634, Japan

Co-authors: Kimihiko Kichikawa, Hajime Oshishi, Hideo Uchida, Kyoichi Hiramatsu, Kiyoshi Ohkuma, Hitoshi Katayama, Masahiro Senmoto, Ryohei Kuwatsuru, Tsutomu Takashima, Toshifumi Gabata, Katuhide Itoh, Kyokyo Hayamizu, Kumiko Naitoh, Syoji Yoshida and Naofumi Hisa

Sabino ILICETO
A.R.C., Via Jacini 18, I-70125 Bari, Italy

Co-authors: Carlo Caiati, Navin C. Nanda and Paolo Rizzon

Sanjiv KAUL
Cardiovascular Division, Box 158, Medical Center, University of Virginia, Charlottesville, VA 22908, U.S.A.

Co-author: Christian Firschke

W. EVANS KEMP, Jr.
Division of Cardiology, Medical Center, Vanderbilt University, Nashville, TN 37209, U.S.A.

Co-author: Benjamin F. Byrd, III

Bijoy K. KHANDHERIA
Department of Cardiology, Mayo Clinic, 200 First Street SW, Rochester, MN 55905, U.S.A.

Jörg P. LANGHOLZ.
Innere Abteilung, Franziskus-Krankenhaus, Budapeststrasse 15/19, D-10787 Berlin, Germany

Tohru MASUYAMA
Echocardiography Laboratory, First Department of Medicine, Osaka University School of Medicine, 2-2 Yamadaoka, Suita 565, Japan

Co-authors: Hiroshi Ito, Young-Jae Lim and Akira Kitabatake

Shoichi MATSUTANI
First Department of Medicine, Chiba University School of Medicine, 1-8-1 Inohana, Chuou-ku, Chiba 260, Japan

Co-authors: Hitoshi Maruyama, Masaaki Ebara, Masaharu Yoshikawa, Hiromitsu Saisho and Masao Ohto

Samuel MEERBAUM
5741 El Canon Avenue, Woodland Hills, CA 91367, U.S.A.

Sharon L. MULVAGH
Cardiovascular Division, Mayo Clinic, W 16B, 200 First Street SW, Rochester, MN 55905, U.S.A.

Co-authors: Hector R. Villaragga, David A. Foley, Sang Man Chung and Navin C. Nanda

Satoshi NAKATANI
The Cardiology Division, National Cardiovascular Center, 5-7-1 Fujishiro-dai, Suita, Osaka 565, Japan

Co-author: Kunio Miyatake

Navin C. NANDA
Division of Cardiovascular Disease, Heart Station, UAB School of Medicine, S102 Spain-Wallace Building, 619 South 19th Street, Birmingham, AL 35233-1924, U.S.A.

Co-author Chapter 6: Edwin L. Carstensen
Co-author Chapter 11: Joel S. Raichlen
Co-authors Chapter 41: Carlos Garcia Del Rio, Gregg W. Taylor, Dipak I. Agrawal, Gopal G. Agrawal, Claudia Carvalho, Miguel Espinal, Sanjay Malhotra, Lance LaMotte and Reinhard Schlief

Natesa G. PANDIAN
Division of Cardiology, Tufts-New England Medical Center, 750 Washington Street, Box 32, Boston, MA 02111, U.S.A.

Co-authors: Jiefen Yao and Navroz D. Masani

Thomas R. PORTER
University of Nebraska Medical Center, 600 S. 42nd Street, Omaha, NE 68198-1165, U.S.A.

Jeffry E. POWERS
Advanced Technology Laboratories, MS 265, P.O. Box 3003, Bothell, WA 98041-3003, U.S.A.

Co-authors: Peter N. Burns and Jacques Souquet

Joel S. RAICHLEN
Division of Cardiology, 763L, Thomas Jefferson University Hospital, Main Building, 132 S. 10th Street, Philadelphia, PA 19107, U.S.A.

Co-author: Navin C. Nanda

Michelle L. ROBBIN
Department of Radiology, Chief Ultrasound, University of Alabama at Birmingham, 619 South 19th Street, Birmingham, AL 35233-1924, U.S.A.

Daniele ROVAI
CNR Clinical Physiology, Via Savi 8, I-56126 Pisa, Italy

Co-authors: Anthony N. DeMaria, Massimo Lombardi, Cecilia Marini, Jan Schneider and Antonio l'Abbate

Reinhard SCHLIEF
Clinical Development Diagnostics, Schering AG, P.O. Box 650311, D-13353 Berlin, Germany

Karl Q. SCHWARZ
Director Echocardiography, University of Rochester Medical Center, Box 679, Cardiology Unit, Rochester, NY 14642, U.S.A.

Co-authors: Xucai Chen and Gian Paolo Bezante

Pravin M. SHAH
Section of Cardiology, Loma Linda University Medical Center, 11234 Anderson Street, P.O. Box 2000, Loma Linda, CA 92354, U.S.A.

Janine R. SHAPIRO
Department of Anesthesiology, Box 604, University of Rochester Medical Center, 601 Elmwood Avenue, Rochester, NY 14642, U.S.A.

Co-author Chapter 12: Richard S. Meltzer

George R. SUTHERLAND
Department of Cardiology, Western General Hospital, Crewe Road, Edinburgh EH4 2XU, U.K.

Helene VON BIBRA
I. Medical Clinic and Policlinic, Klinikum rechts der Isar, Technical University Munich, Ismaninger Strasse 22, D-81675 Munich, Germany

Co-authors: Anja Tuchnitz, Christian Firschke, Harald Becher and Albert Schömig

Samuel A. WICKLINE
Division of Cardiology, Echocardiography, The Jewish Hospital of St. Louis at Washington University Medical Center, 216 South Kingshighway Blvd, St. Louis, MO 63110, U.S.A.

Co-authors: Gregory M. Lanza, Kirk D. Wallace and James G. Miller

PART ONE

Echo-enhancement agents: basic aspects

1. Contrast echocardiography – a historical perspective

PRAVIN M. SHAH

THE BEGINNING YEARS

Nearly 25 years ago, Dr Raymond Gramiak and I embarked on a study, using M-mode echocardiographic recordings of the aortic valve to estimate stroke volume from extent and duration of aortic cusp separation. We proceeded to record the aortic valve and root echo simultaneously with measurements of cardiac outputs using the indicator dilution technique in the cardiac catheterization laboratory. It was then a routine practice at the University of Rochester to place a catheter in the left atrium by the trans-septal technique and measure cardiac output by injection of indocyanine green into the left atrium with peripheral arterial sampling. During these a striking enhancement of echo signals was observed: this was termed contrast echocardiography following an analogy with the term 'contrast angiography' applied to radiographic procedures. We observed this contrast echo effect with indocyanine green injections, as well as with saline or dextrose in water flush of the catheters. This observation reminded Dr Gramiak of a passing comment by Dr Claude Joyner at the First International Course on Diagnostic Ultrasound, that echo enhancement could be observed with saline injections. Although his observation was not in print, our first published report in 1968 [1] dealing with the M-mode echocardiography of the aortic root credited Dr Joyner with the observation. We utilized contrast echocardiography in order to validate the echo recordings of the aortic root and valve.

A systematic prospective study was then undertaken, and the results of 242 injections of echo contrast in 32 patients during diagnostic cardiac catheterization were reported in 1969 [2]. The contrast substances employed were indocyanine green solution, saline, 5% dextrose in water, and the patient's own blood. The injection sites included both atria, both ventricles, and the aortic root. This being the pre-strip chart era, we obtained continuous recordings of the combined ultrasonic and physiologic display on 35 mm film by means of a Fairchild Oscilloscope Record Camera. The echo anatomy and function were illustrated in a number of studies with different transducer positions and in varied pathological conditions.

N. C. Nanda, R. Schlief and B. B. Goldberg (eds.), Advances in Echo Imaging Using Contrast Enhancement, Second Edition, 3–9.
© 1997 *Kluwer Academic Publishers.*

For example, aortic regurgitation was assessed by supravalvular injection of contrast and its appearance in diastole in the left ventricular outflow space anterior to the mitral valve echo. Similarly, contrast injection in the left ventricle in a patient with hypertrophic obstructive cardiomyopathy illustrated the outflow narrowing from systolic anterior motion (SAM) of the mitral valve. Additionally, the contrast injection in the aortic root in this patient confirmed the mid-systolic closure of the aortic valve. Several examples of intracardiac shunts were observed with passage of contrast across the septal, and a case of left ventricle–right atrial shunt (Gerbode defect) was correctly identified prior to angiographic verification. We speculated that the contrast effect represented ultrasonic detection of minia-ture bubbles within the heart produced by gaseous cavitation, which occurs when the contrast agent is injected rapidly, or by miniature bubbles injected in the foam of indocyanine green solutions. Subsequently, Fred Kremkau, then a gradu-ate student in the Department of Electrical Engineering at the University of Rochester, carried out an investigation to determine which of three mechanisms is responsible for the ultrasound contrast effect: (1) acoustic impedance difference between blood and the injected fluid, (2) turbulence resulting from the injection, or (3) production of miniature bubbles. The in vitro studies provided strong evidence that production of microbubbles represents the primary source of con-trast echo [3].

THE EARLY CLINICAL APPLICATIONS

Subsequent to the aforementioned reports, a number of studies reported over the next 15 years the clinical applications of contrast echocardiography. Feigenbaum and colleagues reported the use of indocyanine green injections to identify echoes from the left ventricle in 1970 [4]. Kerber and associates reported in 1974 their experience with use of contrast echo for diagnosis of valvular regurgitation and intracardiac shunts [5]. This was followed by similar reports by Seward et al. [6], Valdes-Cruz et al. [7], Pieroni et al. [8], Kronic et al. [9] and Fraker et al. [10]. Weyman and colleagues reported their observations on negative contrast echocardiography as a new method for detection of left to right shunts [11]. A prolonged clearance time in the right heart chambers was used to quantitate tricuspid regurgitation [12]. The major applications in the 1970s and early 1980s included validation of echo anatomy, detection of intracardiac shunts and detec-tion of valvular regurgitation.

The issue of safety was examined by a survey undertaken by a special task force of the American Society of Echocardiography and reported by Bommer et al. [13]. They found an extremely low level of risk without residual side effects of significant complications. Nevertheless, a potential for blockage of micro-circulation following injection of microbubbles is potentially worrisome.

The advent of Doppler techniques, including color flow imaging, essentially eliminated a need for contrast echocardiography in the diagnosis of valvular regurgitation and most cases of intracardiac shunts. It was a consensus in the early 1980s among several echocardiographers that contrast echocardiography was mostly history.

The more recent reemergence of contrast echocardiography is based on three newer developments, namely development of newer contrast agents, newer applications *vis-à-vis* myocardial opacification, and the growing use of transesophageal echocardiography.

THE NEWER CONTRAST AGENTS

DeMaria and associates [14] and Bommer et al. [15] reported animal experiments using the first manufactured contrast agents composed of saccharide microbubble preparation. They demonstrated a potential for myocardial opacification [16]. However, these studies were abandoned since the macrobubbles blocked arterioles and were poorly tolerated. Armstrong and colleagues employed aortic root injections of hydrogen peroxide in animal studies to obtain myocardial opacification. Although no adverse reactions were reported, the use of hydrogen peroxide has been abandoned in favour of newer contrast agents. Dr Chuwa Tei, working with me as a research associate, brought to my attention in 1981 his observation that hand-agitated saline Renograffin mixture could be used for repeated myocardial opacification when injected directly into the coronary arteries of a dog. This was systematically studied in a series of dog experiments and eventually published in 1983 [17].

Although no adverse effects were noted, transient electrocardiographic changes of ST segment and depression of left ventricular function were observed. Dr Steven Feinstein, as a research fellow in 1983 under my supervision, undertook the development of contrast agents of smaller size and greater stability. It is my recollection that one of the then-cardiology fellows, Dr Peter Lee, working at the Wadsworth VA Medical Center Cardiology Division, suggested the use of a sonicator to develop microbubbles for echocardiographic contrast. Dr Feinstein borrowed a sonicator from a chemistry laboratory and proceeded to sonicate a variety of solutions including sorbitol 70%, sorbitol 70% diluted 1:1 with dextrose 5% in water, dextrose 70%, dextrose 50%, Renograffin-76, Renograffin-75 diluted 1:1 with saline solution. He examined under a microscope the mean diameter and, by A-mode videotaped studies, persistence of the microbubbles produced by sonication [18]. The report of these studies marks the first description of sonicated echo contrast agents. The use of these agents in vivo required a fresh preparation under sterile conditions, owing to their short half life. Having acquired a personal patent for use of sonication in development of contrast agents, Dr Feinstein collaborated in a commercial development of sonicated human albumin marketed under the name Albunex. This agent is supplied as a powder, which on mixing with dextrose in water results in a contrast agent which could be injected in the vascular system. This contrast agent is currently among those most investigated for newer application of myocardial perfusion.

THE DEVELOPING APPLICATIONS

Myocardial opacification

The promise of myocardial opacification by contrast echocardiography, first suggested by the early studies of Bommer et al. [15] and subsequently systematically examined by Armstrong et al. [16] and Tei et al. [17] continues to attract attention from an increasing number of investigators. Armstrong et al. [16] used radiolabelled microspheres to validate the relationship between myocardial echo contrast enhancement and myocardial perfusion. Tei et al. [17] demonstrated the correlation between perfusion bed as delineated by myocardial contrast and distribution of asynergy following arterial occlusion. Intracoronary injections in left main, left anterior descending, and proximal circumflex coronary arteries provided information on the arterial distribution and perfusion. They also examined the extent of asynergy and the extent of negative contrast delineation from the injection in the left main coronary artery after a branch occlusion. In another series of experiments, Tei et al. correlated myocardial echocontrast washout (disappearance rate) with severity of coronary stenosis [19]. Varying grades of coronary stenosis were produced in the intact closed chest of dogs by placement of intraluminal plugs with known reductions in luminal diameters. The contrast disappearance curve was obtained using video densitometric methods in the regions delineated by intramyocardial contrast. The time course of contrast disappearance was shown to be strikingly different between control, and 50% stenosis, and 75% stenosis, and 100% stenosis. Tei further observed that the hyperaemic flow response after Renograffin injection was significantly reduced from induction of coronary stenosis. This was systematically examined by Kondo and associates [20]. The interest generated by these and other studies of contrast echocardiography resulted in publication of a seven-part seminar guest-edited for the *Journal of the American College of Cardiology* by the author along with Drs Eliot Corday and Samuel Meerbaum [21]. The first series of investigations was published in January 1984, and the final paper of the seminar in December 1984 was authored by Kaul et al. [22]. Dr Sanjiv Kaul, previously a cardiology fellow at Wadsworth VA Medical Center, was familiar with the earlier studies by Tei and associates and proceeded to reproduce and extend the observations after leaving Los Angeles.

Although the reproducibility of the technique for identifying regional perfusion deficits and the correlation of echo contrast disappearance rate ('washout') with reductions in flow rate have been repeatedly demonstrated, its wider clinical usefulness is hampered by a need for intracoronary injections for reproducible results. A great deal of the early work has been reproduced in elegant experimental designs by Kaul and associates [23], but in many published reports the early work of Tei and associates has not been recognized [21]. In addition, clinical applicability has been tested in the coronary angiographic suites and during coronary bypass surgery.

Improved endocardial definition

A multicenter study using intravenous injections of sonicated human serum albumin as a contrast agent (Albunex) has reported improved delineation of left ventricular endocardium following its transpulmonary passage. It has been suggested that this application could facilitate echocardiographic assessment of left ventricular global and segmental function. The frequency and accuracy with which this is achieved remains to be established.

Supplement to transesophageal echocardiography (TEE)

A potential for intravenous contrast studies in detection of small or trivial intra-cardiac shunts (e.g., patent foramen ovale) has been demonstrated. The use of contrast studies has found increasing applications in conjunction with TEE when carried out for the indication to rule out intracardiac source of systemic embolism.

THE SECOND GENERATION CONTRAST AGENTS

Attempts to maximize the persistence of the bubbles have employed high density gases with low diffusion and saturation contrast. A number of commercially produced agents are undergoing experimental and clinical investigations. FS069 (Molecular Biosystems, Inc.) consists of perfluoropentane-filled albumin spheres produced by sonication. MRX 115 (Ima Rx$_1$) combines fluorocarbons with microsome. Echogen or QW 3600 (SONUS Pharm.) consists of dodecafluoro-pentane as the active agent. AF0150 or Imagent (Alliance Pharm.) consists of perfluorohexane/nitrogen-filled microspheres. These agents appear to have an enhanced ability to demonstrate myocardial opacification following intravenous injection. The goal of myocardial opacification through echo contrast is likely to be realized with use of the new agents coupled with harmonic and transient imaging approaches.

CONCLUSION

Contrast echocardiography, as discussed above, has clearly come a long way over a span of 25 years. The future development of this technology will be deter-

8   *Pravin M. Shah*

mined by its usefulness to study myocardial perfusion following intravenous injections. The present studies on intracoronary contrast have pretty much run their course, although important areas of interventional and intraoperative applications will continue to be explored.

REFERENCES

1. Gramiak R, Shah PM. Echocardiography of the aortic root. Invest Radiol 1968;3:356–88.
2. Gramiak R, Shah P, Kramer D. Ultrasound cardiography: contrast study in anatomy and function. Radiology 1969;92:939–48.
3. Kremkau FW, Gramiak R, Carstensen EL et al. Ultrasonic detection of cavitation at catheter tips. Am J Roentgenol 1970;110:177.
4. Feigenbaum H, Stone J, Lee D, Nasser W, Chang S. Identification of ultrasound echoes from the left ventricle by use of intracardiac injections of indocyanine green. Circulation 1970;41: 615–21.
5. Kerber R, Kioschos J, Lauer R. Use of an ultrasonic contrast method in the diagnosis of valvular regurgitation and intracardiac shunts. Am J Cardiol 1974;34:722–7.
6. Seward J, Tajik A, Spangler J, Ritter D. Echocardiographic contrast studies. Mayo Clin Proc 1975;50:163–9.
7. Valdes-Cruz L, Pieroni D, Roland J, Varghese P. Echocardiographic detection of intracardiac right-to-left shunts following peripheral vein injections. Circulation 1976;54:558–62.
8. Pieroni D, Varghese P, Freedom RM, Rowe R. The sensitivity of contrast echocardiography in detecting intracardiac shunts. Cathet Cardiovasc Diagn 1979;5:19–29.
9. Kronik G, Slany J, Moesslacher H. Contrast M-mode echocardiography in diagnosis of atrial septal defect in acyanotic patients. Circulation 1979;59:372–8.
10. Fraker T, Harris P, Behar V, Kisslo J. Detection and exclusion of interatrial shunts by two-dimensional echocardiography and peripheral venous injection. Circulation 1979;59:379–84.
11. Weyman A, Wann L, Caldwell R, Hurwitz R, Dillon J, Feigenbaum H. Negative contrast echocardiography: a new method for detecting left-to-right shunts. Circulation 1979;59:498–505.
12. Nanda NC, Shah PM, Gramiak R. Echocardiographic evaluation of tricuspid valve incompetence by contrast injections. Clin Res 1976;24:233A.
13. Bommer WJ, Shah PM, Allen H, Meltzer R, Kisslo J. The safety of contrast echocardiography: report of the committee on contrast echocardiography for the American Society of Echocardiography. J Am Coll Cardiol 1984;3:6–13.
14. DeMaria AN, Bommer WJ, Riggs K et al. Echocardiographic visualization of myocardial perfusion by left heart and intracoronary injections of echo contrast agents (abstr). Circulation 1980;62(suppl II):143.
15. Bommer W, Rasor J, Tickner G et al. Quantitative regional myocardial perfusion scanning with contrast echocardiography (abstr). Am J Cardiol 1981;47:403.
16. Armstrong WF, Mueller TM, Kinney EL, Tickner EG, Dillon JC, Feigenbaum H. Assessment of myocardial perfusion abnormalities with contrast-enhanced two-dimensional echocardiography. Circulation 1982;66:166–73.
17. Tei C, Sakamaki T, Shah PM et al. Myocardial contrast echocardiography. A reproducible technique of myocardial opacification for identifying regional perfusion deficits. Circulation 1983;67:585–93.
18. Feinstein SB, Ten Cate FJ, Zwehl W et al. Two-dimensional contrast echocardiography. 1. In vitro development and quantitative analysis of echo contrast agents. J Am Coll Cardiol 1984; 3:14–20.
19. Tei C, Kondo S, Meerbaum S et al. Correlation of myocardial echo contrast disappearance rate ("washout") and severity of experimental coronary stenosis. J Am Coll Cardiol 1984;3:39–46.

20. Kondo S, Tei C, Meerbaum S, Corday E, Shah PM. Hyperemic response of intracoronary contrast agents during two-dimensional echographic delineation of regional myocardium. J Am Coll Cardiol 1984;4:149–56.
21. Corday E, Shah PM, Meerbaum S. Introduction, seminar on contrast two-dimensional echocardiography: applications and new developments, part I. J Am Coll Cardiol 1984;3:1–6.
22. Kaul S, Pandian NG, Okada RD, Pohost GM, Weyman AE. Contrast echocardiography in acute myocardial ischemia: in vivo determination of total left ventricular "area at risk". J Am Coll Cardiol 1984;4:1 272–82.
23. Sabia PJ, Powers ER, Jayaweera AR, Ragosta M, Kaul S. Functional significance of collateral blood flow in patients with recent acute myocardial infarction. Circulation 1992;85:2080–9.

2. Microbubble fluid dynamics of echocontrast

SAMUEL MEERBAUM

PRINCIPLES

Underlying principles of echocontrast effects are outlined in this section, which aims to reflect the recent advances and detailed reports presented in subsequent chapters. To place microbubble phenomena in context, it was deemed useful to consider parallel multidisciplinary efforts in such areas as hydrodynamic cavitation, diving physiology and bubbly flow processes. Although there are many applications of ultrasound contrast, one difficult but potentially realizable objective was singled out: the quantitation of myocardial perfusion with intravenously administered gaseous bubbles.

The pace of echocontrast developments is accelerating, exemplified by intensive exploration of optimally stabilized microbubbles. Their ability to withstand elevated pressure and other effects is being examined. The recently demonstrated enhancement of the left ventricular chamber is itself a significant advance. Other advances made since the first edition of this book lead one to expect further major progress. Understanding of physical mechanisms is sought to assist achievement of a safe and effective methodology.

Following review of pertinent developments, bubble–fluid dynamics will be discussed. This will comprise the general behavior of a gas bubble in a liquid, unprotected microbubble dissolution, bubble films or encapsulation, pressure effects and sound velocity of a bubbly medium. Ultrasound–microbubble interactions will be addressed briefly, and an attempt will be made to relate the bubble dynamics topics to contrast echo.

ECHOCONTRAST DEVELOPMENTS

Microbubbles are the source

As contrast echocardiography was beginning to be applied [1], the question arose as to the origin of the apparent gas bubbles which were evidently responsible for

11

N. C. Nanda, R. Schlief and B. B. Goldberg (eds.), Advances in Echo Imaging Using Contrast Enhancement, Second Edition, 11–38.
© 1997 *Kluwer Academic Publishers.*

the enhanced images. Considering the substantial experience with cavitation, this process was at first suspected of causing bubble formation within a transiently low pressure region generated by catheter tip injection of a contrast fluid [2]. However, it could be convincingly demonstrated that cavitation should not be expected with commonly practised clinical injection techniques [3]. The overwhelming evidence points to introduced microbubbles being the source of echocontrast enhancement. In the absence of deliberate administration, bubbles could be traced to accidental inclusion of trapped air along with fluid injections. Generally, whenever gaseous bubbles were eliminated in experimental studies, echocontrast effects disappeared. Conversely, controlled introduction of microbubbles led to best contrast echo images.

Research and development

Most of the early studies used agents with surfactant properties (e.g. cardiogreen or renografin), containing air bubbles produced by 'vigorous shaking' of the contrast fluid. Clinical echo contrast evaluations encompassed diagnostic studies of intracardiac shunts, septal defects and valvular insufficiency. Assessment of cardiac output was attempted. The left heart chambers could generally not be opacified with intravenous contrast injections; the bubbles were either too large to pass through the pulmonary microcirculation, or else dissolved during transit. Early on, Roelandt [4] and Meltzer et al. [5] estimated that very small bubbles collapsed rapidly (< 1 s) by virtue of surface tension, whereas normal pulmonary transit alone requires at least 2 s. Identifying the need for microbubble stabilization, several authors reported development of polysaccharide-coated as well as gelatin-encapsulated gaseous bubbles [6, 7].

In the early 1980s, when bubble contrast had been improved through sonication or quasi-standardized hand agitation, there arose enthusiastic efforts to develop effective myocardial contrast echocardiography [8–11]. Most studies applied the contrast agents (mean bubble diameter of the order of $10 \mu m$; smaller in some sonicated media) directly into the coronary arteries or the aortic root. Any toxicity and myocardial effects were critically examined. Satisfactory experimental delineation of regionally ischemic myocardium beyond coronary artery occlusions was achieved using conventional ultrasound scanning. Considering the bubbles as tracers and deriving time–intensity curves, assessment of myocardial blood flow was also envisioned. Introduction of stabilized microbubble agents further broadened development work in a number of laboratories. For example, Keller et al. [12] reported studies of sonicated human serum albumin agent bubbles several microns in diameter, and featuring excellent ultrasound characteristics.

Along with other contributors, a series of myocardial contrast echo investigations was forthcoming from Kaul and his colleagues [13–16]. Most of their studies featured intracoronary administration of sonicated agents (e.g. the commercially available Albunex), but some recent reports described right and left atrial as well as intravenous injections. One particular investigation demonstrated that the intravascular rheology of albumin microspheres closely resembles that of red blood cells [17]. Characterization of myocardial perfusion with echocontrast was elucidated, and the authors validated a reliable delineation of under-perfused myocardial risk zones. Yet other studies concerned coronary flow reserve, evaluation of collateral flow, correlations with ventricular function, and intraoperative applications. Recognizing the need to develop ultrasound equipment specially dedicated to echocontrast imaging, these and other investigators called for systems with greater dynamic range and enhanced data processing capabilities.

Cavitation and sonication

Cavitation [18] involves very rapid bubble motions in the presence of high amplitude forced oscillations. Bubbles change dimensions greatly, even during a single accoustic period, growing by a diffusion process to as much as 2 – 3 times their normal size, and then rapidly collapsing. Many of the basic features of 'stable' acoustic interactions with gaseous bubbles in echocontrasts tend to resemble those of the transient cavitation in a high power ultrasonic field. The cavitation process is characterized by shock wave propagation and forceful bubble collapse following their growth, along with locally extreme pressures and temperatures. With regard to medical imaging, fears of side effects have been largely dispelled as a result of the observed difficulty of generating cavitation in the common lower range of ultrasound intensities and pressure amplitudes. However, a recent paper [19] described direct generation of microbubbles by ultrasonic cavitation in an in vivo canine aorta (1.8 MHz focused sound field, pulse duration 250 ms, applied intensity >4300 W/cm).

It was always recognized that well-characterized bubbles in the micron size region were needed for quantitative echocontrast applications. Hand-agitated preparation was deemed unsatisfactory for microbubble production and was largely supplanted by sonication. The latter exposes the fluid medium to high energy sound waves, which transform the introduced air into microbubbles. The size and number of the generated bubbles depend on the energy and duration of the applied power. Feinstein et al. [20] were the first to apply a commercial sonication device to the preparation of standardized microbubble contrasts capable of microcirculatory transit. The transient ultrasonic cavitation or sonication process consists of successive stages: (1) a pre-initiation phase involving vigorous oscillations of gas microcavities pre-existing within the liquid, caused by the high energy sound field at the tip of the sonication unit's horn, (2) ensuing powerful compressions and expansions of the microcavities, (3) a 'catastrophic' phase during resonance, characterized by shock waves and major pulsations, eventually generating in the sonicated liquid a stable cloud of microbubbles.

Stabilized microbubble contrasts

Echocontrast agents should feature full safety, adequate persistence, unhindered pulmonary capillary passage and ultrasound effectiveness. Having perfected production of micron-sized bubbles, recent attention turned to the stability of microbubbles during their passage from an intravenous administration site via the ventricular chambers to the myocardial microcirculation. The promise of novel agents is linked with some of the bubble dynamics to be discussed, such as their oscillatory response, dissolution tendencies, and bubble interactions with a sound field.

Levovist (SHU 508A, Schering AG, Berlin, Germany) is biodegradable and made up of many crystalline galactose particles, plus a low concentration of a palmitic acid which coats the resulting air microbubbles [21]. Air attached to and within the microparticles forms tiny bubbles, which (after an intravenous injection) produces an increase in ultrasound backscatter from the blood. The diameter of 99% of the bubbles is in the range $1-8\,\mu m$, with an average of $2.7\,\mu m$. Concentrations between 200 and 400 mg particles per ml have been obtained. The physicochemical properties of SHU 508A have been reported [22]. It is said to be stable and its ultrasound characteristics are only moderately compromised at elevated pressures; this is attributed to a thin but highly flexible gas bubble surface coating. Experimental and clinical studies, including enhanced Doppler imaging and excellent opacification of the left ventricle, appeared to validate the potentials of this transpulmonary contrast agent [23].

Albunex (Molecular Biosystems, San Diego, CA) is biodegradable and prepared by a controlled 20 KHz sonication of 5% human serum albumin. This yields a high concentration of rather stable and very echogenic air-filled albumin microspheres. The mean microsphere diameter is $3.8\,\mu m$, and 95% of these albumin-encapsulated bubbles (shell thickness about 15 nm) are less than $10\,\mu m$ in diameter. Concentrations are of the order of $200-400$ million microspheres per ml. Benefits and limiting effects of the relatively stiff proteinaceous shell are being investigated, but the microsphere's effectiveness appears to be adversely affected at high ambient pressures (>200 mmHg). Substantial characterization of the physicochemical properties of Albunex has been reported [24, 25]. Ultrasonic evaluation of Albunex over a range of medical imaging frequencies indicated effective backscatter even at low microsphere concentrations. Albunex served as the echocontrast vehicle for a number of transpulmonary studies [26, 27].

Other microbubble stabilization methods are currently being investigated. One approach aims to prolong microbubble survival by incorporating in the bubble a gas of low blood solubility and diffusivity. Porter et al. investigated perfluoro-propane-exposed sonicated dextrose albumin and claim both safety and superior myocardial contrast imaging following intravenous injections [28]. A new proteinaceous gas-filled microsphere agent, FS069 (Molecular Biosystems Inc., San Diego, CA), is also being studied experimentally for intravenous contrast administration and myocardial imaging [29]. Another new agent, SHU 563A, exemplifies attempts to coat microbubbles with a thin and flexible biodegradable polymer shell which enhances stability and ultrasound effectiveness at low acoustic pressure amplitudes [30]. An ultrasound contrast agent called Echogen (Sonus Pharmaceuticals, WA) is based on a new 'phase-shift' emulsion concept. The agent changes at body temperature from pre-injection non-echogenic stable submicron dodecafluoropentane microparticles or liquid droplets to echogenic vapor microbubbles after intravenous injection [31]. A number of recent articles reported the use of micron-sized nitrogen gas-filled liposomes (aerosomes) as an echocontrast agent. Resulting stabilized yet highly elastic microbubbles are said to facilitate enhancement of the heart's chambers during intravenous contrast injection [32]. Microencapsulation clearly increases bubble stability, but it can impart stiffness, which results in resonance damping.

DYNAMICS OF BUBBLE–FLUID INTERACTIONS

The view and sound of bubbly water in streams or waves or jets are a commonly observed experience, as are the growth of bubbles rising towards a liquid surface, or the generation and collapse of vaporous bubbles during boiling processes. Certain characteristics of gaseous bubbles in a liquid medium motivated inter-disciplinary mathematical modeling, along with experimental validations. The presence of bubbles can substantially change a fluid's properties, such as its sound propagation velocity. The gas bubbles can grow, shrink or assume a quasi-equilibrium state, depending on such properties as surface tension and diffusion. Thus, inside a $5\,\mu m$ radius air bubble in water, surface tension generates an internal overpressure of about 0.3 atmospheres, promoting outward gas diffusion and microbubble shrinkage. A $1\,\mu m$ radius bubble may collapse in as little as 0.07 s. Conversely, gaseous microbubbles in many fluids seem to be enveloped by a very thin film which increases their persistence. In the presence of a sound field and its forcing pressure oscillations, the highly compressible gas bubbles respond by vigorous volume pulsations. Vibrating gas bubbles are extremely effective in absorbing and scattering sound wave energy. Of particular interest are conditions when the driving frequency is equal to the bubble's resonant frequency. The characteristic absorption-scattering cross-section of a resonating air bubble in water appears to exceed, by orders of magnitude, its geometric cross-section.

Behavior of a gas bubble in a liquid medium

Fundamental bubble behavior can be viewed as analogous to the damped oscil-lation of a spring with its natural frequency. The equivalent effective mass is that of the liquid surrounding the bubble, while the compression–expansion of the bubble gas resembles the spring's elastic energy process. The bubble–liquid system also exhibits damping of a thermal and viscous type. In an ultrasound field, bubble volume response peaks when resonance is achieved.

Most theoretical models of bubble behavior are admittedly simplified, but their conceptual usefulness has been reasonably validated. The best known mathematical description of the response of a single gaseous bubble within a large static incompressible fluid volume is the Rayleigh–Plesset equation [33]:

$$R\ddot{R}+\frac{3}{2}\dot{R}^2=\frac{1}{\rho_l}\left[p_g-p_l-\frac{2\sigma}{R}-\frac{4\mu}{R}\dot{R}\right]$$
(1)

R = bubble radius
$\dot{R}$ = velocity of bubble wall motion
$\ddot{R}$ = bubble boundary acceleration
p_l = ambient pressure within liquid (at some distance from the bubble)
p_g = internal bubble gas pressure (partial vapor pressure neglected)
ρ_l = liquid density
μ = liquid viscosity
σ = surface tension

[An attempt will be made to use uniform terminology, but new or differing terms in equations will be introduced as they arise.]

For equilibrium bubble conditions (albeit often unstable because of increased outward gas diffusion due to surface tension):

$$p_{g0}-p_L=\frac{2\sigma}{R_0}$$
(2)

where subscript 0 denotes the initial or equilibrium state.

The gas within the bubble is assumed to follow a polytropic relationship: p_g = constant $\cdot(\rho_g)^k$, where k equals the polytropic constant (adiabatic 1.4, isothermal 1.0). If the mass of gas within the bubble were to remain constant, then

$$p_g=p_{g0}\left(\frac{R_0}{R}\right)^{3k}.$$
(3)

Introducing small pressure perturbations in the liquid pressure, the gas bubble radius may oscillate as follows:

$$R=R_0(1+\varepsilon e^{i\omega t})$$
(4)

where $\varepsilon\ll1$, $\omega=2\pi f$, f = frequency, t = time, i = imaginary number (associated with viscous damping terms $2\mu/R_0^2$).

There have been many formulations of the resonant frequency f_r, including Houghton's mathematical solution [34]:

$$f_r=\frac{1}{2\pi R_0}\left\{\frac{3k}{\rho_l}\left[p_L+\left(1-\frac{1}{3k}\right)\frac{2\sigma}{R_0}\right]-\left(\frac{2\mu}{R_0\rho_L}\right)^2\right\}^{1/2}$$
(5)

Disregarding viscosity and surface tension, one arrives at the simplified Minaert

[35] relation:

$$f_r = \frac{1}{2\pi R_0} \left(\frac{3kp_L}{\rho_L} \right)^{1/2}. \tag{6}$$

For adiabatic air bubbles in water ($\rho_1 = 1$ g/cm^3 at $p_1 \sim 1$ dyne/cm^2), and R_0 in cm, f_r (in Hz) is:

$$f_r \approx 329/R_0. \tag{7}$$

For a microbubble of $R_\sigma = 2 \times 10^{-4}$ cm ($2\,\mu m$), $f_r \approx 1.645$ MHz. For a $5\,\mu m$ bubble radius, the simplified Minnaert calculation deviates only about 11% from the more accurate computations of resonant frequencies. At smaller bubble radii (e.g. $1\,\mu m$), much greater deviations are expected (e.g. 46%), and the more complex Houghton or equivalent relationship has to be used. While still uncertain, it appears that a near isothermal ($k=1$) rather than adiabatic ($k=1.4$) process may apply in the $1-5\,\mu m$ range of bubble radii at diagnostic ultrasound frequencies [36].

Not only must viscosity and surface tension be included as one deals with echo-contrast microbubbles of less than $5\,\mu m$ radius, but one also needs to account for non-Newtonian properties of realistic media. According to Tsujino and Shima [37], the rheology of human blood appears to be well represented by a Casson model incorporating shear stress and shear rate:

$$\tau^{1/2} = S\,\dot{\gamma}^{1/2} + b \qquad \text{for} \qquad \tau^{1/2} > b \tag{8}$$

and

$$\dot{\gamma} = 0 \qquad \text{for} \qquad \tau^{1/2} < b$$

where τ = shear stress, $\dot{\gamma}$ = shear rate, S and b are constants [37]. A more complex relation for bubble resonant frequency was defined for gas bubbles in a Casson fluid:

$$f_r = \frac{1}{2\pi R_0} \left\{ \frac{3k}{\rho_L} \left[p_L + \left(1 - \frac{1}{3k}\right)\frac{2\sigma}{R_0} \right] - \left(\frac{1}{2}\frac{M_0}{R_0\rho_L} \right) \right\}^{1/2} \tag{9}$$

where M_0 is a characteristic of a Casson fluid.

Assuming $p_L = 114.7$ kPa for blood pressure in an artery, $k = 1.4$, and temperature of 310 K, a $10\,\mu m$ radius bubble was computed to have a natural frequency of about 355 kHz. Such a bubble size was employed in numerical computations of characteristic bubble oscillations as the frequency of low amplitude pressure pulsation approaches the resonant bubble frequency. As shown in Figure 1, f is equivalent to our f_r, while f_0 is the pulsation frequency; pressure at a distance from the bubble $= p_L \cdot (1 + A \sin t)$, where A is set to cause relatively large oscillating amplitudes of R/R_0. If the non-Newtonian viscous term was neglected the computed gas bubble oscillation would be more prone to cause bubble collapse. Air bubble oscillations near resonance were generally more damped in blood

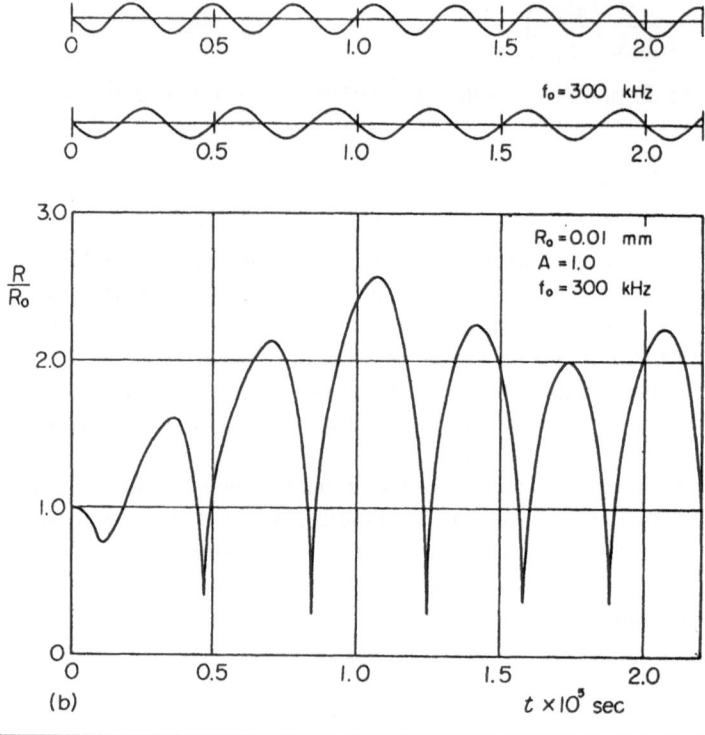

Figure 1. Bubble oscillation in the case where frequency of pulsation is near to natural frequency of a bubble.

than in water, although increasing the imposed pulsation amplitude would theoretically potentiate the bubble collapse tendency.

Lauterborn [38] calculated the bubble frequency response for bubbles of radii $10\,\mu m$ and $1\,\mu m$ in atmospheric water at $20°C$ ($\sigma=72.5\,\text{dyne/cm}$, $\mu=0.01\,\text{g/cm}\cdot\text{s}$, vapor pressure assumed negligible). In Figure 2, R_n ($=R_0$)=bubble radius at rest, p_A ($=A$)=sound pressure amplitude, and v or v_0 represents sound field or resonant frequency, respectively. Frequency response curves for the two bubble radii demonstrated characteristic oscillatory bubble wall excursions $(R_{max}-R_n)/R_\mu$ for differing sound pressure amplitudes. Primary-, sub- and ultraharmonic resonances are indicated.

Non-linear vs. linear bubble dynamics

While the above relationships describe satisfactorily many instances where linearized analysis can be applied, recent mathematical analyses and experi-

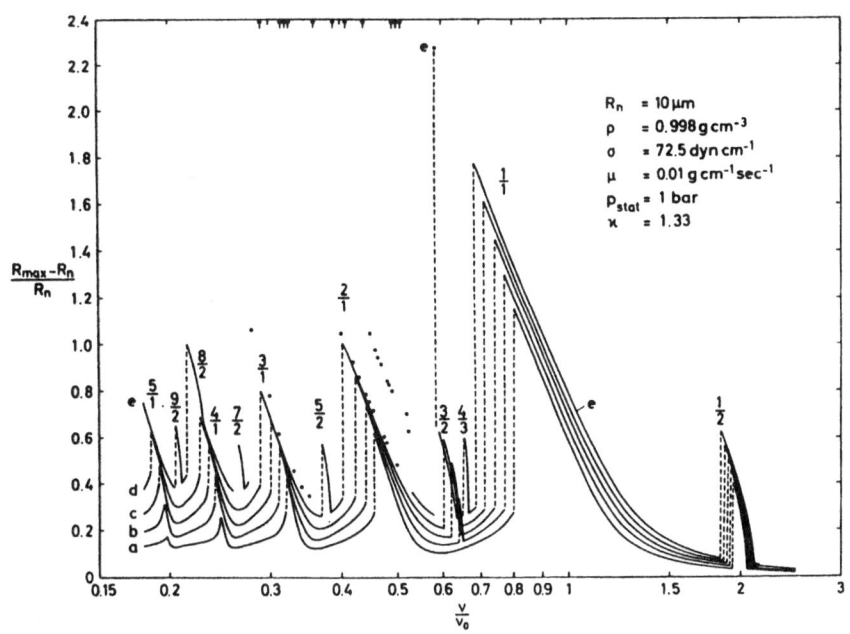

Figure 2. Frequency response curves for a bubble in water with a radius at rest of $R_n = 10\,\mu$m for different sound pressure amplitudes (p_A) of (a) 0.4, (b) 0.5, (c) 0.6, (d) 0.7 and (e) 0.8 bar. The numbers above the peaks of the curves are the orders of the resonances. The dots and arrows belong to curve (e). The arrows indicate that the corresponding stationary solution is out of the range of the diagram, or that no stationary solution could be found. In this case the amplitude values were also very high, oscillating around some value outside the diagram.

mental studies point to the probably more prevalent but unfortunately much more difficult to analyse non-linear behavior of bubbles in liquids. In general, non-linear bubble dynamics can be expected when the amplitude of forced oscillations is not very small, and when frequencies are at or close to resonance. Conversely, Eatock and Nishi [39] concluded that with an acoustic amplitude of 50 kPa, the behavior of bubbles should be essentially linear, as long as the driving frequency is more than 10% away from the bubble resonant frequency. Non-linear behavior is germaine to harmonic bubble phenomena, as examined by Miller [40].

Combining all the above relationships, a fairly general dynamic bubble equation (still assuming polytropic bubble gas behavior, and neglecting thermal as well as acoustic dissipation) is [33]:

$$R\ddot{R}+\frac{3}{2}(\dot{R})^2=\frac{1}{\rho_L}\left[p_{g0}\left(\frac{R_0}{R}\right)^{3k}-\rho_L(1-\eta\cos\omega t)-\frac{2\sigma}{R}-\frac{4\mu\dot{R}}{R}\right] \qquad (10)$$

where η represents a dimensionless pressure amplitude, which is often not very small, and is in part responsible for non-linear bubble behavior. The Lauterborn figure (Figure 2) demonstrates the profound effects on bubble volume pulsations anticipated as the sound amplitude increases. However Prosperetti [41, 42] pointed to non-linear effects even during small-to-moderate amplitude forced oscillations, particularly in the vicinity of the bubble resonant frequency.

Re-examining bubble phenomena in a sound field, Prosperetti et al. [43] modelled radial bubble motions in realistic fluids, which are not totally incompressible. Since the use of polytropic relationships appeared questionable, an alternative mathematical analysis of non-linear bubble dynamics was based upon an evaluation of the bubble interior's thermo-fluid mechanics. For micron-sized bubbles, the alternative numerical computations at moderate acoustic pressure amplitudes and at a frequency ratio of 0.8 demonstrated significant differences (in terms of calculated energy dissipation and bubble oscillatory excursions) from the generally applied polytropic approach. Not surprisingly, there was good agreement between the non-linear numerical computations and polytropic bubble analysis for very low acoustic pressure amplitudes and frequencies remote from resonance.

The normal sound velocity corresponds to infinitesimal oscillatory amplitudes. As the oscillatory pressure amplitude increases, the velocity of sound is higher at the pressure maximum than at its minimum, causing wave distortion and non-linearities. In a 1995 publication, Wu et al. [44] tried to rationalize the frequently markedly nonlinear behavior of bubbly fluids. Their starting point was a Beyer [45] expression for the speed of sound:

$$c^2 = c_0^2 \left[1 + \left(\frac{B}{A} \right) \left(\frac{\rho - \rho_0}{\rho_0} \right) \right] \tag{11}$$

c and c_0 = speed of sound of finite and infinitesimal amplitudes; ρ and ρ_0 = medium's density and its equilibrium value; $(B/A) = \rho_0/(\delta c^2/\delta p)$ = characteristic of a medium's non-linearity [18].

Experiments on very thin polycarbonate sheets with micron-sized pores indicated a B/A as high as $10^4 - 10^5$ for bubbly water, strikingly higher than the $B/A = 5$ for bubble-free water. The presence of bubbles in a medium evidently greatly escalates B/A, particularly, as already indicated, when bubble oscillations approach resonance. The authors carried out only limited analytical modeling, but called attention to the highly dispersive character of a non-linear attenuative bubbly medium, vs. the nearly non-dispersive bubble-free liquid.

Bubble clouds and interactions

It is true that most echocontrast dilutions or moderate concentrations involve few microbubbles per unit blood volume. In contrast to the concentrated bubble interactions observed during cavitation, spatial sparsity of bubbles would tend to justify the common algebraic summing of individual bubble dynamics. There may, however, be intermediate conditions which would require consideration of bubble–bubble interactions, or bubble cloud pressure/velocity gradients. Data are limited, but d'Agostino and Brennon [46] analysed bubble clouds and reported significant differences in natural frequency and acoustic cross-sections of individual bubbles vs. the ensemble (depending on dispersion and interactions). Bubble response (at subresonant frequencies) appeared maximal at the cloud's center.

Recent theoretical work of Pelekasis and Tsampoulos [47] and Doimkov and Zavtra [48] addressed bubble interactions. For even relatively large inter-bubble distances, the authors indicate that pulsating bubbles of equal size tend to attract each other (force inversely proportional to distance squared), occasionally producing shape changes and eventual bubble breakup. In a higher magnitude oscillatory field, two unequal size bubbles moved away from each other whenever the forcing frequency was between the two natural frequencies; otherwise attraction prevailed. For a closer bubble array, numerical calculations modeled the type of multiple scattering and bubble coalescence observed experimentally.

Bubble dissolution and collapse

Interface mass diffusion and surface tension represent major influences on bubble stability. In general, the bubble will either grow or shrink, depending on whether the internal gas concentration is less or greater than the saturation within the surrounding fluid. A $10 \mu m$ diameter free gaseous bubble in an undersaturated fluid would dissolve within seconds, due to surface tension, which increases the pressure within the bubble and promotes outward gas diffusion. Bubble dissolution is particularly rapid for micron-sized free bubbles. Conversely, and of less interest here, when the driving oscillation is of high amplitude, or the bubble is larger, a so-called 'rectified diffusion' may favor bubble growth. Much recent echocontrast agent research has centered on avoiding premature microbubble shrinkage or collapse, without compromising the extremely favorable vigorous bubble oscillations in the ultrasound field.

Epstein and Plesset [49] examined the dissolution of a free bubble (devoid of surface coating or encapsulating shell) in an undersaturated liquid without or with consideration of surface tension. An approximate relation was given as:

$$\left(\frac{R}{R_0}\right)^2 = 1 - \left[\frac{2D(C_s - C_i)}{\rho_g R_0^2}\right] t \tag{12}$$

R_0 = initial bubble radius ($t \leqslant 0$)
R = radius at time t
D = coefficient of diffusivity (2×10^{-5} cm^2/s, air in water)
C_c = dissolved gas (air) concentration in liquid (assumed uniform throughout solution, at constant temperature)
C_s = dissolved gas concentration for a saturated solution, at the same temperature.

Solubility of air at 22°C was assumed as $C_s/\rho_g = 0.02$

ρ_g = density of gas (air) within bubble.

Using the above approximation for an air–water solution at 22°C and $C_i/C_s = 0$, total dissolution time of a bubble with an initial $R_0 = 10\,\mu m$ was calculated as 1.25 s. Inclusion of surface tension in the approximate treatment yielded 1.17 s. Respective values for $C_i/C_s = 0.25$ were 1.67 and 1.46. Epstein and Plesset point out that while a bubble's life may be extended in a saturated solution, dissolution of very small bubbles would still take place by virtue of surface tension (6.63 s for a $10\,\mu m$ bubble).

Rearranging an Epstein–Plesset equation involving surface tension, deJong et al. [50] presented the gas bubble shrinkage rate (dR/dt) as:

$$\frac{dR}{dt} = \frac{D}{R}\frac{BTC_s}{\rho_L}\left[\frac{C_i/C_s - 1}{1 + 4\sigma/3Rp_L}\right]\left\{1 + \frac{R}{(\pi Dt)^{1/2}}\right\}. \tag{13}$$

Terms are as previously defined, and

B = universal gas constant
T = absolute temperature
p_L = pressure in still liquid
t = time.

Assuming an unsaturated setting ($C_i/C_s = 0$) and standard conditions, a $10\,\mu m$ radius air bubble collapsed in water within 1 s, while a $1\,\mu m$ radius bubble dissolved in 0.01 s. Even when C_i/C_s was set at 1, the $10\,\mu m$ bubble collapsed within 6 s and the $1\,\mu m$ bubble still disappeared within 0.01 s. These are theoretical model computations; a number of experiments indicated better microbubble persistence, which might be due to a (difficult to ascertain) surface film or contamination.

Yang [51] studied gas bubble radius vs. time relationships, with special emphasis on plasma and whole blood. Considering gaseous bubble–liquid systems featuring diffusion and surface tension, a theoretical non-dimensional dissolution rate was derived:

$$\frac{dR^*}{dt^*} = -C_\rho^* \frac{1 - C_i^* + \sigma^*/R^*}{1 + \frac{2}{3}(\sigma^*/R^*)}\left(\frac{1}{R^*} + \frac{1}{\sqrt{\pi t^*}}\right) \tag{14}$$

t^* = dimensionless time Dt/R_0
R^* = dimensionless radius R/R_0
$C_\rho^* = C_s/\rho_g$
$C_i^* = C_i/C_s$
$\sigma^* = 2\sigma/R_0 p_{g0}$

subscript 0 always applies to $t \leq 0$.

In the absence of adequate data for plasma and blood, the diffusion coefficient

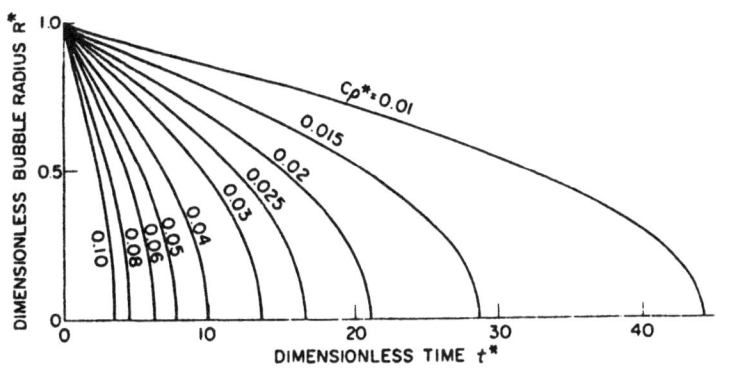

Figure 3. Radius–time relation for dissolving bubbles in plasma ($k^*=0$) for $C^*_x=0$.

of water was used ($D=2\times10^{-5}\,\text{cm}^2\,\text{s}$). Figure 3 illustrates, for bubbles dissolving in plasma, the dimensionless bubble radius R^* vs. dimensionless time t^* results for various C^*_ρ levels. Dissolution is accelerated as C_s is increased and/or ρ_g is decreased. The time to complete bubble dissolution in a saturated liquid ($C^*_i=1$) is approximately

$$t^* = \frac{1+\sigma^*}{3\sigma^*C^*_\rho}. \tag{15}$$

This time to bubble dissolution decreases with higher surface tension.

In terms of factors modifying the dissolution tendencies of free gas bubbles in a liquid, two recent papers can be mentioned. Fyrillas and Szeri [52] analysed the effects of surfactants on mass transfer across the dynamic interface of a bubble undergoing spherical symmetrical oscillations. According to the authors, the presence of surfactants negates Henry's law relating the concentration of the gas in a liquid to the partial pressure of the bubble gas. The factor that seemed dominant in their mathematical analysis was the bubble–liquid interfacial resistance to mass transfer, which is partly determined by the surfactant molecule concentration on the interface. While most of this study deals with bubble growth, a bubble dissolution trend was envisioned in some range of interfacial resistance vs. surfactant effects.

In a more directly applicable study, van Liew and Burkard [53] illustrated the successive dimensional alterations of microbubbles encountered in blood during transit from systemic veins, via pulmonary artery and microvessels, to systemic artery and microvasculatures. The echocontrast microbubbles were stabilized by using a slowly permeating gas (permeation coefficient 1/10 of nitrogen or oxygen, 1/50 of CO_2). The authors modelled the progressive bubble shrinkage or rebound at different bubble sites in the circulation, starting from an intravenous injection, and considering local blood pressures and permeating gas partial pressures.

Ultrasound signal intensities were also estimated. Significant dimensional stabilization of the bubbles was evident when applying a slowly vs. rapidly permeating gas.

Gas bubble surface films or encapsulation

Lieberman [54] hypothesized that, in reality, many gas bubbles in liquids
inherently depart from the theoretical equations due to contamination at the
bubble surface. Considering a thin bubble surface film of thickness δ and a
modified diffusion coefficient D', Lieberman proposed a modification of the
Epstein–Plesset equation as follows:

$$R^2 = R_0^2 - \left(\frac{2D}{\rho}\right)(C_s - C_i)t + \left(\frac{2\delta D'}{D}\right)(R_0 - R). \tag{16}$$

Unfortunately, the film diffusion properties are generally not known, and they,
along with δ, will change as the bubble diameter decreases.

Fox and Herzfeld [55] also raised the question of how micron-sized gaseous
nuclei persist when surface tension, internal bubble gas pressure, and diffusion
across the boundary, theoretically mandate rapid bubble dissolution. Recalling
reports that seafoam is actually stabilized by an organic skin formed of proteins,
they hypothesized a very thin elastic shell surrounding the gaseous bubble, significantly lowering surface tension and hindering diffusion. A capsule would resist
microbubble collapse, unless and until pressures and circumferential tension
became sufficiently high to cause buckling, tearing and destruction of the thin
shell. Although the latter's material properties can vary, an example listed the
tearing strength of a 15 Å thick organic film as 3.3×10^8 dynes/cm^2.

Calling the outer radius R and the thin shell thickness δ, analysis based on
Love's theory [56] allowed computation of the radial extension U of the bubble
shell (apparently limited to lower pulsation frequencies):

$$U = -\frac{P}{3(kp_g)}\frac{R^2}{a}\left(1 + \frac{R}{a}\right)^{-1} \tag{17}$$

$$a = \frac{2}{3}\left(\frac{1}{kp_g}\right)\left(\frac{\delta E}{1 - v}\right)$$

v = Poisson's constant
E = Young's modulus
P = excess (e.g. acoustic) pressure
p_g = pressure in the bubble in the absence of sound wave

$(kp)^{-1}$ = adiabatic gas compressibility.

Assuming atmospheric pressure in the bubble, Fox and Herzberg reported
$\alpha = 2.7 \times 10^{-3}$ cm, or 3×10^{-5} cm, respectively, for fatty or protein skin surrounding
the air bubble. Experimental data were quoted as evidence that the more rigid
encapsulated microbubbles exhibited significantly higher resonance frequencies

than would be expected for free gas bubbles. The modified bubble resonant frequency equation, in its simplified form, was listed as:

$$f_r = \frac{1}{2\pi R}\left[\frac{3kp_g}{\rho_L}\left(1+\frac{a}{R}\right)\right]^{1/2}. \tag{18}$$

The inner bracket constitutes a correction due to the presence of the thin skin.

Assuming that a shell enclosing the microbubble allows some (but less than normal) gaseous diffusion, a slowing of the bubble contraction would be expected, depending on the shell's configuration and other physical properties. Imposed pressure pulsations within the liquid would also be expected to influence the course of the encapsulated gas bubble. Moreover, if the shell surrounding the bubble were to be destroyed, impulsive pressures could further accelerate diffusion and gaseous bubble collapse. Equations have yet to be developed for slow dissolution of encapsulated bubbles.

Yount [57] also considered stabilization by very thin elastic skins, whose compression strength would tend to prevent premature bubble collapse. A modified LaPlace equation was listed as:

$$p_g + \left(\frac{2\Gamma}{R}\right) = p_L + \frac{2\sigma}{R} \tag{19}$$

intrabubble pressure p_g > ambient hydrostatic p_1,

Γ = skin compression (= $\delta E/1 - v$, as per Fox Herzfeld)
δ = film thickness
E = Young's modulus
v = Poisson's ratio.

Although film/shell analyses and measurements are still oversimplified and scarce, there has been recent progress toward assessing the benefits or detriments of microbubble encapsulation. In the case of gaseous bubbles in blood, some denatured proteins may well organize into a presumably very flexible interface film, generating resistance to diffusion and reduced surface tension. Prior studies already indicated that microbubbles are less prone to collapse in blood than in water, although the life of micron-sized bubbles is still too short. The objective of recent efforts is to develop a stable and well characterized contrast agent microbubble, in which the capsule maintains dimension and reduces interfacial diffusion. As already stated, care must be taken to preserve the highly responsive bubble oscillations which underlie the greatly enhanced ultrasound backscatter.

deJong et al. [58] recently analysed the elasticity of a particular encapsulating thin shell (Albunex microsphere). Simplified relations were derived for bubble radius displacement, in terms of pressure and shell dimensions as well as material properties. A characteristic shell stiffness term was derived:

$$S_{shell} = \frac{8\pi Et}{i - v} \tag{20}$$

in which t is the wall thickness, E is the shell's Young modulus of elasticity and v is the Poisson's ratio. S_{shell} is then incorporated as a 'restoring force' into an analytical expression of the encapsulated gas bubble's modified resonance frequency, used to compute the altered damping and ultrasound contrast effectiveness. To improve agreement of analysis with experimental observations on Albunex microspheres in a representative ultrasound field, the computed damping was further adjusted by including an internal shell friction parameter. With these revisions, the Albunex scattering effects could be satisfactorily predicted [59]. The authors' measurements also demonstrated second harmonic scattering.

Given safety, stability and adequate echogenicity, uniformly encapsulated gas bubbles should facilitate contrast delivery from an intravenous site. Apparently, more work needs to be done to provide a rationally tailored bubble coating, in terms of its configuration, strength and interfacial properties. The actual bubble persistence and ultrasound effectiveness of particular coated bubbles must be evaluated, over a representative range of pressures, shear forces and flows.

Effects of ambient fluid pressure

Numerous observations point to accelerated microbubble shrinkage under increased pressure conditions. Motley et al. [60] examined the ultrasonic backscatter of galactose-based Echovist contrast and demonstrated a progressively more rapid decay as pressure in the bubbly water was elevated from atmospheric pressure to 100 mmHg. Similarly, Shandas et al. [61], studying Albunex microbubbles in vitro, established that contrast enhancement and microbubble stability were reduced at pressures exceeding 40 mmHg.

These observations are in line with in vivo experimental results reported by Shapiro et al. [62], who studied intravenous injections for myocardial echocontrast, using sonicated Albunex microspheres or equally small microbubbles produced by adding polysaccharide to a diluent. Myocardial opacification was limited, even at high intravenous microbubble concentrations, and in spite of demonstrated transpulmonary contrast delivery. There was a 60% decrease in left compared with right ventricular contrast intensity. A rapid decrease in left ventricular contrast intensity was observed in early systole, and contrast disappeared almost totally by the end of systole. Could this reflect destruction of the microbubble coatings, or is the rate of bubble shrinkage accelerated at elevated systolic pressures? In an in vitro study of sonicated albumin, a significant relation was found between pressures and video density decay, with rapid contrast disappearance noted at pressures above 60 mmHg. The general implication was that, even though stabilized microbubbles may well be effectively delivered across the pulmonary microcirculation and into the left ventricle, they could subsequently be excessively degraded, to the detriment of quantitative myocardial evaluation.

At the earliest stage of echocontrast development, Tickner and Rasor [63] and

Ooi and Acosta [64] actually proposed using changes in bubble radius and its natural frequency to measure pressure of the fluid medium. More recently, Ran and Katz [65] performed a related analytical and experimental study of gas bubble response to sudden changes in ambient pressure in water. Assuming an isothermal gas process within the bubble, Ran and Katz first re-derived the simplified linear formula for the natural frequency:

$$f_r = \frac{1}{2\pi R}\left[\frac{3}{\rho_L}\left(p_L + \frac{4\sigma}{3R}\right)\right]^{1/2}. \tag{21}$$

Aiming to assess blood pressure from changes in bubble radius, they selected a $50\,\mu m$ radius gas bubble and calculated its natural frequency as $55\,kHz$ (assuming $\sigma = 72\times10^{-3}\,N/m$). They then proceeded to derive the pressure–bubble radius relationship in a stream. Disregarding the liquid vapor pressure, the fluid pressure p_L for fluctuation frequencies less than f_r is given as:

$$p_L = \left(p_{L0} + \frac{2\sigma}{R_0}\right)\left(\frac{R_0}{R}\right)^{3k} - \frac{2\sigma}{R} \tag{22}$$

where p_{L0} and R_0 indicate initial fluid pressure and bubble radius, while R is the bubble dimension resulting from pressure p_L.

In their experiments, the investigators employed bubbles of air, carbon dioxide, helium and hydrogen, providing a range of mass, thermal diffusivities, solubilities in water, and isentropic constants. They used holograms to study changes in the bubbles' diameter. Most of the experiments were performed with pressure changes at a rate of $20\,kPa/ms$. Results indicated that bubbles responded instantaneously to changes in the ambient pressure. The analysis was validated and its isothermal assumption verified.

Figure 4 illustrates, for air in water, the inverse relationship between instantaneous-to-original bubble radius ratio and instantaneous-to-initial ambient pressure ratio. A linear correlation was obtained when data were plotted on a logarithmic scale. These experimental data indicated that for a $P/P_0 = 10$ the corresponding ratio $R/R_0 = 0.4$. Changes in the pressure wave did not significantly alter the bubble size vs. pressure relationship.

Tickner and Rasor [63] derived a simplified relationship between changes in blood pressure P and bubble resonant frequency f_r:

$$\frac{\Delta P}{P_0} = \text{constant}\times\left(\frac{\Delta f_r}{f_r}\right). \tag{23}$$

Based on an experimental constant of 1.25, they predicted that a $P/P_0 = 10$ would cause a reduction in bubble radius to $R/R_0 = 0.16$.

Shima et al. [66] derived complex equations for the impulsive pressure developed at and around a gas bubble by a forcing pressure of given amplitude and frequency. Numerical computations showed that the oscillatory local bubble wall pressure peaked at the primary bubble resonance frequency, its magnitude

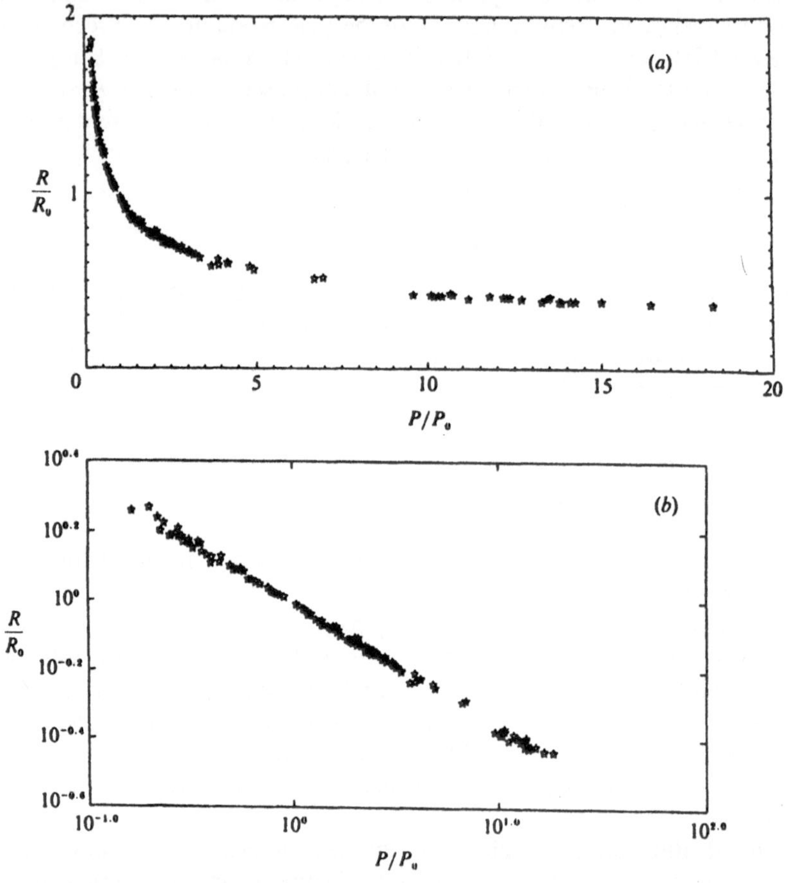

Figure 4. The response of air bubbles to changes in the ambient pressure plotted on both linear (*a*) and logarithmic (*b*) scales. The subscript 0 indicates initial conditions. Air content is 68%.

depending on the imposed pulsation amplitude. This impulsive pressure was distinctly lower in a non-Newtonian fluid, than for a gas bubble in water. The effects of pressure pulsation on the gas bubble–liquid system were noted by Tsujino and Miura [67]. Using the Casson model of blood rheology, a complicated relationship was derived, requiring numerical computations. When a $50\,\mu m$ gas bubble in blood was subjected to a sudden pressure change ($0.3-2\,MPa$), the computed amplitude of the generated forceful bubble pulsation was as shown in Figure 5. Greater damping was noted as the hematocrit increased, and bubble collapse was associated with high impulsive pressure amplitudes.

There is agreement that the higher levels of blood pressure encountered on the left side of the heart and in arteries contribute to microbubble shrinkage, which can influence the concentration of agent required for quality ultrasound contrast

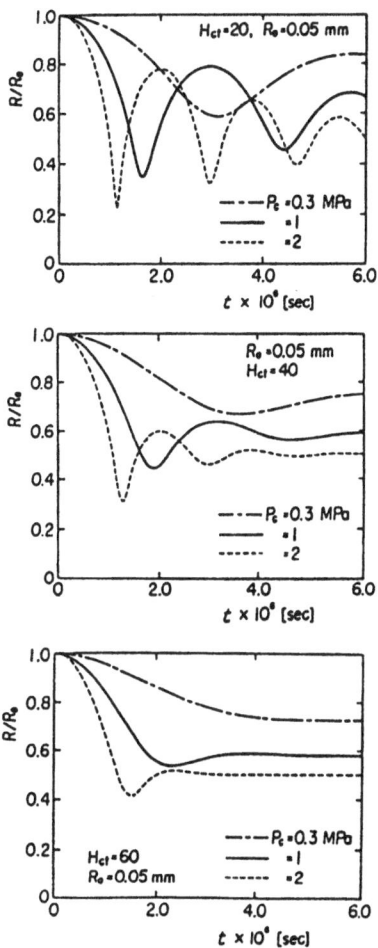

Figure 5. Variation of bubble radius with time. The amplitude of the pulsatile swing in bubble diameter (R/R_0) increases as it is successively exposed to a greater stepped pressure effect ($p = 0.3$, 1, 2 MPa). More wave damping is noted at higher haematocrit levels.

images. Disagreement remains regarding the extent and character of the pressure effects upon the microbubble or its encapsulating shell. Mor Avi et al. [68] carried out a study in an isolated beating rabbit heart model, with a latex balloon placed in the left ventricular chamber. Albunex microbubble contrast was used during 7.5 MHz ultrasound imaging. Experimental left ventricular systolic pressures ranged up to 155 mmHg. The end-diastolic videodensity decreased $8 \pm 6\%$ over 25 consecutive heart beats, and systolic peak videodensity was about 20% less than in diastole. Moderate cyclic compression of the gas-filled microspheres, rather than their destruction, was postulated based on these observations. The relatively slow permanent changes observed were associated with altered permeability of the bubble shell, or factors related to shear forces.

Effects of elevated pressures on Albunex, Levovist, and other echocontrast agents were recently examined in in vitro models by Vuille et al. [69] and Padial et al. [70]. A high static pressure (up to 200 mmHg) was shown to increase the peak backscatter decay rate, strikingly so for Albunex and to a lesser extent with Levovist. In the presence of pulsatile pressures, Albunex again exhibited a rapid, though cyclic, video intensity decay. The gas-containing microspheres were thought to be significantly deformed but probably not destroyed in any quantity at the imposed pressure amplitudes and durations. The observed decreased effectiveness of encapsulated vs. free bubbles was associated with the greater liquid undersaturation at higher pressures, an altered concentration gradient and greater amount of bubble gas that could be dissolved.

Velocity of sound in a bubble medium

The normal speed of sound in air (at standard conditions) is 330 m/s, in water 1480 m/s, in blood 1570 m/s, in muscle 1580 m/s, and in bone 3500 m/s. The presence of highly compressible gas bubbles in a liquid can profoundly affect the velocity of sound in a bubbly mixture. Gibson [71] measured the effects of air bubbles on the speed of sound c in water and obtained a very good correlation with theory. Calling x the volume fraction of air in the air–water mixture (m, l, g subscripts signify mixture, liquid and gas, respectively), the mean density ρ_m is:

$$\rho_m = (1-x)\rho_l + x\rho_g \tag{24}$$

and mean compressibility K_m in terms of liquid K_l and gas K_g is:

$$K_m = (1-x)K_l + xKg. \tag{25}$$

The mixture's speed of sound is then:

$$c = (\rho_m K_m)^{-1/2}. \tag{26}$$

The presence of 1% air decreased the medium's velocity of sound by 90%.

Both Ophir and Parker [72] and deJong et al. [50] presented similar bubble–liquid mixture relations for the mean speed of sound. Gas volume concentrations as low as 0.1% significantly reduced the mixture's velocity of sound transmission. It was pointed out that, in principle, bubble gas concentrations could be assessed from measurements of the speed of sound.

Two illustrations from Anderson and Hampton [36] indicate the effects of frequency, air volume concentrations and pressure levels on the speed of sound in bubbly water (Figures 6 and 7). The various derived formulae can be applied to microbubbles in fluids, up to relatively high frequencies. Above the resonant frequency bubble concentration seems to have little effect.

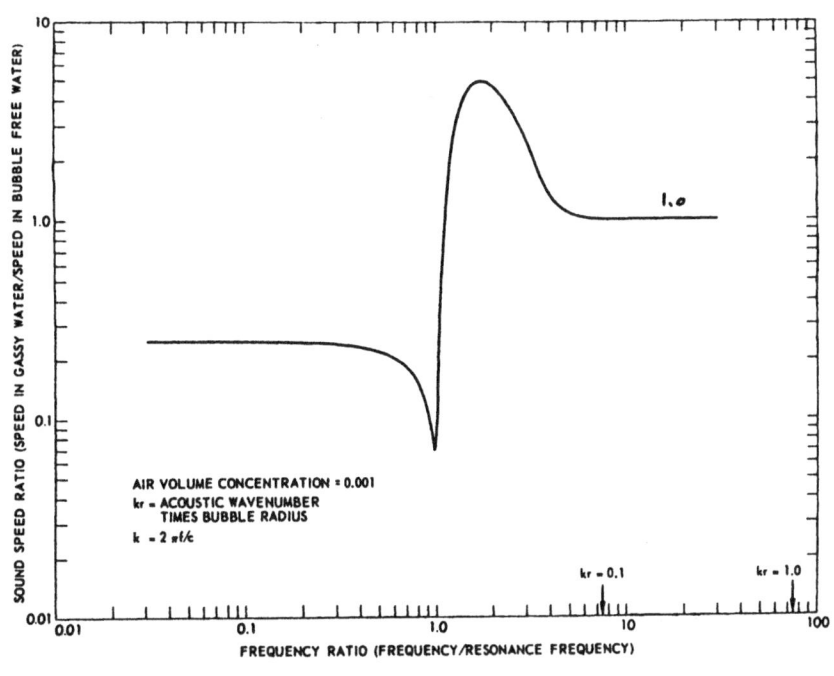

Figure 6. Sound speed ratio vs. frequency ratio for water plus a collection of single-sized bubbles.

EFFECTIVENESS OF ULTRASOUND–MICROBUBBLE INTERACTIONS

The ultimate objective here is to exploit the distinct microbubble behavior to help achieve an optimal echocontrast performance. To minimize redundancies, this chapter will not address to any depth the many-faceted and complex ultrasound issues of echocontrast. Rather, a few topics were selected because of their connection with the aforementioned dynamics of the highly compressible gaseous bubbles, and their vigorous oscillatory response in a sound field. Contemporaneous reviews by Ophir and Parker [72] and deJong et al. [50, 58, 59] offer insights into most of the factors of import to ultrasound effectiveness.

Scattering

Ultrasound scattering is generated from bubble interfaces due to greatly altered accoustic impedance, brought about primarily by the media's changed compressibility. Calling incident wave intensity I and scattered wave intensity S, $Q_s = S/I =$ scattering cross-section (often associated with 'contrast echogenicity'). The interaction of diagnostic ultrasound (1–10 MHz) with microbubbles follows

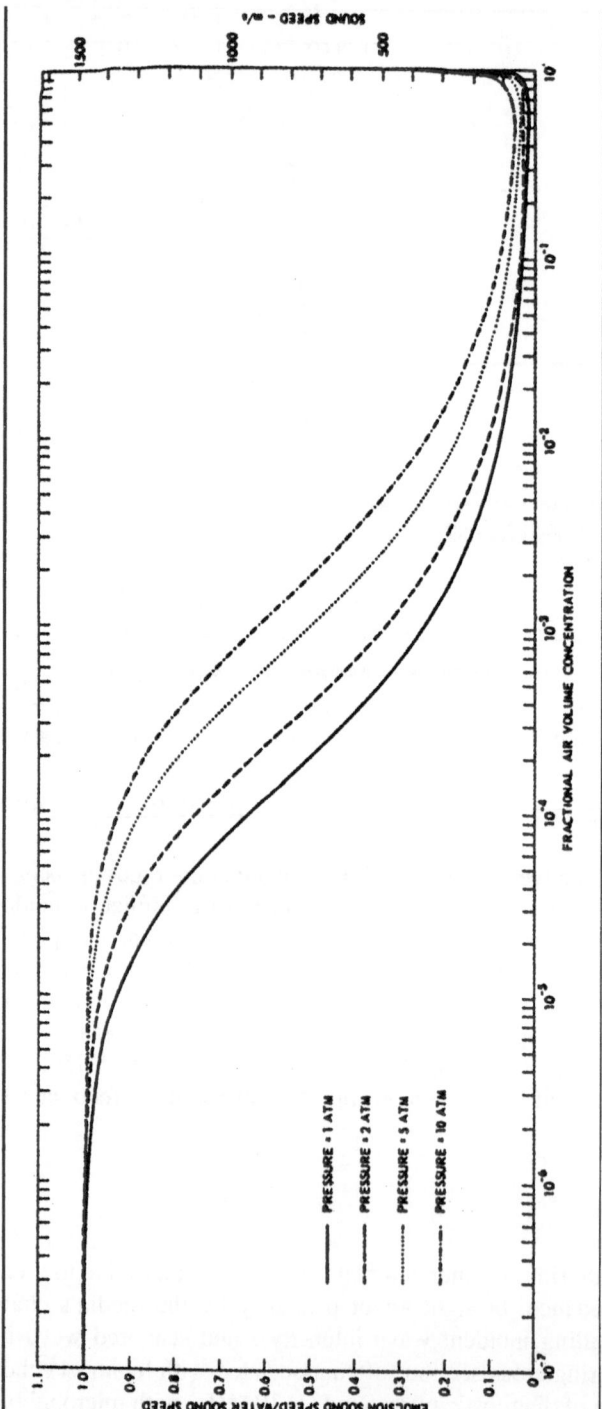

Figure 7. Emulsion sound speed vs. air volume concentration.

in general the rules of Rayleigh scattering. The scattering cross-section is greatly dependent on the bubble radius R, the frequency up to resonance, and on differences of compressibility (as well as density).

As has been reiterated, microbubbles are extremely effective scatterers in an ultrasound field. The gaseous bubbles undergo strong volume pulsations in response to an applied sound pressure of given frequency and amplitude. Assuming a relatively long wave length (compared with the bubble radius R), Morse and Ingard [73] originally and deJong [50] more recently, presented the following simplified relation for the important scattering cross-section Q_s:

$$Q_s = \left[\frac{4}{9}\pi R^2 \left(\frac{2\pi}{\lambda}R\right)^4\right] \left\{\left[\frac{K_g - K_L}{K_L}\right]^2 + \frac{1}{3}\left[\frac{3(\rho_g - \rho_L)}{2\rho_g - \rho_L}\right]^2\right\} \tag{27}$$

where

Q_s = scattering cross-section (m^2)
R = microbubble radius (m)
λ = ultrasonic wave length = c/f (m) (wave number = $2\pi/\lambda$)
K_g = compressibility of bubble gas (m^2/N)
K_L = compressibility of liquid (m^2/N)
ρ_g = density of bubble gas (kg/m^3)
ρ_l = density of liquid (kg/m^3)
c = sound velocity (m/s)
f = frequency (s^{-1}).

The most striking influence on bubble scattering is due to the difference between bubble gas vs. liquid medium compressibility (K_g vs. K_L). Thus, for air at atmospheric pressure, the compressibility K_g is 7.65×10^{-6} m^2/N, which is about 16000 times greater than that for water K_L. Also, Q_s is a 6th power function of R, and a 4th power function of the frequency. deJong [50] compared air bubbles 1 μm in diameter to solid scatterers in water: the calculated scattering coefficient of the former was 2×10^8 higher than Q_s for an iron particle, largely due to the $(K_g/K_L)^2$ factor being 4×10^8 for the air bubble, 10^{-2} for metal and about 0 for liquid scatterers. Additional large potentiation of Q_s results when the gaseous microbubbles pulsate at their resonant frequency. At frequencies higher than resonance frequencies, the bubble's Q_s levels out at its physical reflection cross-section. Fortuitously, echocontrast microbubbles (of the order of 5 μm in diameter) resonate in the medical ultrasound imaging band of 1–10 MHz. Unfortunately, as indicated, such small bubbles tend to dissolve rapidly and need to be protected to assure their stability.

One of the early contributors in the field was Nishi [74], who extended theoretical modelling to include air bubbles in blood, and reported on computations of bubble response in the diagnostic ultrasound range. Recognizing that prior modelling approximations (disregarding viscous/thermal losses and surface tension) were inaccurate for microbubbles, e.g. radii $< 10 \mu$m, Nishi presented more realistic though complicated relations for calculating scattering and

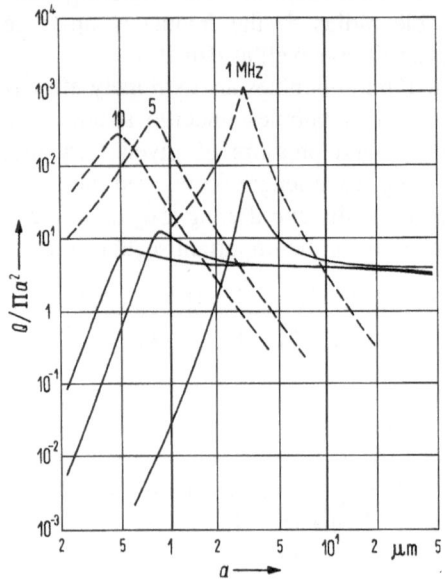

Figure 8. Scattering and absorption cross-sections as a function of bubble radius (a) for an air bubble in blood at 1, 5 and 10 MHz. ---- $Q_a/\pi a^2$; —— $Q_s/\pi a^2$.

absorption. Numerical computations (including thermal, viscous and radiation damping terms) were carried out for an air bubble in either water or in blood. The extraordinarily high ultrasound scattering and absorption with gaseous microbubbles, compared to other microparticles, was corroborated. In Figure 8 two indices of gas bubble effectiveness, i.e. scattering and absorption cross-sections, are plotted against equilibrium radii over a range of pressures. Higher damping effects were noted with blood than with water, and the effects of different frequencies (1, 5 and 10 MHz) were shown.

Harmonic effects

Miller's [40] comparison of second vs. first harmonic effects demonstrated significant differences in the relationship between scattering cross-section Q_s and bubble radius R. Both first and second harmonics exhibited a sharp peak in Q_s at a resonant bubble radius of about $2\,\mu m$ (at 1.64 and 3.28 MHz, respectively). However, for larger radii (more than $4\,\mu m$), the second harmonic response exhibited extremely low scattering, whereas the first harmonic Q_s remained elevated, eventually levelling out at the bubble's physical cross-section scattering. These results point to potential future discriminatory assessment of small vs. large bubbles, or between tissue vs. blood in vessels. Harmonic contrast ultrasound imaging has recently received much attention [75, 76]: for instance, where fundamental frequency echocontrast enhancement in a Doppler study of a small vessel flow might actually make it more difficult to distinguish between tissue and blood, harmonic imaging can apparently overcome this limitation.

Effects of bubble concentration

It is recognized that the usual approach based on analyzing ultrasound effects of single scatterers in a liquid can be inadequate for appropriate modeling of a bubble cloud. Nonetheless, for very small and widely separated scatterers of individual Q_s, it is admissible to formulate an effective bubbly medium's scattering cross-section $Q_{eff}=nQ$, where n represents the number of bubbles in the insonified liquid volume. Raising the bubble concentration increases directly the total scattering cross-section (one should recall, however, that scattering depends even more strongly on bubble diameters or frequencies). As noted in the discussion of bubble dynamics, computations of bubble cloud configurations are not readily accomplished at this time. There also appears to be a generally unfavorable relationship between ultrasound backscatter and depth of view, limiting the use of high concentrations of strongly scattering microbubbles. At high bubble concentrations, shadowing and increased attenuation cause a flattening of the Q_{eff} vs. n curve, and can even lead to a degradation of Q_{eff}.

Effects of bubble encapsulation

Encapsulation of bubbles has already been discussed. For an acoustic wave amplitude of 50 kPa and realistic properties for diagnostic contrast ultrasound, deJong et al. [50, 59] performed calculations comparing an air bubble vs. an Albunex microsphere in water at 20°C. Compared with the unprotected air bubble, ultrasound scattering results with encapsulated bubbles were found to be less spectacular. The shell increases stiffness, causing the diameter at which resonance occurs to be larger than that for an air bubble. It was established that gas-filled Albunex microspheres $<3\,\mu$m in diameter contributed relatively little over a range of current diagnostic frequencies, and those in the $3-5\,\mu$m range provided only moderate effectiveness. Albunex microspheres with diameters between 5 and $8\,\mu$m appeared most effective, while microspheres larger than $8\,\mu$m delivered a substantial contribution at all frequencies. Absorption was generally about 10 times the bubble's scattering.

REFERENCES

1. Gramiak R, Shah PM, Kramer DH. Ultrasound cardiography. Contrast studies in anatomy and function. Radiology 1968;92:939.

2. Ziskin MC, Bonakdarpour A, Weinstein DP et al. Contrast agents for diagnostic ultrasound. Invest Radiol 1972;7:500.
3. Meltzer RS, Tickner EG, Salines TP, Popp RL. The source of ultrasound contrast effect. J Clin Ultrasound 1980;8:121–7.
4. Roelandt J. Review Paper. Contrast echocardiography. Ultrasound Med Biol 1982;8:471–92.
5. Meltzer RS, Sartorius OEH, Lancee GT et al. Transmission of echocardiographic contrast through the lungs. Ultrasound Med Biol 1981;7:377–84.
6. Meltzer RS, Vermeulen HWJ, Valk NK, Verdouw PD, Lancee CT, Roelandt J. New echocardiographic contrast agents: transmission through the lungs and myocardial perfusion imaging. J Cardiovasc Ultrason 1982;1:277–82.
7. Carroll BA, Turner RJ, Tickner EG, Boyle DB, Young SW. Gelatin encapsulated nitrogen microbubbles as ultrasonic contrast agents. Invest Radiol 1980;15:260–6.
8. Tei C, Sakamaki T, Shah PM et al. Myocardial contrast echocardiography. A reproducible technique of myocardial opacification for identifying regional perfusion deficits. Circulation 1983;67:585–93.
9. Armstrong WF, West SR, Mueller EM, Dillon JC, Feigenbaum H. Assessment of location and size of myocardial infarction with contrast-enhanced echocardiography. J Am Coll Cardiol 1983;2:63–9.
10. Kaul S, Pandian NG, Okada RD, Pohost GM, Weyman AE. Contrast echocardiography in acute myocardial ischemia: I. In vivo determination of total left ventricular 'area at risk'. J Am Coll Cardiol 1984;4:1272–82.
11. Kemper AJ, O'Boyle JE, Sharmer S et al. Hydrogen peroxide contrast-enhanced 2-dimensional echocardiography: Real time in vivo delineation of regional myocardial perfusion. Circulation 1983;68:603–11
12. Keller MW, Glasheen W, Kaul S. Albunex: a safe and effective commercially produced agent for myocardial contrast echocardiography. J Am Soc Echo 1989;2:48–52.
13. Kaul S. Quantitation of myocardial perfusion with contrast echocardiography. Am J Card Imag 1991;5:200–16.
14. Porter TR, D'Sa A, Turner C et al. Myocardial contrast echocardiography for the assessment of coronary blood flow reserve: Validations in humans. J Am Coll Cardiol 1993;21:349–55.
15. Villanueva FS, Glasheen WP, Sklenar J, Kaul S. Assessment of risk area during coronary occlusion and infarct size after reperfusion with myocardial contrast echocardiography using left and right atrial injections of contrast. Circulation 1993;88:596–604.
16. Skyba DM, Jayaweera AIR, Goodman NC, Ismail S, Camarano G, Kaul S. Quantification of myocardial perfusion with myocardial contrast echocardiography during left atrial injection of contrast: Implications for venous injection. Circulation 1994;90:1513–21.
17. Keller MW, Segal SS, Kaul S, Duling BR. The behavior of sonicated albumin microbubbles in the microcirculation: A basis for their use as myocardial echo contrast agents. J Am Coll Cardiol 1988;10:75A.
18. Apfel RE. Accoustic cavitation. Methods of experimental physics. 1981;69:355–411.
19. Ivey JA, Gardner EA, Fowlkes JB, Rubin JM, Carson PL. Accoustic generation of intra-arterial contrast boluses. Ultrasound Med Biol 1995;21:757–767.
20. Feinstein SB, Ten Cate FJ, Zwehl W et al. Two-dimensional contrast echocardiography. I. In vitro development and quantitative analysis of echo contrast agents. J Am Coll Cardiol 1984; 3(1):14–21.
21. Schlief R, Schurmann R, Balzer T, Zomack M, Niendorf HP. Saccharide based contrast agents. In: Nanda NC, Schlief R, editors. Advances in echo imaging using contrast enhancement. Dordrecht, The Netherlands: Kluwer, 1993:71–96.
22. Schlief R, Schurmann R, Balzer T et al. Saccharide-based contrast agents and their application in vascular Doppler ultrasound. Adv Echo-contrast 1994;3:60–76.
23. Von Bibra H, Sutherland G, Becher H, Neudert J, Nihoyannopoulos P. Clinical evaluation of left heart Doppler contrast enhancement by saccharide-based transpulmonary contrast agent. J Am Coll Cardiol 1995;25:500–8.

24. Christiansen C, Kryvi H, Sontum PC, Skotland T. Physical and biochemical characterization of Albunex, a new ultrasound contrast agent consisting of air-filled albumin microspheres suspended in a solution of human albumin. Biotechnol Appl Biochem 1994;19:307–20.

25. Bleeker HJ, Shung KK, Barnhart JL. Ultrasonic characterization of Albunex, a new contrast agent. J Accoust Soc Am 1990;87:1792–7.

26. deJong N, Ten Cate FJ, Vletter WB, Roelandt JRTC. Quantification of transpulmonary echocontrast effects. Ultrasound Med Biol 1993;19:279–88.

27. Sonne HS, Christensen PD, Muan B, Assentoft J, Haider T, Kristensen BO. Left ventricular opacification after intravenous injection of Albunex. Int J Cardiac Imag 1995;11:47–53.

28. Porter TR, Xie F, Cricsfeld A, Kilzer K. Noninvasive identification of acute myocardial ischemia and reperfusion with contrast ultrasound using intravenous Perfluoropropane – exposed sonicated Dextrose Albumin. J Am Coll Cardiol 1985;26:33–40.

29. Dittrich HC, Bales GL, Kuvelas T, Hunt RM, McFerran BA, Greener Y. Myocardial contrast echocardiography in experimental coronary artery occlusion with a new intravenously administered contrast agent. J Am Soc Echocardiogr 1995;8:465–74.

30. Uhlendorf V, Hoffmann C. Non-linear accoustical response of coated microbubbles in diagnostic ultrasound. Ultrasonics Symposium 1994. IEEE; 1559–62.

31. Forsberg F, Liu JB, Merton A, Rawool NM, Goldberg BB. In vivo evaluation of a new ultrasound contrast agent. Ultrasonics Symposium. 1994. IEEE; 1555–8.

32. Unger E, Shen DK, Fritz T et al. Gas-filled liposomes as echocardiographic contrast agents in rabbits with myocardial infarcts. Invest Radiol 1993;28:1155–9.

33. Plesset MS, Prosperetti A. Bubble dynamics and cavitation. Am Rev Fluid Mech 1977;9:145–85.

34. Houghton G. Theory of bubble pulsation and cavitation. J Accoust Soc Am 1963;35:1387–93.

35. Minnaertk M. On musical air bubbles and the sound of running water. Phil Mag 1993;16:235.

36. Anderson AL, Hampton LD. Accoustics of gas-bearing sediments. (1) Background. J Accoust Soc Am 1980;67:1865–89.

37. Tsujino T, Shima A. The behavior of gas bubbles in blood subjected to an oscillating pressure. J Biol Mech 1980;13:407–16.

38. Lauterborn W. Numerical investigation of non-linear oscillation of gas bubble in liquids. J Accoust Soc Am 1976;59:283–93.

39. Eatock VC, Nishi RY. Numerical studies of the spectrum of low-intensity ultrasound scattered by bubbles. J Accoust Soc Am 1985;77:1692–701.

40. Miller DL. Ultrasonic detection of resonant cavitation bubbles in a flow tube by their second-harmonic emission. Ultrasonics 1981;19:217–24.

41. Prosperetti A. Bubble phenomena in sound fields: Part I. Ultrasonics 1984;22:69–77.

42. Prosperetti A. Bubble dynamics: some things we did not know 10 years ago. In: Bubble Dynamics and Interface Phenomena, Proceedings of an IUTAN Symposium. Blake JR, Boulton-Stone JM, Thomas NH, editors. Dordrecht: Kluwer Academic Publishers, 1994:3–16.

43. Prosperetti A, Crum LA, Commander KW. Non-linear bubble dynamics. J Accoust Soc Am 1988;83:502–14.

44. Wu J, Zhu Z, Du G. Non-linear behavior of a liquid containing uniform bubbles: comparison between theory and experiments. Ultrasound Med Biol 1995;21:545–52.

45. Beyer RT. Non-linear accoustics. The Naval Ship Systems Command. Washington DC: Naval Department, 1974.

46. D'Agostino L, Brennen CE. Accoustical absorption and scattering cross sections of spherical bubble clouds. J Accoust Soc Am 1988;84:2126–33.

47. Pelekasis NA, Tsanopoulos JA. Bjerkenes forces between two bubbles. Response to a step change in pressure. J Fluid Mech 1993;254:467–99.

48. Doinikov AA, Zavtrak ST. On the mutual interaction of two gas bubbles in a sound field. Phys Fluids 1995;7:1923–30.

49. Epstein PS, Plesset MS. On the stability of gas bubbles in liquid-gas solutions. J Chem Phys 1950;18:1505–9.

50. deJong N, Ten Cate F, Lancee CT et al. Principles and recent developments in ultrasound contrast agents. Ultrasonics 1991;29:324–30.
51. Yang WJ. Dynamics of gas bubbles in whole blood and plasma. J Bio Med 1971;4:119–23.
52. Fyrillas MM, Szeri AJ. Dissolution or growth of soluble spherical oscillating bubbles: the effect of surfactants. J Fluid Mech 1995;289:295–314.
53. van Liew HD, Burkhard ME. Behavior of bubbles of slowly permeating gas used for ultrasonic imaging contrast. Invest Radiol 1995;30:315–21.
54. Lieberman L. Air bubbles in water. J Appl Phys 1957;28:205–11.
55. Fox FE, Herzfeld KF. Gas bubbles with organic skin as cavitation nuclei. J Accoust Soc Am 1954;26:984–9.
56. Love AEH. The mathematical theory of elasticity. New York: Dover Publications, 1944.
57. Yount DE. Bubble nucleation in aqueous media: implications for diving physiology. Appl Sci Res 1982;38:37–44.
58. deJong N, Hoff L, Skotland T, Bom B. Absorption and scatter of encapsulated gas-filled microspheres: Theoretical considerations and some measurements. Ultrasonics 1992;30:95–103.
59. deJong N, Cornet R, Lancee CT. Higher harmonics of vibrating gas-filled microspheres. Part I: simulations. Ultrasonics 1994;32:447–53.
60. Mottley J, Everbach EC, Schwarz KQ et al. Decay of ultrasound integrated back scatter from a saccharide contrast agent is accelerated by increased pressure. Circulation 1990;82(Suppl. 3):28.
61. Shandas R, Sahn DJ, Bales G et al. Persistence of Albunex ultrasound contrast agent: in vitro study of the effects of pressure and accoustic power on particle size and the duration of contrast and Doppler enhancement. Circulation 1990;82(Suppl. 3):95.
62. Shapiro JR, Reisner SA, Lichtenberg GS, Meltzer RS. Intravenous contrast echocardiography with use of sonicated albumin in humans: Systolic disappearance of left ventricular contrast after transpulmonary transmission. J Am Coll Cardiol 1990;6:1607–17.
63. Tickner EC, Rasor NS. An instrument for the noninvasive assessment of pulmonary hypertension. Adv Bioeng. ASME 1978;101–3.
64. Ooi KK, Acosta AJ. The utilization of specially tailored air bubbles as statis pressure sensors in a jet. Trans ASME IJ Fluid Eng 1983;106:459–65.
65. Ran B, Katz J. The response of microscopic bubbles to sudden change in ambient pressure. J Fluid Mech 1991;224:91–115.
66. Shima A, Rajvanjhi SC, Tsujino T. Study of non-linear oscillations of bubbles in Powell Eiring fluids. J Accoust Soc Am 1985;77:1702–9.
67. Tsujino T, Miura M. The motion of bubbles in blood as related to some medical problems. Nippon Kikai Gakai Ronbunshu. B Hen Trans of the Jap Soc of Mech Eng Part B 1989;56: 74–8.
68. Mor-avi V, Shroff SJ, Robinson KA et al. Effects of left ventricular pressure on sonicated albumin microbubbles: Evaluation using an isolated rabbit heart model. J Am Coll Cardiol 1994;24:1779–85.
69. Vuille C, Nidorf M, Morrissey RL, Newell JB, Weyman AE, Picard MH. Effect of static pressure on the disappearance rate of specific echocardiographic contrast agents. J Am Soc Echocardiogr 1994;7:347–54.
70. Padial LR, Chen MH, Vuille C, Guerero JL, Weyman AE, Picard MH. Pulsatile pressure affects the disappearance of echocardiographic contrast agents. J Am Soc Echocardiogr 1995;8:285–92.
71. Gibson FW. Measurement of the effect of air bubbles on the speed of sound in water. J Accoust Soc Am 1970;48:1195–7.
72. Ophir J, Parker J. Contrast agents in diagnostic ultrasound. Ultrasound Med Biol 1989;15: 319–33.
73. Morse PM, Ingard KU. Theoretical accoustics. New York: McGraw Hill, 1968:418–27.
74. Nishi RY. The scattering and absorption of sound waves by a gas bubble in a viscous liquid. Acustica 1975;33:65–74.
75. Schrope V, Newhouse VL, Uhlendorf V. Simulated capillary blood flow measurements using a non-linear ultrasonic contrast agent. Ultrasonic Imag 1992;14:134–58.

3. Physics of microbubble scattering

NICO DE JONG

INTRODUCTION

The earliest reference to bubbles as sound sources was made by Bragg [1] who attributed the murmuring of a brook and the plonk of droplets falling into water to entrained air bubbles. Minnaert [2] has since shown that the sound generated by gas bubbles in liquids is associated with simple volume pulsation of the bubble without changing shape. The bubble behaves as a simple damped oscillating system with one degree of freedom. Therefore, the differential equation of motion is of the same form as the classical mass-spring system. He derived for this system the frequency at which resonance occurs, assuming an adiabatic equation of state for the gas in the bubble and neglecting surface tension and damping factors. At that time experiments showed that liquids containing gases possess higher sound damping characteristics than those which are gas free. Sörensen [3] concluded that just a few widely dispersed bubbles, which are so small as to be invisible, can have an appreciable acoustic effect. Fox et al. [4] also carried out attenuation measurements on bubbly liquids and came to a similar conclusion.

The first theoretical description of the behavior of gas bubbles exposed to an acoustic field, together with some aspects of collapsing bubble, was provided in 1950 by Noltingk and Neppiras [5]. During collapse the walls of the bubble rush inwards until the cushioning action of the gas within the bubble stops the radial motion. At the end of the collapse the contents of the bubble are highly compressed and high instantaneous temperatures may occur within the bubble, depending on the degree to which adiabatic conditions still apply. Assuming strictly adiabatic conditions they calculated that, at room temperature and an ambient pressure of 1 atm, the internal temperature of a collapsed bubble would reach the order of 10000°C. In 1951 Poritsky [6] investigated the effect of several types of damping connected with the motion of cavitation bubbles and concluded that the viscous damping component dominates for very small bubbles. In 1959, Devin [7] thoroughly surveyed the fundamental processes by which pulsating

39

N. C. Nanda, R. Schlief and B. B. Goldberg (eds.), Advances in Echo Imaging Using Contrast Enhancement, Second Edition, 39–64.
© 1997 *Kluwer Academic Publishers.*

bubbles dissipate their energy. This included the portion of energy radiated in the form of spherical sound waves, that part which is transformed into heat energy during the polytropic compression and expansion of the enclosed gas, and that part of the energy lost in viscous dissipation.

In 1954 Fox [8] concluded that the cavitation nuclei consist of very small gas bubbles stabilized by an organic skin, which mechanically prevents loss by diffusion. He also calculated the higher resonance frequency of such bubbles. In 1955 Strasberg [9] described gas bubbles in a liquid as sources of sound, referring to the experiments of Bragg. He considered that the gas bubbles initiate volume pulsation when subjected to a force, e.g. gas bubbles entrained in a liquid passing a fixed body or bubbles formed at a nozzle when they separate. In 1964 Flynn developed a thorough description of cavitation phenomena, including aspects of transient and stable cavitation, tensile strength of fluids and the influence of nuclei in the fluid, equation of motion and possible non-linear effects, as well as results from several experimenters. In 1968 Tucker and Welsby [10] suggested that the second harmonic, which is the result of the non-linear behavior of bubbles, is a very sensitive indicator for the presence of bubbles. One year later, Welsby and Sajar [11] developed an analytical method to describe the first as well as the second harmonic of the bubble motion. Lauterborn [12] performed numerical studies on the near resonance behavior of gas bubbles in 1976 for which only the viscous damping was included in the sub- and superharmonics. In 1977 Medwin [13] described the scattering, absorption and extinction cross-section for the linear case including all the damping factors for all bubble sizes. Miller [14] described in 1981 the ultrasonic detection of resonant cavitation bubbles in a flow tube by their second harmonic emission. The experimentally determined value for small bubbles of $4\,\mu m$ diameter agreed well with theoretical values, which were analytical and comparable with the derivation given by Welsby. Eatock et al. [15] studied the magnitude of the non-linear effect in the scattering of ultrasound by nitrogen bubbles in water using a numerical approach, for ultrasonic frequencies and at amplitudes typical for diagnostic medical devices. He was interested in possible bubble detection in blood or tissue for application in decompression research. He included all the damping mechanisms, viscous, thermal and reradiation, as well as the polytropic exponent and concluded that the application of the non-linear effect to the detection of bubbles would be limited to bubbles smaller than $10\,\mu m$. Chapelon et al. [16] described in 1987 a system for detection and sizing of moving bubbles. They noted Doppler shifts on the sidebands generated at the sum and difference of two impinging frequencies. In 1988 Tamura et al. [17] described acoustic measurements on elastic microcapsules. They used the dependency of the resonance frequency on the ambient pressure to calculate this ambient pressure, out of attenuation measurements, using a double frequency Doppler shift.

Cavitation and contrast echography have been shown to be very closely related, and some methods for the production of ultrasound contrast agents are based on this cavitation principle. Theoretical expressions describing cavitation have proved to be very useful for understanding the mechanism of scattering of ultrasound of gas bubbles. Atchley and Crum [18] have given an excellent overview of the historical background of the cavitation research. Ophir and Parker [19] have reviewed the different types of ultrasound contrast agents currently employed. They concluded that the most promising ultrasound contrast agents are free or encapsulated air bubbles.

SCATTERING

Free or encapsulated gas bubbles can be detected by ultrasound because they possess acoustic characteristics which differ from the surrounding medium. This means a difference in speed of propagation, density, or absorption. In general, the physical size of the free or encapsulated gas bubbles in the contrast agents is much smaller than the wavelength of the acoustic field (at 3 MHz the wave-length in water is 0.5 mm). An important property of such particles is the scattering cross-section (Σ_s), which is defined as:

$$\Sigma_s = \frac{\text{scattered power}}{\text{incident intensity}} \qquad (1)$$

The magnitude of the scattered power is linearly related to the intensity of the sound field. For illustration the intensity distribution of a sound field generated by a 2-dimensional phased array system is given in Figure 1. For a steering angle of $0°$ the generated sound beam is well defined without spurious side lobes. For a $40°$ steering angle a grating lobe at $-45°$ appears with a magnitude of -20 to -30 dB below the main lobe. When strong scatterers such as free or encapsulated gas bubbles are located in this grating lobe direction the resulting image can be easily misinterpreted.

When the generated ultrasound wave hits a free or encapsulated air bubble, a volume pulsation is initiated. Depending on the magnitude of the ultrasound wave the pulsation will be: linear to the applied pressure or non-linear to the applied pressure. The latter can be split into stationary and transient. For the linear case the scattering cross-section can be derived according to Medwin [13]. For the non-linear non-transient case the bubble motion has to be determined by modelling. For the non-linear transient case a qualitative description will be given.

Linear scattering

The scattering, absorption and the extinction cross-sections are important acoustic properties of gas encapsulated bubbles in echocardiographic applications. The scattering cross-section of a particle is defined as the quotient of the scattered

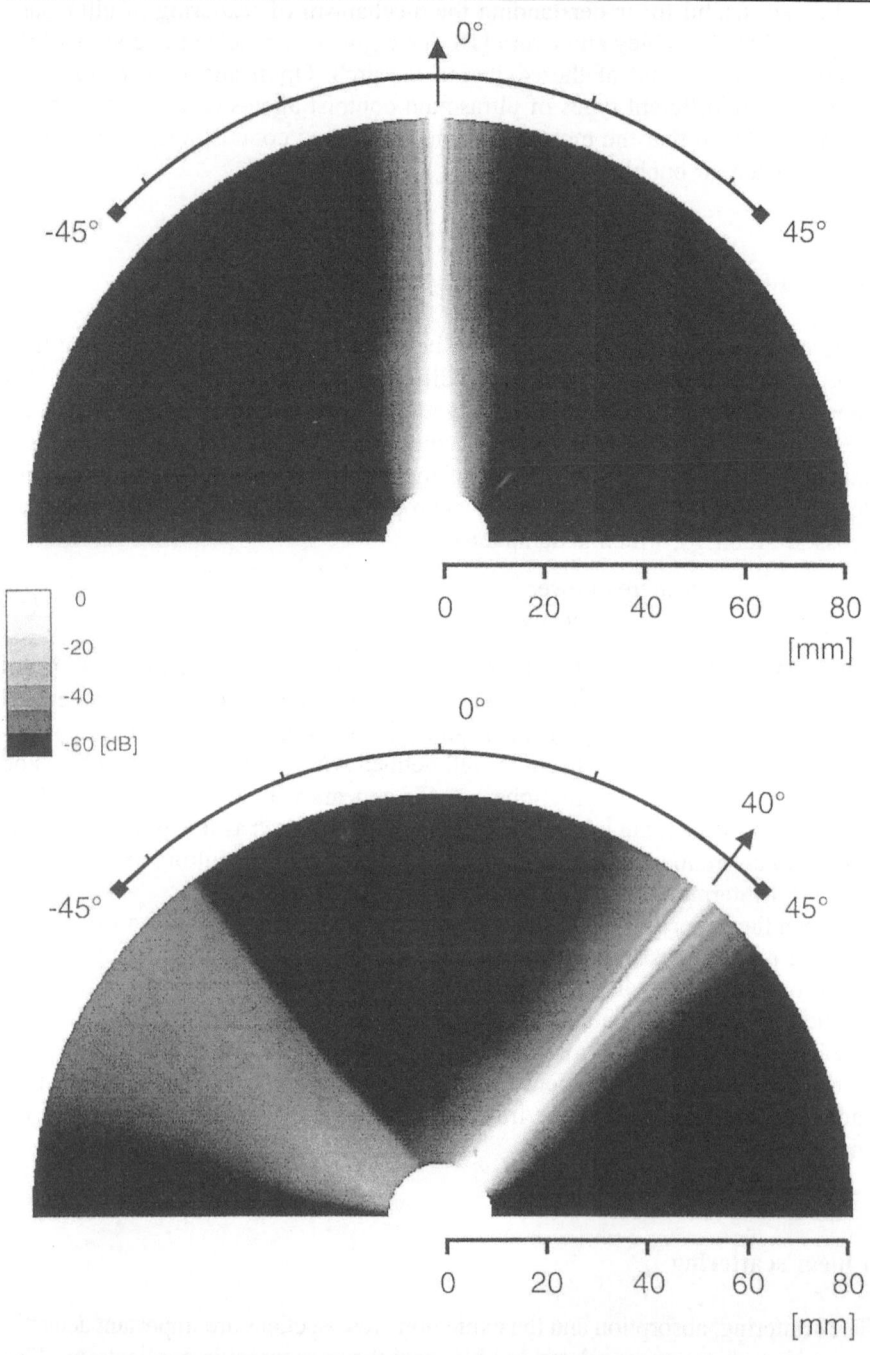

Figure 1. Simulated lateral beam profiles of 5 MHz, 64 element transducer. Pitch 160 μm, focus 70 mm. Top panel: steering angle 0°, bottom panel steering angle 40°.

power and the incident intensity. Medwin [13] has shown that for an air bubble in a liquid this can be expressed as:

$$\Sigma_s = \frac{4\pi R^2}{\left(\frac{f_r^2}{f^2}-1\right)^2 + \delta^2} \tag{2}$$

Σ_s = scattering cross-section
R = radius of the bubble
f_r = resonance frequency
f = frequency of applied ultrasound field
δ = total damping.

The resonance frequency, f_r, and the total damping, δ, in this expression are described in the following section. The resonance frequency of an air bubble in water can be determined assuming the following: (1) the wavelength of the acoustic field is much larger than the bubble diameter, (2) the radial displacement of the bubble is small compared to its radius and (3) the surrounding fluid is incompressible. These conditions are easily fulfilled for bubbles used as contrast agents in echocardiography. The diameter of the bubbles ranges from 1 to 10 μm, while the wavelength of the acoustic field used for medical diagnosis ranges from 0.2 to 0.75 mm (frequency between 7.5 and 2 MHz in water). The fluid (water, blood) is virtually incompressible. Also, the acoustic pressures used in the medical diagnostic field cause a radial displacement, which is small compared to the equilibrium diameter. According to Medwin [13] the angular resonance frequency (ω_{Rg}) for an ideal gas bubble is given by:

$$\omega_{Rg}^2 = \frac{S_a}{m} b\beta \tag{3}$$

S_a = stiffness of the bubble–liquid system
m = effective mass of the system
b = $1/\Gamma$, Γ = polytropic exponent
β = surface tension coefficient.

The equivalent mass m is given by Medwin [13] together with the polytropic exponent b^{-1} and the surface tension coefficient β. In a system consisting of a bubble surrounded by a shell, the shell causes an additional restoring force. Equation (3) thus becomes:

$$\omega_r^2 = \omega_{rg}^2 + \frac{S_{shell}}{m} \tag{4}$$

where ω_r is the angular resonance frequency of the system and S_{shell} the stiffness of the shell. The determination of S_{shell} is provided later in this chapter.

The damping coefficient δ is determined by the following mechanisms: (1) re-radiation, (2) damping due to the viscosity of the surrounding fluid and (3) thermal damping as shown in the following expression:

$$\delta = \delta_{rad} + \delta_{vis} + \delta_{th}$$

$$\delta_{rad} = kR$$

$$\delta_{vis} = 4\frac{\eta}{\rho\omega R^2} \tag{5}$$

$$\delta_{th} = B(\omega,R)\frac{\omega_r^2}{\omega^2}$$

δ_{rad} = damping due to re-radiation
δ_{vis} = damping due to viscosity of the surrounding liquid
δ_{th} = damping due to heat conduction
k = wave number ($2\pi/\lambda$, λ=wavelength)
η = viscosity of the surrounding liquid
ρ = density of the surrounding liquid
R = radius of the bubble
ω_r = resonant angular frequency
ω = applied angular frequency.

The expression for $B(\omega,R)$ is given by Anderson and Hampton [20] and Medwin [13] (called d/b by Medwin).

   The polytropic exponent includes this factor B and is 1 for adiabatic conditions and γ for isothermic conditions (γ=ratio of heat capacity at constant pressure and constant volume, C_p/C_v). The resonance frequency can be determined within 5% of the exact value by taking the polytropic exponent equal to γ for bubbles with a radius smaller than $10\,\mu$m. Taking $b=1$, the calculated resonance frequency remains within 5% of the exact values for bubbles with a radius $>100\,\mu$m. For the frequency range 1–10 MHz, the adiabatic case is valid for bubbles with radii $>50\,\mu$m and the isothermic case for radii $<1\,\mu$m. Thus, the full expression for b given by Medwin [13] has to be used for bubble radii between these two values.

Absorption cross-section
The energy lost in an ultrasonic beam, i.e. that portion which is converted into heat, is referred to as the absorption cross-section. Coackly and Nyborg [21] derived the following expression for Σ_a:

$$\Sigma_a = \Sigma_s\left(\frac{\delta}{\delta_{rad}} - 1\right). \tag{6}$$

Extinction cross-section
The total energy loss for an acoustic beam travelling through a screen of bubbles is given by the sum of the absorption cross-section and the scattering cross-section. This sum is called the extinction cross-section Σ_e:

$$\Sigma_e = \Sigma_s + \Sigma_a. \tag{7}$$

Non-linear (stationary) scattering

The bubble model developed by Rayleigh [22] provides the theoretical basis: the bubble contains gas and vapor, is considered as spherical and is surrounded by an incompressible liquid of infinite extent. The volume is defined by a single variable, the radius, and the motion is assumed to be spherically symmetrical. The wavelength of the ultrasound is much larger than the radius of the bubble and only the behavior of the bubble surface is of interest. The liquid is Newtonian, so its viscosity is constant. It is assumed that the vapor pressure remains constant during the compression/expansion and that there is no rectified diffusion during the short period of exposure to the ultrasound. The gas in the bubble is compressed/expanded according to the gas law with the polytropic exponent Γ remaining constant during the vibration. It is necessary to solve the equations of the conservation of mass and momentum for the gas and the liquid phase to account adequately for the vibration of the bubble. The following expression is given by Eatock [15] for an ideal gas bubble and is called the RPNNP equation, after its developers Rayleigh, Plesset, Noltingk, Neppiras and Poritsky.

$$\rho R\ddot{R} + \frac{3}{2}\rho\dot{R}^2 = p_{go}\left(\frac{R_0}{R}\right)^{3\Gamma} + p_v - p_{lo} - \frac{2\sigma}{R} - \delta_t\omega\rho R\dot{R} - p_{ac}(t) \qquad (8)$$

R = instantaneous bubble radius
$\dot{R}$ = first time derivative of the radius
$\ddot{R}$ = second time derivative of the radius
ρ = density of the surrounding medium
R_0 = initial bubble radius
p_{go} = initial gas pressure inside the bubble
Γ = polytropic exponent of the gas
p_v = vapor pressure
p_{lo} = hydrostatic pressure
σ = surface tension coefficient
δ_t = total damping coefficient
ω = angular frequency of incident acoustic field
$p_{ac}(t)$ = time varying acoustic pressure.

The initial gas pressure inside the bubble, p_{go}, and the total damping coefficient, δ_t, are given by:

$$p_{go} = \frac{2\sigma}{R_0} + p_{lo} - p_v \qquad (9)$$

$$\delta_t = \delta_{rad} + \delta_{vis} + \delta_{th} + \delta_f \qquad (10)$$

$$\delta_f = \frac{S_f}{m\omega} \qquad (11)$$

δ_{rad} = damping coefficient due to re-radiation

δ_{vis} = damping coefficient due to viscosity of the surrounding liquid
δ_{th} = damping coefficient due to heat conduction
δ_f = damping coefficient due to the internal friction inside the shell
S_f = shell friction parameter
m = effective mass of bubble-liquid system.

Expressions for the thermal, re-radiation and viscous damping coefficients together with the effective mass of the bubble-liquid system are the same as for the linear case. S_f represents a shell friction parameter for encapsulated bubbles.

For encapsulated bubbles there is an additional restoring force due to the shell stiffness and additional loss term due to the internal friction inside the shell. The internal friction can be expressed as in equation (11) [23], while the restoring force is equivalent to an additional pressure difference equal to $S_p(1/R - 1/R_0)$ [24]. Equation (8) then becomes:

$$\rho R \ddot{R} + \frac{3}{2}\rho \dot{R}^2 = P_{g0}\left(\frac{R_0}{R}\right)^{3\Gamma} + p_v - p_{lo} - \frac{2\sigma}{R} - S_p\left(\frac{1}{R_0} - \frac{1}{R}\right) - \delta_t \omega \rho R \dot{R} - p_{ac}(t). \quad (12)$$

Albunex is a contrast agent which consists of encapsulated air bubbles with a shell thickness of about 15 nm [25]. The values of the shell elasticity parameter S_p and the shell friction S_f for Albunex are 8 N/m and 4×10^{-6} N.s/m respectively [23]. The damping coefficients in equations (5) and (10) are derived from the linear theory and are therefore assumed to be independent of the applied pressure. For high pressures this may not hold, but the pressure is assumed to be within the limits associated with the linearized theory. In equation (12) all variables, except the radius and its time derivatives, are determined by their initial conditions and are considered as constant during the bubble vibration (the lumped constant approach).

Scattering, absorption and extinction cross-section
The solution of equation (12) results in an instantaneous radius, velocity and acceleration of the bubble wall. The scattered sound pressure at the bubble surface, p_s is defined in equation (13) [21]:

$$|p_s(n\omega)| = |\rho n \omega R_0 \dot{R}(n\omega)| \quad (13)$$

p_s = scattered pressure
ρ = density of the surrounding liquid
ω = angular frequency
n = harmonic number (1,2,3,...1/2, 1/3...)
R_0 = initial radius
$\dot{R}$ = first derivative of the radius.

The scattering cross-section, Σ_s^n, of the nth harmonic of a scatterer, defined as the quotient of the scattered power and the incident acoustic intensity of the first harmonic, is used as the parameter defining the acoustic behavior.

$$\Sigma_s^n = 4\pi R_0^2 \frac{|p_s(n\omega)|^2}{|p_{ac}(\omega)|^2} \qquad (14)$$

$p_{ac}(\omega)$ = applied acoustic pressure.

Based on the scattering cross-section, the absorption and extinction cross-section are determined by equations (6) and (7).

Non-linear (transient) scattering

Non-linear transient scattering has been reported for different ultrasound contrast agents. SHU 563A (Schering, Berlin, Germany) consists of microbubbles coated with a thin biodegradable shell [26]. For low acoustic pressure these particles are durable linear scatters. For high acoustic pressures they will immediately rupture causing a transient non-linear response termed 'acoustically stimulated acoustic emission'. The received scatter amplitude and the corresponding scattering cross-section is much higher than expected with conventional modelling. The spectrum of the scatter signal contains higher frequency components which can be used for harmonic imaging. The described effect is transient. Van der Wouw et al. [27] have reported the extremely sensitive behavior of another ultrasound contrast agent, Quantison (Andaris, Nottingham, UK). During imaging with a commercial 2D echomachine he noticed that below a certain acoustic output power level the scatter magnitude amplitude was too low to be imaged and could not be compensated with increasing the gain setting. Other reports [28] show that intermittent imaging of ultrasound contrast agents after intravenous injection improves the appearance of the agent in the left ventricle cavity and myocardium, suggesting that the scatter effect is transient and is optimal when the ultrasound wave impinges the agent for the first time.

Simulations

Linear acoustic behavior
The presence of a shell around a bubble increases the resonance frequency compared with that for an ideal bubble. This is caused by the stiffness of the shell which decreases the total compressibility. The shell elasticity parameter S_p (which equals $E.t_w/(1-v)$, where E is the Young modulus of the shell, t_w the shell thickness and v the Poisson ratio) is the determinant factor for this change in compressibility. A value of zero for this parameter means that there is no contribution of the shell or that there is no shell present. Measurements, described later in this chapter, on Albunex microspheres show that the value of this shell parameter is about $10\,\mathrm{N\,m^{-1}}$. In the simulation studies it is assumed that the damping parameters are not changed by the addition of a shell. The exact

48   *Nico de Jong*

Table 1. Resonance frequency in MHz for different values of the shell elasticity parameter S_p and bubble diameter.

Bubble diameter (μm)	S_p			
	0	5	10	20
1	9.5	45.4	64.0	90.3
3	2.4	8.9	12.4	17.5
5	1.3	4.2	5.9	8.2
8	0.8	2.1	2.9	4.1
10	0.6	1.5	2.1	2.9
20	0.3	0.6	0.8	1.1

damping coefficient as given in equations (5) and (10) and described by Anderson [2] and Medwin [13] have been used. The surface tension is set to zero except when calculations without the shell are performed. In this case the surface tension value of the water–gas transition is taken. The simulation studies are all carried out for a bubble in water.

The resonance frequencies for several bubble diameters (range 1–20 μm) and shell parameters (range 0–20), are listed in Table 1. These data show that an increase in bubble diameter results in a decrease in the resonance frequency and that the addition of the shell results in an increase in the calculated resonance frequency. Furthermore, it is shown that for diagnostic frequencies (2–7.5 MHz) and for bubbles without a shell the bubble diameters of interest range from ~1 to 5 μm, while this shifts clearly towards higher diameters for high values of the shell parameter, i.e. for a shell parameter value of 10 the bubble diameters of interest range ~4 to 10 μm.

The predicted scattering cross-section as a function of frequency for a single bubble with a diameter of 6 μm is shown for several values of the shell parameter in Figure 2. It is clear from this figure that the greater the shell parameter the higher the resonance frequency. All the curves demonstrate the well known fourth power relationship at low frequencies [29], while at high frequencies the scattering cross-section is equal to the geometrical cross-section. The maximum value of the scattering cross-section decreases for higher values of the shell parameter. The variation of the different damping parameters as a function of the diameter, calculated at the resonance frequency is shown in Figure 3. Thermal damping clearly dominates for bubbles with a diameter >10 μm, while for bubbles <5 μm in diameter damping due to fluid viscosity dominates. If surface tension is neglected, the re-radiation damping at resonance is constant with diameter and its magnitude is small compared with the thermal and viscous damping for bubbles in this size range. The minor increase in small values of diameter is due to the more important influence of the surface tension. The value of all these damping coefficients taken together is about 0.15, and there is little variation in this value for resonating bubbles with diameters between 4 and 50 μm.

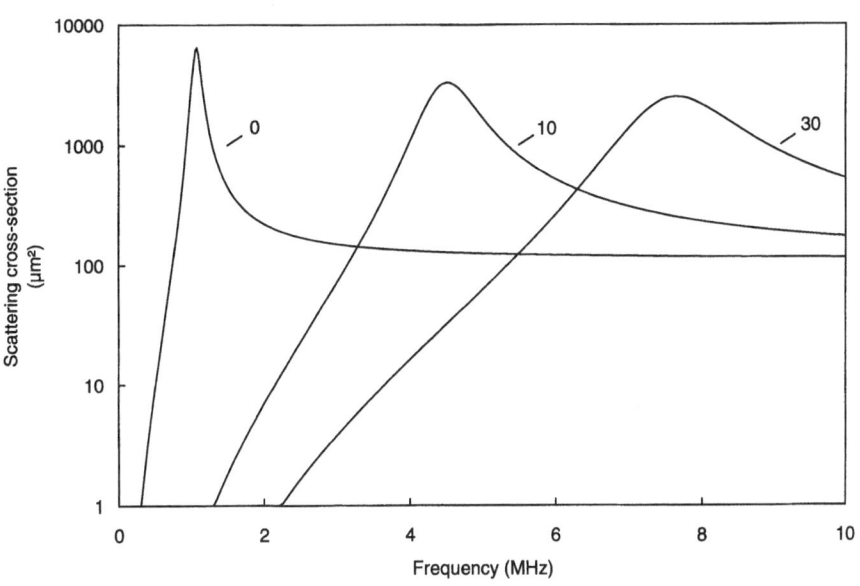

Figure 2. Theoretical scattering cross-section as a function of frequency for three different values of the shell parameter: 0; 10; 30. The simulations are performed assuming a bubble with a diameter of $6\,\mu m$ in water.

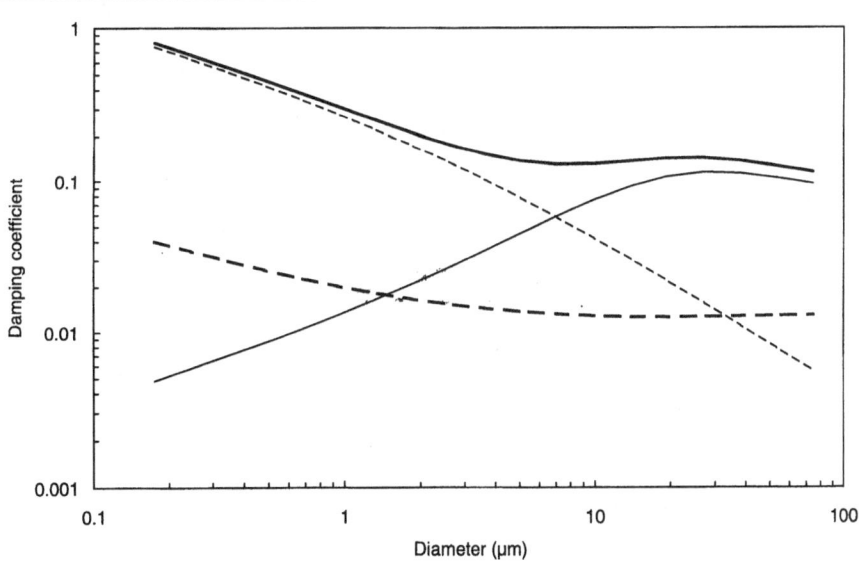

Figure 3. Damping coefficients for air bubbles in water as function of the diameter, calculated at their resonance frequency. –, total damping; –, damping due to thermal conduction; – –, damping due to the viscosity of the surrounding liquid; – –, damping due to re-radiation.

Table 2. Physical constants used in simulations.

Parameter	Symbol	Value
Water		
Density	ρ	$998\,\text{kg/m}^3$
Surface tension	σ	$0.072\,\text{N/m}$
Viscosity	η	$0.001\,\text{N.s/m}^2$
Acoustic velocity	c	$1500\,\text{m/s}$
Vapour pressure	p_v	$2.33 \times 10^3\,\text{N/m}^2$
Air		
Ratio specific heat	γ	1.4
Density	ρ_g	$1.3\,\text{kg/m}^3$
Specific heat at constant pressure	C_{pg}	$1000\,\text{J/kg/K}$
Thermal conductivity	C_g	$0.024\,\text{W/m/K}$
Albunex		
Shell elasticity	S_p	$8\,\text{N/m}$
Shell friction	S_f	$4 \times 10^{-6}\,\text{kg/s}$
General		
Atmospheric pressure	p_{I0}	$1.01 \times 10^5\,\text{N/m}$

Non-linear (stationary) acoustic behaviour

Calculations are performed for an air bubble in water at 20°C. Physical constants (summarized in Table 2) include viscosity, acoustic velocity, specific heat and thermal conductivity which are used to calculate the damping coefficient.

The second-order differential equation (12) can be solved numerically using the fourth-order Runga Kutta method. A sinusoidal wave is taken as driving acoustic pressure which starts at $t=0$ and is cosine windowed for the first five periods to quickly reach a steady state condition. The amplitude of the acoustic wave is 50 kPa unless stated otherwise. The initial conditions at $t=0$ are those of a bubble at rest: $R=R_0$ and $\dot{R}=0$. Values for the instantaneous radius, velocity and acceleration of the bubble are taken after a steady state is reached. The angular resonance frequency ω_0 is determined according to Eatock [15].

Simulations were carried out with driving frequencies, bubble sizes and pressure amplitude which are realistic in the field of diagnostic ultrasound. The following categories are considered:

(a) The scattering cross-section of ideal gas bubbles, with diameters of 2, 3.4 and 6 μm, driven at a frequency of 2 MHz and amplitude of 50 kPa is calculated. The resonance frequency of a 2 μm bubble is 4 MHz; for 3.4 μm, 2 MHz and for 6 μm, 1.1 MHz. The bubbles are, therefore, driven at half resonance, resonance and above their resonance frequencies.

(b) The scattering cross-section of encapsulated bubble, with diameters of 6, 9.6 and 20 μm, driven at a frequency of 2 MHz and an amplitude of 50 kPa is calculated. The resonance frequency of a 6 μm microsphere is 4 MHz, for 9.6 μm, 2 MHz and for 20 μm, 700 kHz. The microspheres are therefore driven at half resonance, resonance and above their resonance frequencies.

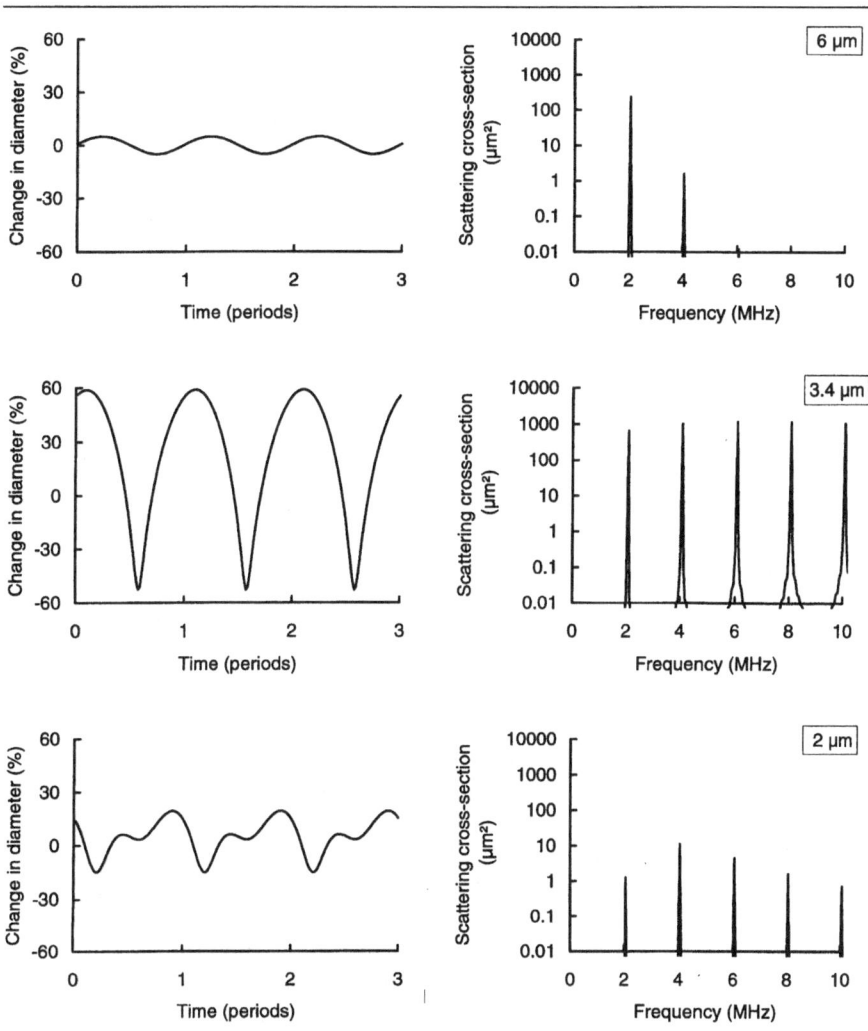

Figure 4. Calculated behavior of an ideal gas bubble in water for an acoustic driving frequency of 2 MHz at a pressure amplitude of 50 kPa. Bubble diameters of 6 μm (above resonance), 3.4 μm (resonance) and 2 μm (half resonance). The left panels show the change in diameter as a function of time, the right panels the corresponding frequency domain components.

Ideal gas bubble: The results of the simulation studies for ideal gas bubbles, with diameters of 2, 3.4 and 6 μm, driven at a fixed frequency of 2 MHz, and an acoustic pressure of 50 kPa are shown in Figure 4. The left panel shows three periods of relative diameter change of the bubble after 15 periods ($t=0$ in the figure), while the right panels present the corresponding Fourier transforms expressed in the scattering cross-section using equations (13) and (14). The scattering cross-section of the bubble with a diameter of $6\,\mu m$ at $2\,MHz$ is $200\,\mu m^2$, which is about twice as high as its physical scattering cross-section, which is equal to $4\pi R^2$, due to the fact that $2\,MHz$ is not far above resonance. The vibration can be considered as only slightly non-linear as shown by the $20\,dB$ drop in the $4\,MHz$ component. The driving pressure and the change in diameter have a phase difference of approximately $180°$, which is typical for a mass-spring system acting above its resonance frequency [8]. A bubble with a diameter of $3.4\,\mu m$ driven at its resonance frequency shows a change in diameter which varies between 60% of its initial diameter in the expanding phase and about 40% in the compression phase. The driving pressure and the diameter are $100°$ out of phase for this amplitude of the driving pressure. For low amplitudes (linear case) the phase difference is $90°$. The resulting scattering cross-section contains higher harmonics, which even exceed that of the first harmonic, which has a value of $670\,\mu m^2$. When such highly non-linear behavior occurs, the level of the first harmonic is no longer independent of the applied pressure, but actually decreases as pressure increases because energy flows into the higher harmonics. A bubble with a diameter of $2\,\mu m$ exposed to an acoustic field of half its resonance frequency generates higher harmonics with a magnitude which is $20\,dB$ higher than the first harmonic.

Encapsulated bubble: Simulation studies for encapsulated bubbles with diameters of 6, 9.6 and $20\,\mu m$, driven at a fixed frequency of $2\,MHz$, and an acoustic pressure of $50\,kPa$ are shown in Figure 5. The encapsulation is equal to that of Albunex microspheres with a shell elasticity parameter of $10\,N/m$ and a shell friction of $4 \times 10^{-6}\,N/s/m$. Compared with an ideal air bubble these results are less spectacular. Since the shell surrounding the Albunex microsphere increases its stiffnesses, the diameter at which resonance occurs is larger than that for an air bubble. For a microsphere with a diameter of $20\,\mu m$, the relative change in diameter is very small: the scattering cross-section is $1200\,\mu m^2$, which is close to its physical scattering cross-section. The second harmonic is $>40\,dB$ below the first harmonic. For the microsphere at resonance ($9.6\,\mu m$ diameter), the diameter and driving pressure are out of phase as shown in the middle panel of Figure 5. The scattering cross-section at $2\,MHz$ is $1200\,\mu m^2$, which is more than four times the physical scattering cross-section. The second harmonic at $4\,MHz$ has a value which is about $20\,dB$ below the first harmonic. The lower panel ($6\,\mu m$ diameter) shows the response for a microsphere driven at half its resonance frequency. The second harmonic is $20\,dB$ below the first harmonic. The absolute value of the scattering cross-section at the first harmonic is $20\,dB$ below the physical scattering cross-section due to the steep frequency dependency of the scattering cross-section below resonance. The absolute value of the second harmonic is of the same order of magnitude as that of the microsphere with a diameter of $20\,\mu m$.

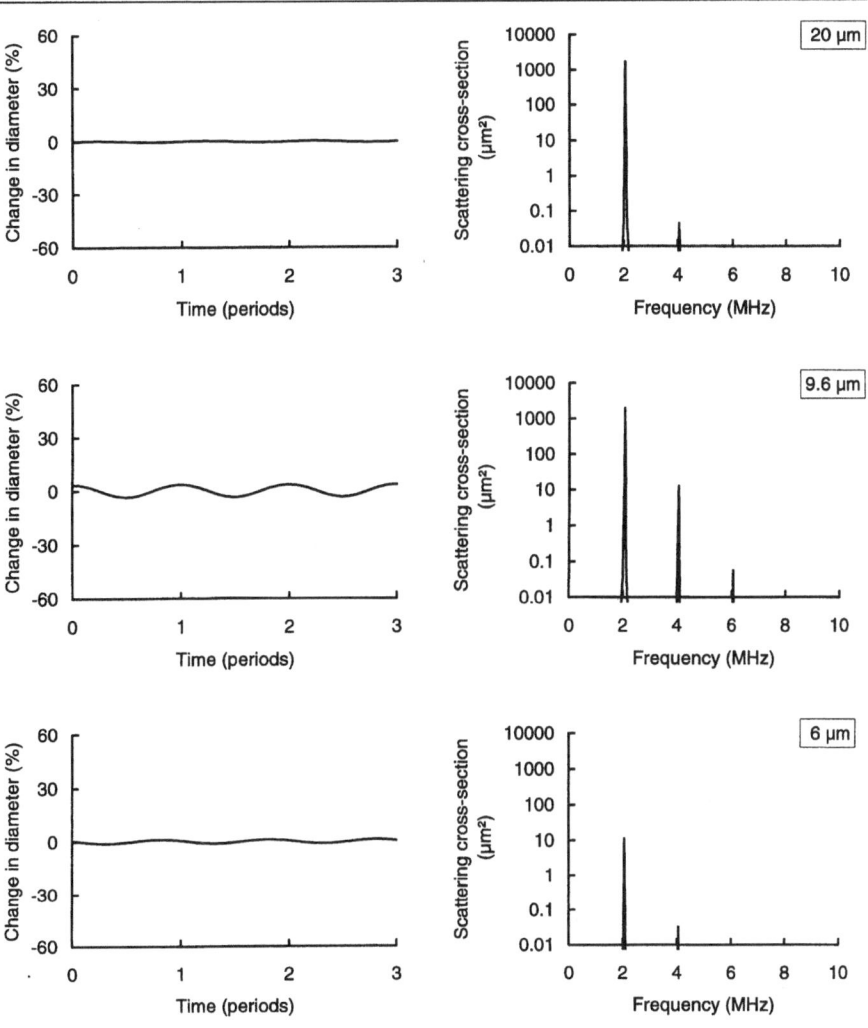

Figure 5. Calculated behavior of an Albunex microsphere in water for an acoustic driving frequency of 2 MHz at a pressure amplitude of 50 kPa. Microsphere diameters of 20 μm (above resonance), 9.6 μm (at resonance) and 6 μm (half resonance). The left panels show the change in diameter as a function of time, the right panels the corresponding frequency components.

ACOUSTIC MEASUREMENTS

When the concentration of the scatterers is low, the scattered power is proportional to the number of scatterers [29], N. The total scattering then becomes, according to equation (1):

$$P_s(R,\omega) = N I_0(\omega)\, \Sigma_s(R,\omega). \tag{15}$$

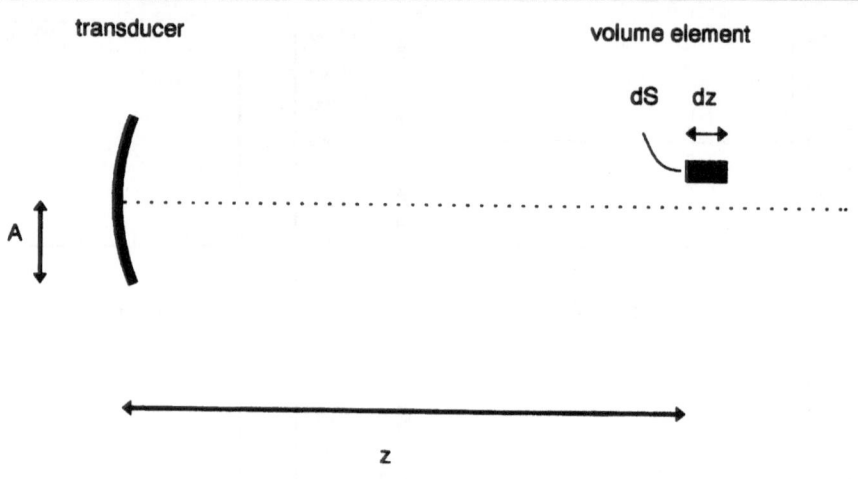

Figure 6. Transducer with small volume element with scatterers. *A* radius of the transducer, *z* focal distance of the transducer, d*z* length of volume element, d*S* area of volume element.

Consider a small volume element containing scatterers homogenously distributed with number concentration $n(R)$ and radii R, as shown in Figure 6. The scattered power from the volume element will be:

$$dP_{sV}(\omega)=I_0(\omega)dV \int_0^\infty n(R)\Sigma_s(R,\omega)\ dR \qquad (16)$$

which can be written as:

$$dP_{sV}(\omega)=I_0(\omega)dS\ dz \int_0^\infty n(R)\Sigma_s(R,\omega)\ dR \qquad (17)$$

$dP_{sV}(\omega)$ = scattered power of the small volume element dV
dS = cross-sectional area of volume element
dz = length of volume element
$I_0(\omega)$ = incident intensity
$n(R)$ = concentration of scatterers with radius R
$\Sigma_s(R,\omega)$ = scattering cross-section of a scatterer with radius R
R = radius of the scatterer
ω = angular frequency.

Integrating over a surface normal to the acoustic beam produces:

$$dP_s(\omega)=dz \iint_S I_0(\omega)dS \int_0^\infty n(R)\Sigma_s(R,\omega)\ dR. \qquad (18)$$

The first integral of this equation denotes the total power, P_T, passing through surface S. The total scattered power then becomes:

$$\frac{dP_s(\omega)}{dz}=P_T \int_0^\infty n(R)\Sigma_s(R,\omega)dR. \qquad (19)$$

For a focused transducer the total transmitted power P_T can be approximated by the reflected power of a perfect flat plate reflector in the focus of the transducer [30].

Scattered power measured with focused transducers

Equation (19) gives the total scattered power over a small distance dz. As the microsphere diameters are much smaller than the wavelength of the acoustic field, their scatter is spherically symmetrical. Having a low concentration, the scattered power from an ensemble of such scatterers, averaged over all configurations of scatterers within the volume element, will also be spherically symmetrical. The power received by the transducer with radius A at a distance z from the scattering volume in the far field is then:

$$P_R = \frac{A^2}{4z^2} P_s \tag{20}$$

P_R = power received by the transducer
P_s = scattered power
A = radius of the transducer
z = distance from scattering volume to transducer.

For curved transducers equation (20) holds for a scatter volume placed in focus, where the far field conditions are valid.

The incident and scattered power of the scatterers in the scatter volume dV are attenuated by the scatterers themselves. The amplitude of the time domain signal can be corrected for this attenuation according to:

$$A_{corr}(t) = A(t) 10^{\frac{\alpha(\omega)}{20} ct} \tag{21}$$

$A_{corr}(t)$ = corrected amplitude of the time domain signal
$A(t)$ = measured amplitude of the time domain signal
c = sound velocity
t = time of flight in suspension with scatterers
$\alpha(\omega)$ = attenuation at angular frequency ω, measured in dB per unit length.

Linear acoustic measurements
Measurements on the linear acoustic behavior have been carried out using four broadband acoustic transducers covering a frequency band from 700 kHz to 12 MHz. The transducers are used both as transmitter and receiver.

Scattered power vs. concentration: The measured scattered power as a function of the concentration of encapsulated bubbles is shown in Figure 7, for the 5 MHz transducer. The concentration of these bubbles (Albunex) is expressed as a volume (μl) of Albunex diluted in 400 ml of Isoton. The scattered power, averaged over the frequency range 4–6 MHz is expressed in dB/cm and measured over a

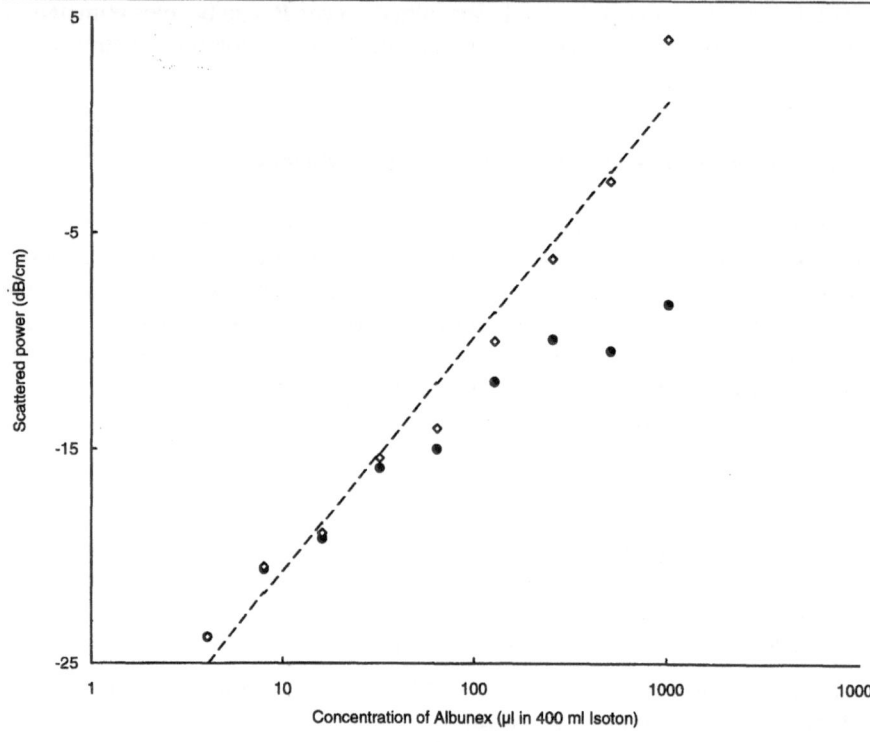

Figure 7. Scattered power as a function of concentration of Albunex microspheres measured at 5 MHz. •, measurement over a window length of 3.75 mm; ◇, corrected for attenuation; --, line fitted to the corrected values.

distance of 3.75 mm in the Albunex suspension. The measured values are shown together with the results corrected for the attenuation. The dotted line denotes a least square fit through these calculated values. The slope of this curve is 3.2±0.2 dB/cm per doubling in concentration. The theoretical value is a 3 dB increase for each doubling of the concentration. It is concluded that attenuation affects the results over the considered window length of 3.75 mm for concentrations of Albunex >50 μl in 400 ml Isoton, but when corrected the results do not differ greatly from the theoretical prediction, even for concentrations as high as 1 ml Albunex in 400 ml Isoton. This confirms the assumption that the scattered power is proportional to the concentration of microspheres.

Scattering and transmission as a function of frequency: Scattering and transmission were measured on Albunex samples. After performing the acoustic measurements, microsphere concentration and size distribution were measured on the same sample. The results were used to calculate the theoretical values for both transmission and the scattering shown in Figures 8 and 9. There is a clear

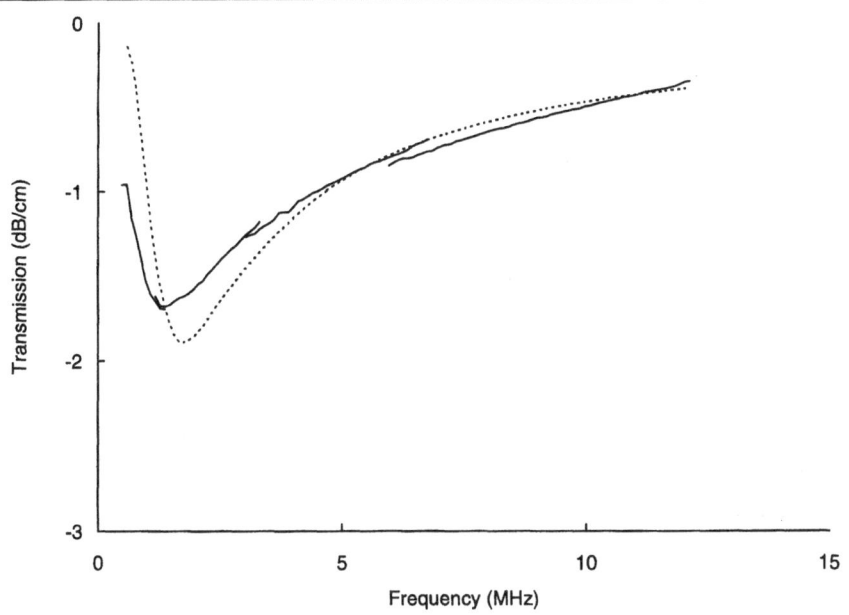

Figure 8. Transmission vs. frequency. —, measured spectrum obtained after dilution of $35\,\mu l$ Albunex in 700 ml Isoton; ---, simulated spectrum obtained with measured size distribution, a shell parameter of 8 N/m and a shell friction of 4×10^{-6} N.s/m.

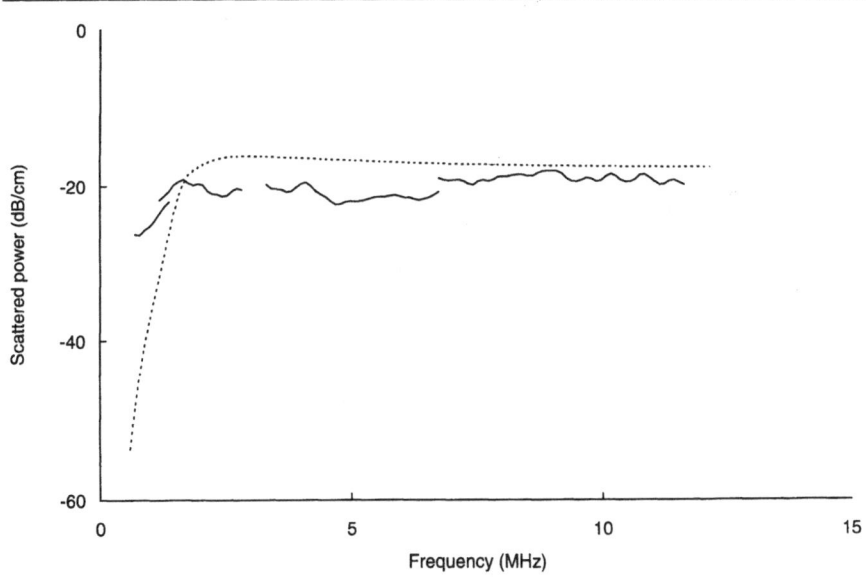

Figure 9. Scattered power vs. frequency, —, measured spectrum obtained after dilution of $35\,\mu l$ Albunex in 700 ml Isoton, ---, simulated spectrum obtained with measured size distribution, a shell parameter of 8 N/m and a shell friction of 4×10^{-6} N.s/m.

minimum of the transmission at 1.5 MHz. The response calculated with the Albunex microsphere model, with a shell elasticity parameter of 8 N/m and shell friction of 4×10^{-6} N.s/m shows the same pattern as the measurements. The total scattered power is virtually independent of the frequency in the range from 1.5 to 12.5 MHz (see Figure 9). For lower frequencies the scattered power increases with frequency. The calculated response according to equations (18) and (19) is shown in the figure. The difference between the calculations and measurements is about 3 dB. The level of the scattered power is about -20 dB/cm, meaning that 1% of the incident energy is scattered away.

Non-linear (stationary) acoustic behavior
An example of an A-mode recording from a medium containing Eccospheres, Albunex and a flat plate reflector is shown in Figure 10. The driving frequency is 2.2 MHz as transmitted by the 3.5 MHz transducer. The time trace as received by the same transducer is shown in the top panel. The flat plate response can be clearly recognized, while the Albunex and the Eccospheres scatter more or less equally for this specially chosen concentration. The received signal, filtered with a Gaussian bandpass filter (center-frequency 2.2 MHz, bandwidth 0.5 MHz), extracts the first harmonic and is shown in the middle panel. The second harmonic, obtained by filtering the original signal with a bandpass filter (center frequency 4.4 MHz, bandwidth 0.5 MHz), is shown in the bottom panel. The scale of the y-axis is different from the other panels in order to magnify the effect. The Albunex response at 4.4 MHz (second harmonic) is much higher than that of the Eccospheres and the flat plate, clearly demonstrating the non-linear behavior of Albunex.

Non-linear (transient) acoustic behavior
According to conventional linear theory the scattering cross-section is a property of the particle and is independent of the applied acoustic pressure. For non-linear stationary scattering the scattering cross-section is approximately independent for relatively low acoustic pressures. For high pressures higher harmonics occur and the effective scattering cross-section for the ground frequency decreases according to equation (14). For non-linear transient scattering the scattering cross-section has been measured for Quantison (Andaris, Nottingham, UK) and is given in Figure 11. Acoustic pressure was varied between 0.2 and 1.8 MPa. For the highest acoustic pressure the measured scattering is 20 dB higher, as expected from the linear theory (this is equal to a factor of 10 in amplitude or a factor of 100 in concentration). Figure 12 shows the result of imaging Quantison with a commercial 2D system. The left panel is for normal acoustic pressures as used in medical diagnosis, the right panel for the highest power output as produced by the Toshiba 160 phased array system. The increase in scattering shown in the right panel cannot be achieved by increasing the gain setting and low power output.

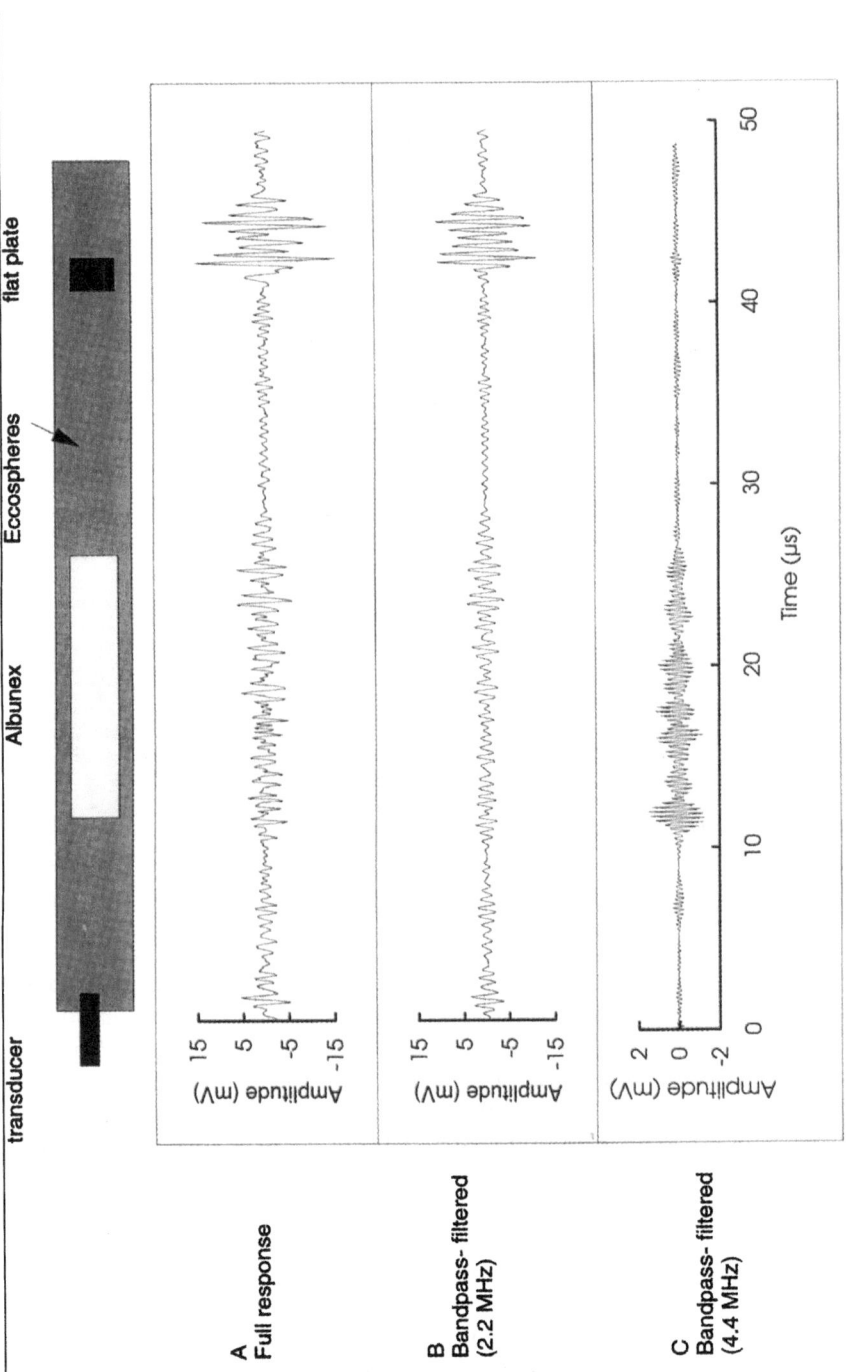

Figure 10. Time signal received by the 3.5 MHz transducer. (A) Time signal as received by the transducer. (B) Time signal after bandpass filtering at 2.2 MHz. (C) Time signal after bandpass filtering at 4.4 MHz.

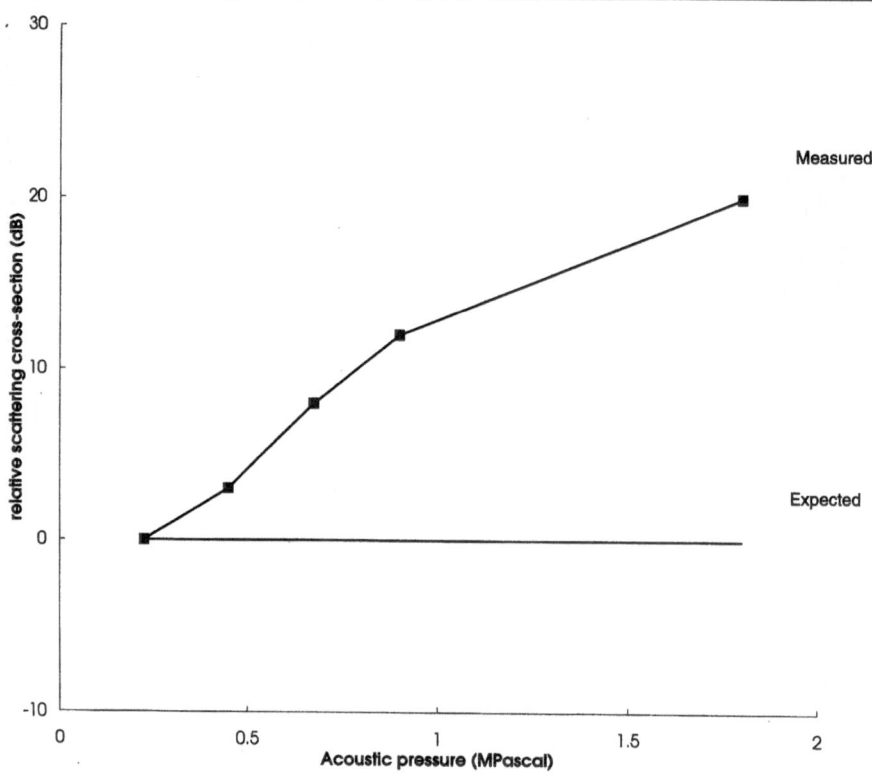

Figure 11. Scattering cross-section as function of the applied acoustic pressure.

CONCLUSIONS

The scattering property of ultrasonic contrast agents remains the most important characteristic and has been employed in combination with conventional two-dimensional imaging to create images of greater clarity. It is anticipated that this property will continue to play an important role in contrast echocardiography for the foreseeable future. Nevertheless, other properties of ultrasonic contrast agents can provide additional information related to the physical properties and characteristics of the fluid carrier. Examples are the change in acoustic velocity for frequencies below the resonance frequency and the resonance phenomenon itself, which depends for example on the ambient pressure. The absorption properties of ultrasonic contrast agents are currently not used. The attenuation properties (= scatter + absorption) are dominant to the scatter properties. For Albunex the quotient of attenuation and scatter magnitude has been measured [23] for different frequencies and different size ranges and the result is given in Figure 13. Microspheres smaller than 5 μm have a low quotient, and their scattered

low pressure
≈ 0.1 MPascal

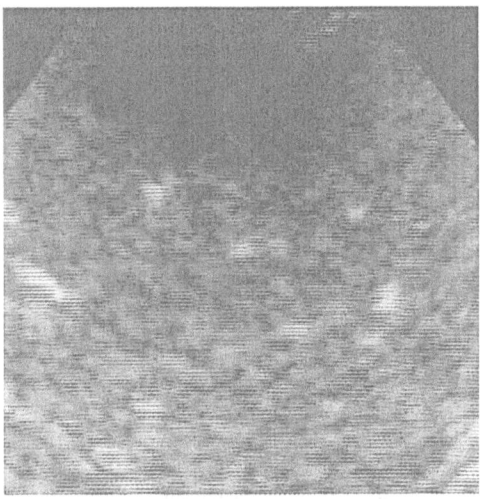

high pressure
≈ 0.6 MPascal

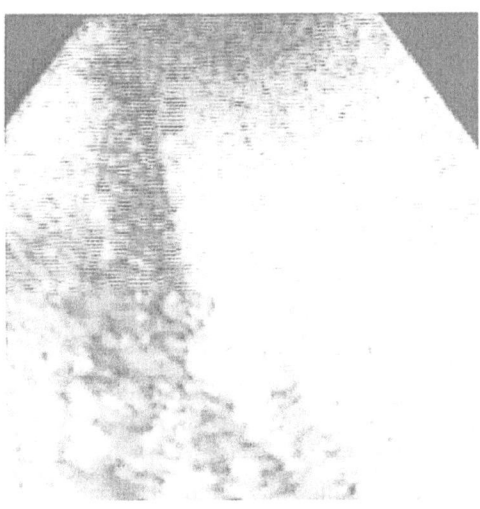

Figure 12. Two-dimensional image of Quantison with a Thosiba 160 phased array system. Top power setting=8, bottom power setting=16.

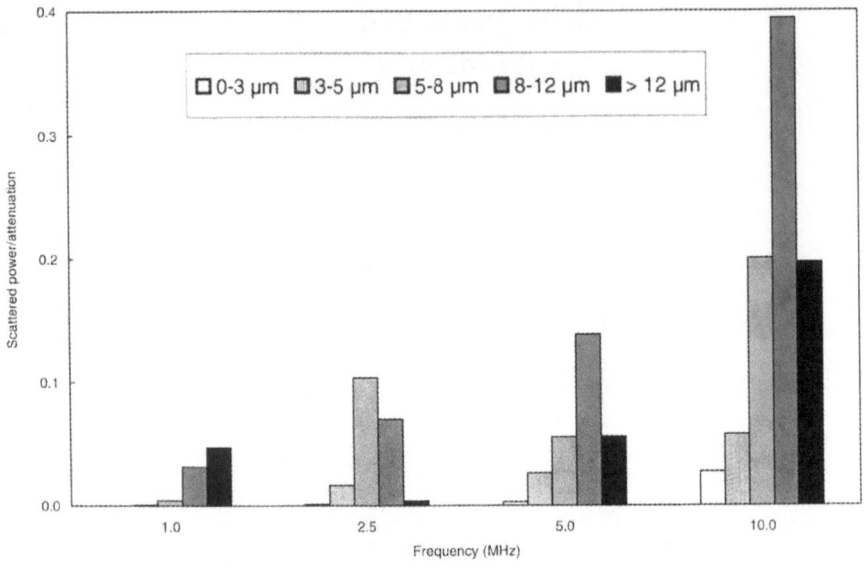

Figure 13. Scattered power of Albunex divided by the attenuation for the frequencies 1, 2.5, 5 and 10 MHz, and five different microsphere size ranges.

power is not high. For both 2.5 and 5 MHz and for size ranges $5-8\,\mu m$ and $8-12\,\mu m$ the quotient is around 0.1, where for the lower frequency the smaller size range dominates and for the higher frequency the larger size range. For 10 MHz the values of the quotient increases to 0.4, but the absorption is still 1.5 times the scattering. For frequencies between 2.5 and 5 MHz, which are often used in medical diagnosis, the absorption is around 10 times the scattering for microspheres with diameters between 5 and $8\,\mu m$.

The small (encapsulated) gas bubbles in a contrast medium react to an external oscillating pressure field with volume pulsations. For an acoustic field of low amplitudes the instantaneous radius of the bubbles is linearly related to the amplitude of the applied external pressure field; as the amplitudes of the external field increase the pulsation of the bubbles becomes non-linear. The spectrum of the scattered ultrasound wave also contains higher harmonics of the ground frequency. For even higher amplitudes of the acoustic field the scatter properties of encapsulated bubbles in a contrast medium increase dramatically and become transient. The scattered frequency spectrum broadens to contain second and third harmonics. The properties of the encapsulated bubbles changes due to rupture, disappearance, change of gas content etc. The whole process can be closely related to microcavitation.

REFERENCES

1. Bragg W. The world of sound. London: G. Bell and Sons Ltd, 1920.
2. Minnaert M. Phil Mag 1933;26:236.
3. Sörensen C. Annu Phys 1936;26:121.
4. Fox FE, Curley SR, Larson GS. Phase velocity and absorption measurements in water containing air bubbles. J Acoust Soc Am 1954;27:534–9.
5. Noltingk BE, Neppiras EA. Cavitation produced by ultrasonics. Proc Phys Soc B 1950;63: 674–85.
6. Poritsky H. The collapse or growth of a spherical bubble or cavity in a viscous fluid. Proc First US Natl Congress Appl Mech E. New York: Sternberg, 1952:813–21.
7. Devin C. Survey of thermal, radiation, and viscous damping of pulsating air bubbles in water. J Acoust Soc Am 1959;31:1654–67.
8. Fox F. Gas bubbles with organic skin as cavitation nuclei. J Acoust Soc Am 1954;26:984–9.
9. Strasberg M. Gas bubbles as sources of sound in liquids. J Acoust Soc Am 1956:281.
10. Tucker DG, Welsby VG. Ultrasonic monitoring of decompression. Lancet 1968;1:1253.
11. Welsby VG, Safar MH. Acoustic non-linearity due to micro-bubbles in water. Acustica 1969; 22:177–82.
12. Lauterborn W. Numerical investigation of nonlinear oscillations of gas bubbles in liquids. J Acoust Soc Am 1976;59:283–93.
13. Medwin H. Counting bubbles acoustically: a review. Ultrasonics 1977;15:7–13.
14. Miller DL. Ultrasonic detection of resonant cavitation bubbles in a flow tube by their second-harmonic emissions. Ultrasonics 1981:217–24.
15. Eatock BC et al. Numerical studies of the spectrum of low-intensity ultrasound scattered by bubbles. J Acoust Soc Am 1985;77:1692–701.
16. Chapelon JY, Newhouse VL, Cathignol D, Shankar PM. Bubble detection and sizing with a double frequency Doppler system. Ultrasonics 1988;26:148–54.
17. Tamura T, Chihara K, Shirae K, Ishihara K, Nagakura T, Tanouchi J, Kitabatake A. Dynamic pressure measurements using two-frequency ultrasound. Jap J Appl Phys 1989;28(Suppl 1): 211–3.
18. Atchley A, Crum A. Acoustic cavitation and bubble dynamics. In: Ultrasound. Its Chemical, Physical and Biological Effects. Suslick S, ed. Weinheim: Verlaggesellschaft, 1988.
19. Ophir J, Parker KJ. Contrast agents in diagnostic ultrasound. Ultrasound Med Biol 1989;15: 319–33.
20. Anderson AL, Hampton LD. Acoustics of gas-bearing sediments: 1. Background. J Acoust Soc Am 1980;67:1865–89.
21. Coackly WT, Nyborg LN. Cavitation: dynamics of gas bubbles: applications. In: Ultrasound: its applications in medicine and biology: part one. Frey FJ, ed. Amsterdam: Elsevier, 1978: 77–159.
22. Rayleigh Lord. Phil Mag 1917;34:94.
23. De Jong N, Hoff L. Ultrasound scatter properties of Albunex microspheres. Ultrasonics 1993; 31:175–81.
24. De Jong N, Hoff L, Skotland T, Born N. Absorption and scatter of encapsulated gas filled microspheres: theoretical considerations and some measurements. Ultrasonics 1992;30:95–103.
25. Christiansen C, Kryvi H, Sontum PC, Skotland T. Physical and biochemical characterization of Albunex, a new ultrasound contrast agent consisting of air-filled albumin microspheres suspended in a solution of human albumin. Biotechnol Appl Biochem 1994;19:307–20.
26. Uhlendorf V, Hoffman C. Nonlinear acoustical response of coated microbubbles in diagnostic ultrasound. Proceedings IEEE Ultrasonics Symposium 1994, Cannes, France.
27. Van der Wouw PA, Van Royen N, de Jong N, TenCate FJ. Optimal ultrasound machine settings for visualising Quantison[R], a new ultrasound contrast agent, in humans. To be published.
28. Porter TR, Xie F. Transient myocardial contrast after initial exposure to diagnostic ultrasound

pressures with minute doses of intravenously injected microbubbles. Demonstration and potential mechanisms. Circulation 1995;92:2391–5.

29. Morse PM, Ingard KU. Theoretical Acoustics. New York: McGraw Hill 1968:418–41.
30. Lizzi L, Greenebaum M, Feleppa EJ, Elbaum M. Theoretical framework for spectrum analysis in ultrasonic tissue characterization. J Acoustic Soc Am 1983;73:1365–73.

4. Spontaneous echocontrast: etiology, technology dependence and clinical implications

ARIC A. AIAZIAN, MEINE A. TAAMS,
FOLKERT J. TEN CATE & JOS R.T.C. ROELANDT

INTRODUCTION

Spontaneous echocardiographic contrast (SEC) is a phenomenon of discrete reflections appearing in the blood inside the cardiac chambers, cavities or vessels without previous injection of echocontrast media or fluids containing microbubbles. SEC can be divided into two categories, based on its appearance. Smoke-like SEC is described as an amorphous, swirling, light gray haze [1]. Its configuration and acoustic density change when observed over several cardiac cycles (Figure 1). Smoke-like SEC can be observed in the left and right heart chambers, great vessels and veins, and is believed to be caused by blood stasis [2–4]. It is most commonly observed in patients with dilated left atrium (Figure 2), mitral stenosis [5, 6], left ventricular dysfunction [7] and aortic aneurysm or dissection [8]. Smoke-like SEC has been associated with stroke and thrombo-embolic events [9–14]. Non-smoke SEC appears either as a 'snowstorm' or as discrete scattered reflections in normal physiological conditions (Figure 3). Such SEC in the left atrium can be enhanced by respiratory maneuvers [15]. Its intensity is mild to moderate and is explained by transient stasis, particularly in the pulmonary circulation [8]. Non-smoke SEC has been also noticed in the inferior vena cava [16].

Because higher ultrasound frequencies (>5 MHz) and unrestricted access to the heart are important for diagnosis of SEC, the introduction of transeso-phageal echocardiography (TEE) provided a powerful tool for both the detection and follow up of SEC [5, 6]. The technique also increased the proportion of patients in whom SEC was observed [17, 18]. Although a number of investigators currently state that SEC mainly is due to blood stasis associated with thrombo-embolic events, the etiology of SEC and therapeutic approaches are still under investigation.

N. C. Nanda, R. Schlief and B. B. Goldberg (eds.), Advances in Echo Imaging Using Contrast Enhancement, Second Edition, 65–83.
© 1997 *Kluwer Academic Publishers.*

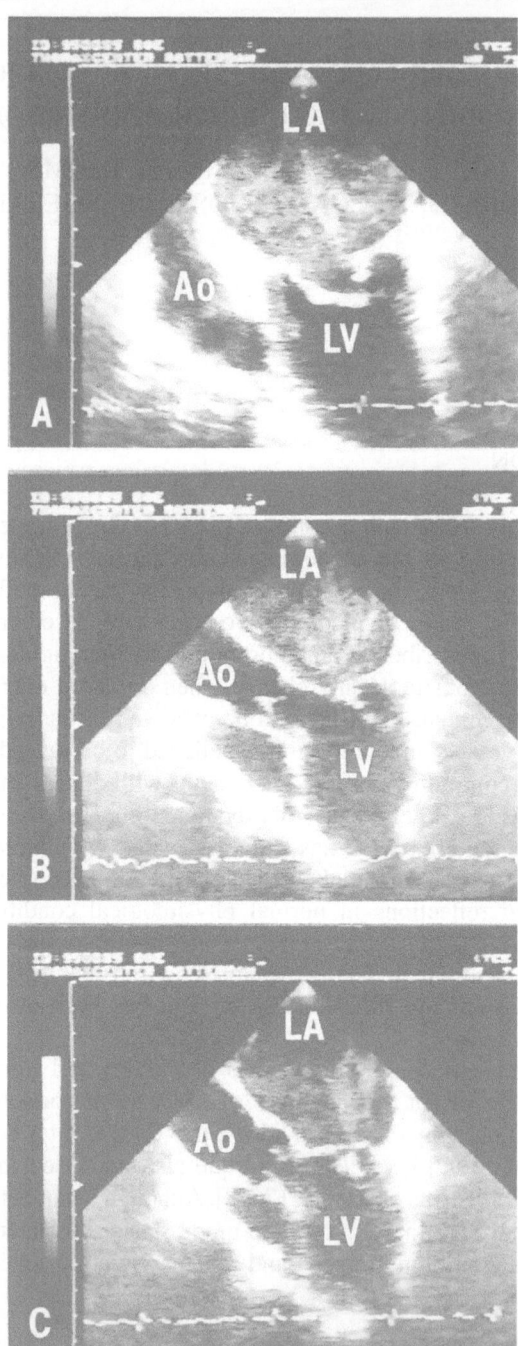

Figure 1. In this patient with mitral stenosis SEC has a typical smoke-like appearance with swirling gray haze changing its configuration over several cardiac cycles (A → B → C). LA, left atrium; LV, left ventricle; Ao, aorta.

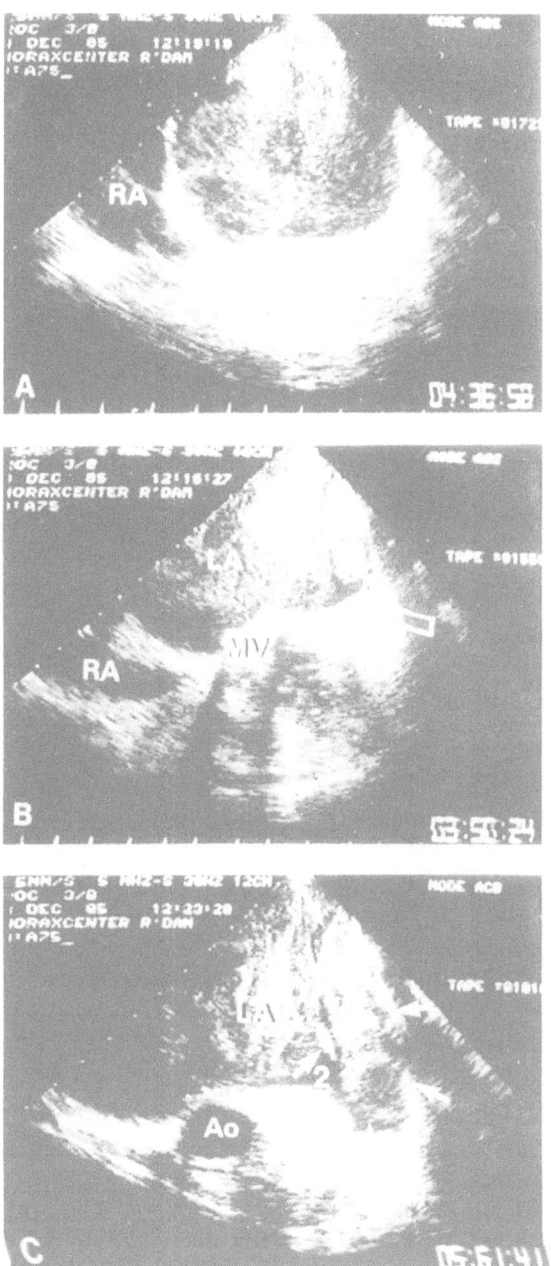

Figure 2. Transesophageal echocardiograms in a patient with dilated left atrium (LA) and obstructed mechanical prosthesis of the mitral valve (MV). SEC is swirling in the LA reflecting the flow from the pulmonary veins (A). A slight tilt of the transducer revealed a thrombus attached to the left atrial wall (B; open arrow) which emerged into the left atrial appendage (C; arrows). Ao, aorta; RA, right atrium.

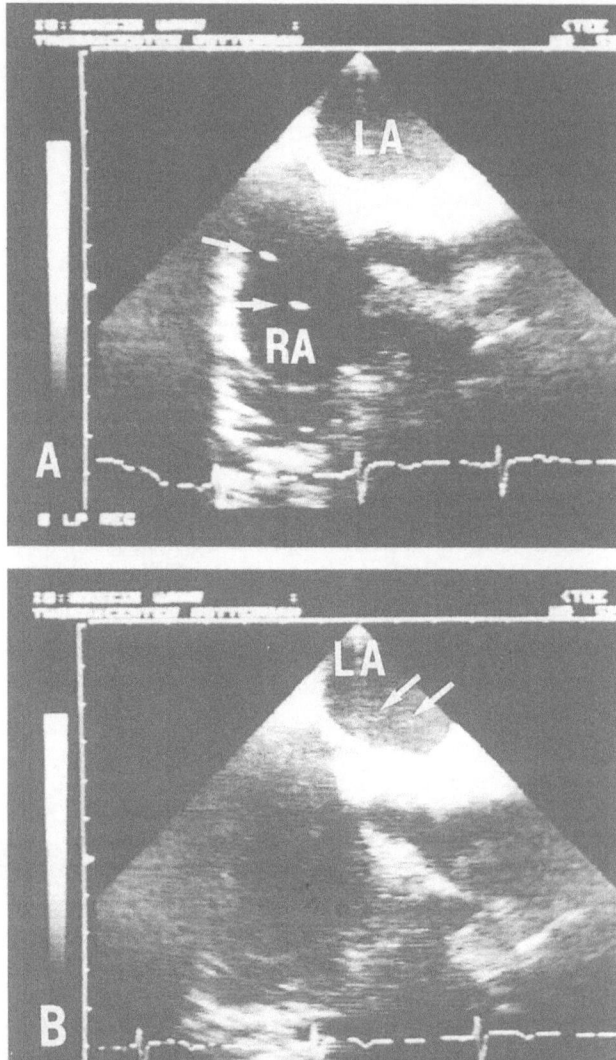

Figure 3. (A) Discrete reflecting particles (arrows) in the right atrium (RA) due to persisting microbubbles following remote intravenous injection. LA, left atrium. (B) Non-smoke SEC (arrows) in the left atrium. These faint discrete reflectances were observed during few cardiac cycles following Valsalva maneuver and were explained by release of blood aggregates formed in the pulmonary circulation during transient stasis.

ETIOLOGY

Different theories have been put forward to explain the phenomenon of SEC. SEC was first observed by Gramiak and Shah in 1971, and explained by the appearance of gas bubbles resulting from cavitations [19]. In 1976, using M-mode echocardiography, Feigenbaum described SEC in the left ventricle next to dyskinetic myocardial segments and associated it with sluggish blood flow [20]. Subsequently, several in vitro and in vivo studies have been carried out to investigate the etiology of SEC.

Several authors reported an association between smoke-like SEC and enlarged cardiac cavities (left atrium [11], right atrium [21], left ventricle [22]), vessels (aorta [8], inferior cava [16]) as well as decreased myocardial function (including atrial fibrillation [17, 23]) leading to an abnormal and decreased blood flow. Similar effects were observed in patients with mitral stenosis [24], prosthetic valves [17] and severe aortic regurgitation resulting in the obstruction to mitral inflow [25]. Mitral regurgitation was not associated with the appearance of SEC in the left atrium in patients with mitral stenosis due to agitation of blood in the left atrium [26]. Hjemdahl-Monsen et al. [27] described SEC in the inferior vena cava in a patient with constrictive pericarditis; the SEC effect disappeared after pericardiectomy. The common feature in all these conditions appears to be transient stagnation of blood. SEC occurs at the site of stagnation in the absence of an external source of echogenic substances. After the SEC is generated it flows with the blood stream and persists during the passage of flow into the next cardiac chamber or vessel until it dissipates.

Non-smoke SEC with discrete reflecting particles can appear in the blood flowing from the pulmonary veins into the left atrium; such SEC can be induced by respiratory maneuvers, particularly the Valsalva maneuvere [15]. The reflectance of this respiratory contrast is lower than that produced by the injection of contrast material containing microbubbles of air. Van Camp et al. explained this phenomenon by the transient stasis in pulmonary circulation induced by respiratory maneuvers. Intravascular ultrasound studies have shown that the Valsalva maneuver immediately induces contrast in the internal jugular vein [28]. Blood aggregates ('rouleaux') caused by transient stasis in the pulmonary veins are suddenly released into the left atrium creating SEC [15].

Other sources of SEC were assumed for brighter non-smoke contrast in the inferior vena cava and right heart [29, 30]. Meltzer et al. [30] assumed that these bubbles were produced by gas absorbed from the intestine and transported via porta-systemic shunting. According to other investigators similar SEC appeared due to microbubbles associated with a recent or remote intravenous injection which persist in the circulation as a result of delayed clearance [29].

In vitro and in vivo studies have been conducted to unravel the mechanism of SEC development. Several factors can affect blood echogenicity, including hematocrit, sedimentation rate and presence of blood cell aggregates. Smoke-like SEC is caused when the blood transiently becomes echogenic at a specific location where there is a decreased flow.

Erythrocytes and platelets were considered to form aggregates which are visualized as SEC. SEC is observed at the initial stage of thrombus formation and platelet aggregates were thought to be the main source of SEC [31, 32]. This theory was supported by the resolution of SEC in patients receiving anticoagulant therapy [7, 33]. However, the first study reported only one case; the second study evaluated SEC in the brachial vein in normal volunteers, making it difficult to extrapolate the findings to SEC in cardiac chambers and great vessels under different flow conditions. Experimentally induced stasis was shown to cause red blood cells to aggregate and to form rouleaux with a smoke-like SEC appearance; following release of stasis the rouleaux disappear to form erythrocytes again [34].

In an important experimental study, Merino et al [1] studied different blood elements and flow conditions. Smoke-like SEC appeared to be primarily caused by the interaction of red blood cells with plasma proteins at low flow and low shear rate conditions. Platelets became echogenic only in dense clumps and did not present as a smoke-like image; the appearance was more characteristic of a non-smoke SEC, as observed by other authors [33]. The reversibility of SEC was explained by a weak nature of the protein binding. Shear forces rather than blood flow was the factor related to 'smoke' production. SEC was visualized only in whole blood and not in platelets alone, plasma alone, platelets and plasma or in the presence of very high red blood cell concentrations [11].

TECHNOLOGY DEPENDENCE

Using real time imaging, Sigel et al. [35, 36] demonstrated that blood echogenicity was related to the shear rate and the frequency of the transducer used. For example, shear rate was estimated as $2-9/s$ in the left atrium in severe mitral stenosis; this permitted red cell aggregation and rouleaux formation [3]. At shear rates $>40\,s$ SEC disappeared (when the ultrasound system is used with $2-5\,MHz$ transducers).

Interaction of ultrasound with the particles is determined by Rayleigh scattering. The ability to visualize SEC depends on the ratio between the size of the blood particles or aggregates and the wavelength of the ultrasound: increase in size of the particles or decrease in ultrasound wavelength lead to visualization of SEC. Ultrasound wavelength is inversely related to the transducer's frequency. The practical consequence of a high ultrasound frequency for the quality of the image is not only improved resolution, but also an increase in sensitivity because of enhanced backscattering. A backscatter signal intensity (I) is proportional to the 4th power of the transducer's frequency and to the 6th power of the particle's radius. The higher frequencies used with TEE transducers ($5\,MHz$) compared with the conventional transthoracic (TTE) probe ($2.0-3.75\,MHz$) is the main reason for the greater sensitivity of TEE in detecting SEC. According to de Belder et al. [17], who studied 80 patients with atrial fibrillation using both

TTE and TEE, left atrial SEC was observed in one patient (1%) by TTE and in 26 patients (33%) by TEE. In the study by Castello et al. [18] the prevalence of the atrial SEC in patients undergoing TEE was 19%. A further increase in the transducer frequency (to 7 MHz) showed spontaneously echogenic blood without stasis and aggregation in the left ventricular cavity during epicardial echocardiography, while a 3 MHz probe in this setting did not visualize SEC [37]. In a study of 10 normal volunteers Kearney et al. [5] were able to see non-smoke SEC in the peripheral venous blood using a 10 MHz transducer. Frequency dependent SEC is observed with intravascular ultrasound. Using a 30 MHz transducer normal blood flow inside the vessel shows SEC.

SITES OF SEC APPEARANCE

Left atrium

Cormier et al. [24] observed SEC in the left atrium by TEE in 65% of 82 patients with mitral stenosis referred for percutaneous mitral comissurotomy. After the intervention, following a two-fold increase in mitral valve area, the incidence of SEC was 50% and dropped further to 28% at 6 months. In this study sinus rhythm was the only independent factor which predicted the disappearance of SEC. In the absence of mitral stenosis, stasis of blood in the left atrium is mainly caused by impaired diastolic function of the left ventricle [38].

Atrial fibrillation plays also an important role in the development of left atrial SEC. Of the patients with SEC in the left atrium studied by Black et al. [2], 87% were in atrial fibrillation; in the study by Castello et al. [18] the proportion was 64%. Among the patients with mitral stenosis evaluated by Kronzon et al. [39] 74% were in atrial fibrillation and left atrial SEC was present in 74%. Several studies of the left atrial appendage Doppler flow patterns [10, 40, 41] showed absence of flow waves in combination with dilated left atrial appendage to be associated with the presence of SEC. Pop et al. [38] studied the relationship between left atrial size, mitral and left atrial appendage flows and appearance of SEC in patients with chronic non-valvular atrial fibrillation. According to this study all those with SEC in the left atrium had a corrected base index left atrial diameter of ≥ 24 mm, corrected time velocity integral of mitral flow < 10 cm and mean peak velocity in the left atrial appendage of 0.12 m/s (vs 0.21 m/s in patients without SEC). Another study stated that although forward left atrial appendage velocities were significantly reduced in patients with SEC by univariate analysis, multivariate analysis demonstrated that atrial fibrillation was the most significant predictor of SEC [23].

SEC has been found to be highly associated with previous stroke or peripheral embolism in patients with atrial fibrillation or mitral stenosis [13]. Among patients with atrial fibrillation, the finding of atrial SEC and risk for thrombus

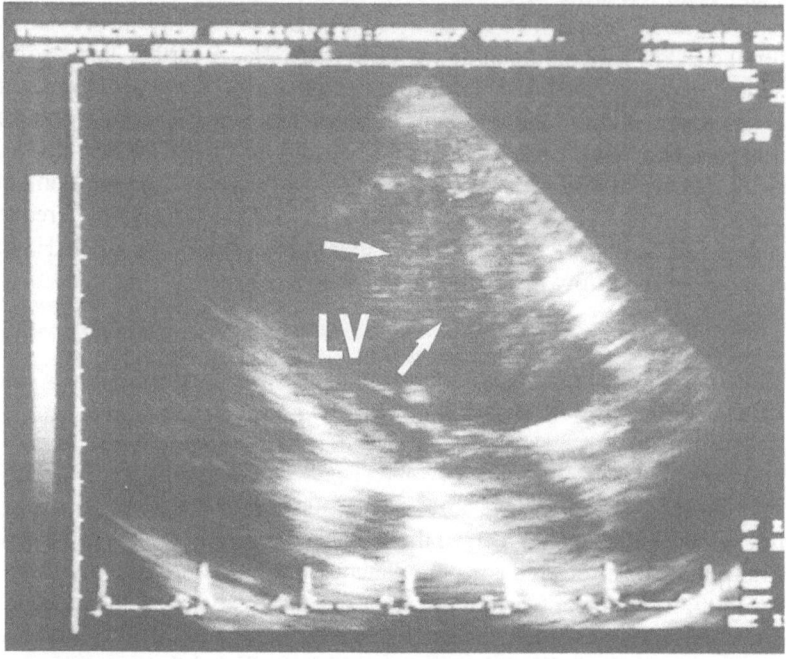

Figure 4. Example of a smoke-like SEC visualized in the left ventricular cavity of a 21-year-old patient with diffuse left ventricular hypokinesis and dilatation due to a long standing aortic insufficiency. This patient had a history of systemic embolism.

formation was more frequent among those with reduced left ventricular function (78% vs 37%) [42].

SEC developing in the left atrium after respiratory maneuvers, as described above [15], is not a predictor of thrombus formation in the left atrium.

Left ventricle

SEC is less frequently observed in the left ventricle than in the left atrium (Figure 4). Feigenbaum, using M-mode echocardiography first described SEC in the left ventricle of the patient with coronary artery disease and regional wall motion abnormalities [20]. Doud et al. [22] studied 16 patients with SEC in the left ventricle visualized by two-dimensional echocardiography: nine had coronary artery disease, six presented with dilated cardiomyopathy and one with hypertensive cardiomyopathy; the mean ejection fraction was 17.6%. SEC disappeared in those who had subsequent improvement of left ventricular function or underwent aneurysmectomy. A rare condition of a cavity filled by SEC is seen in a left ventricular pseudoaneurysm (Figure 5).

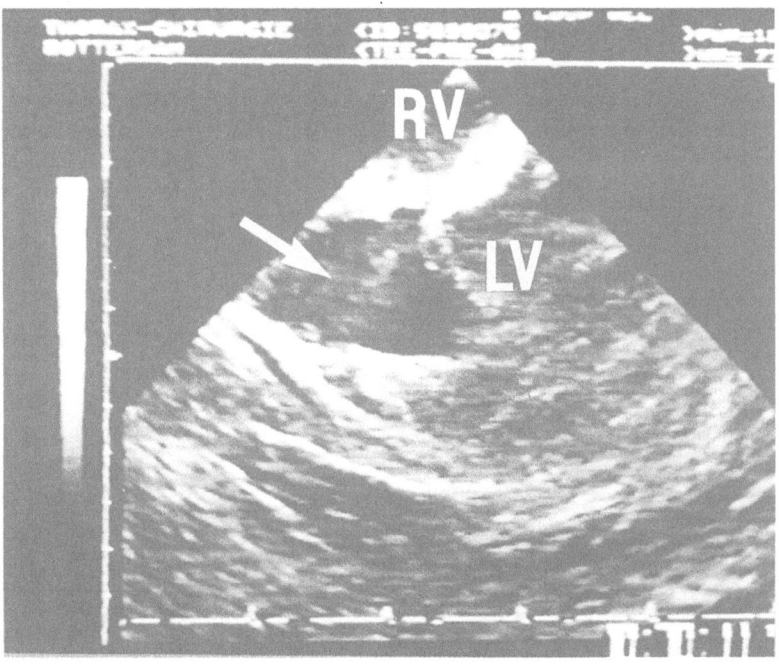

Figure 5. Pseudoaneurysm of the left ventricle (LV) in a patient with recent inferior wall myocardial infarction. SEC is indicated by arrow. RV, right ventricle.

In an experimental setting, Mikell et al. [44] were able to produce SEC in the left ventricle of a dog by ligating the anterior descending artery, which caused dyskinetic motion of the myocardial region supplied by this artery. The SEC disappeared when left ventricular contractility was increased by infusion of dopamine. Asystole was then induced by intravenous injection of potassium chloride and the blood inside the left ventricle became highly echogenic and showed a 'sludge-like' motion during external compression of the heart.

Thoracic aorta

The introduction of TEE has allowed the evaluation of SEC in thoracic aorta. SEC is observed in both dilatation and dissection of the thoracic aorta. Dilated aorta in combination with a low cardiac output represents the condition of a vessel with slow and swirling blood flow. Aortic aneurysm is another condition predisposing to the appearance of SEC and thrombus formation (Figure 6).

In patients with dissection of thoracic aorta TEE usually shows SEC in the false lumen. This is related to slow or reversed blood flow, often in combination with thrombus formation [44–46]. In patients with noncommunicating retro-

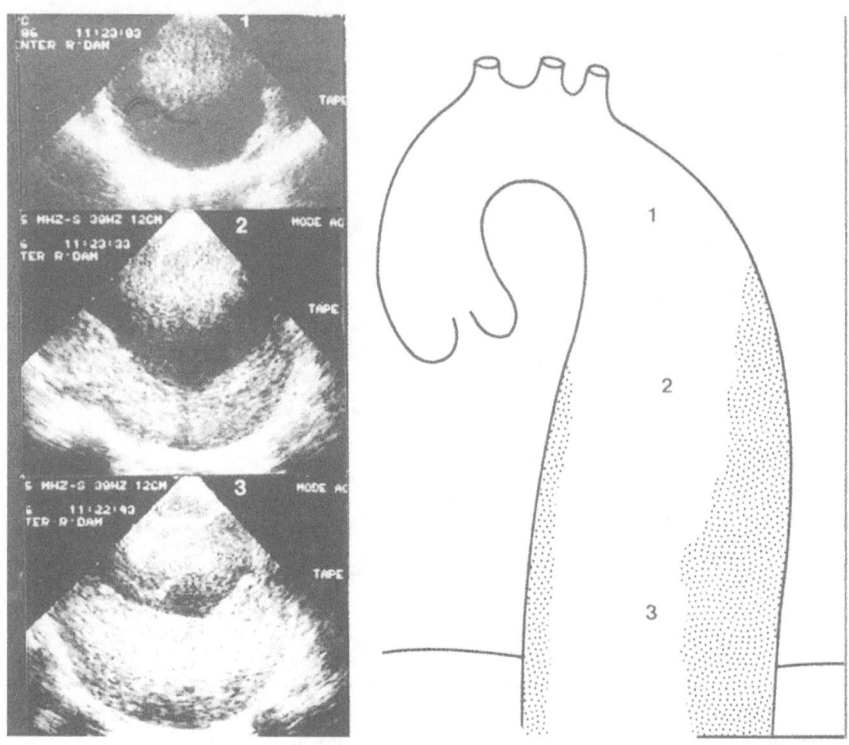

Figure 6. Diagram and corresponding cross-sections (1–3) of an aneurysmatic descending aorta in a 70-year-old patient. Gradual increase in size of the aneurysm distally is associated with increase in the intensity of SEC and thickness of the thrombus due to decrease in the blood flow velocity.

grade dissections or dissections with small tears the flow inside the false lumen is absent or slow and SEC is common. A healing thrombosis of the false lumen eventually occurs [46, 47]. The absence of SEC in the false lumen implies high blood flow due to large tears, and thrombus formation in the false lumen seems not to occur in these individuals [36, 48]. According to Taams et al. [8] thrombus formation occurring proximally and SEC more distally in the false lumen of type III aortic dissection indirectly suggests a distal entry tear (Figure 7). In type I dissection the intensity of SEC in the false lumen decreases towards the entry tear (Figures 8, 9). Type I dissection with a thrombus inside the false lumen of the proximal ascending aorta and SEC filling the false lumen of the descending aorta suggests that the entry tear is localized in the 'blind' zone (distal ascending aorta) or in the aortic arch [49].

One should always be aware of the possibility of SEC appearing in the descending aorta as a result of the high frequency of the TEE transducer in patients without structural abnormalities of the aorta [50].

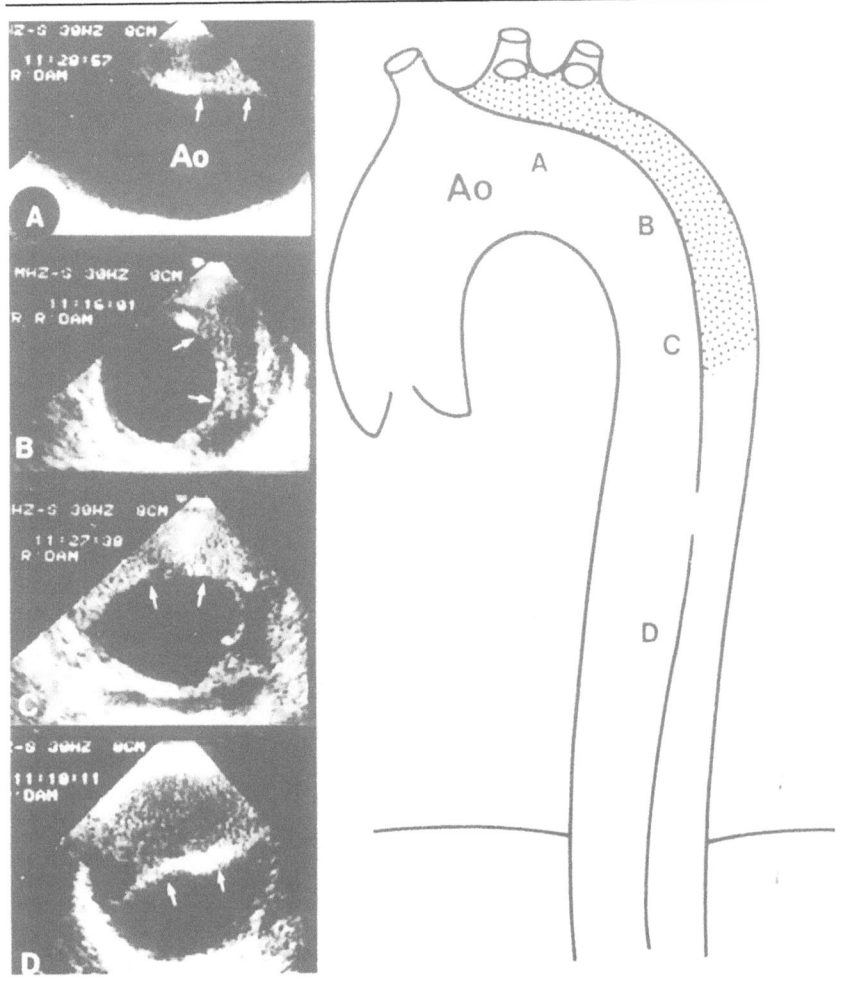

Figure 7. TEE images and corresponding diagram (A–D) of a DeBakey type III aortic dissection with a thrombus formation in the proximal part of the false lumen; SEC appears in the distal part of the false lumen and suggests that the entry tear is in the lower part of the descending aorta. Arrows indicate the intimal flap.

OTHER LOCATIONS OF SEC

Right heart

Smoke-like SEC is relatively rare in the right heart and is usually associated with severe right heart dysfunction. De Georgia et al. [21] found eight patients with atrial SEC among 648 consecutive patients undergoing TEE. Common findings

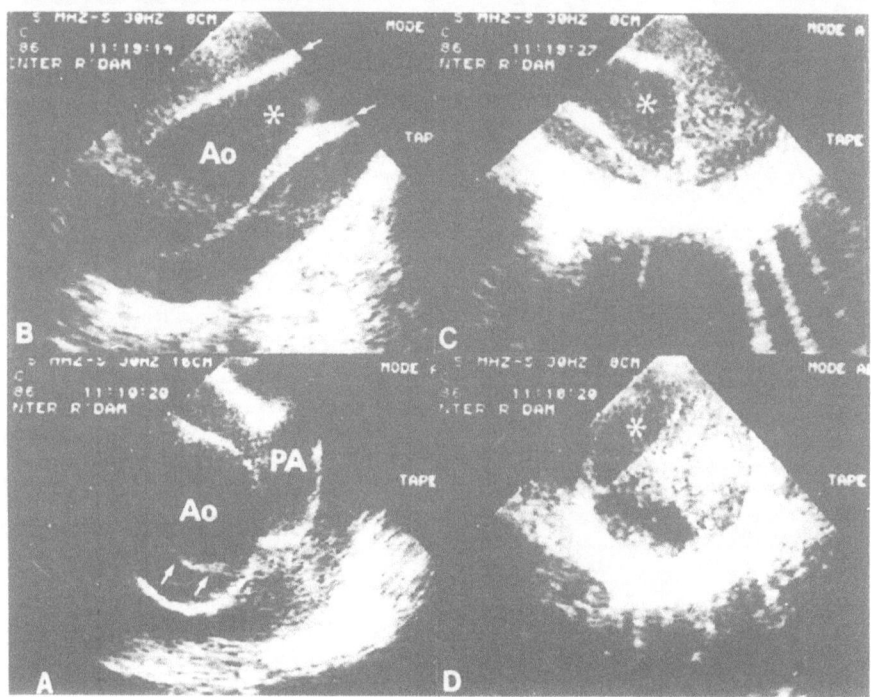

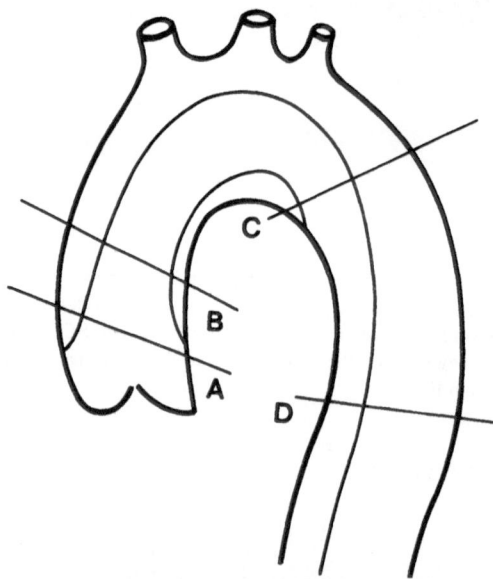

Figure 8. TEE images of the ascending (A–C) and descending (D) aorta (Ao) and corresponding diagram with visualization of an aortic dissection type I. The false lumen is echo-free proximally and filled by SEC in the descending aorta. This suggests an entry tear in the false lumen of the proximal ascending aorta. PA, pulmonary artery; asterisk indicates the true lumen, arrows show the intimal flap.

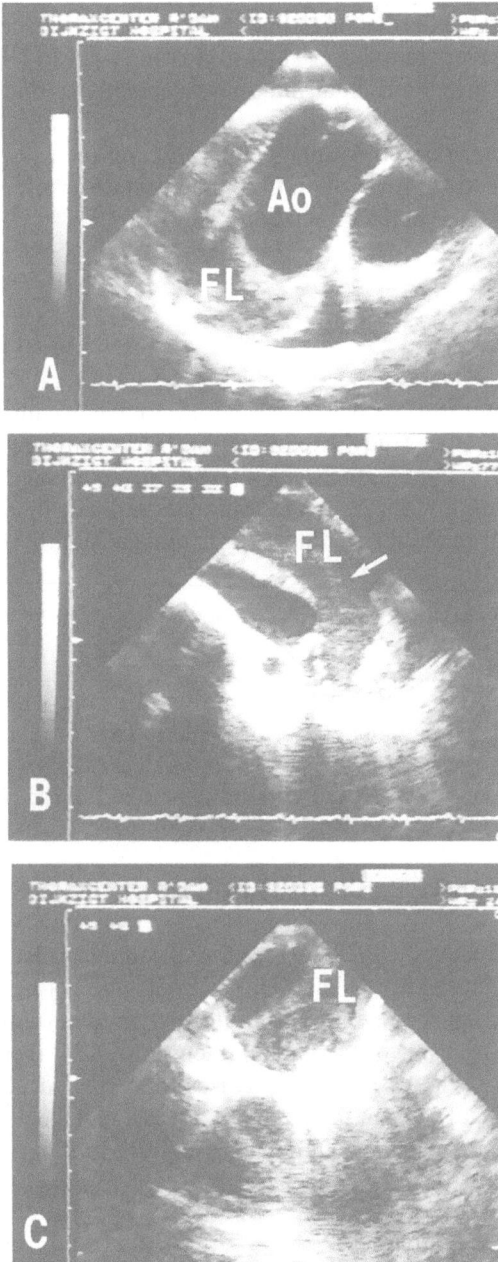

Figure 9. Type I 'spiral' dissection with false lumen (FL) filled by dense SEC proximally (A) and in the descending aorta (C). Smoke-like SEC (arrow) was observed in the proximal descending aorta (B). The entry tear was localized in the arch.

in these eight individuals were right atrial enlargement (8 patients), tricupsid regurgitation (7), atrial fibrillation or flutter (6), elevated right ventricular pressure (5), moderate or severe mitral valve disease (5) and right-to-left interatrial shunt (3). Persisting SEC in the right atrium may eventually lead to the atrial thrombosis [51].

Other authors have described non-smoke SEC on the right side of the heart assumed to be due to gas absorption from the intestine or intravenously injected solutions which persist in the circulation [29, 30].

Inferior vena cava

SEC in the inferior vena cava has been described by several authors. An intense slow-moving contrast effect was described in a patient with constrictive pericarditis [27]. In another study, Meltzer et al. [16] examined the inferior vena cava by TTE in 106 consecutive patients. Ten of these had definite and seven equivocal SEC in the inferior vena cava. The authors did not find a specific cardiac condition and considered the finding of little clinical significance.

Pericardial effusion

SEC can be found in patients with pericardial effusion, when it is caused by clots or red blood cell aggregates (for example in case of malignant pericardial effusion [52]). A rare case of SEC in pneumopericardium was described due to gas production from infected fluid within the pericardial sac [53].

DETECTION OF SEC BY INTRAVASCULAR ULTRASOUND

Due to the high frequency (30 MHz) used with intravascular ultrasound transducers blood particles of normal size and without aggregation can be visualized (Figure 10). Their appearance provides information about intravascular hemodynamics and the true size of the lumen. Intensity of SEC increases proximally to a tight stenosis of the vessel and decreases following successful angioplasty [54]. Other sources of SEC in the coronary arteries include debris from atherosclerotic plaques produced by rotational angioplasty [55].

CLINICAL IMPLICATIONS

SEC has been recognized as a predictor of thrombus formation and a potential risk factor for embolism [9–14]. These types of SEC have a smoke-like appearance and are usually located within the left atrium and left atrial appendage.

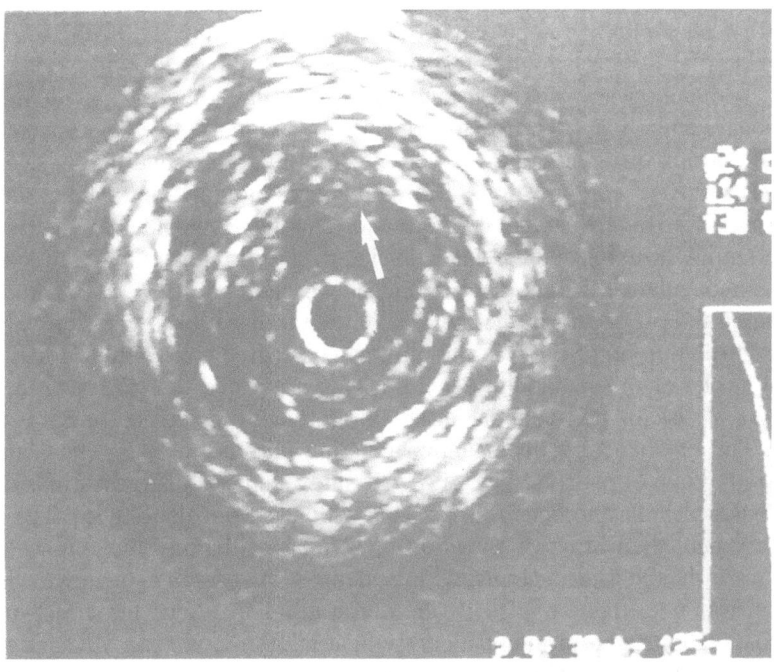

Figure 10. SEC in the coronary artery observed due to a high frequency (30 MHz) of the intravascular ultrasound transducer.

Leung et al. [14] found left atrial SEC in 161 (59%) of 272 consecutive patients with non-valvular atrial fibrillation undergoing TEE. During the follow up period (mean 17.5 months) multivariate analysis showed SEC to be the only positive predictor ($p = 0.03$) of subsequent stroke or other embolic events, with a significantly reduced survival ($p = 0.025$). According to Mitusch et al. [11] dense SEC and left atrial thrombi are associated with thromboembolic complications in patients with non-rheumatic atrial fibrillation. In another study by Chimowitz et al. [13] six of 12 patients with SEC had had a stroke or peripheral embolism, compared with one of 28 patients without SEC ($p < 0.001$; relative risk, 27.0).

Kamp et al. [10] compared groups of patients with and without documented systemic embolism and found that the occurrence of left atrial SEC on TEE was significantly ($p = 0.01$) higher in the group with embolism. They suggested that left atrial SEC could be used to identify patients with atrial fibrillation prone to systemic embolism and to define high- and low-risk groups to be randomized for treatment with warfarin versus aspirin. Transthoracic echocardiography with Doppler assessment of the mitral flow has been proposed as a screening technique for the prediction of the presence of left atrial SEC and the identification of patients with non-valvular atrial fibrillation with increased thromboembolic risk

[38]. In the study by Zotz et al. [9] only left atrial SEC was statistically associated with thromboembolism in patients with left atrial thrombi ($p = 0.038$).

Right atrial SEC was found on several occasions in patients with severe right heart dysfunction and was mentioned as a potential source of paradoxical or pulmonary embolism [21]. Due to the dangerous sequelae of SEC, especially in the left atrium, different attempts were made to treat this condition or decrease the possibility of complications. Both anticoagulation and antiplatelet therapy have been tried to prevent the SEC formation and the development of thrombosis.

Left atrial thrombi can be reduced in size by the administration of antiplatelet and anticoagulative agents (aspirin and phenprocoumon) [9]. In the study by Leung et al. [14], warfarin therapy on follow up of patients with non-valvular atrial fibrillation and SEC was the only negative predictor ($p = 0.02$) of subsequent stroke or other embolic events. It was suggested that TEE can identify patients at risk of non-valvular atrial fibrillation by demonstrating left atrial SEC. The risk/benefit ratio of anticoagulation may be most favorable in this high-risk group of patients. The AFASAK study suggests that warfarin is useful and that aspirin has no benefit in patients with atrial fibrillation [56]. The more recently published Stroke Prevention in Atrial Fibrillation Study showed that both warfarin and aspirin are useful in patients <75 years of age; above this age aspirin was not effective [57]. De Belder et al. [17] studied the effect of oral anticoagulation and aspirin in patients with atrial fibrillation and SEC and found that the presence of SEC was not related to therapy. Indeed, the main cause of SEC is the abnormal flow pattern which could not be changed by either therapy. Instead, therapy is effective by preventing thrombus formation. Although anticoagulation can prevent thrombosis, it has no effect on 'smoke' formation [1]. Seven days aspirin administration was shown to be effective to reduce non-smoke SEC in the peripheral veins of normal volunteers but the clinical implications of this finding require further investigation.

SUMMARY

SEC is a phenomenon of discrete reflections appearing in the blood of cardiac chambers, cavities or vessels without previous injection of the echocontrast media or fluids containing microbubbles of air. The currently available data suggest that smoke-like SEC is due primarily to the interaction of red blood cells and plasma proteins at low flow and low shear rate conditions. This type of SEC has been associated with thromboembolic disease.

Higher frequency ultrasound transducers facilitate the detection of SEC. The introduction of TEE has offered a reliable tool of risk stratification in patients with suspected SEC, especially in the left atrium. In this group of patients with a high risk of thromboembolic complications, anticoagulation therapy is recommended.

REFERENCES

1. Merino A, Hauptman P, Badimon L et al. Echocardiographic 'smoke' is produced by an interaction of erythrocytes and plasma proteins modulated by shear forces. J Am Coll Cardiol 1992; 20:1661–8.
2. Black IW, Hopkins AP, Lee LCL, Walsh WF. Left atrial spontaneous echo contrast: a clinical and echocardiographical analysis. J Am Coll Cardiol 1991;18:398–404.
3. Beppu S, Nimura Y, Sakakibara H, Nagata S, Park Y, Izumi S. Smoke-like echo in the left atrial cavity in mitral valve disease: its features and significance. J Am Coll Cardiol 1985;6:744–9.
4. Iliceto S, Antonelli G, Sorino M, Biasco G, Rizzon P. Dynamic intracavitary left atrial echoes in mitral stenosis. Am J Cardiol 1985;55:603–6.
5. Erbel R, Stern H, Ehrenthal W et al. Detection of spontaneous echocardiographic contrast within the left atrium by transesophageal echocardiography: spontaneous echocardiographic contrast. Clin Cardiol 1986;9:245–52.
6. Daniel WG, Nellessen V, Schroder E et al. Left atrial spontaneous contrast in mitral valve disease: an indicator for an increased thromboembolic risk. J Am Coll Cardiol 1988;11: 1204–11.
7. Mahony C, Sublett KL, Harrison MR. Resolution of spontaneous contrast with platelet disaggregatory therapy (Trifluoperazine). Am J Cardiol 1989;63:1009–10.
8. Taams MA, Gussenhoven EJ, Schippers LA et al. The value of transesophageal echocardiography for the diagnosis of thoracic aorta pathology. Eur Heart J 1988;9:1308–16.
9. Zotz RJ, Pinnau U, Genth S, Erbel R, Meyer J. Left atrial thrombi despite anticoagulant and antiplatelet therapy. Clin Cardiol 1994;17:375–82.
10. Verhorst PM, Kamp O, Visser CA, Verheugt FW. Left atrial appendage flow velocity assessment using transesophageal echocardiography in nonrheumatic atrial fibrillation and systemic embolism. Am J Cardiol 1993;71:192–6.
11. Mitusch R, Lange V, Stierle U, Maurer B, Scheikhzadeh A. Transesophageal echocardiographic determinants of embolism in nonrheumatic atrial fibrillation. Int J Cardiac Imaging 1995;11: 27–34.
12. Gueret P, Vignon P, Fournier P et al. Transesophageal echocardiography for the diagnosis and management of nonobstructive thrombosis of mechanical mitral valve prosthesis. Circulation 1995;91:103–10.
13. Chimowitz MI, De Georgia MA, Poole RM, Hepner A, Armstrong WM. Left atrial spontaneous echo contrast is highly associated with previous stroke in patients with atrial fibrillation or mitral stenosis. Stroke 1994;25:1295–305.
14. Leung DY, Black IW, Cranney GB, Hopkins AP, Walsh WF. Prognostic implications of left atrial spontaneous echo contrast in nonvalvular atrial fibrillation. J Am Coll Cardiol 1994;24: 755–62.
15. Van Camp G, Cosyns B, Vandenbossche JL. Non-smoke spontaneous contrast in left atrium intensified by respiratory manoeuvres: a new transesophageal echocardiographic observation. Br Heart J 1994;72:446–51.
16. Meltzer RS, Klig V, Visser CA, Teichholz LE. Spontaneous echographic contrast in the inferior vena cava. Am Heart J 1985;110:826.
17. de Belder MA, Lovat LB, Tourikis L, Leech G, Gamm A. Left atrial spontaneous contrast echoes – markers of thromboembolic risk in patients with atrial fibrillation. Eur Heart J 1993; 14:326–35.
18. Castello R, Pearson AC, Labovitz AJ. Prevalence and clinical implications of atrial spontaneous contrast in patients undergoing transesophageal echocardiography. Am J Cardiol 1990;65: 1149–53.
19. Gramiak R, Shah PM. Detection of intracardiac blood flow by pulsed echo-ranging ultrasound. Radiology 1972;100:415–18.
20. Feigenbaum H. Echocardiography. 2nd edn. Philadelphia: Lea and Febiger, 1976:350.

21. De Georgia MA, Chimowitz MI, Hepner A, Armstrong WF. Right atrial spontaneous contrast: echocardiographic and clinical features. Int J Cardiac Imaging 1994;10:227–32.
22. Doud DN, Jacobs WR, Moran JF, Scanlon PJ. The natural history of left ventricular spontaneous contrast. J Am Soc Echocardiogr 1990;3:465–70.
23. Jue J, Winslow T, Fazio G, Redberg RF, Foster E, Shiller NB. Pulsed Doppler characterization of left atrial appendage flow. J Am Soc Echocardiogr 1993;6:237–44.
24. Cormier B, Vahanian A, Lung B et al. Influence of percutaneous mitral comissurotomy on left atrial spontaneous contrast of mitral stenosis. Am J Cardiol 1993;71:842–7.
25. Evangelista A, Ganzalez T, Garcia del Castillo H, Candell J, Soler Soler J. Contraste echocardiografico espontaneo en el ventriculo izquierdo en relacion con insuficiencia aortica. Rev Esp Cardiol 1990;43:581–3.
26. Karatasakis GT, Gotsis AC, Cokkinos DV. Influence of mitral regurgitation on left atrial thrombus and spontaneous echocardiographic contrast in patients with rheumatic mitral valve disease. Am J Cardiol 1995;76:279–81.
27. Hjemdahl-Monsen CE, Daneils J, Kaufman D, Stern EH, Teichholz LE, Meltzer RS. Spontaneous contrast in the inferior vena cava in a patient with constrictive pericarditis. J Am Coll Cardiol 1984;4:165–7.
28. Attubato MJ, Katz ES, Feit F, Bernstein N, Schwartzman D, Kronzon I. Venous charges occurring during Valsalva manoeuvre: an intravascular ultrasound study [abstract]. Circulation 1992;86:1–87.
29. Segal R, Baltazar RF, Mower MM, Stewart C. Spontaneous contrast visualization on the right side of the heart during echocardiography. Am J Med Sci 1986;292:363–6.
30. Meltzer RS, Lancee CT, Stewart GR, Roelandt J. Spontaneous echocardiographic contrast on the right side of the heart. J Clin Ultrasound 1982;10:240–2.
31. Mahony C, Ferguson J. The effect of heparin versus citrate on blood echogenicity in vitro: the role of platelet and platelet–neutrophil aggregates. Ultrasound Med Biol 1992;18:851–9.
32. Mahony C, Ferguson J, Fischer PL. Red cell aggregation and the echogenicity of whole blood. Ultrasound Med Biol 1992;18:579–86.
33. Kearney K, Mahony C. Effect of aspirin on spontaneous contrast in the brachial veins of normal subjects. Am J Cardiol 1995;75:924–8.
34. Wang X, Liu L, Cheng TO et al. The relationship between intravascular smoke-like echo and erythrocyte rouleaux formation. Am Heart J 1992;124:961.
35. Sigel B, Coelho JCH, Spidos DG et al. Ultrasonography of blood during stasis and coagulation. Invest Radiol 1981;16:71–6.
36. Sigel B, Machi J, Beitler JC, Justin JR, Coelho JC. Variable ultrasound echogenicity in flowing blood. Science 1982;218:1321–3.
37. Garcia-Fernandez MA, Lopez-Sandoru J, Corna-Canela I et al. Echocardiographic detection of circulating blood in normal canine hearts. Am J Cardiol 1985;56:834–6.
38. Pop GAM, Meeder HJ, Roelandt JRTC et al. Transthoracic echo/Doppler in the identification of patients with chronic non-valvular atrial fibrillation at risk for thromboembolic events. Eur Heart J 1994;15:1545–51.
39. Kronzon I, Tunick PA, Colossman E et al. Transesophageal echocardiography to detect atrial clots in candidates for percutaneous transseptal mitral balloon valvuloplasty. J Am Coll Cardiol 1990;16:1320–2.
40. Garcia-Fernandez MA, Torrecilla EG, San Roman D et al. Left atrial appendage Doppler flow patterns: implications on thrombus formation. Am Heart J 1992;124:955–61.
41. Pollick C, Taylor D. Assessment of left atrial appendage function by transesophageal echocardiography. Implications for the development of thrombus. Circulation 1991;84:223–31.
42. Brown J, Sadler DB. Left atrial thrombi in non-rheumatic atrial fibrillation: assessment of prevalence by transesophageal echocardiography. Int J Cardiac Imaging 1993;9:65–72.
43. D'Cruz IA, Jain M, Jain A. Dynamic intracavitary echoes in left ventricular pseudoaneurysm. Am Heart J 1986;112:418–20.

44. Mikell FL, Asinger RW, Elsperger KJ, Anderson WR, Hodges M. Regional stasis of blood in the dysfunctional left ventricle: echocardiographic detection and differentiation from early thrombosis. Circulation 1982;66:755–63.

45. Erbel R, Borner N, Steller D et al. Detection of aortic dissection by transesophageal echocardiography. Br Heart J 1987;58:45–51.

46. Erbel R, Mohr-Kahaly S, Oelert H et al. Diagnostic strategies in suspected aortic dissection: comparison of computed tomography, aortography, and transesophageal echocardiography. Am J Cardiac Imaging 1990;4:157–72.

47. Mohr-Kahaly S, Erbel R, Rennollet H et al. Ambulatory follow-up of aortic dissection by transesophageal two-dimensional and color-coded Doppler echocardiography. Circulation 1989; 80:24–33.

48. Erbel R, Bednarczyk I, Pop T et al. Detection of dissection of the aortic intima and media after angioplasty of coarctation of the aorta: An angiographic, computer tomographic and echocardiographic comparative study. Circulation 1990;81:805–14.

49. Erbel R, Oelert H, Meyer J et al. New classification of aortic dissection by transesophageal echocardiography. Circulation 1991;84:11–23.

50. Castello R, Pearson AC, Fagan L, Labovitz AJ. Spontaneous echocardiographic contrast in the descending aorta. Am Heart J 1990;120:915–9.

51. Barron JV, Neibla MC, Keirns C et al. Echocardiographic detection of dynamic intracavitary echoes and right atrial thrombus. Am Heart J 1987;113:829–31.

52. D'Cruz IA, Holman MS, Childers LS. Spontaneous mobile contrast echoes in pericardial effusion. Am Heart J 1990;120:1472–5.

53. Hsu TL, Chen CC, Lee GW, Huang ST, Hsiung MC, Chiang BN. Intrapericardial spontaneous contrast echoes in pneumopericardium due to a gas-forming organism. Am J Cardiol 1986;58: 1143–4.

54. Coy KM, Park JC, Scheffer A et al. Spontaneous echo contrast effect as an indicator of flow-limiting coronary stenosis. Am J Cardiol 1991;68:276–7.

55. Zotz R, Stahz P, Erbel R, Auth D, Meyer J. Analysis of high frequency rotational angioplasty-induced echo contrast. Cath Cardiovasc Diagn 1991;22:137–44.

56. Petersen P, Godtfredsen J, Boysen G, Andersen ED, Andersen B. Placebo-controlled, randomised trial of warfarin and aspirin for prevention of thromboembolic complications in chronic atrial fibrillation: the Copenhagen AFASAK study. Lancet 1989;1:175–8.

57. Stroke Prevention in Atrial Fibrillation Study Group Investigators. Preliminary report of the Stroke Prevention in Atrial Fibrillation Study. N Engl J Med 1990;322:863–8.

5. Echo-enhancing agents: their physics and pharmacology

REINHARD SCHLIEF

INTRODUCTION

The contrast media used to create or enhance contrast in radiographic and magnetic resonance images exploit the physical characteristics of specific elements to achieve their objective. Iodine is radio-opaque and the powerfully paramagnetic rare earths such as gadolinium decrease proton relaxation time, shortening T1 and T2 times and brightening the tissues that take them up. Superparamagnetic iron oxide particles alter magnetic susceptibility, reducing T2 and T2* times, lowering signal strength and darkening the tissues. The complex molecules that deliver the elements play no part in the process of image enhancement. Ultrasound echo-enhancers are very different entities whose function is to reflect incident ultrasound energy. This effect depends on the physical structure of the echo enhancer, not the properties of any specific element. Ultrasound echo enhancement is a far grosser process than the sub-molecular interactions of gadolinium or iodine.

Physics dictates that the most efficient reflectors of ultrasound are gas-filled microbubbles and all ultrasound echo-enhancers are variations on a microbubble theme. However, between that simple statement of principle and its realization as an aid to clinical practice lies a host of practical problems. The microbubbles must be the right size, they must stay in the circulation for long enough to be useful, they must not coalesce into bigger, potentially occlusive aggregates and they must be well tolerated. The fundamental similarity of all the echo-enhancers distinguishes them from X-ray and MRI enhancers because the microbubbles are confined to the vascular bed. A bubble formulation that is sufficiently stable will enhance the acoustic signal from the entire blood pool, while X-ray and MRI enhancers leak from the blood vessels into the extravascular space. An X-ray or magnetic resonance contrast enhancer that stays in the blood pool is still a fairly distant prospect.

N. C. Nanda, R. Schlief and B. B. Goldberg (eds.), Advances in Echo Imaging Using Contrast Enhancement, Second Edition, 85–113.

Table 1. Milestones in microbubble evolution.

Right heart enhancers	Not stable enough to cross the pulmonary barrier	1980
Left heart enhancers	Cross the pulmonary barrier intact but do not survive in the arterial tree	1990
Blood pool enhancers	Cross the pulmonary barrier and survive in the entire blood pool intact	1995
More persistent enhancers, organ/tissue specific enhancers and enhancers with specialized acoustic properties	Expand the diagnostic applications of echo-enhancement	1999?

EVOLUTIONARY STAGES

The first problem to be overcome in the evolution of echo-enhancers that produce their effect after an intravenous injection was that of bubble stability. The problem has been approached from several directions but the two that have proved most productive to date use galactose microparticles as nuclei for bubble formation or employ sonication to create gas-filled microspheres with shells of denatured human albumin. The first successful echo-enhancer was albumin-based but, more recently, albumin has been replaced by a variety of biodegradeable polymers. With increasing stability the bubbles were able to cross the pulmonary bed intact and enhance the echo from left heart cavity and the entire blood pool. Future developments will include the production of microbubbles with an affinity for specific organs or tissues and others with a much longer duration of action or a very characteristic acoustic response. Within the next 5 years we will probably see the first attempts to use microbubble formulations therapeutically rather than diagnostically, in the treatment of tumors and the delivery of gene therapy. Microbubble echo-enhancers have evolved steadily over the last 15 years (Table 1).

Echo-enhancers: approaches to stability

The main problem with gas microbubbles is ensuring that they last long enough to be useful. Ideally they should be stable enough to pass through the lungs into the left chambers of the heart to increase the echogenicity of the entire vascular bed. The pulmonary capillaries are extremely efficient filters, and producing bubbles stable enough to pass through them intact is a demanding task. Protecting the bubbles against destruction by the pressure changes within the chambers of the heart and the major arteries is equally difficult, but this stability is essential if a microbubble formulation is to provide echo-enhancement within the entire blood pool. There are three distinct approaches to the stabilization of gas microbubble suspensions (Figure 1). The simplest preparations are suspensions of free

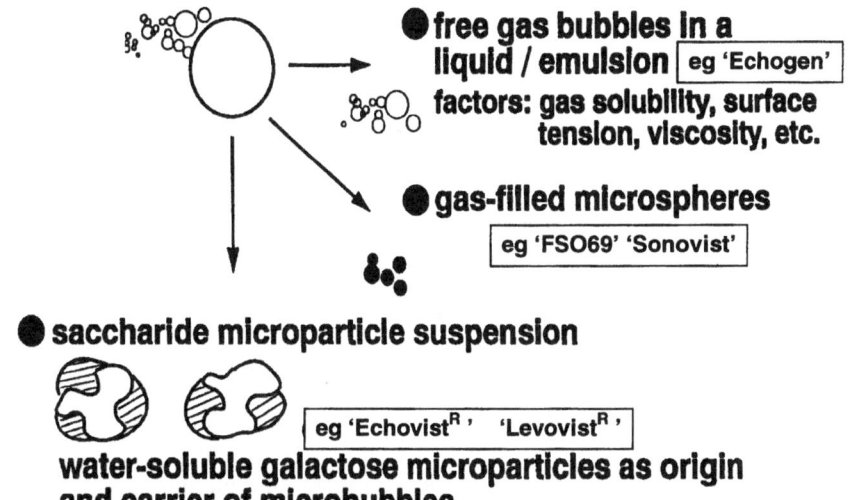

Figure 1. Pharmaceutical techniques for microbubble stabilization.

bubbles made by agitating solutions of saline or Renografin. The things that determine bubble duration in these ad hoc suspensions are the rate of diffusion of the gas from the bubbles into the fluid, the viscosity of the fluid and its surface tension. Gas-filled microspheres protected against gas loss by an albumin shell (Albunex) provide a more effective suspension, but this approach did not ensure transpulmonary stability until the advent of the perfluorocarbon gases with very low blood solubility. Water-soluble saccharide microparticles that originate and carry the gas microbubbles provide a different solution to the problem of bubble stability. When the particles are mixed with an aqueous vehicle microbubbles form at the originating and stabilizing sites. After injection into the plasma, the microparticles dissolve, liberating stable echogenic gas bubbles. There are two types of saccharide microparticle contrast enhancers, Echovist which does not cross the pulmonary barrier, and Levovist which contains a trace of palmitic acid to create a thin film of fatty acid that surrounds the bubble. The fatty acid is not easily recognized by the reticuloendothelial cells, so the bubbles escape destruction to cross the lungs and act as a blood pool echo-enhancer.

Other agents at an earlier stage are all variants of a gas filled polymer with a physical structure similar to that of the albumin-stabilized bubbles of Albunex. In one formulation, BR1, the microbubbles contain sulphur hexafluoride, a dense gas that diffuses from the bubbles more slowly than air, which should extend their life and prolong their echogenic effect. Liposomal preparations are also under development as ultrasound echo-enhancers and drug delivery systems. A conceptually novel approach was offered by the phase-shift enhancer Echogen, a fluorocarbon emulsion at room temperature that changes phase to produce a suspension of echogenic gas microbubbles at blood temperature. Recent clinical trials have shown that Echogen demands a special activation procedure during preparation to produce an effective and reproducible echogenic suspension. This procedure helps the phase change by initiating gas formation in the syringe immediately before injection. Microspheres of human albumin, like Albunex, become much more effective when they are filled with a fluorocarbon gas instead of air. Sonovist, a suspension of microbubbles encapsulated in polymeric cyanoacrylate, is probably the most stable formulation yet developed. The bubbles do not dissolve spontaneously but they can be destroyed by a high power ultrasound beam, with the emission of a characteristic sound (stimulated acoustic emission). If they remain in the circulation they are removed by phagocytosis, predominantly by the hepatic macrophages of the Kuppfer cells.

Three echo-enhancers have reached the market at the time of writing. Echovist and Levovist are based on galactose microparticles, while Albunex is a suspension of microbubbles stabilized with human albumin. According to the best available information, 14 agents are in clinical development and another six are undergoing preclinical evaluation.

MARKETED PRODUCTS

Galactose-based echo-enhancers

The first galactose microparticle echo-enhancer, SHU 454 (Echovist®, Schering AG, Berlin), was marketed in 1991 and the second, SHU 508A (Levovist, Schering AG, Berlin) has been licensed in several European countries (Germany, Austria, Switzerland, Italy and Spain) and others will soon follow. Echovist is used outside echocardiography to assess tubal patency during the investigation of infertility. Hysteringosalpingocontrast sonography (HyCoSy) using the galactose-based enhancer gives a clear image of flow through the Fallopian tubes, demonstrating tubal patency and avoiding the risks of radiation and the necessity for anesthesia [1, 2]. Levovist has a much wider range of applications, the most important of which depend on the enhancement of the vascular Doppler signal throughout the entire blood pool.

Physical characteristics
Under moderate magnification the granules of microparticle aggregates look rather like snowballs. When these snowballs are vigorously shaken in an aqueous solvent they break down into their constituent microparticles, each of which presents a multitude of stabilizing sites on which microbubbles can form. Echovist and Levovist are chemically very similar; Echovist contains galactose alone whereas Levovist includes palmitic acid (0.1%) as an additive. This minute amount of fatty acid makes Levovist microbubbles stable enough to transit the lung and enter the arterial circulation. Soon after intravenous injection the stabil-

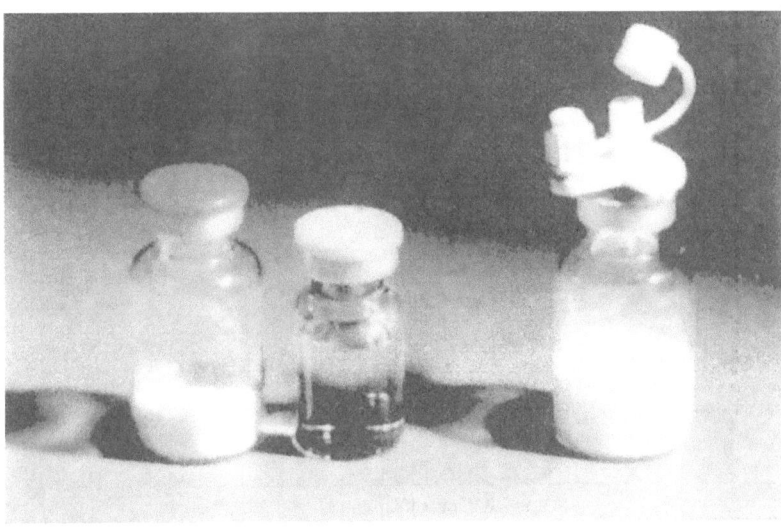

Figure 2. The Levovist pack.

izing sites within the galactose microparticles, at which the bubbles formed, start to dissolve, releasing fatty acid-shielded microbubbles into the bloodstream (Figure 2).

The size of the galactose particles is closely related to the size of the micro-bubbles; the bubbles of Echovist and Levovist have a mean diameter of 3 μm. In practice the granules are energetically shaken (by hand) with sterile water (Levovist) or galactose solution (Echovist) for 5–10 s immediately before use, producing a milk white suspension containing 200, 300 or 400 mg of micro-particles per ml. After standing for 1–2 min to reach equilibrium the suspension retains its echogenicity for 10–15 min. At concentrations <200 mg/ml the micro-bubble suspension becomes rather less stable and at concentrations >400 mg/ml the viscosity increases rapidly.

The physicochemical characteristics of Echovist and Levovist are very similar to those of a saccharide solution. Their viscosity depends on concentration and temperature. At the concentrations used clinically Echovist and Levovist are rather less viscous than typical X-ray contrast media (Figure 3). Figure 4 shows the osmolality of the liquid compartments of Echovist and Levovist after stirring and centrifuging. At a concentration of 300 mg/ml, the osmolality of Levovist is lower than that of Magnevist, a remarkably well tolerated MRI contrast enhancer, and a little higher than that of a non-ionic X-ray contrast medium (Figure 3). The total osmotic load which these echo-enhancers would produce if all the microparticles dissolved is their potential osmolality. These values are much higher than the osmolality of the liquid compartment (at 300 mg/ml, Levovist would have a potential osmolality of 2.060 mosm/kg H_2O), but they have no

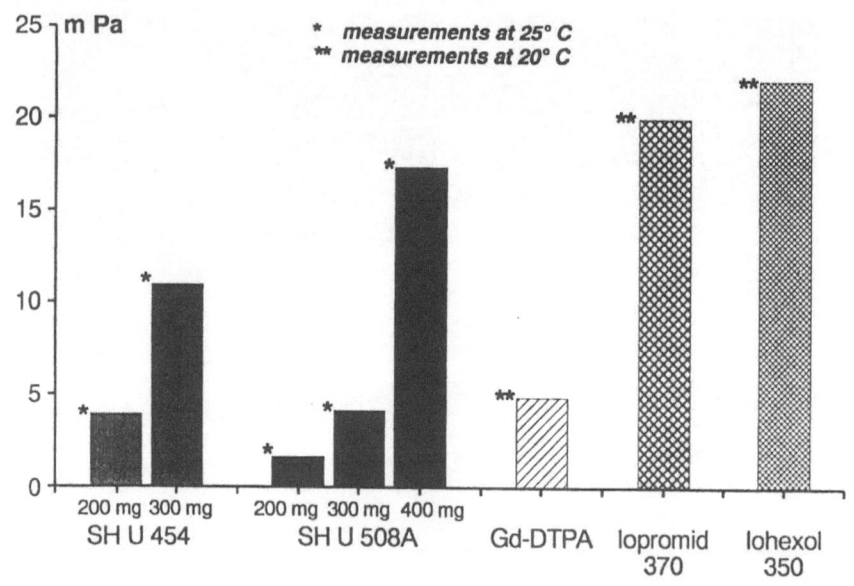

Figure 3. Levovist and Echovist: comparative viscosity.

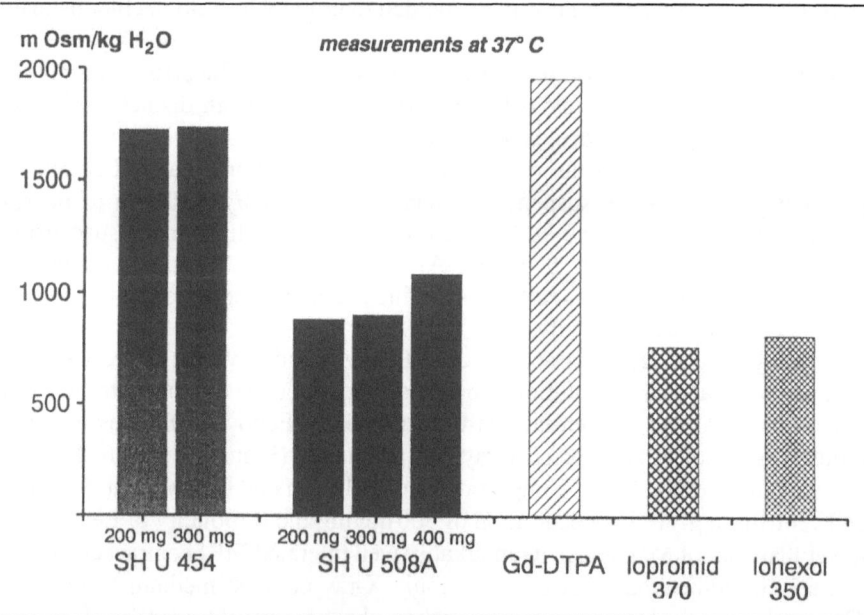

Figure 4. Levovist and Echovist: comparative osmolality.

Table 2. Sound velocity in Levovist at different concentrations (5 MHz, atmospheric pressure, 22°C).

Medium	Sound velocity (m/s)
Sterile water	1473
Levovist 100 mg/ml	1487
Levovist 150 mg/ml	1517
Levovist 200 mg/ml	1527
20% w/v galactose solution	1540
Soft tissue (average)	1540

clinical importance, because they never occur in vivo. The only application of this potential osmolality is in the calculation of the total osmotic load of each injection. Because the injection volume is small the total osmotic load is also rather small.

Acoustic properties
Both agents produce a predictable, dose-dependent increase in ultrasound back-scatter. In the case of Echovist this increased echogenicity is confined to the venous blood. Levovist microbubbles pass through and opacify the left heart and decay later in the arterial bed. Levovist increases the echogenicity of the entire vascular bed, making it the first echo-enhancer to enhance backscatter from the entire blood pool.

Speed of sound: The speed of sound within a microbubble suspension is governed mainly by the liquid compartment. The speed of sound in Levovist suspensions differs very little from the speed in water, a solution of galactose or a range of body tissues (Table 2). The very slight reduction in velocity in the microbubble suspensions is probably due to the air content of the bubbles. For comparison, sound moves through living tissue at between 1450 (fat) and 1540 (soft tissue) m/s.

Attenuation and frequency: The tissue attenuation of ultrasound rises with increasing frequency. Galactose-stabilized microbubble suspensions behave quite differently, and ultrasound attenuation by Levovist is strongly frequency-dependent, rising to a maximum at around 2.5 MHz and falling off above that level [3] (Figure 5). The reason for this frequency dependence is that galactose-stabilized microbubbles are flexible and oscillate within the ultrasound field. They resonate at the frequencies used diagnostically, which raises the interesting possibility of ultrasound imaging using the backscattered energy in the second harmonic returned by the microbubbles [3].

Doppler shift: In some cases, the maximum flow velocities in Levovist-enhanced Doppler spectra seemed to be rather higher than those in the non-enhanced baseline spectrum, raising the suspicion that echo-enhancement could

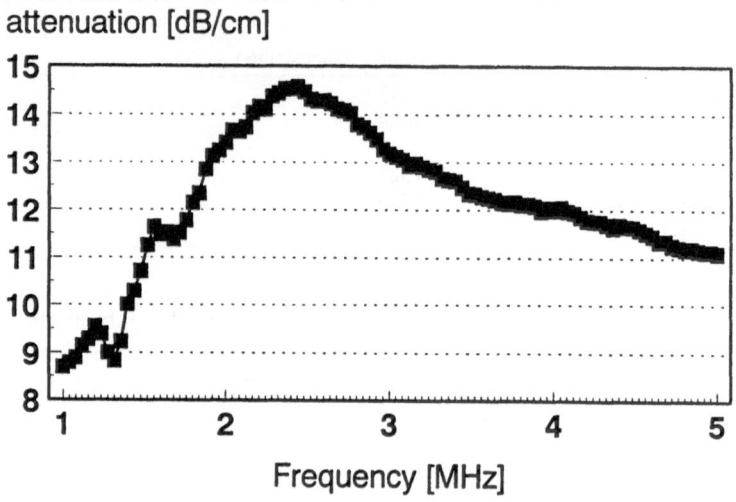

Figure 5. Levovist: attenuation vs. frequency.

give a false indication of maximum flow velocity. Petrick has investigated the effect of Levovist on the Doppler shift in great detail, using a flow model that reproduces the behavior of the circulation in vivo very closely [4]. His pulsatile circulatory system differs from other models in two important respects: first, it uses an elastic tube so that the wall distends during the cycle, reproducing the hydrodynamic damping that occurs in large vessels such as the carotid artery during the cardiac cycle. Second, it simulates the variable acoustic attenuation of the extravascular tissue by surrounding the elastic tube with oil. The technique employed a specially developed pulsatile flow model that made it possible to record the Doppler signals with great accuracy and to analyse the recorded data very extensively. It was also possible to change the elasticity and the bore of the tube to simulate physiological changes in vascular elasticity and resistance and reproduce these typical flow shapes:

- Monophasic velocity pattern with parabolic flow profile
- Monophasic velocity pattern with plug flow
- Triphasic velocity pattern with parabolic flow
- Triphasic velocity pattern with plug flow.

Each series of measurements consisted of a baseline unenhanced value and five readings made after injections of 1.25–20 ml Levovist. Levovist enhanced the Doppler signal by 16–31 dB and the various graphical representations of the data showed that the enhanced and unenhanced Doppler power spectra were identical under all the flow conditions that the model allowed. Petrick's results provide no evidence that echo-enhancement with Levovist falsifies the Doppler

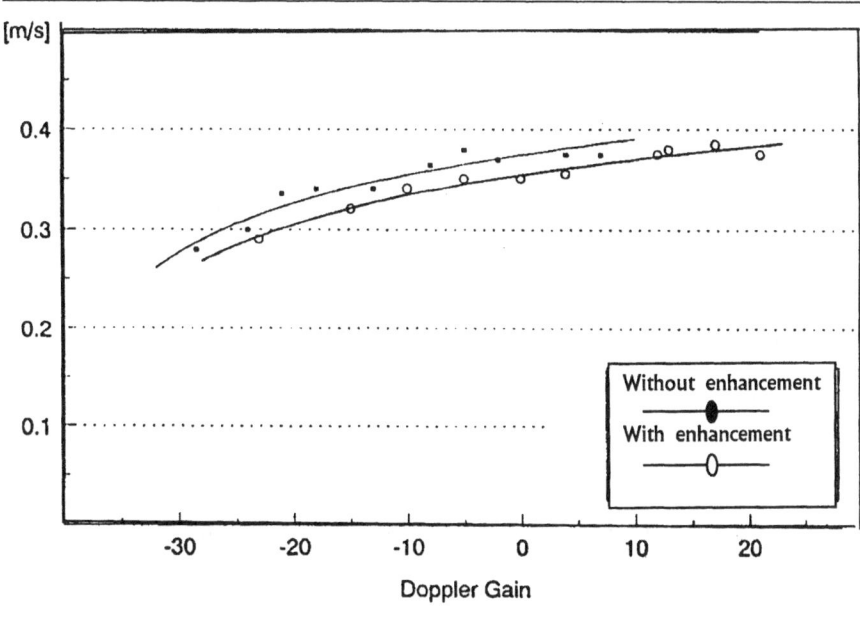

Figure 6. Mean and maximum Doppler shift before and after Levovist.

measurement of flow velocity. In his model the frequency distribution of the Doppler signals were unaffected by the echo-enhancer. Low signal intensity can lead to the incorrect interpretation of Doppler spectra but echo-enhancement, provided that excessive gain and oversaturation is avoided, amplifies the Doppler signal and leaves its spectral distribution unchanged (Figure 6).

Effects of pressure

Static pressure: Levovist retains its echogenicity at relatively high pressures for several minutes: at a pressure of 100 mmHg it loses less than 1 dB of its echogenicity per minute. Reports suggest that, at these pressures, the echogenicity of sonicated human albumin decays within a few seconds. The greater stability of Levovist is probably due to the effect of the fatty acid on gaseous diffusion. High ambient pressures can be produced accidentally within the syringe during a bolus injection but, if we assume an injection rate of 2 ml/s, this high pressure will persist for only 5 s. This is too short a period to produce any significant bubble decay in the saccharide-based agents but it may reduce echogenicity appreciably in suspensions of sonicated albumin [5–8].

Pressures below atmospheric can be produced during the preparation of the injection. When the suspension is transferred to a syringe from the 20 ml vial pressures between 50 and 200 mmHg have been measured and this could lead to some degassing. If the products are handled properly however, these low pressures are not generated. No practical problems have been reported with either Echovist or Levovist.

Cyclic pressure variations: The circulating microbubbles are subjected to arterial pressure swings with a frequency of 1–2 Hz and an amplitude within the normal blood pressure range. Measurements of the echogenicity of Levovist within the left ventricular cavity show no significant difference between systole and diastole [9]. Quantitative videodensitometry shows a decrease in intensity during systole but the change barely exceeds the visibility threshold. Elastic microbubbles, like those of Levovist, should respond very quickly to pressure changes with a change in diameter inversely proportional to the change in pressure (provided that we ignore the effects of gas diffusion and varying temperature). Holographic measurements confirm this expectation and show almost instantaneous changes in diameter in response to rapid pressure changes [10]. This suggests that microbubbles could provide a direct measurement of changing intravascular pressure. Other in vitro investigations show that the relationship between the intensity of the backscattered signals produced by Levovist microbubbles and the ambient pressure is reproducible [6, 9], showing that the change in diameter is reversible. At a frequency of 1 Hz, varying the pressure between 100 and 120 mmHg produces a cyclic decrease in backscatter of 2–3 dB. Before this pressure response can be used to measure intravascular pressure changes we must devise some means of detecting the small changes in the backscattered signals caused by the varying pressure.

Pharmacokinetics
The pharmacokinetics of the galactose-based echo-enhancers reflect the metabolism and excretion of their saccharide and fatty acid components.

Metabolism and excretion: After intravenous injection, the galactose microparticles dissolve in the blood stream. Galactose is predominantly metabolized in the liver first to glucose-1-phosphate and then, via the glucose metabolic pathway, to CO_2. If the plasma galactose level rises above 50 mg/100 ml galactose is eliminated via the kidneys rather than the liver. The plasma half-life is 10–11 min in healthy adults and about 7–9 min in children [12]. Elimination is considerably prolonged by alcohol and in patients with impaired liver function, but there is no evidence that this produces any adverse effect. The palmitic acid component of Levovist represents <5% of the free fatty acids already present in the body. Its half-life is about 2–4 min. Intravenous injections of high doses of galactose (>28 g, equivalent to 0.5 g/kg) have been used as a liver function test for many years [13]. Twenty-eight grams of galactose is contained in 70 ml of the highest recommended Levovist concentration of 400 mg/ml. That is about eight times the usual diagnostic dose. These high dose injections are well tolerated, even in patients with liver disease.

Toxicology: Standard tests for assessing acute and subchronic toxicity, including local tolerance, embryotoxicity and mutagenicity were performed in several species with entirely negative results. Furthermore, specially designed models were used to identify any potential risk from the injection of suspensions of microparticles and microbubbles [14]. There was no evidence of any increased risk associated with Echovist or Levovist. Repeated administration of the diagnostic doses of Levovist or Echovist at short time intervals carries no risk of acute toxicity. Long-term studies in beagle dogs showed no important differences between the physiological effects of Levovist or saline and detailed histological investigations showed no sign of any abnormalities due to the contrast agent.

Effects on hemodynamics and the microcirculation: In dogs, injections of Levovist and injections of plain galactose solution in the same concentration and volume produced the same mild and transient changes in blood pressure and cardiac output. Other studies in animals with coronary stenoses or acute heart failure showed that Levovist is safe even in patients with severe right-to-left shunts. Because of the transient hemodynamic changes seen in animals given the highest concentration of 400 mg/ml it is recommended that this rather viscous concentration of Levovist should not be injected directly into the coronary arteries. Microscopic examination of the microcirculation of rabbits, cats and hamsters given injections of Levovist showed no sign of any emboli or of any change in the diameter of the arterioles or capillaries.

An open-chest dog model has been used to compare the acute effect of Levovist on hemodynamics [15]. Three injections were given: Levovist prepared according to the manufacturer's recommendations; bubble-free Levovist, prepared by making up the suspension and leaving it to stand for 2 days. Before injecting it the suspension was subjected to overpressure within the syringe to destroy any remaining microbubbles. Control animals received saline. Previous studies investigated the hemodynamic changes from 5 min to several hours after the injection and showed no acute effects that the microbubbles might produce. This study investigated the changes that took place within the first 2 min after the injection and used an ultrasonic flow probe that gave an instantaneous reading of volumetric blood flow. Left atrial and ventricular pressure were measured with an intracardiac pressure transducer within the cardiac chambers, and central arterial and pulmonary arterial pressures were recorded conventionally. A large range of doses were given (Table 3). Because of the wide range of doses and animal size the results were analyzed according to the weight-adjusted dose. The hemodynamics were measured at four time intervals: T0 (10s before the appearance of the echo-enhancement); T15 (15s after peak enhancement); T40 (mid-effect) and T90 (late effect). The results are summarized below.

Following injection of Levovist prepared according to the manufacturer's instructions, marked increases in LA pressure (up to 60%) were seen, but only at the higher doses. Osmotic loading resulted in a fall in systemic vascular resistance and an increase in cardiac output. Arterial pulse pressure rose,

Table 3. Weight adjusted dosage.

Dose range	Multiple of human dose	Volume of dose (ml/kg)	Weight of dose (g/kg)
A	1	≥0.266	≥0.076
B	2.4	0.267–0.482	0.077–0.142
C	3.8	0.483–0.628	0.143–0.208
D	5.3	0.699–0.914	0.209–0.274
EF	6.7	0.915–1.143	0.275–0.341

Adapted from Ref. 15.

producing a modest increase in mean BP and a slight decrease in heart rate. There was a dose-dependent increase in pulmonary vascular resistance and this, combined with the increase in cardiac output, produced a marked increase in PA pressure (>25%) at the highest doses. The systemic vascular changes caused by bubble-free Levovist were essentially the same as those seen with Levovist. There was a slight increase in PA pressure. This was due to an increase in cardiac output that was greater than the fall in pulmonary vascular resistance. Volume loading due to saline injection produced a slight increase in LA pressure (<20%) at the maximum dose. This produced a slight increase in BP with a fall in heart rate that was probably a reflex response. Systemic vascular resistance fell slightly and overall mean cardiac output increased. There was a modest increase in PA pressure despite the slight fall in pulmonary vascular resistance produced by the increase in cardiac output.

The only difference between the effects of conventional and bubble-free Levovist was that PA pressure rose to a greater extent after the echogenic formulation than after injections of the bubble-free galactose vehicle. These changes only appeared after the highest dose. When the dose used in normal clinical practice was given, all the hemodynamic changes were extremely small (<5%) and the larger changes produced by higher volume injections of Levovist were (with the possible exception of the changes in the pulmonary circulation) a direct consequence of volume and osmotic loading. These results confirm the good tolerance of Levovist within and above the normal dose range.

Cavitation: When microbubbles start to resonate in a high amplitude therapeutic ultrasound field they can collapse violently (cavitation) and, theoretically, release enough energy to cause local effects on tissue. This possibility has been investigated in an isolated perfused guinea pig heart exposed to an ultrasound field more intense than that used during echocardiographic investigations. Coronary flow was measured after an injection of acetylcholine and the procedure was repeated after an injection of Levovist during continuous insonation. There was no difference in the flow response to acetylcholine before and during the injection of the echo-enhancer. This suggests very strongly that Levovist, even under intense insonation, has no influence on the endothelial response to acetylcholine which, in turn, suggests that Levovist-associated cavitation, if it occurs, has no detectable physiological effect.

During Phase I trials of Levovist, biochemical and hematological variables were measured in patients given the maximum recommended dose of Levovist before, during and after insonation at high energy levels. There was no evidence of any changes that could have been associated with microbubble cavitation. The earlier galactose-based enhancer Echovist has been used to confirm tubal patency in more than 1000 women as part of the investigation of their infertility using HyCoSy [1, 2]. Again there was no evidence of any cavitation-induced tissue damage. Some of the women became pregnant in the cycle during which they had received the Echovist-enhanced tubal investigation. This suggests that, even if microbubble cavitation had occurred during the tubal investigation, it produced no clinically relevant effects. To date there is no evidence, either clinical or experimental, that microbubble cavitation during diagnostic ultrasound investigations poses any danger to patients.

Albumin-stabilized microbubbles

Sonicating a solution of albumin produces small, relatively stable, albumin-encapsulated microbubbles. The high temperatures reached during the cavitation process that creates the bubbles produce superoxide free radicals that oxidize the cysteine residues of the albumin chains [16]. The intra-chain crosslinking which results from this oxidation creates an albumin shell between 30 and 50 nm thick around each microbubble [17]. These encapsulated bubbles are rather less effective as ultrasound scatterers than free gas bubbles of the same size but they are much more stable [17].

Physical characteristics
Albunex contains sonicated albumin microbubbles with a mean diameter of 4 μm; 95% of the bubbles are < 10 μm in diameter [17]. The microbubbles are supplied in ready-to-use vials containing $300-500 \times 10^6$/ml. The product has a shelf life of more than 6 months.

Albunex contains three protein fractions [17]: a carrier protein fraction that contains 5% human serum albumin and accounts for > 99% of the total protein content, a water-soluble fraction (shell protein) with the same peptide chain and surface-charge as the carrier protein, and a water-insoluble fraction (shell protein) with the same peptide chain as the carrier protein but a different surface charge.

Sonicated albumin microbubbles are sensitive to pressure in vitro and in vivo. As pressure rises, backscatter intensity falls and the rate of backscatter decay is inversely related to microbubble concentration [18, 19]. This suggests that the albumin shells remain intact despite the loss of echogenicity, suggesting that their air content has been replaced by liquid. This acoustic instability has also been demonstrated in vivo as a systolic fall in videointensity following intravenous injection of sonicated Albunex [20]. This decay was reduced when greater concentrations of albumin microbubbles were delivered to the left ventricle

[21]. This reduction may be due to longer bubble survival at left ventricular systolic pressures or to increased bubble stability at high concentrations. Further investigation of the in vitro and in vivo stability of albumin-stabilized microbubble contrast agents is needed before these questions can be answered.

Acoustic properties
The attenuation and backscattering of Albunex microbubbles in vitro are related linearly to microbubble concentration [22].

Pharmacokinetics
The plasma half-life of radiolabelled Albunex microbubbles after intravenous injection is < 1 min [17]. Six minutes after the injection 80% of the radioactivity was detected in the liver with only 10% remaining in the liver after 6 h.

Safety and toxicology
Albunex microbubbles have no significant effects on systemic, pulmonary or coronary hemodynamics, ECG or respiration in animals [22, 23] or in humans [24–26]. Albumin microbubbles produce no adverse effects after intracoronary injection in humans and no adverse effects or hemodynamic alterations were noted in patients receiving injections of Albunex through saphenous vein grafts while undergoing coronary artery bypass surgery [27]. In early 1996 Albunex was withdrawn from the market in Europe and Japan, partly for commercial reasons and partly due to an apparent absence of acoustic response in some 30% of patients. This fall in efficacy may have been due to problems in maintaining the stability of the microspheres. The Albunex package leaflet issued by Mallinkrodt in the USA warns that mishandling the product by shaking the vial or placing it on a benchtop too roughly can impair the stability of the microbubble suspension.

ULTRASOUND ECHO-ENHANCERS IN LATE CLINICAL DEVELOPMENT

Echogen

Physical characteristics
The persistence of gas microbubbles in water depends upon three factors: the density of the gas inside the microbubbles, the rate at which the gas diffuses out of the microbubbles, and the water solubility of the gas. These factors can be combined to give a numerical indicator of persistence, the 'Q factor'. This 'Q factor' provided the theoretical underpinning for the development of Echogen. The gas used in Echogen, dodecafluoropentane, has a density of 1.66 g/ml and boils at 28.5°C. These properties led to the elegant concept of a phase–shift formulation of dodecafluoropentane as an emulsion that would transform itself

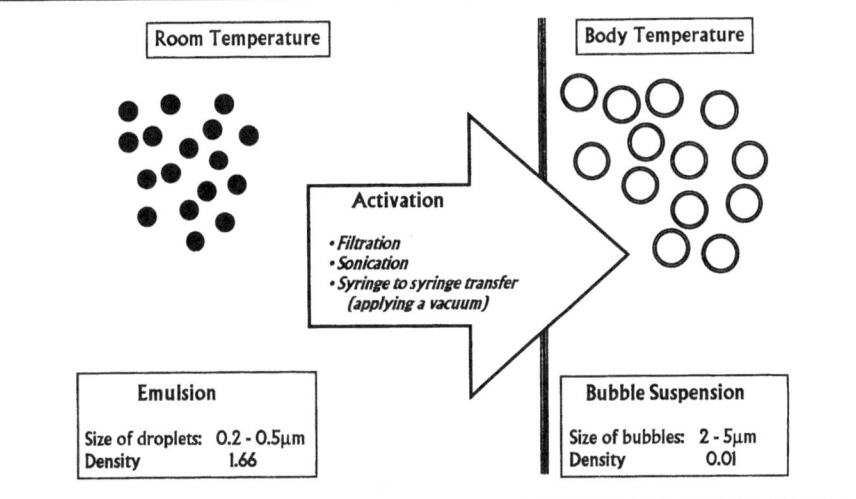

Figure 7. Echogen: assisted phase-shift (adapted from Ref. 23).

into an echogenic suspension of dodecafluoropentane gas microbubbles as the liquid particles changed to the gas phase at body temperature. Experience has shown that the phase change is incomplete and that injecting the emulsion does not produce diagnostically adequate enhancement [28]. The manufacturers (Sonus Pharmaceuticals, Bothell, WA, USA) developed a filter with a pore size of 1.2 μm that exploits the Bernoulli effect to accelerate the phase shift as the emulsion is pumped across it. Later in the development process other methods, including sonication and activation by pumping the suspension from one syringe to another via a cavitation chamber, were introduced. A more elegant alternative involves a combination of low pressure and the creation of a transient pressure wave within the syringe immediately before injection (Figure 7). High doses of filter-activated Echogen (0.7 ml/kg) gave good gray-scale enhancement that persisted for up to 15 minutes. When the emulsion was activated by sonication it produced the same effect at a dose 20 times lower than that needed after filter activation [28]. Studies in monkeys show appreciable myocardial opacification at a dose as low as 0.02 ml/kg.

Acoustic properties
Echogen, 0.25, 0.35, 0.45 and 0.65 ml/kg i.v., was used to enhance the gray-scale image of the renal cortex in three dogs [29]. There was a noticeable increase in image brightness as early as 10 s after the injection. Brightness reached a peak 70–100 s after the injection and gradually decreased thereafter. The detailed results are summarized in Table 4. The relationship between videodensity and dose was complex and non-linear. On occasions the density curves drifted below

Table 4. Acoustic properties of Echogen.

Dose (ml/kg)	Time to peak (sec)	T½	Time to baseline
0.25	70	2.5	5.5
0.35	75	2.0	5.5
0.45	60	2.8	17.5*
0.65	55	3.5	18.6*
Mean ± SD	65 ± 9	2.7 ± 0.6	11.8 ± 7.3

*Curves did not return to baseline within 15 mins. These values have been extrapolated from the figures at 15 mins (adapted from Ref. 29).

the pre-injection values, which probably reflected the attenuation produced by the microbubbles.

Doppler enhancement was studied in three animal species (woodchucks, rabbits and dogs) after intravenous injections of 0.005–0.8 ml/kg [30, 31]. Quantitative analysis showed a non-linear dose response relationship with a maximum enhancement of 18 dB after a dose of 0.6 ml/kg in a dog. Nine hepatomas were imaged in the woodchucks used in this study. Flow was detectable in two of the tumors before enhancement and in eight after Echogen. The mean enhancement, for doses ≥ 0.3 ml/kg, was 10–14 dB and enhancement persisted for 2–3 min.

Pharmacokinetics

The pharmacokinetics of Echogen were investigated during the Phase I clinical trials [32]. The fluorocarbon content of expired air and blood was measured by gas chromatography in samples taken from 49 subjects after a bolus i.v. injection of Echogen at doses of 0.01, 0.02, 0.05, 0.1, 0.15, 0.25 and 0.35 ml/kg. The results fitted an open single compartment model. The area under the curve was closely related to the dose ($r^2 = 0.96$) except at the highest dose level. Mean blood clearance was 59.3 ml/min/kg and the blood apparent distribution volume ranged from 0.08 to 0.45 l/kg. All of the injected dose (±13%) was excreted in the expired breath within 2 h and the mean time for 50% excretion was 3.7 min. There were no extraneous peaks in the gas chromatograph traces which suggested that dodecafluoropentane was not metabolized.

Dodecafluoropentane bubbles are much more persistent than bubbles containing air, lasting for up to 20 min after a single injection [32]. This persistence is largely due to the low water solubility and slow diffusion of dodecafluoropentane, but other mechanisms are probably also involved. Dodecafluoropentane could, like other fluorocarbons, increase vascular permeability and enter the extravascular space. Another possibility is that the microbubbles expand in vivo and become trapped in the microvasculature. A study of the behaviour of dodecafluoropentane microbubbles in a cat mesenteric preparation showed that in vivo the bubbles, which were considerably larger than erythrocytes, appeared to adhere to the endothelial surfaces of the precapillary arterioles [33]. This effect could have been due to the surfactants used to stabilize the emulsion. Other studies of mesenteric microcirculation and radiolabelled microsphere analysis

Table 5. Adverse events associated with Echogen.

Vasodilatation	3/59
Dyspnoea	2/59
Tachycardia	1/59
Hypertension	1/59
Chest pain	1/59
Back pain	2/59
Urticaria	1/59

Adapted from Ref. 27.

show that the persistent effect of Echogen is not due to capillary blockage. Details of these investigations are as yet unpublished.

Safety/toxicology
Echogen was used to investigate myocardial opacification in a dog study in which the same dose (0.5 ml/kg) was given intravenously [34]. Two protocols were used: four doses of 0.5 ml/kg at 30 min intervals and eight doses of 0.5 ml/kg at 10 min intervals. There were no significant hemodynamic or blood gas changes in the animals given four doses of Echogen. The dogs receiving eight doses showed an increase in pulmonary artery pressure and a progressive decrease in arterial oxygen saturation, mean arterial pressure, stroke volume and cardiac output. Two animals developed severe respiratory distress and bradycardia after the seventh injection and were withdrawn from the study. There were no alterations in regional wall motion after injections at 30 min intervals but there was marked paradoxical septal motion in all the animals at 10 min intervals. These changes appeared after the third or fourth injection and appeared to be related to the increased pulmonary artery pressure and hypoxemia. Later injections of Echogen produced moderate to severe global left ventricular hypokinesis. Post mortem examinations of the lungs in the dogs injected at 10 min intervals showed marked accumulation of dodecafluoropentane.

Clinical tolerance
Echogen produced adverse events in about 10% of the 59 patients investigated in the completed Phase II study [28] (Table 5). The manufacturer has now changed the formulation, with a concomitant reduction in frequency of adverse events of some 50% [28].

BY 963

Physical characteristics
BY 963 is a suspension of air-containing microbubbles (spherisomes) stabilized with a soya-based phospholipid (3-SN-phosphatidyl-DL glycerostearyl-Na) [35]. The suspension is reconstituted using a syringe that contains a cavitation chamber

in which the lyophilized phospholipid and the vehicle are mixed before injection. The bubbles have a mean diameter of $3.9\,\mu m$, and 95% have a diameter $<8.0\,\mu m$. The molecular weight of the phospholipid is $301\,g/mol$ and the suspension has an osmolality of $300\,mOsm$, containing 170×10^6 bubbles/ml. It remains stable for long periods of time.

Acoustic properties
Microbubbles of BY 963 opacify the left ventricle after intravenous injection [35]. It causes no obvious attenuation in the left ventricle.

Pharmacokinetics
The bubbles have a half life of 10 min [35]. At the time of writing there is no published information on the detailed pharmacokinetics of BY 963.

Safety/toxicology
At the time of writing there is no published information on the safety and toxicology of BY 963.

FSO 69

Physical properties
This agent is described by its manufacturers (Molecular Biosystems Inc., San Diego, CA) as a suspension of proteinaceous gas-filled microspheres. FSO 69 appears to have similarities to the original albumin-stabilized microbubble suspension, Albunex. The microbubbles range from 3.6 to $5.4\,\mu m$ in diameter. They are metabolized in the liver and, although there is some recirculation, their myocardial enhancement is essentially a first-pass effect [36].

Acoustic properties
The available information suggests that FSO 69 is a derivative of Albunex with rather more persistence in the circulation [37]. In an animal model using coronary artery occlusion FSO 69 produced well visualized myocardial opacification [37]. In a Phase III study FSO 69 improved endocardial border definition in at least 90% of 203 patients at 14 USA centres, compared with 59% of those patients given Albunex [38]. Complete left chamber filling was seen in 90% of the patients imaged with FSO 69 and only 37% of those given Albunex.

Pharmacokinetics
At the time of writing there is no published information on the pharmacokinetics of FSO 69.

Safety/toxicology
There is some suggestion that FSO 69 is slightly better tolerated than Albunex. In the Phase III study referred to earlier the incidence of adverse events in patients

given FSO 69 or Albunex was 7% and 9%, respectively [38]. All these events were described as mild and transient. At the time of writing there is no other published information on the safety and toxicology of FSO 69.

ULTRASOUND ECHO-ENHANCERS IN EARLY CLINICAL DEVELOPMENT

Imagent (AFO145)

Physical characteristics
Imagent (AFO145; Alliance Pharmaceutical Corporation, San Diego, CA) is composed of unspecified structural components, surfactants, sodium chloride and phosphate buffers, presented as a dry powder in a vial that also contains gaseous perfluorohexane [39]. The suspension is reconstituted with sterile water to give a preparation containing 90 μl/ml of perfluorohexane. In the reconstituted suspension the bubbles which both contain and are stablilized by the fluoro-carbon, have a mean diameter of 6 μm and are present in a concentration of 5×10^8/ml.

Acoustic properties
Conventional imaging showed homogenous left ventricular contrast and good border delineation [39]. The contrast was very persistent and lasted long enough to allow the probe to be changed to give different views. There is some evidence that increasing the power of the ultrasound beam reduces peak enhancement and shortens the duration of Imagent echo enhancement in an animal model of hepatic carcinoma [40]. Pixel intensity altered by around 10% between systole and diastole and second harmonic imaging showed bright opacification of the entire left ventricular cavity [38]. In an animal model of renal infarction, second harmonic imaging doubled the intensity of the image contrast after Imagent enhancement [41].

Safety/toxicology
There were no significant changes in blood pressure or heart rate after a single injection or a series of repeated injections in dogs [39]. Likewise, there was no significant change in cardiac output, pulmonary artery pressure or coronary flow reserve or changes in blood gases after intravenous doses of Imagent in doses of 0.15, 0.3 or 0.6 mg/kg.

Pharmacokinetics
At the time of writing there is no published information on the pharmacokinetics of Imagent.

Table 6. Pressure resistance of gas bubbles.

Gas	Solubility in water*	Mol wt (g)	P (50% absorbance) (mmHg)
CH_4	0.032	16	51
Air	0.0167	29	43
C_2H_6	0.042	30.1	57
Ar	0.031	40	61
Kr	0.059	83.8	67
CF_4	0.0038	86	94
Xe	0.106	131.3	50
SF_6	0.006	140.1	127
C_4F_8	0.016	200	266

Adapted from Ref. 43.
*Bunsen coefficient (25°C).

BR1

Physical characteristics
BR1 (Bracco Research SA, Carouge-Geneva) is a suspension of microbubbles filled with the dense gas sulphur hexafluoride, a compound used by the semiconductor industry to etch silicon films [42]. Sulphur hexafluoride was chosen as the gas component of this novel echo-enhancer because of its high pressure resistance [43]. The manufacturers measured the resistance of BR1 to pressure changes with a simple photometric method that recorded the light absorbancy of BR1 suspension as the pressure rises. Using this system they were able to compare pressure responses by using the point at which the absorbance of the suspension was reduced by 50% (P 50% absorbance; Table 6). Microbubbles containing sulphur hexafluoride resist increasing pressure more effectively than bubbles containing air.

BR1 is presented as lyophilized particles that contain phospholipid and polyethylene glycol 4000. The particles are contained in a vial filled with the gas sulphur hexafluoride and the echo-enhancer is prepared by adding 5 ml saline to the vial and shaking. The suspension thus created contains between 1 and 5×10^8 bubbles/ml with a mean diameter of $2-3 \mu m$. More than 90% of the bubbles are $<8 \mu m$ in diameter. The SF_6 content of the suspension varies from 2 to $10 \mu l/ml$. The reconstituted suspension is iso-osmolar with saline and its viscosity at 25°C is $<2.0 MPa/cm$. It remains stable for 6 h and the bubble diameter is reported to remain constant throughout that period.

Acoustic properties
The backscattering coefficient reaches a peak between 3 and 5 MHz but the peak is small and there is no significant difference in backscatter coefficient between 1 and 10 MHz [44]. Attenuation reaches a peak at 4.5 MHz, suggesting that most of the bubbles resonate at this frequency (Figure 8). BR1 remains highly echogenic over the entire diagnostic frequency range (1–10 MHz) [44].

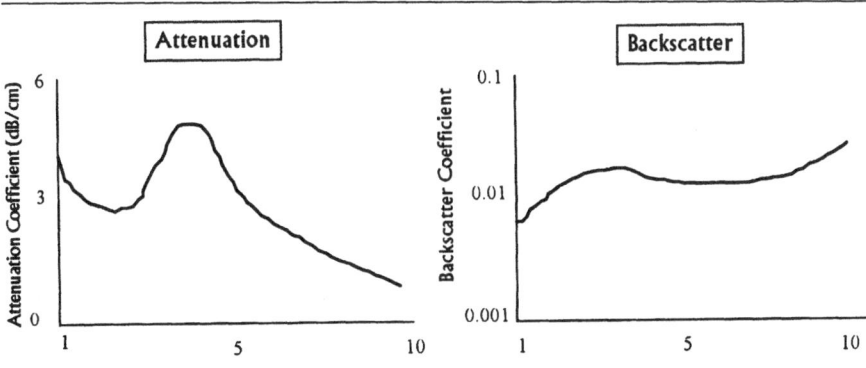

Figure 8. Attenuation and backscatter coefficient of BR1 as a function of frequency (adapted from Ref. 43).

Pharmacokinetics
There is no published information on the pharmacokinetics of BR1.

Safety/toxicology
Sulphur hexafluoride is conventionally regarded as non-toxic although there have been very few formal investigations of its systemic effects. It is administered as an intraocular injection in the treatment of retinal detachment [45]. The gas has been widely used as an electrical insulator since 1937 and the only problems associated with it have been occasional instances of suffocation caused by breathing the gas in high concentration [42]. The manufacturers of BR1 have investigated its toxicology at doses 30 times higher than those used for imaging and have seen no evidence of any cardiovascular or CNS toxicity [44]. At the time of writing details of those investigations remain unpublished.

SHU 563A

Physical characteristics
This new agent is the first of a new generation of ultrasound contrast enhancers with specific acoustic properties. It consists of air-filled microspheres, mean diameter 1 μm, the shells of which are formed by a thin layer of a biodegradeable cyanacrylate polymer. The microspheres are preformed and presented as a lyophilized powder which is suspended in physiological saline before injection. This suspension is isotonic and remains stable for several hours.

SHU 563A is extremely persistent and the only mechanisms by which the microbubble are eliminated are phagocytosis (mainly by the hepatic macrophages of the Kupffer cells) and the effects of ultrasound irradiation. Consequently, the agent has the potential to differentiate normal from abnormal tissue in the liver

Table 7. Effects of stimulated acoustic emission of Sonovist.

Spectral Doppler	Single bright spikes
Color Doppler	Color pixels with random velocity coding that give valid information on spatial position
B-mode	Grey/white pixels
Power Doppler	Pixel Doppler signal

and spleen. It also raises the possibility of new targeted approaches to diagnosis and therapy. Another exciting possibility is that SHU 563A could be used to image tissue perfusion as effectively, and much more conveniently, than radionucleide scintigraphy.

Acoustic properties
At low signal amplitudes the microbubbles behave as passive backscatterers. As the amplitude of the ultrasound signal increases the microbubbles begin to resonate, emitting the second and third harmonics of the insonating fundamental. At the same time, their average backscattering cross-section increases and their backscattering efficiency rises. As the amplitude rises still further the micro-bubbles respond with a sudden and very distinctive signal collapse. This whole sequence of responses is termed *stimulated acoustic emission* [48]. The effect is produced to a limited extent by other microbubble agents, including Levovist, but with SHU 563A the effect is more pronounced and reproducible and may provide clinically useful quantitative information. *Stimulated acoustic emission* produces a variety of effects in imaging displays (Table 7). The response during harmonic imaging is particularly useful because the random mosaic pattern of the SHU 563A signals is easy to distinguish from the signals showing blood flow. In B-mode and power Doppler imaging the SHU 563A-enhanced display is not similar to the usual display pattern.

Pharmacokinetics
SHU 563A microparticles are extremely stable and circulate throughout the blood pool unaffected by pressure changes or local turbulence until they are taken up by the splenic macrophages of the Kupffer cells. They remain intact for some time even after phagocytosis, and retain their structural integrity and gas content for many days, unless they are exposed to high amplitude ultrasound.

Safety/toxicology
The cyanoacrylate polymer of the microbubble shells has been used for many years as a tissue adhesive and in the treatment of varices [49, 50]. The total mass of cyanoacrylate at diagnostic doses is very low ($\sim \mu g/kg$). Routine toxicological studies showed good tolerance and no evidence of any untoward effects at doses more than 100 times greater than those used diagnostically. The phase I trials of

SHU 563A have shown no evidence of any adverse events or local discomfort [46]. These studies included extensive scanning with high amplitude ultrasound.

Quantison® and Quantison Depot®

Physical characteristics
The technology used to manufacture Quantison (Andaris, Nottingham) allows the production of human albumin microcapsules with diameters ranging from 0.1 to 70 µm [51]. The albumin is cross-linked but resembles native soluble human albumin extremely closely; as a consequence the particles are very hydrophilic and remain in the circulation for ≥1 h. The technology also allows a great deal of freedom to manipulate the size distribution, the shell properties and the rate of degradation. The batch-to-batch variation is reported to be <40 nm [51]. The early formulations contained a large number of particles >6.5 µm in diameter; when smaller particles were produced their shells were thinner and their echogenicity was enhanced. The resonant frequency of the particles can be altered by adjusting the degree of cross linking within the particle shells.

Quantison: The microparticles of Quantison have a mean diameter of 3.2 µm and virtually none are >7 µm in diameter. The particles are highly pressure resistant: the manufacturer's internal quality control standard demands that the particles should withstand a pressure of 300 mmHg and they will probably withstand 600 mmHg without loss of gas [49].

Quantison Depot: Quantison Depot is designed to measured myocardial perfusion. The particles have a mean diameter of 12 µm and none have a diameter >20 µm [49]. The suspension must be given by intra-arterial injection because the particles are far too large to cross the pulmonary circulation. In animal models it produces a depot effect within the myocardium that gives prolonged enhancement. It probably produces this sustained effect by blocking enough of the capillaries within the coronary microcirculation to allow the microparticles to remain in situ within the myocardium. In nuclear medicine a similar methodology, microembolization, uses solid radiolabeled particles of human albumin to achieve the same effect.

Acoustic properties
Results from animal investigations show that Quantison Depot gives very effective myocardial gray scale opacification.

Pharmacokinetics
There is no published information on the pharmacokinetics of Quantison or Quantison Depot.

Safety/toxicology
There is no published information on the toxicology of Quantison or Quantison Depot.

Perfluorocarbon-exposed sonicated dextrose albumin (PESDA)

Physical characteristics
The microbubbles of PESDA are prepared ad hoc by sonicating solutions of glucose and human albumin with a fluorocarbon [52]. There is no information on bubble size or bubble size distribution.

Acoustic properties
In a dog model, PESDA, used together with pulsed ultrasound, gave a transient but significant increase in myocardial contrast intensity. The technique used, transient response imaging (TRI), was to delay real time ultrasound transmission until after the microbubbles had entered the myocardium [52]. As the frequency of the ultrasound pulses was reduced (to less than one per cardiac cycle) the myocardial contrast increased very markedly [52]. This increase was followed by the abrupt disappearance of the enhancement, probably as a result of bubble collapse [53, 54]. The mechanism of this effect is not well established but it could conceivably be a variant of the stimulated acoustic emission seen with Sonovist (SHU 563A) [48].

Pharmacokinetics
There is no published information on the pharmacokinetics of PESDA.

Toxicology
There is no published information on the toxicology of PESDA.

Aerosomes

Physical characteristics
The multilamellar structure of Aerosomes (Imarex, Tucson, AZ) has been con-firmed by electron microscopy [55] (Figure 9). The lipid bilayers are composed of dipalmitoylphosphatidylcholine and the structures differ from classical lipo-somes in that the internal space is filled with gas rather than fluid. In this for-mulation 1 mg of lipid stabilizes 1 ml of gas. The vesicles have a mean diameter of $7-8\,\mu m$ and they are reasonably pressure resistant, retaining approximately 70% of their initial gas content after 60 pressure cycles to 130 mmHg [56].

Acoustic properties
As the concentration of the vesicles rises ultrasound attenuation increases and the speed of sound through the suspension falls. The acoustic impedance of the

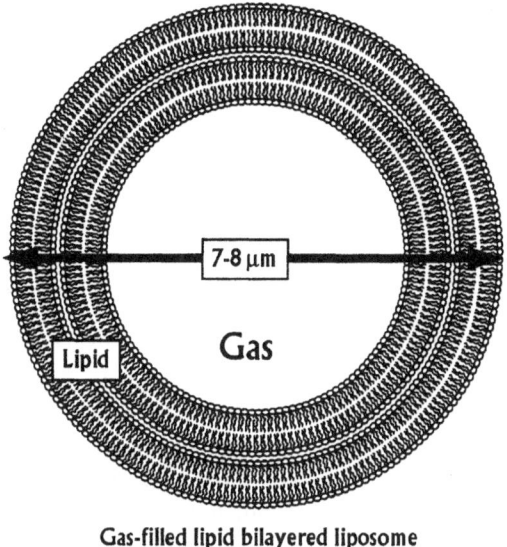

Gas-filled lipid bilayered liposome

Figure 9. Liposomal enhancers.

suspension, on the other hand, remains constant at 1.5 Rayls [56, 57]. In contrast, water-filled vesicles had very little influence on acoustics. In rabbits, a dose of 0.5 ml (0.14 ml/kg) produced four-chamber cardiac enhancement that lasted for 30–40 s; after a dose of 1 ml (0.29 ml/kg) the enhancement lasted for 60 s [20]. In a rabbit infarct model the liposomes improved the visualization of ventricular wall motion, endocardial border definition and the images of mitral valves and papillary muscles. In pigs a dose of 0.2 ml/kg produced four-chamber enhancement that lasted for 2.5 min and intra-aortic injections of 0.1 ml/kg visualized myocardial perfusion.

Pharmacokinetics
There is no published information on the pharmacokinetics of these liposomal enhancers.

Safety
A 1 ml bolus (≥40 ml/kg) killed none of the test mice [55]. The dose that produced adequate enhancement in pigs and rabbits (0.1 ml/kg) contains roughly 0.1 mg lipid/kg, a fairly low dose. These results suggest that the therapeutic index of this liposome suspension may be > 100.

Table 8. Diagnostic imaging and the use of contrast enhancers (USA 1994).

Imaging procedure	Number of procedures (millions)	Percent using contrast
X-ray angiography	4.1	100
Computerized tomography	21.3	44
Nuclear medicine	9.1	100
Magnetic resonance	7.7	26
Sub total	42.2	58
Ultrasound	50.9	0

Adapted from Ref. 55.

FUTURE PROSPECTS

The technology that enables the production of stable microbubble formulations is now well understood and this understanding will certainly lead to the production of new agents with specific properties. The gas-filled polymer microparticles, typified by SHU 563A, represent one of the most interesting of these new developments. The phenomenon of stimulated acoustic emission combined with new scanning techniques could lead to a conceptual shift from the echo-enhancement with which we are familiar to an entirely novel acoustic equivalent of radionucleide scintigraphy. The more persistent agents will provide quantitative information on myocardial, hepatic, renal and cerebral perfusion and microbubbles with extended half lives will allow the use of very low doses to target organs like the liver or the lymph nodes. Another promising approach will be the fusion of antibody technology with microbubble production to give agents with affinities for individual organs or cell types.

The use of echo-enhancers may expand enormously during the next few years. The size of the potential demand for effective agents is obvious from a survey of the current situation in the USA (Table 8).

By the beginning of the next century the figures for ultrasound contrast will have begun to approach that for the other imaging procedures such as magnetic resonance, as the first generation echo enhancers become more widely used and new agents designed to address specific problems make their appearance.

REFERENCES

1. Holz K, Becker R, Schürmann R. Ultrasound in the investigation of tubal patency in infertile women. A meta-analysis of three comparative studies of Echovist 200. In press.
2. Degenhardt F, Jibril S, Eisenhauer B. Hysteringo-contrast sonography (HyCoSy) for determining tubal patency. Clin Radiol 1996;51(Suppl 1):15–18.
3. Burns PB. Potential advantages of harmonic imaging. Advances in Echo-Contrast 1994;4: 17–19.
4. Petrick J, Zomack M, Schlief R. An investigation of the relationship between ultrasound echo-enhancement and Doppler frequency shift using a pulsatile arterial flow phantom. Invest Radiology, in press.

5. Schwarz KQ, Bezante GP, Chen X et al. Quantitative echo contrast measurement by Doppler sonography. Ultrasound Med Biol 1993;19:289–97.
6. Shapiro JR, Reisner SA, Lichtenberg GS et al. Intravenous contrast echocardiography with use of sonicated albumin in humans. Systolic disappearance of left ventricular contrast after transpulmonary transmission. J Am Coll Cardiol 1990;16:1603–7.
7. Shandas R, Shan DJ, Bales G et al. Persistence of Albunex ultrasound contrast agent: in vitro study of the effects of pressure and acoustic power on particle size and duration of contrast and Doppler enhancement (Abstract). Circulation 1990;82(Suppl III):Abstr. 372.
8. Wiencek JG, Aronson S, Walker R et al. In vitro and in vivo stability of Albunex contrast agent (Abstract). J Am Coll Cardiol 1992;19:175A.
9. Schlief R, Cramer E, Poland H et al. Cyclical variations in echointensity of the transpulmonary contrast agent SHU 508 are correlated with pressure changes. Abstract for 63rd Scientific Sessions, American Heart Association, Dallas, Texas, November 12–15, 1990.
10. Ran B, Katz J. The response of microscopic bubbles to sudden changes in ambient pressure. J Fluid Mech 1991;224:91–115.
11. Schlief R, Cramer E, Poland H et al. Cyclical variations in echointensity of the transpulmonary contrast agent SHU 508 are correlated with pressure changes. Abstract for 63rd Scientific Sessions, American Heart Association, Dallas, Texas 1990 November 12–15.
12. Vink CLJ. The 'half-life' concept in the determination of the functional capacity of the liver. Clin Chim Acta 1959;4:583.
13. Marchesini G, Fabbri A, Bugianesi E et al. Analysis of the deterioration rates of liver function in cirrhosis, based on galactose elimination capacity. Liver 1990;10:65–71.
14. Fritzsch Th, Maass B, Müller B et al. Composition and tolerance of galactose-based echo contrast media. New Dimensions of Contrast Media. Hitoshi, Katayama, Brash RC, ed. Tokyo: Excerpta Medica, 1991:156–62.
15. Schwarz KQ, Bezante GP, Chen X et al. Hemodynamic effects of microbubble echo contrast. J Am Soc Echocardiogr 1996;9:795–804.
16. Grinstaff MW, Suslick KS. Air-filled proteinaceous microbubbles: synthesis of an echo contrast agent. Proc Natl Acad Sci 1991;88:7708–10.
17. Barnhart J, Leven H, Villapando E et al. Characteristics of Albunex: air-filled albumin microspheres for echocardiography contrast enhancement. Invest Radiol 1990;25:S162–4.
18. de Jong N, Hoff L, Skotland T et al. Absorption and scatter of encapsulated gas filled microspheres: theoretical considerations and some measurements. Ultrasonics 1992;30:95–103.
19. Shandas R, Sahn DJ, Bales G et al. Persistence of Albunex ultrasound contrast agent in vitro. Study of the effects of pressure and acoustic power on particle size and duration of contrast enhancement (Abstract). Circulation 1990;82:111–95.
20. Wiencek JG, Willker RC, Gretler D et al. In vitro and in vivo stability of Albunex contrast agent (Abstract). J Am Coll Cardiol 1992;19:175A.
21. Shapiro JR, Reisner SA, Lichtenberg GS et al. Intravenous contrast echocardiography with use of sonicated albumin in humans. Systolic disappearance of left ventricular contrast after transpulmonary transmission. J Am Coll Cardiol 1990;16:1603–7.
22. Powsner SM, Keller MW, Sanie J et al. Quantitation of echo-contrast effects. Am J Physiol Imaging 1986;1:124–8.
23. Walker R, Wiencek JG, Aronson S et al. The influence of intravenous Albunex injection on pulmonary artery pressure, gas exchange and left ventricular peak intensity. J Am Soc Echocardiogr 1992;5:460–72.
24. Keller MW, Glasheen W, Kaul S. Albunex: A safe and effective commercially produced agent for myocardial contrast echocardiography. J Am Soc Echocardiogr 1989;2:48–52.
25. Keller MW, Feinstein SB, Watson DD. Successful left ventricular opacification following peripheral intravenous injection of sonicated contrast agents: an experimental evaluation. Am Heart J 1987;114:570–5.
26. Reisner SA, Ong LS, Lichtenberg GS et al. Myocardial perfusion imaging by contrast echo-

cardiography with use of intracoronary sonicated albumin humans. J Am Coll Cardiol 1989; 14:660–5.

27. Ismail S, Johnson SH, Utsunomiya H et al. Safety and efficacy of sonicated albumin microspheres in perfusion and vein graft patency assessments. Clin Cardiol 1991;14:V29–V32.

28. Correas J-M, Kessler D, Worah D et al. Echogen emulsion: Current clinical status in the development of the first fluorocarbon gas-based contrast agent. Presented at the First European Symposium on ultrasound contrast imaging. Rotterdam, Jan 25–26th 1996.

29. Corres JM, Quay SD. EchoGen™ emulsion: A new ultrasound contrast agent based on phase shift colloids. Clin Radiol 1996;51(Suppl 1):11–14.

30. Sehgal CM, Arger PH, Pugh CR. Sonographic enhancement of renal cortex by contrast media. J Ultrasound Med 1995;14:741–8.

31. Forsberg F, Liu J-B, Merton DA et al. Parenchymal enhancement and tumor visualisation using a new sonographic contrast agent. J Ultrasound Med 1995;14:949–57.

32. Correas J, Bothwell WA, Meuter AR et al. Human pharmacokinetics of a US contrast agent studied with gas chromatography. Http://www.rsna.org/sci_program/scisess/y002.016.008.html.

33. Gong Z, Giraud G, Pantely G et al. Time course of chamber and myocardial opacification with EchoGen: a videodensitometric study in monkeys with microvascular visualization studies in a cat mesentery model (Abstract). Circulation 1994;90(Suppl I):1–556.

34. Grayburn PA, Erickson JM, Escobar J et al. Peripheral intravenous myocardial contrast echocardiography using a 2% dodecafluoropentane emulsion: Identification of myocardial risk area and infarct size in the canine model of ischemia. J Am Coll Cardiol 1995;26:1340–7.

35. Kamp JGF. Right and left ventricular opacification and myocardial contrast echocardiography by intravenous administration of BY 963. Presented at the First European Symposium on ultrasound contrast imaging. Rotterdam, Jan 25–26th 1996.

36. Dittrich HC, Bales GL, Kuvelas T et al. Myocardial contrast echocardiography in experimental coronary artery occlusion with a new intravenously administered contrast agent. J Am Soc Echocardiogr 1995;8:465–74.

37. Tejpal Y. First clinical studies with FSO 69. Presented at the First European Symposium on ultrasound contrast imaging. Rotterdam, Jan 25–26th 1996.

38. Molecular Biosystems' contrast agent demonstrates improved delineation. PJB Publications PHIND (Daily and Current). Ref No 00502536 dated 18/07/96.

39. Aeschbacher BC. Experience with Imagent US, a new ultrasound contrast agent. Presented at the First European Symposium on ultrasound contrast imaging. Rotterdam, Jan 25–26th 1996.

40. Mattrey RF, Steinbach GC, Baker KG et al. Effect of ultrasound transmit power on liver enhancement following the administration of Imagent US, a perfluorochemical-stabilized microbubble contrast agent. Http://www.rsna.org/sci_program/scisess/Y005.020.005.html.

41. Mattrey RF, Girard MS, Steinbach GC. Improvement in tissue contrast enhancement at the second harmonic frequency in a renal infarct model following the intravenous injection of a perfluorochemical-stabilized microbubble contrast agent. Http://www.rsna.org/sci_program/scisess/y005.020.006.html.

42. Sulfur hexafluoride (SF6). A colourless, odorless, nontoxic, nonflammable liquefied gas. US Technical Data Sheets. Http://www.boc.com/gases/solution/electron/data/sf6.htm.

43. Schneider M, Arditi M, Barrau M-B et al. BR1: A new ultrasonographic contrast-agent based on sulfur hexafluoride-filled microbubbles. Invest Radiol 1995;30:451–7.

44. Broillet J. BR1: a new ultrasonographic contrast agent based on sulfur hexafluoride-filled microbubbles. Presented at the First European Symposium on ultrasound contrast imaging. Rotterdam, Jan 25–26th 1996.

45. van Rooyen MM, van Biljon G, du Toit AM. Outpatient treatment of retinal detachment. S Afr Med J 1989;75:12–15.

46. Bauer A, Mahler M, Urbauk A. Microvascular imaging – results from a phase I study of the novel polymeric ultrasound contrast agent SHU 563A. This book, Chapter 43.

47. Hoffmann C. Mechanische charakterisierung von ultraschallkontrast mitteln. Ultraschall Klin Prax 1995;9:208–10.

48. Burns PN, Fritzsch T, Weitschies W et al. Psuedo Doppler shifts from stationary tissue due to the stimulated emission of ultrasound from a new microsphere contrast agent. Radiology 1995; 197(P).
49. Fischer G, Sackmann M. Stellenwert der endoskopischen Farbdopplersonographie zur Diagnostik von Magenvarizen und zur Erfolgskontrolle einer therapeutischen Cyanoacrylate-Verklebung. [Value of endoscopic color Doppler ultrasound in the diagnosis of gastric varices and follow-up of therapeutic cyanoacrylate obliteration.] Z Gastroenterol 1995;33:727–8.
50. Koukoubis TD, Glisson RR, Feagin JA et al. Augmentation of menisca repairs with cyano-acrylate glue. J Biomed Mat Res 1995;29:715–20.
51. Sutton A. Deposit echocontrast agents. Presented at the First European Symposium on ultra-sound contrast imaging. Rotterdam, Jan 25–26th 1996.
52. Porter TR, Xie F, Klizer K. Intravenous perfluoropropane-exposed sonicated dextrose albumin produces myocardial contrast which correlates with coronary blood flow. J Am Soc Echo-cardiogr 1995;8:710–18.
53. Porter TR. Transient myocardial contrast after initial exposure to diagnostic ultrasound pressures with minute doses of intravenously injected microbubbles. Demonstration and potential mech-anisms. Circulation 1995;92:2391–5.
54. Porter TR, Xie F, Armbruster R et al. Improved myocardial echocardiographic contrast with second harmonic imaging in humans using intravenous perfluorocarbon-exposed sonicated dextrose albumin. Presented at the First European Symposium on ultrasound contrast imaging. Rotterdam, Jan 25–26th 1996.
55. Unger E, Shen D, Fritz T et al. Gas-filled liposomes as ultrasound contrast agents. Invest Radiol 1994;29(Suppl 2):S134–S136.
56. Unger E, Shen D, Fritz T et al. Gas-filled liposomes as echocardiographic contrast agents in rabbits with myocardial infarcts. Invest Radiol 1993;28:1155–9.
57. Arai A, Kenny A, Shiota T et al. Transpulmonary passage of Aerosomes, a pressure-stable, lipid-based echocardiographic contrast agent: studies in pigs. ACC 43rd Scientific Session, 1994; (Abstract).
58. Theta Corporation, AMR, Klein Biomedical. Http://www.researchmag.com/company/fs/fssnus.html.

6. Echo-enhancing agents: safety

NAVIN C. NANDA & EDWIN L. CARSTENSEN

BACKGROUND

The enhanced echocontrast associated with the gas microbubbles produced when a liquid is rapidly injected into a large blood vessel was first reported by Gramiak and Shah in 1968 [1]. Twenty-eight years later, microbubbles remain the standard echocontrast enhancers. They have been used clinically to verify the presence of intracardiac shunts [2] and valvular regurgitation [3, 4] and for the measurement of cardiac output [5]. Contrast echocardiography (CE) has also been used as a means of defining myocardial perfusion non-invasively [6]. In recent times this rapidly developing diagnostic technique has been more accurately named myocardial contrast two-dimensional echocardiography (MC-2DE). Contrast enhancers have extended the applications of cardiac color Doppler imaging, and gas bubbles have also found applications outside the cardiovascular system, in establishing tubal patency during investigations of infertility and in the diagnostic evaluation of potential malignancies [7, 8]. In the early days of CE, clinical investigators speculated that the gaseous microbubbles present in the echocontrast agents (ECA) were probably the source of the observed contrast [6, 9]. This discovery was confirmed [10–13] and it is now accepted that the ultrasonic targets in contrast-enhanced ultrasonography are the microbubbles themselves.

Since the first description of CE the applications of diagnostic ultrasound have expanded greatly, largely because of the enhanced spatial resolution of the tomographic gray-scale image and the general acceptance of two-dimensional color Doppler imaging. The main limitation of ultrasound is high sound attenuation and low signal-to-noise ratio: echo-enhancement overcomes that limitation very effectively. One problem with the original microbubble enhancers was their inability to cross the pulmonary barrier and opacify the left side of the heart and the arterial tree. The most recent saccharide-stabilized enhancer, SHU 508A and, to a lesser extent, the sonicated albumin enhancer Albunex or Infoson, have overcome this limitation. SHU 508A, a galactose stabilized agent, contains a

115

N. C. Nanda, R. Schlief and B. B. Goldberg (eds.), Advances in Echo Imaging Using Contrast Enhancement,
Second Edition, 115–131.
© 1997 *Kluwer Academic Publishers.*

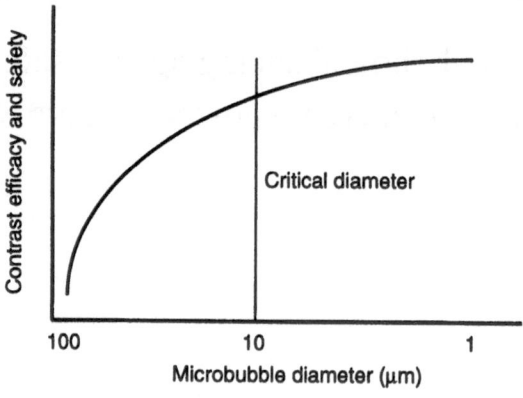

Figure 1. Schematic representation of relationship between microbubble diameter and magnitude of contrast effect and safety.

trace of palmitic acid which enables it to cross the lungs and enhance the signal throughout the blood pool, significantly expanding the role of the ultrasound echo enhancers.

MICROBUBBLE FORMATION AND SIZE

By the late 1970s the ultrasound contrast effect had been observed in a variety of unrelated injected solutions, and a relationship was shown to exist between the microbubble content and the magnitude of the contrast effect. Many of the ECAs provided microbubbles in variable quantity and size and with unknown stability. An early paper by Kort et al. [13] showed that large diameter microbubbles (100 μm) lodged in rat arterioles, obstructing blood flow for up to 200 s. Eventually, the microbubbles shrank and microcirculation was restored. This was the first indication that myocardial echocontrast suspensions would need to contain very small microbubbles to ensure safety and efficacy and freedom from the dangerous side-effect of gas embolization. Confirmation that control of microbubble size and uniformity were of prime importance in assuring the non-toxicity of echocontrast agents came from Xie et al. [14] and Dick et al. [15]. Both groups concluded that echocontrast preparations containing air microbubbles must ensure a diameter of <10μm in order to prevent trapping the capillary vasculature and to avoid the risk of significant but transient toxicity (see Figure 1).

By 1984, Feinstein's group [10, 16] had described an alternative to the traditional hand-agitation method of preparing small microbubbles. The new procedure involved the use of ultrasonic energy derived from a piezoelectric crystal which was introduced into the echocontrast medium (ECM) to produce agitation

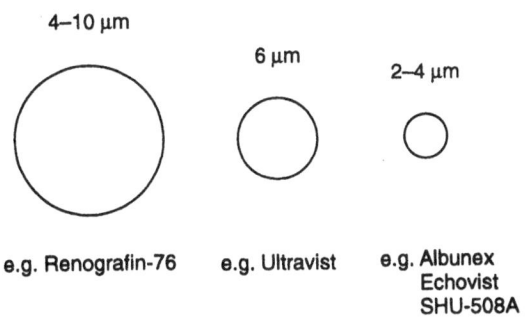

Figure 2. Newer, more advanced agents contain stable microbubbles of a smaller, more uniform size.

Table 1. Factors affecting the safety of echocontrast media.

Microbubbles	Carrier solution
Size of bubbles	Osmolality
Stability of bubbles	Viscosity
Uniformity of size	Surfactant properties

and cavitation. The first generation cavitation bubbles ultimately collapsed to second generation 'by-product' microbubbles, which were then used as sonicated ECM. These resulting microbubbles were significantly smaller and more stable than earlier hand-agitated microbubbles and were more uniform in size (Figure 2).

It is now very clear that, while size, stability and uniformity of the microbubbles generated in the ECM are vitally important for safety, the characteristics of the carrier solution must also be carefully selected. The most important physicochemical factors are osmolality and viscosity and the surfactant properties of the carrier medium (Table 1).

Osmolality

The osmolality of echocontrast suspensions has received far too little attention from investigators. Hyperosmolality can lead to direct cardiodepressive effects, such as lowering of coronary blood flow. Hadjuczki et al. [17] showed a significant and direct relationship between observed reactive hyperaemia and the osmolality of carrier solutions. Most of the transient regional contractions observed in their study in dogs occurred in animals given the solutions of highest osmolality, such as 70% sorbitol. Iopamidol, an agent with low osmolality, demonstrated a markedly lower hyperaemic response.

The osmolality of the saccharide-stabilized enhancers has been exhaustively investigated. The osmolality of suspensions of SHU 508A are similar to those of the commonly used ionic and non-ionic X-ray contrast media but the injected volumes of the echo-enhancer, and hence the total osmotic load, are lower than those of the radiographic agents. The osmolality of the liquid compartment of the saccharide suspension (910–980 mmol/kg) is much lower than that of the standard MRI contrast enhancer gadopenetate (1980 mmol/kg). The soluble microparticles of SHU 508A could conceivably enhance the hypertonicity of the blood downstream from the injection site, unlike the effect of true solutions which become progressively more dilute. The well-known effects of hyper-osmolar injections (warmth, pain and local paraesthesia) are rather less localized after SHU 508A than after an X-ray or MRI contrast enhancer. The excess osmotic load contributed by SHU 508A is similar to that associated with other widely used intravenous formulations. It should be taken into consideration when considering the use of the product in patients in whom severely impaired cardiac function verges upon cardiac decompensation [18].

Viscosity

The viscosity of carrier solutions has a profound influence on the size and stability of microbubbles and yet it is only recently that its importance has been recognized. Koenig and Meltzer [19] compared the relative viscosity of a number of ECAs and found an inverse correlation between microbubble size and viscosity. This may be caused by a 'shell' of viscous material that surrounds the microbubble and prevents if from coalescing with its neighbours. Highly viscous echocontrast carrier solutions are associated with adverse effects such as transient changes in coronary flow, left ventricular function and left ventricular hemodynamics.

Surfactant properties

Many of the early successful echocontrast solutions were good surfactants. The pressure inside a small microbubble is greater than that inside a larger one, so bubbles in an ECM containing a good surfactant will be less likely to coalesce to reduce their internal pressure. As a consequence, the ECM will retain a high number of smaller bubbles and those small bubbles will decay more slowly.

SAFETY AND TOXICITY STUDIES

Early contrast enhancers

From 1968, freshly agitated indocyanine green dye was widely used as an echo-contrast agent (Table 2) [1]. Solutions of the green dye have very favorable

Table 2. Echocontrast enhancers.

Early media	New media
Indocyanine green	Gelatin-stabilized
Saline (alone or with 5% dextrose)	Viscosity-stabilized ionic (e.g. Renografin-76)
Carbonated water + 5% dextrose	Non-ionic (e.g. Ultravist-370)
	Albumin-stabilized (e.g. Albunex)
	Saccharide-stabilized (e.g. Echovist [SHU-454], SHU 508A)
	Phase-shift liquid/gas (Echogen)

surfactant properties capable of stabilizing microbubbles. An equally popular early ECA was saline, either alone or mixed with 5% dextrose in water [20] and some investigators also used carbonated water with 5% dextrose [21]. Early attempts to use hydrogen peroxide led to the formation of microbubbles with a large and variable size range [20, 21], and the procedure was not developed for clinical use. In 1984, the first survey of the safety of echocontrast enhancers was made by Bommer et al. [22]. They reported, on behalf of the American Society of Echocardiography, on the relative safety of CE in patients, and evaluated a retrospective survey of 363 physicians in detail. The major ECAs under review were indocyanine green and certain saline solutions, some of which contained preservatives. Only 32 significant cases of side-effects were reported from an estimated exposure of 51 180 patients. The authors suggested that the most serious complication following the injection of ECAs was gas embolization. This could account for some of the transient neurological side-effects observed. At about the same time, a case report appeared of a transient cerebrovascular defect in a patient with a right-to-left shunt who received an echocontrast enhancer [23]. The observed defect improved significantly within 24 h and the event was attributed to the possibility of a microbubble air embolism produced by the injected ECA.

More recent echocontrast enhancers

The possibility of embolization has probably been most thoroughly investigated in connection with the saccharide-stabilized agent SHU 508A. Neither the early preclinical studies of SHU 508A nor the later extensive clinical investigations showed any evidence of embolization. Experience with lung perfusion scintigraphy which uses radiolabelled macroaggregates 15–40 μm in diameter to occlude lung capillaries, is relevant here. The scintigraphic aggregates obstruct roughly one in every 1000 lung capillaries (out of a total of around 10^8) and the occlusions are cleared by enzymatic degradation and the reticuloendothelial system with a half-life of 3–6 h [24, 25]. Pulmonary scintigraphy is well tolerated, even in patients with severe pre-existing lung embolization. The microbubbles and microparticles of SHU 508A have a diameter 10 times less than that of the scintigraphic macroaggregates, well within the size range of the cellular blood elements. In addition, the microparticles and microbubbles dissolve rapidly in the blood. The risk of microembolism associated with SHU 508A is extremely small and any embolization that did occur would be minor and short-lived because of the high solubility of galactose and palmitic acid.

THE INTERACTION OF ULTRASOUND WITH CONTRAST ENHANCERS

The gas microbubbles of an ultrasound contrast enhancer can interact with incident ultrasound in two ways:

1. The ultrasound energy can destroy the contrast agent.
2. The microbubbles can act as nuclei for inertial cavitation, a phenomenon that occurs when exposure to ultrasound drives the bubbles into oscillation. The oscillation can become very large and inertial cavitation occurs when the bubbles expand to more than twice their equilibrium radius before collapsing violently.

Glass et al. [26] were the first to report a decrease in scattering of the contrast agent Albunex under exposure to diagnostic levels of ultrasound. They found that the reduction was greater at frequencies around 1 MHz than at those >5 MHz. It is reasonable to expect that large bubbles would be more fragile than small bubbles but whether that is an explanation of the selective effects of ultrasound remains to be investigated. The process seems to be very rapid.

Inertial cavitation has the potential to damage tissue and it is probably responsible for the lysis seen when dilute cell suspensions are exposed to ultrasound [27]. Microbubble contrast agents enhance this cell lysis [28–30]. The first studies of ultrasound lysis used cell concentrations of 1 or 2% by volume but microbubble contrast agents can produce cell lysis at physiological hematocrits [28]. Cell lysis has been also demonstrated in mice injected with Albunex [31]. The exposures used pulse lengths and repetition frequencies similar to those employed diagnostically. The acoustic pressure threshold for hemolysis was approximately 3.5 MPa for the peak positive pressure (~2 MPa, negative pressure; ~200 W/cm^2, pulse average intensity) at 1 MHz. The thresholds are greater at higher frequencies. Few if any commercial ultrasound machines have output levels high enough to produce hemolysis in patients during diagnostic procedures. There is evidence to suggest that commercially available diagnostic ultrasound equipment can produce lung hemorrhage in experimental animals [32–37]. The threshold for this effect is approximately 1 MPa at 2 MHz. Intestinal hemorrhage has also been demonstrated with the pulsing regimens used in diagnostic ultrasound [38]. The mechanism for this effect is probably inertial cavitation of bubbles within the lumen. Since the thresholds for this phenomenon are near the upper limits of commercially available ultrasound systems, it is unlikely to have any clinical relevance. The acoustic wave emitted by a piezoelectric lithotripter causes lung and intestinal hemorrhage in mice, but the lung hemorrhage is not enhanced by ultrasound contrast agents [39]. This and other characteristics of the phenomenon leave the underlying physical mechanism undefined. Other soft tissues do not appear to be affected even with pressure pulses as great as 40 MPa. Once a microbubble contrast agent is injected into the blood of the animal, however, almost all of the tissues are subject to hemorrhage and the thresholds are quite low – of the order of 2 MPa [39]. Earlier observations showed that it is not possible to extrapolate from the biological effects of lithotripter exposures to the possible effects of pulsed ultrasound [40, 41]. Hence, the implications of these findings for diagnostic ultrasound remain to be evaluated.

Ultrasound can induce hemolysis and hemorrhage and ultrasound contrast agents can magnify these effects. The mechanism underlying the additive effect of the contrast enhancer is probably inertial cavitation and the pressure threshold for hemolysis in vivo is probably near to the maximum that can be achieved with current diagnostic equipment. There is no direct evidence that any of these interactions cause problems during the normal use of ultrasound contrast agents in clinical practice.

CONCLUSIONS

Bommer et al. [22] recommended that all investigators using CE should take precautions similar to those observed during routine catheterizations, namely, the removal of all visible air from the injection apparatus prior to injection. This was especially important when dealing with patients with right-to-left shunts. Side-effects associated with indocyanine green dye, used by the majority of investigators, were also highlighted. Idiosyncratic reactions had already been reported following its use in catheterization. There were no reports of anaphylactic reactions in this survey but the authors cautioned against the use of repeated injections of the dye in patients with prior sensitivity. Side-effects of auditory hallucinations, paraesthesia and abdominal pain were recorded with ECA injections containing bacteriostatic saline solution, and no side-effects of this kind occurred with pure saline solution. Once again, the authors warned against the repeated injection of contrast agents containing preservatives.

Overall, the survey showed that the risk of side-effects from these early ECAs was very low and the side-effects that were recorded were transient. During the ensuing decade the high benefit-to-risk ratio for the CE technique stimulated research into the development of safe echocontrast enhancers that could opacify the left heart. The more recently introduced agents are at least as safe as the original formulations and, in general, their manufacture is much more tightly controlled.

ECHOCONTRAST ENHANCERS IN CLINICAL USE

Gelatin-stabilized media

Animal studies
Several investigators [42–44] have studied gelatin-encapsulated microbubbles as potential ECAs, using a preparation from Rasor Associates (Sunnyvale, CA), which was later withdrawn by the manufacturers. Meltzer et al. [45] found commercially available gelatin to be a suitable medium for the generation of precision microbubbles for experimental contrast echocardiographic use. Bommer et al. [46] reported excellent left heart contrast in animal studies when gelatin-encapsulated bubbles were injected into the distal pulmonary artery. No major vital changes were recorded.

Human studies
Gelatin has been used for years as a plasma expander. It has surfactant properties and can stabilize microbubbles and prevent coalescence, but gelatin plasma expanders are no longer commercially available in the USA, having been withdrawn because of occasional allergic reactions. Haemaccel (Hoechst) is a gelatin-based solution, still commercially available in Europe, that is effective clinically as an echocontrast enhancer, giving contrast that lasts for as long as 60 s [47, 48]. Unfortunately, mild transient side-effects (transient second-degree atrioventricular block, QRS axis deviation, transient T wave changes) occur in a few patients, although the majority (21 of 25), experienced no significant electrocardiographic changes [48]. Berwing et al. [49–50] tested a new ECM based on a commercially available agent Gelifundol (Biotest Pharma ERG), consisting of 5.4% oxypolygelatin, widely used outside the USA as a plasma substitute. Gelifundol was stabilized with lecithin, oleum sojae and sodium-iron-3-gluconate complex. Gelifundol opacified the left ventricle after peripheral intravenous injection in 95% of patients who had no shunt connection. Only three patients reported a slight taste sensation with the treatment.

This new agent offers relatively good physiological ranges of osmolality and viscosity compared with 70% dextrose. It does not appear to give rise to coronary flow changes and subsequent reactive hyperaemia, but a much fuller evaluation is still required in clinical practice to establish its overall safety profile and to assess its transcapillary stability.

Viscosity-stabilized media: ionic vehicles

Animal studies
Highly viscous suspensions of echocontrast enhancers have been studied because of the small size and stability of the microbubbles they contain, compared with those found in saline or 5% dextrose contrast solutions. Feinstein et

al. [16] used viscous solutions to develop high intensity sonication for micro-bubble creation. They first examined in vitro a range of sonicated viscous echo-contrast enhancers such as Renografin-76 (ER Squibb and Sons Inc.: meglumine sodium diatrizoate), 50% or 70% sorbitol or 70% dextrose. The most viscous solution, 70% sorbitol, yielded the smallest range of microbubble size ($6 \pm 2 \mu$m) and the bubbles persisted for the longest time. Renografin-76 yielded micro-bubbles of size $10 \pm 4 \mu$m. A second study, in cats [10], demonstrated the poten-tial ability of the microbubbles in sonicated Renografin-76 to flow with the red blood cells in the mesenteric circulation. Feinstein's group [16] concluded that sonicated viscosity-stabilized ECAs could show the extent of regional under-perfusion. Satisfactory quantitation, however, had still to be demonstrated. Further studies of the effects of sonicated Renografin-76 on left ventricular contractility in dogs showed that it was the carrier rather than the microbubbles that had an effect upon left ventricular function [51]. Gillam et al. [52] used Renografin-76 via intracoronary injection in dogs to study the functional and pathological effects of the repeated use of echocontrast enhancers. They found no gross evidence of neurological impairment and an absence of major patho-logical ill effects, but they saw consistent short-lived depression of ventricular wall motion within the contrast-enhanced segments. They concluded that the direct depressant effect on myocardial contractility appeared to be partly related to the hyperosmolality of Renografin-76, as well as to its ionic content.

Human studies
Sonicated Renografin-76 microbubbles were used as an ECM to evaluate blood flow in 60 patients undergoing cardiac catheterization [53]. The microbubbles measured $3.8 \pm 2.1 \mu$m and no complications were reported. In a further study in 14 patients, Feinstein et al. [54] used MC-2DE with sonicated Renografin-76 (bubble size $4.52 \pm 2.8 \mu$m) to determine myocardial perfusion patterns in patients without significant coronary artery disease. No major complications were reported, although there were transient changes in heart rate, blood pres-sure and ECG during the intracoronary injections. The results showed that intra-myocardial blood flow varied between individuals, and the authors suggested that the highly viscous and hyperosmolar solution of Renografin-76 may have produced an autoregulatory vascular response. Reisner et al. [55] used sonicated meglumine to obtain immediate quantitative assessment of the results of coron-ary angioplasty in 19 patients. Three patients experienced mild side-effects. One had transient precordial pain and ECG changes, a second experienced brief AV block, while a third developed a symptomatic sinus bradycardia. All highly viscosity-stabilized ionic ECA vehicles studied to date produce some degree of intrinsic myocardial depression. For these reasons alone they are far from ideal for the clinical application of MC-2DE.

Viscosity-stabilized media: non-ionic vehicles

Human studies
Some of the problems associated with the use of radiographic contrast agents as vehicles for echogenic microbubbles are due to their highly ionized radio-opaque content. Non-ionic vehicles, such as iopramid (Ultravist-370; Schering AG, Berlin) might overcome this problem. Mudra et al. [56] used a sonicated suspension of microbubbles ($6 \pm 4\,\mu$m) in iopramid to produce echocontrast enhancement in 31 patients. Satisfactory MC-2DE images were obtained following intracoronary injection of the ECA and iopramid was well tolerated, producing no serious side-effects. In a second study in 11 patients with coronary vessel disease, sonicated iopramid was injected directly into the bypass graft during surgery [57]. An effective echocontrast picture of bypass flow rate was obtained and there were no appreciable changes in ECG, pulmonary pressure or aortic pressure during the procedure.

Albumin-stabilized media

Animal studies
Bommer et al. [46] described an echocontrast medium containing albumin as early as 1979 but 7 more years passed before Feinstein [58] recognized its potential in MC-2DE. He found that sonicated albumin microspheres could pass through the capillary vasculature almost as readily as red blood cells. The microspheres ($2-4\,\mu$m) were iso-osmotic to human serum and echogenic for several days. The work was continued with sonicated albumin microspheres ($3\,\mu$m) in dogs [59]. The investigators obtained good opacification of myocardium with no significant changes in coronary blood flow or reduced left ventricular function and hemodynamics after repeated injections. In another series of intracoronary injections in rabbits, Keller et al. [60] compared the echocontrast properties of sonicated albumin stabilized bubbles ($5.8\,\mu$m) with those of sonicated Renografin-76 and noted that the small size of the microbubbles in the albumin preparation allowed them to pass through capillary beds. Sonicated human albumin did not produce significant changes in coronary blood flow or regional systolic left ventricular function. In comparison, both sonicated and non-sonicated Renografin-76 caused significant changes in these variables. Results from pathological studies showed no ill effects after a 10 ml injection of sonicated albumin but, after 20 ml, significant changes, especially microinfarcts within the myocardium, were observed. The authors concluded that sonicated albumin possessed minimal toxic properties and that, in clinically applicable quantities, it produces no pathological effects. This ECM, they said, showed promise as a tracer for blood flow during MC-2DE.

Human studies
The potential of sonicated albumin as a carrier in MC-2DE myocardial per-fusion imaging has been demonstrated in humans [61]. In a subgroup of nine patients the investigators injected sonicated albumin or sonicated Renografin-76, using the same technique. Sonicated albumin caused neither arrhythmias nor any depression in hemodynamic function, and there were no subjective symp-toms during intracoronary injection. More recently, two groups have studied the safety and clinical efficacy of sonicated human serum albumin (Albunex, Molecular Biosystems Inc., San Diego) [62, 63]. Feinstein et al. [62] admin-istered Albunex as an i.v. bolus in 71 patients. The efficacy of the injections was assessed by two independent blinded observers. All injections were well toler-ated and no clinically significant ECG, physical or neurological abnormalities were seen. A small number of transient reactions to Albunex were observed, but all resolved spontaneously. The authors concluded that Albunex was non-toxic, safe and effective but that additional clinical experience was still required. There was some variability in the degree of ventricular cavity opacification which appeared to be related to the volume of Albunex injected. Similar variability in left ventricular myocardial opacification was recorded by Sanders et al. [63] in a study in 30 mildly hypertensive patients, who received Albunex by continuous i.v. infusion in a forearm vein. The investigators found this ECM to be well tolerated in all patients in their study and found no changes in vital signs or hemodynamic variables.

More recently Muan has reviewed the results of the European Albunex clinical trial program in which several hundred patients were investigated [64]. The trial concentrated on three applications, echo ventriculography, cardiac Doppler flow and myocardial contrast echocardiography. The results showed that Albunex was well tolerated and showed no evidence of any neurological, immunological or hemodynamic toxicity. Albunex was withdrawn from the European market early in 1996.

Saccharide-stabilized media

Animal studies
Saccharide-stabilized microbubbles suspended in carrier medium provide a very promising new approach to echocontrast enhancement in MC-2DE. One poten-tial agent is SHU-454 (Echovist: Schering AG, Berlin), a preparation of galac-tose microcrystals penetrated by air passages with very precisely controlled dimensions. When the microcrystals are dispersed in a saturated solution of galactose, the solution enters the air passages which collapse to produce a stable microbubble suspension. In animal studies [65, 66] SHU-454 gave an excellent reproducible opacification of the right heart when compared with other agents. It was well tolerated with minimal side-effects. SHU-508A is a development of SHU-454 in which the surface characteristics of the microbubbles are subtly

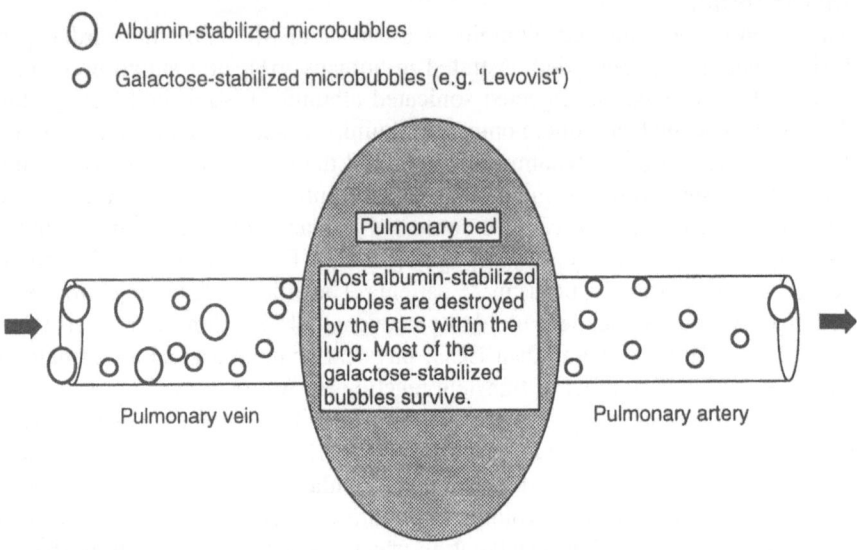

Figure 3. Schematic representation of the passage of stabilized microbubbles through the pulmonary bed.

altered by the addition of a trace of palmitic acid. Because of this alteration, the bubbles do not attract the attention of the lungs' reticuloendothelial system. Consequently they pass through the pulmonary bed and opacify the left heart and the entire blood pool (Figure 3). In a comparative study in eight dogs, SHU-508A produced significantly greater and more homogeneous left heart contrast than any of the other agents (indocyanine green, saline and SHU-454) [67] with no detectable cardiodepressant effect.

Human studies
Rovai et al. [67] have recently described fully the safety, efficacy and reproducibility of SHU-454 in a clinical trial involving 103 patients. The ECM was administered i.v., and high quality and reproducible echocontrast enhancement was obtained in the majority of patients. Only six patients reported any symptoms, always of mild intensity and short lasting, none of which required the study to be interrupted. The authors concluded that SHU-454 was effective and well tolerated.

The first study of SHU-508A in B-mode enhancement used four-step dosage increments starting with placebo (0.9% saline) and ending with 16 ml of Levovist 400 mg/ml. There was no detectable effect on systolic blood pressure, ECG trace (PQ interval, QRS complex, ST time), cardiac output, diastolic blood pressure, impedance cardiography (thoracic fluid index, ventricular ejection time, ejection velocity index, stroke volume) or heart rate [18]. The second B-mode study used higher volume injections (35 and 70 ml, 400 mg/ml) given over 3 min, 1 min or 30 s but did not use a saline control. In four of the five volunteer subjects 70 ml of SHU-508A given over 30 s produced a transient fall in systolic and diastolic blood pressure of between 6 and 47 mmHg, but impedance cardiography showed no changes. In the absence of any control injection the reasons for the fall in blood pressure are hypothetical. The changes could be a consequence of the injection volume and the osmolality and viscosity of the suspension. In any event this 70 ml dose at which hemodynamic effects first appeared, is more than eight times higher than the maximum recommended dose of 8 ml of a suspension containing 400 mg/ml.

The Phase II and III trials of SHU-508A included a total of 1819 patients [18]. Most experience was gained during the European Phase III study which included 1255 patients. Another 590 patients were investigated in studies conducted in the USA and Japan. Adverse events were recorded over the 24 h after the injection in all the studies and biochemical and hematological measurements were made at baseline and 30 m and 24 h after the injection. Four trials have been conducted in the USA in a total of 157 patients, using protocols almost identical to those used for the European studies. The Japanese trials included 427 patients, again with protocols similar to those used in the European trial. SHU-508A was extremely well tolerated by the vast majority of patients. Any adverse events that did occur were almost always related to the intravenous injection of a hyperosmolar fluid, rather than to any specific properties of this formulation. To minimize discomfort to the patient during and immediately after the injection the number of doses should be minimized. None of the adverse events reported during the clinical trials suggest that Levovist poses any specific risk to the patients with the various cardiac and vascular diseases likely to be referred for diagnostic ultrasound.

Phase-shift agents

One of the most elegant concepts to be applied to the development of an ultrasound echoenhancer is that of the shift in phase from liquid to gas. The agent is an emulsion of the fluorocarbon dodecafluoropentane in an aqueous vehicle containing two surface-active agents. The fluorocarbon boils at body temperature, automatically creating a suspension of echogenic microbubbles. In animal studies the formulation was very effective, producing bubbles from 3 to 5 μm in diameter and giving both left chamber opacification and some degree of myocardial enhancement [68]. More recent information shows that the phase change alone is not sufficient to produce a uniform microbubble suspension [69]. The emulsion must first be activated by energetic agitation. The fluorocarbon-based agent has been investigated in Phase I, II and III trials in around 500 patients with promising results. One unanswered question is the reason for the rather prolonged enhancement that it produces – up to 20 min. The agent is cleared from the blood in 30–48 min and leaves unchanged, via the lungs.

CONCLUSIONS

The search for the ideal ECM continues. Uppermost in the minds of investigators is the need for a highly reproducible echocontrast enhancer which is free from toxicity. There are two important safety aspects to be considered: the toxicity of the microbubbles and the toxicity of the carrier medium. With respect to the microbubble toxicity, early studies showed the potential of echogenic substances to bring about gas embolization, probably because the microbubble suspensions contained bubbles of varying sizes, some of which had a diameter greater than the apparently critical value of $10\,\mu m$. Fortunately, the advent of sonication has allowed the reliable preparation of microbubbles $< 10\,\mu m$ in diameter. Although transient mechanical obstruction of capillaries by microbubbles will always be a potential hazard with any new agents, sonication has substantially reduced the risk. The satisfactory transpulmonary passage of an ECM is a reasonable guarantee of freedom from microbubble toxicity.

There is currently intense activity surrounding the manufacture of a readily available and reproducible ECM. Central to the research programme is the toxicity of the carrier medium. The side-effects of the carrier will be minimized if its ionic composition, osmolality and viscosity are similar to those of blood. Several investigators have examined naturally occurring, biodegradable substances such as gelatin and albumin as potential carriers. Protein-based carriers such as these always raise the possibility (albeit a remote one) of immunological or allergenic reactions. Two new carriers are currently under evaluation commercially as potential transpulmonary echocontrast enhancers. The first, Albunex, is an albumin-based microbubble enhancer, which is well tolerated and has been reported in one study to opacify the left heart in 63% of patients [57]. Opacification was most satisfactory in patients given the highest doses. SHU-508A crosses the lungs to opacify the left heart and enhance the signal throughout the blood pool. This increases the scope of echocontrast enhancement quite significantly. SHU-508A has probably been investigated more exhaustively than any other contrast enhancer to date. The results have confirmed the value of this saccharide based enhancer and comprehensively validated the whole concept of echoenhancement.

The risk of side-effects from echocontrast enhancers used in MC-2DE or Doppler flow mapping compares favorably with that of any other cardiac investigative procedure. The goal of reproducible and safe myocardial perfusion imaging with a peripheral injection of a contrast agent is almost within our grasp. The development of echocontrast agents, unlike that of the radiographic contrast agents, is likely to be limited by the problems of formulation, microbubble stability and image processing rather than toxicity.

REFERENCES

1. Gramiak R, Shah PM. Echocardiography of the aortic root. Invest Radiol 1968;3:356–66.
2. Valdes-Cruz LM, Sahn DJ. Ultransonic contrast studies for the detection of cardiac shunts. J Am Coll Cardiol 1984;3:978–85.
3. Kerber RE, Kioschos JM, Lauer RM. Use of an ultrasonic contrast method in the diagnosis of valvular regurgitation and intracardiac shunts. Am J Cardiol 1974;34:722–7.
4. Meltzer RS, van Hoogenhuyze DCA, Serruys PW et al. The diagnosis of tricuspid regurgitation by contrast echocardiography. Circulation 1981;63:1093.
5. De Maria AN, Bommer JW, Rasor J et al. Determination of cardiac output by two-dimensional contrast echocardiography. In: Meltzer RS, Roelandt J, editors. Contrast Echocardiography. The Hague: Martinus Nijhoff Publishers 1982;289–97.
6. Santoso J, Roelandt J, Mansyoer H et al. Myocardial perfusion imaging in humans by contrast echocardiography using polygelin colloid solution. J Am Coll Cardiol 1985;6:612–20.
7. Degenhardt F. Echo-enhanced salpingography. An investigation of Echovist in 224 patients. Adv Echo-Contrast 1995;4:58–9.
8. Lees W. Doppler ultrasound in tumour diagnosis. Adv Echo-Contrast 1995;4:59–61.
9. Kremkau EP, Gramiak R, Carstensen EL. Ultrasonic detection of cavitation at catheter tips. Am J Roentgenol 1979;110:177–83.
10. Feinstein SB, Shah PM, Bing RJ et al. Microbubble dynamics visualized in the intact capillary circulation. J Am Coll Cardiol 1984;4:595–600.
11. Barrera JG, Fulkerson PK, Rittgers SE et al. The nature of contrast echocardiographic 'targets' (abst). Circulation 1978;58(Suppl II):233.
12. Meltzer RS, Tickner EG, Sahines TP et al. The source of ultrasound contrast effect. J Clin Ultrasound 1980;8:121–7.
13. Kort A, Krougon I. Microbubble formation: In vitro and in vivo observations. J Clin Ultrasound 1982;10:117–20.
14. Xie F, Shapiro J, Meltzer R. Toxicity of intracoronary microbubbles in contrast echo. Circulation 1987;76(Suppl IV):505.
15. Dick CD, Feinstein SR, Peterson EM et al. Biodistribution of transpulmonary echo-cardiographic contrast agents. Circulation 1987;76(Suppl IV):506.
16. Feinstein SB, Ten Cate FJ, Zwehl W et al. Two-dimensional contrast echocardiography. I. In vitro development and quantitative analysis of echo contrast agents. J Am Coll Cardiol 1984; 3:14–20.
17. Hajduczki I, Rajagopalam RE, Meerbaum S et al. Effects of intracoronary administered echo-contrast agents on epicardial coronary flow, ECG and global and regional hemodynamics. J Cardiovasc Ultrason 1987;6:85–93.
18. Schlief R. The safety and tolerance of SHU-508 A (Levovist) a galactose-based blood pool enhancer. In Press.
19. Koenig K, Meltzer RS. Effect of viscosity on the size of microbubbles generated for use as echocardiographic contrast agents. J Cardiovasc Ultrason 1986;5:3–4.
20. Roelandt J. Contrast echocardiography. Ultrasound Med Biol 1982;8:471–92.
21. Meltzer RS, Serruys PW, Hugenholtz PG et al. Intravenous carbon dioxide as an echo-cardiographic contrast agent. J Clin Ultrasound 1981;9:127–31.
20. Wang X, Wand J, Huang Y et al. Contrast echocardiography with hydrogen peroxide I: Experimental study. Chinese Med 1979;92:595–9.
21. Gross CM, Wann LS, Hurley SE. Evaluation of myocardial perfusion by contrast echo-cardiography using hydrogen peroxide. Circulation 1982;66(Suppl II):28.
22. Bommer WJ, Shah PM, Allen H et al. The safety of contrast echocardiography. Report of the Committee on Contrast Echocardiography for the American Society of Echocardiography. J Am Coll Cardiol 1984;3:6–13.
23. Lee F, Ginzton L. A central system complication of contrast echocardiography. J Clin Ultrasound 1983;11:292–94.

24. Büll U, Hör G (eds). Klinische Nuklearmedezin. Edition Medizin. 1987;135.
25. Hermann HJ. Nuklearmedezin. München: Urban and Schwarzenberg, 1982;200.
26. Glass JM, Yao LX, Rashka PS et al. Frequency dependence of the half life of Albunex. J Am Soc Echocardiography 1991;4:300.
27. Carstensen EL, Kelly P, Church CC et al. Lysis of erythrocytes by exposure to CW ultrasound. Ultrasound Med Biol 1993;19:147–65.
28. Brayman AA, Azadniv M, Makin IRS et al. Effect of a stabilized microbubble echo contrast agent on hemolysis of human erythrocytes exposed to high intensity pulsed ultrasound. Echocardiography 1995;12:13–21.
29. Miller DL, Thomas RM. Ultrasound contrast agents nucleate inertial cavitation in vitro. Ultrasound Med Biol 1995;21:1059–65.
30. Miller MW, Azadniv M, Doida Y et al. Effect of a stabilized microbubble contrast agent on CW ultrasound induced red blood cell lysis in vitro. Echocardiography 1995;12:1–12.
31. Dalecki D, Child SZ, Raeman CH et al. Hemolysis in vivo from exposure to pulsed ultrasound. Ultrasound Med Biol 1996 (in preparation).
32. Raeman CH, Child SZ, Dalecki D et al. Exposure time dependence of the threshold for ultrasonically induced murine lung hemorrhage. Ultrasound Med Biol 1996;22:139–41.
33. Baggs R, Penney DP, Cox C et al. Thresholds for ultrasonically induced lung hemorrhage in neonatal swine. Ultrasound Med Biol 1996;22:119–28.
34. Harrison GH, Eddy HA, Wang JP et al. Microscopic lung alterations and reduction of respiration rate in insonated anesthetized swine. Ultrasound Med Biol 1995;21:981–3.
35. Penney DP, Schenk EA, Maltby K et al. Morphologic effects of pulsed ultrasound in the lung. Ultrasound Med Biol 1993;19:127–35.
36. Raeman CH, Child SZ, Carstensen EL. Timing of exposures in ultrasonic hemorrhage of murine lung. Ultrasound Med Biol 1993;19:507–12.
37. Zachary JF, O'Brien WD Jr. Lung hemorrhage induced by continuous and pulsed wave (diagnostic) ultrasound in mice, rabbits, and pigs. Vet Pathol 1995;32:43–54.
38. Dalecki D, Carol Raeman CH, Child SZ et al. Intestinal hemorrhage from exposure to pulsed ultrasound. Ultrasound Med Biol 1995;21:1067–72.
39. Dalecki D, Child SZ, Raeman CH et al. The influence of contrast agents on hemorrhage produced by lithotripter flields. Urology (in preparation).
40. Carstensen EL, Hartman C, Child SZ et al. Test for kidney hemorrhage following exposure to intense, pulsed ultrasound. Ultrasound Med Biol 1990;16:681–5.
41. Raeman CH, Child SZ, Dalecki D et al. Damage to murine kidney and intestine from exposure to the fields of a piezoelectric lithotripter. Ultrasound Med Biol 1994;20:589–94.
42. Meltzer RS, Vermeulen HW, Valk N et al. New echocardiographic contrast agents: transmission through the lungs and myocardial perfusion imaging. J Cardiovasc Ultrason 1982;1:277–82.
43. Meltzer RS, Tickner EG, Popp RL. The source of ultrasound contrast effect. J Clin Ultrasound 1980;8:121–7.
44. Carroll BA, Turner RJ, Tickner EG et al. Gelatin encapsulated nitrogen microbubbles as ultrasonic contrast agents. Invest Radiol 1980;15:260–6.
45. Meltzer RS, Klig V, Teichholz LE. Generating precision microbubbles for use as an echocardiographic contrast agent. J Am Coll Cardiol 1985;5:978–82.
46. Bommer WJ, Mason DT, DeMaria AN. Studies in contrast echocardiography: Development of new agents with superior reproducibility and transmission through the lungs. Circulation 1979;60(Suppl II):17.
47. Ernst A, Cikes I, Cistovic F. Polygelin colloid solution as an echocardiographic contrast agent. J Cardiovasc Ultrason 1984;3:143–5.
48. Santoso T, Roelandt J, Mansyoer H et al. Myocardial perfusion imaging in humans by contrast echocardiography using polygelin colloid solution. J Am Coll Cardiol 1985;6:612–20.
49. Berwing K, Schlepper M. Kontrast-Echokardiographische Darstellung des linken Ventrikels bei peripher venoser Kontrastmittelingiektion bie Patienten. Z Kardiol 1985;74(Suppl 5):15.

50. Berwing K, Schlepper M. Echocardiographic imaging of the left ventricle by peripheral intravenous injection of echo-contrast agent in patients. Am Heart J 1988;115:399.
51. Lang RM, Borow KM, Neumann A et al. Echocardiographic contrast agents: effects of sonicated microbubbles and carrier solutions on left ventricular contractility. J Am Coll Cardiol 1987;9:910–9.
52. Gillam LD, Kaul S, Fallon JT et al. Functional and pathologic effects of multiple echocardiographic contrast injections on the myocardium, brain and kidney. J Am Coll Cardiol 1985;6:687–94.
53. Feinstein SB, Lang RM, Neumann A et al. Intracoronary contrast echocardiography in humans: perfusion and anatomic correlates. Circulation 1985;72(Suppl III):57.
54. Feinstein SB, Lang RB, Dick C et al. Contrast echocardiography during coronary arteriography in humans: Perfusion and anatomic studies. J Am Coll Cardiol 1988;11:59–65.
55. Reisner SA, Ong LS, Lichtenberg GS et al. Quantitative assessment of the immediate results of coronary angioplasty by myocardial contrast echocardiography. J Am Coll Cardiol 1989; 13:852–6.
56. Mudra H, Zwehl W, Klauss V et al. Myocardial contrast echocardiography using sonicated iopramid (Ultravist 370) before and after coronary angioplasty. Z Kardiol 1991;80:367–72.
57. Ten Cate KJ, Drury JK, Meerbaum S. Myocardial contrast two-dimensional echocardiography. Experimental examination at different coronary flow levels. J Am Coll Cardiol 1984;3: 1219–26.
58. Feinstein SB. Myocardial perfusion imaging: contrast echocardiography today and tomorrow. J Am Coll Cardiol 1986;8:251–3.
59. Keller MW, Feinstein SB, Watson DD. Successful left ventricular opacification following a peripheral venous injection of a sonicated contrast agent: an experimental evaluation. Am Heart J 1987;114:570–5.
60. Keller MW, Glasheen W, Teja K et al. Myocardial contrast echocardiography without significant hemodynamic effects or reactive hyperemia: a major advantage in the imaging of regional myocardial perfusion. J Am Coll Cardiol 1988;12:1039–47.
61. Reisner SA, Ong LS, Shapiro JR et al. Efficacy and safety of myocardial perfusion imaging using intracoronary sonicated albumin in humans (abst). Circulation 1988;78:565.
62. Feinstein SB, Cheirif J, Folkert J et al. Safety and efficacy of a new transpulmonary ultrasound contrast agent: Initial multicenter clinical results. J Am Coll Cardiol 1990;16:316–24.
63. Sanders WE, Cheirif J, Desir R et al. Contrast opacification of left ventricular myocardium following intravenous administration of sonicated albumin microspheres. Am Heart J 1991; 122:1660–5.
64. Muan B. A survey of the results from clinical trials of Infoson. Presented at the First European Symposium on Ultrasound Contrast Imaging, Rotterdam, Jan 25–26 1996.
65. Smith MD, Kwan OL, Reiser HJ et al. Superior intensity and reproducibility of SHU-454, a new right heart contrast agent. J Am Coll Cardiol 1984;3:992–8.
66. Smith MD, Elion JL, McLure RR. Left heart opacification with peripheral venous injection of a new saccharide echocontrast agent in dogs. J Am Coll Cardiol 1989;13:1622–8.
67. Rovai D, L'Abbate A, Lombardi M et al. Non-uniformity of the transmural distribution of coronary blood flow during the cardiac cycle. Circulation 1989;79:179–87.
68. Careas J-M. Phase change agents. Adv Echo-contrast 1995;4:52–3.
69. Careas J-M. Echogen emulsion: current clinical status in the development of the first fluorocarbon gas-based contrast agent. First European symposium on ultrasound contrast imaging. Jan 25–26 Rotterdam.

7. Behavior of echocontrast microbubbles in the microvasculature

GEORGE D. GIRAUD & DAVID J. SAHN

Echocardiographic (echo) contrast material has been used for a broad range of applications in cardiology including identification of cardiac structures, detection of intracardiac shunts, visualization of blood flow, detection of valvular regurgitation, analysis of complex congenital heart disease and determination of cardiac output using indicator dilution techniques. Most recently the availability of transpulmonary echocontrast agents which persist in the circulation long enough to give myocardial signal has raised questions about these microbubble agents which traverse both the pulmonary and systemic vasculature [1]. Despite the broad use of and current enthusiasm about echocontrast agents, little is known about the behavior of echocontrast microbubbles in the microvasculature. Early echocontrast agents were limited to study of the right side of the circulation and detection of right-to-left shunts. Initially, development of transpulmonary echocontrast agents was limited by available techniques to produce and to stabilize bubbles that would transverse the pulmonary microcirculation. This has involved the use of stabilizing agents such as gelatin, albumin and saccharides which has allowed production of echocontrast agents with small bubble size ($2-5\,\mu$m) and echocontrast agents that decay more slowly in storage as well as in the circulation. Examples of such echocontrast agents are Albunex and Levovist. Albunex consists of encapsulated gas-filled microspheres suspended in human serum albumin. Albunex microbubbles have a mean diameter of $4\,\mu$m with 95% of the microspheres smaller than $10\,\mu$m. Levovist is an echocontrast agent which is formed immediately after mixing specially manufactured microparticles and water-based galactose solution. Levovist microbubbles have a mean diameter of $3\,\mu$m with 95% of the microspheres $<6\,\mu$m. Both these agents have undergone extensive clinical testing and have been shown to be both safe and effective agents that pass through the lung and allow imaging of left-side cardiac structures. The most recent generation of intravenous myocardial echocontrast agents have achieved further prolongation of bubble persistence by using gases such as perfluorocarbons, with the aim of filling these bubbles with substances that diffuse very poorly into the blood. The physics of bubble

133

N. C. Nanda, R. Schlief and B. B. Goldberg (eds.), Advances in Echo Imaging Using Contrast Enhancement,
Second Edition, 133–137.
© 1997 *Kluwer Academic Publishers.*

passage through the circulation is not directly affected by the gas in the bubble, however.

Despite these advances, little is known about the behavior of these ultrasonic contrast bubbles in the pulmonary microvasculature. Direct visualization of ultrasonic contrast bubble transit through the pulmonary microvasculature is not possible. There are no techniques which allow visualization of echocontrast material as it passes through the pulmonary microvasculature. We can make only indirect observations regarding the behavior of these agents at a microvascular level. Much of what we know about the transit of microparticles to and through the pulmonary microvasculature is from studies using solid microspheres. Although the lungs' primary function is that of gas exchange, the lungs also perform a secondary function as a blood filter. A number of studies have investigated the filtering capabilities of the lung. In order for an echocontrast agent to be a useful left-side ultrasonic contrast agent, the agent must pass the pulmonary circulation. To do this the bubbles must be smaller than the capillaries of the lung. In addition to microbubble size, deformability of the echocontrast envelope or bubble may also play an important role in passing through the microcirculation, as is the case with red blood cells. The cut-off diameter for the filtering function of the pulmonary circulation varies somewhat depending on the particle used. Early investigators studying the filtering function of the lung using glass beads found cut-off diameters of around $15 \mu m$ diameter in rabbits and rats [2]. Similarly, investigators using polystyrene spheres to study the filtering function of the dog lung found cut-off diameters of around $15 \mu m$ [3]. It must be kept in mind, however, that the cut-off for solid particles may not apply for gas particles dispersed in solution since gas particles can be deformed as they pass through the microvasculature [4]. It is also important to note that microparticles injected into the peripheral venous circulation have effects on the pulmonary vasculature that are not a direct result of obstruction of the pulmonary capillary. Interactions with non-inert particles or bubbles might involve capillary surface–bubble surface or even cell-mediated interactions. Researchers studying the effect of intravenous administration of liposomes (200–300 nm diameter) on the pulmonary circulation of sheep found a close relationship between injection of liposomes and increases in pulmonary artery pressure following the injections. These investigators found a close correlation between liposome retention in the lung and the intravascular macrophages [5]. Liposomes were associated with mononuclear cells in the lumen of the alveolar wall microvessels. These investigators suggested that elevated pulmonary artery pressure was likely due to a mechanism involving the arachidonic acid cascade, principally thromboxane. Smaller air-filled albumin microspheres, such as Albunex microspheres, do pass through the pulmonary circulation, but not without consequence. In pigs, injection of 400×10^6 air-filled albumin microspheres per ml (mean diameter $4 \pm 1 \mu m$) caused a dose-dependent increase in mean pulmonary artery pressure [6]. This increase was abolished by pretreatment with indomethacin, suggesting that the pressure response in pigs may be explained by release of thromboxane A_2 from pulmonary intravascular macrophages during phagocytosis of the microspheres. In the same study, these investigators found that this response to Albunex was absent in rabbits and monkeys, suggesting species specificity of the response. Dick and associates [7] found that only 0.07% of albumin microbubbles (99% < 10 μm) were trapped in the lung of the rabbit. In dogs, rats and humans, the clearance of blood-borne particles from the blood occurs predominantly in hepatic Kupffer cells and splenic macrophages [8]. In contrast, removal of blood-borne particles in calves, sheep, goats, cats and pigs occurs predominantly in pulmonary intravascular macrophages [8, 9]. These observations suggest a clear species difference in the response of the pulmonary vasculature to injected microparticles. In some species, microbubbles produce a response that is not a result of obstruction of the pulmonary capillary.

We are no better off regarding our understanding of the behavior of microbubbles in the systemic microvasculature. Few experimental models allow direct visualization of the peripheral microvasculature, and these models are technically very difficult to set up and use. As a result, there are few studies that have looked at microvascular transit of materials through systemic capillary beds. An early study investigated the mechanism governing the creation and disappearance of microbubbles in vitro as well as in the rat mesenteric arterial circulation [10]. The researchers [10] injected what they called 'large hand-agitated microbubbles' into the rat aorta. They found that these microbubbles blocked mesenteric arterioles for a significant period of time, varying from a few seconds up to 200 s, before decreasing in size enough to allow passage downstream.

The development of sonication techniques to produce microbubble preparations with diameters < 10 μm significantly advanced the manufacture of echocontrast agents for use in left-side imaging. One of the most thorough studies of small bubble size ultrasonic contrast in the systemic microvasculature was undertaken by Feinstein and his associates [11]. These investigators used direct microscopic examination of a cat mesenteric preparation to determine the behavior and fate of three different microbubble preparations, spanning a broad range of sizes, in the microcirculation. They used sonicated sorbitol 70% to produce microbubbles 10 μm in diameter, sonicated 70% dextrose to produce microbubbles 10–15 μm in diameter and hand-agitated Renografin-76/saline solution to produce microbubbles ≥ 15 μm in diameter. The results appear to indicate that all 10 μm diameter microbubbles counted passed through the microvessels under examination. Microbubbles of 10–15 μm diameter generally passed through the vessels, while 15 μm microbubbles showed transient blockage followed by shrinkage or subsequent collapse of the microbubble. These investigators found that all 25 microbubbles ≥ 15 μm in diameter blocked the capillary vessels during passage for a brief period, varying from a few seconds to > 1 min.

Other workers have studied a new phase-shift fluorocarbon echocontrast agent, Echogen, in the cat mesenteric preparation [12]. This echocontrast agent has the unique feature of producing robust myocardial opacification lasting up to 35–40 min after peripheral venous injection. Following Echogen injection, capillary flow velocity was noted to decrease transiently, returning to the pre-injection level within 40–70 s. Also of interest was margination and endothelial adhesion of microbubbles or white blood cells without slowing of central RBC flow velocity following injection of Echogen. This phenomenon could account for the long persistence of echocontrast opacification with this agent.

There is indirect evidence against long-term effects at a microvascular level. Despite the large size of microbubbles in hand-agitated echocontrast solutions, no histological abnormalities attributable to vascular occlusion in the brain, heart or kidney have been noted. Animal studies by Gillam and co-workers [13] used indirect methods to study the effects of intra-arterial injections of hand-agitated Renografin-76 (microbubble size range $2-25\,\mu$m, mean $11.5\pm6.6\,\mu$m). Intra-arterial injection of hand-agitated Renografin-76 in dogs produced significant acute depression of myocardial function. Light microscopic examination of tissue 24 h after echocontrast injection revealed no myocardial or cerebral changes attributable to these injections. Findings in the kidneys suggested non-specific changes consistent with osmotic diuresis.

SUMMARY

Many factors affect the behavior of microbubbles in the circulation. There are factors unique to the vascular bed which affect microbubble behavior. Some factors appear to be species-specific, while others appear to be related to the endothelial surface, chemicals or cells. The manner in which microbubbles themselves behave in the circulation depends on a number of factors. Physical factors related to the microbubble behavior include microbubble size, uniformity of microbubble size and microbubble stability. Some bubbles become larger while others diminish as gas diffuses into or out of the microbubbles during their stay in the circulation. Carrier solutions could have an effect, but this should be diluted with mixing and passage through the first vascular bed, the pulmonary vascular bed. Nonetheless, factors such as osmolarity, viscosity and surfactant properties of the carrier solutions could alter bubble behavior. Many of these considerations will require active investigation and close and open communication between investigators and companies developing new echocontrast materials.

REFERENCES

1. Unger E, Fritz T, Shen D et al. Gas filled lipid bilayers as imaging contrast agents. J Liposome Res 1994;4:861–74.
2. Gordon DB, Flasher BJ, Dbrury DR. Size of the largest arteriovenous vessels in various organs. Am J Physiol 1953;173:275–81.

3. Ring GC, Blum AS, Kurbatov T, Moss WD, Smith W. Size of microspheres passing through the pulmonary circuit in the dog. Am J Physiol 1961;200:1191–6.
4. Butler BD, Hills BA. The lung as a filter for microbubbles. J Appl Physiol 1979;47:537–43.
5. Miyamoto K, Schultz E, Heath T, Mitchell MD, Albertine KH, Staub NC. Pulmonary intravascular macrophages and hemodynamic effects of liposomes in sheep. J Appl Physiol 1988; 64:1143–52.
6. Ostensen J, Hede R, Myreng Y, Ege T, Holtz E. Intravenous injecting of Albunex® microspheres causes thromboxane mediated pulmonary hypertension in pigs but not monkeys or rabbits. Acta Physiol Scand 1992;144:307–15.
7. Dick CD, Feinstein SR, Peterson EM, Stark VJ, Ryan JW, Harper PV. Biodistribution of transpulmonary echocardiographic contrast agent. Circulation 1987;76:IV–506.
8. Winkler GC. Pulmonary intravascular macrophages in domestic animal species: review of structural and functional properties. Am J Anat 1988;181:217–34.
9. Warner AE, Brain JD. Intravascular pulmonary macrophages: a novel cell removes particles from the blood. Am J Physiol 1986;250:R728–32.
10. Kort A, Kronzon I. Microbubble formation: in vitro and in vivo observations. J Clin Ultrasound 1982;10:95–101.
11. Feinstein SR, Shah PM, Bing RJ et al. Microbubble dynamics in the intact capillary circulation. J Am Coll Cardiol 1984;4:595–600.
12. Ge S, Giraud GD, Shiota T et al. Microcirculatory flow dynamics during peripheral intravenous injection of Echogen®. Microscopic visualization of mesenteric microcirculatory flow with simultaneous transthoracic echo imaging in cats. J Am Coll Cardiol Special Issue 1995;277A.
13. Gillam LD, Kaul S, Fallon JT et al. Functional and pathologic effects of multiple echocardiographic contrast injections on the myocardium, brain and kidney. J Am Coll Cardiol 1985;6:687–94.

8. Imaging instrumentation for ultrasound contrast agents

JEFFRY E. POWERS, PETER N. BURNS & JACQUES SOUQUET

INTRODUCTION

The past decade has seen dramatic improvements in ultrasound imaging system performance. The introduction of color Doppler imaging has added a new dimension to blood flow measurement, displaying blood flow as a real-time map over a two-dimensional image rather than as a spectrum from the single point of pulsed Doppler or the single line of continuous wave Doppler. The more recent refinement of power Doppler (also known as Color Power Angiography™) has increased the imaging sensitivity of color Doppler to the point where images such as that shown in Figure 1 have become common place. If the function of a contrast agent is to enhance the echo from blood, is such help needed with state of the art ultrasound instruments? We believe that it is. First, not all patients yield images like that shown in Figure 1, nor can such images be obtained in all anatomical locations. Contrast agents can extend the anatomical scope, and hence the clinical utility, of conventional ultrasound imaging. Second, as we shall see, contrast agents allow renegotiation of some of the fundamental compromises inherent in a blood flow image and allow system design changes which result in improvements that go far beyond Doppler signal enhancement.

There are two major reasons why images like Figure 1 cannot be obtained in all patients or in all anatomical sites. First, tissue attenuation due to absorption reduces the amplitude of the echo and tissue inhomogeneity distorts the ultrasound beam. This makes some examinations, such as transcranial Doppler, difficult to perform on any patient. These factors are also responsible for the variable success of abdominal and cardiac ultrasound in different patients. Second, tissue motion obscures the blood flow signal from Doppler detection. One obvious example is attempting to image flow in the coronary arteries. The heart muscle itself moves much faster than the blood flows in the microcirculation, making it very difficult for Doppler techniques to distinguish the motion of blood from that of the tissue.

139

N. C. Nanda, R. Schlief and B. B. Goldberg (eds.), Advances in Echo Imaging Using Contrast Enhancement, Second Edition, 139–170.
© 1997 *Kluwer Academic Publishers*.

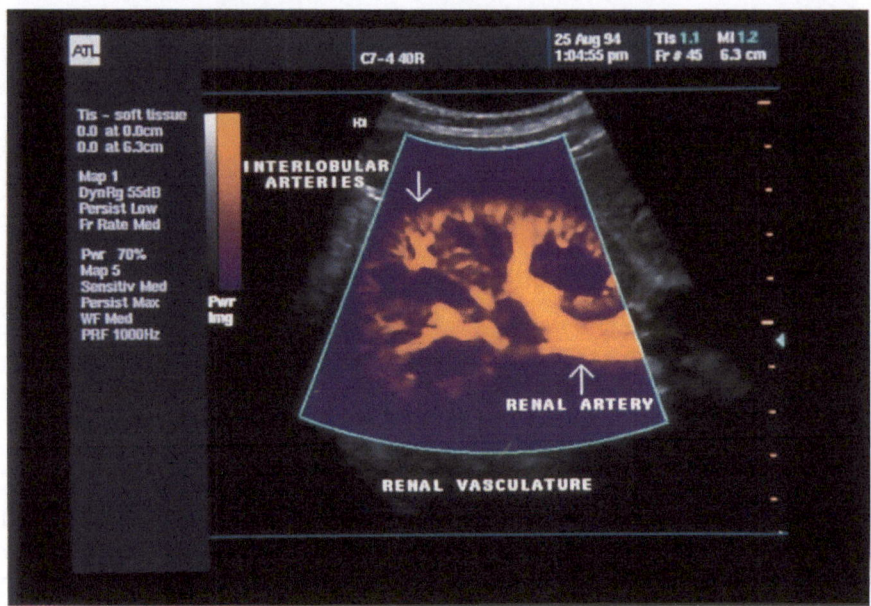

Figure 1. Color power angiography of native kidney, no contrast enhancement.

One obvious use for contrast agents is as a means to improve the signal-to-noise ratio in color flow and other Doppler examinations. This will find application in many circumstances in which Doppler is routinely performed today, but where the weakness of the echo from blood is a problem. As the field matures, more applications of contrast agents will be developed which allow entirely new kinds of examinations not currently possible without contrast agents. In this chapter we discuss the impact that contrast agents will have on ultrasound system design and how the equipment can best be optimized to exploit their capabilities.

We will first describe the characteristics of blood flow and tissue which allow the ultrasound system to be able to differentiate their echoes without the use of a contrast agent. These characteristics will be referred to as the 'signature' of a particular blood flow regimen. This will lead to a description of the signal processing used to detect that particular signature and how it is optimized, both with and without contrast agents. The more distinct the signature of the blood echo, the better the system is able to differentiate flow from moving tissue. We shall then describe some signatures of contrast agents which exploit the non-linear properties of the microbubbles themselves and how this leads to strategies by which the system can detect the echo from blood with even better separation from that of tissue.

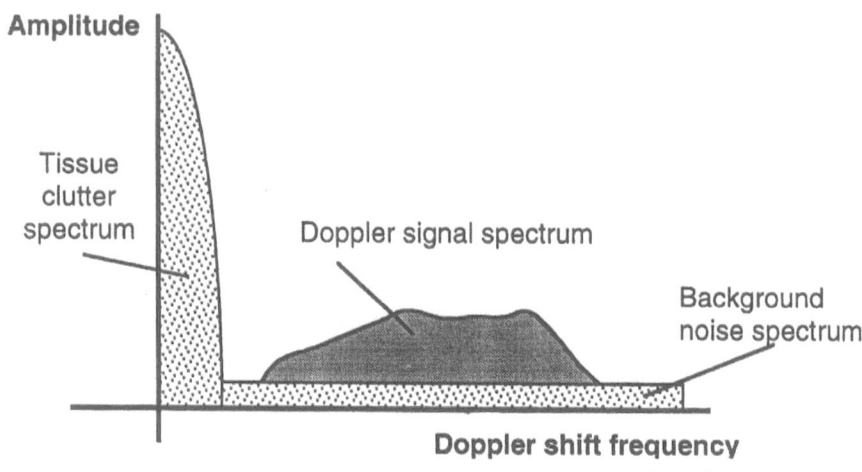

Figure 2. Schematic representation of the received signal spectrum showing clutter, noise and Doppler signal.

THE CHARACTERISTIC SIGNATURE OF THE ECHO FROM FLOWING BLOOD

The detection of blood flow raises a number of difficulties which can be overcome by careful analysis of the signature of the echo returned by the red blood cells. The primary difficulty is the small amplitude of the signal returned from blood compared to the large amplitude (perhaps 1000 times greater) of that returned from tissue.

The concept of noise

As no electronic or acoustic detection process for detecting an echo is perfect, there are unwanted components in the signal, referred to as 'noise'. It is useful to distinguish two types of noise in a Doppler system. First there is the system noise, which may come from electrical noise in the receiver or from thermal noise in the transducer. This is a low amplitude random noise signal spread across the entire frequency band. The second kind of noise, called clutter, comprises the echo from stationary or slowly moving tissue. The term *clutter* has been borrowed from the radar field where it signifies the signals received from surrounding structures which must be eliminated in order to detect the target. This is equally unwanted, but comes from real targets in tissue and therefore has a predictable structure, typically with a high amplitude and low velocity (Figure 2).

Thus the blood flow signature is that of an echo whose amplitude is greater than that of the system noise, but which moves more rapidly than the surrounding tissue. What is the best method to detect it? It would be very difficult to identify the signature using a single pulse since the blood moves only a tiny amount during the time it is being interrogated with the short burst of sound. By sending out multiple bursts and comparing the echo received from consecutive bursts, it is possible to identify the component which has moved and to separate it from the accompanying clutter. It is also possible to perform averaging which reduces the effect of the random, low amplitude system noise. These two steps are fundamental to any blood flow detection scheme, including all forms of spectral Doppler, color Doppler, and Doppler power imaging. The differences between these flow imaging methods lie only in how many pulses are sent, how they are compared so as to eliminate clutter, how much they are averaged and which attribute of the blood echo finally is displayed.

The simplest method for eliminating clutter simply subtracts the echoes from consecutive pulses and is known from its use in radar as the *moving target indicator* (MTI). More complex methods have been developed which combine several pulses simultaneously, but the basic principle is still the same. These techniques enhance the system's ability to separate the slowly moving clutter from the blood flow, improving the detection of the blood on the basis of one of its specific signatures, its motion. This part of the system is called the wall filter and is present in all forms of Doppler flow detectors, including power Doppler. Once the clutter has been eliminated, some attribute of the signal must be displayed. Spectral Doppler displays a velocity spectrum, showing the variation in time of the velocity components within a single small region of space. Color Doppler displays a single velocity value for every location on the screen, thus showing the variation in space of the velocity within a small interval of time. Power Doppler is a variation of color Doppler in which the amplitude of the signal from the moving blood is mapped to a color at each location, rather than its mean Doppler shift frequency. Each of these characteristics of the blood flow is useful for different clinical situations and each allows different trade offs in the number of pulses used and how they are averaged.

SPECTRAL DOPPLER

In order to gain the most velocity information from a single point in the image, the system must interrogate that location continuously. However, such a continuous wave (CW) technique provides no range resolution, simply superimposing the Doppler signals from all echoes received along the sensitive volume of the beam. It also means that tissue signals coming from everywhere along the sound beam contribute to the clutter problem, making clutter elimination even more difficult. The system is thus required to be able to detect very small signals from the echo of interest along with the huge clutter signals generated by nearby, shallow structures. The ability to maintain very small signals along with very large

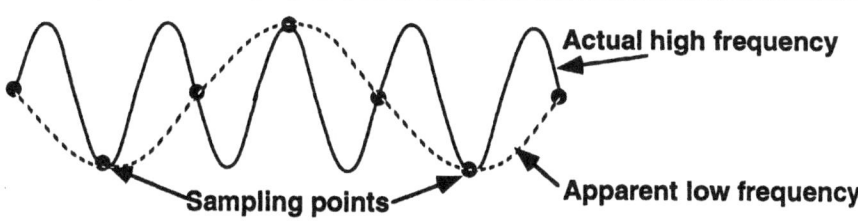

Figure 3. Aliasing seen with pulsed Doppler and color flow. High velocities sampled too infrequently look like low velocities.

ones is referred to as dynamic range, and is a key attribute of any blood flow detection system.

If some compromise can be allowed in the highest velocity which needs to be detected, the transducer can be pulsed in the direction of interest to provide range resolution. With pulsed wave (PW) Doppler, the position of the volume of red blood cell scatterers is only measured once every pulse repetition interval (PRI). The (PRI) is the reciprocal of the pulse repetition frequency (PRF), a more commonly used term. The small change in position is actually measured by the shift in phase of the echo with respect to the sound transmitted. It can be seen that if the volume of scatterers has moved more than one half of a wavelength of sound between pulses, there is no way of telling whether it has moved a long distance forward or a short distance backwards. This artifact, illustrated in Figure 3, is known as aliasing and can lead to high velocity flow appearing to go backwards at a lower velocity.

COLOR FLOW

PW and CW Doppler displays provide a spectral output showing all the different velocity components present at one location, but they require an entire two-dimensional display because three variables – frequency, amplitude and time – are being shown simultaneously. If blood flow information is required over the entire field of view, further compromises must be made in the amount of information displayed at each location. If only an estimate of the mean velocity, corresponding to the average of all the Doppler shift frequency components present at a given time, is adequate, this can be encoded in color and displayed at the location in space where that velocity was detected, leading to a two-dimensional image of the flow. Color flow imaging uses one set of colors to depict flow moving towards the transducer and a different set to depict flow away from the transducer. A range of hues of these colors is chosen to correspond to a range of velocities. The fact that only the mean velocity in each sample volume is displayed also allows a less complete detection of all velocities to be

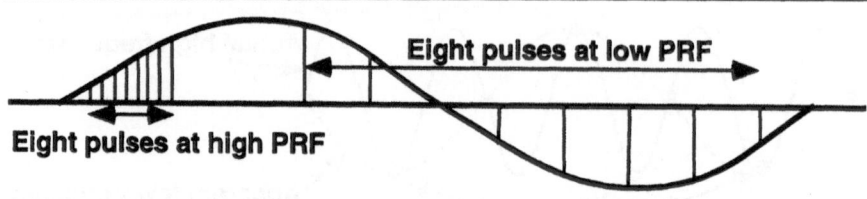

Figure 4. Eight pulses at a high PRF only capture a short segment of a low frequency, making it difficult to eliminate clutter or capture flow. Eight pulses at a low PRF capture more of the signal, improving both clutter rejection and slow flow detection.

made at each point: this compromise is necessary to allow the beam to dwell for the minimum time at each location and thus complete the image within a usefully short time.

With pulsed Doppler, the same location is interrogated with a steady stream of pulses. In order to detect flow throughout the entire image, the interrogating beam must be stepped from location to location rapidly, firing a number of pulses in each direction just sufficient to remove the clutter and obtain a satisfactory estimate of the mean velocity. The number and rate of these pulses has a significant impact on the range of velocities that can be detected as well as on the frame rate. In general, the more pulses fired along a given scan line, the better the detection process works, but the slower the frame rate. As discussed above, the pulse repetition frequency (PRF) determines the highest velocity that can be resolved unambiguously without aliasing. The PRF, along with the total number of pulses transmitted per line (also known as the *ensemble length*), also helps determine the *lowest* frequency that can be detected. The longer the interval between pulses (PRI) and the longer the ensemble length, the longer the time interval over which the signal is actually being detected and analyzed. Since low frequencies have longer periods, they require a greater duration of observation to estimate their frequency and either reject them as clutter or accept them as flow. This is illustrated in Figure 4. Thus, knowing approximately the velocity or frequency range that is expected from both the flow and the tissue enables the system to be better optimized for a particular flow regimen.

One critical aspect of a color Doppler imaging system is the method by which the final decision is made as to whether or not there exists flow in a given location in the image. As with any decision making process, a priori knowledge of the expected flow and clutter characteristics can help in optimizing the final image. In other words, if the properties of the received signal do resemble the expected signature for blood, the system displays it as flow, otherwise it displays the B-mode image at that location, which it has acquired concurrently. For this process such characteristics of the signal as its amplitude, frequency and randomness are all taken into account. Figure 5 illustrates how different flow

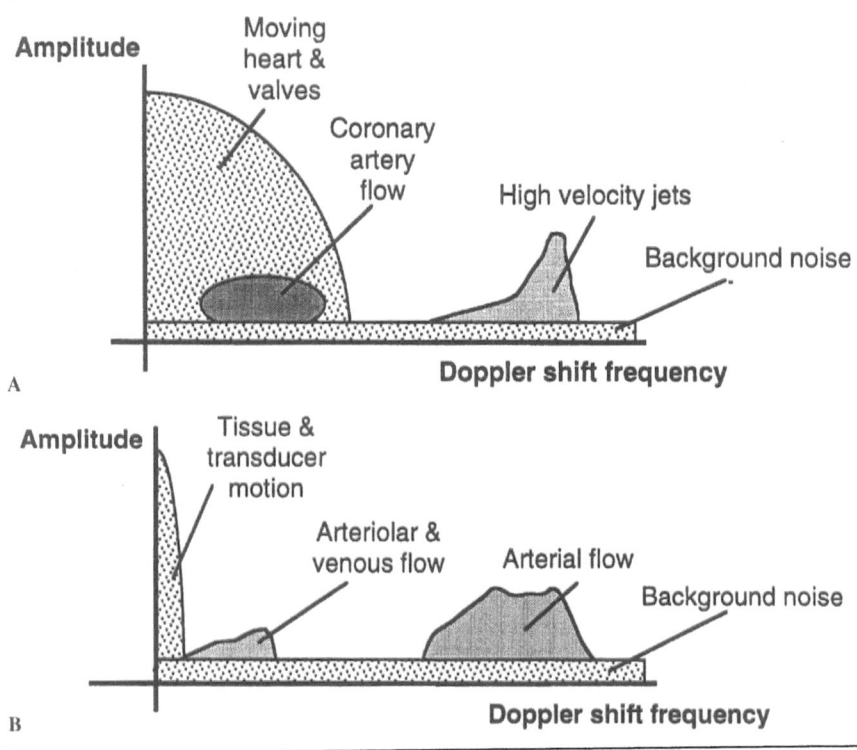

Figure 5. Diagramatic representation of tissue and blood flow 'signatures' in typical clinical situations. (A) Cardiac Doppler; (B) a radiological application.

regimens in differing anatomic levels of the circulation lead to different characteristic signatures.

COLOR POWER DOPPLER

Suppose that we do not care how fast or in what direction blood is flowing at a particular location, but instead are interested only in detecting the existence of flow with the greatest sensitivity. We still need to look at the signal for a component that is moving and that is stronger than the system noise and smaller than tissue clutter signals, but we can throw away all information about its actual velocity. By displaying the power in the Doppler signal (Figure 6A) we can indicate whether there is flow in each location and how much there is, but not its direction or speed. How does that change the situation?

In practice, this rather simple modification to the display method brings about several advantages for flow detection, which are particularly pertinent for contrast

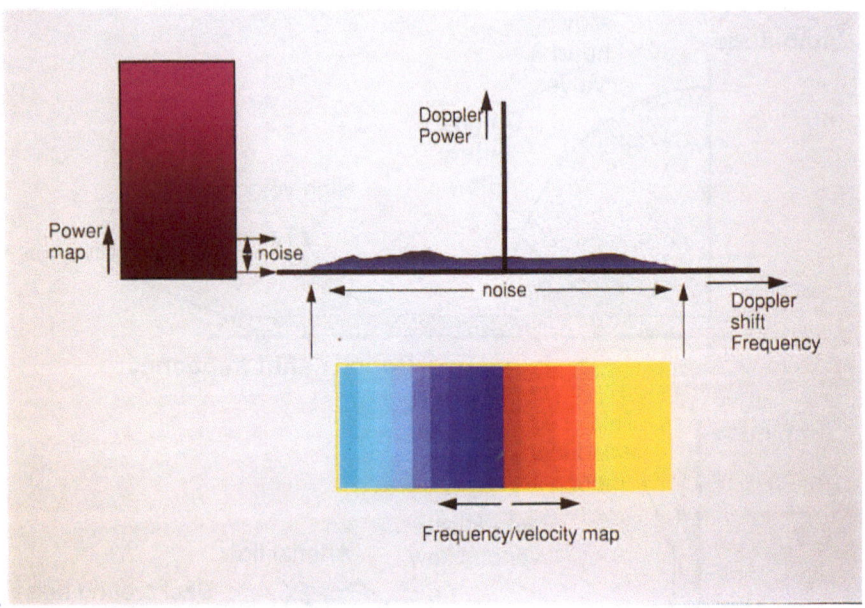

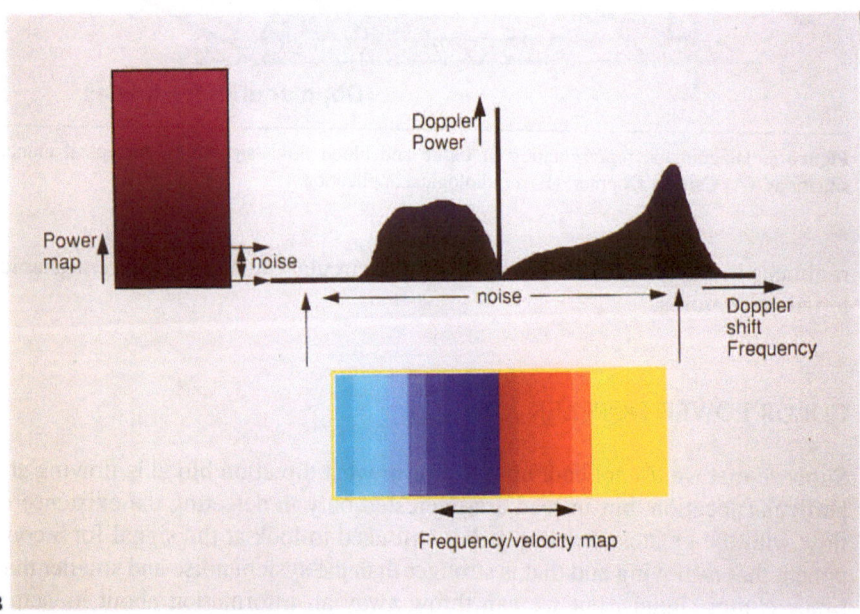

Figure 6. Power (energy) mode. (A) In the velocity map, colors are mapped to the estimated frequency of the Doppler signal. In power mode, they are mapped to the power of the Doppler signal. (B) Noise from the Doppler receiver at high gain settings is mapped to many different colors in velocity mode, but to a single color in power mode. This forms a background 'wash', effectively increasing the usable dynamic range of the Doppler image.

agent detection. Since directional information is not displayed, all flow is displayed with the same basic color hue so that aliasing is not shown. Referring again to Figure 4, since aliasing is no longer an issue, the PRF can be made very low and thus allow the detection of very low velocities. There is an additional important benefit for low level signal detection related to the background noise characteristics. Any location in the image has a frequency associated with it, even the low level background noise. In fact, as noise comprises an even mixture of all frequencies, its mean frequency is a quantity which varies at random, leading to the typical red/blue mottled appearance seen when the gain is turned up too high on a color flow system. However, it typically has a low amplitude which is fairly uniform over the entire image (Figure 6B). When the Doppler power signal is displayed, the noise then becomes a uniform low background level instead of random bright color dots. This is much less distracting and allows the gain to be turned up significantly, effectively increasing the dynamic range. Another advantage is the amount of temporal smoothing that can be tolerated. With velocity-encoded color images, averaging a few frames improves sensitivity but it degrades the real-time responsiveness of the color display. With power Doppler imaging, there is no attempt to image the pulsatility of flow, so the signals can be averaged over a longer period of time, allowing reduction of the noise and thus greater sensitivity.

THE ADDITION OF A CONTRAST AGENT

How does the addition of a contrast agent affect the strategies described above for detecting blood flow? The most obvious way to describe the action of a contrast agent is that it simply increases the amplitude of the echo returned from blood. Of itself, this has a significant effect on the optimization of the ultrasound system and this will be discussed in the next section. However, we have already seen that this is too narrow a classification by which to understand the true potential of a new scattering target in the blood. What contrast agents really do is to modify the characteristic signature of the echo from blood. Amplitude is only one aspect of this characteristic signature, and other characteristics of the contrast agent signature allow dramatic changes in the design and performance of ultrasound systems.

Color flow optimization for contrast-enhanced Doppler studies

Contrast agents greatly increase the signal received from blood. Enhancements of the backscattered echo from arterial blood of between 10 and 25 dB (a factor of about 3–20) have been measured [1] following injection of microbubble contrast agents in very small quantities (0.01–0.1 ml/kg of Levovist, for example) into a peripheral vein. Signal enhancements of up to 17 dB have been recorded in the vasculature of experimental liver tumors [1]. The effect on spectral and color Doppler is dramatic, with signals from flow in vessels too small or too deep, and, therefore, hitherto undetectable, being rendered visible on color Doppler imaging [2, 3]. While this enhancement increases detectability of weak flow states, it can also present problems when there is too great an echo amplitude. If the system has been optimized to detect signal levels from unenhanced blood and they increase by 20 dB, the image can often become filled with color, with the result that detail resolution is compromised. In addition, the methods used to discriminate between blood and tissue need to take into account the increased signal level from the contrast agent mixed with the blood. If, for example, the system is programmed to reject very high amplitude signals on the assumption that they are clutter, flow signals can actually be rejected because they are too strong.

Color saturation and blooming can come about in a number of ways. First, on a color image relatively small vessels always appear larger than they actually are due to the bandwidths of both the transmitted waveform and the receive signal processing. Relatively narrow bandwidths are often used for color flow imaging to increase sensitivity. When the received signal amplitude is higher than expected, this can smear the resulting image (Figure 7). This is exacerbated by the fact that, since color flow displays a frequency rather than an amplitude, a decision threshold is used to decide whether or not to display flow at a particular location; this decision threshold is usually based on the amplitude of the signal. This thresholding effect can lead to smearing of small vessels. Furthermore, if flow is displayed, it is typically displayed at full brightness, making this artifact of contrast enhanced color images very prominent (Figure 8). This effect typically results in symmetrical smearing both proximal and distal to the vessel.

A second artifact can result in asymmetrical smearing which appears only distal to the vessel. In a large vessel the distal wall acts as a strong specular reflector of the sound used for color flow imaging. Some of the sound reflected from the distal wall also scatters from the contrast agent within the vessel which again reflects off the distal wall and is received by the transducer, as illustrated in Figure 9. There is no way for the system to distinguish these secondary reflections from signals received from actual flow distal to the vessel, an effect often referred to as the 'pulse-echo mirror image' artifact [4]. The natural response of the sonographer when these artifacts are observed is to reduce the color gain. This is often done prior to injecting the contrast because the bloom-ing artifact is so common. Reducing the gain, however, is the equivalent of throwing away potentially useful signals. Other controls can be changed which can decrease the sensitivity to these artifacts, but instead of throwing away the additional signal it can be used to improve other aspects of the image quality.

Most color flow systems are optimized to image flow signals whose strength is very close to the noise floor (the amplitude at which a signal is obscured by the random noise of the system). To ensure maximum sensitivity during use, the gain is typically increased by the operator until the screen shows random noise,

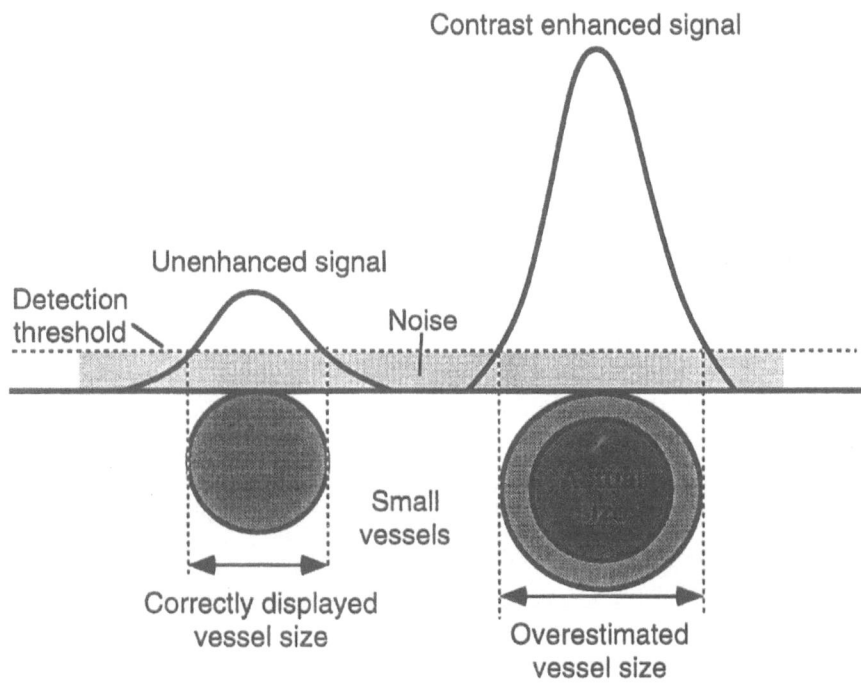

Figure 7. The 'blooming' artifact. The enhanced echo from the contrast agent can cause color smearing and exaggerated vessel size.

and then turned down until it goes away. During system design and optimization, various parameters such as ultrasound center frequency, bandwidth, line density, ensemble length, wall filter settings, etc. are adjusted to provide the best resolution and frame rate possible to give good sensitivity for the given clinical application. These settings are then stored as presets to allow rapid optimization with a minimum of user interaction.

Unfortunately, the objectives of maximizing sensitivity and resolution (both spatial and temporal) are mutually exclusive and controls that increase one almost always decrease the other. System optimization consists of finding the best balance between these two performance aspects for all parameters in the system, a balance that depends quite strongly on the expected flow and clutter characteristics. When a contrast is injected that balance can change significantly, requiring reoptimization of all of these parameters to form the best image. Table 1 shows some of the more common system parameters and how they affect sensitivity and resolution. If the sensitivity is found to be too high, producing color blooming, the parameters can be adjusted to reduce the sensitivity but increase resolution or frame rate. For example, to increase lateral and axial resolution without affecting frame rate, the system bandwidth could be increased,

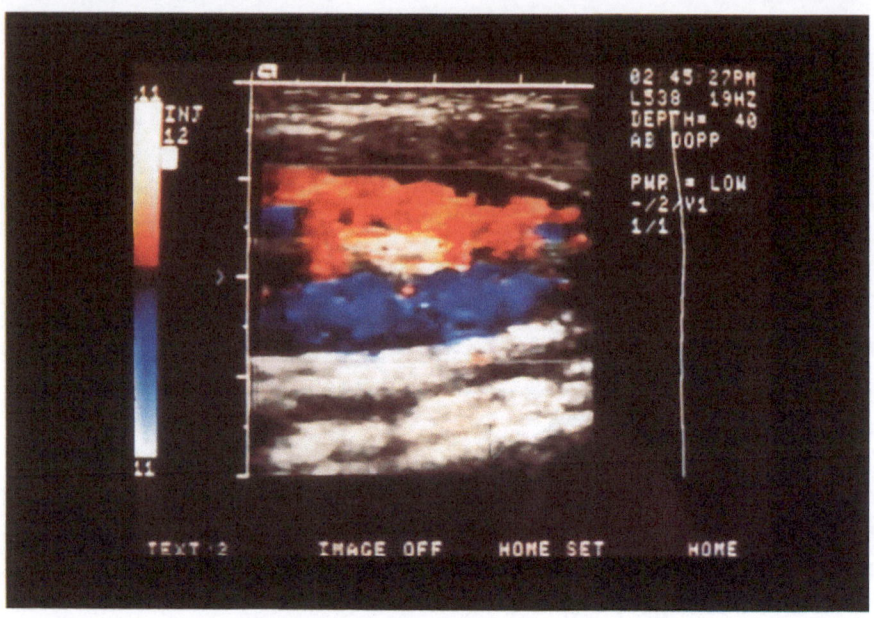

Figure 8. An example of contrast agent-induced blooming.

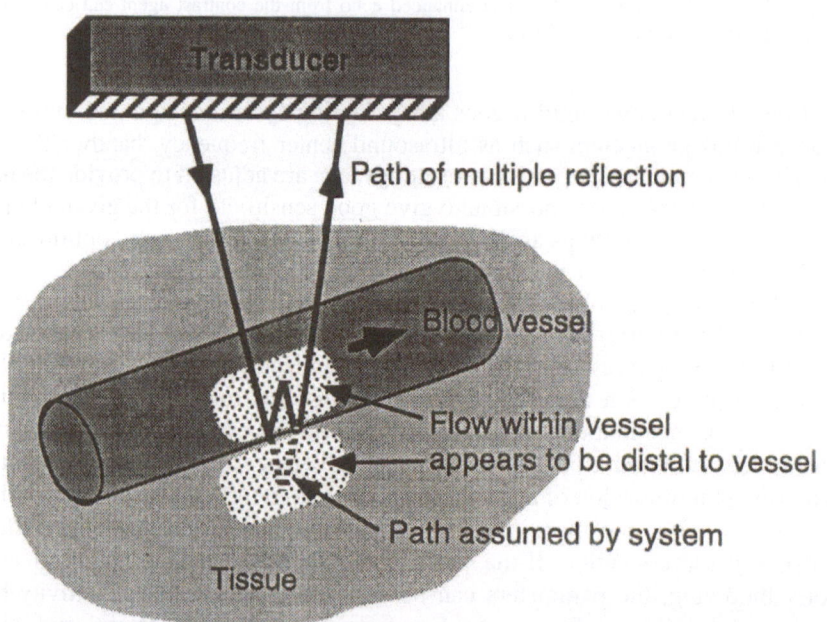

Figure 9. A strong reflection from agent in a vessel can lead to the distal 'mirror image artifact', another kind of blooming.

Table 1. Image quality/frame rate compromises forced by some common color flow parameters.

Parameter increased	Sensitivity	Resolution	Frame rate
Pulse repetition frequency	Decreased	Same	Increased
Ensemble length	Increased	Same	Decreased
Bandwidth	Decreased	Increased	Same
Line density	Same	Increased	Decreased
Persistence	Increased	Same	Decreased

the ensemble length reduced and the line density increased. To increase frame rate, ensemble length and persistence can both be reduced. Both of these maneuvers will reduce the overall sensitivity much as if the gain were reduced, but at the same time improving resolution. In this way, the tendency of the agents to cause artifacts can be turned into improvements in other aspects of performance.

Contrast power Doppler imaging

The power map is a natural choice for contrast enhanced color Doppler studies because it maps a parameter directly related to the acoustic quantity that is enhanced by the contrast agent, . In cases where velocity measurements are required, color flow is still necessary of course, possibly combined with spectral Doppler. However, when it is only necessary to determine the location of a vascular bed, Doppler power imaging works well with contrast agents. Doppler power imaging has the advantage in that it can significantly reduce the color blooming discussed above. Individual pixels displayed in color flow are typically on or off, with the color representing the velocity at that location. Wherever the flow crosses the threshold, it is displayed. With the Doppler power display the gradations in returned power are displayed which makes the color blooming less objectionable.

Most color flow controls also affect Doppler power imaging, but an important additional control is display dynamic range. The dynamic range of the display has a meaning similar to that used above when discussing the system signal processing. It is the ratio of the maximum signal that can be displayed without saturating to the minimum signal that can be seen above the background noise. While a large dynamic range is always good in a signal processing system, the display dynamic range should be tailored to the range of signal that is to be expected. If the display dynamic range is too great and the signal never exceeds a low level, only a small portion of the color map is used, reducing image contrast. If the display dynamic range is too small, strong signals will saturate, reducing image contrast in areas of strong signals (Figure 10). This is important to the use of contrast agents. Prior to injection the system is typically set up to obtain the best image of the anatomy, with the dynamic range set for the signal

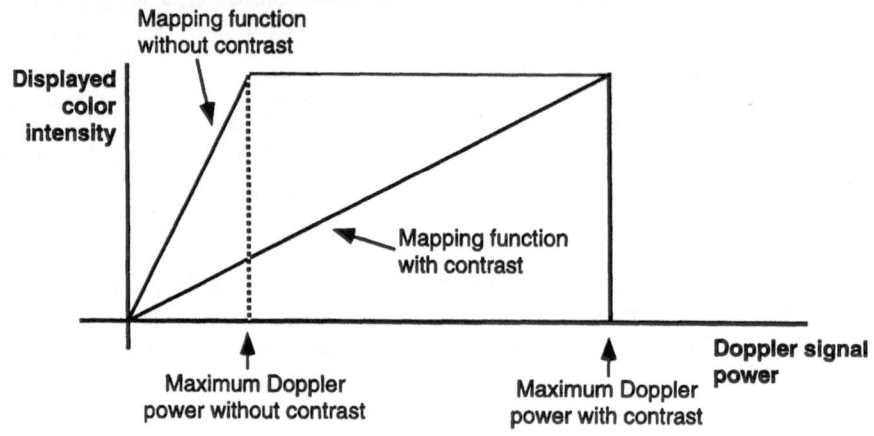

Figure 10. The display mapping function determines the effective dynamic range. If input signal is greater than expected, color saturates at the top of the color map. By changing the display dynamic range when contrast is injected, saturation is eliminated.

levels expected from unenhanced blood. When the contrast is injected, the signal level increases greatly and much of the image saturates. By simply increasing the display dynamic range, much of the saturation can be eliminated.

Well suited though the mode may be to contrast studies, however, the advantages of the power map for contrast-enhanced detection of small vessel flow are offset by a significant shortcoming: its increased susceptibility to interference from clutter. Clutter is both detected more readily, because of the power mode's increased sensitivity to low velocities, and displayed more prominently, because the high amplitude is displayed rather than the low velocity. Typical deep abdominal or cardiac examinations are plagued by the high intensity flash that corresponds to the high amplitude signals from slowly moving tissue. Furthermore, frame averaging has the additional effect of sustaining and blurring the flash over the cardiac cycle, exacerbating its effect on the image. Surprisingly, perhaps, contrast agents are capable of offering a solution to this problem too.

THE MODE OF ACTION OF MICROBUBBLE CONTRAST AGENTS

So far we have discussed the signature of the flow to be detected only in terms of its amplitude and velocity. Other features of the contrast agent echo signature can also be used to improve their detection. To understand these it is helpful to consider more completely the mode of action of a microbubble agent.

The first observation to make is that the echo enhancement produced by most agents in very high dilution is unexpectedly strong. For example, a peripheral venous injection of as little as $10\,\mu$l/kg of Levovist (Schering AG) produces a 7 dB enhancement in a systemic artery [5]. This is much greater than would be expected from such sparse scatterers of this size in blood. Second, investigations of the acoustic characteristics of several agents [6] have demonstrated an approximately linear dependence of backscatter coefficient on numerical density of the agent at low concentrations, as would be expected from a collection of particles, but a dependence of attenuation on ultrasound frequency different to that predicted by the Rayleigh law. Instead, peaks exist which are dependent on both ultrasound frequency and the size of the microbubbles, suggestive of resonance phenomena. This important observation suggests that the bubbles undergo radial oscillation in the ultrasound field, and have a natural frequency of oscillation at which they will both absorb and scatter ultrasound with a peculiarly high efficiency. Such resonance in a bubble would account for its exceptionally strong echo: the bubble behaves as an active source of sound rather than the passive reflector conceived of in the classical Rayleigh scattering of sound by red blood cells [7].

Considering the linear oscillation of a free bubble of air in water, simple theory [8] predicts the resonant frequency of radial oscillation of a bubble of $3\,\mu$m diameter (the median diameter of a typical transpulmonary microbubble agent) to be about 3 MHz, precisely the center frequency of ultrasound used in a typical abdominal scan. One consequence of this extraordinary coincidence is that bubbles undergoing resonant oscillation in an ultrasound field can be induced to non-linear motion. It has long been recognized [9] that if bubbles are 'driven' by the ultrasound field at sufficiently high acoustic pressures, the oscillatory excursions of the bubble reach a point where its alternate expansions and contractions are not equal. Lord Rayleigh, the originator of the theoretical understanding of sound upon which ultrasound imaging is based, was first led in 1917 to investigate this by his curiosity over the creaking noises made by his tea-kettle as the water came to the boil.

The consequence of such non-linear motion is that the sound emitted by the bubble, and detected by the transducer, contains harmonics, just as the resonant strings of a musical instrument, if plucked too vigorously, will produce a 'harsh' timbre containing overtones. Figure 11A shows the frequency spectrum of an echo produced by a microbubble contrast agent following a 3.75 MHz burst. This particular agent is Levovist, for which the manufacturer estimates the median bubble diameter as approximately $2\,\mu$m, with a 97th centile value of approximately $6\,\mu$m [10]. Ultrasound frequency is on the horizontal axis, with the relative amplitude on the vertical axis. A strong echo, at -13 dB with respect to the fundamental, is seen at twice the transmitted frequency, known as the second harmonic. Figure 11B shows that a group of scatterers that do not undergo non-linear oscillation, such as red blood cells, do not produce the second harmonic echo. This is, in fact, the echo from another ultrasound contrast agent, showing that not all agents respond in the same way to the ultrasound field. Key factors in the harmonic response of an agent are

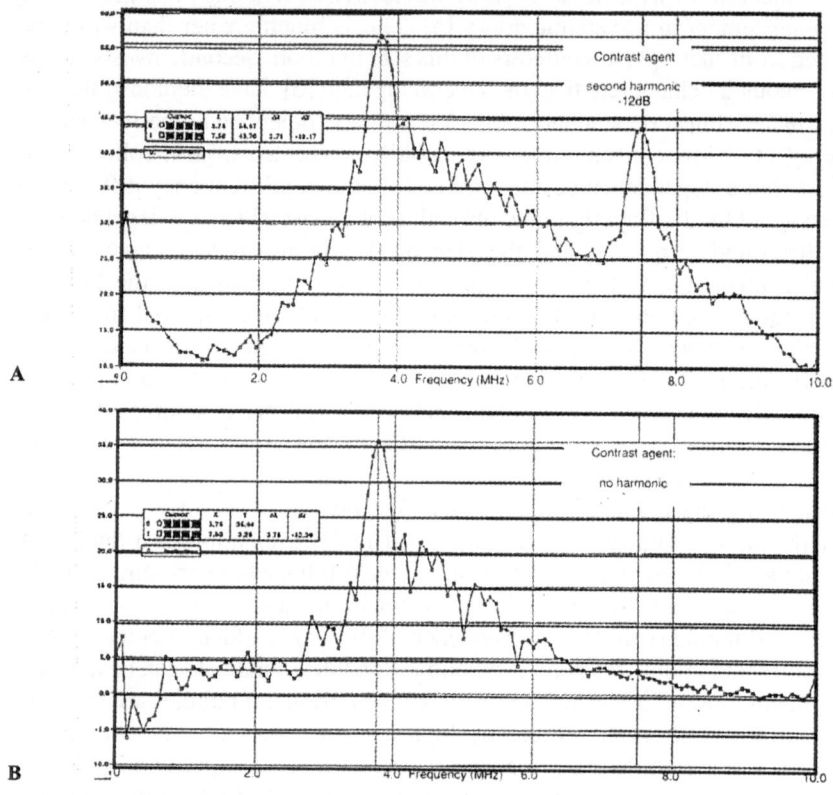

Figure 11. Harmonic emission from microbubble contrast agents. (A) A sample of Levovist is insonated at 3.75 MHz and the echo analysed for its frequency content. It is seen that most of the energy in the echo is at 3.75 MHz, but that there is a clear second peak in the spectrum at 7.5 MHz. This echo is only 13 dB less than that of the main, or fundamental echo. Harmonic imaging and Doppler aims to separate and process this signal alone. (B) Not all agents display a similar non-linear response. Here, although the echo is increased in amplitude at the fundamental frequency, there is no harmonic response.

likely to be the size distribution of the bubbles, the mechanical properties of the bubble capsule (a stiff capsule, for example, will dampen the oscillations and attenuate the non-linear response) and, most of all, the peak incident pressure.

The importance of incident pressure is illustrated by Figure 12, in which it can be seen that there is a regimen for which the incident sound is sufficiently weak to cause a bubble to behave as a linear, though perhaps resonant scatterer. At higher incident pressure the bubble oscillates in a non-linear way and harmonics, subharmonics and ultra-harmonics are seen. Finally, at a critical point (which depends on many conditions, some of which are unknown) a bubble in such resonant oscillation will be disrupted and destroyed. Measurements indicate

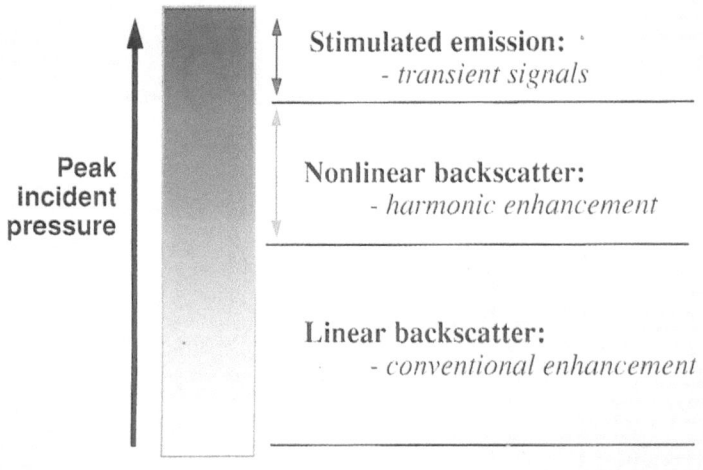

Figure 12. The acoustic mode of action of a microbubble contrast agent depends on the incident pressure. At low pressures, the bubble is a linear resonant scatterer which enhances the blood echo. At higher levels, the scattering becomes non-linear and harmonics can be detected. As the point is reached at which the bubble is disrupted by the incident sound, a transient, non-linear signal is emitted.

that at the point of disruption, an agent such as Levovist emits a transient, broad-band signal; a distinct – and final – signature in its echo. Of course, such borders in these regimens of acoustic behaviour are not precisely defined; the bubbles comprise a population with a spectrum of size and physical properties, and characteristics of the medium, such as its viscosity, are also likely to be important.

Harmonic imaging

These properties have led to a new real-time imaging and Doppler method based on the phenomenon of non-linear oscillation of resonant microbubbles, which we call harmonic imaging [11]. The system transmits normally at one frequency, but when in harmonic mode selectively receives echoes only at double that frequency. A commercially available digital color flow system (UM9 HDI and HDI 3000, ATL Inc., Bothell WA) has been modified to transmit at between 1.5 and 4 MHz and receive at the second harmonic, between 3 and 8 MHz. The principle is illustrated in Figure 13. In harmonic mode, echoes from the contrast agent are received preferentially by means of a bandpass filter whose center frequency is at the second harmonic. Echoes from solid tissue, as well as red blood cells themselves, are suppressed (Figure 13B). Real-time harmonic imaging is now commercially available (HDI 3000, ATL Inc., Bothell, WA) and has also

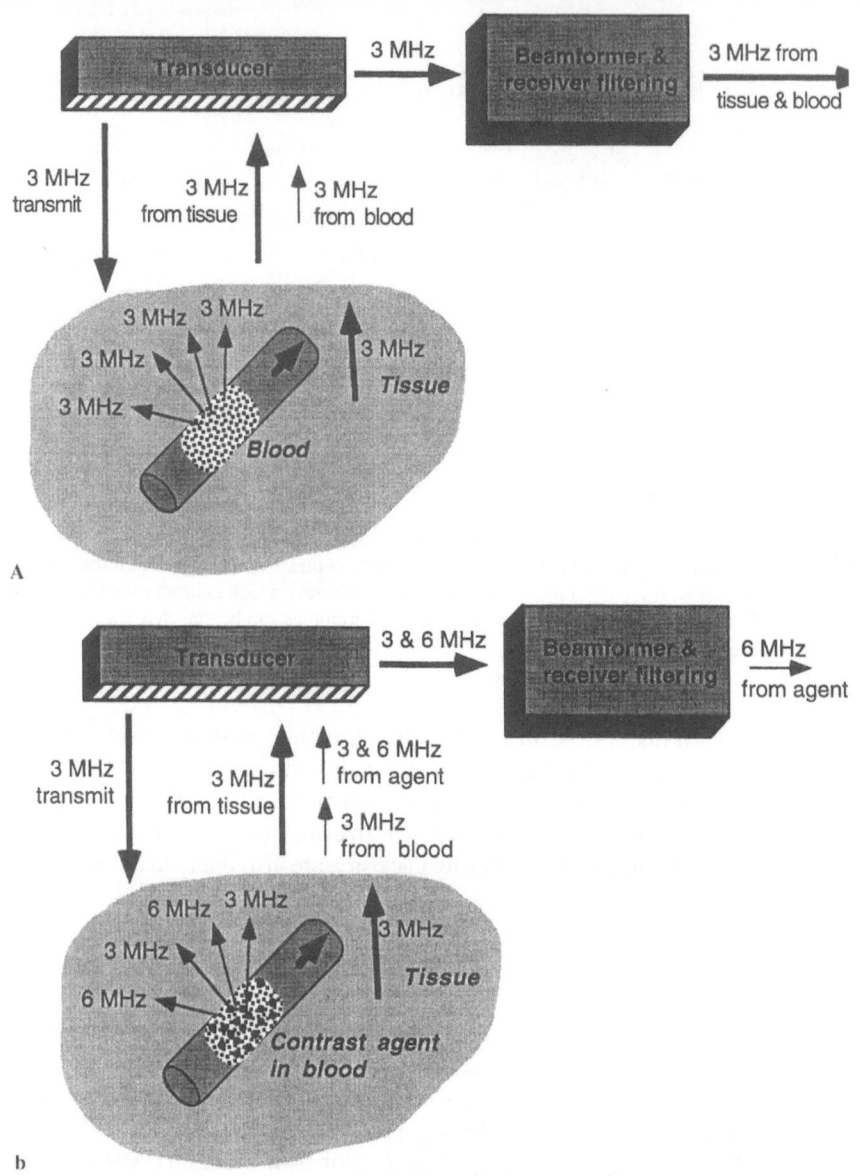

Figure 13. (A) Conventional imaging. The system transmits and receives at 3 MHz. The beamformer is unable to differentiate the echoes from blood and tissue. (B) Harmonic imaging. The system transmits at 3 MHz and receives at 6 MHz. The beamformer can separate the contrast agent echo from that of tissue.

been implemented experimentally on a number of commercially available systems. Transmitting sound at one frequency and receiving signals at twice that frequency requires a combination of a broadband transducer array and a broadband, flexible beamformer. The ability to use different frequencies simultaneously for 2D, color, and Doppler is also important. Fortunately much effort has been directed in recent years towards increasing the bandwidth of transducer arrays because of its significant bearing on conventional imaging performance. The new generation of all digital systems combining digital beamforming with flexible digital signal control allow such wide bandwidth and frequency agility.

In harmonic images, the echo from tissue mimicking material is reduced but not eliminated, reversing the contrast between the agent and its surroundings (Figure 14). In principle, harmonic images should show no tissue and render visible only the contrast agent. So why, then, is tissue still visible in a harmonic image? The residual image from normal, non-contrast bearing tissue may have its origin in number of factors relating to the ultrasound machine itself, the tissue target or the intervening acoustic path.

First, the transmitted pulse of ultrasound unintentionally may contain some second harmonic components, which is scattered by tissue and displayed on the harmonic image. In practice, the requirement of low second harmonic 'distortion' of the transmitted pulse (and of the receiver itself) depends on the system design but can be difficult to attain. Second, the receiver may display energy from echoes which are not exactly at the second harmonic but are slightly higher or lower than that frequency. This might be regarded as an inadequacy of the receiving filter, but is in fact an inevitable consequence of the bandwidth required to form an image with an acceptable axial resolution. This overlap is illustrated in Figure 15. Finally, the passage of ultrasound through tissue itself generates some second harmonic energy. Since this higher frequency energy is preferentially absorbed by tissue it has been assumed to be undetectable. Imaging this very small signal requires exceptional system bandwidth and dynamic range, only recently possible with broadband digital beamforming. Nonetheless, it has been demonstrated that this is indeed a source of the 'residual' image of tissue in harmonic mode.

In Doppler modes, the effect of the harmonic method is to reduce the clutter component of the echo. Figure 16 shows harmonic spectral Doppler: it can be seen that as the system switches from fundamental to harmonic mode, the Doppler shift frequency increases. It can also be seen that the signal intensity decreases in relation to that at the fundamental, because the scattered signal is weaker at the harmonic than at the fundamental. When spectral Doppler is applied to a region in which there is tissue motion as well as blood flow, however, its striking ability to suppress clutter is evident. In Figure 17A a large sample volume has been positioned so as to include a signal from the anterior wall of the aorta. The spectrum from conventional Doppler shows a large solid tissue clutter ('thump') component. In harmonic mode, this is rejected without the loss of flow data caused by filters (Figure 17B). Harmonic color images demonstrate flow in

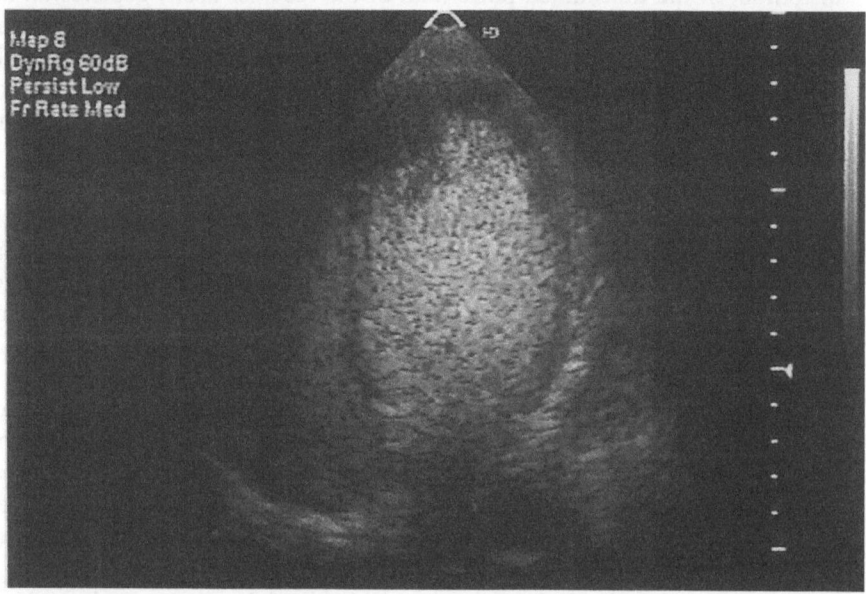

Figure 14. Harmonic imaging. Images taken at peak contrast enhancement with Albunex in an adult human showing the contrast barely visible in conventional mode (A). Harmonic imaging (B) is seen to reverse the contrast between the microbubble agent and the surrounding myocardium.

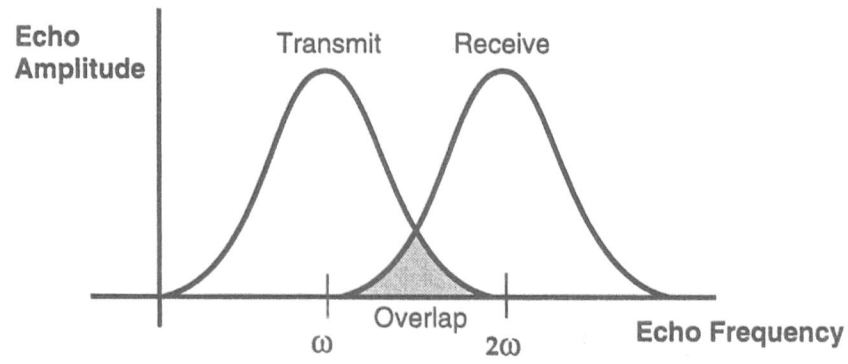

Figure 15. System frequency response in harmonic mode showing area of overlap between transmit and receive frequencies, resulting in residual tissue image.

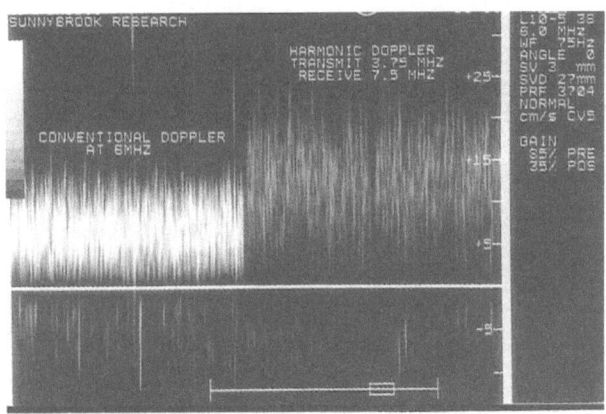

Figure 16. Harmonic spectral Doppler. As the system is switched from conventional to harmonic mode, the Doppler shift frequency doubles.

moving tissue without the flash artifact (Figure 18). In vivo measurements from spectral Doppler show signal-to-clutter ratios improved by a total of 35 dB [12] (Figure 19). This represents the combined effect of the enhancement of the echo pro-duced by the contrast agent and the reduction of the clutter produced by the harmonic method.

Harmonic power imaging

At the small expense of some sensitivity, amply compensated by the enhancement caused by the agent, harmonic mode helps to mitigate the clutter problem

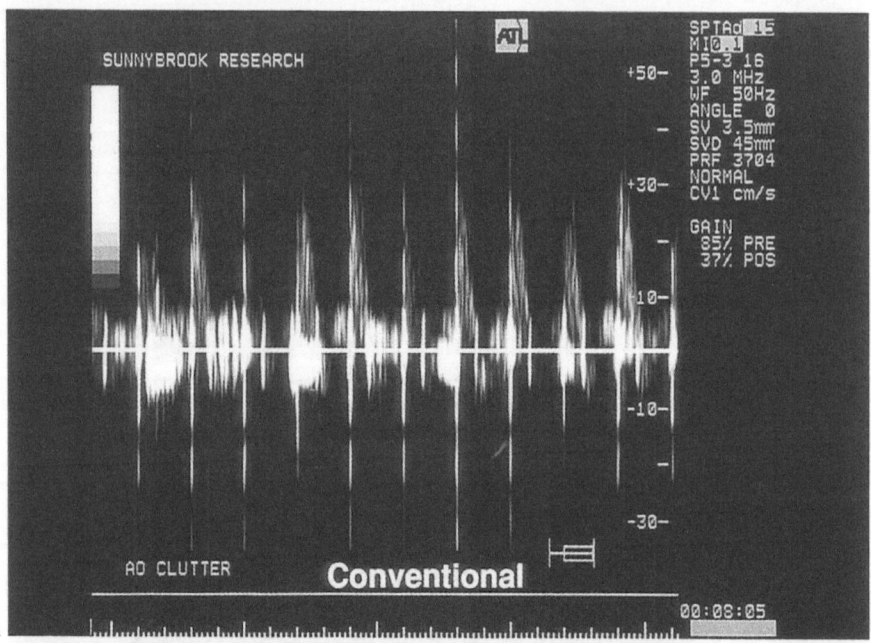

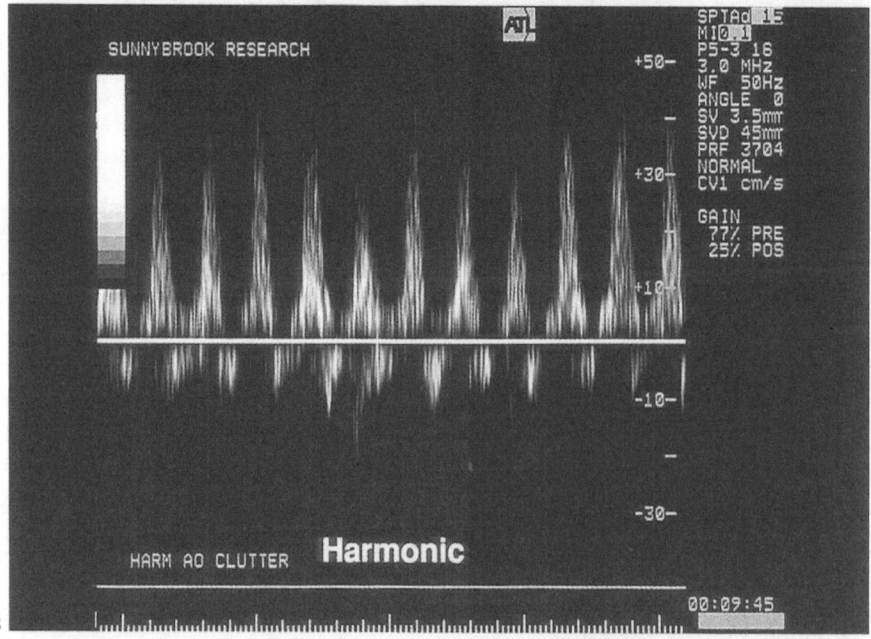

Figure 17. Clutter rejection with harmonic spectral Doppler. (A) The abdominal aorta of a rabbit is examined with harmonic spectral Doppler. In conventional mode, clutter from the wall (arrow) requires a 200 Hz filter, which also eliminates diastolic flow. (B) In harmonic mode, the clutter is still visible, but greatly reduced so that flow can be resolved (i.e. the signal-to-clutter ratio is within the dynamic range of the instrument). Note that the filter setting has been reduced to 25% of the previous level.

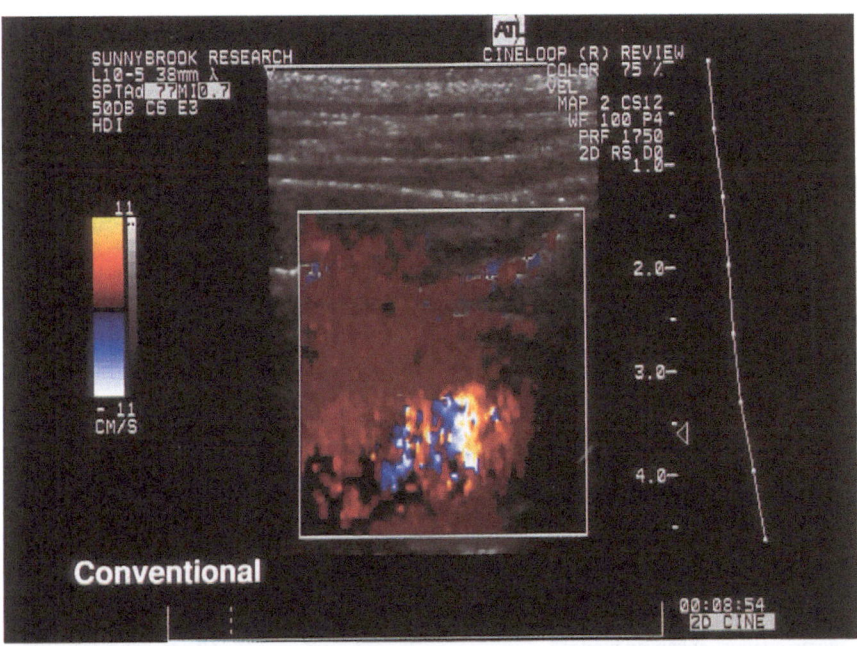

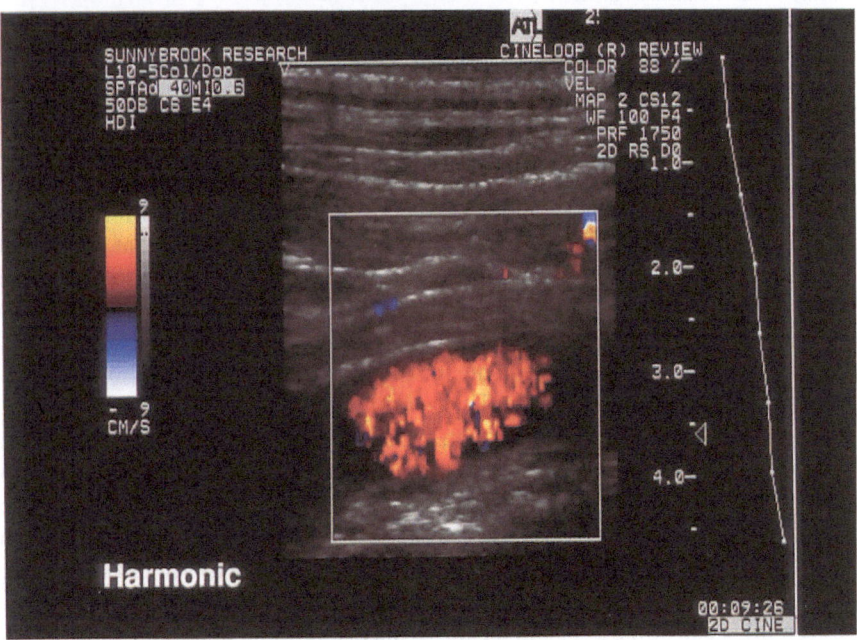

Figure 18. Clutter rejection with harmonic color Doppler. (A) Flow in the abdominal aorta of an experimental animal is superimposed on respiratory motion, producing severe flash artifact. (B) In harmonic mode at the same point in the respiratory cycle, the flash artifact disappears. Flow from a smaller vessel (the cranial mesenteric artery) is visualized. All instrument settings are the same.

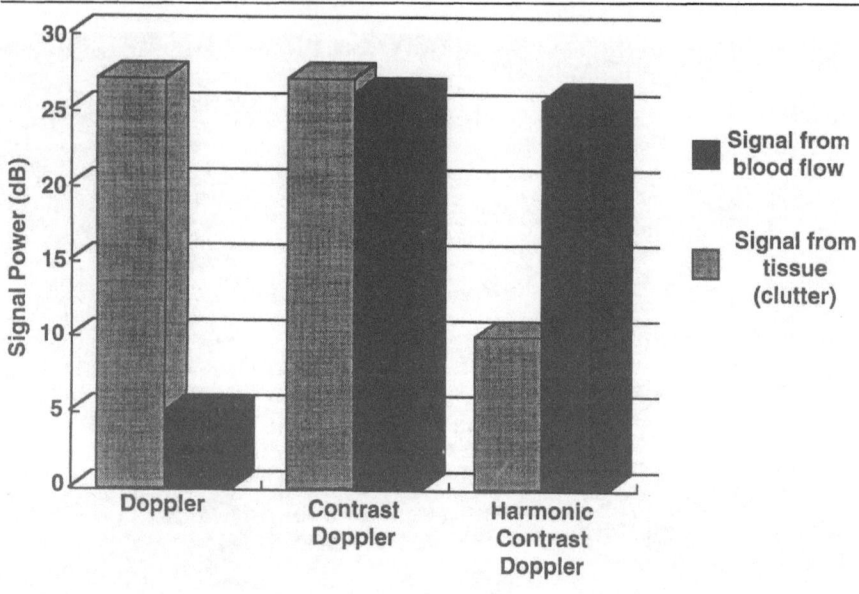

Figure 19. Clutter and spectral Doppler signal levels in conventional, contrast-enhanced and harmonic contrast-enhanced modes.

in power Doppler imaging (Figure 20). Combining the harmonic method with power Doppler produces an especially effective tool for the detection of flow in the small vessels of the organs of the abdomen which may be moving with cardiac pulsation or respiration. In a study in which flow imaged on contrast enhanced power harmonic images was compared with histologically sized arterioles in the corresponding regions of the renal cortex [5], it was concluded that the method is capable of demonstrating flow in vessels <40μm diameter; about 10 times smaller than the corresponding imaging resolution limit, even as the organ was moving with normal respiration.

An interesting benefit of harmonic imaging is that the agents tend to last significantly longer in harmonic mode. This is primarily due to the decreased threshold of detection caused by the reduction in tissue clutter signal. All agents have a finite lifetime once injected, lasting from a few seconds to 10 min or so. The decay time and process depends greatly on the agent. Some agents are exhaled through the lungs, some removed by the liver, and some destroyed by the ultrasound insonation itself. In all cases there is a rapid wash-in phase followed by a much slower wash-out phase. The time of the wash-out phase during which the contrast can still be visualized above the background noise and clutter signal is extended with harmonic imaging (Figure 21).

Harmonic imaging demands exceptional performance from the transducer array and system beamformer. Its implementation forces implicit compromises

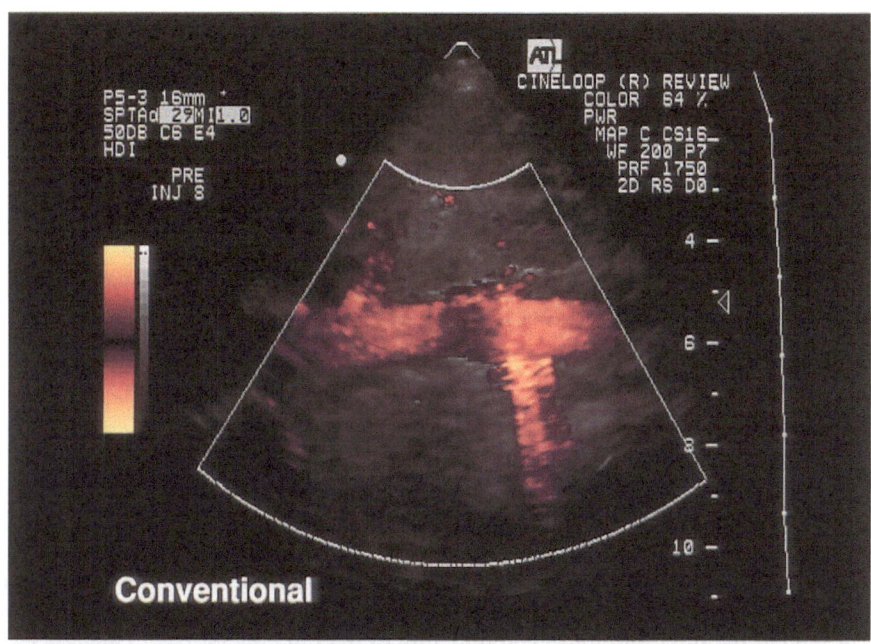

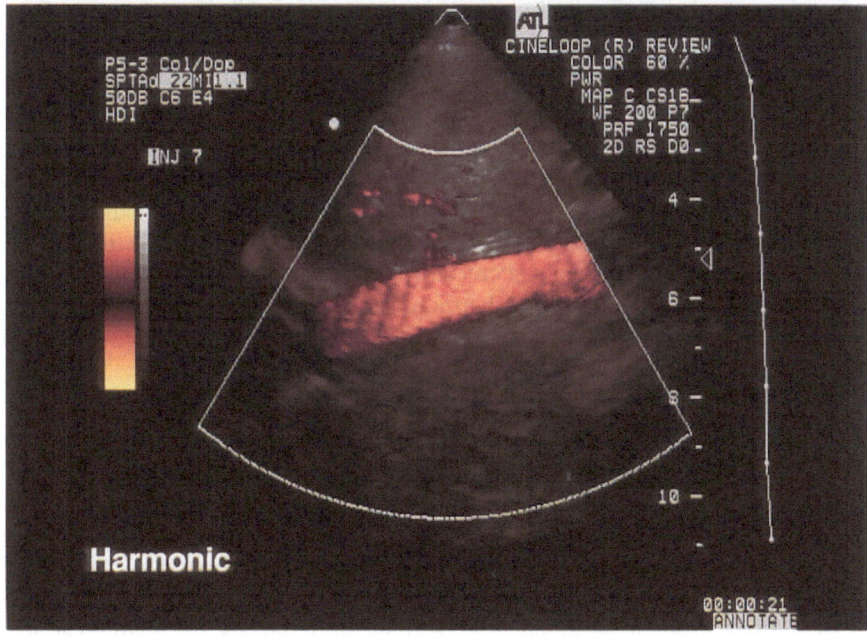

Figure 20. Reduction of the flash artifact in harmonic power Doppler. The harmonic contrast method helps overcome one of the principal shortcomings of power Doppler, its increased susceptibility to tissue motion. (A) Aortic flow in power mode with flash artifact from cardiac wall motion. (B) In harmonic mode at the same point in the cardiac cycle, the flash is largely suppressed. All instrument settings are the same.

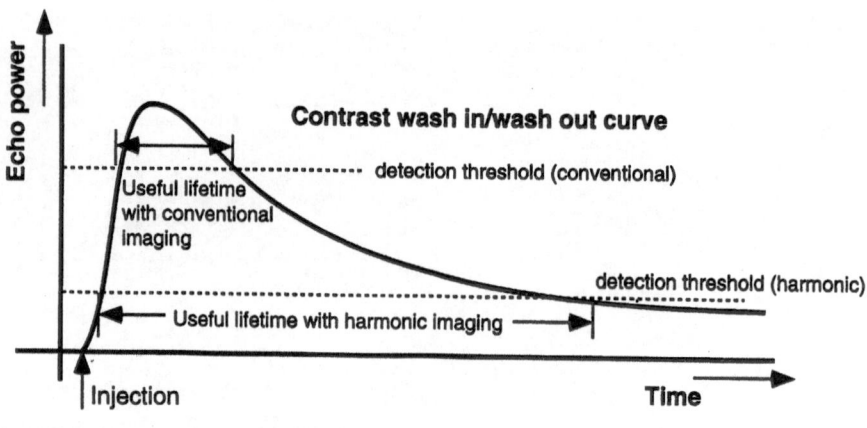

Figure 21. Time–enhancement curve caused by circulation of contrast agent bolus showing how harmonic imaging, by reducing the detection threshold, increases the imaging lifetime of the agent.

between, for example, image resolution and rejection of the non-contrast agent echo. It places unusual demands on the bandwidth performance of transducers, as well as the flexibility of the architecture of the imaging system. The ease with which it has been developed on modern instruments, however, reflects this flexibility and augers well for the future of the method. More significantly, contrast agents are now being developed specifically with non-linear response as a design criterion. Entirely new agents present opportunities for entirely new detection strategies. Our laboratory measurements show that some new contrast agents are capable of creating an echo with more energy in the second harmonic than at the fundamental, that is, they are more efficient in harmonic than conventional mode [10]. We have already produced harmonic B-mode images that demonstrate flow-related enhancement in the parenchyma of moving abdominal organs [13]. Recently, the first harmonic color images from the flow in the moving myocardium were reported by Fritzsch et al. using the Schering agent Sonovist [14].

Other contrast signatures

Other properties of microbubbles may also be used to enhance their detection. It has been shown, for example, that Sonovist, exhibits the unique property of producing a strong color Doppler signal even when it is stationary. The echo is essentially a strong noise signal, with zero mean frequency and a large variance. This appears to be due to the disruption of the microbubbles by the ultrasound field itself, generating a transient broadband acoustic signal, or ultrasonic 'pop'. This phenomenon has been labelled 'stimulated acoustic emission' [10] and it

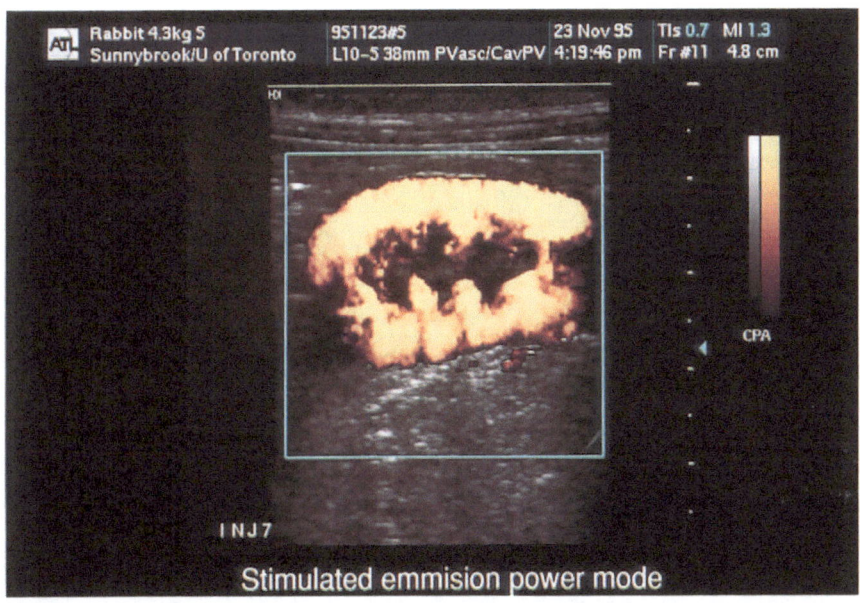

Figure 22. This power Doppler image of the kidney was produced entirely by the transient destruction of the agent Sonovist [15].

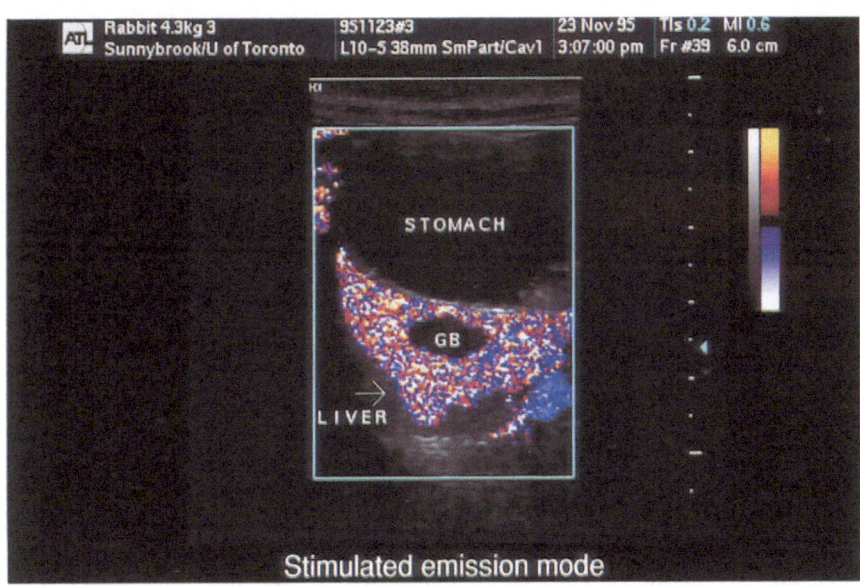

Figure 23. 30 min after injection of Sonovist, a rabbit liver with normal Kupffer cells is imaged in color mode as a random pattern of Doppler shifts. Note the absence of such shifts from the surrounding tissue or from the gall bladder (GB).

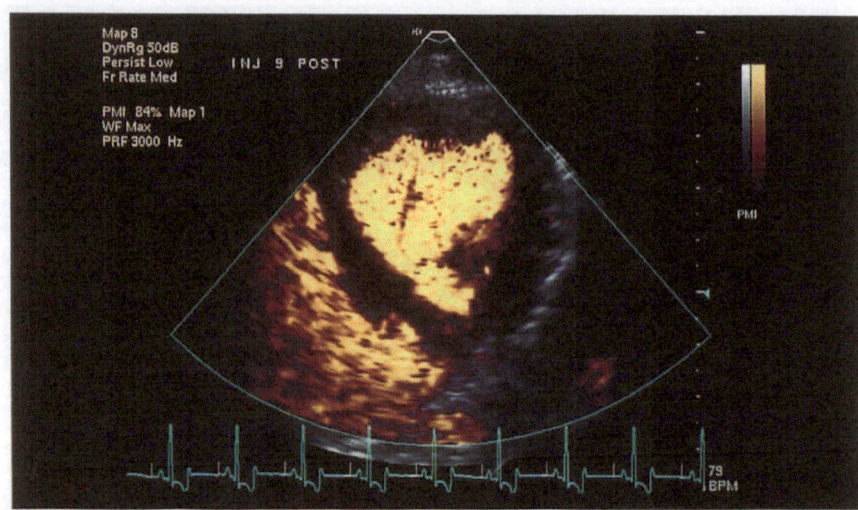

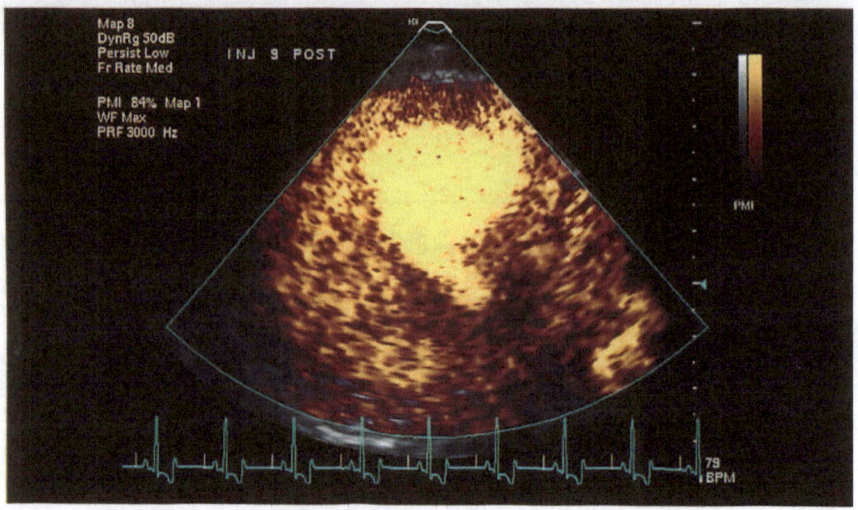

Figure 24. Short axis cardiac image following injection of Levovist. (A) Agent seen filling right ventricle, has just entered left ventricle, none visible in the myocardium. (B) Agent has filled left ventricle and is clearly visible in myocardium.

shows up in color flow as the random noise pattern shown in Figure 22. This may solve the problem of the ultrasound detection of particles that are selectively targeted to an organ site. Sonovist, after its blood pool phase, is phagocytosed and appears in the reticuloendothelial system of the liver and spleen about 30 min after injection. Color Doppler and power imaging can be used to detect this agent in both phases, disrupting it during the scanning process. In the blood pool phase, images such as that shown in Figure 22 can be acquired very rapidly. In the reticuloendothelial phase scanning the liver produces images such as Figure 23. Tumors of the liver, which contain a modified density of Kupffer cells, are seen as areas devoid of these transient 'pseudo-Doppler' signals [15].

In the same way that frequency and amplitude are both used as part of the signature for normal blood flow, so combinations of other contrast signatures can be used to enhance further the system's ability to separate the contrast from the moving tissue. This is especially important in the heart. The combination of stimulated acoustic emission detection along with harmonic power mode can be used to image myocardial blood flow with unprecedented separation from tissue. Figure 24A shows the agent filling the right ventricle and beginning to enter the left ventricle. No contrast is visible in the myocardium. In Figure 24B the agent has completely filled the left ventricle and can clearly be seen in the myocardium.

Intermittent contrast imaging

Although Sonovist is the only agent which has been shown to have such strong acoustic emission due to bubble disruption, it is becoming clear that most agents are destroyed, at least partially, by the interrogating ultrasound field [16]. This effect may not be apparent in large vessels or in the cardiac chambers because there is a constant supply of fresh blood containing new contrast. It is, however, a potential problem when trying to image flow in smaller vessels or veins. It is hypothesized that when the agent is moving slowly, as it does in a capillary bed, the agent is destroyed by the real time scanning before it has a chance to completely fill the vascular bed. By imaging at periodic intervals of every second or more, more blood has a chance to enter the region of interest before it is imaged and subsequently destroyed.

Calculation of wash-in/wash-out characteristics

Another potentially fertile area of development for contrast agent use is that of measuring the dynamics of contrast agent passage through tissue. When a bolus of contrast agent is injected in a peripheral vein, it takes several heartbeats to arrive at the location of interest in the left side of the circulatory system. The signal in that area rapidly increases to a peak value and then slowly decays back to the original baseline value (Figure 21). The time to peak and the decay time might be modified by the presence of a cancer, which contains arteriovenous shunts [17], or by the myocardial perfusion rate [18]. To measure these enhancement transit times effectively the system must be able to store a very large amount of data internally to allow post-processing, since the total transit can be in excess of 10 min. The alternative is that the system has the computational power to perform the necessary calculations in real time during the examination,

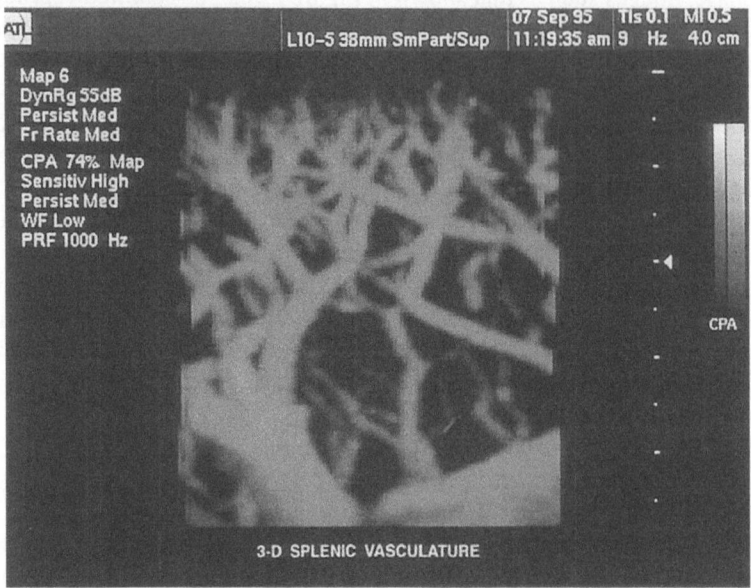

Figure 25. 3D image of splenic vasculature.

and that the user select the regions of interest prior to injection of the contrast agent.

THE FUTURE: 3D AND BEYOND

As the agents mature and are subject to more clinical investigation, and as ultrasound system capabilities increase, there is potential for a great variety of new applications. One example is the recent development of integrated 3D Color Power Angio™ (power Doppler). Complex vascular structures can be difficult to visualize using a single scan plane cross-section since the vessels often go in and out of the scan plane, producing a number of small vessel cross-sections. Often the operator must spend a great deal of time rocking the scanhead back and forth to build up a mental image of what the vascular bed really looks like. When imaged in 3D the complex nature of the vascular structure can be more easily appreciated, as shown in Figure 25.

The combination of contrast agents with 3D imaging is natural. Since the contrast increases the power of the Doppler signal it will have a direct impact on the amount of vascularity which can be seen with 3D. As the scanhead must be moved to obtain a 3D data set, it may be subject to motion artifact. Harmonic imaging promises to reduce the artifact from tissue or scanhead motion. Since the duration of contrast enhancement is limited, acquisition of a 3D data set during peak enhancement will allow more detailed analysis of the vascular bed long after the contrast effect has diminished. Perhaps the combination of contrast agents, power Doppler and this method will allow the production of 3D vascular maps of, for example, human breast tumors.

Fritzsch [19] has described novel ideas for new contrast agent development involving agents targeted to specific organs and therapeutic agents, such as cytotoxic drugs or even vectors carrying genes, encapsulated within the micro-particles themselves. These would then be released under ultrasonic interrogation at the region of interest. The first such site-specific particulate agent is Sonovist, which, following its blood pool phase, is selectively taken up by the Kupffer cells of the liver and spleen. Clearly these developments will bring new challenges to the sophistication and flexibility of ultrasound instrument design.

CONCLUSION

Modern agents for the production of ultrasound contrast offer the prospect that extremely small, harmless injections of a solution into a peripheral vein will produce spectacular improvements in the detection of blood-filled structures and blood flow velocity in the arterial system. New uses are likely to include the detection of smaller vessels, the detection of neovascular flow associated with tumors and wound healing and possibly the measurement of relative blood flow rates in such tissues as the myocardium. Optimization of ultrasound instruments to the inclusion of contrast agents in moving blood presents a set of exciting challenges which have not yet been completely met. Harmonic imaging will give new capability to imaging and Doppler systems, effectively allowing real-time 'subtraction' imaging of flowing blood with ultrasound and the detection of flow in smaller vessels than hitherto possible. The production of transient and non-linear acoustic phenomena by the latest generation of contrast agents promises to stimulate a new set of imaging solutions which will allow ultrasound to probe deeper into the microcirculation than ever before.

REFERENCES

1. Burns PN, Hilpert P, Goldberg BB. Intravenous contrast agent for ultrasound Doppler: in vivo measurement of small vessel dose-response. IEEE Eng Med Biol Soc 1990;12:322–4.
2. Goldberg BB, Hilpert PL, Burns PN et al. Hepatic tumors: signal enhancement at Doppler US after intravenous injection of a contrast agent. Radiology 1990;177:713–7.
3. Goldberg BB, Liu J-B, Burns PN, Merton D, Forsberg F. Galactose-based intravenous sono-graphic contrast agent: experimental studies. J Ultrasound Med 1993;12:463–70.
4. Burns PN. Doppler artifacts. In: Taylor KJW, Burns PN, Wells PNT, eds. Clinical Applications of Doppler Ultrasound, Vol. 2. New York: Raven Press, 1995:46–76.
5. Burns PN, Powers JE, Hope Simpson D et al. Harmonic power mode Doppler using micro-

bubble contrast agents: an improved method for small vessel flow imaging. Proc IEEE UFFC 1995:1547–50.

6. Bleeker H, Shung K, Barnhart J. On the application of ultrasonic contrast agents for blood flowmetry and assessment of cardiac perfusion. J Ultrasound Med 1990;9:461–71.

7. Mo LYL, Cobbold RSC. Theoretical models of ultrasonic scattering Doppler ultrasound in blood. In: Shung KK, Thieme GA, editors. Ultrasonic scattering in biological tissues. Boca Raton: CRC Press, 1993:125–71.

8. Ophir J, Parker KJ. Contrast agents in diagnostic ultrasound. Ultrasound Med Biol 1989;15: 319–33. [Published erratum appears in Ultrasound Med Biol 1990;16:209.]

9. Neppiras EA, Nyborg WL, Miller PL. Nonlinear behavior and stability of trapped micron-sized cylindrical gas bubbles in an ultrasound field. Ultrasonics 1983;21:109–15.

10. Fritzsch T, Hauff P, Heldmann F, Lüders F, Uhlendorf V, Weitschies W. Preliminary results with a new liver specific ultrasound contrast agent. Ultrasound Med Biol 1994;20:137.

11. Burns PN, Powers JE, Fritzsch T. Harmonic imaging: a new imaging and Doppler method for contrast enhanced ultrasound. Radiology 1992;185(P):142.

12. Burns PN, Powers JE, Hope Simpson D, Uhlendorf V, Fritzsch T. Harmonic contrast enhanced Doppler as a method for the elimination of clutter – in vivo duplex and color studies. Radiology 1993;189:285.

13. Burns PN. Contrast-enhanced studies: how we can use present instruments? J Ultrasound Med 1995;14:S50.

14. Fritzsch T, Hauff P, Scholle FD, Uhlendorf V. Second harmonic response of a polymer coated microbubble suspension. J Ultrasound Med 1996;15:S57.

15. Burns PN, Fritzsch T, Weitsches W et al. Pseudo Doppler shifts from stationary tissue due to the stimulated emission of ultrasound from a new ultrasound contrast agent. Radiology 1995; 197(P):577.

16. Porter T, Xie F. Transient myocardial contrast following initial exposure to diagnostic ultrasound pressures with minute doses of intravenously injected microbubbles: demonstration and potential mechanisms. In: Nanda N, Schlief R, Goldberg B, editors. Advances in Echo Imaging Using Contrast Enhancement. 2nd edn. Dordrecht: Kluwer Academic Publishers, 1997.

17. Kedar RP, Cosgrove DO, Smith IE, Mansi JL, Bamber JC. Breast carcinoma: measurement of tumor response to primary medical therapy with color Doppler flow imaging. Radiology 1994; 190:825–30.

18. Skyba D, Jayaweera A, Goodman N, Ismail S, Camarano G, Kaul S. Quantification of myocardial perfusion with myocardial contrast echocardiography during left atrial injection of contrast: implication for venous injection. Circulation 1994;90:1513–21.

19. Fritzsch T. Contrast-enhanced studies: the future of contrast agent technology. J Ultrasound Med 1995;14:S50.

9. Doppler myocardial imaging

GEORGE R. SUTHERLAND

INTRODUCTION

Current standard cardiac ultrasound techniques derive their information on myocardial function indirectly either from parameters measured from the endo- and epicardial specular reflections or from blood pool Doppler indices.

Doppler myocardial imaging (DMI) is a new cardiac ultrasound technique in which Doppler signal processing is applied to the reflected ultrasound signals originating within the myocardium [1, 2]. Information on myocardial function can be derived either from the use of color Doppler principles or by performing spectral or power analysis of the pulsed Doppler signal derived from a sample volume placed within the myocardium. In the color Doppler approach, mean velocity estimation is based on the autocorrelation technique and quantifies regional intramural velocities by detecting consecutive phase shifts of the reflected echoes from myocardium. The use of Doppler techniques to interrogate myocardial motion is not a new concept. Pulsed Doppler recordings of myocardial motion using a single sample volume technique were first described by Isaaz et al. in 1989 [3]. Although deemed to be interesting, little clinical value was ascribed at that time to this approach. It was not until 1991 that McDicken et al. [4] first described the modifications required to a standard cardiac ultrasound which would allow visualization of the motion of a tissue-equivalent phantom that the potential clinical value of the color Doppler processing of myocardial signals began to be appreciated.

Doppler signal processing of myocardial ultrasound information has several theoretical advantages over standard gray-scale imaging. Doppler processing has an intrinsically lower noise floor and higher signal-to-clutter ratio, and allows the accurate measurement of differing velocities at a large number of neighbouring intramural sites. This is simpler, quicker and more reliable than the alternative technique of tracking the speckle pattern within conventional B and M-mode images. In addition, as the Doppler technique measures frequency shift rather than signal amplitude, the velocity image information is less affected by

171

N. C. Nanda, R. Schlief and B. B. Goldberg (eds.), Advances in Echo Imaging Using Contrast Enhancement, Second Edition, 171–186.
© 1997 *Kluwer Academic Publishers.*

tissue attenuation than gray-scale imaging. It is, therefore frequently possible to obtain diagnostic image quality DMI images from patients who are 'poorly echogenic' using standard gray-scale imaging. Although gray-scale image processing is superior to autocorrelation Doppler signal processing in terms of maximal achievable frame rate and spatial resolution, color Doppler technique can still provide clinical diagnostic information. Current generation ultrasound processing techniques allied to digital signal processing within the new generation of ultrasound machines allow color Doppler images to be displayed with a frame rate of 50–80/s. While image spatial resolution is inferior to gray-scale imaging, this approaches 1×1 mm at low color Doppler frame rates and 2×2 mm at frame rates of 40–50/s.

MYOCARDIAL VELOCITY AND ACCELERATION ESTIMATION TECHNOLOGY

In a conventional ultrasound machine, prior to blood pool Doppler signal processing taking place, the reflected Doppler ultrasound signal is passed through a high-pass (clutter) filter which will reject the high intensity, low-frequency components which arise from the myocardium. In order to visualize the myocardium an ultrasound machine must therefore be adapted by changing the thresholding and filtering algorithms to reject the low amplitude echoes from the blood pool and to allow the high amplitude, low velocity information from the myocardium to pass to subsequent determination of the mean Doppler shift, and hence mean velocity, using standard autocorrelation methodology (Figure 1). The velocity information may be displayed in either the two-dimensional (Figure 2) or M-mode format (Figure 3). Using the mean velocity information, the machine can also be programmed to calculate and display regional myocardial accelerations in a two-dimensional format (although with a reduced frame rate). In addition, a further myocardial Doppler parameter can be measured and displayed – the power of the Doppler signal (Table 1).

A series of in vitro phantom studies has confirmed the accuracy of DMI velocity encoding over the range of velocities at which normal and abnormal myocardium would be expected to move [5]. Velocity estimation has been shown to be affected by target velocity, target material, system receive gain and the pulse train size, but the inherent error is at worst $\pm 10\%$ of the true mean velocity [5]. The spatial resolution of both the two-dimensional velocity and power maps is at best 1×1 mm and at worst 3×3 mm, with a slightly inferior axial resolution compared to standard gray-scale imaging but a similar lateral resolution [2]. This means that the DMI technique is a better real-time spatial discriminator than real time magnetic resonance imaging, position emission tomography or current nuclear perfusion techniques.

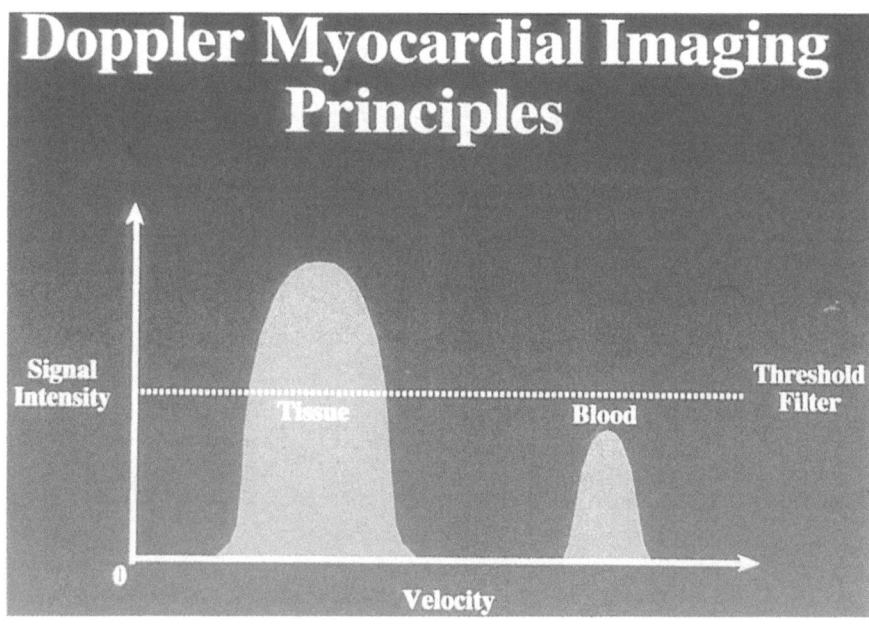

Figure 1. Principles underlying the implementation of Doppler myocardial imaging. To obtain tissue information as opposed to blood pool Doppler information, the signal intensity threshold filter has to be altered. In addition, algorithms have to be changed to allow encoding of the low velocities at which the myocardium moves with concomitant rejection of high velocity blood pool information.

DOPPLER POWER INTERROGATION OF THE MYOCARDIUM

The power (or signal strength) of reflected myocardial Doppler ultrasound information is directly related to the number of scatterers within the tissue block insonated. Thus measurement of regional Doppler power levels could provide information on tissue characterization. The Doppler power signal is both velocity independent and angle independent as this measurement is independent of the phase-shift of the signal. Regional Doppler power can be displayed in two formats: as a two-dimensional (with a dynamic range of $0-40\,\text{dB}$) or M-mode image (Figure 4), or as the temporal variation in signal strength of the signal derived from a pulsed Doppler sample volume placed within the myocardium (but excluding the endocardial and pericardial specular reflectors). This latter measurement can be obtained by measuring the power of the raw audio data (which is available on all ultrasound machines) using a specially developed on-line power meter. The temporal variations in power levels can then be fed back into the machine via an auxillary channel and displayed as a trace in real time simultaneously with the Doppler myocardial velocity information. Raw audio data are relatively unprocessed ultrasound information obtained early in the

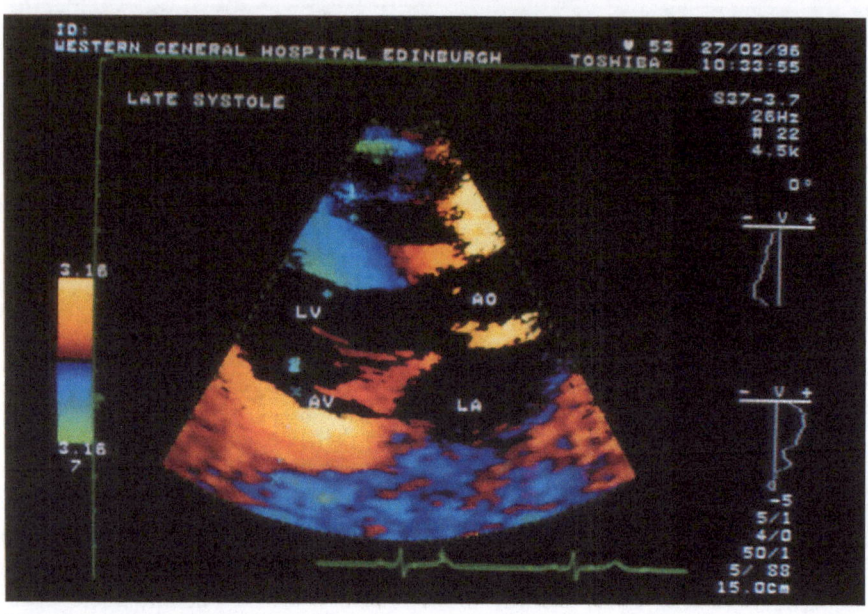

Figure 2. A left ventricular long axis Doppler myocardial image. Note that within the posterior wall and septum, a gradation of velocities is seen with the highest recorded subendocardially and the lowest subepicardially.

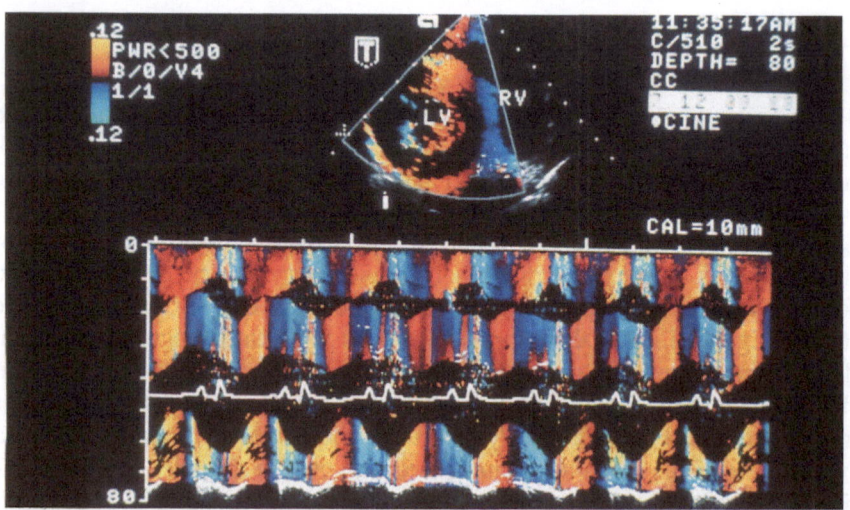

Figure 3. An M-mode short axis Doppler myocardial imaging study. Note again the velocity gradients both in early systole and in early and late diastole which are present within all cardiac walls.

Table 1. Currently available formats in which Doppler myocardial imaging information can be obtained.

Format	Parameter measured
2-D color maps	Velocity, acceleration, energy
M-mode[a]	Velocity, energy
Pulsed Doppler	Velocity, energy[b]

[a]M-mode of acceleration is simply velocity information, and thus an M-mode of acceleration data is not available.
[b]Not currently available as standard technique.

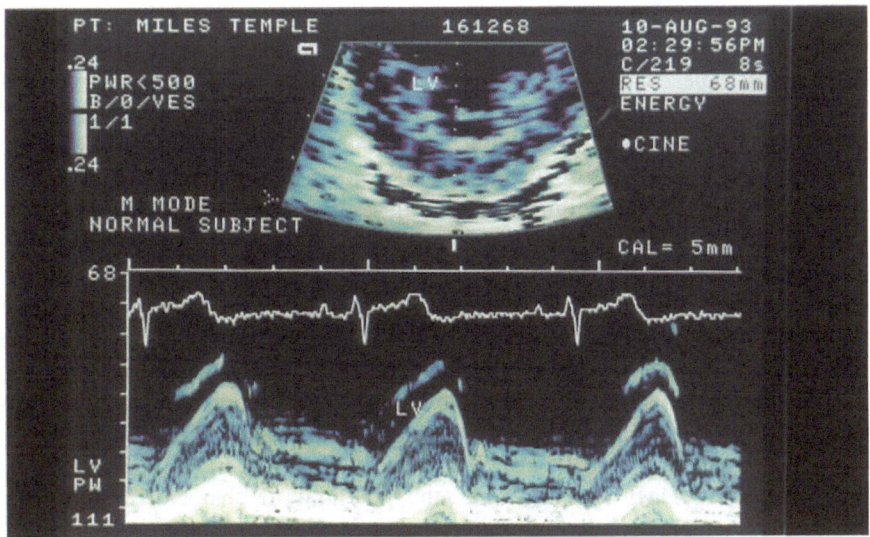

Figure 4. Combined image illustrating the Doppler power (energy) mode. The upper image is a short-axis view of the left ventricular posterior wall. The lower image is the corresponding M-mode Doppler power image. There is no gray-scale information retained within this image. This is a pure Doppler power image.

signal processing path with the ultrasound machine. These data have a wide dynamic range (typically $0-70\,$dB) and are linearly processed. They should be the direct equivalent of integrated backscatter data derived from the myocardium. Using a similar Doppler power measurement technique, Schwartz et al. [6] have already demonstrated in a series of in vitro studies that single pulsed Doppler power measurement is the direct equivalent of integrated backscatter measurement in terms of changes in blood pool reflectivity produced by varying the concentrations of echo contrast agent.

DMI STUDIES OF NORMAL MYOCARDIAL FUNCTION

To circumvent the temporal resolution problems inherent in measuring intra-mural velocities using the 2-dimensional approach, DMI M-mode interrogation of intramural velocities was developed. This technique allows myocardial sampling at 4 ms intervals with high spatial resolution but as with all M-mode techniques, should only be carried out where the myocardium can be insonated at 90°. Such M-mode velocity studies in normal patients have confirmed the presence of transmural systolic and diastolic velocity gradients across the left and right ventricular walls during the cardiac cycle (Figure 3) [7–9]. These velocity gradients are presumed to represent short-axis contractile function only and are in accord with prior reports on intramural velocity gradients recorded by placement of a series of ultrasonic crystals within the myocardium during animal experiments. These latter studies have also recorded an early systolic transmural velocity gradient with higher velocities encoded in the subendocardial region and lowest velocities in the subepicardial region. A series of DMI M-mode velocity studies has confirmed the normal velocity distribution and timing of peak velocity gradients within the left ventricular posterior wall and interventricular septum during the cardiac cycle. These studies have also identified predictable changes in intramural velocities which occur associated with ageing. DMI M-mode studies have also been used to interrogate long-axis shortening of the heart by aligning the M-mode beam along the apex–base axis of the interventricular septum or lateral left and right ventricular walls. These studies have demonstrated a base–apex longitudinal velocity gradient in both the septum and ventricular free walls, with the highest long-axis shortening velocities recorded at the cardiac base and with almost zero velocities recorded at the apex. Thus the DMI technique has the potential to characterize long-axis function of the heart, including regional abnormalities in long-axis contraction and relaxation. DMI velocities have also been shown to change in a predictable manner in response to both positive and negative inotropes in an animal model. In addition, single pulsed Doppler DMI velocity sample volume interrogation of the myocardium allows the determination of regional myocardial peak velocities and their temporal relationship to regional mechanical and electrical events in the heart (Figure 5).

Left ventricular diastolic and left atrial function

Current ultrasound parameters used to evaluate left ventricular diastolic function are indirect indices derived either from digitized M-mode traces of the endocardial surfaces of the left ventricle or from Doppler indices derived from blood flow through the left heart. DMI provides a direct measure of intramural velocities during left ventricular relaxation and hence may provide a better and more clinically relevant measure of diastolic function. Early studies showed that

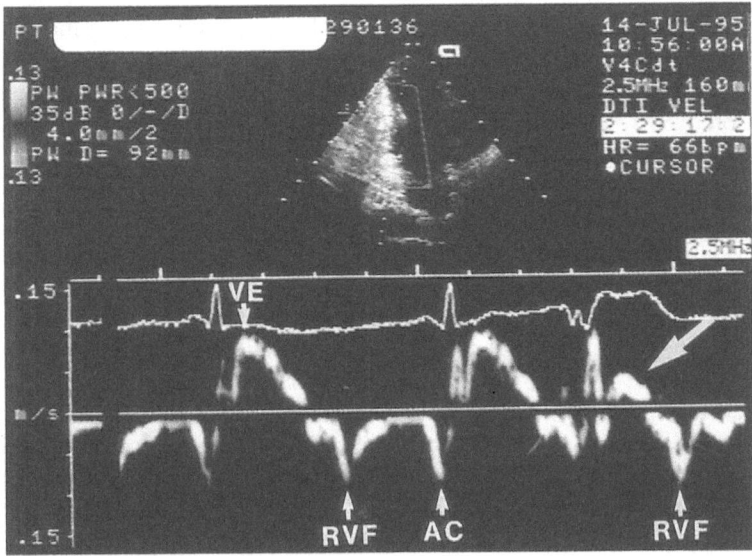

Figure 5. A pulsed Doppler myocardial imaging study in a normal patient. The image is acquired from the apical 2-chamber view with a sample volume placed in the inferior wall. The first beat recorded is in normal sinus rhythm, the third beat being a ventricular premature beat. Note the changes in long-axis systolic function induced by the premature beat.

DMI indices parallel transmitral Doppler blood flow measurements and normal changes in these latter indices associated with ageing [8]. Other studies have shown that measured DMI diastolic velocities of long axis lengthening to correlate with atrioventricular valve ring velocities during diastole [10, 11]. In addition, DMI indices have been shown to correlate well with both left atrial wall function indices and indices of left atrial appendage function [12, 13]. The extent to which these new indices provide more clinically relevant information remains to be determined but it is likely that such direct measurements may supplant current indirect measurements of left ventricular diastolic function.

DMI STUDIES IN CARDIOMYOPATHIES

A series of clinical studies in patients with dilated cardiomyopathies, hypertrophic cardiomyopathies, and patients with concentric left ventricular hypertrophy determined a range of abnormalities in intramural velocities which could not be predicted from either standard gray-scale 2-dimensional or M-mode studies. The results of these studies showed a significant decrease in mean velocities and velocity gradients during all systolic phases during atrial contraction in patients with dilated cardiomyopathy [14]. In those with hypertrophic cardiomyopathy velocity gradients were significantly decreased or reversed in all systolic cardiac phases, despite apparently normal M-mode fractional thickening indices [15]. These abnormalities are likely to be due to the abnormal myocardial architecture in HCM patients and may prove to be a new ultrasound marker specific for this condition.

DMI STUDIES IN ISCHEMIC HEART DISEASE

DMI can determine changes in myocardial contractile function induced by ischemia [16, 17], as has been shown both in animal models and in patients. The earliest ischemia-induced changes have been recorded using the DMI pulsed Doppler technique to identify regional velocity changes. Significant early changes in regional diastolic velocities in the ischemic zone were found after only 15 s of ischemia in dogs [18]. Later changes in systolic peak velocities were also observed which paralleled changes in the transmural velocity gradient obtained by 2-dimensional DMI reduction in wall thickening on M-mode gray-scale imaging. Encouraging results have also been reported with regard to detection of reversible ischemic changes using the DMI single sample volume pulsed Doppler technique in conjunction with low-dose dobutamine infusions (Figure 6) [19]. Predictable changes in transmural velocities associated with a marked reduction in Doppler power signal strength have also been noted in patients with acute myocardial infarction (Figure 7A–G).

PULSED DOPPLER SIGNAL STRENGTH – A DIRECT EQUIVALENT TO INTEGRATED BACKSCATTER MEASUREMENTS?

Normal and ischemic studies

Measurements of variation in the power of the pulsed Doppler myocardial signal from normally contracting myocardium have shown this to demonstrate a cyclic variation in signal strength which parallels the temporal variation in standard integrated backscatter measurements. The largest variation in myocardial pulsed Doppler signal strength and highest signal intensities are recorded at the base of the heart, while the lowest levels are recorded at the apex. These findings again parallel integrated backscatter parameters. Clinical studies have also demonstrated that ischemia blunts both the absolute peak signal levels and the cyclic variation of power of the myocardial pulsed Doppler signal (Figure 7A–G). There is no measurable cyclic variation in pulsed Doppler signal strength in infarct zone, although the absolute level of the signal may vary depending on the relative amounts of oedema or fibrosis present. Furthermore,

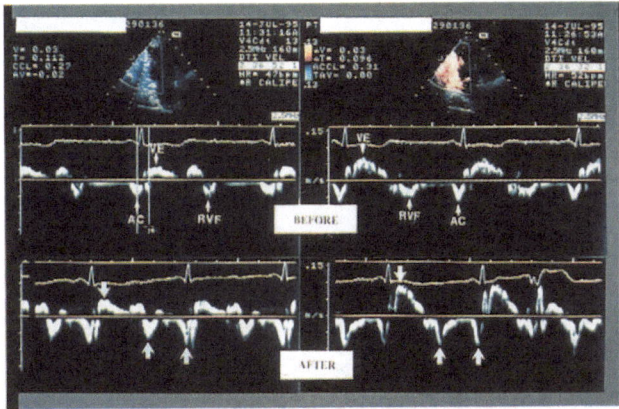

Figure 6. Images taken during a low dose dobutamine study for myocardial viability early after myocardial infarction. In the upper left panel, the pulsed Doppler sample volume is placed in the apical portion of the inferior left ventricular wall. The apparent double abnormal low velocity contraction pattern seen in systole here is pathognomonic of established infarction. In the lower left panel, during low dose dobutamine, there is no significant change in this myocardial segment during systole but there is an increase in diastolic filling velocities. In the upper right panel, an area which appeared equally acontractile on 2-dimensional gray-scale imaging is shown to be hypocontractile (long-axis contractility) prior to low dose dobutamine but function returns to near normal values (lower right panel) following 5 mg/kg dobutamine. This is clearly viable myocardium.

the intensity of the pulsed Doppler myocardial signal can be increased by the presence of a myocardial contrast agent.

DMI stress echocardiography

One initial hope was that with appropriate development, Doppler myocardial imaging would allow quantification of two-dimensional stress echo images by determining ischemia-induced intramural velocity changes. The initial low frame rates present in the first generation of DMI two-dimensional imaging systems made such measurements unlikely to be of clinical value. Even with the development of a second generation of DMI instruments with frame rates equal to or greater than those of current standard 2-D gray-scale imaging, the problems associated with the Doppler angle of insonation of the myocardium continue to preclude off-line measurement of absolute velocity changes during stress echo studies. However, subsequent initial DMI studies during low and high dose dobutamine stress echocardiography using the left parasternal window approach have demonstrated a predictable dose dependent increase in mean systolic velocities in both the inferior and anterior left ventricular walls in normal patients while patients with coronary artery disease were shown to have impaired augmentation of systolic velocities [20].

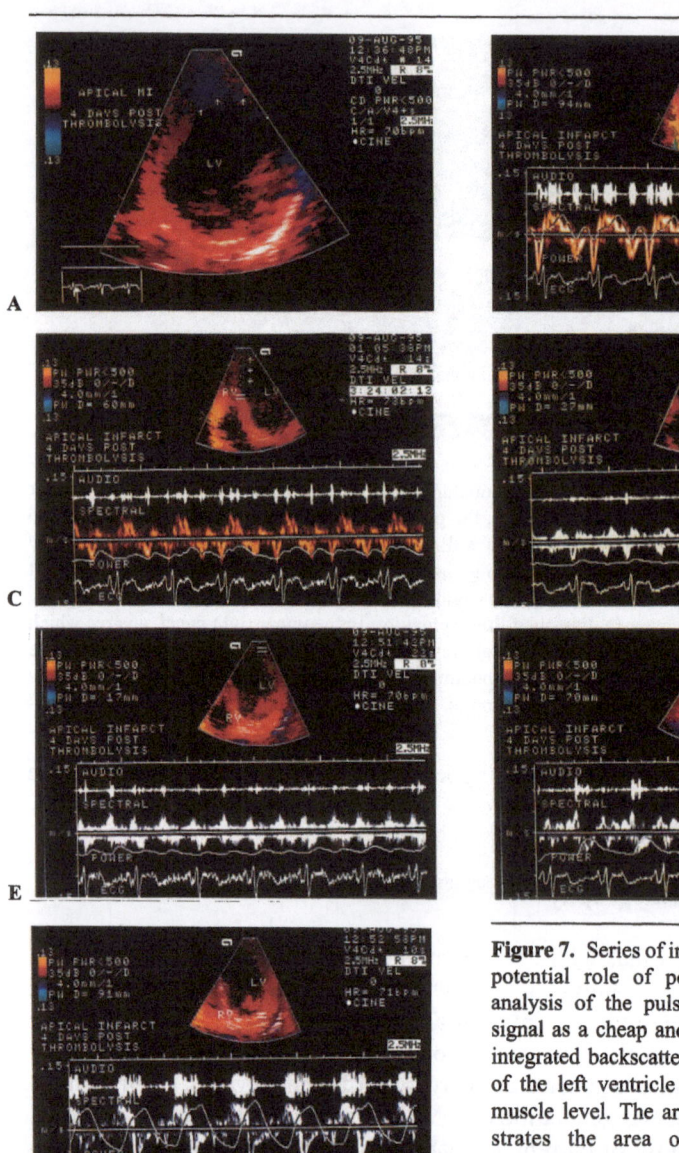

Figure 7. Series of images to demonstrate the potential role of power (energy) Doppler analysis of the pulsed Doppler myocardial signal as a cheap and effective alternative to integrated backscatter. (A) A short-axis view of the left ventricle at just below papillary muscle level. The arrowed indicator demonstrates the area of anterior myocardial infarction. In panels B–G the pulsed Doppler sample volume is tracked around the myocardium from the normally functioning segments, through the peri-infarct ischemic zone and into the infarct. Subsequently, the sample volume is tracked through the lateral aspect of the myocardium. (B) The normal basal zone; (C) the peri-infarct ischemic zone; (D) the infarct; (E) the viable lateral wall; (F) the normally functioning lateral walls; (G) the normally functioning basal wall. The amplitude of the power signal is seen to blunt as the infarct area is approached with a flat response in the infarct zone and a return to normal in the lateral wall as tissue viability becomes normal. This paralleled the changes recorded in integrated backscatter levels.

DOPPLER POWER (ENERGY) EVALUATION OF MYOCARDIAL PERFUSION

Current opinion suggests that the successful detection of regional myocardial perfusion by cardiac ultrasound requires the combination of an effective left heart contrast agent and an appropriate ultrasound detection technique. Gray-scale videodensitometric analysis is increasingly being shown to be inappropriate for myocardial contrast detection due to the inherent gray-scale processing and compression algorithms within the current generation of ultrasound machines. Radiofrequency data acquisition, second harmonic imaging and Doppler power (energy) data acquisition are all more appropriate imaging modalities because of their individual unique properties and linear processing algorithms. Doppler power has been shown to be effective in detecting the presence of a galactose-based left heart contrast agent within the myocardium in both a closed-chest animal model and in normal subjects (Figure 8) [21]. Two-dimensional Doppler power imaging has also been shown to detect regional perfusion abnormalities caused by concomitant dipyridamole infusion and myocardial reperfusion in an animal model [22]. Why should the Doppler power mode offer advantages over gray-scale imaging? Firstly, the mode of insonation is different. Doppler power is based on information derived from color Doppler signal processing. To produce an image, the transducer is programmed to send out up to six impulses per line and then waits to receive the returning impulses for a longer period than for a corresponding gray-scale image. This has the effect of delivering more average acoustic power to the contrast agent and in a more intermittent manner than with gray-scale imaging. As acoustic power is important in creating a non-linear response in contrast agent reflectivity, this may account in part for the greater sensitivity of Doppler power mode imaging. Another factor which may be important is the intrinsically lower noise floor of the color Doppler system. It is thus possible that any signal enhancement brought about by the contrast agent may not be identified if it lies within the noise floor of gray-scale imaging, but that the same level of signal enhancement would be detected by the Doppler system as it lies above the noise floor. The likely role of any of the Doppler power modalities in a clinically effective ultrasound technique remains to be determined, but currently it shows equal promise when compared to either radiofrequency data or second harmonic imaging and has a major advantage in its relative ease of implementation within an ultrasound system.

DMI EVALUATION OF ARRHYTHMIAS

As DMI can identify regional myocardial velocities and regional variations in myocardial acceleration, theoretically both normal and abnormal patterns of myocardial velocities and accelerations could be identified on a 2-dimensional

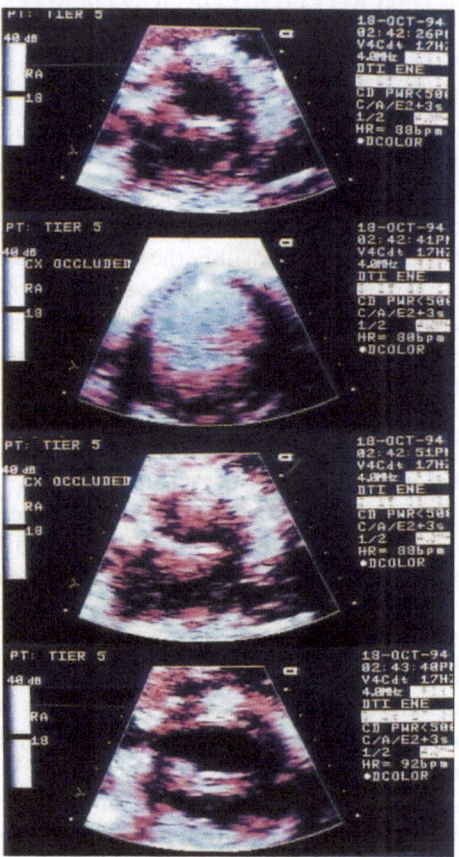

Figure 8. A Doppler energy study in an animal model using a left heart contrast agent (Levovist: Schering AG). In the upper panel, the resting LV short-axis Doppler energy image is visualized. The second panel shows the contrast agent entering the right and left heart. In the third panel during acute circumflex occlusion, an increase in echo intensity in the normal anterior and lateral zone is seen where no increase in echogenicity is seen in the circumflex (posterior left ventricular wall) territory. In the bottom panel, the clip has been released on the circumflex coronary artery and immediately a return in signal intensity is seen as reperfusion occurs within this wall. It is clear from this series of images that Doppler energy can identify the presence of a left heart contrast agent in a transthoracic animal model and can determine an absolute lack of perfusion and reperfusion.

image, as could the presence and precise location of foci of onset of abnormal ventricular contraction [22]. The higher temporal resolution of pulsed Doppler myocardial imaging could also be used to derive instantaneous regional myocardial contraction and relaxation velocities [23]. The normal right and left ventricular myocardial velocity and acceleration sequences during the cardiac

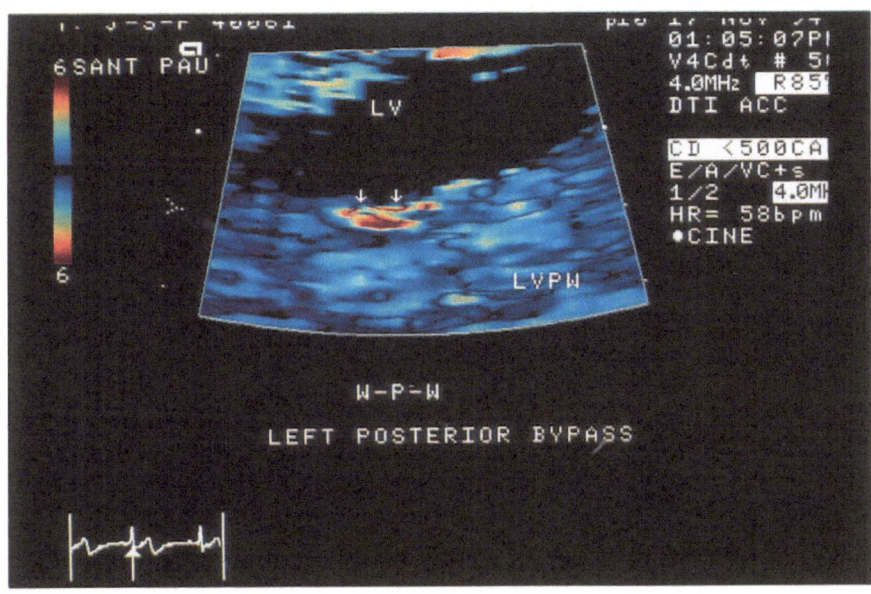

Figure 9. A late diastolic frame in a patient with a left-sided Wolff–Parkinson–White bypass tract. The localized early onset of activation (arrowed) of the myocardium is clearly seen in the left ventricular posterior wall situated in a subendocardial position.

cycle have been determined by DMI (2-D, pulsed and Doppler and M-mode) modalities. These data have been compared with myocardial velocity and acceleration information from patients with abnormal ventricular depolarization including patients with pacemakers and bundle branch block. Patients with ventricular arrhythmias and patients with functioning accessory bypass tracts have also been studied. The combination of DMI velocity and acceleration mapping using 2-D, M-mode and pulsed DMI techniques can identify specific normal and abnormal sequences of myocardial acceleration which accurately reflected the mode and timing of electrical depolarization. Abnormal early regional changes in velocity or acceleration associated with either WPW bypass tracts, unifocal VPBs or sustained ventricular tachycardias have been accurately located, as has the immediate normalization in depolarization induced during the monitoring of radiofrequency ablation (Figure 9) [23]. This information was only in part available from standard gray-scale imaging. Thus DMI would appear to be an important new adjunct to the non-invasive investigation of abnormal myocardial depolarization, including the real-time monitoring of radiofrequency ablation.

DOPPLER MYOCARDIAL IMAGING – 3-D RECONSTRUCTION

Three-dimensional echocardiography (3-D), has been validated as an accurate technique for determining left ventricular volume (LVV). However, the major limitation of transthoracic 3-D data acquisition is often the poor standard of gray-scale image quality available for reconstruction. In contrast, DMI is relatively independent of the amplitude of the ultrasonic signal returning from the interrogated myocardium and is less affected by the attenuating effect of the chest wall. In addition, the DMI algorithm is a powerful boundary detection technique, and hence can potentially provide a more complete data set for 3-D reconstruction than that obtained by gray-scale imaging. LVV measurements have been validated by comparing transthoracic 3-D DMI with standard gray-scale imaging techniques [24]. Such studies have shown that, despite the slightly inferior spatial resolution of the DMI technique, 3-D LVV estimation was equally well effected by both. However, the superior boundary detection and relative lack of chest wall signal attenuation inherent in DMI allowed for accurate volume measurements and successful 3-D volume reconstruction in 'difficult-to-image' patients in whom gray-scale imaging failed to acquire a data set [25].

DOPPLER IMAGING OF SKELETAL MUSCLE

Potential applications of DMI are not confined to cardiac imaging. Early validation work demonstrated that DMI could be used to identify contracting skeletal muscle groups. Since then, a systematic examination of skeletal muscle contraction in healthy volunteers has been undertaken [26]. This study demonstrated the capability of DMI to identify tetanic skeletal muscle contraction, to differentiate between active contraction and passive muscle movement, and to characterize isotonic, isometric and reflex skeletal muscle contraction profiles. Thus, there may be considerable potential for further evaluation of DMI in neurological and musculoskeletal applications. In particular, DMI may have practical application in the design and validation of biomechanical assist systems such as latissimus dorsi cardiomyoplasty [27, 28].

CONCLUSIONS

DMI is a valuable addition to the clinical ultrasound investigative modalities providing new information not available from current standard echo techniques. The possible areas of clinical use of aspects of the technique are shown in Figure 10. At present, it remains a technique in its early stages of development. However, with sufficient frame rates, resolution improvement, the development of off-line angle-independent velocity estimation and the implementation of multiple pulsed Doppler sample volumes its diagnostic range should be further extended.

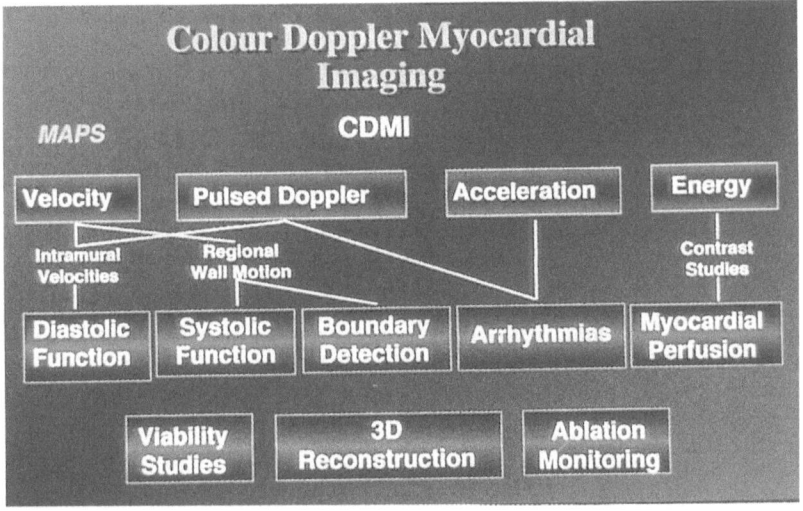

Figure 10. The potential clinical roles of Doppler myocardial imaging.

REFERENCES

1. Sutherland GR, Steward MJ, Grounstroem KWE et al. Colour Doppler myocardial imaging: a new technique for the assessment of myocardial function. J Am Soc Echocardiogr 1994;7: 441–58.
2. Miyatake K, Yamagishi M, Tanaka N et al. A new method for the evaluation of left ventricular wall motion by color-coded tissue Doppler imaging: in vitro and in vivo studies. J Am Coll Cardiol 1995;25:717–24.
3. Isaaz K, Thomson A, Ethevenot G, Cloez JL, Brembella B, Pernot C. Doppler echo-cardiographic measurement of low velocity motion of the left ventricular posterior wall. Am J Cardiol 1989;64:66–75.
4. McDicken WN, Sutherland GR, Moran CM, Gordon L. Colour Doppler velocity imaging of the myocardium. Ultrasound Med Biol 1992;18:651–4.
5. Fleming AD, McDicken WN, Sutherland GR et al. Assessment of color Doppler tissue imaging using test phantoms. Ultrasound Med Biol 1994;20:937–51.
6. Schwarz KQ, Bezante GP, Chen X. When can Doppler be used in place of integrated back-scatter as a measure of scattered ultrasound intensity? Ultrasound Med Biol 1995;21:231–42.
7. Fleming AD, Xia X, McDicken WN, Sutherland GR, Fenn L. Myocardial velocity gradient by Doppler imaging. Br J Radiol 1994;67:679–88.
8. Palka P, Lange A, Fleming AD, Sutherland GR, Fenn LN, McDicken WN. Doppler tissue imaging: myocardial wall motion velocities in normal subjects. J Am Soc Echocardiogr 1995;8(in press).
9. Palka P, Fleming AD, Lange A, Fenn LN, Sutherland GR, McDicken WN. Doppler myocardial imaging: mean myocardial velocity and velocity gradient in normal subjects. Br Heart J 1995;73(Suppl 3):85.
10. Rodriguez L, Garcia M, Ares M, Leung D, Thomas JD, Griffin B. Is mitral annulus motion during early diastole active or passive? Clinical evidence of elastic recoil. J Am Coll Cardiol 1995;February, Special Issue:P57A.

11. Rodriguez L, Garcia M, Nakatani S, Ares M, Griffin B, Thomas JD. Longitudinal axis diastolic dynamics in patients with left ventricular hypertrophy: a Doppler tissue imaging study. J Am Soc Echocardiogr 1995;8:391.
12. Rodriguez L, Garcia M, Nakatani S, Leung D, Grimm R, Griffin B, Thomas JD. Quantitation of left atrial systolic function by Doppler tissue imaging: clinical validation. J Am Soc Echocardiogr 1995;8:349.
13. Rodriguez L, Nakatani S, Leung D, Griffin B, Stewart W, Thomas JD, Grimm R. Measurement of atrial appendage cycle length by transthoracic echocardiography using Doppler tissue imaging. J Am Soc Echocardiogr 1995;8:391.
14. Lange A, Palka P, Sutherland GR et al. Doppler tissue imaging assessment of systolic and diastolic transmural velocity gradients in dilated cardiomyopathy: a new diagnostic index. Eur Heart J 1995;16(Suppl):298.
15. Palka P, Lange A, Sutherland GR et al. Doppler tissue imaging: assessment of systolic and diastolic transmural velocity gradients in hypertrophic cardiomyopathy. A new diagnostic index. Eur Heart J 1995;16(Suppl):105.
16. Masaaki Uematsu, Kunio Miyatake, Norio Tanaka et al. Myocardial velocity gradient as a new indicator of regional left ventricular contracton: detection by a two-dimensional tissue Doppler imaging technique. J Am Coll Cardiol 1995;26:217–23.
17. Stewart MJ, Sutherland GR, Moran CN, Fleming AD, Fenn LN, McDicken WN. Imaging of ischemic and infarcted myocardium by Doppler tissue imaging. Circulation 1993;88:I-47.
18. Garcia-Fernandez MA, Azevedo J, Puerta M, Moreno E, Torrecilla E, San Roman D. Quantitative analysis of segmental left ventricular wall dysfunction by pulsed Doppler tissue imaging. A new insight into diastolic performance. Eur Heart J 1995;16(Suppl):451.
19. von Bibra H, Tuchnitz A, Firschke C, Schühlen H, Schömig A. Doppler tissue imaging of left ventricular myocardium – initial results during pharmacologic stress. J Am Coll Cardiol 1995;February, Special Issue:P57A.
20. Fontavet HL, Puleo JA, Davis MG, Lockely M. Quantative dobutamine stress echocardiography utilising Doppler tissue imaging. J Am Coll Cardiol 1996;27:65A.
21. Sutherland GR, von Bibra H, Tuchnitz A, Henke J, Schönig A. Transthoracic detection of regional myocardial perfusion abnormalities using a pervenous contrast agent – a comparative study of Doppler energy and grey scale imaging. Br Heart J 1995;73(Suppl 3):57.
22. Yamagishi M, Tanaka N, Itoh S, Miyatake K, Yamayaki N, Hirama M. An enhanced method for detection of early contraction site of ventricles in Wolff-Parkinson-White syndrome using color coded tissue Doppler echocardiography. J Am Soc Echocardiogr 1993;6:30.
23. Sutherland GR, Pons Llado GP, Carreras F et al. Doppler myocardial imaging in the evaluation of normal and abnormal ventricular depolarisation. Br Heart J 1995;73(Suppl 3):85.
24. Sutherland GR, Caso P, Palka P, Fenn LN, Lange A, McDicken WN. Doppler myocardial imaging assessment of left ventricular volume and area: a comparison with grey-scale imaging. Eur Heart J 1995;16(Suppl):392.
25. Lange A, Bouki K, Fenn LN, Palka P, McDicken WN, Sutherland WN. A comparison study of grey scale versus Doppler tissue imaging left ventricular volume measurement using three dimensional reconstruction. Eur Heart J 1995;16(Suppl):266.
26. Grubb NR, Fleming A, Fox KAA, Sutherland GR. Evaluation of Doppler tissue imaging for the assessment of skeletal muscle function in healthy volunteers. Radiology 1995;194:837–42.
27. Grubb NR, Sutherland GR, Campanella C. Optimisation of myostimulation in dynamic cardiomyoplasty. Eur J Cardiothoracic Surg 1995;9:45–9.
28. Grubb NR, Fleming A, Sutherland GR, Campanella C, Fox KAA, Sinclair C. Assessment of latissimus dorsi muscle function after cardiomyoplasty using Doppler tissue imaging. Br Heart J 1994;71(Suppl):254,302,334.

10. Quantitative Doppler intensitometry

KARL Q. SCHWARZ, XUCAI CHEN & GIAN PAOLO BEZANTE

INTRODUCTION

All ultrasound imaging techniques rely on the display of information contained in backscattered ultrasound signals. The backscattered signals can be thought to contain two types of information: frequency or phase and amplitude. The way in which backscattered ultrasound signals are processed determines the type of image displayed. There are two basic forms of ultrasound imaging in wide use today: amplitude or intensity mapping (traditional B-mode and M-mode imaging) and Doppler velocity mapping (spectral Doppler and two-dimensional color flow mapping). The ability of intensity mapping techniques to distinguish structures is determined by the size of the structures relative to one another and their size relative to the ultrasound scanning frequency, ultrasonic characteristics of the scatterers, and other factors, such as transmit power, attenuation, gain and signal processing. This contrasts with Doppler velocity mapping, where the most important feature that allows adjacent structures to be distinguished is their velocity relative to one another and to the ultrasound transducer. This 'velocity filter' allows the differentiation of structures that may be too small to be distinguished from background artifact with intensity mapping.

While Doppler processing converts raw high-frequency ultrasound signals (2–7 MHz) to much lower frequency audio signals (0.1–10 kHz), the intensity component of the original backscattered signal is preserved. Quantitative Doppler intensitometry refers to the processing of Doppler signals for the intensity component as a substitute for more traditional methods of quantifying ultrasonic backscatter and is a technique in the early stages of development. However, it is already clear that Doppler signals have several distinct advantages over all other types of ultrasound for the assessment of backscatter intensity. The most important advantage (and limitation) is the requirement that the subject of interest is in motion with respect to the ultrasound transducer. This requirement allows Doppler ultrasound to 'see' objects that are either too small or too clouded in echo arti-

187

N. C. Nanda, R. Schlief and B. B. Goldberg (eds.), Advances in Echo Imaging Using Contrast Enhancement,
Second Edition, 187–206.
© 1997 *Kluwer Academic Publishers.*

fact to be imaged by any other ultrasound technique. It also allows Doppler to separate the subject of interest from a multiple of surrounding and overlapping echoes simply by applying a velocity criterion.

In the past decade there has been substantial interest in methods by which the intensity of backscattered ultrasound signals from tissue can be quantified, to allow differentiation between healthy and disease states. Examples include: the detection of myocardial ischemia [1, 2], the evaluation of viable myocardium after infarction [3–7], monitoring of cardiac ageing [8], assessment of myocardial hypertrophy [8–10], quantification of echocontrast agents in macroscopic vessels and cardiac chambers [11–15] and documentation of tissue microvasculature [16, 17]. In the case of echocontrast agents, determining backscattered intensity over time holds the promise of non-invasively assessing global blood flow and regional perfusion and quantifying shunts and embolic risk.

Integrated backscatter refers to measuring the energy of radio frequency (RF) signals received by the ultrasound transducer directly and before extensive signal processing. This technique is considered the most precise method for measuring the intensity of scattered ultrasound signals. The amount of signal processing prior to integration is in part dependent on the scanner type (phased array vs. mechanical sector scanner) and the characteristics of the individual scanner and ultrasound transducer. The integrated backscatter technique requires an ultrasonic scanner which has been modified to output the RF ultrasound signal (the 'raw RF') directly from the transducer and a set of sophisticated instruments for measuring the intensity of these high frequency signals. The combination of modified ultrasound scanner and associated instrumentation exists in only a few research laboratories.

Interest in quantifying scattered ultrasound intensity is currently increasing with the introduction of microbubble-based echocontrast agents into clinical practice. Our hypothesis is that the intensity of Doppler signals, assessed by unmodified clinical instruments, can be used as an audio frequency equivalent to ultrasonic-integrated backscatter to quantify scattered ultrasound intensity in systems with motion, such as the evaluation of echocontrast in vessels [18]. Further, we propose that the Doppler signal intensity is largely independent of the usual constraints of the Doppler frequency measurements, namely the dependence on the target velocity and direction of travel [19].

DOPPLER BASICS

Like all forms of ultrasonic imaging, Doppler machines consists of an ultrasonic pulser and receiver. High-frequency, brief ultrasonic pulses are transmitted, followed by much longer periods when the ultrasound transducer receives the backscattered signals. Ultrasound machines create images based on the intensity of the backscattered signal and its timing with respect to the original transmitted pulse. When in Doppler mode, ultrasound machines compare the backscattered signals with the transmitted pulse for differences in frequency. Only structures which are in motion relative to the ultrasound transducer will create back-scattered signals that are shifted in frequency compared with the transmitted pulse. The magnitude of the frequency shift (f_d) can be calculated using the Doppler equation and depends on this relative velocity difference (v), the ultra-sound imaging frequency (f_o), the speed of sound (c) and the angle between the direction of travel and the ultrasound beam (Θ).

$$f_d = \frac{2f_o v \cos(\Theta)}{c}$$

At clinical ultrasound frequencies (2–7 MHz) and typical blood flow velocities (10–500 cm/s), the Doppler shift is in the audio frequency range (20–10 000 Hz). On the other hand, the intensity of Doppler signals is determined by the intensity of the backscattered signal and is largely independent of velocity. If a strong echo is received, then a strong Doppler signal will be output from the Doppler processor as long as the Doppler frequency is greater than the low velocity filter. Backscattered signals from structures with little or no motion relative to the ultrasound transducer are eliminated by low velocity filters in the Doppler processing unit. Backscattered signals from structures with sufficient motion pass through the Doppler processing unit with their intensity signature essentially intact.

When in spectral Doppler mode, ultrasound machines broadcast the Doppler signals from a loudspeaker while simultaneously showing their frequency spectra on a video display. In a clinical setting, the audio output is provided primarily as an alignment tool for the operator. When a clear signal is heard containing the appropriate frequency (or pitch) components, the operator assumes the ultrasound beam is properly aligned with the mean flow velocity vector. However the audio Doppler signal is more than just an alignment tool, as it represents the true 'raw' or unprocessed Doppler signal, much as the RF backscatter signal. Like back-scattered ultrasound, the Doppler energy depends on the intensity of the original transmitted wave packet, the efficiency of the scatterers and the degree of signal attenuation. When the backscattered signals are processed for the Doppler com-ponent, the processing preserves the energy information contained in the original backscattered signals. Commercial Doppler instruments display Doppler intensity in variable ways, depending on the 'mode' of presentation and the manu-facturer's choice of algorithm to implement the display. The standard display mode for spectral Doppler (pulsed or continuous wave) outputs the intensity of the received signal as brightness for the spectral display and volume for the audio signal. There is no equivalent standard display for color flow mapping, but most machines have a velocity mode (or velocity-variance mode) where the Doppler intensity is displayed as the color brightness and most manufacturers also include an intensity-only mode.

CAN DOPPLER BE USED AS A REPLACEMENT FOR MORE TRADITIONAL MEASURES OF ULTRASONIC BACKSCATTER?

It is commonly assumed that Doppler processing removes much of the energy information contained in backscattered ultrasound signals or that the intensity of Doppler signals relates more to the velocity of the scatterers than to their scattering properties. To illustrate that neither of these two assumptions are correct, we used phantoms to study the relationship of the scattering characteristics of the target, target motion, and ultrasound scanning techniques on the resultant Doppler energy and frequency [19]. As a 'gold standard' measure of backscatter energy, we used RF integrated backscatter. All backscatter measurements were made with a commercial scanner (XP128c, Acuson, Mountain View, CA) which was modified to allow access to the RF backscatter signal in all imaging modes. To compare RF backscatter with Doppler-derived backscatter intensity, both the RF signal and pulsed wave Doppler audio signal were simultaneously digitized and analyzed for energy and frequency content while the scattering characteristics of the target, the target's motion or the scanning technique were altered (Figure 1).

To assess the effects of target motion and scanning technique on the RF back-scatter and Doppler signals, a rotating rubber disk of uniform scatterer concentration submersed in a water tank was used (4.5×10^3 particles/mm^3, ATS Laboratories, Bridgeport, CT). A solid Doppler target was chosen to avoid fluctuations in backscatter energy created by bubble contaminants found in most fluid-based Doppler phantoms. To assess how the ultrasonic characteristics of the target itself effect the RF backscatter and Doppler signals, a liquid flow Doppler phantom with echocontrast (Levovist, Schering, Berlin, Germany) was used.

How does target velocity effect the backscatter signals?

In the first protocol, the rubber disk phantom was used to dispel the notion that Doppler energy is dependent on target velocity. In this protocol, the rotational speed of the rubber disk phantom was increased in a stepwise fashion while holding all other aspects of imaging constant. The corresponding linear velocity at the Doppler target area was 13.3–68.1 cm/s (pulsed wave sample volume near the disk edge). Three different Doppler pulse repetition frequencies (Nyquist limits) were used (60, 75 and 150 cm/s) to determine the effect of the Doppler scale setting on the relationship between the target velocity and the Doppler audio frequency and energy and RF integrated backscatter energy. The target velocity range and 3 Nyquist limits were chosen such that the maximum target velocity (68.1 cm/s) was significantly higher that the lowest Nyquist limit setting (60 cm/s) and lower than the highest Nyquist limit (150 cm/s). The RF integrated backscatter and audio Doppler energy/frequency were calculated at each rotational speed and compared with the linear velocity of the Doppler target area and linear velocity as a percentage of the Nyquist limit by linear

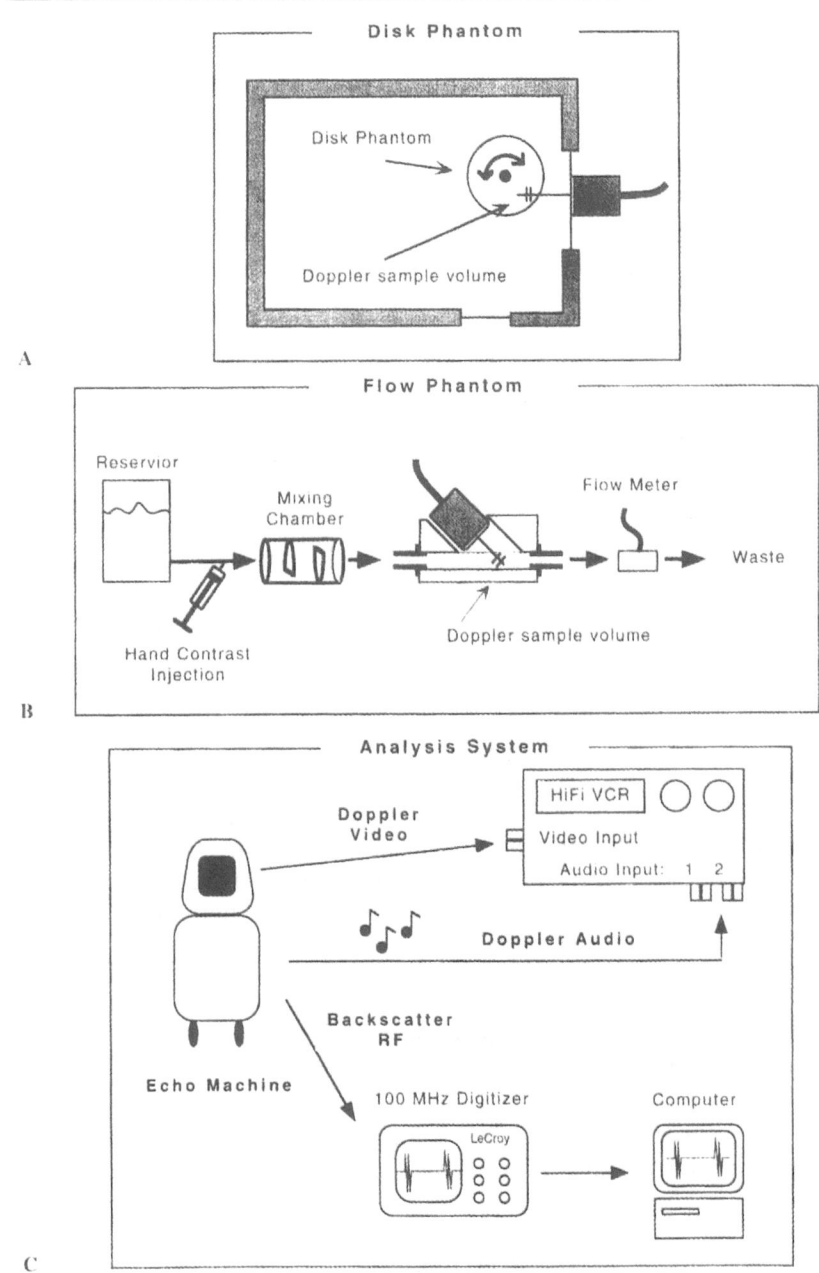

Figure 1. The experimental model. (A) A solid rubber disk Doppler phantom. (B) A flow Doppler phantom. The Doppler signal has three components that are acquired differently (C). The video image of the pulsed wave Doppler spectral display is recorded along with the Doppler audio signal on a HiFi VCR for later intensity analysis. The backscatter RF ultrasound signal is immediately digitized and stored to computer disk. Like the Doppler audio signal, time-intensity analysis is completed later using digital signal processing techniques described later in the text.

regression (Figure 2). As can be seen in the Figures, and as expected by the Doppler equation, the Doppler frequency increased linearly with the velocity of the target. At velocities above the Nyquist limit (V_{max}, Figure 2A) the Doppler frequency failed to track the actual linear velocity due to aliasing. The Doppler intensity and RF intensity were both constant throughout the velocity range, but there were differences between the two measures at the higher velocity levels. The audio Doppler intensity started to drop as the target velocity exceeded $0.75 V_{max}$ (Figure 2G, H), while the RF intensity was constant at all velocities. This is due to the gradual shifting of the audio Doppler signal from one channel to another as more and more of the spectral signal begins to alias. The shift in audio signals between the two stereo channels (corresponding to forward and reverse flow) started to occur at a modal frequency below V_{max} because all Doppler signals have some degree of spectral broadening. The higher frequency components therefore start shifting to the other channel before the modal frequency reaches the level predicted by the Nyquist calculation.

How does the direction of travel affect the backscatter signal?

The disk phantom was set at a constant velocity below the Nyquist V_{max} limit while the Doppler and RF signals were measured at various angles. Care was taken to avoid changing the amount of attenuation or distance between the Doppler targeting area and the ultrasound transducer. A total of 12 different angles were analyzed, from 10 to 68°. As expected from the Doppler equation, the Doppler frequency varied as the cosine of the angle between the direction of travel and the direction of ultrasound interrogation (Figure 3). On the other hand, the RF and Doppler intensity were independent of direction of travel. The moderate degree of scatter in the intensity curves is probably due to unavoidable changes in attenuation as the angle of attack between the ultrasound beam and the edge of the disk phantom alters.

How do acoustic attenuation, transmit power and gain affect the backscatter signal?

The effects of acoustic attenuation, transmit power and receive gain were studied with the disk phantom. In the case of attenuation, the RF backscatter and Doppler

---

Figure 2. (Opposite and following pages) How does target velocity affect backscatter signals? Displayed are three groups of graphs depicting the relationship between: (1) Doppler frequency and target velocity (A–C), (2) Doppler audio intensity and RF intensity as a function of target velocity (D–F), and (3) Doppler audio intensity as a function of target velocity, when velocity is expressed in terms of the Nyquist limit ($\%V_{max}$, G–I). In each group, the target velocity range is the same (13–68 cm/s), but the Doppler Nyquist is different (60, 75, and 150 cm/s).

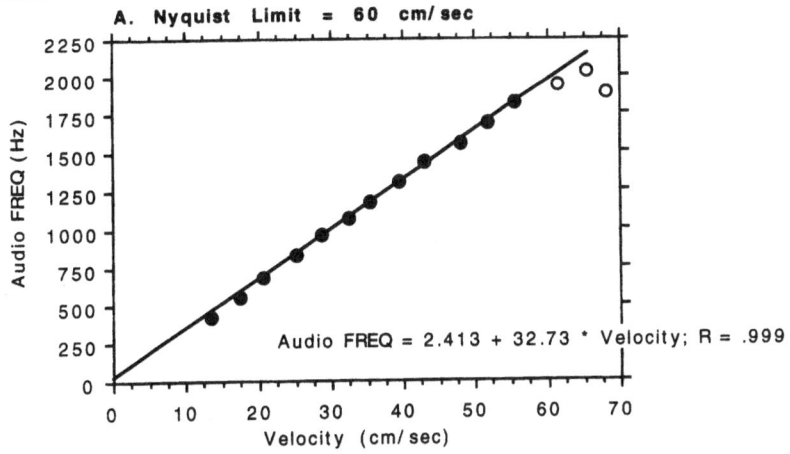

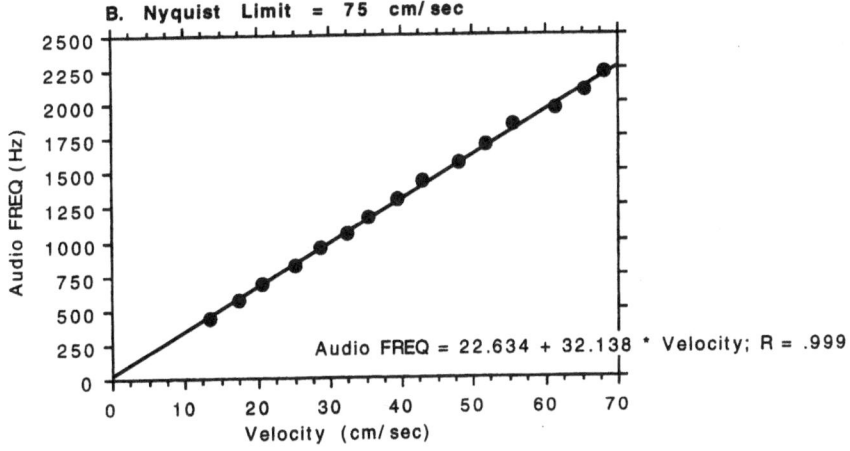

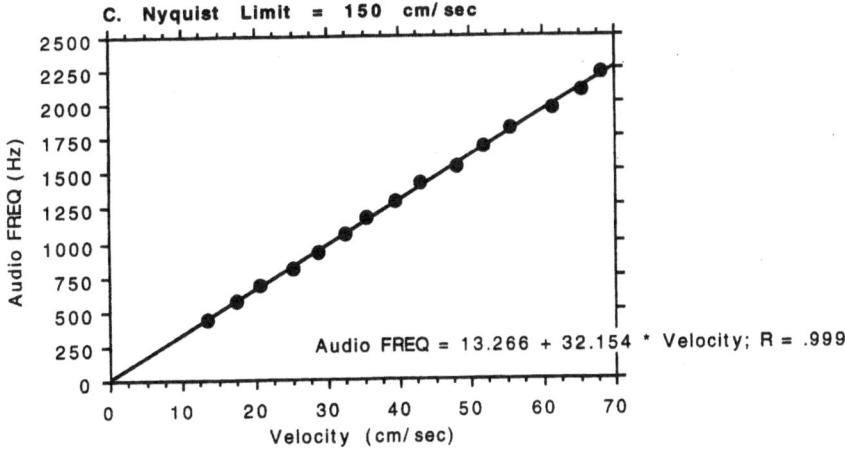

Figure 2A.

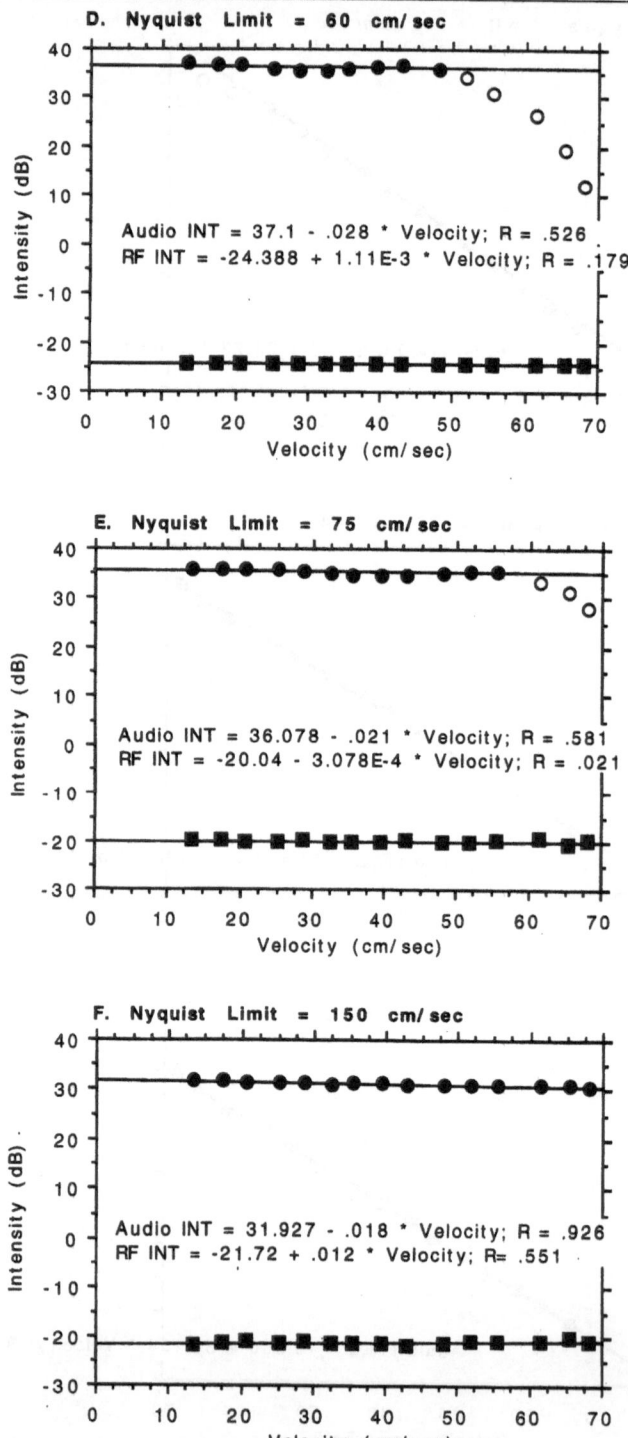

D. Nyquist Limit = 60 cm/sec

Audio INT = 37.1 - .028 * Velocity; R = .526
RF INT = -24.388 + 1.11E-3 * Velocity; R = .179

E. Nyquist Limit = 75 cm/sec

Audio INT = 36.078 - .021 * Velocity; R = .581
RF INT = -20.04 - 3.078E-4 * Velocity; R = .021

F. Nyquist Limit = 150 cm/sec

Audio INT = 31.927 - .018 * Velocity; R = .926
RF INT = -21.72 + .012 * Velocity; R= .551

Figure 2B.

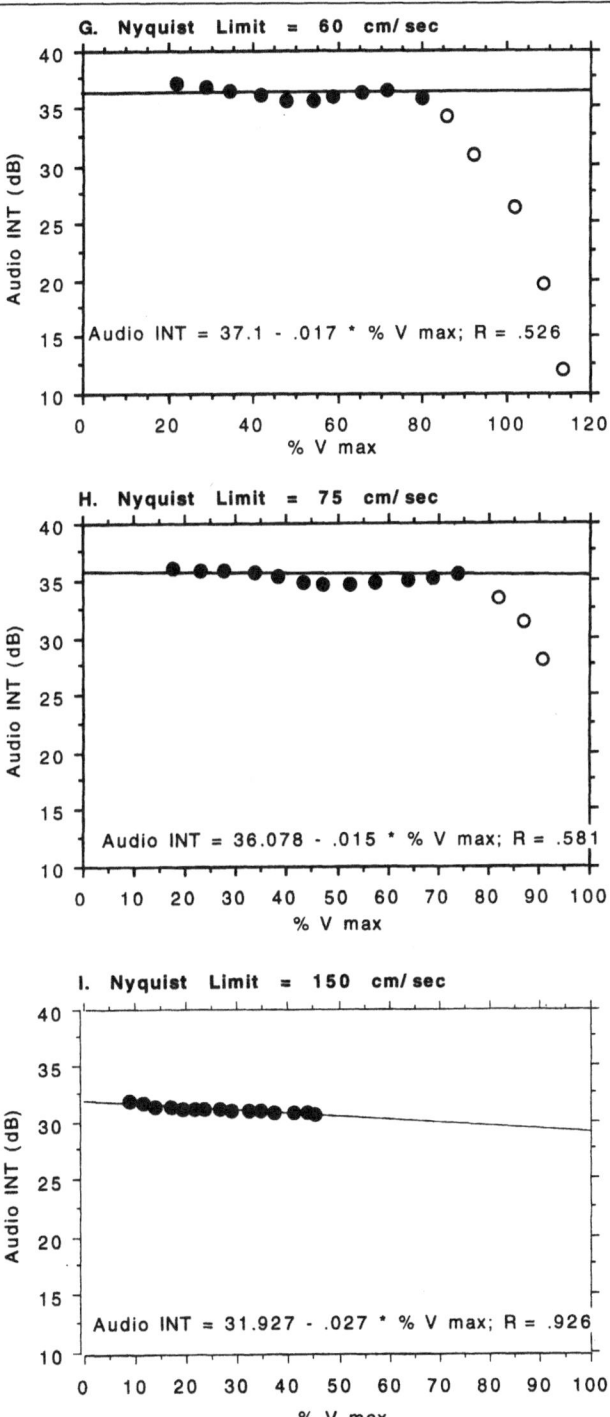

Figure 2C.

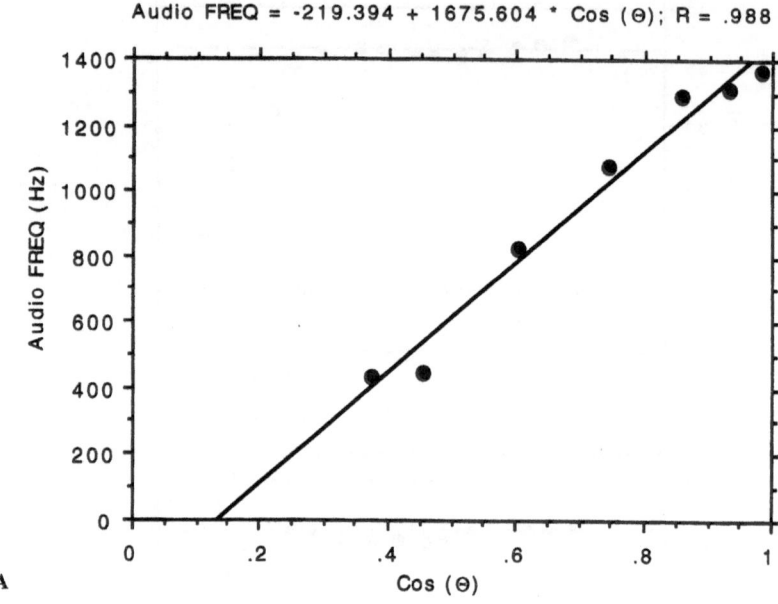

A

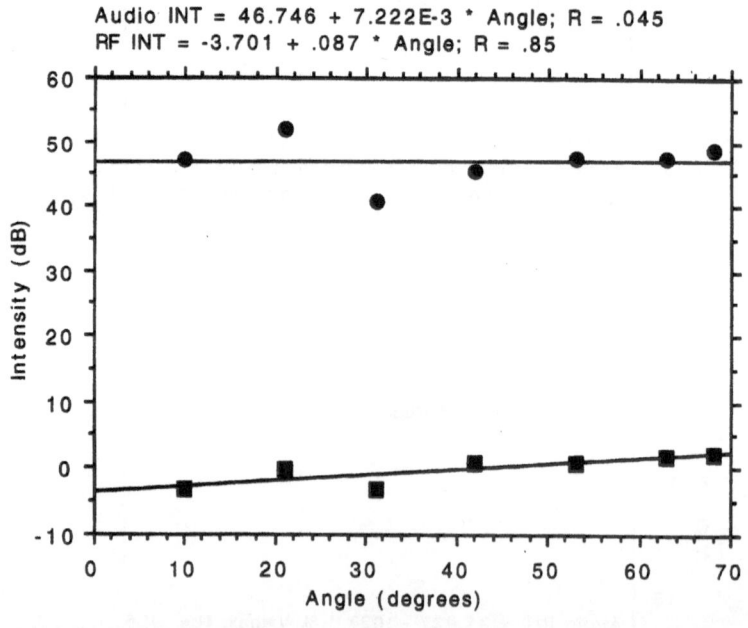

B

Figure 3. How does the direction of travel affect the backscatter signal? (A) Doppler audio frequency is compared with the angle between the direction of travel and the ultrasound beam (cos(Θ)). (B) Both the backscatter RF intensity and audio Doppler intensity are shown as a function of angle.

signals were analyzed as different Lexan® disks (acoustic attenuators) of varying thicknesses (1/8, 1/4, 3/8 and 1/2 inch) were added between the transducer and the Doppler target. The Doppler frequency was independent of backscatter intensity, but the Doppler intensity tracked the RF intensity for all levels of attenuation (Figure 4). The transmit power and receive gain similarly affected Doppler and RF intensity, but had no effect on Doppler frequency (data not shown).

How do the acoustic characteristics of the target affect the backscatter signal?

When considering Doppler as a replacement for traditional RF backscatter, the most important part of the proof is to determine how the acoustic characteristics of the target area affect the RF backscatter and Doppler energy. For this, we used a flow system and echocontrast to vary the acoustic characteristics of the target area. Unfortunately, it is not possible to determine precisely the number and size distribution of bubbles in any individual vial of echocontrast, even with the current commercially available agents. Further, because bubbles are effervescent, it is not possible to know with certainty how many bubbles are present once they are diluted into another medium (such as a flow system). However, using indicator dilution principles, it is possible to know with great certainty how the concentration of bubbles will change with time following equal dose bolus injections into a flow system of constant geometric volume and flow rate. We used such a system and measured the Doppler intensity/frequency and RF intensity over the time. As expected from indicator dilution principles, a rapid increase in Doppler and RF intensity was noted, followed by a gradual washout of the indicator from the system (Figure 5). During the period of wash-out there was an orderly (exponential) decrease in the number of bubbles in the targeting area. During this period the Doppler intensity (or frequency) was compared with the simultaneously acquired RF intensity (dashed lines, Figure 5). A one-to-one correlation was found between Doppler and RF intensity as the 95% confidence interval for the linear regression coefficient contained unity. The offset between Doppler and RF intensity was due to different levels of receiver gain between the two imaging modalities (51 dB in the example shown in Figure 5). Because all intensity measurements are relative, this offset has no physical importance. The conclusion of this portion of the proof is that changes in RF intensity (integrated backscatter) are parallelled by equal changes in Doppler intensity.

A clinical example

Consider the example of femoral artery flow assessed by pulsed wave Doppler imaging. A triphasic spectral Doppler waveform is typically seen in the common

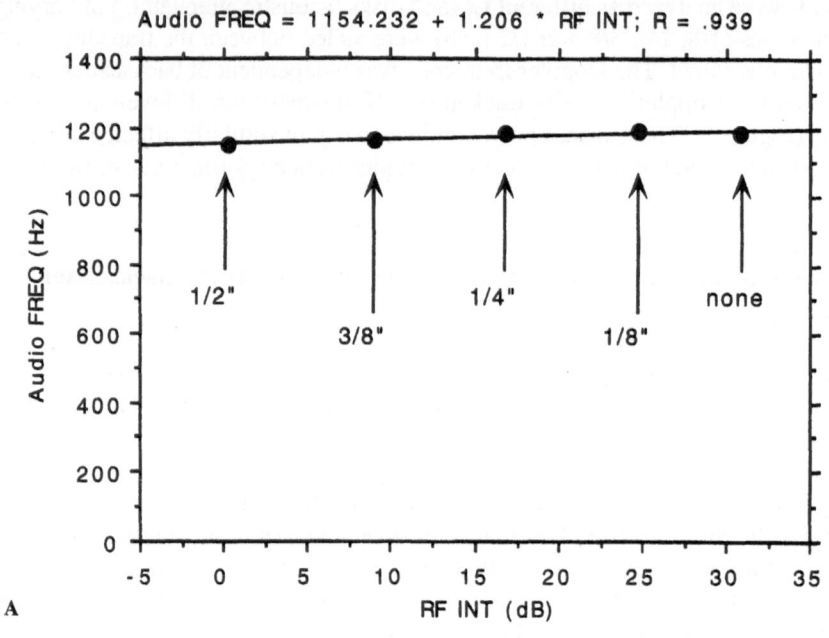

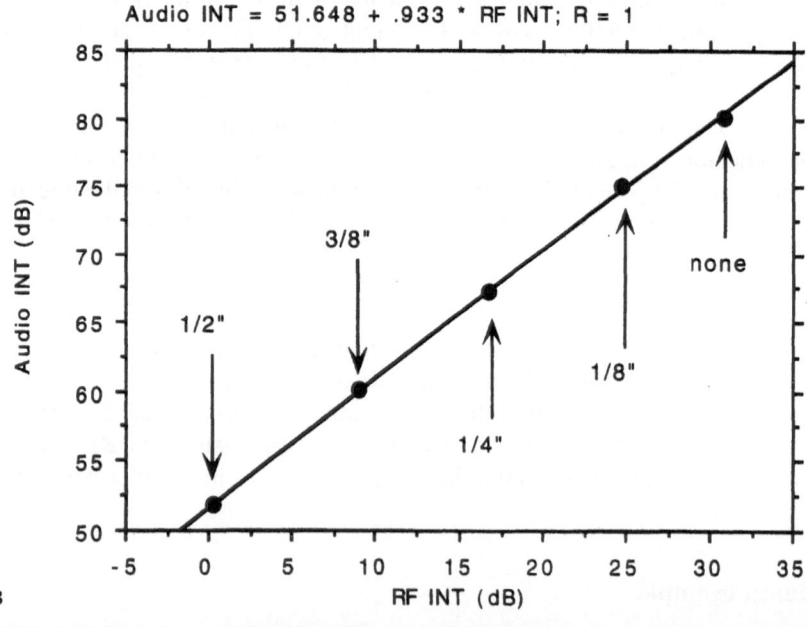

Figure 4. How does acoustic attenuation affect the backscatter signal? The audio Doppler frequency (A) and audio Doppler intensity (B) are both shown as a function of the backscatter RF intensity as acoustic attenuators of varying thicknesses are positioned between the ultrasound transducer and the Doppler target.

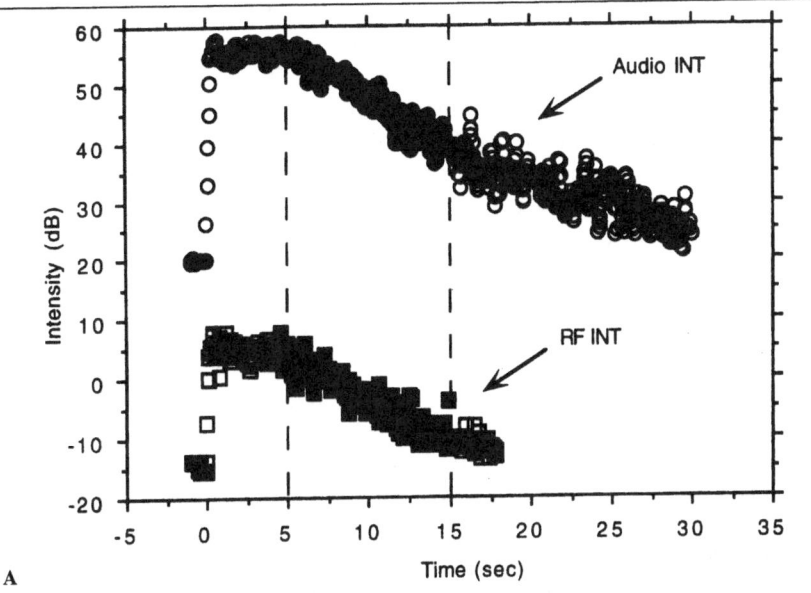

A

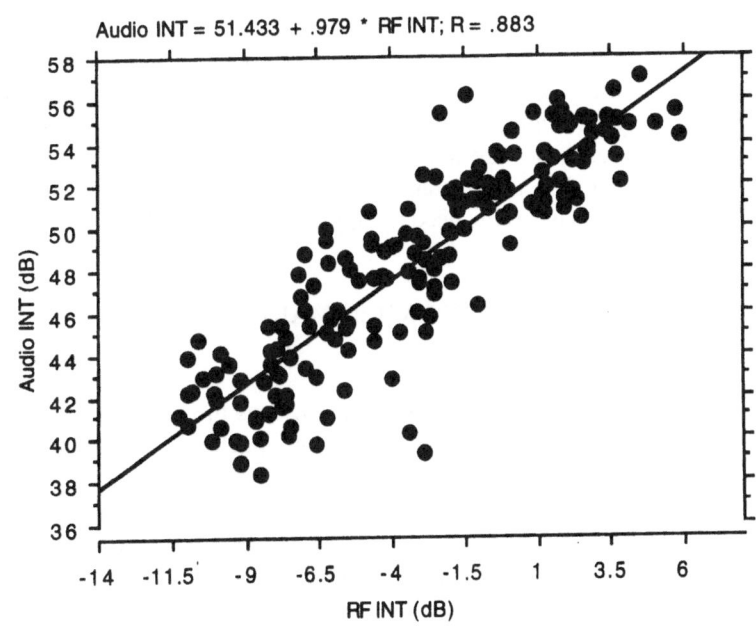

B

Figure 5. How do the acoustic characteristics of the target affect the backscatter signal? (A) RF backscatter and Doppler audio energy response to a bolus injection of echocontrast in a flow system is depicted as a time–intensity graph. During the washout phase, there is an orderly decrease in the concentration of microbubbles and a corresponding decrease in the RF and audio intensity (dashed vertical lines). (B) Relationship between the audio and RF intensity.

femoral artery (Figure 6). The first component is the high velocity wave associated with left ventricular systole (dashed line 1, Figure 6). This is followed by a second brief phase characterized by no flow or even flow reversal (dashed line 2, Figure 6). Finally, a third lower velocity end diastolic forward flow wave is observed (dashed line 3, Figure 6). The audio Doppler signal from this cardiac cycle was analyzed for energy and spectral content. The Doppler frequency mirrored the spectral Doppler waveform, exhibiting the rich spectral content of both forward flow waves and a brief period of no flow immediately following the systolic wave. On the other hand, the Doppler intensity waveform appeared more like a series of equal-amplitude square waves. As a first approximation, the blood flowing through the femoral artery can be thought of as a target with uniform scattering characteristics, but travelling at variable rates of speed. Based on the concepts presented above, one would therefore expect the Doppler intensity to be a constant value whenever the Doppler velocity is above the low velocity filter setting and significantly below the Nyquist limit. This is exactly what is seen in the femoral Doppler intensity tracing, as both the systolic and diastolic forward flow waves are of equal intensity (dashed lines 1 and 3, Figure 6), and the Doppler intensity falls to the noise level during the brief period of stasis immediately following the systolic forward wave (dashed line 2, Figure 6). The backscatter intensity can therefore be determined from upper envelope of the Doppler time–intensity graph. The Doppler noise level can be determined from the the lower envelope of the same graph if there are periods of no-flow, as shown in Figure 6. If the Doppler signal does not contain periods with no flow (external carotid artery, middle cerebral artery, etc), then Doppler noise level cannot be directly determined. The intermediate values for Doppler intensity between these two extremes are largely an artifact of digital signal processing (windowing Doppler data at the transition zones between flow and no-flow). This DSP artifact emphasizes the need for appropriate data grouping (see below).

TECHNICAL CONSIDERATIONS WHEN MAKING DOPPLER INTENSITY MEASUREMENTS

The audio output of the ultrasound machine provides the most direct access to the 'raw' Doppler signal. Most modern ultrasound machines have a stereo audio output with one channel dedicated for flow towards and the other for flow away from the transducer. When making quantitative pulsed Doppler intensity measurements the first task is to set the proper values for Doppler velocity scale and baseline. As was noted above, the pulse repetition frequency (PRF) needs to be adjusted so that the maximal measured velocities are less than 75% of the Nyquist limit. Signals with mean velocities above this limit may begin to have spectral components that start to alias before reaching the Nyquist limit. The position of the Doppler baseline must be centered in the spectral display. While shifting the Doppler baseline from the center of the display allows most modern

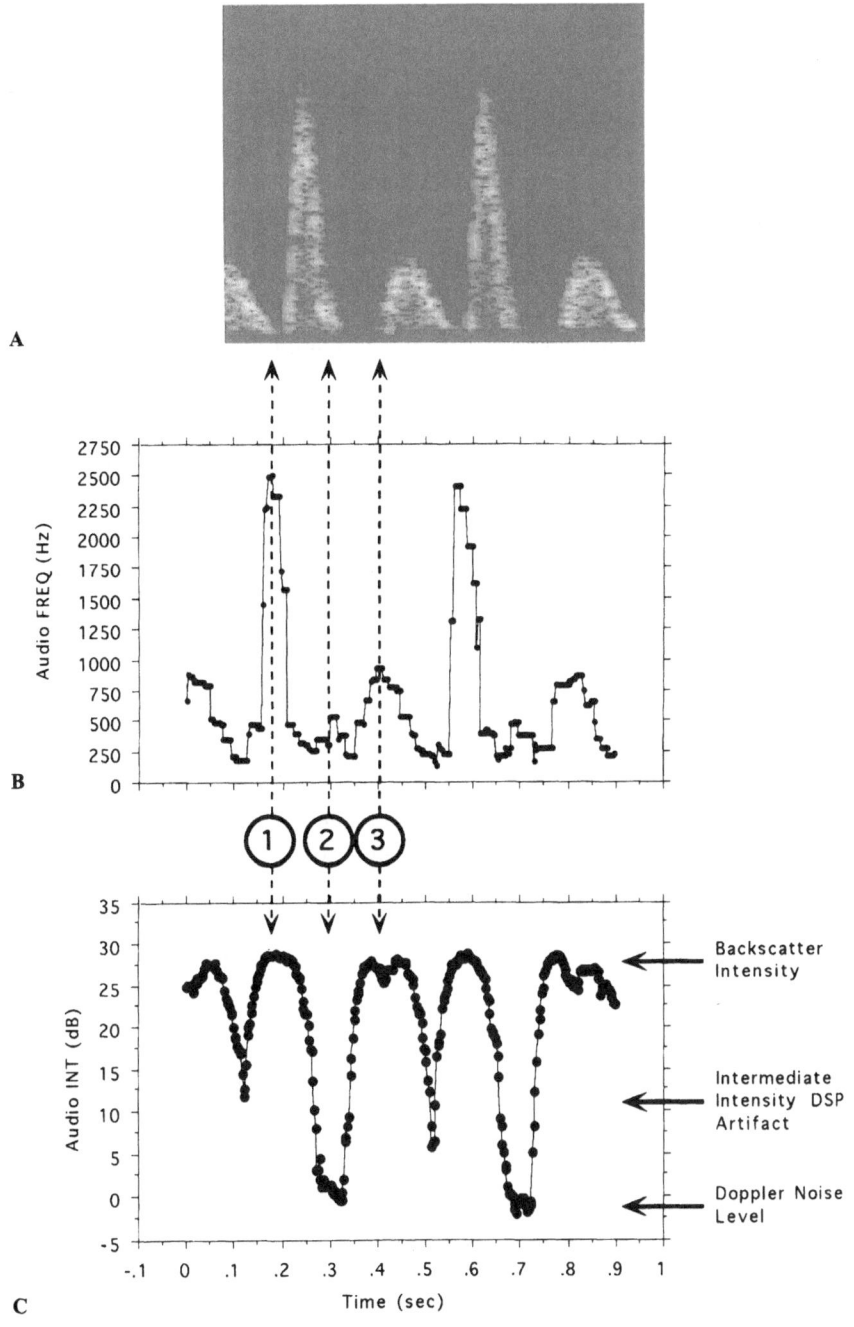

Figure 6. (A) A common femoral artery spectral Doppler trace. Corresponding Doppler time–frequency (B) and time–intensity (C) graphs. Dashed line 1 corresponds to the systolic forward flow wave, dashed line 2 identifies the period of flow stasis, and dashed line 3 is the low velocity late-diastolic forward flow wave.

ultrasound machines to measure velocities up to twice the Nyquist limit, the audio output is unaffected by these changes.

The low velocity or 'wall' filter gives Doppler ultrasound a fundamental advantage in signal-to-noise ratio over all other types of ultrasound imaging. When it is properly adjusted, unwanted low velocity echoes from structural components or from signal noise can be completely eliminated. If it is set too low, these unwanted and often very intense echoes contaminate the Doppler signal. If it is set too high, some of the flow signal will be eliminated. The best method for adjusting the low velocity filter is for the operator to listen to the audio output of the echo machine and use the lowest possible filter setting that permits a clear Doppler signal without artifact.

The transmit power and receiver gain must be adjusted so that the signal in question is within the dynamic range of the Doppler instrument and the recording system. In the case of echocontrast, the Doppler signal intensity may increase 20 dB or more following a bolus injection. The operator must adjust the Doppler instrument before the injection so that clear Doppler signals of the lowest intensity possible are heard before the injection to ensure that all of the contrast effect will be within the dynamic range of the instrument. Changes in receiver gain or transmit power during the contrast effect will have a direct effect on the Doppler signal intensity. Likewise, automatic control in Doppler gain or recorder gain present in some ultrasound systems and VCRs must be turned off before acquiring Doppler data for quantitative intensity measurements.

Once the audio Doppler signal is acquired and recorded on videotape, it can be analyzed for intensity or frequency content. This may be done using analogue circuitry or through digital signal processing. However, the constraints of pulsatile flow which make up all physiological signals is the same for both methods. Like pulsatile blood flow, physiological Doppler signals alternate between signal and no signal with cycle lengths of approximately 330–1000 ms. An instantaneous value for Doppler intensity would be inaccurate unless gated for the cardiac cycle. An average of the Doppler signal intensity over several seconds would also be inaccurate due to the effects of systolic/diastolic cycle length and the effects of heart rate. Taking the femoral artery Doppler signal shown in Figure 6 as an example, the entire cardiac cycle length is only approximately 400 ms (HR = 150) and the smallest component (the systolic forward wave) is only about 100 ms in duration. The constraint of pulsatile flow on Doppler signals requires any analysis of signal intensity be performed with a short time constant (less than 50 ms) and to detect the upper envelope of the Doppler intensity waveform. To accomplish this goal using digital signal processing, the audio signal should be digitized at least 24 000 Hz, which is sufficient to detect almost all physiological Doppler signals which are below 10 000 Hz. A 16 bit digitizer is preferred, with the dynamic range adjusted to the output of the recording device (typically ± 1.6 V). Once acquired, the Doppler intensity can be calculated by squaring the RMS voltage using data groups at the recommended < 50 ms data length. The data may then be converted to decibels by taking the logarithm of the Doppler intensity (base 10) and multiplying by ten. To improve precision, overlapping data groups with windowing can be applied. The result will be a Doppler intensity waveform plot similar to that shown in Figure 6.

BACKSCATTER (DOPPLER) INTENSITY AND ECHOCONTRAST

The determination of microbubble concentration in vivo is the main goal of Doppler intensity measurements when used in conjunction with echocontrast. RF-integrated backscatter and video intensitometry have been utilized for this purpose for years. We have already shown that Doppler intensity can be used as a substitute for RF-integrated backscatter measurements, and so one can assume that Doppler can be used to assess echocontrast concentration. Doppler ultrasound is ideal for this purpose because echocontrast is always contained within the vascular space. However, one fundamental piece of information has been lacking: the relationship between scattered ultrasound intensity and echocontrast concentration. It has been assumed that the relationship is linear, but that was not shown until recently [14].

The intensity of ultrasound reflected from microbubbles depends on the incident ultrasound intensity and the scattering characteristics of the bubbles. If the microbubbles are freely suspended in a fluid medium and the microbubble concentration is low enough to allow each bubble to participate in the scattering process (bubbles must be separated by roughly one ultrasound wavelength), then the scattering characteristics of the contrast agent will depend on the number of bubbles and their scattering cross-section [20, 21]. If the bubbles are encapsulated (such as Albunex, San Diego, CA), then the scattering characteristics will be modified by encapsulant. An 'effective' scattering cross-section is used to describe the microbubbles contained in commercial contrast agents, because it may be affected by the encapsulant (if any), the diluent medium, a distribution of geometric sizes and the insonification frequency.

To test the hypothesis that ultrasound could be used to measure the concentration of microbubbles, we infused Levovist at various controlled rates into a constant velocity, constant pressure flow system. The backscatter intensity was determined by Doppler intensity at each infusion rate. We found that backscatter intensity could not be used to measure the absolute echocontrast concentration, but that relative changes in contrast concentration could be measured. When the relative change in contrast concentration (Δ[SHU 508A]) was compared with the corresponding change in the audio Doppler intensity (Δ Audio INT), a close linear correlation was found ($r^2 = 0.9183$, Figure 7). After correcting for the effects of microbubble wastage (decay), a 1:1 relationship between changes in echocontrast concentration and backscatter intensity was documented.

The absolute concentration of microbubbles is never known in either in vitro flow systems or in vivo situations. This is because the absolute number and size distribution of bubbles in the original injectate is unknown and because of

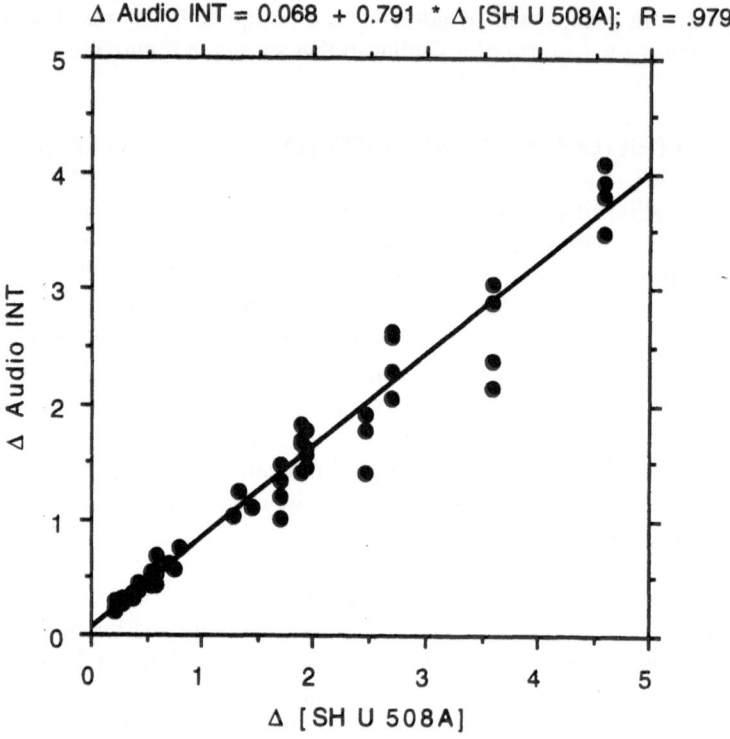

Figure 7. Backscatter intensity and echocontrast concentration. Relationship between relative change in audio Doppler intensity and the corresponding relative change in echocontrast concentration.

unpredictable changes in those parameters during and after injection. However, backscatter ultrasound intensity (Doppler intensity) can be used to measure quantitative changes in echocontrast concentration, even when the absolute concentration of contrast is unknown. In fact, in in vivo situations it is the relative difference in backscatter intensity that will permit perfusion imaging with echocontrast.

SUMMARY

Contrary to popular belief, Doppler ultrasound signals contain the same overall backscatter energy information present in the original RF signal. Thus, Doppler signals are capable of quantifying backscatter energy, as well as determining the target's velocity and direction of travel. The low velocity filter, which is part of all Doppler systems, provides a powerful mechanism for improving the signal-to-noise ratio of low energy backscatter signals and for isolating a region of interest based on velocity criteria. This makes Doppler ultrasound an ideal technique for backscatter energy quantification in systems that have low signal-to-noise ratio, but which are in motion with respect to the ultrasound transducer. Audio pulsed wave Doppler signals are an ideal source of quantitative ultrasound backscatter information because: the audio signal is not modified by changes in the screen display (log compression, post-processing algorithm, velocity baseline, etc), the frequency range of audio signals permits low cost data storage (HiFi VCRs) and data processing, and all pulsed wave Doppler systems are equipped with a 2-D spatial data gate (the Doppler sample volume) and a velocity gate (the low velocity filter). Color Doppler energy algorithms permit the 2-D display of backscatter energy and may be quite useful for real-time qualitative image interpretation. Examples include: tumor localization, vessel identification and organ perfusion. Unfortunately, color Doppler energy systems are more limited than audio spectral Doppler for backscatter quantification due to lower temporal and spatial resolution, lower dynamic range, complex image processing algorithms and more complex data analysis (real-time video images rather than audio signals). In the future, pulsed and color Doppler intensitometry is expected to find its greatest utility in image enhancement, tissue characterization and perfusion assessment, particularly when coupled with echo-contrast.

REFERENCES

1. Milunski MR, Mohr GA, Wear KA, Sobel BE, Miller JG, Wickline SA. Early identification with ultrasonic integrated backscatter of viable but stunned myocardium in dogs. J Am Coll Cardiol 1989;14:462-71.
2. Sagar KB, Pelc LR, Rhyne TL, Komorowski RA, Wann S, Warltier DC. Role of ultrasonic tissue characterization to distinguish reversible from irreversible myocardial injury. J Am Soc Echocardiogr 1990;3:471-7.
3. Sagar KB, Pelc LR, Rhyne TL, Howard J, Warltier DC. Estimation of myocardial infarct size with ultrasonic tissue characterization. Circulation 1991;83;1419-28.
4. Milunski MR, Mohr GA, Perez JE et al. Ultrasonic tissue characterization with integrated backscatter: acute myocardial ischemia, reperfusion, and stunned myocardium in patients. Circulation 1989;80;491-503.
5. Hajduczki I, Jaffe M, Areea J, Kar S, Nordlander R, Haendchen RV. Preservation of regional myocardial ultrasonic backscatter and systolic function during brief periods of ischemia by synchronized coronary venous retroperfusion. Am Heart J 1991;122:1300-7.
6. Vandenberg BF, Stuhlmuller JE, Rath L et al. Diagnosis of recent myocardial infarction with quantitative backscatter imaging: preliminary studies. J Am Soc Echocardiogr 1991;4:10-8.
7. Vered Z, Mohr GA, Barzilai B et al. Ultrasound integrated backscatter tissue characterization of remote myocardial infarction in human subjects. J Am Coll Cardiol 1989;13:84-91.
8. Masuyama T, Nellessen U, Schnittger I, Tye TL, Haskell WL, Popp RL. Ultrasonic tissue characterization with a real time integrated backscatter. J Am Coll Cardiol 1989;14:1702-8.
9. Lattanzi F, Di Bello V, Picano E et al. Normal ultrasonic myocardial reflectivity in athletes with increased left ventricular mass. A tissue characterization study. Circulation 1992;85:1828-34.
10. Wong AK, Verdonk ED, Hoffmeister BK, Miller JG, Wickline SA. Detection of unique trans-

mural architecture of human idiopathic cardiomyopathy by ultrasonic tissue characterization. Circulation 1992;86:1108–15.

11. Rovai D, Lombardi M, Ghelardini G et al. Discordance between responses of contrast echo intensity to increased flow rate in human coronary circulation and in vitro. Am Heart J 1992; 124:398–404.

12. Rovai D, Lombardi M, Mazzarisi A et al. Flow quantification by radio frequency analysis of contrast echocardiography. Int J Cardiac Imag 1993;9:7–19.

13. Rovai D, Ghelardini G, Trivella MG et al. Intracoronary air-filled albumin microspheres for myocardial blood flow measurement. J Am Coll Cardiol 1993;22:2014–21.

14. Schwarz KQ, Bezante GP, Chen X, Schlief R. Quantitative echo contrast concentration measurement by Doppler sonography. Ultrasound Med Biol 1993;19:289–97.

15. Schwarz KQ, Bezante GP, Chen X, Mottley JG, Schlief R. Volumetric arterial flow quantification using echo contrast. An in vitro comparison of three ultrasonic intensity methods: radio frequency, video and Doppler. Ultrasound Med Biol 1993;19:447–60.

16. Reisner SA, Ong LS, Shapiro JR et al. Efficacy and safety of myocardial perfusion imaging using intracoronary sonicated albumin in humans. Circulation 1988;78:565.

17. Wilson B, Shung KK, Hete B, Levene H, Barnhart JL. A feasibility study on quantitating myocardial perfusion with Albunex®, an ultrasonic contrast agent. Ultrasound Med Biol 1993; 19:181–91.

18. Schwarz KQ, Becher H, Schimpfky C, Vorwerk D, Bogdahn U, Schlief R. Doppler enhancement with SH U 508A in multiple vascular regions. Radiology 1994;193:195–201.

19. Schwarz KQ, Bezante GP, Chen X. When can Doppler be used in place of integrated backscatter as a measure of scattered ultrasound intensity? Ultrasound Med Biol 1995;21:231–42.

20. Ophir J, Parker K. Contrast agent in diagnostic ultrasound. Ultrasound Med Biol 1989;4: 319–33.

21. Tuthill T, Baggs R, Violante M, Parker K. Ultrasound properties of liver with and without particulate contrast agent. Ultrasound Med Biol 1991;17:231–7.

Contrast echocardiography: clinical aspects

11. Contrast echocardiography today

NAVIN C. NANDA & JOEL S. RAICHLEN

INTRODUCTION

Contrast echocardiography has moved an enormous distance since ad hoc suspensions of air microbubbles were first injected into the aortic root [1], but these simple techniques can still give useful information on the state of affairs within the right chambers of the heart. While the echoenhancers were confined to the venous circulation by their inability to cross the lungs, their applications were inevitably restricted. The development of more robust microbubble formulations able to cross the pulmonary barrier and opacify the left heart and arterial tree has transformed the field of contrast echocardiography. Nevertheless, simple air bubble suspensions, prepared at the point of use, can provide useful echo-enhancement in the right cardiac chambers.

THE RIGHT SIDE OF THE HEART

Suspensions of microbubbles prepared by hand agitation and injected into an arm help to identify the structures of the right heart, the superior vena cava, the right atrium, the inferior vena cava and the right ventricle. Opacification of the right atrium highlights the atrial septum and defines its surface. Common abnormalities such as dilated coronary sinus are easily detected using the transthoracic or transesophageal (TEE) approach. Atrial septal aneurysms, which are often associated with small left to right septal shunts, are readily visible using ultrasound contrast enhancers and TEE. The classic application of contrast-enhanced TEE in the right heart is probably the detection of intra-atrial shunts through a patent foramen ovale. The technique can detect very small defects, and color Doppler visualization of the flow through the shunt into the left atrium is often possible. Contrast enhancement can also visualize left-to-right shunts, which appear as areas of negative enhancement as the contrast-free blood from the left chamber swirls through the shunt into the right cardiac cavity. The

209

N. C. Nanda, R. Schlief and B. B. Goldberg (eds.), Advances in Echo Imaging Using Contrast Enhancement,
Second Edition, 209–216.
© 1997 *Kluwer Academic Publishers.*

Doppler flow velocity of the regurgitant tricuspid jet allows the calculation of pulmonary artery pressure, but in a significant number of patients the Doppler signal is too weak and poorly defined to provide dependable information on flow velocity. Contrast enhancement often improves the signal-to-noise ratio sufficiently to allow an accurate estimate of flow velocity and to confirm or refute the diagnosis of pulmonary hypertension. Contrast enhancement can also help in the echocardiographic investigation of the pulmonary hemodynamic response to exercise, which demands that the Doppler signals give an adequate description of peak flow velocity. In these more demanding applications pharmaceutical microbubble formulations such as Levovist given more predictable results than ad hoc microbubble suspensions [2, 3].

A rather unusual application of echocontrast enhancement in the cardiopulmonary circulation is in the diagnosis of the hepatopulmonary syndrome, a condition associated with cirrhosis and characterized by a failure of the mechanisms that regulate pulmonary blood flow. Normally the pulmonary vasculature reacts to alveolar hypoxia by constricting the smaller vessels. In the hepatopulmonary syndrome this response is reversed and alveolar hypoxia produces pulmonary vasodilatation. This vasodilatation produces an intrapulmonary shunt that is pathophysiological rather than anatomical, and the shunt can be demonstrated quite easily by giving an intravenous injection of a suspension of microbubbles that are too large to cross the intact pulmonary barrier. The appearance of echo-enhancement in the left ventricle under these circumstances is absolutely diagnostic of the hepatopulmonary syndrome [4].

CARDIAC OUTPUT

In principle, contrast echocardiography should allow the determination of cardiac output by applying classic indicator dilution theory to contrast appearance and decay within the cardiac chambers. In practice, the essential relationship between microbubble concentration and observed videointensity depends on a host of variables, none of which is constant from one subject to another. This makes the determination of absolute flow extremely difficult but suggests that contrast behavior could show small changes within the same subject quite readily [2].

MITRAL VALVE INCOMPETENCE

Attempts to use the color Doppler display to classify mitral regurgitation in transthoracic echocardiography have often been disappointing. A transpulmonary echocontrast enhancer such as Levovist augments the flow display during systole and diastole, increasing the area of the regurgitant jet by as much as 270% and showing a late systolic flow component not readily apparent in unenhanced views. In transthoracic scans, echo-enhanced color Doppler maps of regurgitant flow are as informative as the better accepted images from the transesophageal view. These enhanced transthoracic images give a reliable indication of the severity of the valvular defect even in technically difficult patients. Echo-enhancement could avoid the necessity for transesophageal imaging in the majority of patients with mitral valve incompetence [5].

AORTIC STENOSIS

Colour guided conventional Doppler assessment of the peak flow velocity through an aortic stenosis, together with the simplified Bemoulli equation [6] allows the calculation of the transaortic pressure gradient which gives a quantitative indication of the severity of the stenosis. To measure flow velocity accurately the Doppler beam must be positioned in the center of the stream but, in tight stenoses, the stream is narrow and difficult to detect. Echo-enhancement increases the signal to noise ratio, allowing precise direction of the ultrasound beam and accurate assessment of peak flow velocity. Using Albunex, the duration of the enhancement fell as left ventricular systolic pressure rose, suggesting that microbubble destruction had accelerated. The combination of echo-enhancement and color Doppler (preferably with an enhancer that can withstand high intraventricular pressure) could provide a useful non-invasive technique for the assessment of the severity of aortic stenosis [7].

CORONARY STENOSES

Blood flow velocity in the left anterior descending coronary artery (LAD) can be measured by Doppler TEE but in the majority of patients the Doppler signal is inadequate. Echo-enhancement with Levovist boosts the signal intensity, increasing the color flow-mapped area in the LAD and extending the length and width of the flow-mapped arterial segment. This more complete record of coronary flow shows the local acceleration due to relatively slight vessel narrowing and has been shown to improve the detection of stenoses in the proximal left circumflex coronary artery and the LAD. It also makes the Doppler-based assessment of coronary flow reserve a feasible proposition [8].

LEFT VENTRICULAR FUNCTION

If echo-enhancers such as Levovist or Albunex that cross the pulmonary barrier can opacify the left heart chambers completely at end-diastole they could allow the measurement of left ventricular end-diastolic volume and, by enhancing the delineation of the endocardial border, allow the assessment of left ventricular wall thickness. Attempts to increase the effect of echo-enhancers in conventional 2D imaging have not been entirely successful and, in many patients, complete opacification is difficult to achieve with this imaging technique. Colour Doppler enhancement has proved much more rewarding, particularly in patients with incomplete ventricular filling. Enhanced color Doppler shows flow throughout the ventricular chamber, but the very slow frame rates involved prevent detailed examination of wall motion. However, placing a relatively narrow color sector around the region of interest can accelerate the frame rate so that wall motion can be visualized clearly throughout systole. The prolonged enhancement produced by Levovist during color Doppler interrogation could allow the assessment of left ventricular function in several image planes after a single injection [9].

MYOCARDIAL VIABILITY

The identification of those areas of the myocardium that remain viable after an infarction is, potentially, one of the most valuable applications of echocontrast enhancement. Positron emission tomography and magnetic resonance spectroscopy detect the presence of metabolic activity, thallium scintigraphy identifies intact cell membranes and technetium imaging assesses mitochondrial function. Contrast echocardiography, on the other hand, shows the detailed topography of the microvascular perfusion that maintains the integrity of the myocytes. In regions with flow rates below 0.15 ml/min/kg, the myocytes start to die within 60 min. Because very few microbubbles enter the microvasculature, regions with flow rates lower than 0.15 ml/min/kg show no echo enhancement. Because of this fortunate coincidence, echo-enhancement within the infarct zone indicates the presence of viable tissue and collateral perfusion and might give a good estimate of infarct size. The shift in the distribution of transmural blood flow towards the epicardium during myocardial ischemia can be seen in animal studies using radiolabelled microspheres. Myocardial contrast echocardiography is the only imaging technique that has shown this transmural shift in man. Enhanced two-dimensional images before and after rapid atrial pacing show a decrease in endocardial enhancement and a corresponding increase in epicardial opacification.

The pattern of enhancement immediately after an infarction is complicated by the effects of reactive hyperaemia which is much more pronounced (and potentially more confusing) when the microvascular damage is limited. Echo-enhancement after administration of a coronary vasodilator defines the infarcted tissue as a zone in which flow does not increase. This simple picture may be complicated by the existence of a functional border zone, an area of non-ischemic but akinetic myocardium that surrounds the infarcted tissue. This border zone has so far been shown only in animal models. If this functional border zone occurs in humans it could limit the ability of echo-enhancement to delineate infarcted tissue. A more immediate practical limitation is the necessity to give the echo-enhancer by intracoronary or intra-aortic injection. This limitation may be over-come by the evolution of new more persistent enhancers, improved image pro-cessing procedures (such as harmonic imaging) or the development of transient response imaging, which limits the destructive effect of the ultrasound on the microbubbles and maximizes enhancement using short pulses of ultrasound. Myocardial contrast echocardiography is superior to thallium scintigraphy in its ability to show small shifts in regional perfusion but is much more dependent on the experience of the operator. The lowest cost of ultrasound imaging is an increasingly important factor and, in the near future, echo-enhanced Doppler ultrasound will probably become one of the standard methods for assessing myocardial perfusion [10–12].

THREE-DIMENSIONAL IMAGING

Short-axis views give several transverse images of the left ventricle but fail to image large portions of the myocardium, especially the apex. Three-dimen-sional views constructed from sequentially acquired 2D echo-enhanced images gated to the cardiac and respiratory cycles could provide anatomically realistic images of the entire myocardium. In experimental animals the left ventricle can be sequentially imaged from the apical window, rotating the transducer between scans by steps of 1–3° in a fan-shaped arc. The resulting images give a clear indication of the size and topography of ischemic and infarcted myocardial territories. The technique has some limitations, the most important of which are the attenuation of signals from posterior cardiac structures produced by intense opacification of the left heart cavity, and the relatively long period (1–3 min) needed for complete data acquisition. Perhaps the most tempting possibility offered by 3D imaging is densitometric processing of all the data from the 3D image space. This could allow the assessment of regional blood volume and flow throughout the left ventricle [13].

INTRAOPERATIVE ULTRASOUND

Echo-enhanced TEE provides an effective technique for the intraoperative evaluation of myocardial structure and function. One of its most promising appli-cations is to trace the progress of the cardioplegic infusion delivered by antero-grade or retrograde infusion. Initial studies, in which the enhancer was given with the cardioplegic solution, showed that incomplete myocardial delivery of the solution, often because of aortic valve incompetence or intra-atrial shunting, which is quite common in patients undergoing coronary surgery. Identification of regions with poor delivery of cardioplegia can direct surgeons to place their first grafts and then to instill cardioplegia solution down that graft prior to placement of the other grafts. Intraoperative infusion of the enhancer into the proximal end of a saphenous vein coronary graft can provide useful information on the area of reperfusion supplied by the graft. A videodensitometric assessment can be performed in real time and the results used to assist intraoperative decision making. Another important application of intraoperative myocardial contrast echocardiography is in the evaluation of left ventricular function after the cardiopulmonary bypass has been stopped. The technique can usefully differentiate between areas of stunned or hibernating myocardium and other areas that have been damaged irreversibly, a distinction that can have important implications for postoperative patient management [14, 15].

STRESS ECHOCARDIOGRAPHY

Exercise testing is not well suited to echocardiography because of the difficulty of obtaining good images from a moving patient. Pharmacological stress testing provides an alternative approach, using either dobutamine, a powerful β-agonist that increases resting heart rate, myocardial contractility and myocardial oxygen demand or a coronary vasodilator like adenosine or dipyridamole that increases blood flow through non-stenotic vessels, and in some cases may steal blood from the stenosed vessels, precipitating local ischemia. Conventional 2D imaging produces multiple tomographic images of the left ventricle during rest and at peak stress and these images allow a detailed assessment of regional wall motion and myocardial thickening. In 10–20% of patients the conventional images are inadequate for a variety of anatomical and functional reasons. In these individuals administration of an intravenous echo-enhancer, by opacifying the left ventricle, can produce a significant improvement in endocardial border definition and show wall motion abnormalities that were previously invisible. The routine use of an echo-enhancer during stress echocardiography has been shown to reduce the need for expensive radionuclide investigations and cardiac catheterization studies. The associated cost reduction could be appreciable [16].

MYOCARDIAL PERFUSION

The ability of an intravenous echo-enhancer to visualize myocardial perfusion is governed by the basic principles of coronary anatomy and physiology. The coronary circulation during rest demands between 5 and 10% of the cardiac output, giving a flow rate of 250–500 ml/min. The number of microbubbles that reach the myocardium is a very small fraction of the number injected into a vein: only 5–10 of every 100 bubbles that reach the aortic root will traverse the coronary circulation. The concentration of bubbles in the muscle and blood of the resting myocardium is roughly 7% of their concentration in the blood. The bubbles that enter the myocardium must be very persistent and extremely echogenic.

Increasing the dose is one way of tilting the balance in favour of the micro-bubbles. Analysing the raw radiofrequency signal from the transducer avoids the distortions of non-linear processing and improves the detection of very low concentrations of echo-enhancers, but demands enormous processing power and vast amounts of computer memory. Another approach is to use an echo-enhancer such as the dodecafluoropentane emulsion EchoGen, which stays in the myocardium for long periods of time. Several other potentially persistent myocardial opacifiers are under development. They include perfluorocarbon-exposed sonicated dextrose albumin (PESDA), liposomal Aerosomes (MRX Arizona) and BR1 (Bracco Research, Switzerland), a formulation of micro-bubbles that contain a gas, sulphur hexafluoride, that diffuses across the bubble shell very slowly and is virtually insoluble in blood [17, 18].

An alternative approach is to maximize the echogenicity of the microbubbles by exploiting their acoustic properties. Harmonic imaging, which involves transmitting at the bubbles' resonant frequency and constructing the image from the harmonic component of the returning ultrasound eliminates the noise of the ultrasound reflections from static tissue, in principle, gives an enormous increase in the signal to noise ratio. The combination of transient response intermittent ultrasound with harmonic imaging, using PESDA microbubbles, is another approach that offers enormous potential. Power Doppler imaging exploits the energy content of the Doppler signal rather than its frequency shift. A very recent and extremely promising development combines power Doppler with harmonic imaging [19].

Several unsolved problems still complicate the imaging of myocardial flow. They include the differing acoustic reflectivity of the ventricular walls (aniso-tropy), signal attenuation caused by the echo-enhancer itself, which can elimin-ate the echocardiographic image altogether (shadowing) and the necessity to reach a critical myocardial bubble concentration before visualization becomes possible at all (thresholding) [17].

CONCLUSION

Over the next 5 years, contrast echocardiography will begin to make significant inroads into some of the areas previously dominated by radionuclide imaging. Continued pressure for cost containment will almost certainly favour the relatively low cost procedures of ultrasound imaging.

REFERENCES

1. Gramiak R, Shah PM. Echocardiography of the aortic root. Invest Radiol 1968;3:356–66.
2. Khandheria BK. Identification of right sided structures by contrast transeophageal echo-cardiography with emphasis on patent foramen ovale. Chapter 17, this volume.

3. Kemp WE, Byrd BF. Right heart echo enhancement in the assessment of pulmonary artery pressures and right ventricular function. Chapter 13, this volume.
4. Abrams GA, Fallon MB, Gomez CR, Nanda NC. Detection of intrapulmonary vasodilatation and diagnosis of hepatopulmonary syndrome using contrast echocardiography. Chapter 16, this volume.
5. Von Bibra H, Tuchnitz A, Firschke C, Becher A, Schömig A. Contrast enhanced color Doppler in the assessment of mitral regurgitation. Chapter 14, this volume.
6. Stamm RB, Martin RP. Quantification of pressure gradients across stenotic valves by Doppler ultrasound. J Am Coll Cardiol 1983;2:707–18.
7. Nakatani S, Miyatake K. Contrast enhanced Doppler in the assessment of aortic stenosis. Chapter 15, this volume.
8. Caiati C, Iliceto S, Rizzon P. Transesophageal echocardiographic assessment of coronary flow reserve and coronary artery stenosis using echo enhancement. Chapter 19, this volume.
9. Raichlen JS, Nanda NC. Contrast color flow Doppler evaluation of the left ventricle. Chapter 21, this volume.
10. Firschke C, Kaul S. Assessment of myocardial viability post-infarction with myocardial contrast echocardiography. Chapter 25, this volume.
11. Beppu S. Can myocardial contrast echo replace thallium? Chapter 26, this volume.
12. Winkelmann JW, Block RJ, Feinstein SB. Usefulness of echo-enhancement in stress echocardiography (USA experience). Chapter 22, this volume.
13. Yao J, Masani ND, Pandian NG. Three-dimensional echocardiographic assessment of myocardial perfusion abnormalities. Chapter 30, this volume.
14. Shapiro JR. Intraoperative applications of contrast echocardiography. Chapter 27, this volume.
15. Masuyama T, Ito H, Lim Y-J, Kitabatake A. Myocardial contrast echocardiography in ischemic heart disease: Osaka experience. Chapter 24, this volume.
16. Winkelmann JW, Block RJ, Feinstein SB. Usefulness of echo-enhancement in stress echocardiography (USA experience). Chapter 22, this volume.
17. Porter TR. Transient response imaging during contrast echocardiography. Chapter 29, this volume.
18. Schlief R. Echo-enhancing agents: their physics and pharmacology. Chapter 5, this volume.
19. Villarraga HR, Foley DA, Chung SM, Nanda NC, Mulvagh SL. Harmonic imaging during contrast echocardiography: basic principles and potential clinical value. Chapter 28, this volume.

12. Clinical use of contrast agents: technical (practical) considerations

JANINE R. SHAPIRO & RICHARD S. MELTZER

INTRODUCTION

The contrast echocardiography effect was first described at the University of Rochester in 1968 by Gramiak and Shah [1], who observed a cloud of echoes in the catheterization laboratory during injection of indocyanin green dye for construction of a dye curve. Since then, contrast echocardiography has been used for the identification of structure, intracardiac and intrapulmonary shunts, valvular regurgitation, Doppler enhancement and, most recently, for myocardial perfusion imaging. Contrast echocardiography remains a rapidly developing field which continues to hold great promise for the real-time assessment of myocardial perfusion in humans [2–8]. With the continued development of new echocardiographic contrast agents capable of pulmonary transmission and improvement in measurement techniques, the goal of non-invasive left heart contrast echocardiography can be achieved.

ECHOCARDIOGRAPHIC CONTRAST AGENTS

The source of the ultrasound contrast effect was found to be related to the presence of gaseous microbubbles [9]. Most of the initial research in contrast echocardiography has focused on the development and evaluation of contrast agents. The ideal contrast agent for clinical use should be non-toxic, have negligible hemodynamic or myocardial effects, and feature highly uniform and echogenic microbubbles capable of transcapillary passage. A number of echogenic contrast agents have been used. Microbubbles have been produced in solutions of saline, glucose, X-ray contrast media and various other carriers [10]. In the earliest studies, hand agitation was used to introduce microbubbles into various solutions. However, most microbubbles produced by this technique were trapped in the capillary vasculature because of their relatively large size [11]. Feinstein et al.

N. C. Nanda, R. Schlief and B. B. Goldberg (eds.), Advances in Echo Imaging Using Contrast Enhancement, Second Edition, 217–225.

introduced the use of ultrasonic energy (sonication) for microbubble preparation [12]. Microbubbles produced by sonication were significantly smaller, more uniform and more stable than those produced by hand agitation. Sonicated albumin suspensions have been shown to contain extremely small microbubbles which provide good opacification of the myocardium in two-dimensional echocardiographic images, without producing significant changes in coronary blood flow, left ventricular function or hemodynamics [13]. First generation echocardiographic contrast agents are now on the market. Albunex and Levovist are safe after intravenous injection and can cause left heart cavity opacification, but do not provide adequate enhancement of the myocardium to enable unaided visual assessment of myocardial perfusion imaging. Several newer agents are under development which hold some potential to attain this goal. They include Echogen (Sonus Pharmaceuticals, Bothell, WA), FS069 (Molecular Biosystems, San Diego, CA), Aerosomes (ImaRx Pharmaceuticals, Tucson, AZ), BY 963 (Bracco Byk Gulden, Konstanz, Germany), Imagent (Alliance Pharmaceuticals, San Diego, CA), and agents from Acusphere (Cambridge, MA) and Schering (Berlin). Many of these agents have been shown to yield good myocardial opacification after intravenous injection in animal studies, and some are in early clinical trials. Many are based on fluorocarbon preparations which dissolve poorly in aqueous solution and thus remain stable in the circulation longer than most air-containing microbubble contrast agents. With the entry of multiple companies into second generation contrast agent development, the field has progressed markedly in the past 2 years. The goal of clinical echocardiographic myocardial perfusion imaging in real time, after intravenous injection, now seems attainable. This development might have a major impact on the diagnosis and management of coronary disease.

CLINICAL APPLICATIONS OF CONTRAST ECHOCARDIOGRAPHY

Contrast echocardiography for shunt detection

One of the original uses of contrast echocardiography was the detection of intracardiac and intrapulmonary shunts; this remains by far the most common application of contrast echocardiography. Typically, agitated saline or dextrose is used as the contrast agent, and an intravenous injection is made to search for right-to-left shunting. This is probably the most sensitive method of searching for patent foramen ovale and it is, therefore, usually undertaken during every study performed in patients looking for a cardiac source of emboli. Contrast echocardiography is particularly important in pediatric cardiology, when looking for shunting in congenital heart disease.

'Geographic' versus 'kinetic' studies of myocardial perfusion

Myocardial perfusion imaging by contrast echocardiography was shown to be feasible in animals in 1982 (Figure 1) [14, 15]. Investigators initially noted the 'geographic' distribution of contrast in the myocardium and related this to the coronary distribution area. Many subsequent studies of myocardial contrast focused on the 'kinetics' of contrast wash-in and wash-out from the myocardium or in the cardiac chambers. An example of a kinetic study is shown in Figure 2, looking at the time course of myocardial contrast videodensity after contrast injections before and after intracoronary papaverine [16]. The videodensity curve is similar to a more traditional indicator dilution curve, and seems to promise the possibility of obtaining important hemodynamic data in individual patients. Although there seems to be some continuing promise for the kinetic studies (looking at various parameters of videodensity vs. time curves), the pendulum in the last few years has been swinging away from these sorts of studies. Many aspects of the technology which limit videodensity–time curves from being true indicator dilution curves have been explored and emphasized by many investigators. Furthermore, the vitally important issues of relative and absolute myocardial blood flow seem particularly difficult to analyze by contrast echo kinetic studies. The kinetics of the various contrast parameters are affected in different directions by changes in myocardial blood flow holding myocardial blood volume constant, or by changes in myocardial blood volume holding myocardial flow constant: contrast wash-out is more rapid for a greater blood flow, but slower when examining the kinetics of wash-out from a larger myocardial blood volume. In practice, both myocardial blood flow and myocardial blood volume rise and fall together in most physiological situations, making kinetic analysis very difficult.

Delineation of area at risk, collateral circulation, myocardial viability and the 'no-reflow' phenomenon

Data from geographic studies of contrast distribution within the myocardium are more promising than those from many myocardial kinetic contrast studies. For more than a decade it has been realized that myocardial echocontrast can help to delineate areas at risk within coronary territories – i.e. the area likely to infarct following the occlusion of a coronary artery in experimental animal models. More recent studies have shown that contrast echocardiography can be used to delineate the extent of myocardium that has an effective collateral circulation [17, 18]. That is, the area of myocardium opacified by each of two separate injections, one into the right and one into the left coronary artery, is presumably functionally perfused by flow from both coronary arteries. Such myocardium is less likely to infarct and more likely to be viable after an infarction than myocardium not perfused by dual circulation [19, 20]. Another important

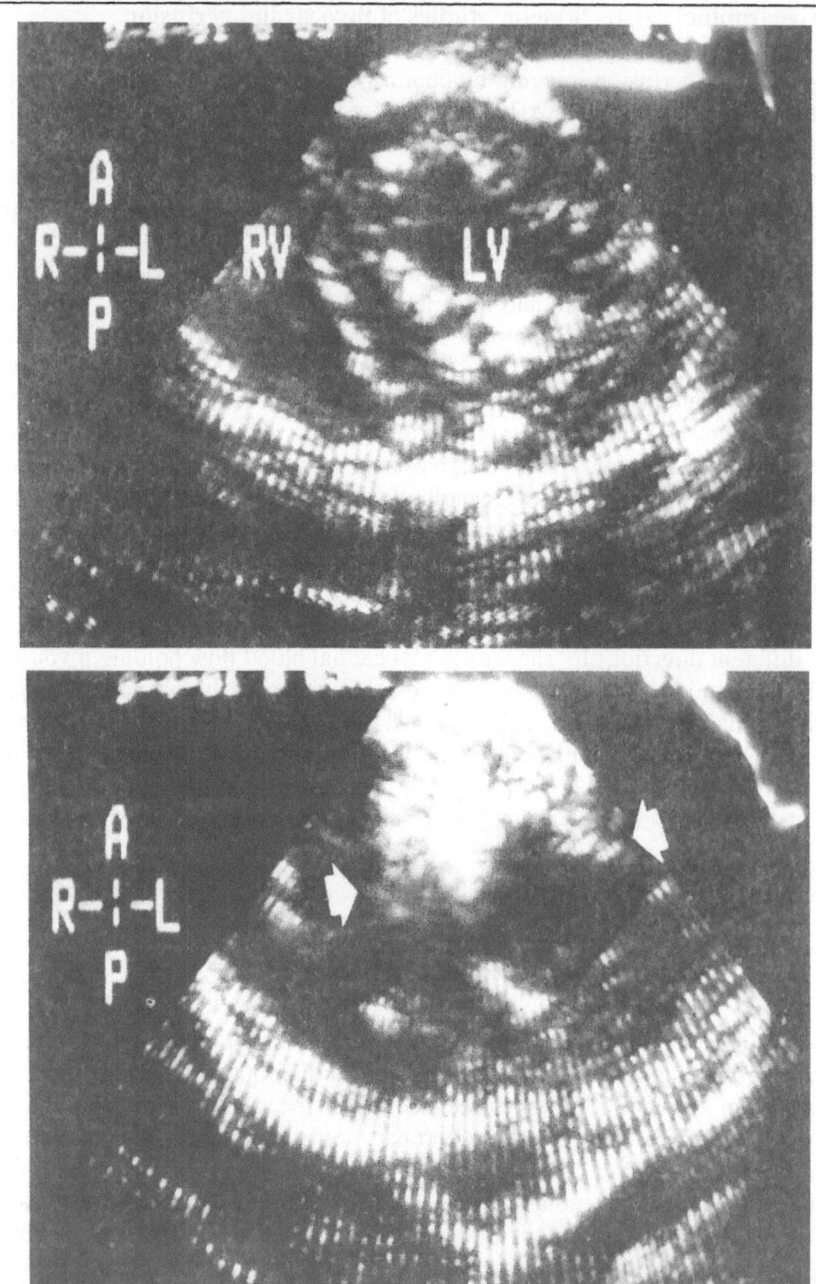

Figure 1. Stop frame photographs from one of the first published studies of myocardial perfusion imaging by contrast echocardiography. Epicardial short-axis images in an open-chest pig are shown at baseline (A) and after injection of contrast into the LAD coronary artery (B). Note the 'geographical' distribution of contrast in the LAD perfusion bed (arrows). (Reproduced with permission from: Meltzer RS et al. J Cardiovasc Ultrasonogr 1982;1:277–82).

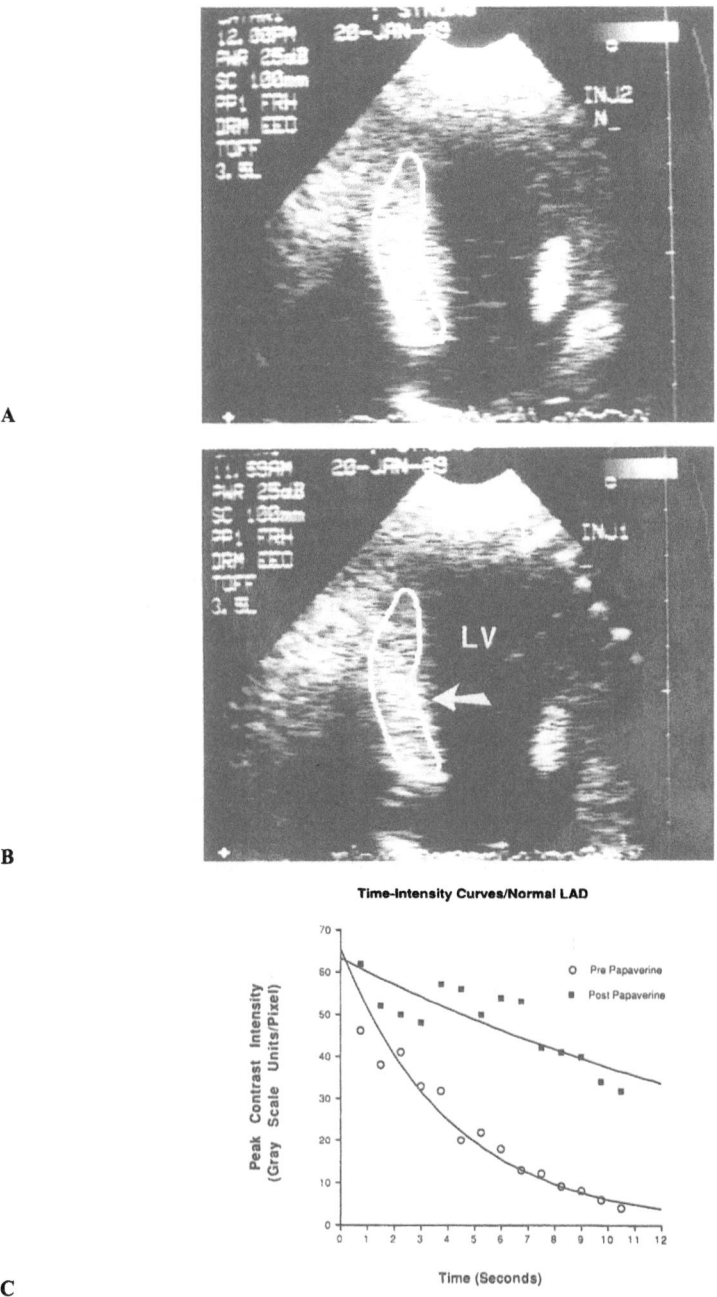

Figure 2. Two-dimensional echocardiographic images during contrast injection before (A) and after (B) intracoronary papaverine in a patient without significant left anterior descending coronary artery disease. Significantly higher contrast is seen in the septal area after papaverine resulting in higher peak contrast intensity and area under the curve in the corresponding time–intensity curves (C). The arrow points to the large area of interest chosen. LV=left ventricle; o=pre-papaverine. (Reproduced with permission from: Reisner et al. Eur Heart J 1992;13:389–94).

effect of microvascular physiology on gross left ventricular function relates to the 'no reflow' phenomenon seen after reperfusion following an acute myocardial infarction. Ito et al. showed that immediately after reperfusion which is successful by angiographic criteria, a small area of myocardium in the distal perfusion bed of the occluded coronary artery will not show a contrast echocardiographic blush [21]. This is probably due to microvascular swelling and is similar to the 'no-reflow' phenomenon which has been described in animal models. Ito et al. and others have shown that the prognosis for full return to function of myocardium showing the contrast echocardiographic 'no reflow' phenomenon is guarded and worse than that of dysfunctional myocardium, which perfuses as expected by contrast. Immediately after angioplasty and reperfusion, improved wall motion is often seen, so that stunned myocardium is not necessarily the initial finding after opening a closed vessel [22].

Intraoperative uses of myocardial perfusion imaging

Intraoperative myocardial perfusion imaging represents an exciting application of contrast echocardiography [23]. There are probably two main potential uses for intraoperative myocardial perfusion imaging. The first is to identify the distribution of cardioplegic solution delivered to the heart at the start of cardiopulmonary bypass. If there are areas of underperfusion, increased myocardial protection can be attempted by retrograde cardioplegia delivery and bypass of the underperfused area first, with immediate delivery of cardioplegia antegrade through the newly established bypass. The second area of intraoperative application for myocardial perfusion imaging by contrast echocardiography is in the assessment of bypass function. Lack of myocardial contrast in the expected coronary distribution area of a newly established bypass could alert the surgeon to potential problems, before ceasing cardiopulmonary bypass or during the period of weaning from bypass with the heart beating, with the patient still cannulated, when re-establishing bypass can be more easily accomplished. Lack of contrast blush could be due to multiple factors, such as distal runoff, the 'no-reflow' phenomenon, as well as to surgically correctable anastomotic problems.

Other clinical uses of contrast echocardiography

Contrast echocardiography is currently being used to enhance endocardial border definition during stress echocardiography and in technically difficult cases. It can be used to bring out Doppler signals as well as for 2-D and M-mode imaging. Nelson Schiller at the University of California San Franscico has reported that the yield of being able to measure pulmonary artery pressure by the tricuspid regurgitation method is significantly increased with the addition of intravenous contrast injections, and this also aids the measurement of the response of pulmonary artery pressures to exercise. Contrast has been used to bring out weak color Doppler signals, such as those in diastole and in the left ventricular apex. Contrast might be used to measure ejection fraction, though technical problems exist, such as the need for bolus delivery, and many of the problems that limit the use of videodensitometry for indicator dilution type studies apply here. In fact, the ultimate 'market' for contrast echocardiography is still unclear and it can be anticipated that new applications and contrast agents designed for the new applications will continue to evolve during the next decade.

FUTURE PROSPECTS

There will almost certainly be a significant increase in the interest and use of contrast echocardiography over the next few years, with the availability and marketing of first generation contrast agents and with the development and introduction of second generation agents that may allow myocardial perfusion imaging after intravenous injections. Quantitative analysis of contrast echocardiography and myocardial perfusion imaging is critically dependent on the intravascular stability of the microbubble agent used. Since agents are being developed capable of pulmonary transmission by different mechanisms, future research should continue to focus on methods needed to achieve denser opacification, consistent and quantifiable myocardial contrast and to assure safety, as well as on the development of sophisticated ultrasound systems for data acquisition.

There is also an exciting possibility that contrast agents may enhance therapeutic ultrasound applications involving acoustic cavitation, such as accelerating thrombolysis and perhaps ultrasound angioplasty [24–26]. A significant part of the current market for nuclear cardiology myocardial perfusion studies may be displaced eventually by contrast echocardiographic myocardial perfusion imaging, due to the lower cost, lack of ionizing radiation, wider availability of equipment and expertise to perform and interpret echocardiographic studies, and better spatial and temporal resolution of echocardiography compared with radioisotope imaging.

REFERENCES

1. Gramiak R, Shah PM. Echocardiography of the aortic root. Invest Radiol 1968;3:356–66.
2. Santoso T, Roelandt J, Mansyoer H, Abdurahamn N, Meltzer RS, Hugenholtz PG. Myocardial perfusion imaging in humans by contrast echocardiography using polygelin colloid solution. J Am Coll Cardiol 1985;6:612–20.
3. Moore CA, Smucker ML, Kaul S. Myocardial contrast echocardiography in humans. I. Safety – a comparison with routine coronary angiography. J Am Coll Cardiol 1986;8:1066–72.
4. Lang RM, Feinstein SB, Feldman T, Neumann A, Chua KG, Borow KM. Contrast echocardiography for evaluation of myocardial perfusion: Effects of coronary angioplasty. J Am Coll Cardiol 1986;8:232–5.

5. Cheirif J, Zoghbi WA, Raizner AE et al. Assessment of myocardial perfusion in humans by contrast echocardiography. I. Evaluation of regional coronary reserve by peak contrast intensity. J Am Coll Cardiol 1988;11:735–43.

6. Keller MW, Glasheen W, Smucker ML, Burwell LR, Watson DD, Kaul S. Myocardial contrast echocardiography in humans. II. Assessment of coronary blood flow reserve. J Am Coll Cardiol 1988;12:925–34.

7. Reisner SA, Ong LS, Lichtenberg GS et al. Quantitative assessment of the immediate results of coronary angioplasty by myocardial contrast echocardiography. J Am Coll Cardiol 1989; 13:852–6.

8. Lim YJ, Nanto S, Masuyama T et al. Visualization of subendocardial ischemia with myocardial contrast echocardiography in humans. Circulation 1989;79:233–44.

9. Meltzer RS, Tickner EG, Sahines TP, Popp RL. The source of ultrasound contrast effect. J Clin Ultrasound 1980;8:121–7.

10. Reisner SA, Shapiro JR, Amico AF, Meltzer RS. Contrast agents for myocardial perfusion studies: mechanisms, state of the art, and future prospects. In: Meerbaum S and Meltzer RS, editors. Myocardial Contrast Two-Dimensional Echocardiography. Dordrecht: Kluwer, 1989:45–59.

11. Meltzer RS, Tickner EG, Popp RL. Why do the lungs clear ultrasonic contrast? Ultrasound Med Biol 1980;6:263–9.

12. Feinstein SB, Ten Cate FJ, Zwehl W et al. Two-dimensional contrast echocardiography. I. In vitro development and quantitative analysis of echo contrast agents. J Am Coll Cardiol 1984;3:14–20.

13. Keller M, Glasheen W, Teja K, Gear A, Kaul S. Myocardial contrast echocardiography without hyperemic response or hemodynamic effects. J Am Coll Cardiol 1988;12:1039–47.

14. Meltzer RS, Vermeulen HWJ, Valk NK, Verdouw PD, Lancee CT, Roelandt J. New echocardiographic contrast agents: Transmission through the lungs and myocardial perfusion imaging. J Cardiovasc Ultrasonogr 1982;1:277–82.

15. Armstrong WF, Mueller TM, Kinney EL, Tickner EG, Dillon JC, Feigenbaum H. Assessment of myocardial perfusion abnormalities with contrast-enhanced two-dimensional echocardiography. Circulation 1982;66:166–72.

16. Reisner SA, Ong LS, Fitzpatrick PG et al. Evaluation of coronary flow reserve using myocardial contrast echocardiography in humans. Eur Heart J 1992;13:389–94.

17. Sabia PJ, Powers ER, Ragosta M, Sarembock IJ, Burwell LR, Kaul S. An association between collateral blood flow and myocardial viability in patients with recent myocardial infarction. N Engl J Med 1992;327:1825–31.

18. Sabia PJ, Powers ER, Jayaweera AR, Ragosta M, Kaul S. Functional significance of collateral blood flow in patients with recent acute myocardial infarction: A study using myocardial contrast echocardiography. Circulation 1992;85:2080–9.

19. Camarano G, Ragosta M, Gimple LW, Powers ER, Kaul S. Identification of viable myocardium with contrast echocardiography in patients with poor left ventricular systolic function caused by recent or remote myocardial infarction. Am J Cardiol 1995;75:215–9.

20. Galiuto L, Marchese A, Cavallari D, Iliceto S, Rizzon P. Evaluation of postinfarction viable myocardium at jeopardy by dobutamine echocardiography and myocardial contrast echocardiography. Echocardiography 1994;11:337–42.

21. Ito H, Tomooka T, Sakai N et al. Lack of myocardial perfusion immediately after successful thrombolysis: A predictor of poor recovery of left ventricular function in anterior myocardial infarction. Circulation 1992;85:1699–705.

22. Reisner SA, Shapiro JR, Ong LS, Lichtenberg GS, Amico AF, Meltzer RS. Immediate improvement of segmental left ventricular function following coronary angioplasty. J Cardiovasc Technol 1991;10:11–9.

23. Villaneuva FS, Spotnitz WD, Jayaweera AR, Dent J, Gimple LW, Kaul S. On-line intraoperative quantitation of myocardial perfusion during coronary bypass graft operations with

myocardial contrast two-dimensional echocardiography. J Thorac Cardiovasc Surg 1992; 104:1524–31.

24. Meltzer RS, Schwarz KQ, Mottley JG, Everbach EC. Therapeutic cardiac ultrasound. Am J Cardiol 1991;67:422–4.

25. Kornowski R, Meltzer RS, Chernine A, Vered Z, Battler A. Does external ultrasound accelerate thrombolysis? Results from a rabbit model. Circulation 1994;89:339–44.

26. Meltzer RS, Porder JB, Porder K. Ultrasound bioeffects, mechanisms, and safety. In: Siegel R, editor. Ultrasound Angioplasty. Boston: Kluwer Academic Publishers;1996:55–68.

13. Right heart echo-enhancement in the assessment of pulmonary artery pressures and right ventricular function

W. EVANS KEMP, Jr & BENJAMIN F. BYRD, III

INTRODUCTION

Echocardiographic contrast has several applications in the evaluation of right heart function as well as anatomy. In addition to enhancing endocardial borders and thus improving 2D echocardiographic assessment of right ventricular function, intravenous contrast can be used with various techniques to determine cardiac output as well as enhance spectral Doppler signals used to determine right ventricular and pulmonary artery pressures. Spectral Doppler measurement of tricuspid and pulmonary valve regurgitation velocities can provide accurate estimations of pulmonary artery systolic [1–3] and diastolic [3–7] pressures, respectively. Doppler determination of right-sided pressures is essential in the diagnosis of pulmonary hypertension [2], may be used to follow the response to therapy [4] and may be used to measure the pulmonary artery pressure response to exercise [7].

This chapter will review the use of contrast agents in the assessment of pulmonary artery pressure and cardiac output at rest and during exercise. Diagnostic quality Doppler tricuspid and pulmonary valve regurgitation signals are present in a significant number of patients, particularly those with pulmonary hypertension. Estimates of the incidence of tricuspid regurgitation vary from 61–96% in patients with pulmonary hypertension and 0–44% in patients with normal pulmonary artery pressures [2, 3, 7–9]. Pulmonary insufficiency is found in 5–78% of patients with normal hearts [10, 11] and is certainly present more often in those with pulmonary hypertension. By adding estimated central venous pressure to the transvalvular Doppler gradient determined by using the modified Bernoulli equation, $P=4V^2$, systolic and diastolic pulmonary artery pressures may be calculated from the tricuspid and pulmonary regurgitant velocities, respectively.

227

N. C. Nanda, R. Schlief and B. B. Goldberg (eds.), Advances in Echo Imaging Using Contrast Enhancement,
Second Edition, 227–237.
© 1997 *Kluwer Academic Publishers.*

TRICUSPID REGURGITATION

A significant number of patients undergoing echocardiographic examination do not have tricuspid regurgitation demonstrated by Doppler or have trivial, incomplete signals with a poor envelope and poorly described peak velocity, inadequate for accurate estimation of systolic pulmonary artery pressure. Berger et al. [2] suggested that patients with tricuspid regurgitation which is undetectable by Doppler are unlikely to have a significant elevation in pulmonary artery pressure. However, in their group of patients with pulmonary artery catheter-documented pulmonary hypertension, 20% had no detectable tricuspid regurgitation by pulsed or continuous-wave Doppler, primarily those with systolic pulmonary artery pressure <50 mmHg. These patients may have had tricuspid regurgitation which could have been detected by Doppler techniques after contrast enhancement. It is critical to obtain a diagnostic quality tricuspid regurgitation signal whenever clinicians raise the question of pulmonary hypertension, and contrast Doppler enhancement will play a major role in this area.

Before the development of Doppler spectral and color flow imaging, detection of a regurgitant pulmonary or tricuspid valve relied on the use of contrast-enhanced techniques. M-mode and two-dimensional contrast-enhanced echocardiography of the inferior vena cava, right atrium and ventricle have been used to detect tricuspid valve regurgitation [12–15]. Initially, indocyanine green dye was used as a contrast agent [14, 16]. In 1969 Bove et al. demonstrated that rapid injection of intravenous solutions created a microcavitation phenomenon that provided ultrasound contrast [17]. Others have described useful ultrasonic contrast with CO_2 in D_5W [18] and agitated saline [19, 20]. In 1978 DeMaria et al. [13] demonstrated the safety and utility of combining the microcavitation technique with 2D echocardiographic imaging to demonstrate tricuspid regurgitation. Using 10 cm^3 injections of intravenous D_5W, they demonstrated retrograde contrast flow from the right ventricle to right atrium or right atrium to inferior vena cava in 88% of patients with known tricuspid valve insufficiency. They also noted an increased clearance time of the contrast in patients with tricuspid regurgitation or congestive heart failure vs. those without these conditions. In a review of the use of contrast echocardiograpy, Meltzer et al. suggested that tricuspid regurgitation is severe when the clearance time of contrast in the inferior vena cava exceeds 20 cardiac cycles [21]. Thus, contrast echocardiography not only allowed the detection of tricuspid regurgitation but also indicated its severity.

Lieppe et al. [14], using rapid injection of intravenous saline, reported the first use of contrast-enhanced 2D echocardiographic imaging to demonstrate the presence of tricuspid regurgitation in patients with no physical or phonocardiographic evidence of tricuspid insufficiency. Subsequently, saline solutions were agitated prior to injection to improve the contrast effect. In 1982, Ogawa et al. [19] reported that i.v. administration of 20 cm^3 of 'shaken saline' demonstrated microbubbles consistent with tricuspid regurgitation by 2D echocardiography in eight of 15 patients with no clinical evidence of valvular insufficiency. In 1988 we reported our initial experience using i.v. injection of agitated saline to enhance trivial or incomplete Doppler tricuspid regurgitation signals (signals without well defined peaks or borders) in patients with normal and elevated pulmonary artery pressures [20]. Thirty-eight patients referred for echocardiography and seven normal volunteers were studied. Fourteen patients had indwelling pulmonary artery catheters, 50% of whom had a systolic pulmonary artery pressure >35 mmHg. The presence or absence of tricuspid regurgitation was initially determined by pulsed-wave Doppler examination of the tricuspid valve in all subjects. Using a three-way stopcock, two sequential boluses of 2 cm^3 of agitated saline were injected into a right antecubital vein or pulmonary artery catheter side port during continuous-wave Doppler spectral analysis of the tricuspid valve. Signals were considered complete when a well-defined smooth line outlined the spectral signal at peak systole. In 17 of 19 patients (89%) with previously incomplete tricuspid regurgitation signals, a complete Doppler spectrum was seen with saline contrast enhancement. After saline enhancement continuous-wave Doppler detected tricuspid regurgitation in six of 16 patients (38%) without previously identified tricuspid regurgitation, but the spectra were trivial and incomplete in four of these subjects. Both patients with diagnostic signals had indwelling pulmonary artery catheters which may have caused trivial tricuspid insufficiency. Doppler and catheter-determined pulmonary artery pressures correlated closely in 10 of 14 patients with indwelling pulmonary artery catheters who also had adequately enhanced Doppler signals (r = 0.99, SEE = 1.5 mmHg). Thus, agitated saline can be used to enhance incomplete Doppler tricuspid regurgitation signals in many patients, providing accurate estimates of pulmonary artery pressure. However, in patients without regurgitation detectable by careful pulsed-wave Doppler examination, agitated saline contrast enhancement is unlikely to provide diagnostic signals.

In 1991 we studied the use of albumin microspheres for contrast enhancement of tricuspid regurgitation spectral Doppler signals [22]. Using a 1 cm^3 injection of Albunex, a commercially prepared 5% solution of sonicated albumin microspheres 4–10 μm in diameter, we demonstrated optimal enhancement in 73% of 15 patients with previously incomplete tricuspid regurgitation Doppler spectra. In all eight patients in whom agitated saline was used there was adequate enhancement of these signals. Unlike saline contrast, which enhanced Doppler signals for only 2–3 s, sonicated albumin showed enhancement for up to 20–30 s.

To determine the optimal dose of Albunex for enhancing incomplete tricuspid regurgitation signals we have recently studied 21 patients with non-diagnostic spectral Doppler tricuspid regurgitation signals [23]. The mean age was 52 ± 13 (range 25–81) years and there were 11 female and 10 male patients. The reason for examination included evaluation of coronary artery disease/chest pain (9), congestive heart failure/cardiomyopathy (4), known pulmonary hypertension (3) and other conditions (4). Each patient was given one set of three varying doses of both Albunex and agitated saline in a random order. Each agent was prepared by standard techniques and delivered via a three way stopcock into the right

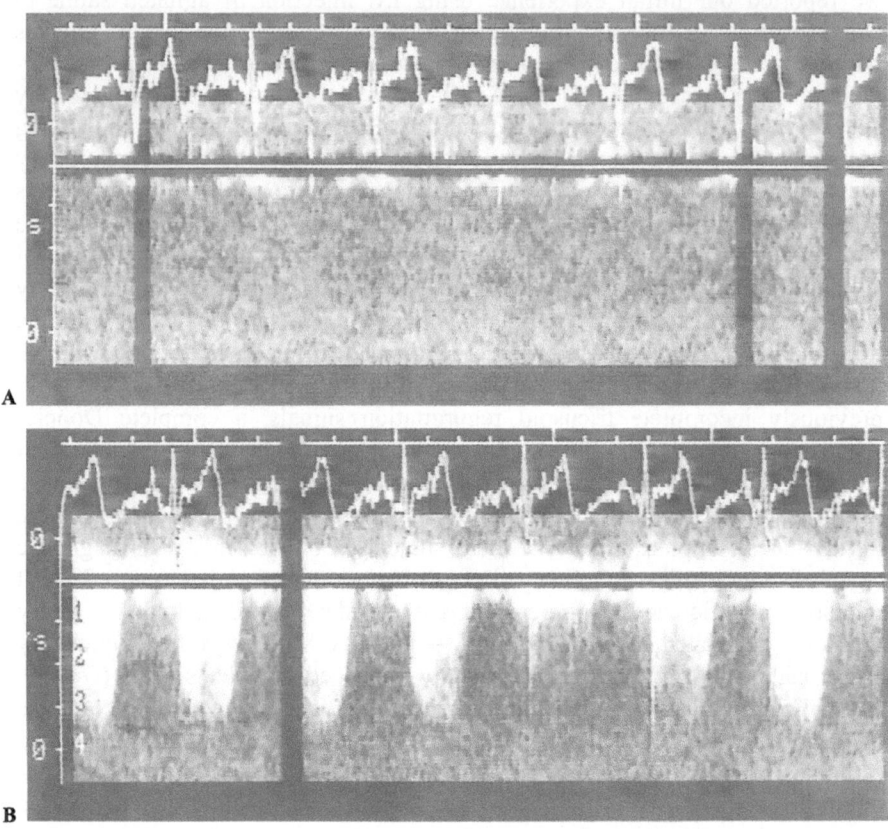

Figure 1. Spectral continuous wave Doppler tricuspid regurgitation signals at baseline (A) and after 1 cm³ of intravenous Albunex (B). There is dramatic Doppler signal enhancement, with a complete spectral envelope and an easily determined peak velocity of 3.7 m/s, green arrow. Scale markers are in m/s.

antecubital vein (whenever possible) through a 20 gauge or larger catheter. Undiluted Albunex (1 cm³) was injected initially, with subsequent doses adjusted up to 4 cm³ or down to 0.25 cm³, depending on the degree of spectral distortion (white-out due to increased noise) or inadequate spectral enhancement. All patients received 1, 2 and 4 cm³ of agitated saline. Color Doppler was used to guide continuous wave spectral Doppler alignment for optimal signal recording. Continuous wave spectral Doppler settings were optimized prior to each set of injections and kept constant. A diagnostically enhanced tricuspid regurgitation Doppler signal was defined as having well-defined borders and an easily determined peak velocity (Figure 1). Enhanced tricuspid valve insufficiency was distinguished from left ventricular outflow or mitral regurgitation by location of the Doppler cursor on duplex 2D imaging.

Saline injection yielded diagnostic tricuspid regurgitation spectra in 10 of 21 patients (48%). Diagnostic tricuspid regurgitation spectra could not be obtained with either agent in three patients. In 18 of 21 patients (86%) injection of Albunex led to a diagnostic spectral Doppler signal. An Albunex dose of $1-4\,cm^3$ demonstrated a diagnostic signal in 13 of the 18 patients (72%). Albunex requires some operator judgment in dose selection but appears to be a better agent for contrast enhancement than agitated saline. Similar findings were reported by Beppu et al. [24], who studied 20 patients with incomplete spectral Doppler tricuspid regurgitation signals using $2-4\,cm^3$ of both agitated saline and sonicated 5% human albumin prepared in their laboratory. All patients underwent right heart catheterization within 5 days of study. They noted that sonicated albumin more often provided adequate spectral enhancement and that these enhanced signals more accurately estimated transtricuspid pressure gradients than agitated saline, as determined at catheterization. They suggested that sonicated albumin has more numerous microbubbles of uniform size, providing a more reliable Doppler ultrasound contrast effect than the larger and fewer bubbles generated by agitation of saline. We had also noted in our 1991 study [22] that the tricuspid regurgitation spectral Doppler envelope was more sharply defined with sonicated albumin than with agitated saline.

In addition to improving the reliability of Doppler estimates of systolic pulmonary artery pressure in patients at rest, contrast Doppler echocardiography can assist the study of the pulmonary hemodynamic response to exercise. Himmelman et al. [7] investigated the use of agitated saline contrast Doppler echocardiography for estimating pulmonary artery pressure at rest and during exercise in patients with tricuspid regurgitation and chronic lung disease. Patients with chronic lung disease often have hemodynamics obtainable only by right heart catheterization due to their technically difficult echocardiographic examinations. Thirty-six patients with chronic lung disease and 12 normal subjects were studied at rest and during exercise with a variable-load supine bicycle ergonometer. No patients had pulmonary valve disease. All patients were given $2-3\,cm^3$ of agitated saline at rest, at each 2 min incremental stage of exercise, and during recovery. Continuous full screen monitoring of tricuspid insufficiency was performed using duplex continuous wave Doppler. Diagnostic quality Doppler signals, as defined previously, were seen in 56% of patients at rest and 39% during exercise, improving to 93% and 89% with contrast, respectively. Using this contrast Doppler technique right ventricular systolic pressure increased during exercise to a greater extent in patients with chronic lung disease than in normal controls ($p < 0.001$). Correlation of Doppler-estimated pulmonary artery systolic pressures with catheter-measured pressures in the 10 patients with indwelling catheters was excellent ($r = 0.98$). The authors noted, as did Beppu et al. [24], that contrast-enhanced tricuspid regurgitation Doppler signals routinely underestimated catheter-determined pressures, especially during exercise. They used the following formula to estimate right ventricular (and thus pulmonary artery) systolic pressure: $PASP_{catheter} = (RVSP_{Doppler}/0.88) - 4$.

One factor which caused underestimation of pulmonary artery pressures by Doppler was malalignment of the ultrasound beam parallel to the maximal regurgitant jet, especially during exercise. Another was failure to account for changes in right atrial pressure with exercise. The authors assumed a constant right atrial pressure estimated from 2D imaging of the inferior vena cava at rest, whereas patients with indwelling catheters were found to increase mean right atrial pressure from an average of 7 mmHg at rest to 14 mmHg during exercise. Some authors have noted less efficacy with saline contrast-enhanced Doppler evaluation of tricuspid regurgitation in patients with chronic obstructive pulmonary disease [25], which may be related to their use of instruments without automatic Doppler gain control [26]. Working without automatic gain control requires careful Doppler gain adjustments to minimize spectral distortion. In our experience, saline contrast-enhanced echocardiography significantly improves the non-invasive assessment of systolic pulmonary artery pressure at rest and during exercise in patients whose symptoms are primarily exertional.

PULMONARY INSUFFICIENCY

While tricuspid regurgitation velocities reflect systolic right ventricular and thus pulmonary artery pressures, pulmonary regurgitation velocities provide accurate estimates of mean [5] and diastolic [3–7] pulmonary artery pressure. Pulmonary insufficiency is often detected by Doppler echocardiography in patients with normal or minimally elevated pulmonary artery pressures. While pulmonary insufficiency is often of limited hemodynamic significance, accurate display of its velocity profile allows non-invasive assessment of right-sided diastolic pressures. Early pulmonic valve M-mode echocardiographic indicators of pulmonary hypertension, such as an absent 'a' wave and early closure, have numerous possible causes and do not accurately estimate pulmonary artery pressure. In 1979 Koizumi et al. [27] presented the first report using intravenous contrast-enhanced M-mode echocardiography of the pulmonary valve to demonstrate linear contrast echoes which were thought to represent pulmonary valve insufficiency. In 1981, Gullace et al. [28] studied this phenomenon with 2–10 cm³ D_5W injected i.v. in patients undergoing M-mode echocardiography. They confirmed that the linear contrast echoes noted proximal to the pulmonary valve in diastole represented regurgitation of blood from the pulmonary artery into the right ventricular outflow tract. They also used contrast M-mode echocardiography to demonstrate that forward flow across the pulmonary valve stopped at the time of mid-systolic notching in patients with pulmonary hypertension. Meltzer et al. [29] used i.v. injection of 5–10 cm³ of D_5W in patients with known pulmonary insufficiency undergoing M-mode examination and found the technique highly sensitive and specific, as well as safe. As with tricuspid insufficiency the development of color Doppler simplified the detection of pulmonary insufficiency. However, the technical limitations which plague reliable spectral Doppler assessment of tricuspid regurgitation velocities also effect pulmonary insufficiency study, and the techniques for contrast enhancement of tricuspid regurgitation certainly apply to the pulmonary valve. Albunex has demonstrated reliable contrast enhancement of pulmonary insufficiency spectral Doppler. In 11 of 12 patients with baseline incomplete pulmonary insufficiency spectral Doppler signals (92%) studied by Tanabe et al. [30], Albunex provided reliable and accurate diagnostic enhancement. The initial dose was 0.5 cm^3 and was increased to 5 cm^3 or decreased to 0.25 cm^3 depending on the initial result. Doppler estimation of diastolic pulmonary artery pressure using this technique showed a strong correlation with simultaneous pulmonary artery catheter pressure determination ($r=0.93$).

CARDIAC OUTPUT

Contrast 2D echocardiographic imaging of the right ventricle provides a baseline tomographic image which initially shows the rapid entry of ultrasound contrast, followed by eventual dissolution of the agent and return to baseline. This technique has been shown to aid in the diagnosis of cardiac tamponade by demonstrating right ventricular collapse when poor image quality prevents baseline detection of myocardial and pericardial borders [31]. Measuring changes in the intensity of echocardiographic contrast effect over time, has also provided one approach to non-invasive measurement of cardiac output.

Microbubbles created by microcavitation have intraventricular velocities similar to red blood cells [32], and their velocities should reflect cardiac output. De Maria et al. [14] noted that the persistence of contrast effect was significantly longer in patients with congestive heart failure than in those with normal systolic function. Thus, the wash-out rate of contrast effect to some degree reflects forward flow. Hypothesizing that cardiac output could be determined using this principle, Bommer et al. [33] derived reliable indicator dilution type curves from photometric sampling in the right ventricle during rapid injection of D_5W. The upstroke of such a curve is related to the appearance of contrast in the right ventricle, and its downslope is related to disappearance of the agent (Figure 2). Using the time from peak contrast to 50% (DT/50) and 90% (DT/90) decreases in amplitude, they reproducibly distinguished decreased intracardiac flows for patients with tricuspid regurgitation or congestive heart failure from that of normal controls. This technique was then investigated for its use in determining cardiac output.

The calculation of cardiac output from indicator dilution curves is based on the Hamilton equation, where cardiac output$=I/(c \times t)$, where $I=$quantity of indicator injected, $c=$mean concentration of indicator detected and $t=$total duration of the curve obtained. To determine the mass of injectate, it is necessary to use an agent with predictable and constant microbubble size. Calculating the mean concentration of indicator present during detection requires a factor which relates contrast concentration to observed videointensity. Unfortunately, this factor depends on many variables including microbubble scattering and attenuation characteristics, instrument power and gain settings, transducer frequency, blood viscosity, and impedance characteristics of the chest wall. Another factor only recently recognized is the alteration [34, 35] and even destruction [36–38] of microbubbles by the imaging ultrasound waves themselves. Since these variables are not constant between subjects they make absolute flow quantitation very difficult. However, using this principle to detect change within the same subject may be useful as these variables should remain constant. DeMaria et al. [39] injected 30 μm plastic microbubbles into the left atrium of mongrel dogs to derive indicator dilution type curves from videodensitometric analysis of the left ventricle. As expected, the technique proved poor in predicting absolute cardiac output (r=0.61–0.65). However, the technique was very precise for measuring changes in cardiac output within each dog in response to pharmacological alterations (r=0.90–0.94).

Plastic microbubbles provide a contrast agent of known, constant size and concentration, but they cannot be used in humans. Since the bubble size range and concentration of Albunex (and several other experimental microbubble preparations) are known, it may be useful in indicator dilution curve cardiac output determinations. At present, there are no studies demonstrating this use for Albunex, although encouraging preliminary studies using galactose microbubbles have been performed (Figure 2A) [40].

One different approach taken by Schwarz et al. [41] used audiodensitometric analysis of Doppler detected flow rather than videodensitometry. Again, this method accurately measured changes in flow, but not absolute flows. Another technique was utilized by Galanti et al. [42], who measured transpulmonary transit times of intravenous Albunex in dogs as an indicator of cardiac output. They noted an excellent correlation between the pulmonary transit rate, or the time from first 2D echocardiographic contrast appearance in the right ventricle to its first appearance in the left ventricle, and thermodilution cardiac output (r=0.92). This technique has not been validated in humans. At present, problems related to the multiple factors listed above, as well as variations in microbubble passage through diseased pulmonary vessels, appear daunting.

CONCLUSION

Contrast echocardiography has demonstrated many applications in evaluating right-sided cardiac function. Contrast enhancement of tricuspid and pulmonic valve Doppler flow spectra allows accurate estimation of pulmonary artery pressures. In the vast majority of patients, saline or Albunex contrast enhancement can render inadequate Doppler signals diagnostic, often when no regurgitation is detected by standard techniques. Even in patients with technically limited echocardiographic examinations, such as those with chronic lung disease,

SHU-454   NaCl

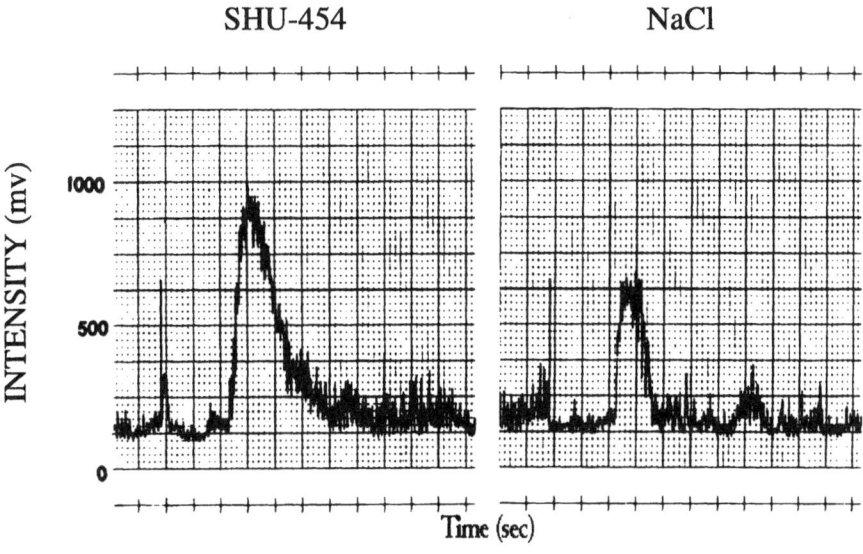

Figure 2. Videodensitometric curve obtained after injection of galactose microbubbles (SHU-454) (A) and agitated saline (B) with the detector in the right ventricle. From Smith MD, Kwan OL, Reiser HJ, DeMaria AN. Superior intensity and reproducibility of SHU-454, a new right heart contrast agent. J Am Coll Cardiol 1984;3:992–8, with permission.

contrast-enhanced Doppler study of tricuspid regurgitation has allowed evaluation of pulmonary hemodynamics both at rest and during exercise. Accurate assessment of cardiac output remains a theoretical but possible application of contrast echocardiography. As new contrast agents are developed and more is learned about the effects of instrument power, transducer frequency and second-harmonic imaging on microbubble performance, further advances in assessing right heart function will occur.

REFERENCES

1. Yock PG, Popp RL. Noninvasive estimation of right ventricular systolic pressure by Doppler ultrasound in patients with tricuspid regurgitation. Circulation 1984;70:657–62.
2. Berger M, Haimowitz A, Van Tosh A, Berdoff RL, Goldberg E. Quantitative assessment of pulmonary hypertension in patients with tricuspid regurgitation using continuous wave Doppler ultrasound. J Am Coll Cardiol 1985;6:359–65.
3. Hatle L, Angelsen B. Doppler Ultrasound in Cardiology: Physical Principles and Clinical Applications. 2nd edn. Philadelphia: Lea and Febinger, 1985.
4. Masuyama T, Uematsu M, Nakatani S, Sato H, Kodama K. Doppler echocardiographic assessment of changes in pulmonary artery pressure associated with vasodilating therapy in patients with congestive heart failure. J Am Soc Echocardiogr 1991;4:35–42.
5. Masuyama T, Kodama K, Kitabatake A, Sato H, Nanto S, Inoue M. Continuous-wave Doppler

echocardiographic detection of pulmonary regurgitation and its application to noninvasive estimation of pulmonary artery pressure. Circulation 1986;74:484–92.

6. Lee RT, Lord CP, Plappert T, St John Sutton M. Prospective Doppler echocardiographic evaluation of pulmonary artery diastolic pressure in the medical intensive care unit. Am J Cardiol 1989;64:1366–70.

7. Himelman RB, Stulbarg M, Kircher B et al. Noninvasive evaluation of pulmonary artery pressure during exercise by saline-enhanced Doppler echocardiography in chronic pulmonary disease. Circulation 1989;79:863–71.

8. Kostucki W, Vandenbossche H, Friart A, Englert M. Pulsed Doppler regurgitant flow patterns of normal valves. Am J Cardiol 1986;58:309–13.

9. Waggoner A, Quinones M, Young J et al. Pulsed Doppler echocardiographic detection of right-sided valve regurgitation. Am J Cardiol 1981;47:279–86.

10. Choong CY, Abascal CM, Weyman J et al. Prevalence of valvular regurgitation by Doppler echocardiography in patients with structurally normal hearts by two-dimensional echocardiography. Am Heart J 1989;117:636–42.

11. Seiichi T, Miyatake K, Izumi S, Kinoshita N, Sakakibara H, Yasuharu N. Physiological pulmonary regurgitation detected by the Doppler technique and its differential diagnosis. J Am Coll Cardiol 1985;5:499.

12. Wise NK, Myers S, Fraker TD, Stewart JA, Kisslo JA. Contrast M-mode ultrasonography of the inferior vena cava. Circulation 1981;63:1100–3.

13. DeMaria AN, Bommer W, George L, Neumann A, Weinert L, Mason DT. Combined peripheral venous injection and cross-sectional echocardiography in the evaluation of cardiac disease. Am J Cardiol 1978;41:370.

14. Lieppe W, Behar VS, Scallion R, Kisslo JA. Detection of tricuspid regurgitation with two-dimensional echocardiography and peripheral vein injections. Circulation 1978;57:128–32.

15. Meltzer RS, McGhie J, Roelandt J. Inferior vena cava echocardiography. J Cardiac Ultrasound 1982;10:47–51.

16. Feigenbaum H, Stone JM, Lee DA, Nasser WK, Chang S. Identification of ultrasound echoes from the left ventricle by use of intracardiac injections of indocyanine green. Circulation 1970; 41:615–21.

17. Bove AA, Ziskin MC, Mulchin WL. Ultrasonic detection of intracavitation and pressure effects of high-speed injections through catheters. Invest Radiol 1969;4:236–40.

18. Meltzer RS, Serruys PW, Hugenholtz PG, Roelandt J. Intravenous carbon dioxide as an echo-cardiographic contrast agent. J Cardiac Ultrasound 1981;9:127–31.

19. Ogawa S, Hayashi J, Sasaki H et al. Evaluation of valvular prolapse syndrome by two-dimensional echocardiography. Circulation 1982;65:174–80.

20. Beard JT, Byrd BF. Saline contrast enhancement of trivial Doppler tricuspid regurgitation signals for estimating pulmonary artery pressure. Am J Cardiol 1988;62:486–8.

21. Meltzer RS, Vered Z, Roelandt J, Heufeld HN. Systematic analysis of contrast echocardiograms. Am J Cardiol 1983;52:375–80.

22. Byrd BF, O'Kelly BF, Schiller NB. Contrast echocardiography enhances tricuspid but not mitral regurgitation. Clin Cardiol 1991;14(suppl 5):V-10–V-14.

23. Kemp Jr WE, Kerins D, Shyr Y, Byrd III BF. Optimal Albunex dosing for enhancement of Doppler tricuspid regurgitation spectra. Am J Cardiol 1997;79:232–4.

24. Beppu S, Tanabe K, Shimizu T et al. Contrast enhancement of Doppler signals by sonicated albumin estimating right ventricular systolic pressure. Am J Cardiol 1991;67:1148–50.

25. Sajkov D, Cowie RJ, Bradley JA, Mahar L, McEvoy RD. Validation of new pulsed Doppler echocardiography techniques for assessment of pulmonary hemodynamics. Chest 1993;103: 1348–53.

26. Schiller NB. Letter to the editor. Chest 1994;107:1902–3.

27. Koizumi K, Umeda T, Machii K. Estimation of the flow pattern and velocity in the main pulmonary artery studied by simultaneous cross-sectional and M-mode echocardiography combined

with contrast technique. Fourth World Congress of the World Federation of Ultrasound in Medicine and Biology, Miyazaki, Japan, 1979:71.

28. Gullace G, Savoia MT, Ravizza P, Locatelli V, Addamiano P, Ranzi C. Contrast echocardiographic features of pulmonary hypertension and regurgitation. Br Heart J 1981;46:369–73.

29. Meltzer R, Vered Z, Hegesh T, et al. Diagnosis of pulmonic regurgitation by contrast echocardiography. Am Heart J 1984;107:102–7.

30. Tanabe K, Asanuma T, Yoshitomi H et al. Doppler estimation of pulmonary artery end-diastolic pressure using contrast enhancement of pulmonary regurgitation signals. Am J Cardiol 1996; 78:1145–8.

31. Stratienko AA, Pollock SG, Keller MW, Sarembock IJ. Use of venous contrast echocardiography for diagnosis of cardiac tamponade. Am J Cardiol 1989;64:691–2.

32. Levine RA, Teichholz LE, Goldman ME, Steinmetz MY, Baker M, Meltzer RS. Microbubbles have intracardiac velocities similar to those of red blood cells. J Am Coll Cardiol 1984;3: 28–33.

33. Bommer W, Neef J, Neumann A et al. Indicator-dilution curves obtained by photometric analysis of two-dimension echo-contrast studies. Am J Cardiol 1978;41:370.

34. Holt RG, Crum LA. Acoustically forced oscillations of air bubbles in water: experimental results. J Acoust Soc Am 1992;91:1924–32.

35. Crum LA, Roy RA, Dinno MA, Church CC. Acoustic cavitation produced by microsecond pulses of ultrasound: a discussion of some selected results. J Acoust Soc Am 1992;91:1113–19.

36. Mor-Avi V, Robinson K, Shroff S, Lang RM. Stability of Albunex microspheres under ultrasonic irradiation: an in vitro study. J Am Soc Echocardiogr 1994;7:S29.

37. Vandenberg GR, Melton HE. Acoustic lability of albumin microspheres. J Am Soc Echocardiogr 1994;7:582–9.

38. Wray RA, Zoghbi WA, Quinones MA, Cheirif J. Contrast echocardiography: relation of acoustic power and time gain compensation to contrast intensity duration. J Am Soc Echocardiogr 1991;4:286.

39. DeMaria AN, Bommer W, Kwan OL, Riggs K, Smith M, Waters J. In vivo correlation of thermodilution cardiac output and videodensitometric indicator-dilution curves obtained from contrast two-dimensional echocardiograms. J Am Coll Cardiol 1984;3:999–1004.

40. Smith M, Kwan OL, Reiser HJ, DeMaria AN. Superior intensity and reproducibility of SHU-454, a new right heart contrast agent. J Am Coll Cardiol 1984;3:992–8.

41. Schwartz KQ, Bezante GP. Can the intensity of audio Doppler signals be used to assess echo-contrast concentration? Adv Echo Contr 1993;3:22–9.

42. Galanti G, Jayaweera AR, Villaneuva FS, Glasheen WP, Ismail S, Kaul S. Transpulmonary transit of microbubbles during contrast echocardiography: implications for estimating cardiac output and pulmonary blood volume. J Am Soc Echocardiogr 1993;6:272–8.

14. Contrast-enhanced color Doppler in the assessment of mitral regurgitation

HELENE VON BIBRA, ANJA TUCHNITZ, CHRISTIAN FIRSCHKE, HARALD BECHER & ALBERT SCHÖMIG

CLINICAL DILEMMA

The correct evaluation of mitral regurgitation still poses a considerable dilemma. Clinical assessment relies on a combination of auscultation and palpation, the drawbacks of which are only too frequently encountered. Severe mitral regurgitation may not be associated with a clinically audible murmur, and may co-exist with other lesions which also give rise to systolic murmurs. The auscultatory features of severe regurgitation are due to an increased volume of diastolic flow across the mitral valve and may be missed in the failing heart, as will the apical heave. The advent of M-mode and two-dimensional echocardiography has not solved these problems, as both of these techniques can provide only indirect clues related to valvular morphology and left atrial and ventricular chamber size or function during early diastolic filling [1, 2]. The introduction of conventional Doppler, however, permitted the detection of mitral regurgitation with a level of sensitivity hitherto unknown. In fact, the false-positive diagnosis of pathologic mitral regurgitation by this technique should be avoided by distinguishing the pattern of physiological mitral regurgitation (i.e. brief early systolic flow disturbance close to the mitral valve leaflets at the time of valve closure) from the mid- or pansystolic jet of pathologic regurgitation. Color Doppler flow imaging promised considerable potential for quantifying mitral regurgitation. The application of this technique to valvular regurgitation has increased our understanding of jet formation and behavior under different hemodynamic circumstances. However clinical studies that measured the spatial distribution of the regurgitant jets demonstrated that the technique provided only a semiquantitative estimate of the severity of mitral regurgitation [3–5].

CLASSIFICATION OF MITRAL REGURGITATION

The classification of mitral regurgitation into mild, moderate or severe according to jet area visualized on color Doppler flow imaging has been widely adopted in

N. C. Nanda, R. Schlief and B. B. Goldberg (eds.), Advances in Echo Imaging Using Contrast Enhancement,
Second Edition, 239–251.
© 1997 *Kluwer Academic Publishers.*

clinical practice, as it has proved to be feasible with the information being rapidly acquired. This method of classification has, however, been questioned as being unreliable because of poor quality images of the mitral regurgitant jet, due to either an impaired signal to noise ratio or areas of image drop-out, particularly when using the apical window [6–8].

The advent of transesophageal Doppler echocardiography allowed the study of mitral regurgitant jets with fewer intervening tissue interfaces and from shallower depths (range 5–15 cm) when compared to the transthoracic approach (range 13–22 cm). This technique has provided a greater sensitivity for the detection of regurgitation from both the native [9] and, in particular, from prosthetic mitral valves [10]. It can be used for the classification of mitral valve regurgitation if the transthoracic approach does not provide adequate images [11] or during surgical valve repair. However, a recent systematic study compared regurgitant jet areas measured from the transesophageal window with those from the apical transthoracic approach and demonstrated them to be significantly larger (172%) [12].

The unique esophageal ultrasound window can also allow more comprehensive studies of pulmonary venous flow patterns than the precordial window. Transesophageal studies have demonstrated a reduced velocity of forward flow with moderate mitral regurgitation and reversed systolic flow in severe regurgitation [13, 14]. These signs are now well established as useful additional criteria when classifying mitral regurgitation from the transesophageal approach.

Acceptance of transesophageal studies

The somewhat disappointing correlative clinical results in the semiquantitative evaluation of mitral regurgitation from an apical transducer position have prompted two reactions. In clinical routine they have led to the acceptance of the need for transesophageal studies in the conscious patient, although transesophageal ultrasound is more unpleasant for the patient and cannot be completely free of risk [15]. They have also prompted extensive experimental work in order to analyze the mechanisms for the display of regurgitant jet area using color Doppler techniques and in order to discover truly quantitative methods. Thus, the proximal flow convergence method has been proposed to quantify the regurgitant flow rate [16, 17].

The relationship between the color Doppler dimensions of the regurgitant jet and regurgitant volume has been found to be influenced by both instrumental factors and hemodynamic variables. The former include color flow gain, wall filter and transducer frequency as well as pulse repetition frequency [18, 19]; the latter include transvalve pressure gradient, orifice size and chamber compliance and size [20, 21]. In vitro and animal studies have demonstrated the major dependence of the spatial jet area on these factors. However, all this experimental work has been carried out in an ideal environment in which the problems of ultrasound absorption and attenuation were minimized by using few tissue interfaces and by having short ultrasound penetration depths. In these idealized situations, the hemodynamic variables were easily shown to be the most prominent factors influencing jet size. However, the patient combines the problems both of marked attenuation of ultrasound due to chest wall structures and a variable hemodynamic situation. No clinical data are available which indicates the extent to which the latter influences jet size, once the attenuation problem has been resolved. To clarify this, the essential first step is to evaluate the correlation between the degree of mitral regurgitation and the jet size using an approach which can display the complete regurgitant jet. Only then can the influence of hemodynamic factors on jet size be evaluated in the clinical setting.

POOR ECHOGENICITY

With regard to the many prior attempts to define the problems associated with the color Doppler display of regurgitant jets, it is surprising how little attention has been paid to the fact that moving blood reflects only minute amounts of ultrasound: it has a poor echogenicity. This small amount of reflected ultrasound is, however, the only substrate available for Doppler interrogation. No wonder therefore, that the Doppler information thus derived has a low signal intensity. The weak signal, however, may not reach the intensity threshold of flow display. Instead it will be misinterpreted as noise by the ultrasound device and be eliminated by the cut-off filter. This limitation often impedes flow detection and flow display when using color Doppler, as has been shown in experimental work [22] and as is obvious from the high incidence of intracardiac areas which contain no flow signal.

These sensitivity problems are encountered in virtually every transthoracic color Doppler study but they are hardly ever addressed in publications. They may well be a major reason for the disappointing results obtained when using regurgitant jet display to classify mitral regurgitation. This theory would appear to be confirmed by examining the corresponding transesophageal images, which usually reveal gross underestimation of mitral regurgitation jet dimensions by transthoracic echocardiography [10–12]. There is considerable concern over how to compensate for these sensitivity problems. Theoretically, an increase of the echogenicity of the moving blood by the use of echocontrast agents should enhance low intensity color Doppler signals.

INCREASED ECHOGENICITY

Earlier in vitro and human studies showed enhancement of color Doppler flow signals by contrast agents [23–29], described previously. This article reviews the results of clinical trials using Levovist (Schering AG), with particular emphasis on the improved evaluation of mitral regurgitation [30, 31].

Dose dependency

Dose dependency was demonstrated for the signal intensity of normal anterograde flow in color Doppler flow imaging and also for the color-coded jet area in mitral regurgitation or normal anterograde flow [30]. Thus, the optimal dose for reproducible color Doppler contrast enhancement needs to be clarified.

The dose producing the optimal contrast effect should be the maximal dose which can be given without producing artefacts in color Doppler imaging and which is just below the level at which the tissue priority algorithm would wrongly encode the contrast-enhanced Doppler information as tissue information. Normal subjects show a high degree of interindividual variation in the echogenicity of flowing blood and in the amount of ultrasound absorption by precordial tissue. Thus, it is not surprising that in a multicenter study of 152 patients, the concentration of SHU 508A required for good quality Doppler enhancement varied between 200 and 400 mg/ml [31]. The data from these patients indicate that the initial dose of SHU 508A should be 16 ml (200 mg/ml). If there is still insufficient color coding of normal anterograde atrial flow (< 70% of the left atrial area), an increase in the concentration of subsequent injections is warranted to 300 mg/ml and then to 400 mg/ml. If artefacts are noted, a reduction of dosage to 10 ml of 200 mg/ml coupled with a slower injection rate is advisable.

Color Doppler enhancement of flow display

SHU 508A provided a significant increase in left heart color Doppler signal intensity throughout systole and diastole [30]. Normal anterograde flow and regurgitant jets were, therefore, modified by contrast enhancement. The increase of normal flow color-coded areas was in the range of 50% for ventricular flow and of about 300% for anterograde atrial flow (Figure 1). If we assume that the cardiac cavities are normally full of moving blood, then flow detection is somewhat closer to reality after contrast enhancement, when about 40–80% of any cavity is color coded. Improvement in flow display was demonstrated for low flow velocities (diastolic atrial flow), moderate flow velocities (diastolic ventricular flow) and high flow velocities (regurgitant jets of mitral and tricuspid regurgitation) to a comparable extent [30]. Thickened mitral valve structures or bioprothesis did not reduce color Doppler enhancement of LA flow in a subset of patients when compared with the whole group [30]. Finally, attenuation from the use of greater ultrasound penetration depths may be considered the major cause of poor sensitivity in flow detection by color Doppler, as demonstrated by the significant difference of displayed flow areas between the left atrium

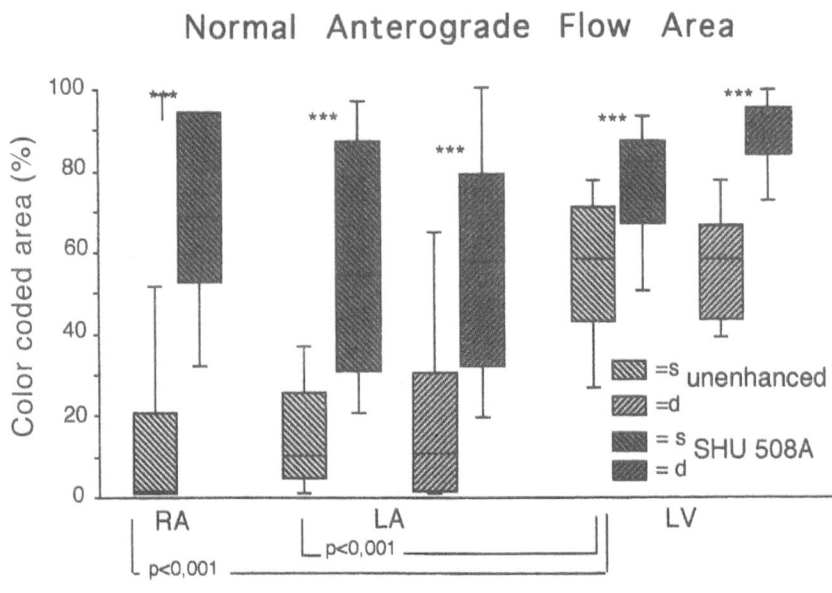

Figure 1. Contrast-induced increase of color-coded areas of normal, anterograde systolic (s) and diastolic (d) flow expressed as a percentage of the respective cardiac cavity. Note that reality, i.e. flow display in 70–100% of each cardiac chamber, is demonstrated only after contrast enhancement. Median with 75 and 90 percentiles. ***$p < 0.001$.

($16 \pm 17\%$) and left ventricle ($52 \pm 20\%$) when using unenhanced color Doppler (Figure 1). Increased echogenicity considerably improves the signal-to-noise ratio in color Doppler imaging, thus increasing the sensitivity of the color Doppler system for flow detection. Augmentation of diagnostic sensitivity and specificity can therefore be expected when using contrast-enhanced color Doppler. This hypothesis was thus tested and supported for that most problematic valvular lesion, mitral regurgitation.

Evaluation of mitral regurgitation

The mean increase in mitral regurgitation jet area as displayed by contrast-enhanced color Doppler was 270%, with a high level of interindividual variation. These results have been described in 33 patients of a two-center study [30] and in another 61 patients from a multicenter study [31]. With contrast enhancement, there was an increased sensitivity for the detection of mitral regurgitant jets: in a control group of 10 patients with no mitral regurgitation visualized on the left ventricular angiogram, two patients had early systolic regurgitation visible before and eight after administration of SHU 508A. These jets had relatively small dimensions and represented trivial or physiological regurgitation. This

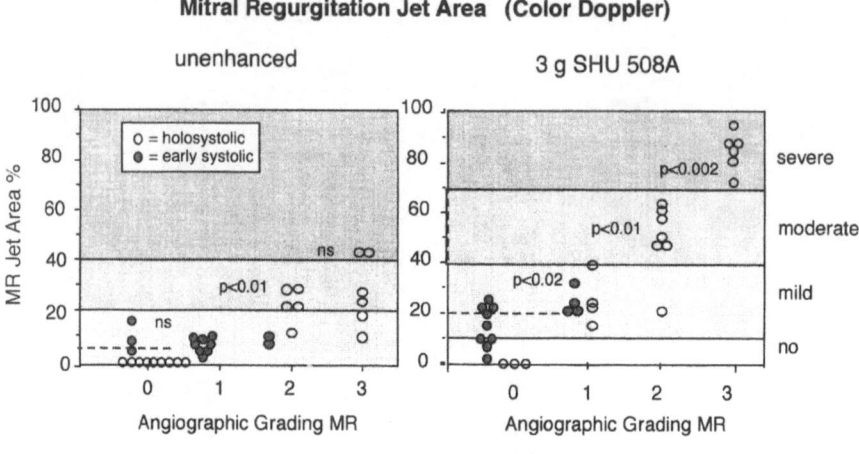

Figure 2. Correlation of reference grading of severity of mitral regurgitation and individual values of color Doppler jet area as a percentage of left atrial area in 29 patients with suspected mitral regurgitation and 10 normal individuals. Reprinted with permission from J Am Coll Cardiol 1993; 22:521–8.

entity can be found more frequently after contrast enhancement. These jets have specific time course and area dimensions, and can easily be differentiated from pathologic mitral regurgitation (Figure 2).

In patients with pathological mitral regurgitation, the measured increase in the regurgitant jet area by 270% (Figure 3) due to increased echogenicity was not unexpected. It is noteworthy that eccentric jets from flail mitral valve could be similarly enhanced (Figure 4), as well as the more centrally placed jets, so that the Koanda effect was not an obvious limitation to evaluation of severity. However, the increased sensitivity of flow detection was accompanied by an alteration in the displayed time course of the regurgitation, which changed from early systolic to holosystolic in eight of 29 patients. Obviously, the late systolic component to this flow had been camouflaged in the unenhanced color Doppler study. It is likely that the extent of regurgitation decreased so much late in systole that the weak flow signals were not of sufficient strength to be displayed. Both the unrestricted display of the jet area and the time course should provide useful additional information when evaluating the degree of mitral regurgitation [33]. The non-artifactual nature of these changes was evident when comparing these findings to the reference grading (Figure 2).

Classification of mitral regurgitation

An essential question is, whether contrast enhancement is useful for the assessment of severity of mitral regurgitation. The relationship between the regurgitant

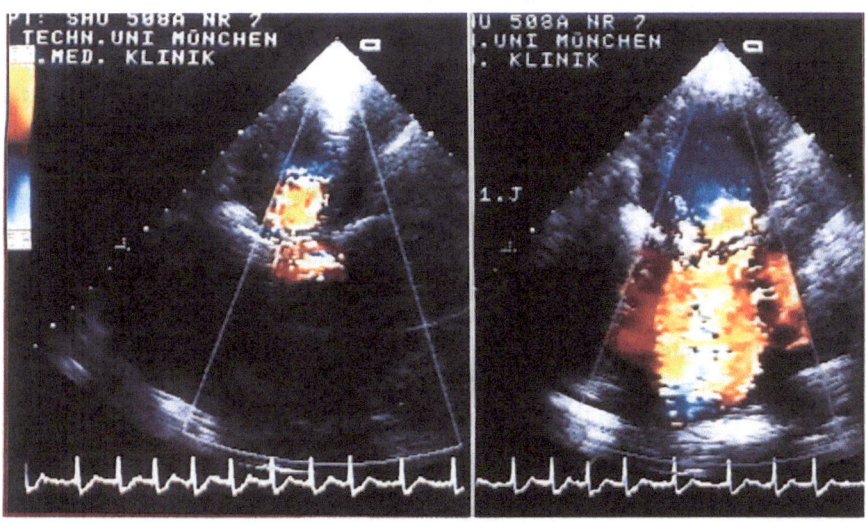

Figure 3. Original color Doppler recording in the 4 chamber view: left, prior to; and right, after 3 g intravenous SHU 508A in a patient with severe mitral regurgitation, which was not displayed in the unenhanced color Doppler.

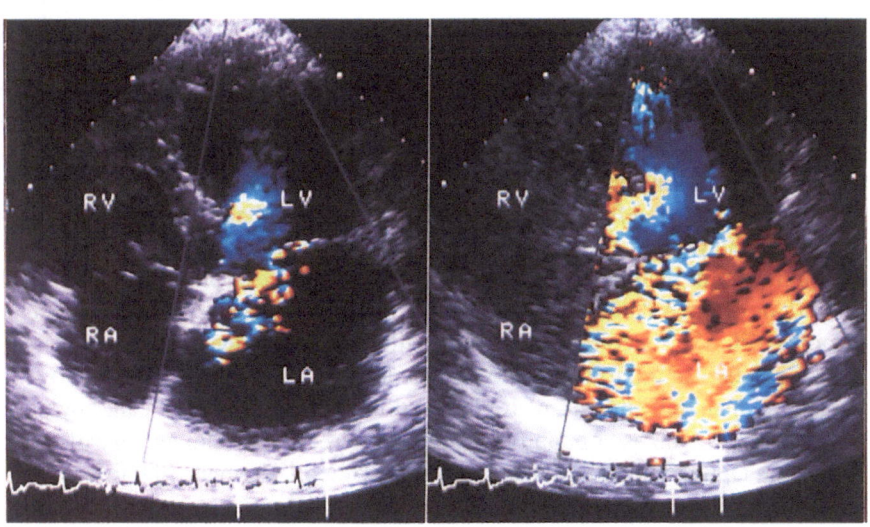

Figure 4. Same setting as Figure 3 in a patient with severe mitral insufficiency from mitral valve prolapse with an eccentric jet.

jet area and the left atrial area was used to grade mitral regurgitation, which was then compared with the angiographic classification (0–3). As shown in previous studies, measuring jet area by color Doppler is not reliable for the classification of mitral regurgitation because of sensitivity problems due to the patient or the imaging device [10–12]. We demonstrated a wide scatter of individual jet areas in each angiographic class, with a large overlap between data obtained from individuals with severe and mild regurgitation in the unenhanced color Doppler images (Figure 2). After injection of SHU 508A the correlation was much improved, indicating a better ultrasound differentiation of the severity of mitral regurgitation.

Nevertheless, we consider that there is persistent reluctance among cardiologists to accept the contrast-enhanced display as a clinical tool simply because many years of low signal intensity transthoracic imaging have made the image of a mainly black-coded left atrium so familiar. We therefore compared transthoracic jet areas in 14 patients before and after contrast enhancement with the better accepted images obtained using the transesophageal approach (which is well-known for its higher sensitivity in the detection of flow). Recent studies have shown that jet areas displayed in the transesophageal setting are considerably larger than those seen using the transthoracic approach [12, 19] and semiquantification has been proposed using transesophageal imaging [34]. The data in this study confirmed gross underestimation of jet areas displayed using the unenhanced transthoracic approach when compared with transesophageal imaging (Figure 5). Contrast-enhanced jet areas, however, were virtually the same size as those displayed with transesophageal imaging. This complete agreement is surprising since the two imaging modalities differ with respect to factors known to influence the sensitivity of flow detection by color Doppler flow imaging, including attenuation due to ultrasound penetration depth and absorption from precordial tissues, both of which are severe handicaps in transthoracic color Doppler display. The higher ultrasound frequency used with the transesophageal approach reduces jet area in the presence of low Doppler signal intensity [35]. Our data suggest that ultrasound penetration depth and precordial absorption are dominant limitations of unpredictable extent for the transthoracic display of regurgitant jet velocities in the unenhanced setting with low Doppler signal intensities. Doppler signals of too low intensity are obviously misinterpreted as noise by the ultrasound machine and cut off by the wall filter. This limitation can, however, be overcome by the use of contrast agents to increase the echogenicity of blood and hence Doppler signal intensity, thus allowing all Doppler flow information to be displayed. It follows therefore that using this compensational technique improves the correlation between jet area and angiographic grading of severity in mitral regurgitation to the level we have hitherto attributed to the transesophageal technique alone as we have shown with these present data.

The improvement brought about by the use of contrast-enhanced color Doppler compensated for the sensitivity problems of the ultrasound machine. These sensitivity problems are fairly independent of the type of ultrasound machine

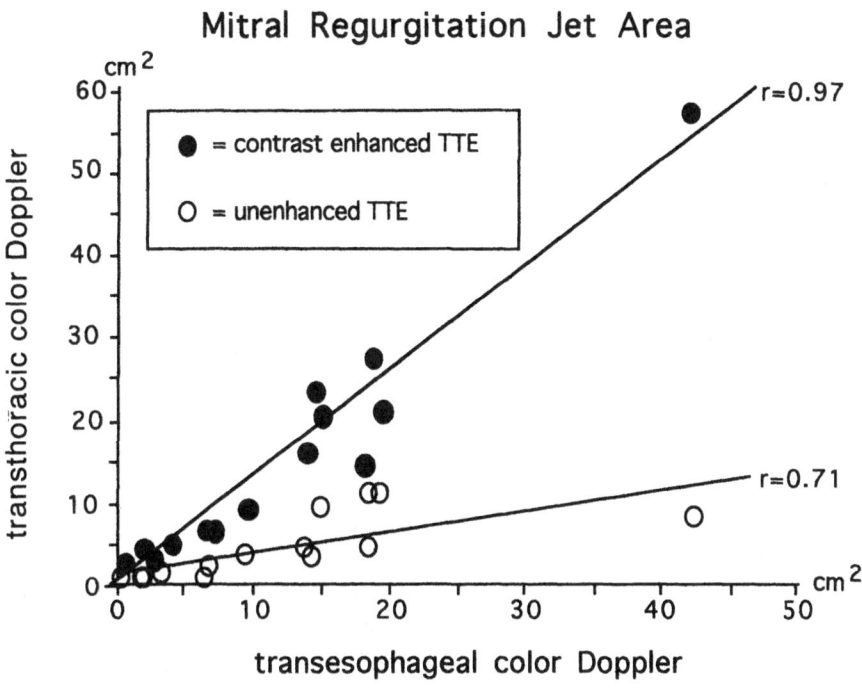

Figure 5. Maximal jet area of mitral regurgitation as obtained when the transthoracic (TTE) color Doppler (vertical) before (open circles) and after contrast enhancement (closed circles) are plotted against maximal jet area obtained at unenhanced transesophageal (TEE) color Doppler (horizontal). Note the substantial underestimation by unenhanced TEE. Reprinted with permission from J Am Coll Cardiol 1995;25:500–8.

used [30] or the center where the study was performed [31]. It is possible to judge from the color Doppler image the degree of any inherent sensitivity problem within the machine: normally, the left atrium is full of moving blood during systole and diastole. A color-coded area >80% of the left atrial cross-sectional area therefore indicates sufficient sensitivity of flow detection, whereas display of normal anterograde flow as <75% of the atrial area indicates loss of flow information for normal anterograde as well as for pathological retrograde flow display [36].

The results reflect the importance of improved flow detection and display. They also emphasize the basic relationship between regurgitant volume and jet area, which is influenced by many variables, including instrumental factors, patient factors and hemodynamic factors. Most of these variables, however, may be dealt with in the clinical setting. Recommendations have been given as to the optimal machine settings [35]. Within patient factors, we have shown echogenicity of the blood to be an important key for visualization of Doppler information, being capable of compensating for the sensitivity problems due to attenuation and penetration depth. The hemodynamic factors associated with the valvular lesion may be roughly estimated by the experienced cardiologist, who observes all information available from echocardiography, ECG, blood pressure and physical signs. An improved detection and display of flow requires new criteria for the classification of jets. The present data from 94 patients indicate the following discrimination of severity:

contrast-enhanced jet area expressed as % of LA area:
    0–10=no pathologic MR,
    11–32=mild MR,
    33–64=moderate MR,
     ≥65=severe MR.

Effects of other echocardiographic contrast agents

The principles of contrast enhancement for Doppler sonography apply to all left heart contrast agents. The specific composition of the particular microbubbles will determine acoustic emission and persistence of the contrast agent in blood and hence localization and duration of the enhancement effect.

ENHANCED PULMONARY VENOUS FLOW PATTERN

The assessment of pulmonary venous flow patterns in order to differentiate between different degrees of mitral regurgitation has as yet been confined to transesophageal echocardiography [38, 39]. However, color Doppler enhancement of left atrial flow display by SHU 508A is usually accompanied by an associated improved display of pulmonary venous inflow, so that the pulsed Doppler sample may be correctly placed in the right upper pulmonary vein [37]. Figure 6 shows the effect of contrast enhancement on pulsed Doppler recordings from the right upper pulmonary vein in a patient with severe mitral regurgitation. In the unenhanced registration, a poor signal to noise ratio had excluded quantitative analysis in three of four patients. The recordings after SHU 508A demonstrated the typical feature of reversed systolic flow in all four patients with severe mitral regurgitation [31].

Evaluation of pulmonary venous flow patterns with regard to their diagnostic and hemodynamic relevance has mainly been performed with transesophageal echocardiography [13–15]. We therefore compared measurements of peak flow velocities obtained from transesophageal with contrast-enhanced transthoracic velocities and found good agreement [31]. Our results indicate that contrast-enhanced Doppler provides similar information, thus obviating the need for transesophageal echocardiography in the majority of patients.

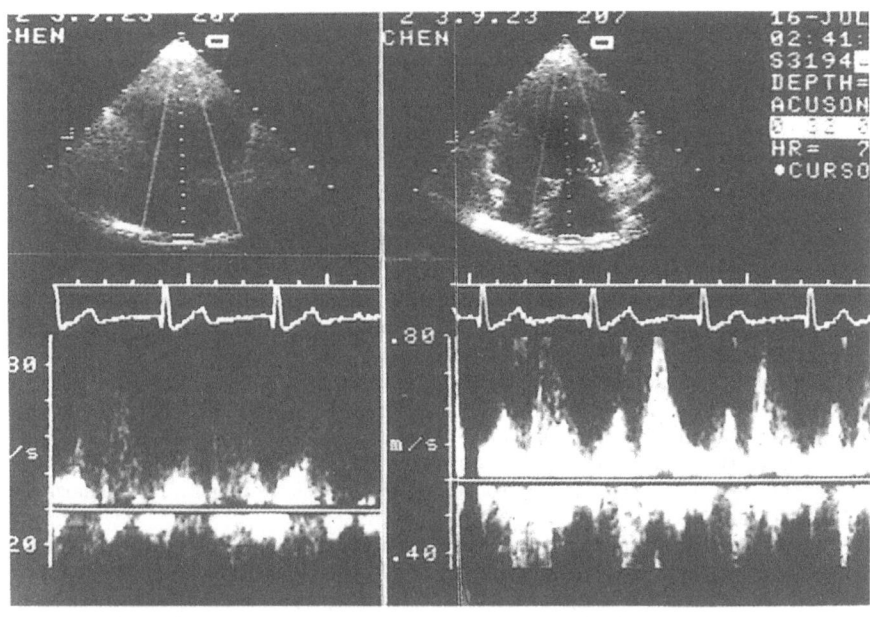

Figure 6. Original pulsed Doppler recording of flow pattern in the right upper pulmonary vein in a patient with severe mitral regurgitation (left panel prior to and right panel after contrast enhancement). The diagnostic systolic reflux is displayed after contrast enhancement.

CLINICAL IMPLICATIONS

Contrast-enhanced Doppler allows confident evaluation of severity of mitral regurgitation even in technically difficult patients. This technique facilitates further the analysis of pulmonary venous flow pattern and thus provides another valuable aid for the semiquantification of mitral regurgitation, which has so far been a domain of transesophageal echocardiography. Taking into account the patient's discomfort associated with the transesophageal technique, Doppler contrast enhancement would now appear to be an important new alternative in the sequence of diagnostic tools used to assess mitral regurgitation. Furthermore, contrast enhancement facilitates better recordings of the maximal mitral regurgitant jet velocity by continuous wave Doppler. This may have potential in the calculation of max dp/dt for the additional evaluation of LV function [40]. Thus contrast enhancement can improve evaluation of valve lesions and their functional hemodynamic consequences and may obviate the use of transesophageal echocardiography in a significant number of cases.

REFERENCES

1. Mintz GS, Kotler MN, Segal BL et al. Two-dimensional echocardiographic evaluation of patients with mitral insufficiency. Am J Cardiol 1979;44:670–8.
2. Gibson DG, Brown D. Measurement of left ventricular dimension and filling rate in man using echocardiography. Br Heart J 1973;35:1141–9.
3. Miyatake K, Izumi S, Okamoto M et al. Semiquantitative grading of the severity of mitral regurgitation by real-time two-dimensional Doppler flow imaging technique. J Am Coll Cardiol 1986;7:82–8.
4. Helmcke F, Nanda NC, Hsiun MC et al. Color Doppler assessment of mitral regurgitation with orthogonal planes. Circulation 1987;75:175–83.
5. Spain MG, Smith MD, Grayburn PA, Harlamert EA, De Maria AN. Quantitative assessment of mitral regurgitation by Doppler color flow imaging: angiographic and hemodynamic correlations. J Am Coll Cardiol 1989;13:585–90.
6. Shah PM. Quantitative assessment of mitral regurgitation. J Am Coll Cardiol 1989;13:591–3.
7. Becher H, Mintert C, Grube E, Lüderitz B. Classification of mitral regurgitation by color flow mapping. Z Kardiol 1989;78:764–70.
8. von Bibra H, Stempfle HU, Poll A, Scherer M, Blüml G, Blömer H. Limitations of flow detection by color Doppler: in vitro comparison to conventional Doppler. Echocardiography 1991;8:633–42.
9. Schlüter M, Langenstein BA, Hanrath P et al. Assessment of transesophageal pulsed Doppler echocardiography in the detection of mitral regurgitation. Circulation 1982;66:784–9.
10. Nellessen U, Schnittger I, Appleton CP et al. Transesophageal two-dimensional echocardiography and color Doppler flow velocity mapping in the evaluation of cardiac valve prothesis. Circulation 1988;78:848–55.
11. Seward JB, Khanderia BK, Oh JK, Abel MD, Freeman WK, Tajik AJ. Transesophageal echocardiography: technique, anatomic correlations, implementation and clinical applications. Mayo Clin Proc 1988;63:649–77.
12. Smith MD, Harrison MR, Pinton R, Kandil H, Kwan OL, De Maria AN. Regurgitant jet size by transesophageal compared with transthoracic Doppler color flow imaging. Circulation 1991;83:79–86.
13. Dennig K, Henneke KH, Dacian S, Rudolph W. Estimation of the severity of mitral regurgitation by parameters derived from the velocity profile of pulmonary venous flow using transesophageal Doppler technique. J Am Coll Cardiol 1990;15:91a.
14. Kreis A, Lamberz H, Gerich N, Hanrath P. Value of the transesophageal echocardiographic pulmonary venous flow in classification of mitral insufficiency. Circulation 1989;80:II-557.
15. Daniel WG, Erbel R, Kasper W et al. Safety of transesophageal echocardiography. A multi-center survey of 10 419 examinations. Circulation 1991;83:817–21.
16. Recusani F, Bargiggia GS, Yoganathan AP et al. A new method for quantification of regurgitant flow rate using color Doppler flow imaging of the flow convergence region proximal to a discrete orifice. An in vitro study. Circulation 1991;83:594–604.
17. Rodriguez L, Anconia J, Flachskampf FA, Weyman AE, Levine RA, Thomas JD. Impact of finite orifice size on proximal flow convergence. Implications for Doppler quantification of valvular regurgitation. Circ Res 1992;70:923–30.
18. Tamura T, Yoghanathan A, Sahn D. In vitro methods for studying the accuracy of velocity determination and spatial resolution of a color Doppler flow mapping system. Am Heart J 1987;114:152–8.
19. Sahn D. Instrumentation and physical factors related to visualization of stenotic and regurgitant jets by Doppler color flow mapping. J Am Coll Cardiol 1988;12:1354–65.
20. Simpson IA, Valdes-Cruz LM, Sahn DJ, Murillo A, Tamura T, Chung KJ. Doppler color flow mapping of simulated in vitro regurgitant jets: evaluation of the effects of orifice size and hemodynamic variables. J Am Coll Cardiol 1989;13:1195–207.

21. Hoit BD, Jones M, Eidbo EE, Elias W, Sahn DJ. Sources of variability for Doppler color flow mapping of regurgitant jets in an animal model of mitral regurgitation. J Am Coll Cardiol 1989;13:1631–6.
22. von Bibra H, Stempfle HU, Poll A, Scherer M, Blüml G, Blömer H. Limitations of flow detection by color Doppler: in vitro comparison to conventional Doppler. Echocardiography 1991;8: 633–42.
23. von Bibra H, Hartmann F, Petrik M, Schlief R, Renner U, Blömer H. Contrast color-coded Doppler flow imaging – improved diagnosis of right heart disease after intravenous injection of Echovist. Z Kardiol 1989;78:101–8.
24. Beard JT, Byrd BF. Saline contrast enhancement of trivial Doppler tricuspid regurgitation signals for estimating pulmonary artery pressure. Am J Cardiol 1988;62:486–8.
25. Wilkenshoff UM, Gast D, Kruck I. Contrast-echo: a comparison of color Doppler, contrast echo and enhanced color Doppler in the detection of cardiac shunts in adults. J Am Coll Cardiol 1990;15:195A.
26. Becher H, Schlief R. Improved sensitivity of color Doppler by SHU 454. Am J Cardiol 1989; 64:374–7.
27. von Bibra H, Stempfle HU, Poll A, Schlief R, Emslander HP. Echocontrast agents improve display of color Doppler – in vitro studies. Echocardiography 1991;8:533–40.
28. Feinstein SB, Cheirift J, Ten Cate F et al. Safety and efficacy of a new transpulmonary contrast agent: initial multicenter clinical results. J Am Coll Cardiol 1990;16:316–24.
29. Schlief R, Staks T, Mahler M, Rufer M, Fritzsch T, Seifert W. Successful opacification of the left heart chambers on echocardiographic examination after intravenous injection of a new saccharide based contrast agent. Echocardiography 1990;7:61–4.
30. von Bibra H, Becher H, Firschke C, Schlief R, Emslander H, Schömig A. Enhancement of mitral regurgitation and normal left atrial color Doppler flow signals with peripheral venous injection of a saccharide based contrast agent. J Am Coll Cardiol 1993;22:521–8.
31. von Bibra H, Sutherland G, Becher H, Neudert J, Nihoyannopoulos P. Clinical evaluation of left heart Doppler contrast enhancement by a saccharide based trans-pulmonary contrast agent. The Levovist Cardiac Working Group. J Am Coll Cardiol 1995;25:500–8.
32. Beppu S, Tanabe K, Shimizu T et al. Contrast enhancement of Doppler signals by sonicated albumin for estimating right ventricular systolic pressure. Am J Cardiol 1991;67:1148–50.
33. Mohr-Kahaly S, Erbel R, Zenker G et al. Semiquantitative grading of mitral regurgitation by color coded Doppler echocardiography. Int J Cardiol 1989;23:223–30.
34. Castello R, Lenzen P, Aguirre F, Labovitz A. Quantitation of mitral regurgitation by trans-esophageal echocardiography with color Doppler flow mapping: correlation with cardiac catheterization. J Am Coll Cardiol 1992;19:1516–21.
35. Sahn D. Instrumentation and physical factors related to visualization of stenotic and regurgitant jets by Doppler color flow mapping. J Am Coll Cardiol 1988;12:1354–65.
36. Tuchnitz A, von Bibra H, Becher H, Firschke C, Schömig A. Wie viel Farbdoppler Kontrast-verstärkung? Z Kardiol 1995;84(Suppl 1):172.
37. Becher H, von Bibra H, Glänzer K, Vetter H. Verbesserter transthorakaler Nachweis des Lungenvenenflusses mittels Kontrast verstärkter Farbdoppler-Echokardiographie. Z Kardiol 1992;81(Suppl A):147.
38. Nishimura RA, Abel MD, Hatle LK, Tajik AJ. Relation of pulmonary vein to mitral flow velocities by transesophageal Doppler echocardiography. Circulation 1990;81:1488–97.
39. Tuccillo B, Fraser AG. Pulmonary venous flow. In: Sutherland G, Roeland J, Fraser A, Anderson R, editors. Transesophageal Echocardiography in Clinical Practice. London, New York: Gower Medical Publishing, 1991:5.1.–5.18.
40. Chen C, Rodriguez L, Guerrero JL et al. Noninvasive estimation of the instantaneous first derivative of left ventricular pressure using continuous wave Doppler echocardiography. Circulation 1991;83:2101–10.

15. Contrast-enhanced Doppler in the assessment of aortic stenosis

SATOSHI NAKATANI & KUNIO MIYATAKE

INTRODUCTION

The severity of aortic stenosis has been assessed by measuring the transaortic pressure gradient and/or the stenotic aortic valve area [1, 2]. Although these measurements are usually performed by cardiac catheterization, this is not suitable for bedside and serial assessment of the severity because of its invasiveness and cost. M-mode and two-dimensional echocardiography may be useful to detect the narrowing of aortic valve. However, they lack the specificity for evaluating the severity of aortic stenosis [3, 4]. Doppler echocardiography provides accurate and non-invasive measurement of transvalvular velocity [5]. Theoretically, transvalvular maximal velocity can provide accurate determination of transvalvular maximal pressure gradient for most stenotic orifices by applying the simplified Bernoulli equation [6–8]. Recently, transvalvular velocity has also been used to determine the stenotic valve area based on the continuity equation [9–12]. Thus, Doppler echocardiography is clearly a potential tool for assessing severity of valvular stenosis. Accurate determination of transaortic velocity is indispensable for non-invasive assessment of the severity of aortic stenosis. Pulsed wave Doppler echocardiography, however, has a major limitation of limited sampling rate to detect transaortic velocity in significant aortic stenosis. Consequently, continuous wave Doppler echocardiography has been used for the determination of transaortic velocity. However, optimal recordings of the velocity may not always be possible [13, 14]. Significant expertise is occasionally required to obtain the clear velocity envelope in technically difficult patients in whom the intensity of Doppler signals is weak. To overcome this limitation of conventional continuous wave Doppler echocardiography, a method of enhancing Doppler signal intensity would be desirable.

Two-dimensional echocardiography performed during the injection of a contrast agent, called contrast echocardiography, has been used in clinical settings to visualize blood flow [15, 16]. Agents used for contrast echocardiography contain tiny suspended microbubbles or particles that scatter sound waves to a greater

253

N. C. Nanda, R. Schlief and B. B. Goldberg (eds.), *Advances in Echo Imaging Using Contrast Enhancement, Second Edition,* 253–263.
© 1997 *Kluwer Academic Publishers.*

extent than the red blood cells [17]; they can, therefore, also be used as Doppler signal enhancers [18]. Several investigators have shown that the peripheral venous injection of contrast agents can enhance the Doppler signal associated with trivial tricuspid regurgitation [19–21]. However, studies have been limited to the right heart chamber. Recently, left heart echocardiography using contrast agents capable of transpulmonary passage has been reported [22–26]. We have shown that sonicated albumin microbubbles can pass through the pulmonary circulation, following which they not only opacify the left heart chamber but also enhance the left heart Doppler signals following peripheral venous injection [27]. If sonicated albumin could enhance the weak Doppler signal of aortic stenotic flow it would help in the assessment of the severity of aortic stenosis [28]. In this chapter, we discuss the usefulness of transpulmonary contrast enhanced Doppler echocardiography using sonicated albumin in the assessment of severity of aortic stenosis.

PREPARATION OF CONTRAST AGENT

We used albumin microbubbles as contrast agents [22, 25–28]. These are as small as erythrocytes [29], capable of transpulmonary passage and have already been used safely in myocardial contrast echocardiography [30, 31]. Albumin microbubbles were made by sonication of albumin solution as reported elsewhere [21]. Briefly, 5 ml of 5% human serum albumin was sonicated with a commercially available sonicating system (Sonifier 250, Branson Inc.) with a 3 mm titanium horn at 20 kHz and 20 W for 30 s. Immediately after sonication, 2 ml of the sonicated solution was injected into an antecubital vein via a 20-gauge needle followed by a 5 ml saline flush.

DOPPLER ECHOCARDIOGRAPHY

Doppler echocardiography was performed using a combined two-dimensional and Doppler ultrasonograph with a 2.5 MHz transducer (Toshiba SSH 160A, Toshiba Corp., Tokyo, Japan). Continuous wave Doppler signals of transaortic flow velocity were obtained from an apical window, directing the Doppler team to the aortic flow with the assistance of audio signals and the color Doppler mode. We usually detect the aortic flow from various approaches including a suprasternal and an apical approach. However, in this study, we only used the apical approach because we aimed to demonstrate the effectiveness of the contrast enhanced Doppler echocardiography. Velocity waveforms were recorded on a strip chart continuously before and after intravenous injection of sonicated albumin. Transaortic pressure gradient was calculated from the peak aortic flow velocity when determined, using the simplified Bernoulli equation [5]. When the aortic velocity signal was enhanced following sonicated albumin injection, the

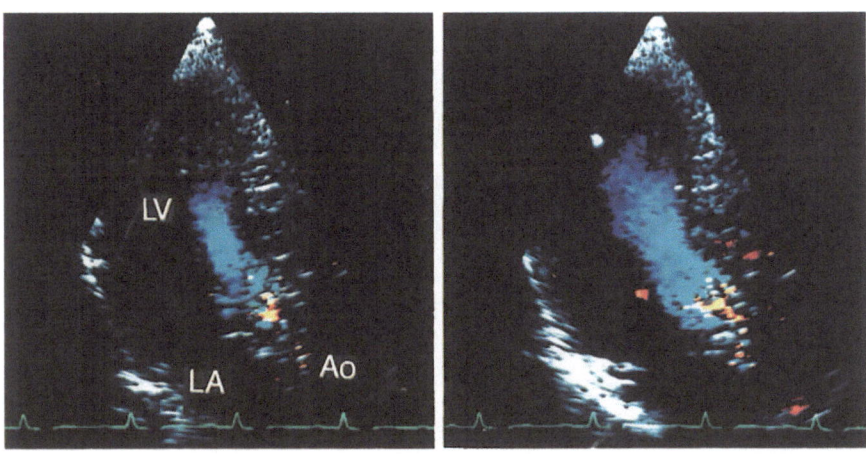

Figure 1. Color Doppler signal of aortic flow in mid-systole before (left) and after (right) intravenous sonicated albumin injection. The signal area is increased after injection. LV, left ventricle; LA, left atrium; Ao, aorta.

duration of the enhancement was measured as the time during which the velocity envelope was clearer than before injection throughout the ejection period. Doppler gain and filter setting were held constant throughout the study. Sonicated albumin injection and Doppler recordings were performed at least twice with an interval of at least 7 min. All reported values represent the average of at least two measurements.

FINDINGS

Color Doppler signal enhancement

Figure 1 shows color Doppler signals of the aortic stenotic flow before and after intravenous sonicated albumin injection. Color Doppler signal of the flow after injection was enhanced and the signal area was larger than before, making the positioning of the Doppler beam just to the stenotic flow easier [26]. It is important to position the Doppler beam in the center of the flow and to optimize the angle of incidence when determining accurate flow velocity. In significant aortic stenosis, however, the streamline of the aortic flow is narrow and is occasionally hard to detect even with the aid of color Doppler signal. Enhancement of the color Doppler signal of aortic stenotic flow with sonicated albumin facilitates the detection of the narrow streamline of significant aortic stenosis. Fan et al. [32] reported that the width of the color Doppler signal of aortic stenotic flow was useful to assess the severity of aortic stenosis. Clinical usefulness of

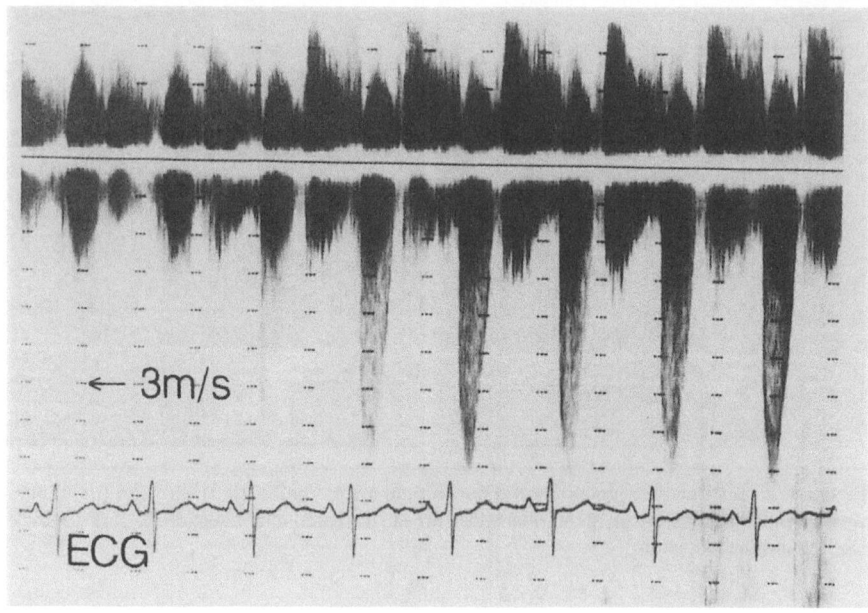

Figure 2. Serial change of aortic flow velocity envelopes determined by continuous wave Doppler technique accompanied by sonicated albumin injection. Before sonicated albumin injection, the velocity envelope was too indistinct to determine the peak velocity and therefore, transaortic pressure gradient. After injection, intensity of Doppler signal was increased and the velocity envelope became clearer, enabling the calculation of transaortic pressure gradient. ECG, electrocardiogram.

their method would be much enhanced with this contrast enhanced Doppler technique because this technique improves sensitivity of the Doppler system to detect the weak color Doppler signal.

Continuous wave Doppler signal enhancement

Figure 2 shows the serial change of transaortic velocity profile determined by continuous wave Doppler echocardiography during intravenous injection of sonicated albumin. Before injection the velocity profile was too indistinct to determine peak velocity, and no transaortic pressure gradient could be obtained. After injection, the velocity profile became clear enough, enabling determination of peak velocity and transaortic pressure gradient. We investigated 22 patients with aortic stenosis with continuous wave Doppler echocardiography from the apical approach during sonicated albumin injection. In 10 patients with aortic stenosis, the transaortic velocity envelope was too indistinct to determine peak velocity before sonicated albumin injection. After injection the envelope became

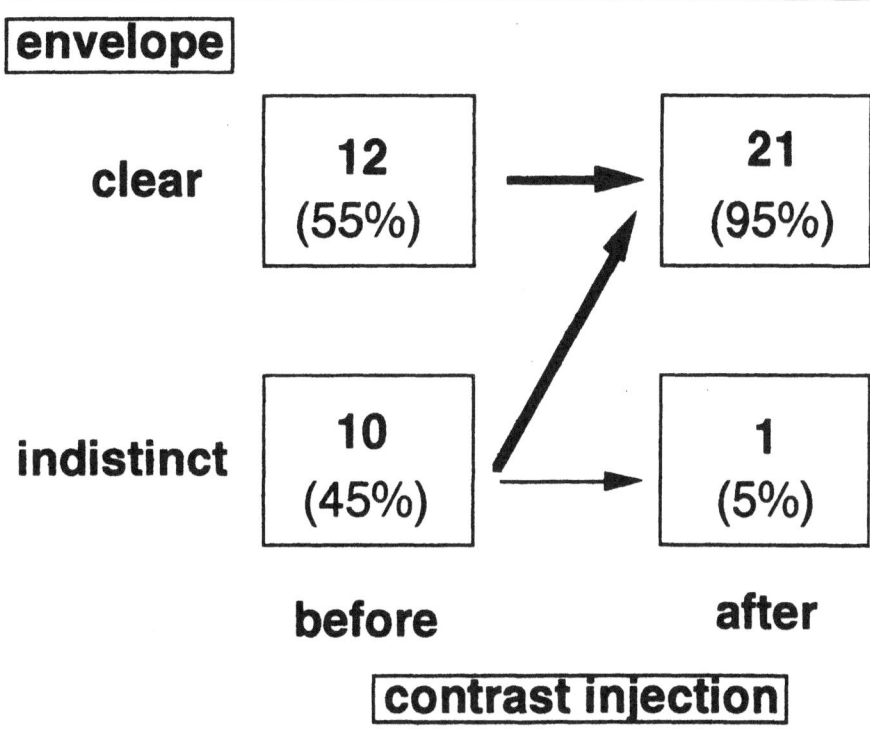

Figure 3. Enhancement of Doppler flow signals in aortic stenosis. The numbers of patients are shown in rectangles.

clear enough to determine peak velocity in nine of the 10 patients, thus enabling calculation of transaortic pressure gradient. The remaining 12 patients showed clear velocity envelope before injection, which became much clearer after injection of sonicated albumin. In these cases, peak transaortic flow velocity and, therefore, transaortic pressure gradient were not altered by administration of sonicated albumin. This is reasonable because intracardiac velocities of albumin microbubbles are similar to those of red blood cells as reported [33]. Thus, enhancement of transaortic flow velocity following intravenous injection of sonicated albumin was achieved in 21 out of 22 patients with aortic stenosis (95%) (Figure 3).

Accuracy of measured velocity after enhancement

To assess the accuracy of contrast-enhanced Doppler echocardiography, we compared the transaortic pressure gradient measured by cardiac catheterization and the contrast-enhanced Doppler technique. The transaortic pressure gradient

cath PG (mmHg)

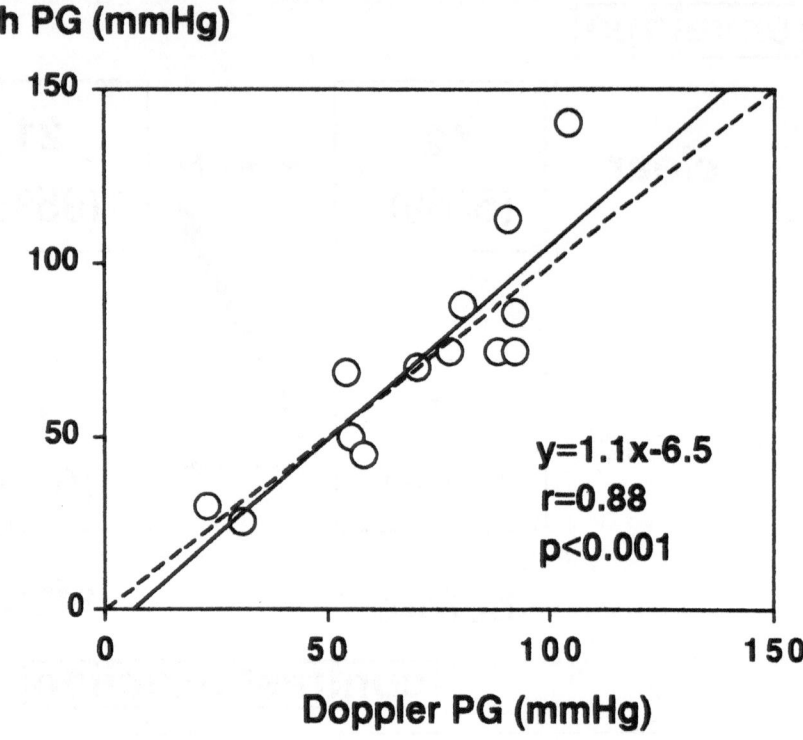

Figure 4. Relationship between transaortic pressure gradients determined by contrast-enhanced Doppler and catheterization measurements. A good agreement between them was obtained ($y=1.1x-6.5$, $r=0.88$, $p<0.001$). Dotted line shows the line of identity. Doppler PG, transaortic pressure gradient determined by contrast-enhanced Doppler technique; cath PG, instantaneous peak transaortic pressure gradient determined by catheterization.

at cardiac catheterization was measured by superimposing pressure tracings of the left ventricle and the ascending aorta obtained by the pull-back technique. Figure 4 shows the relationship between transaortic pressure gradient obtained by both measurements. There was a good agreement, with a correlation coefficient of 0.88. Thus, contrast-enhanced Doppler is an accurate method for estimating the instantaneous transaortic pressure gradient.

Duration of enhancement

As shown in Figure 2, the Doppler signal enhancement effect was stable for a while. In our initial experience we noticed that the enhancement lasted for a shorter time in patients with aortic stenosis than in normal volunteers. Accordingly,

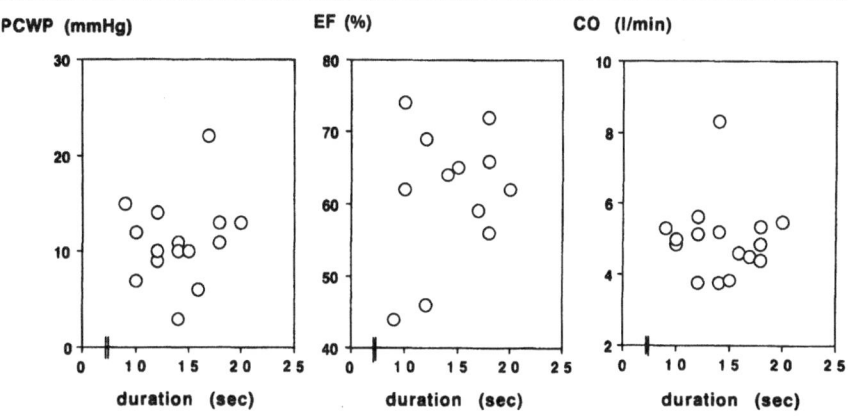

Figure 5. Relationship between duration of Doppler signal enhancement and hemodynamic parameters obtained by catheterization. Duration, duration of Doppler signal enhancement; PCWP, mean pulmonary capillary wedge pressure; EF, ejection fraction; CO, cardiac output.

we analyzed the relationship between hemodynamic parameters obtained by cardiac catheterization and duration of the Doppler signal enhancement in aortic stenosis (Figure 5). Recently, left heart echo opacification with a transpulmonary contrast agent has been reported to be dependent on pulmonary arterial pressure [34]. However, we found no significant correlation between pulmonary capillary wedge pressure and enhancement duration. Furthermore, there were no significant correlations between ejection fraction and cardiac output and enhancement duration. This might be due to the narrow range of these parameters in the patients studied. However, a significant correlation was found between left ventricular systolic pressure and enhancement duration (Figure 6). Albumin microbubbles have been reported to be susceptible to high pressure [35, 36]. The rapid disappearance of the enhancement effect with high chamber pressure may be explained in part by this pressure sensitivity of albumin microbubbles. Aortic regurgitation, which is often associated with aortic stenosis, would also affect the enhancement duration. If the associated aortic regurgitation is severe, a relatively large part of aortic flow would reverse into the left ventricle during diastole and therefore, the enhancement duration would be prolonged. Although half of our patients had significant associated aortic regurgitation (grade 3+ and 4+ demonstrated by aortic root angiography), its association seemed to have little influence on the enhancement duration.

CLINICAL IMPLICATIONS

Advances in technology have been remarkable and the sensitivity of the Doppler system to detect the flow with weak Doppler signal has improved. Using con-

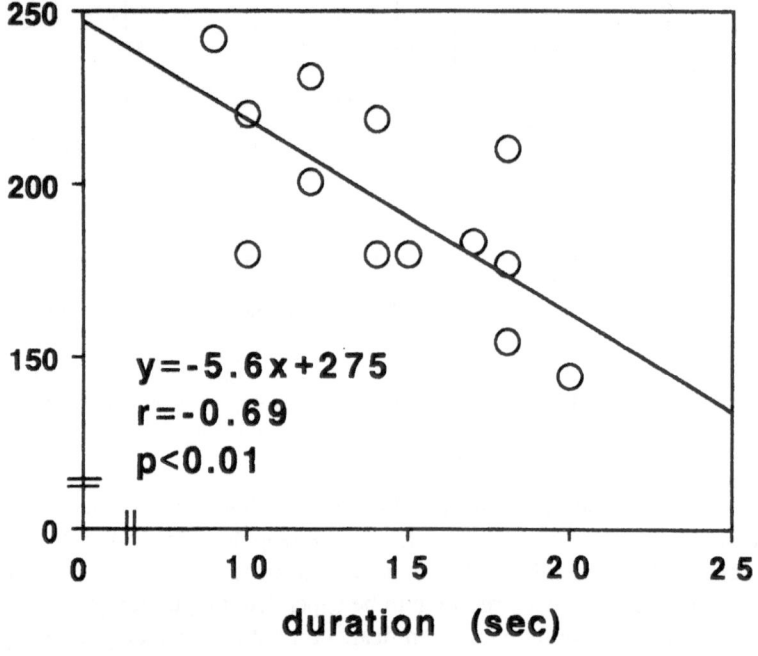

Figure 6. Relationship between duration of Doppler signal enhancement and left ventricular systolic pressure determined by catheterization. A significant correlation between them was found ($y=-5.6x+275, r=-0.69, p<0.01$). Duration, duration of Doppler signal enhancement; LVSP, left ventricular systolic pressure.

tinuous wave Doppler echocardiography the beam direction of the transmitted ultrasound was fixed and only the direction of the received ultrasound beam was movable. Flow detection with continuous wave Doppler echocardiography was limited and correction of the incident angle between the Doppler beam and the flow was often required. Steerable continuous wave Doppler beam has been developed recently. Moreover, the advent of color Doppler echocardiography enables the positioning of the Doppler beam just in the center of the flow. However, in spite of the progress in technology, aortic flow detection by continuous wave Doppler technique is still difficult in certain patients with aortic stenosis because of increased absorption and scattering of sound waves by tissue between the transducer and the heart (which is an inherent flaw of the ultrasound technique), and because of narrow streamline of significant aortic stenosis. In such patients, the measurement is sometimes time-consuming and there is little assurance that maximal velocity has in fact been recorded. Further, in such patients, invasive measurement often fails because of the inability of the catheter to cross the aortic valve during cardiac catheterization. To obtain flow signal from these patients, the Doppler gain should be increased. However, an increase of gain would simultaneously increase noise levels and would therefore not necessarily lead to optimal recording of the flow. Contrast-enhanced Doppler echocardiography is a promising method to improve the signal-to-noise ratio. This method accurately estimates the transaortic flow velocity even in patients with poor Doppler recordings. It enhances diagnostic sensitivity of the Doppler system and facilitates assessment of the severity of aortic stenosis.

Although some contrast agents are considered sensitive to high pressure, there are few data to support this finding [35, 36]. In this study, we showed that albumin microbubbles were sensitive to high pressure in situ. Contrast-enhanced Doppler echocardiography will be used widely in the future as a method of enhancing weak Doppler signals in the left heart or in other organs. However, the sensitivity of albumin microbubbles to pressure might limit clinical application of transpulmonary contrast-enhanced Doppler echocardiography. In our preliminary study, we found that the enhancement duration of pulsed Doppler-determined aortic flow signal in hypertensive patients after intravenous sonicated albumin was shorter than that in normotensive patients. A variety of contrast agents has been developed recently. Contrast agents without pressure sensitivity would overcome this limitation, or pressure sensitivity of the contrast agent might be useful to estimate intracardiac pressure non-invasively.

CONCLUSION

Doppler signal of the aortic stenotic flow was enhanced after peripheral venous injection of sonicated albumin, enabling accurate measurements of transaortic peak velocity and transaortic pressure gradient even in patients with poor Doppler recordings. Interestingly, duration of the enhancement was affected by left ventricular systolic pressure; the higher the left ventricular systolic pressure, the more quickly the Doppler signal enhancement disappeared. Transpulmonary contrast-enhanced Doppler echocardiography using sonicated albumin is useful to assess the severity of aortic stenosis. This technique will enhance the clinical importance of Doppler echocardiography as a method of assessing aortic stenosis non-invasively.

REFERENCES

1. Braunwald E. Valvular heart disease. In: Braunwald E (editor) Heart Disease. 4th edn. Philadelphia: WB Saunders, 1992:1007–77.
2. Carabello BA, Grossman W. Calculation of stenotic valve orifice area. In: Grossman W, Baim DS, editors. Cardiac Catheterization, Angiography and Intervention. 4th edn. Philadelphia: Lea & Febiger, 1992:152–65.

3. Weyman AE, Feigenbaum H, Dillon JC, Chang S. Cross-sectional echocardiography in assessing the severity of valvular aortic stenosis. Circulation 1975;52:828–34.
4. DeMaria AN, Bommer W, Joye J, Lee G, Bouteller J, Mason DT. Value and limitations of cross-sectional echocardiography of the aortic valve in the diagnosis and quantification of valvular aortic stenosis. Circulation 1980;62:304–12.
5. Hatle L, Angelsen B. Doppler Ultrasound in Cardiology. 2nd edn. Philadelphia: Lea and Febiger, 1985.
6. Stamm RB, Martin RP. Quantification of pressure gradients across stenotic valves by Doppler ultrasound. J Am Coll Cardiol 1983;2:707–18.
7. Currie PJ, Seward JB, Reeder GS et al. Continuous-wave Doppler echocardiographic assessment of severity of calcific aortic stenosis: a simultaneous Doppler-catheter correlative study in 100 adult patients. Circulation 1985;71:1162–9.
8. Yeager M, Yock PG, Popp RL. Comparison of Doppler-derived pressure gradient to that determined at cardiac catheterization in adults with aortic valve stenosis: implications for management. Am J Cardiol 1986;57:644–8.
9. Skjaerpe T, Hegrenaes L, Hatle L. Noninvasive estimation of valve area in patients with aortic stenosis by Doppler ultrasound and two-dimensional echocardiography. Circulation 1985;12: 810–8.
10. Zoghbi WA, Farmer KL, Soto JG, Nelson JG, Quinones MA. Accurate noninvasive quantification of stenotic aortic valve area by Doppler echocardiography. Circulation 1986;73;452–9.
11. Come PC, Riby MF, McKay RC, Safian R. Echocardiographic assessment of aortic valve area in elderly patients with aortic stenosis and of changes in valve area after percutaneous balloon valvuloplasty. J Am Coll Cardiol 1987;10:115–24.
12. Nakatani S, Masuyama T, Kodama K, Kitabatake A, Fujii K, Kamada T. Value and limitations of Doppler echocardiography in the quantification of stenotic mitral valve area; Comparison of the pressure half-time and the continuity equation methods. Circulation 1988;77:78–85.
13. Penn IM, Dumesnil JG. A new and simple method to measure maximal aortic valve pressure gradients by Doppler echocardiography. Am J Cardiol 1988;61:382–5.
14. Zoghbi WA, Galan A, Quinones MA. Accurate assessment of aortic stenosis severity by Doppler echocardiography independent of aortic jet velocity. Am Heart J 1988;116:855–63.
15. Gramiak R, Shah PM. Echocardiography of the aortic root. Invest Radiol 1968;3:356–66.
16. Armstrong WF, Kinney EL, Mueller TM, Tickner EG, Dillon JC, Feigenbaum H. Assessment of myocardial perfusion abnormalities with contrast-enhanced two-dimensional echocardiography. Circulation 1982;66:166–73.
17. Meltzer RS, Tickner EG, Popp RL, Roelandt J. The source of echocardiographic contrast. In: Meltzer RS, Roelandt J, editors. Contrast Echocardiography. The Hague: Martinus Nijhoff, 1982:7–16.
18. Becher H, Schlief R. Improved sensitivity of color Doppler by SHU 454. Am J Cardiol 1989; 64:374–7.
19. Himelman RB, Stulbarg M, Kircher B et al. Noninvasive evaluation of pulmonary artery pressure during exercise by saline-enhanced Doppler echocardiography in chronic pulmonary disease. Circulation 1989;79:863–71.
20. Waggoner AD, Barzilai B, Perez JE. Saline contrast enhancement of tricuspid regurgitant jets detected by Doppler color flow imaging. Am J Cardiol 1990;65:1368–71.
21. Beppu S, Tanabe K, Shimizu T et al. Contrast enhancement of Doppler signals by sonicated albumin for estimating right ventricular systolic pressure. Am J Cardiol 1991;67:1148–50.
22. Keller MW, Feinstein SB, Watson DD. Successful left ventricular opacification following peripheral venous injection of sonicated contrast agent: an experimental evaluation. Am Heart J 1987;114:570–5.
23. Berwing K, Schlepper M. Echocardiographic imaging of the left ventricle by peripheral intravenous injection of echo contrast agent. Am Heart J 1988;115:399–408.
24. Smith MD, Elion JL, McClure RR, Kwan OL, DeMaria AN. Left heart opacification with

peripheral venous injection of a new saccharide echo contrast agent in dogs. J Am Coll Cardiol 1989;13:1622–8.

25. Feinstein SB, Cheirif J, Ten Cate FJ et al. Safety and efficacy of a new transpulmonary ultrasound contrast agent: initial multicenter clinical results. J Am Coll Cardiol 1990;16:316–24.

26. Nakatani S, Beppu S, Ohta T, Matsuda H, Miyatake K. Efficacy and adverse effects of transpulmonary contrast echocardiography using sonicated albumin. Intern Med 1994;33:204–9.

27. Terasawa A, Miyatake K, Nakatani S, Yamagishi M, Masuda H, Beppu S. Enhancement of Doppler flow signals in the left heart by intravenous injection of sonicated albumin. J Am Coll Cardiol 1993;21:737–42.

28. Nakatani S, Imanishi T, Terasawa A, Beppu S, Nagata S, Miyatake K. Clinical application of transpulmonary contrast-enhanced Doppler technique in the assessment of severity of aortic stenosis. J Am Coll Cardiol 1992;20:973–8.

29. Feinstein SB, Ten Cate FJ, Zwehl W et al. Two-dimensional contrast echocardiography. I. In vitro development and quantitative analysis of echo contrast agents. J Am Coll Cardiol 1984; 3:14–20.

30. Keller M, Glasheen W, Teja K, Gear A, Kaul S. Myocardial contrast echocardiography without hyperemic response or hemodynamic effects. J Am Coll Cardiol 1988;12:1039–47.

31. Reisner SA, Ong LS, Lichtenberg GS et al. Myocardial perfusion imaging by contrast echocardiography with use of intracoronary sonicated albumin in humans. J Am Coll Cardiol 1989; 14:660–5.

32. Fan PH, Kapur KK, Nanda NC. Color-guided Doppler echocardiographic assessment of aortic valve stenosis. J Am Coll Cardiol 1988;12:441–9.

33. Levine RA, Teichholts LE, Goldman ME, Steinmetz MY, Baker M, Meltzer RS. Microbubbles have intracardiac velocities similar to those of red blood cells. J Am Coll Cardiol 1984;3: 28–33.

34. Zotz RJ, Wagner B. Left heart contrast echocardiography – an independent predictor of pulmonary artery pressure. Circulation 1991;84(Suppl II):II–360.

35. Gottlieb S, Ernst A, Litt L, Schwarz KQ, Meltzer RS. Effect of pressure on echocardiographic videodensity from sonicated albumin: an in-vitro model. J Am Soc Echo 1990;3:238.

36. Shapiro JR, Reisner SA, Lichtenberg GS, Meltzer RS. Intravenous contrast echocardiography in humans: systolic disappearance of left ventricular contrast after transpulmonary transmission. J Am Coll Cardiol 1990;16:1603–7.

16. Detection of intrapulmonary vasodilatation and diagnosis of hepatopulmonary syndrome using contrast echocardiography

GARY A. ABRAMS, MICHAEL B. FALLON,
CAMILO R. GOMEZ & NAVIN C. NANDA

INTRODUCTION

The hepatopulmonary syndrome is defined by a triad including liver disease and/or portal hypertension, gas exchange abnormalities (arterial $pO_2 < 70$ mmHg or alveolar–arterial gradient > 20 mmHg) and intrapulmonary vasodilatation. The mainstay of diagnosis of this syndrome is the detection of intrapulmonary vasodilatation. Contrast echocardiography (CE) and lung perfusion scanning (Tc^{99}-MAA) are the current modalities utilized to detect intrapulmonary vasodilatation and to diagnose hepatopulmonary syndrome (HPS). This chapter will briefly review the HPS and focus on the use of CE in diagnosis as well as potential future diagnostic modalities.

HISTORICAL PERSPECTIVE

The first description of a patient with cirrhosis, cyanosis and clubbing was reported from Germany in 1884 [1]. Seventy-two years later Rydell and Hoffbauer identified intrapulmonary arteriovenous communications near the pulmonary hilum in a 17-year-old juvenile cirrhotic with cyanosis [2]. In 1957, Calabresi et al. found porto-pulmonary anastomoses in 20 post-mortem examinations of patients with cirrhosis [3]. They suggested porto-pulmonary shunts cause veno-arterial admixture resulting in the hypoxemia associated with liver disease. However in 1966, Berthelot et al. were the first to use micropaque gelatin suspension to examine the pulmonary vasculature in autopsy lungs from patients with cirrhosis [4]. In all 13 cases they found an absolute increase in the number of peripheral small branches caused by precapillary arterial vasodilatation rather than arteriovenous malformations. The vessels affected were between 100 to 500 μm in diameter and predominantly located in the lower lobes of the lungs. Also, the gross appearance on the surface of the lung revealed pleural vessels that were similar to cutaneous spider angiomata. This provided the first description

265

N. C. Nanda, R. Schlief and B. B. Goldberg (eds.), Advances in Echo Imaging Using Contrast Enhancement, Second Edition, 265–275.
© 1997 Kluwer Academic Publishers. Printed in Dubai.

of widespread 'intrapulmonary vascular dilatation' (IPVD), a phrase later coined by Krowka and Cortese in 1989 [5]. In the mid 1970s multiple case reports demonstrated hypoxemia in cirrhotics due to dilatation of pulmonary vessels [6–8]. Kennedy and Kundson in 1977 were the first to label this syndrome consisting of hypoxemia, hypocapnia, and orthodeoxia due to intrapulmonary vasodilatation in cirrhosis the hepatopulmonary syndrome (HPS) [9].

PATHOPHYSIOLOGY OF ARTERIAL HYPOXEMIA IN HPS

Two mechanisms have been proposed to explain how intrapulmonary vaso-dilatation causes arterial hypoxemia in the hepatopulmonary syndrome. These include ventilation-perfusion mismatching (V/Q mismatch) and diffusion-perfusion abnormalities.

V/Q mismatch

Dauod and colleagues initially suggested that the cause of gas exchange abnor-malities in patients with cirrhosis was impaired hypoxic pulmonary vaso-constriction [10]. They studied 10 cirrhotics and each patient failed to increase pulmonary pressures when inhaling 10% oxygen. Normally, the vasculature senses alveolar hypoxia and selectively vasoconstricts small vessels in that area. Dauod suggested that a failure in regulation of pulmonary blood flow led to capillary dilatation and gas exchange abnormalities in cirrhosis. These findings were not reproduced in another study [11] and led Rodriguez-Roisin et al. to use the multiple inert gases elimination technique (MIGET), a research tool that quantitatively analyzes the distribution of V/Q ratios, to directly evaluate V/Q ratios in cirrhotics [12]. Their study revealed that cirrhotics with normal pul-monary and systemic hemodynamics had normal vasoconstrictive responses to alveolar hypoxia. However, those cirrhotics with a hyperdynamic circulation had marked dilatation of their pulmonary vessels, V/Q mismatch and gas exchange abnormalities. Therefore, the primary abnormality in these patients was dilatation of the pulmonary capillary bed resulting in V/Q mismatch and an inability to vasoconstrict normally to alveolar hypoxia.

Diffusion-perfusion abnormalities

In 1976 Genovesi et al. suggested that the hypoxemia found in a patient with hereditary hemorrhagic telangiectasia and enlarged pulmonary capillaries was due to a diffusion limitation of oxygen from the alveolus to RBCs in dilated capillaries [13]. They proposed that RBCs in the central stream of dilated capillaries were too far from the alveolus to become saturated while breathing room air. If the patient was allowed to breathe 100% oxygen, the increased driving pressure of oxygen in the alveoli would increase the diffusion of oxygen into the capillary, with saturation of central RBCs and normalization of oxygenation. This physiological shunting was termed a diffusion-perfusion defect and the mechanism was extrapolated to explain the hypoxemia seen in cirrhotics with hepatopulmonary syndrome. Subsequent studies using the MIGET in hepatopulmonary syndrome have demonstrated that V/Q mismatch is the predominant mechanism of hypoxemia in mild to moderate disease. In more severe disease, diffusion-perfusion impairment becomes an important mechanism.

In summary, pulmonary vasodilatation observed in cirrhosis causes a physiologic rather than an anatomic shunt, which can be overcome by the administration of 100% O_2. V/Q mismatch appears to be the major mechanism of arterial hypoxemia in mild to moderate disease while diffusion-perfusion abnormalities become important in more severe disease.

PATHOPHYSIOLOGY OF INTRAPULMONARY VASODILATATION IN HPS

The hepatopulmonary syndrome occurs in approximately 15–20% of cirrhotics and its development appears to be independent of the type or severity of liver disease [14, 15]. Intrapulmonary vasodilatation without evidence of gas exchange abnormalities is more common and may be seen in up to 20–25% of patients [15]. To date, no specific pathophysiologic mechanisms have been identified. However, studies have focused on the possibility that there is a unique susceptibility of the pulmonary microcirculation in affected individuals. Preliminary work in an animal model of HPS [16] has demonstrated that pulmonary levels of vasodilatory mediators are increased in affected animals compared with controls. Continuing investigation using model systems will provide insight into the mechanisms of and treatment for intrapulmonary vasodilatation.

TREATMENT

No effective medical therapy exists. Agents investigated to date, including almitrine [17, 18], somatostatin [19–21] and indomethacin [22] have been unsuccessful. A case report by Caldwell et al. [23] and our preliminary data [24] have demonstrated improvement in gas exchange following intake of garlic, although the mechanism of action is unknown. Liver transplantation provides a surgical option for HPS: recent reports have demonstrated complete resolution of this syndrome after transplantation. The role of liver transplantation in the treatment of HPS, especially in the setting of pulmonary disease with compensated liver function, remains under investigation but some believe that HPS alone should be an indication for liver transplantation [25].

INTRAPULMONARY VASODILATATION AND DIAGNOSIS OF HPS

Gas exchange abnormalities must be accompanied by intrapulmonary vaso-dilatation to fulfill criteria for HPS. Although no gold standard diagnostic test for vasodilatation exists, contrast echocardiography and the lung perfusion scan are the modalities used to detect intrapulmonary 'shunts' or vasodilatation.

The use of contrast echocardiography was first reported in 1969 [26] and the disappearance of contrast material from blood during passage through the lungs was subsequently explained by Meltzer and colleagues [27]. Large micro-bubbles are sorted out or fragmented into smaller bubbles by a normal size pulmonary circulation ($<15\,\mu m$). These small bubbles have high internal pressures secondary to a very high surface tension and completely dissolve in approximately 190 ms. Since the transit time from the pulmonary circulation to the left atrium is at least 2 s the bubbles do not enter the left heart chambers.

In 1976, Shub and colleagues first demonstrated intrapulmonary right-to-left shunting in a patient with diffuse arteriovenous fistulae secondary to hereditary hemorrhagic telangiectasia with the use of M-mode echocardiography and indocyanine green contrast [28]. This report also documented the ability of contrast echocardiography to differentiate intrapulmonary from intracardiac shunting. Intrapulmonary shunting results in delayed (after at least 2 heart beats) visualization of contrast in the left heart chambers due to the transit time through the pulmonary circulation, whereas, immediate visualization is seen in patients with intracardiac shunts. Hind and Wong [29] were the first to use 2-D contrast echocardiography to detect intrapulmonary vasodilatation in a patient with HPS (Figure 1). The addition of transesophageal contrast echocardiography allows identification of contrast in the pulmonary veins entering the left atrium directly confirming intrapulmonary 'shunting' (Figure 2). Contrast echocardio-graphy screening studies [30, 31] and preliminary reports [32, 33] performed in patients undergoing evaluation for liver transplantation have demonstrated a prevalence of positive 2-D contrast echoes of 13–47%. However, many of these patients had normal gas exchange, demonstrating that a mild degree of intra-pulmonary vasodilatation can be present without causing arterial hypoxemia.

We have screened 40 consecutive outpatient cirrhotics using 2-D contrast echocardiography, lung perfusion scan (^{99}TcMAA) and arterial blood gases to determine the sensitivity and specificity of these modalities in detecting intra-pulmonary vasodilatation and diagnosing HPS [15]. The ^{99}TcMAA scan utilizes the intravenous injection of technetium-labeled macroaggregated albumin (particles ranging from $20-90\,\mu m$) which are trapped in the normal pulmonary microcirculation, but will pass through lungs showing capillary and precapillary dilatation and lodge in extrapulmonary sites [34]. This allows quantification of the extrapulmonary shunt fraction based on the relative radioactivity retained in the lung and shunted to extrapulmonary sites. A comparison of radioactivity in the lung and cerebrum, which has a relatively constant blood flow in cirrhotics, provides the most accurate means of calculating the shunt fraction (Figure 3).

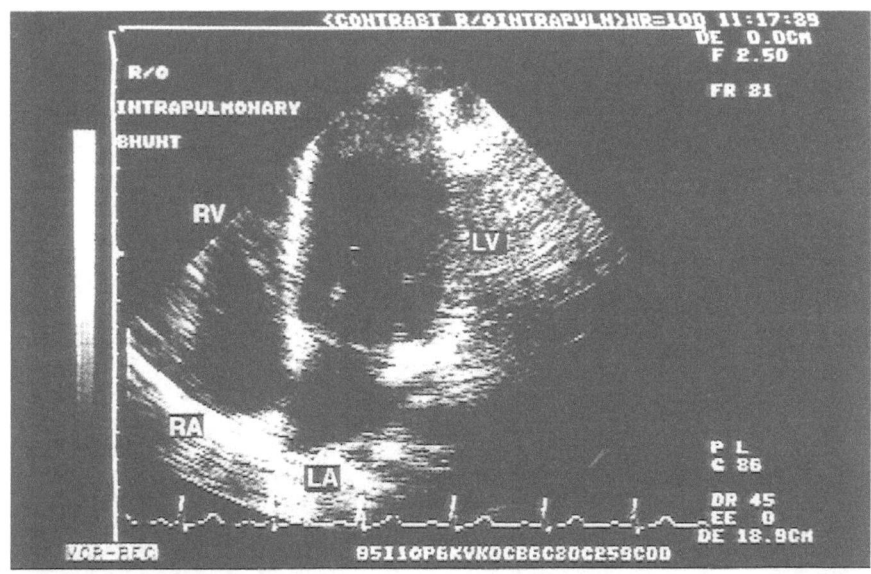

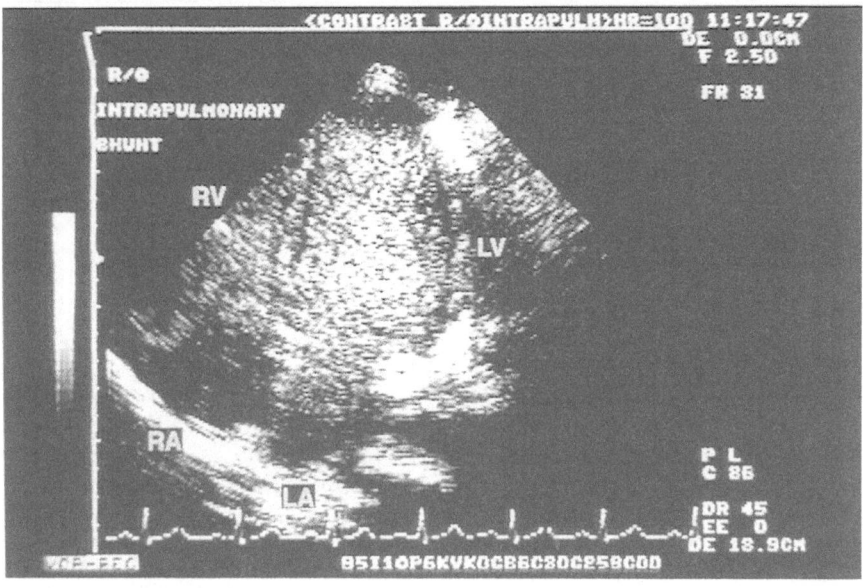

Figure 1. Transthoracic two-dimensional echocardiogram in a patient with hepato-pulmonary shunt. (A) Baseline apical four chamber view. (B) Intravenous injection of $10 \, cm^3$ normal saline resulted in opacification of both right and left sided chambers indicative of a right-to-left shunt. LA = left atrium; LV = left ventricle; RA = right atrium; RV = right ventricle.

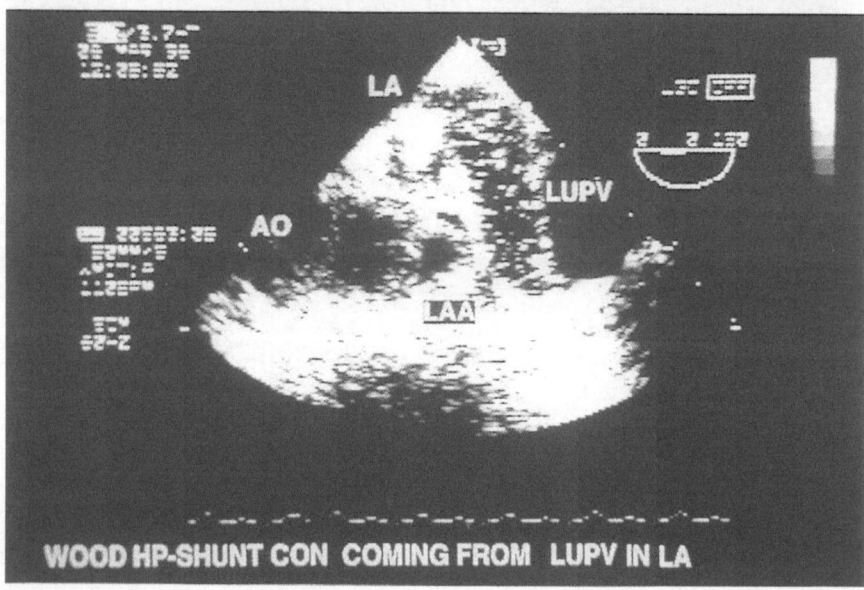

Figure 2. Transesophageal echocardiogram in another adult patient with hepatopulmonary shunt. (A) Baseline aortic short-axis view showing no contrast echoes in the left upper pulmonary vein (LUPV). (B) Intravenous injection of 10 cm³ normal saline resulted in the appearance of contrast echoes (arrow) in the LUPV indicative of a right to left shunt at the pulmonary capillary level. AV = aortic root; LA = left atrium; LAA = left atrial appendage.

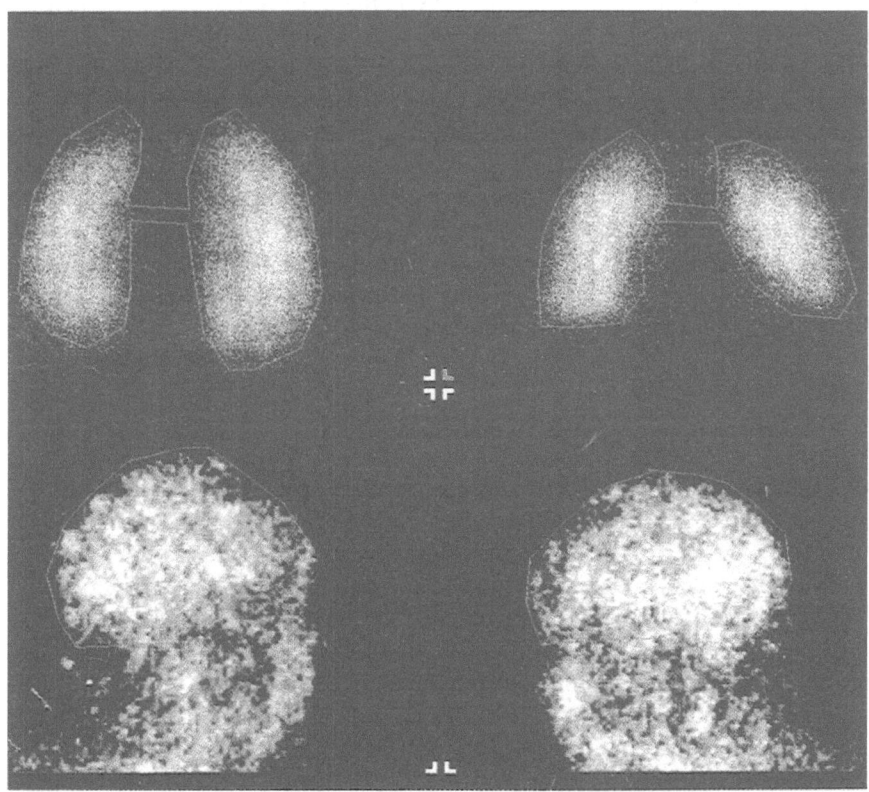

Figure 3. Lung perfusion scan showing views of the anterior and posterior lungs and lateral head. Regions of interest are outlined and demonstrate increased cerebral radioactive counts resulting in an extrapulmonary 'shunt fraction' of 35% (normal ≤6%).

Our study demonstrated that 15 of 40 (38%) cirrhotics studied had positive contrast echocardiograms. Seven of these 15 patients had gas exchange abnormalities and could be considered to have HPS (7/40; 17.5%). Three of these patients were readily diagnosed as having HPS because they were hypoxemic, had no concurrent cardiopulmonary disease, and each had a positive contrast echocardiogram and lung perfusion scan. The remaining four patients (three hypoxemic and one normoxemic with an elevated A-a gradient) had co-existing intrinsic lung disease and/or chest X-ray abnormalities complicating the diagnosis of HPS. In addition, each of these patients had a positive echocardiogram but a negative lung scan. Whether gas exchange abnormalities in these four patients were due to HPS or to intrinsic lung disease is difficult to discern. Eight patients with positive echocardiograms had normal lung scans and normal gas exchange. No patient had a positive lung scan and a negative contrast echocardiogram.

In summary, contrast echocardiography is the most useful screening test for detecting intrapulmonary vasodilatation and, therefore, the most sensitive screening test for diagnosing HPS. However, contrast echocardiography lacks specificity for the diagnosis of HPS because the majority of patients with positive contrast echoes are normoxemic. A negative contrast echocardiogram in cirrhotics rules out HPS. The lung scan appears to be a more specific test for the presence of HPS. It is positive only in hypoxemic cirrhotics who are readily diagnosed with HPS and was normal in a control population of hypoxemic patients with intrinsic lung disease and no liver disease. In hypoxemic cirrhotics with positive intrinsic lung disease, a positive lung perfusion scan suggests that arterial deoxygenation is due, at least in part, to HPS.

FUTURE PERSPECTIVES AND CONCLUSION

An unknown quantity and heterogeneous population of gaseous microbubbles is generated after agitating normal saline for contrast echocardiography. This limitation of saline contrast echocardiography prevents its use in determining the size of the pulmonary microcirculation and from quantifying a 'shunt fraction'. Preliminary work using custom-mixed microspheres (5.5–10 μm) of known size distribution and quantity has allowed sizing of the pulmonary microcirculation and quantification of shunting in an animal model of HPS [35]. A natural extension of this work would be to develop similar techniques in humans using known quantities of contrast materials ranging from 20 to 100 μm. These types of studies could determine whether there is a degree of intrapulmonary vasodilatation necessary to disturb gas exchange. Currently, echocontrast enhancers such as Albunex, Ultravist and Renograffin are smaller than the normal pulmonary microcirculation and will traverse the pulmonary bed, opacifying left heart chambers [36]. The development of larger and safe contrast agents which can be sized and quantified prior to injection will be of significant importance in advancing the use of contrast sonography in the diagnosis and assessment of severity of HPS. In order to quantify intrapulmonary shunting, a non-invasive means of measuring the size and number of contrast components bypassing the lung and entering the arterial circulation is needed. Transcranial Doppler sonography is such a diagnostic modality that is able to detect emboli within the cerebral circulation [37]. Ongoing studies in patients with HPS demonstrate that the number and density of microbubbles can be evaluated after contrast injection (Figure 4). Thus, if the size and number of contrast material injected were known, a quantification of pulmonary shunting and an assessment of the size of the microcirculation could be made. The total amount of emboli could quantify a shunt fraction and the density, which correlates with the size of the embolus [38], could be used as a surrogate to estimate the pulmonary bed size.

The recognition that the hepatopulmonary syndrome is more common than previously appreciated warrants investigators to determine why certain patients

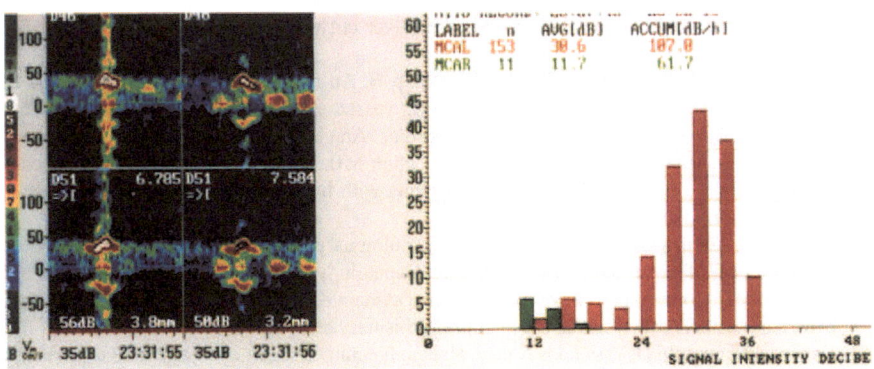

Figure 4. Transcranial Doppler sonography detecting contrast material in the right middle cerebral artery (left panel). Right panel demonstrates the number and size intensity of the emboli passing through the right middle cerebral artery.

develop HPS, its natural history, and the development of safe and efficacious medical therapy. Contrast sonography will undoubtedly play an important role in all of these aspects and will also provide important information regarding the pathophysiology of this syndrome.

REFERENCES

1. Fluckiger M. Vorkommen von trommelschagelformigen Fingerendphalangen ohne chronische Veranderungen an den Lungen oder am Herzen. Wien Med Wchnschr 1884;34:1457.
2. Rydell R, Hoffbauer FW. Multiple pulmonary arteriovenous fistulas in juvenile cirrhosis. Am J Med 1956;21:450.
3. Calabresi P, Abelmann WH. Portocaval and porto-pulmonary anastomoses in Lannec's cirrhosis and in heart failure. J Clin Invest 1957;36:1257–65.
4. Berthelot P, Walter JG, Sherlock S, Reid L. Arterial changes in the lungs in cirrhosis of the liver-lung spider nevi. N Engl J Med 1966;274:291–8.
5. Krowka MJ, Cortese DA. Pulmonary aspects of liver disease and liver transplantation. Clin Chest Med 1989;10:593–616.
6. Stanley NN, Williams AJ, Dewar CA et al. Hypoxia and hydrothorax in a case of liver cirrhosis: correlation of physio-logical, radiographic, scintigraphic and pathological findings. Thorax 1977;32:457.
7. Davis HH, Schwartz DJ, Lefrak SS, Susanan N, Schainker BA. Alveolar-capillary oxygen disequilibrium in hepatic cirrhosis. Chest 1978;73:507–11.
8. Williams A, Trewby P, Williams R, Reid L. Structural alterations to the pulmonary circulation in fulminant hepatic failure. Thorax 1979;34:447–53.
9. Kennedy TC, Knudson RJ. Exercise aggregated hypoxemia and orthodeoxia in cirrhosis. Chest 1977;72:305.
10. Daoud F, Reeves J, Schaefer JW. Failure of hypoxic pulmonary vasoconstriction in patients with liver cirrhosis. J Clin Invest 1972;51:1076–80.
11. Naeije R, Melot C, Hallemans R et al. Pulmonary hemodynamics in liver cirrhosis. Sem Resp Med. 1985;7:164.

274   *Gary A. Abrams et al.*

12. Rodriguez-Roisin R, Roca J, Augusti A, Mastai R, Wagner P, Bosch J. Gas exchange and pulmonary vascular reactivity in patients with liver cirrhosis. Am Rev Resp Dis 1987;135: 1085–92.
13. Genovisi MG, Tierney DF, Taplin GV, Eisenberg H. An intravenous radionuclide method to evaluate hypoxemia caused by abnormal alveolar vessels. Am Rev Resp Dis 1976;114:59–65.
14. Lange P, Stoller J. The hepatopulmonary syndrome. Ann Intern Med 1995;122:521–9.
15. Abrams GA, Jaffe C, Hoffer PB, Binder HJ, Fallon MB. Diagnostic utility of contrast echocardiography and lung perfusion scan in patients with hepatopulmonary syndrome. Gastroenterology 1995;109:1283–8.
16. Fallon MB, Abrams GA, Chen YF et al. Altered pulmonary expression of vasoactive mediators in a rat model of hepatopulmonary syndrome. Hepatology 1995;22:221A.
17. Krowka MJ, Cortese DA. Severe hypoxemia associated with liver disease: Mayo Clinic experience and the experimental use of almitrine bismesylate. Mayo Clin Proc 1987;62:164–73.
18. Nakos G, Evrenoglu D, Vassilakis N et al. Haemodynamics and gas exchange in liver cirrhosis: the effect of orally administered almitrine bismesylate. Resp Med 1993;87:93–8.
19. Krowka MJ, Dickson ER, Cortese DA. Hepatopulmonary syndrome: Clinical observations and lack of therapeutic response to somatostatin analogue. Chest 1993;104:515–21.
20. Schwartz SM, Pound DC. Hepatopulmonary syndrome: failure of response to somatostatin analogue. Gastroenterology 1992;102:882A.
21. Hobeika J, Houssin D, Bernard O et al. Orthotopic liver transplantation in children with chronic liver disease and severe hypoxemia. Transplantation 1994;57:224–8.
22. Shijo H, Sasaki H, Yuh K et al. Effects of indomethacin on hepatogenic pulmonary angiodysplasia. Chest 1991;99:1027–9.
23. Caldwell SH, Jeffers LJ, Narula OS et al. Ancient remedies revisited: does *Allium sativum* (garlic) palliate the hepatopulmonary syndrome? J Clin Gastroenterol 1992;15:248–50.
24. Abrams GA, Krowka MJ, Crapo RO, Fallon MB. Hepatopulmonary syndrome. Open label trial with *Allium sativum* (garlic). Hepatology 1995;22:167A.
25. Van Obbergh L, Carlier M, De Clety S et al. Liver transplantation and pulmonary gas exchange in hypoxemic children. Am Rev Resp Dis 1993;148:1408–10.
26. Gramiak R, Shah P, Kramer D. Ultrasound cardiography. Contrast studies in anatomy and function. Radiology 1968;92:939.
27. Meltzer RS, Tickner EG, Popp RL. Why do lungs clear ultrasonic contrast? Ultrasound Med Biol 1980;6:263–9.
28. Shub C, Tajik AJ, Seward JB, Dines DE. Detecting intrapulmonary right to left shunt with contrast echocardiography. Mayo Clin Proc 1976;51:81–4.
29. Hind C, Wong C. Detection of pulmonary arteriovenous fistulae in patient with cirrhosis by contrast 2D echocardiography. Gut 1981;22:1042–5.
30. Krowka MJ, Tajik AJ, Dickson ER, Wiesner RH, Cortese DA. Intrapulmonary vascular dilatations (IPVD) in liver transplant candidates. Screening by two-dimensional contrast-enhanced echocardiography. Chest 1990;97:1165–70.
31. Hopkins WE, Waggoner BA, Barzilai B. Frequency and significance of intrapulmonary right-to-left shunting in end-stage hepatic disease. Am J Cardiol 1992;70:516–9.
32. Rodriguez L, Lange P, Garcia M, Lombardo H, Stoller JK, Thomas JD. Positive echo-contrast does not correlate with physiologic shunt studies in patients with severe liver disease. J Am Coll Cardiol 1994:25A.
33. Kures PR, Soble JS, Nermann A et al. Frequency of intrapulmonary right to left shunt detected by contrast echocardiography in advanced liver disease (abstr). J Am Soc Echocardiogr 1994; 7:27E.
34. Robin ED, Horn B, Goris ML et al. Detection, quantitation and pathophysiology of lung spiders. Clin Res 1975;23:448.
35. Fallon M, Abrams G, Hou Z, Luo B. Intrapulmonary vasodilatation occurs in a rat model of hepatopulmonary syndrome. Gastroenterology 1996;110:A1188.

36. Nanda N. Echocontrast enhancers – how safe are they? In: Nanda N, Schlief R, eds. Advances in Echo Imaging using Contrast Enhancement. 1st edn. Dordrecht: Kluwer Academic Publishers, 1993:97–110.
37. Brucher R, Russell D. Embolus detection with Doppler ultrasonography – background and principles. In: Tegeler C, Babikian V, Gomez C, eds. Neurosonology. St. Louis: Mosby-Year Book Inc., 1996:231–4.
38. Russell D, Brucher R, Madden K. The intensity of the Doppler signal caused by arterial emboli depends on embolus size. In: Oka M, Reutern GV, Furuhata H, Kodaira K, eds. Recent Advances in Neurosonology: Proceedings of the Fourth Meeting of the Neurosonology Research Group of the World Federation of Neurology. Amsterdam: Exerpta Medica, 1992:57–60.

17. Identification of right-sided structures by contrast transesophageal echocardiography, with emphasis on patent foramen ovale

BIJOY K. KHANDHERIA

INTRODUCTION

The concept of contrast echocardiography, wherein any biologically compatible solution containing microbubbles of air, when injected into the circulation, makes the blood 'echogenic' is not a new one. This has been used extensively with both M-mode echocardiography and 2D echocardiography since the early 1970s [1–6]. The applications of contrast echocardiography include structure identification, diagnosis or exclusion of intracardiac shunt, diagnosis of complex congenital heart disease, quantitation of valvular regurgitation, myocardial perfusion and improved quantitation of the left ventricle including delineation of the walls [2, 3, 6–21]. Although both transthoracic and transesophageal echocardiography are commonly used, much work in contrast echocardiography deals with transthoracic echocardiography. However, the same information could be extrapolated for use in transesophageal echocardiography.

This chapter describes the use of contrast echocardiography in identification of the right-sided structures with special emphasis of shunting across the foramen ovale, particularly with respect to transesophageal echocardiography.

METHODOLOGICAL ASPECTS

The predominant cause of echocardiographic contrast is the visualization of air present in the injectate or the injecting apparatus. A rapid injection of a solution containing microbubbles through a peripheral vein will yield the best result insofar as the right sided structures are concerned. If the injectate crosses the pulmonary bed, the contrast effect is seen in the left side of the heart. This is a subject of intense investigation and will be discussed in subsequent chapters.

In order to perform peripheral venous contrast injection, a large bore (16 or 18 gauge) intravenous cannula is inserted into the antecubital vein, or into a hand vein if no antecubital vein is available. The left arm is preferred in order

N. C. Nanda, R. Schlief and B. B. Goldberg (eds.), Advances in Echo Imaging Using Contrast Enhancement, Second Edition, 277–288.
© 1997 *Kluwer Academic Publishers.*

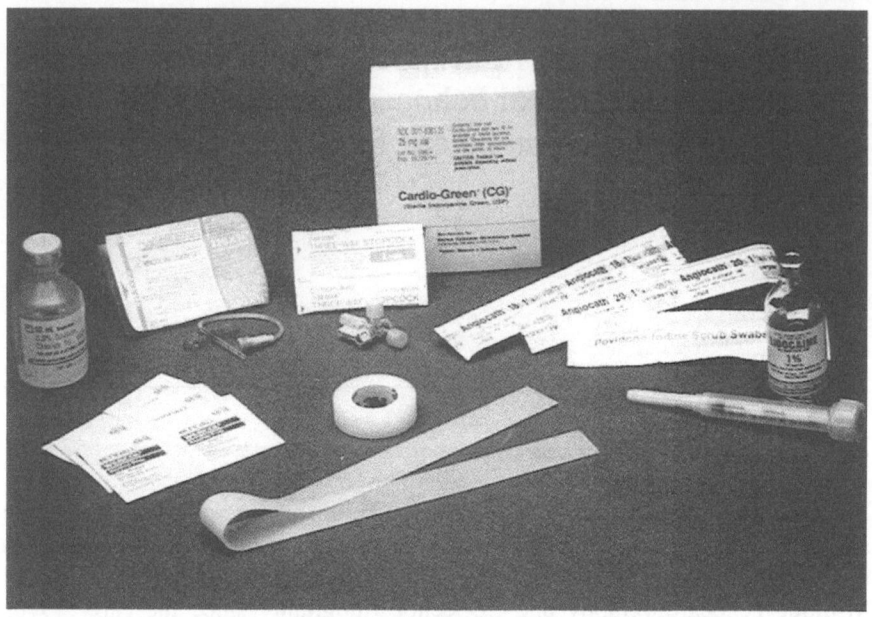

Figure 1. Contrast injection set-up used at our laboratory. Cardio green (indocyanine green) or agitated saline mixed with the patient's own blood are the preferred contrast agents used in our laboratory for identification of right-sided cardiac structures, visualization of right to left shunt, and negative echocontrast for the visualization of left-to-right shunts at the atrial level, in conjunction with transesophageal echocardiography (TEE) or transthoracic echocardiography (TTE). Shown in the figure is the normal saline (extreme left) that is used to reconstitute the cardio green powder, infusion set, three-way stopcock, 20 gauge angiocath, syringe, tourniquet, alcohol swab, adhesive tape and lidocaine solution for local use.

not to miss a persistent left superior vena cava, which usually drains into the coronary sinus. Contrast agents that can be used include 5–10 ml of agitated saline solution. The best results of agitation are obtained with the use of two syringes attached by a two-way stopcock. This agitated solution of saline is rapidly injected into the peripheral vein, and the examiner watches for the appearance of microbubbles in the right side of the heart. Dense opacification can be obtained if the saline is mixed with 2–3 ml of venous blood; this can be withdrawn with the syringe first, and then agitated using the double syringe technique and a three-way stopcock. Alternative agents include indocyanine green dye (1 mg/ml in adult patients and 0.6–1.25 mg/ml in younger patients weighing <30 kg). Indocyanine green is available as powder that needs to be reconstituted with normal saline. The set-up required for performing a peripheral venous contrast injection is shown in Figure 1. Agitated saline or indocyanine green are our methods of choice for identification of right-sided structures and detection of intracardiac shunts.

Dextrose 5% sonicated albumin [22] can also be used for peripheral venous injections. In the USA sonicated albumin (Albunex) injected via a peripheral vein has been approved for certain indications.

Echocardiographic studies are best recorded during held respiration, as well as with the release phase of the Valsalva maneuver: the latter is particularly useful for visualization of a right-to-left shunt through a patent foramen ovale.

IDENTIFICATION OF NORMAL RIGHT-SIDED STRUCTURES

Contrast echocardiography was originally used to identify cardiac structures on M-mode echocardiograms and later on transthoracic 2D echocardiograms [23–28]. Contrast echocardiography can be used to identify right-sided cardiac structures such as the superior vena cava, right atrium, Eustachian valve, inferior vena cava and right ventricle during transesophageal echocardiography. The most common use of contrast echocardiography is for determining the presence or absence of intracardiac shunt across the atrial septum in patients being referred for evaluation of a suspected cardiac source of embolism. The best views that afford visualization of these structures are the four-chamber views, basal short-axis views and longitudinal views obtained with the biplane or multiplane TEE transducer showing the long axis of the superior and inferior vena cava as well as the atrial septum (Figures 2, 3) [41–43]. Upon injection of peripheral contrast via the antecubital vein, the superior vena cava, right atrium and right atrial appendage will become opacified, in that order. This is best visualized in the long-axis view shown in Figure 3. It also allows identification of structures such as Eustachian valve and inferior vena cava, where there will be a negative contrast effect due to blood streaming into the right atrium from the inferior vena cava, and preventing opacification of the right atrium at this level. The Eustachian valve is at the inferior vena cava–right atrial junction and can be delineated well at this level. The opacification of the right atrium also serves to highlight the atrial septum and define the surface of the atrial septum. Other structures that can be identified with this injection include the right ventricular outflow tract as the contrast makes it way to this area. The best view for visualization of this is the right ventricular outflow view in the longitudinal plane, or it can be seen as a continuum of images using multiplane transducers (Figure 3) [43].

IDENTIFICATION OF ABNORMAL RIGHT-SIDED CARDIAC STRUCTURES

Common abnormalities that can be diagnosed with the aid of peripheral contrast injections include the superior vena cava entering the right atrium through a dilated coronary sinus. A contrast injection into the left arm can help delineate this abnormality, with either transthoracic echocardiography (TTE) [27, 30] or

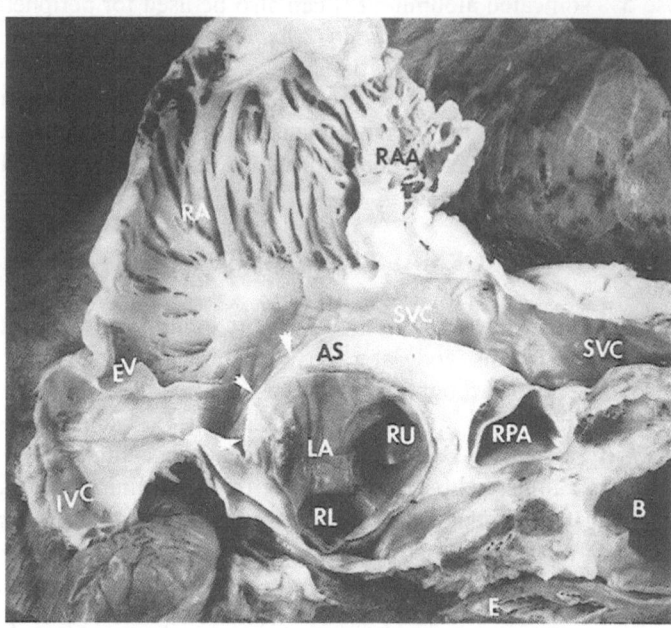

Figure 2. Anatomic specimen cut along the longitudinal plane, cavae view. The esophagus (E) is located posteriorly. Immediately anterior to the esophagus is the left atrium (LA) with the entry of the right upper (RU) and right lower (RL) pulmonary veins, respectively. The fatty limbus of the atrial septum (AS) is seen, as is the valve of the fossa ovalis (arrowheads). The Eustachian valve (EV) is located at the junction of the inferior vena cava (IVC). The right atrium (RA) and the right atrial appendage (RAA) are the anterior most structures. The superior vena cava (SVC) is also seen in its long axis, as it enters the right atrium. B=bronchus, RPA=right pulmonary artery.

transesophageal echocardiography (TEE). The left superior vena cava is best visualized using the longitudinal plane view of the pulmonary vein or the frontal four-chamber view with the coronary sinus.

Atrial septal aneurysm can be well delineated using contrast injection in conjunction with TEE. The two most common views used for this purpose are the frontal four-chamber view and the primary longitudinal plane view with the cavae. Atrial septal aneurysms have been associated with small fenestrations, which result in the right-to-left or left-to-right shunts. These are easily detected using contrast TEE. The mass of echocontrast homogeneously opacifies the right atrium and outlines the right atrial border of the atrial septum, even in those patients in whom echocardiographic dropout may simulate atrial septal defect. Hence, contrast echocardiography is an excellent tool to complement standard 2D imaging for exclusion or diagnosis of atrial septal defects, either alone or in association with atrial septal aneurysm.

There are reports that atrial septal aneurysm may be associated with embolic

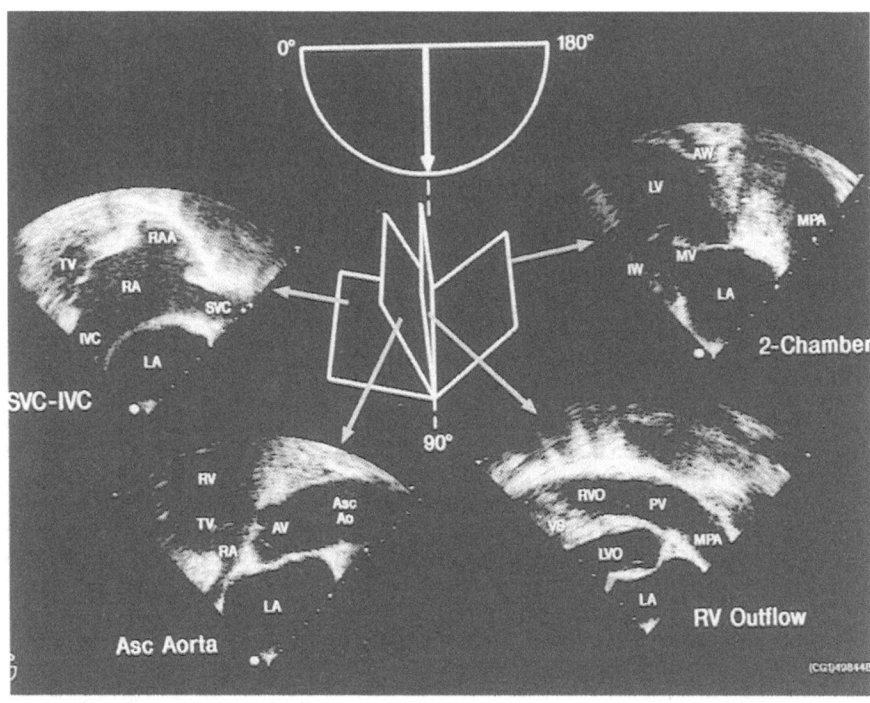

Figure 3. Multiplane examination showing views that can be obtained for visualization of the long axis of the superior vena cava (SVC), inferior vena cava (IVC) as well as other views. RA = right atrium, RAA = right atrial appendage, TV = tricuspid valve, RV = right ventricle, LA = left atrium, Asc Ao = ascending aorta, LVO = left ventricular outflow, VS = ventricular septum, RVO = right ventricular outflow, PV = pulmonary valve, MPA = main pulmonary artery, IW = inferior wall, MV = mitral valve.

events [31, 32]. The numbers of patients in these reports are small; and further confirmation is needed. In addition, some of the studies have major flaws in the methodology, and it is unwise to draw strong conclusions about the association between embolic strokes and atrial septal aneurysm at this time.

Patent foramen ovale (Figures 4–7) can also be detected using contrast TEE. This subject has been most extensively studied [2, 29, 33–37]. Contrast echo-cardiography can detect shunts even as small as 3–5% and is more sensitive than the indicator dilution technique or oximetry. The appearance of one or more microbubbles in the left atrium following injection into the peripheral circulation has been considered to be diagnostic of right-to-left shunt at the atrial level. However, we feel that one must be able to visualize 3–5 microbubbles in the left atrium within three cardiac cycles for a result to be considered a true positive. To improve the sensitivity as well as specificity of contrast TEE in the detection of patent foramen ovale and the accompanying right-to-left shunt, it is

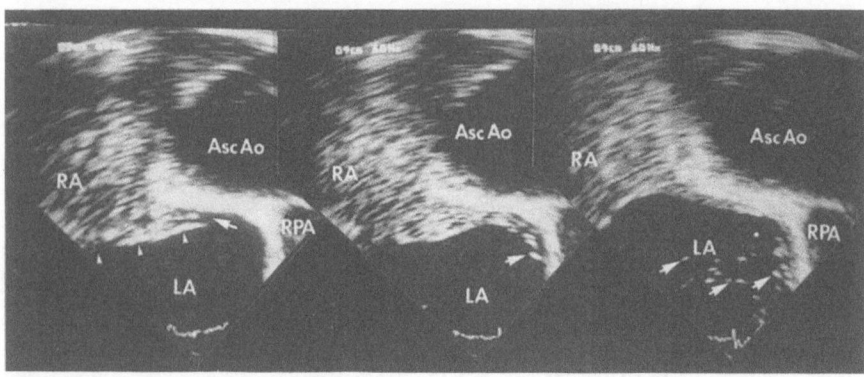

Figure 4. Patent foramen ovale. Still frames of 2D image and contrast TEE image from a patient with patent foramen ovale. Left panel: Longitudinal plane, ascending aorta view showing the left atrium (LA) atrial septum with an echo dropout in the region of the fossa ovalis (arrow), right atrium (RA), right ventricle (RV), tricuspid valve (TV), and the ascending aorta (Ao). Middle panel: Following contrast injection, there is an appearance of microbubbles (arrow) in the left atrium through the patent foramen ovale, indicating and confirming the presence of a patent foramen ovale. Right panel: Microbubbles persist in the left atrium following injection (arrows).

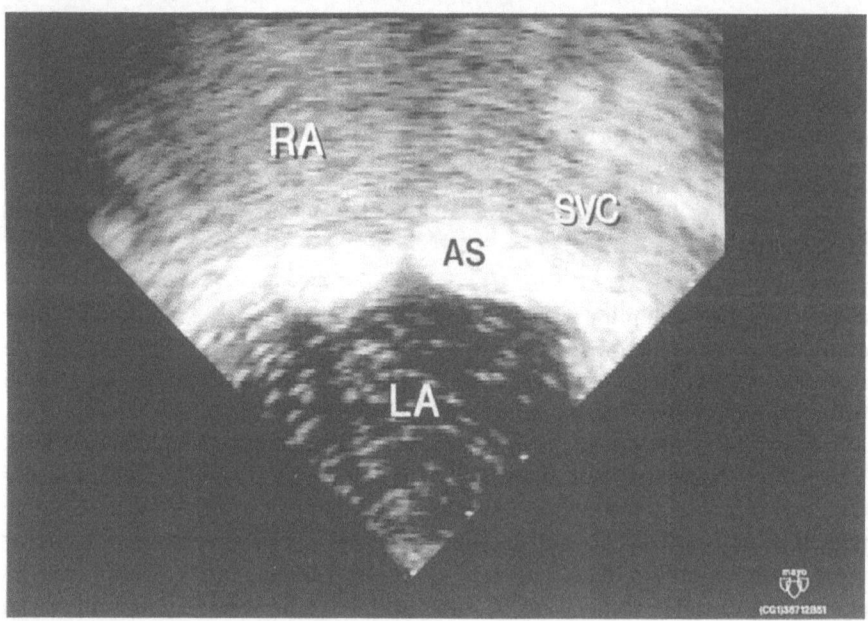

Figure 5. Patent foramen ovale. Still frame of a 2D image from a multiplane TEE examination showing the presence of a Grade III shunt across the atrial septum. Long-axis view shows the superior vena cava (SVC) filled with contrast, right atrium (RA) also showing dense opacification, atrial septum (AS) and opacification of the left atrium (LA), which occurred within the first three cardiac cycles. More than 10 microbubbles can be seen in the left atrium.

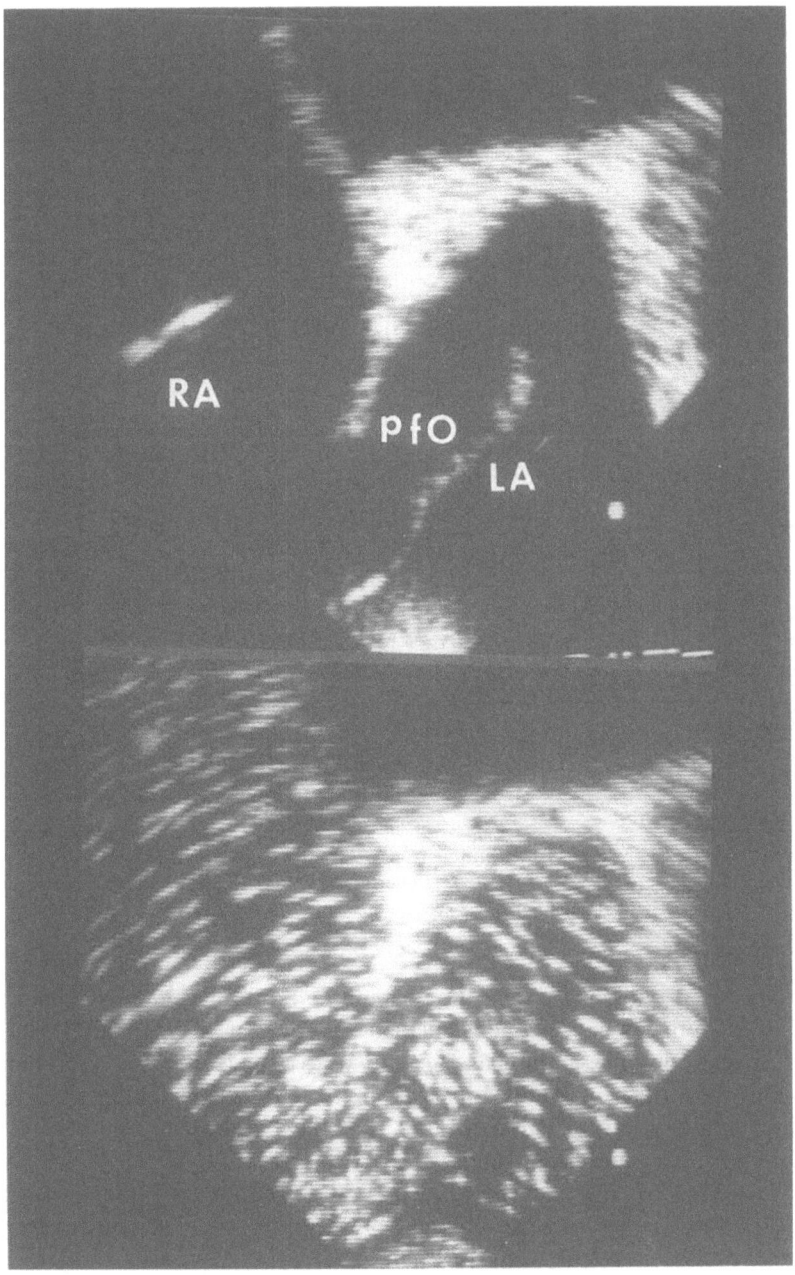

Figure 6. Patent foramen ovale (pfo). Top panel: Still frame of a 2D image showing a separation of the valve of the fossa ovalis indicating an anatomic abnormality. Right atrium (RA) and left atrium (LA) are also shown in this view. Bottom panel: following injection of agitated saline into the left antecubital view, microbubbles appear in the left atrium within the first cardiac cycle and dense opacification of the left atrium is evident.

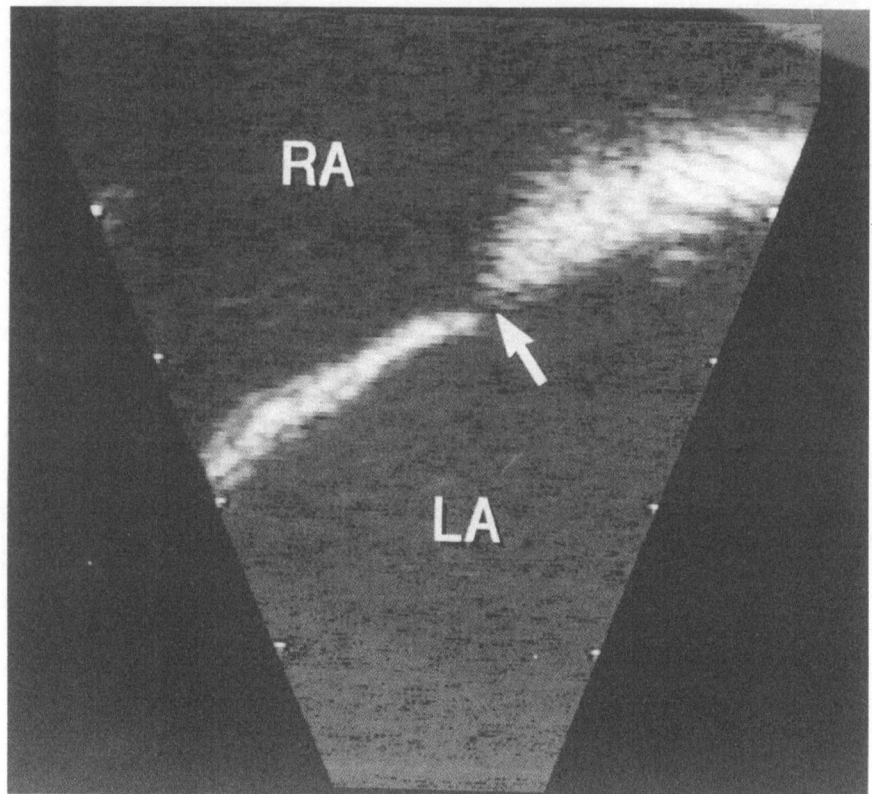

Figure 7. Patent foramen ovale. Still frame from a 2D image using a 7.5 MHz transducer shows the defect which is 5 mm in size (arrow). Higher resolution imaging transducers provide better visualization of the anatomic defect, as is seen in this case. RA = right atrium, LA = left atrium.

recommended that contrast injection should be performed in conjunction with the Valsalva maneuver or Mueller maneuver. Injection of the contrast must be performed during the strain phase of the Valsalva maneuver, and microbubbles crossing the atrial septum and appearing in the left atrium should be looked for following the release of Valsalva. This, we believe, represents the best method of assessing for right-to-left shunt via a patent foramen ovale.

The degree of shunt can be graded semi-quantitatively using an arbitrary scale of the number of microbubbles appearing on the left side within five cardiac cycles of complete opacification of the right atrium following injection of the contrast medium. Grade 1 would be 5 microbubbles or less, and clearly indicates a trivial amount of shunt, perhaps of no clinical significance. Grade II would be 5–10 microbubbles and Grade III would be > 10 microbubbles.

The pattern and appearance time of contrast in the left atrium is influenced by the level of the shunt and the relative pressures in the cardiac chambers. False-positive findings include spontaneous echocontrast in the region of the pulmonary veins as they enter the left atrium, as well as the noise in the near field.

Visualization of the left-to-right shunt is also possible by negative echocontrast: this is diagnostic for the left-to-right shunt seen in patients with atrial septal defects. When a left-to-right shunt exists at the atrial level contrast medium fills the right atrium; however, the non-contrasted, non-opacified left atrial blood passing through the defect may be visualized because it superimposes a negative contrast or a contrast-free area that outlines the defect. This negative contrast effect occurs during end-systole and early diastole. A problem that one may encounter is the masking from the very bright reflections from the positive contrast around the negative contrast area.

Contrast echocardiography has been used for the diagnosis of tricuspid regurgitation. However, with the advent of color flow imaging this has been relegated to a non-existent role. Contrast injection may be used to enhance color Doppler signals. When it is imperative to obtain right ventricular systolic pressure, and the continuous wave Doppler signal is of a poor quality, contrast injection will enhance the signal quality.

COMPLEX CONGENITAL HEART DISEASE

It is possible to determine the position of the great vessels and their ventriculo-atrial connections using contrast TEE. Together with the ability to detect and localize intracardiac shunts, contrast TEE offers an important advantage for diagnosis of complex congenital heart disease. Characteristic contrast echocardiographic patterns have been described in univentricular hearts with either one or two atrioventricular valves, tricuspid atresia, truncus arteriosus, straddling tricuspid valve and double inlet left ventricle [4–6, 38, 39]. Ventricular septal defect can occur as an isolated lesion or as part of a complex cardiac defect. Right-to-left shunting may occur through a ventricular septal defect when the peak right ventricular systolic pressure reaches 50% or more of the systemic pressure. This can be demonstrated with peripheral contrast injection, wherein the flow from right to left is seen as a large bolus, predominantly in diastole. Pulmonary hypertension must be present for the demonstration of this phenomenon. More commonly, one can see negative echocontrast, indicating left-to-right shunt. This modality is of less value since color flow imaging can easily depict ventricular septal defects, as well as outlining the flow patterns.

MISCELLANEOUS APPLICATIONS

Contrast injection into the pericardial space can be used to localize the needle during pericardiocentesis [9]. It has been used to diagnose hemiazygous veins as well as azygous veins.

FUTURE DIRECTIONS

The future of contrast echocardiography lies in the development of newer contrast agents that will cross the circulation and appear in the left side, permitting studies of perfusion, volumes and endocardial border definition via a peripheral injection. TEE, with its potential for excellent image quality, affords the best route to study these newer agents. For the present, contrast TEE finds limited application, especially in those patients referred for source of embolism to detect interatrial shunting, and for those patients with atrial septal defects, anomalous pulmonary venous drainage.

REFERENCES

1. Nanda NC, Gramiak R, Manning JA. Echocardiography of the tricuspid valve in congenital left ventricular-right atrial communication. Circulation 1975;51:1268–72.
2. Shub C, Tajik AJ, Seward JB, Dines DE. Detecting intrapulmonary right-to-left shunt with contrast echocardiography. Observations in a patient with diffuse pulmonary arteriovenous fistulas. Mayo Clin Proc 1976;5:81–4.
3. Seward JB, Tajik AJ, Hagler DJ, Ritter DG. Peripheral venous contrast echocardiography. Am J Cardiol 1977;39:202–12.
4. Seward JB, Tajik AJ, Hagler DJ, Ritter DG. Contrast echocardiography in single or common ventricle. Circulation 1977;55:513–9.
5. Sahn DJ, Allen HD, George W, Mason M, Goldberg SJ. The utility of contrast echocardiographic techniques in the care of critically ill infants with cardiac and pulmonary disease. Circulation 1977;56:959–68.
6. Tajik AJ, Seward JB. Contrast echocardiography. Cardiovasc Clin 1978;9:317–41.
7. Crouse LJ. Sonicated serum albumin in contrast echocardiography: improved segmental wall motion depiction and implications for stress echocardiography. Am J Cardiol 1992;69: 42H–45H.
8. Curtius JM, Thyssen M, Breurer HW, Loogen F. Doppler versus contrast echocardiography for diagnosis of tricuspid regurgitation. Am J Cardiol 1985;56:333–6.
9. Chandraratna PA, Reid CL, Nimalasuriya A, Kawanishi D, Rahimtoola SH. Application of 2-dimensional contrast studies during pericardiocentesis. Am J Cardiol 1983;52:1120–2.
10. Cheirif J, Zoghbi WA, Raizner AE et al. Assessment of myocardial perfusion in humans by contrast echocardiography. I. Evaluation of regional coronary reserve by peak contrast intensity. J Am Coll Cardiol 1988;11:735–43.
11. Cheirif J, Zoghbi WA, Bolli R et al. Assessment of regional myocardial perfusion by contrast echocardiography. II. Detection of changes in transmural and subendocardial perfusion during dipyridamole-induced hyperemia in a model of critical coronary stenosis. J Am Coll Cardiol 1989;14:1555–65.
12. Byrd BF, O'Kelly BF, Schiller NB. Contrast echocardiography enhances tricuspid but not mitral regurgitation. Clin Cardiol 1991;14(Suppl 5):V10–V14.
13. Bell EF. Contrast echocardiography in diagnosis of PDA (letter). J Pediatr 1981;99:668–9.
14. Amano K, Sakamoto T, Hada Y, Yamaguchi T et al. Detection of tricuspid regurgitation by contrast echocardiography. Jpn Circ J 1982;46:395–401.
15. Feinstein SB, Lang RM, Dick C et al. Contrast echocardiography during coronary arteriography in humans: perfusion and anatomic studies. J Am Coll Cardiol 1988;11:59–65.
16. Feinstein SB. Myocardial perfusion imaging: contrast echocardiography today and tomorrow (editorial). J Am Coll Cardiol 1986;8:251–3.

17. Feinstein SB, Ong K, Staniloff HM et al. Myocardial contrast echocardiography: examination of intracoronary injections, micro bubble diameters, and video-intensity decay. Am J Physiol Imaging 1986;1:12–18.
18. Kaul S, Jayaweera AR, Glasheen WP et al. Myocardial contrast echocardiography and the transmural distribution of flow: a critical appraisal during myocardial ischemia not associated with infarction. J Am Coll Cardiol 1992;20:9–16.
19. Kaul S. Clinical applications of myocardial contrast echocardiography. Am J Cardiol 1992; 69:46H–55H.
20. Keidar S, Grenadier E, Binenboim C, Palant A. Transient right to left atrial shunt detected by contrast echocardiography in the acute stage of pulmonary embolism. J Clin Ultrasound 1984; 12:417–9.
21. Keller MW, Feinstein SB, Watson DD. Successful left ventricular opacification following peripheral venous injection of sonicated contrast agent: an experimental evaluation. Am Heart J 1987;114:570–5.
22. Feinstein SB, Cheirif J, Ten CFJ et al. Safety and efficacy of a new transpulmonary ultrasound contrast agent: initial multicenter clinical results. J Am Coll Cardiol 1990;16:316–24.
23. Valdes CLM, Pieroni DR, Roland JM, Shematek JP. Recognition of residual postoperative shunts by contrast echocardiographic techniques. Circulation 1977;55:148–52.
24. Pritchard DA, Maloney JD, Seward JB et al. Peripheral arteriovenous fistula. Detection by contrast echocardiography. Mayo Clin Proc 1977;52:186–90.
25. Serruys PW, Hagemeijer F, Bom AH, Roelandt J. Contrast echocardiography in two dimensions and in real time. 2. Clinical applications. Arch Mal Coeur Vaiss 1978;71:611–26.
26. Hernandez A, Strauss AW, McKnight R, Hartmann AF. Diagnosis of pulmonary arteriovenous fistula by contrast echocardiography. J Pediatr 1978;93:258–61.
27. Snider AR, Ports TA, Silverman NH. Venous anomalies of the coronary sinus: detection by M-mode, two-dimensional and contrast echocardiography. Circulation 1979;60:721–7.
28. Mortera C, Hunter S, Tynan M. Contrast echocardiography and the suprasternal approach in infants and children. Eur J Cardiol 1979;9:437–54.
29. Cheng TO. Contrast echocardiography in cerebral vascular accidents (letter). Presse Med 1991;20:268.
30. Stewart JA, Fraker TDJ, Slosky DA, Wise NK, Kisslo JA. Detection of persistent left superior vena cava by two-dimensional contrast echocardiography. J Clin Ultrasound 1979;7:357–60.
31. Belkin RN, Hurwitz BJ, Kisslo J. Atrial septal aneurysm: association with cerebrovascular and peripheral embolic events. Stroke 1987;18:856–62.
32. Pearson AC, Nagelhout D, Castello R, Gomez CR, Labovitz AJ. Atrial septal aneurysm and stroke: a transesophageal echocardiographic study. J Am Coll Cardiol 1991;18:1223–9.
33. Lechat P, Guggiari M, Lascault G, Fuschiardi M et al. Detection by contrast ultrasonography of patent foramen ovale before neurosurgery. Presse Med 1986;15:1409–10.
34. Pieroni DR, Varghese PJ, Freedom RM, Rowe RD. The sensitivity of contrast echocardiography in detecting intracardiac shunts. Catheter Cardiovasc Diagn 1979;5:19–29.
35. Pieroni DR, Valdes CLM. Atrial right-to-left shunt in infants with respiratory and cardiac distress but without congenital heart disease. Demonstration by contrast echocardiography. Pediatr Cardiol 1982;2:1–5.
36. Sahasakul Y, Chaithiraphan S, Jootar P, Prachuabmoh C. Accuracy of peripheral venous contrast echocardiography in diagnosis of atrial septal defect. J Med Assoc Thai 1986;69:518–24.
37. Weyman AE, Wann LS, Caldwell RL et al. Negative contrast echocardiography: a new method for detecting left-to-right shunts. Circulation 1979;59:498–505.
38. Knight DB, Yu VY. Contrast echocardiographic assessment of the neonatal ductus arteriosus. Arch Dis Child 1986;61:484–8.
39. Kitoh N, Takao A, Satomi G et al. Transposition of the great arteries associated with the straddling tricuspid valve diagnosed preoperatively by two-dimension echocardiography: report of a case. J Cardiogr 1982;12:753–62.

288   *Bijoy K. Khandheria*

40. Khandheria BK, Oh J. Transesophageal echocardiography: state-of-the-art and future directions. Am J Cardiol 1992;69:61H–75H.
41. Seward JB, Khandheria BK, Oh JK et al. Transesophageal echocardiography: technique, anatomic correlations, implementation, and clinical applications. Mayo Clin Proc 1988;63: 649–80.
42. Seward JB, Khandheria BK, Edwards WD et al. Biplanar transesophageal echocardiography: anatomic correlations, image orientation, and clinical applications. Mayo Clin Proc 1990;65: 1193–213.
43. Seward JB, Khandheria BK, Edwards WD et al. Multiplane transesophageal echocardiography: image orientation, examination technique, anatomic correlations and clinical applications. Mayo Clin Proc 1993;68:523–51.

18. Contrast echo-enhancement during transesophageal echocardiography: indications and clinical benefits

DIRK HAUSMANN, ANDREAS MÜGGE,
HENNING KÜHN & WERNER G. DANIEL

INTRODUCTION

During recent years, transesophageal echocardiography (TEE) has evolved as a major adjunct to conventional transthoracic echocardiography. Although TEE is a semi-invasive procedure, carries some discomfort for the patient and has a small, clinical risk [1], it has a clear advantage over transthoracic echocardiography in selected patients. Due to higher transducer frequencies, TEE provides improved image quality for cardiac structures with close proximity to the esophagus. In patients with poor transthoracic image quality (those with emphysema, chest wall abnormalities, prior heart surgery or on artificial ventilation) TEE is also a clinically helpful alternative to the transthoracic technique. Finally, TEE currently offers the only available echocardiographic window during open heart surgery.

As is the case with transthoracic echocardiography, contrast enhancement has been applied during the transesophageal approach. Current applications of contrast-enhanced TEE include detection of patent foramen ovale, color-coded and pulsed wave Doppler examination of coronary blood flow, visualization of myocardial contrast and enhancement of intracardial structures.

DETECTION OF PATENT FORAMEN OVALE (PFO)

Anatomic and pathophysiologic considerations

In the fetal heart the foramen ovale allows unidirectional blood flow from the right to the left atrium. The slit-like foramen ovale is bordered by the limbus of the fossa ovalis (septum secundum) and is guarded by the septum primum. A PFO differs substantially from an atrial septum defect: whereas an atrial septum defect allows both left-to-right and right-to-left shunting, a PFO has a valve-like mechanism which usually allows only right-to-left shunting. The physiologic

289

N. C. Nanda, R. Schlief and B. B. Goldberg (eds.), *Advances in Echo Imaging Using Contrast Enhancement*, *Second Edition*, 289–308.
© 1997 *Kluwer Academic Publishers.*

function of the foramen ovale is conduction of the bloodstream from the umbilical veins to the systemic circulation; thus, right-to-left shunting consists primarily of inferior vena cava blood flow [2, 3]. After birth, the septum primum is pressed against the septum secundum and in most cases, permanent closure of the foramen ovale is achieved by fibrous adhesions. However, the foramen ovale remains patent (PFO) in about one-third of subjects. During 965 autopsies, Hagen et al. [4] found a PFO in 33–36% of otherwise normal humans during the first three decades of life, the width of which ranged from 1 to 10 mm (mean 3–4 mm).

Right-to-left atrial shunting is possible in normal subjects, although mean pressure at rest is lower in the right compared to the left atrium [5]. There are several potential explanations. First, during respiratory maneuvers, such as Valsalva [5–10] or coughing [7–11], central venous return is temporarily reduced; after these maneuvers, venous return is increased and right atrial pressure rises. This may allow interatrial shunting despite normal pressures under resting conditions. These maneuvers might also play a role during paradoxic thromboembolism because thrombotic material in the lower body veins may be dislodged in the presence of an opened window to the systemic circulation. Second, the bloodstream from the lower part of the body is preferentially directed into the region of the foramen ovale [2]. This might be another explanation for interatrial shunting despite normal intracardiac pressures. The direction of blood flow from the inferior vena cava also makes clear why contrast studies with injections into arm veins may be false-negative in some cases with proven PFO [12]. In the presence of abnormal masses in the right atrium, the bloodstream might be directed even more precisely towards the PFO [13]. Third, right-to-left-atrial shunting through a PFO can occur after an increase in right atrial pressure, such as right ventricular infarction [14], tricuspid regurgitation [15], pericardial tamponade [16], pulmonary arterial hypertension due to chronic pulmonary embolism [17], pulmonary vascular disease [18] or chronic obstructive lung disease [19], pulmonary embolism [20], positive endexpiratory airway pressure (PEEP) [21], or rapid ascent to high altitude [22]. Fourth, although mean pressure is lower in the right than in the left atrium, the pressure difference may be reversed transiently during normal respiration. At end-systole, right atrial pressure may be higher than left atrial pressure as shown by simultaneous right and left atrial pressure recordings [5, 6].

Detection of PFO

Patency of the foramen ovale cannot be detected by physical examination, electrocardiogram or roentgenogram; indicator dilution curves performed during resting conditions are also normal in patients with PFO. During right heart catheterization a PFO may only be diagnosed when the catheter incidentally crosses from the right to the left atrium [5, 7, 23].

Transthoracic echocardiography

Since no reliable non-invasive test for interatrial shunting through a PFO was available before the introduction of 2-dimensional transthoracic echocardiography, the overall clinical relevance of paradoxical embolism was unclear and only illustrated by occasional case reports. However, contrast transthoracic echocardiography has become the initial diagnostic tool to test for interatrial right-to-left shunting [10, 24] in patients with otherwise unexplained stroke [8, 9, 25–28].

The accuracy of contrast transthoracic echocardiography depends on the type of contrast material, the site of contrast injection and the respiratory maneuvers for provocation of the right-to-left shunt. Different agents are currently used for peripheral venous injection: agitated saline (with or without some drops of blood), agitated oxypolygelatine or galactose-based ultrasound contrast agents (Echovist, Schering AG, Berlin). During clinical routine, contrast injection is usually performed via the antecubital vein. However, Gin et al. [12] detected a PFO more frequently following vein delivery of contrast compared with upper vein delivery (33% vs. 13%; $p < 0.001$). In some patients, interatrial shunting may occur during normal breathing. However, increasing right atrial pressure by respiratory maneuvers significantly improves the sensitivity of contrast echocardiography [6–12, 23, 29–31]. The incidence of interatrial shunting detected by contrast transthoracic echocardiography in subjects without suspected paradoxical embolism ranges between 5 and 13% during normal breathing and is 7–24% during Valsalva maneuver. These numbers are still significantly lower than the 24–36% PFO incidence found at autopsy [4]. The relatively poor sensitivity of this technique has recently been confirmed by studies using contrast transthoracic echocardiography in patients with PFO proven by cardiac catheterization and/or surgery [23]. Finally, when color Doppler imaging is used during transthoracic echocardiography, the interatrial shunt can be visualized only in extremely rare cases [32, 33].

Transesophageal echocardiography (TEE)

TEE is currently considered as the 'gold standard' for in vivo diagnosis of inter-atrial right-to-left shunts through a PFO. Peripheral venous injection of contrast agents is performed as described for contrast transthoracic echocardiography and respiratory maneuvers are also used to provoke a transient increase in right atrial pressure (Figure 1). In contrast to the transthoracic approach, TEE allows optimal visualization of the atrial septum and the region of the fossa ovalis in virtually all patients under basal conditions and also during respiratory maneuvers. Furthermore, higher transducer frequencies can be applied leading to further improvement in image quality that allows detection of even small numbers of contrast particles crossing the septum.

The incidence of a PFO in patients without suspected cardiac source of embolism ranges between 3 and 39% [33–38] compared with 7–24% by contrast

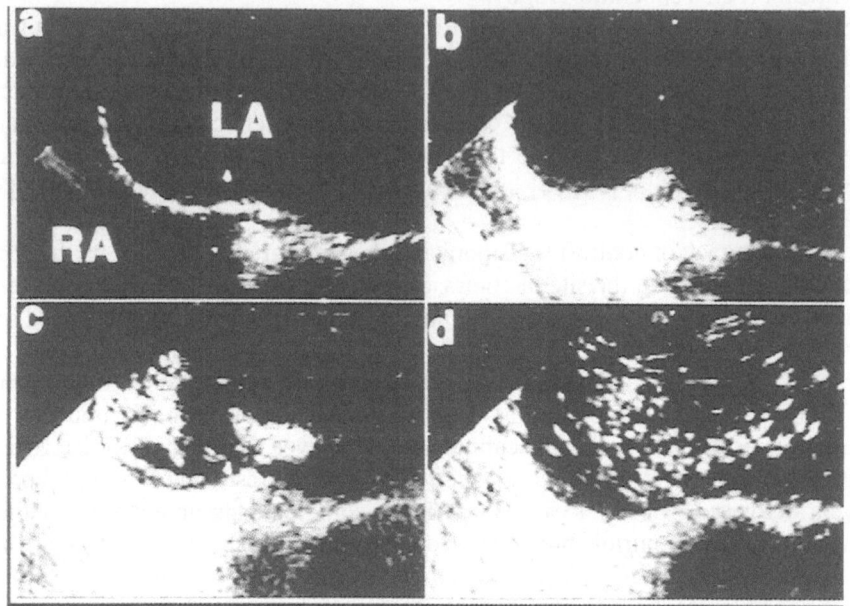

Figure 1. Right-to-left atrial contrast shunt through a patent foramen ovale detected by transesophageal echocardiography (longitudinal plane). (a) Interatrial septum before contrast injection; (b) contrast material appearing in the right atrium (RA) after i.v. injection; (c) right-to-left atrial contrast shunt through the patent foramen ovale after release from Valsalva maneuver; (d) left atrium (LA) filled with contrast bubbles.

transthoracic echocardiography. Since this is in the range of numbers derived from autopsy studies, it might be speculated that nearly all patients with a PFO can be identified using contrast TEE with respiratory maneuver. The accuracy of contrast TEE has recently been proven by studies using anatomic confirmation of the echocardiographic findings [23]. All patients with proven PFO (by cardiac catheterization and/or surgery) were correctly identified using contrast TEE. The additional cardiac axes provided by biplane and multiplane TEE allow a unique and individualized anatomic view of the region of the foramen ovale. In particular, the septum primum and secundum can clearly be differentiated and the valve-like structure of the PFO can be visualized. The direction of the bloodstream from the inferior vena cava to the region of the foramen ovale can also be recognized. Contrast TEE evaluation of a PFO is associated with only a minor intra- and interobserver variability [39].

In our experience, the longitudinal TEE axis is preferred for visualization of the region of the PFO. In nearly all patients, the septum primum and secundum can be differentiated using this view. Two phenomena can usually be observed in patients with PFO: the septum primum and secundum separate after release

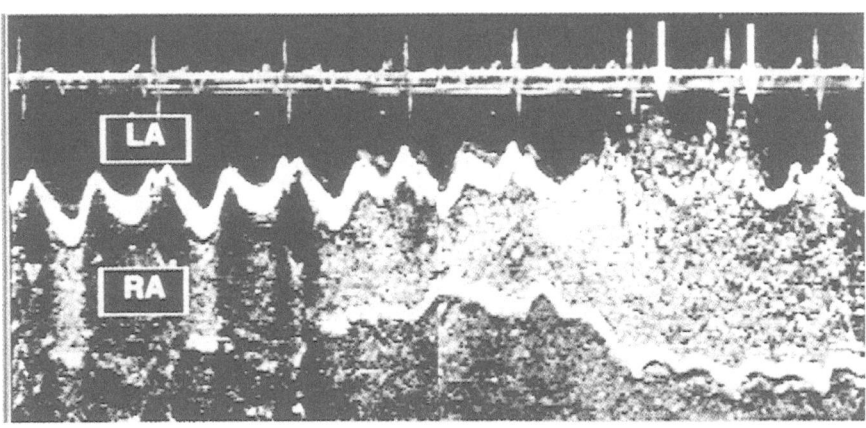

Figure 2. Right-to-left contrast shunt through a patent foramen ovale. M-mode recording during transesophageal echocardiography of the region of the interatrial septum. After i.v. injection of contrast material, bubbles appear in the right atrium (RA). After release from Valsalva maneuver contrast bubbles shunt into the left atrium (LA) (arrows).

from the Valsalva maneuver; this can be seen even without contrast application. Second, transit of contrast bubbles through the interatrial septum can be imaged after the Valsalva maneuver (Figure 2). A PFO is considered to be present when at least three contrast bubbles appear on the left atrial side during the first three cardiac cycles after release from the Valsalva maneuver. Transpulmonary passage of contrast material can be observed in many patients, even when only right-heart contrast agents are used and this can be misinterpreted as indicating the presence of a PFO (Figure 3). Important clues in the differentiation between transseptal and transpulmonary contrast passage are the time course of contrast appearance in the left atrium (early vs. late), the size of the bubbles (large bubbles vs. small, dust-like contrast) and the site of first appearance of the bubbles (region of the PFO vs. pulmonary veins). When a clear diagnosis cannot be achieved using these criteria, contrast injections with and without respiratory maneuvers should be repeated while the pulmonary veins rather than the region of the PFO are visualized by TEE (Figure 2). With this strategy, a definite diagnosis of the presence or absence of a PFO is possible in virtually all patients.

Due to clear visualization of the interatrial septum, detection of interatrial shunting using color Doppler ultrasound during TEE is possible in many patients with PFO [32, 33, 37, 40, 41]. Nevertheless, recognition of the typical color Doppler shunt signal in the left atrium during TEE is more difficult than recognition of contrast bubbles. Evaluation of PFO by this technique requires careful frame-by-frame analysis of the videotape and therefore contrast echocardiography is mostly preferred during clinical routine [42].

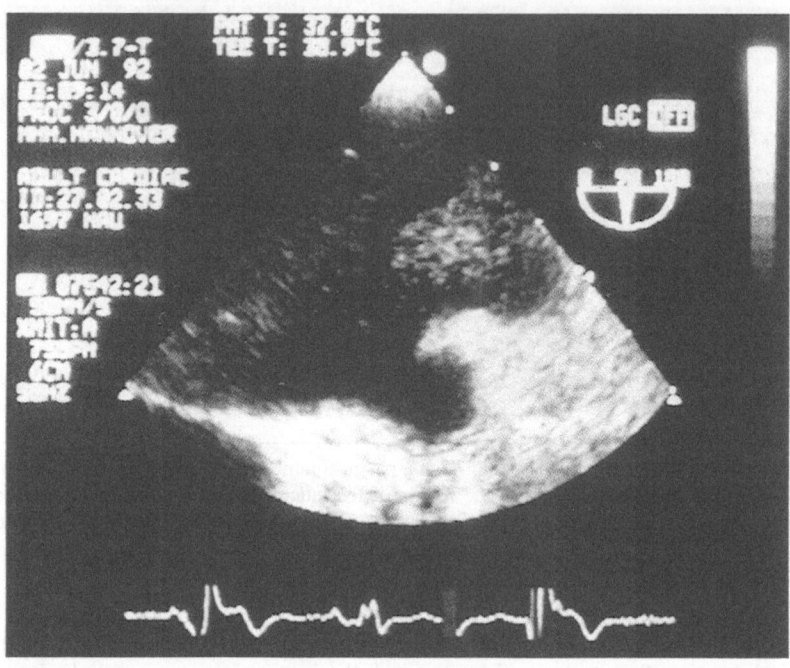

Figure 3. Transpulmonary contrast passage. Visualization of the region of the left upper pulmonary vein and the left atrial appendage by transesophageal echocardiography (longitudinal plane). After i.v. injection of Gelifundol (oxypolygelatine), contrast bubbles appear in the pulmonary vein. Compared with trans-septal contrast passage, transpulmonary shunting results in later appearance of the contrast bubbles and a fine, dust-like contrast.

Transcranial Doppler ultrasound (TCD)

TCD examination of the middle cerebral artery during intravenous contrast injection is a non-invasive technique for the diagnosis of intracardiac right-to-left shunting [43–47]. This technique is more available to the neurologist and has clearly a higher sensitivity for detection of an intracardiac right-to-left shunt than contrast transthoracic echocardiography [44, 45]. It has therefore been proposed as an initial diagnostic tool in the evaluation of cardiac source of embolism in patients with otherwise unexplained stroke. In two recent studies comparing contrast transthoracic echocardiography and TCD, paradoxical shunting was found in 22–26% of patients by transthoracic echocardiography and in 41–50% by TCD [44, 46]. Although the sensitivity of contrast TCD seems to be comparable to contrast TEE [45–47], a significant number of false-positive results can occur with TCD [47]. Since transpulmonary passage of contrast bubbles depends on the type of contrast agent used, the incidence of false-positive TCD studies varies with different agents [48].

Klotzsch et al. [49] compared contrast TEE and TCD for detection of PFO.

Using contrast TEE as a gold standard the sensitivity of TCD was 91%, specificity 94%, and overall accuracy 93%. However, in four of 111 patients TCD studies produced false-positive results. Jauss et al. [50] performed TCD and TEE simultaneously in 50 patients. When contrast TEE was considered as the gold standard, TCD with Valsalva maneuver showed a sensitivity of 93% and a specificity of 100% for detection of right-to-left atrial shunting. DiTullo et al. [51] studied patients with ischemic stroke or transient ischemic attack. Sensitivity of TCD was 68% and specificity was 100% compared with contrast TEE. The patients with false-negative transcranial studies had only very small right-to-left shunts detected by TEE.

In conclusion, TCD may be used as a first screening method in patients with otherwise unexplained embolic events. Patients with positive results as well as patients with negative TCD studies but additional clinical signs for paradoxical embolism (pulmonary embolism, deep venous thrombosis) require further TEE examination.

Clinical significance of PFO

Paradoxical thromboembolism

Since the first description by Cohnheim in 1877 [52], several reports of paradoxical thromboembolism have been published [53, 54]. Paradoxic embolism can be considered as proven when a thrombus crossing the interatrial septum is found. Initially, these observations were restricted to autopsy [55, 56]. With the introduction of echocardiography, detection of these rare events became possible during life (Figure 4) [54, 56–61]. Whether paradoxical embolism is a relevant pathological mechanism in patients with suspected cardiogenic embolism is not known. However, two independent groups reported in 1988 that the incidence of a PFO as detected by transthoracic contrast echocardiography is increased in young patients (<45 and <55 years, respectively) with otherwise unexplained ischemic stroke [46, 47]. The authors concluded that paradoxical embolism may be a more common phenomenon than previously suggested, at least in young patients. These results were subsequently confirmed by others [26, 28], and may also hold true for elderly patients [26].

Using contrast TEE as the superior technique for the detection of PFO, similar results have been reported, i.e. an increased incidence of PFO in patients with otherwise unexplained arterial embolism [33, 34, 36, 63]. In a recent study from our group [33], contrast TEE revealed a PFO in nine (50%) of 18 patients with stroke and age <40 years compared to an incidence of 22% in a control group without embolic events or to an incidence of 21% in a second control group of patients with embolic events but other potential cardiac sources of embolism (e.g. left atrial thrombus emitral and aortic valve disease). It should be pointed out that contrast TEE may not only be more reliable in the detection

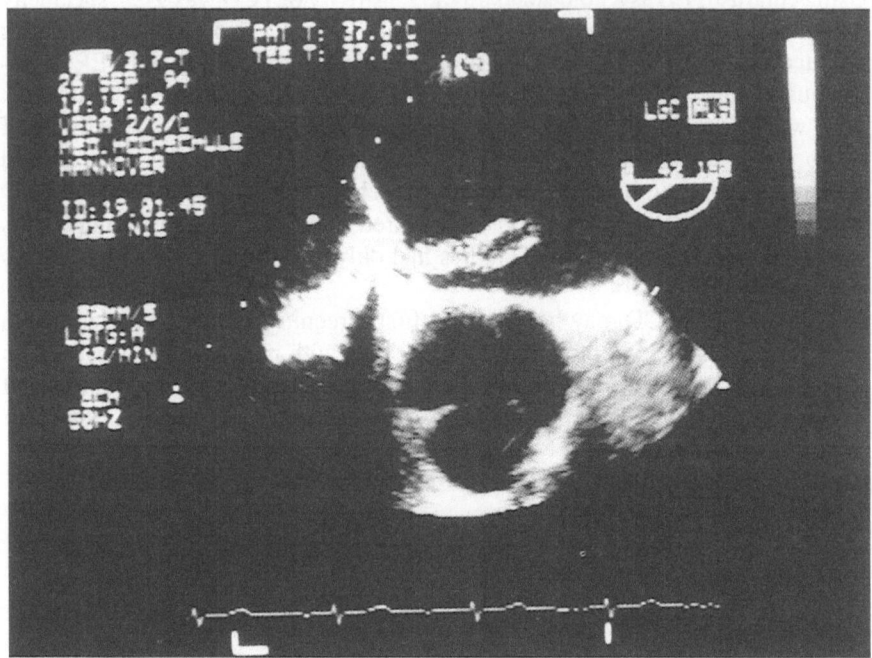

Figure 4. Impending paradoxical embolism. Multiplane transesophageal echocardiography demonstrates an overriding thrombus in a patent foramen ovale. The mobile ends of the thrombus can be seen in the left and right atrium.

of PFO than transthoracic echocardiography, but is also more sensitive in the detection of other potential sources of cardiogenic embolism which could be incidently associated with a PFO.

Arterial hypoxemia
Venous shunting through a PFO can cause significant arterial hypoxemia in the absence of cyanotic heart disease. Interatrial shunting in these cases usually requires the existence of elevated right atrial pressures under normal resting conditions. Provocation of interatrial shunting during respiratory maneuvers is not sufficient to induce arterial desaturation. Right ventricular infarction, severe tricuspid regurgitation, pericardial tamponade, pulmonary arterial hypertension, pulmonary embolism and pulmonary stenosis are possible underlying disorders. In clinical practice, acute episodes of hypoxemia associated with increased right heart pressures can be misinterpreted as pulmonary embolism. It is important to remember that hypoxemia in patients with chronic or acute pulmonary hypertension may not only be caused by the lung disease, but may also be aggravated by significant interatrial shunting.

Paradoxical air embolism
During neurosurgery in the sitting position, 30–40% of patients show venous air embolism with paradoxical air embolism occurring in 6–12%; the inflow from the superior vena cava might play a role, but paradoxical embolism is relatively rare under these circumstances [2, 63]. Pre- or intraoperative contrast echocardiography to exclude a PFO is recommended in patients undergoing these procedures [64]. Guggiari et al. [31] found a PFO in 22 (10%) of 218 patients before neurosurgery; all patients without detectable PFO subsequently underwent uncomplicated surgery. Venous air embolism is a rare but, in the presence of a PFO, a potentially serious complication also during cesarean section; a body position with the right atrium below the uterus aims to reduce the incidence of such events. Venous air embolism seems to occur frequently during liver transplantation; Ellis et al. detected paradoxical embolism of air and/or thrombi in two (13%) of 16 patients during transplantation [65].

Decompression (divers') sickness
During decompression, bubbles of nitrogen gas are liberated from solution in the body tissue and appear in the venous blood. In the presence of intracardiac right-to-left shunts, bubbles may reach the nervous system and cause the neurological symptoms of decompression sickness [30, 66, 67]. Right-to-left atrial shunting via a PFO has been reported in up to 61% of divers with this syndrome, suggesting that a PFO may be important within the pathomechanism of this disorder [30, 66].

Paradoxical tissue embolism
Every tissue entering the venous circulation may potentially cross to the left heart in the presence of a PFO. Previous reports described systemic embolism caused by paradoxical embolism of fat [68], tumor cells [69] and transitional cells desquamated into open periurethral veins [70]. Although reports of such events are extremely rare, they conclusively prove the pathomechanism of paradoxical embolism when tissue is histologically identified in the systemic circulation.

Diagnosis of paradoxical embolism

Identification of a potential window for paradoxical embolism (e.g. PFO) in patients with systemic embolism does not necessarily indicate that paradoxical embolism has occurred. Even when no other cause or risk factor for the embolic event can be found it often remains unclear whether paradoxical thromboembolism is the underlying cause. In clinical practice, patients are often referred for echocardiographic assessment of a potential cardiac source of embolism some time after the acute ischemic event. Thus, extracardiac signs of paradoxic embolism may already have disappeared when the diagnosis of PFO is made. Furthermore, small venous thrombi may go clinically undetected [71]. Ranoux et al. [62] have shown that a PFO is frequently found in patients with stroke (47% of stroke patients <55 years had PFO), but other clinical signs suggestive for paradoxical embolism (deep venous thrombosis, pulmonary embolism, Valsalva maneuver during the event) were not more frequent in these patients than in stroke patients without PFO.

Due to the relatively low sensitivity of clinical signs, objective assessment of venous thrombosis is mandatory in all patients with a suspected cardiac source of embolism and PFO. The discrepancy between clinical and objective signs of venous thrombosis has recently been shown by Stollberger et al. [72] in a study of 49 patients with PFO and clinically suspected embolic events. Among the 42 patients undergoing venographic study, 24 (57%) had venous thrombosis, only six of which had been suspected clinically. Among the 17 patients undergoing venographic examination early after the ischemic events (<7 days), 15 (88%) had venous thrombosis. More patients with venous thrombosis than without thrombosis had a history of previous thromboembolism. The implications of this important study are twofold: objective signs of venous thrombosis should be evaluated and these examinations should be performed early in the course of the disease [72].

Relationship of PFO characteristics to the risk of paradoxical embolism

Recent studies indicated that, in addition to the presence or absence of a PFO, other functional and morphologic characteristics of the PFO should also be evaluated by TEE [8, 34, 51, 73, 74]. According to these initial observations, small PFOs are considered to indicate a lower risk for paradoxical embolism. However, before these PFO characteristics can be used for clinical decision making, further prospective studies are required.

The size of PFOs measured during autopsy shows a wide range between 1 and 19 mm [2]. Using balloon sizing in patients prior to transcatheter PFO closure, Bridges et al. [75] compared the diameter of the stretched PFO with the predicted PFO size based on age-adjusted autopsy data. In 67% of their patients, the actual PFO size was at least 2 standard deviations larger than the predicted mean size for the corresponding age group. In 40 patients with otherwise unexplained stroke, Webster et al. [8], found a severe right-to-left shunt (>20 bubbles) in 20%, a moderate shunt (6–20 bubbles) by contrast TTE in 20%, and a small shunt (1–5 bubbles) in 10% of the patients, compared with 0%, 5% and 10% in normal controls. Similarly, van Camp and co-workers [34] found a significant difference in incidence of PFO between 29 patients with unexplained stroke and 28 normal controls only if severe shunts were considered (six of 29 vs. none of 28 patients; $p<0.05$). In patients with ischemic stroke or transient ischemia, DiTullo et al. [51] found that false-negative contrast transthoracic echocardiography was associated with small right-to-left shunts (as defined by contrast TEE). Contrast transthoracic echocardiography was more sensitive in patients with cryptogenic stroke than in patients with stroke of determined origin. Therefore, patients with cryptogenic stroke (and possibly paradoxical embolism) may be those with larger, more easily detectable shunts.

In a recent study using contrast TEE, we compared three groups of patients with PFO: those with a high likelihood of paradoxical embolism, patients with an unexplained embolic event but with a low likelihood of paradoxic embolism (no additional signs of deep venous thrombosis or associated pulmonary embolism), and those without a history of embolic events [74]. Severe contrast shunting and wide opening of the PFO were seen significantly more often in patients with a high likelihood of paradoxical embolism than in the two other groups. We suggested that the likelihood for paradoxical embolism is low in patients with a small PFO [73]. Homma et al. [76] performed contrast TEE in 74 consecutive patients referred for evaluation of ischemic stroke. The PFO dimension (separation of septum primum from secundum) was significantly larger in patients with cryptogenic stroke than in patients with an identifiable cause of stroke. The number of microbubbles was also greater in those with cryptogenic stroke.

In addition to the size of the PFO, other anatomic characteristics may play a role in the patency of the foramen ovale and the risk of paradoxical embolism. Schneider et al. [74] observed Chiari's network in 2% of 1436 consecutive patients undergoing TEE. The presence of a network was significantly associated with right-to-left atrial shunting, atrial septal aneurysm and a history of suspected cardiac source of arterial embolism. These authors concluded that the Chiari's network may favor persistence of a patent foramen ovale, formation of an atrial septal aneurysm and may also facilitate paradoxical embolism.

ASSESSMENT OF CORONARY BLOOD FLOW

Coronary flow reserve, i.e. the difference between maximal and baseline flow, is a valuable indicator for the functional significance of coronary stenoses. Measurement of coronary flow reserve, however, usually relies on invasive techniques or cost-intensive techniques such as positron emission imaging. Recent studies have shown that TEE allows visualization of proximal portion of epicardial coronary arteries [77]. In most patients, the left main stem, the proximal portion of the left anterior descending and the circumflex artery can be imaged. Using color Doppler flow mapping the coronary artery blood flow can be visualized and coronary blood flow velocities can be measured by pulsed wave Doppler techniques [78]. Initial clinical studies have shown that measurement of coronary flow reserve by TEE is a feasible, safe and rapid approach [79].

For evaluation of coronary blood flow, the TEE transducer should be located at the level of the aortic root. The origin of the left main coronary artery can be visualized in virtually all patients. As the next step during the examination, the bifurcation of the left main coronary artery should be imaged using color Doppler flow mapping. Minor adjustments of the image plane and rotation angles (if multiplane TEE is used) should then be performed to visualize the further course of the left anterior descending artery and the circumflex artery. The color Doppler signal of the proximal portion of the vessel should clearly be seen in the image plane during diastole (when most of the coronary blood flow occurs). Finally, the pulsed wave sample volume should be placed in the proximal portion of the vessel to measure flow velocities.

Major limitations of TEE examination of coronary blood flow velocities are the precise localization of the sample volume in relation to the coronary vessel and the complex movements of the vessel in relation to the transducer (systolic–diastolic, respiratory). Most important, Doppler assessment of coronary flow velocities by TEE is hampered by a low signal to noise ratio. An attempt to overcome these limitations is contrast enhancement of Doppler flow signals. Previous studies have shown that intravenous contrast injections produce significant improvements in Doppler flow signals in the right heart [80]. Recently, sonographic contrast agents that permit transit of the pulmonary capillary bed ('left-heart' contrast agents) have been developed. The safety and efficacy of transpulmonary echocontrast agents have been analyzed in various studies using transthoracic echocardiography [81–85]. These studies have shown that these agents improve regional wall motion analysis in the left ventricle during intravenous injection [82–84] Bibra et al. [85] clearly showed that SHU 508A, a saccharide-based transpulmonary echocontrast agent, enhances both spectral and color Doppler signals in the left heart. In this study using transthoracic echocardiography, the increase in signal to noise ratio after contrast application improved the assessment of aortic stenosis, pulmonary venous flow and mitral regurgitation.

Enhancement of left heart Doppler flow signals by contrast agents may also facilitate the assessment of coronary flow velocities by TEE. Iliceto et al. [86] performed TEE Doppler studies of the left coronary artery in 35 patients. SHU 508A, a new, saccharide-based lung-crossing contrast agent, was injected i.v. and used for contrast-enhancement of coronary flow. In 105 coronary segments (left main, left anterior descending and circumflex artery), Doppler flow mapping studies were optimal in 11% before and 67% after contrast injection ($p < 0.0001$); pulsed wave Doppler measurements were optimal in 23% before and 66% after contrast injection ($p < 0.0001$). No adverse effects of the contrast material were observed. The duration of the contrast effects was sufficient (> 100 s) to allow careful Doppler interrogation of coronary flow. SHU 508A was injected in 200 mg/ml and 300 mg/ml concentrations (5 and 10 ml bolus, respectively). A higher proportion of optimal Doppler recordings was obtained with higher concentrations; however, the 10 ml bolus resulted in excessive enhancement of color Doppler signals in some cases [86].

Aggarwal et al. [87] performed TEE evaluation of coronary blood flow velocities in 10 patients and compared the findings with those of coronary angiography. SHU 508A was injected intravenously at 200 and 300 mg/ml concentrations (5 and 10 ml bolus) during TEE imaging. With contrast application, color flow signals in the left main coronary artery were improved in all patients. Using this approach the absence of significant angiographic vessel narrowings was correctly identified in all patients. Five of six angiographically proven stenoses (>50% narrowing) in the left anterior descending or circumflex artery were correctly diagnosed by detecting a localized area of color flow aliasing and acceleration during TEE imaging after contrast injection. Kozakova et al. [88] also used SHU 508A injections during TEE to enhance the color Doppler and pulsed wave signal in coronary arteries. After contrast injection, the Doppler signals were enhanced for approximately 100 s, and both length and diameter of color-coded flow signals increased significantly in the left main, left anterior descending and left circumflex coronary artery. During pulsed wave Doppler measurements, systolic and diastolic velocities increased significantly.

ASSESSMENT OF MYOCARDIAL PERFUSION

Contrast enhancement of the myocardium during echocardiography is used successfully to delineate myocardial areas at risk. Both experimental models of coronary occlusion [89] and clinical studies of myocardial ischemia [90] have shown the feasibility of this approach. The area at risk defined by myocardial contrast echocardiography clearly correlates with the ultimate infarct size in animal models of prolonged coronary occlusion [91].

Although TEE uses ideal transducer frequencies and an optimal acoustic window for in vivo visualization of myocardial contrast, it can only be used when image quality is significantly reduced during transthoracic echocardiography. Therefore the experience of TEE imaging of myocardial contrast is limited. Voci et al. [92] used sonicated albumin microbubbles before and during dipyridamole infusion in 11 patients. During i.v. contrast injection mean signal intensity in the myocardium increased significantly. During dipyridamole infusion there was a further increase in signal intensity, probably reflecting increased myocardial blood flow. In another study, TEE imaging of the left ventricular myocardium was performed after i.v. injection of Albunex (0.08 and 0.22 mg/kg); measurements were performed by analysis of video images and of radiofrequency data [93]. Contrast injection resulted in visible opacification of the left ventricular cavity but not of the myocardium; radiofrequency analysis, however, facilitated detection of myocardial contrast enhancement. Again, since the myocardium can be adequately imaged in the majority of patients using transthoracic echocardiography, the utility of TEE seems to be limited for this application of contrast enhancement.

On the other hand, TEE is an ideal tool for intraoperative evaluation of myocardial structures. When contrast agents are used, myocardial perfusion can be studied with this approach. Adequate myocardial perfusion is a main factor for immediate clinical outcome after heart surgery and is maintained with antegrade or retrograde perfusion with cardioplegic solution. Voci et al. [94] used contrast TEE to image myocardial perfusion during coronary bypass surgery in 40 patients. The goal of the study was to uncover causes of inadequate retrograde or antegrade cardioplegia. For contrast enhancement, 5% human albumin micro-bubbles were used. During antegrade cardioplegia (coronary perfusion main-tained via coronary arteries), the contrast material was injected into the aortic root; aortic regurgitation was mild in 17%, moderate in 10% and severe in 15% of the patients without preoperative regurgitation. The degree of myocardial con-trast opacification obtained during antegrade cardioplegia was inversely related to the degree of valvular leakage. With retrograde cardioplegia (myocardial per-fusion maintained via coronary veins), contrast was injected in the right atrium; the right-to-left atrial shunt was mild in 32% and moderate or severe in 18% of the patients. The authors concluded that incomplete myocardial distribution of cardioplegia (secondary to transient aortic valve incompetence or shunting through a PFO) is not uncommon in patients undergoing coronary surgery. Allen et al. [95] also examined intraoperative delivery of cardioplegic solutions; contrast TEE was used to analyze regional myocardial perfusion. Contrast media (sonicated Isovue) was injected retrograde via a coronary sinus catheter. Using TEE imaging, the authors found that retrograde cardioplegia provides only poor right ventricular myocardial perfusion. Zaroff et al. [96] performed intra-operative contrast TEE to visualize the quality of cardioplegia and its relation to post-operative outcome. All patients with low preoperative ejection fraction and low intraoperative myocardial perfusion (assessed by contrast TEE) required post-operative inotropic support. However, low preoperative ejection fraction with adequate intraoperative cardioplegia was not associated with postoperative exogenous circulatory support requirements. Aronson et al. [97] studied 15 patients during aorto-coronary bypass surgery. Contrast injections into the free proximal end of the bypass vein provided information about the magnitude and geometric distribution of coronary artery – vein bypass run-off and enabled identification of poorly perfused myocardial regions. Furthermore, intraoperative contrast was injected in the aortic root and TEE was used to assess regional myocardial per-fusion pattern. The myocardial perfusion pattern based on preoperative coronary angiography could be predicted by intraoperative contrast TEE.

Contrast enhancement of the myocardium may also improve delineation of cardiac structures. In a preliminary study, Asamuna et al. [98] injected sonicated albumin and observed opacification of the muscular trabeculae in the left atrial appendage. In this study, myocardial contrast enhancement during TEE allowed better differentiation of trabeculae from thrombotic material in the left atrial appendage. Lindert et al. [99] showed that the i.v. application of left-heart con-trast agents (BY 963, Byk-Gulden, Konstanz, Germany) during TEE improved visualization of left ventricular wall motion in the majority of patients: complete opacification of the left ventricle was achieved in >70% of the patients.

CONCLUSIONS

TEE is a useful alternative to the transthoracic approach in patients in whom the transthoracic approach is limited by impaired image quality (emphysema, chest wall abnormalities, prior heart surgery, artificial ventilation) or when the proximity of cardiac structures to the esophagus allows dramatic improvements in image resolution. Contrast enhancement further increases the clinical and scientific utility of TEE.

Contrast TEE has evolved already as the gold standard for in vivo detection of PFO by visualizing the right-to-left atrial shunting; in addition to the presence of a PFO, shunt severity assessed by contrast TEE may be a helpful marker in patients with suspected paradoxical embolism. Initial studies have shown that the quality of coronary flow measurements during TEE can be augmented by contrast enhancement of color and pulsed wave Doppler signals. Contrast imaging of myocardial perfusion during bypass surgery using interoperative TEE may be another future application of this technique.

REFERENCES

1. Daniel WG, Erbel R, Kasper W et al. Safety of transesophageal echocardiography. A multicenter survey of 10,419 examinations. Circulation 1991;83:817–21.
2. Daly JJ. Venoarterial shunting in obstructive pulmonary disease. N Engl J Med 1968;278: 952–3.
3. Meister SG, Grossman W, Dexter L, Dalen JE. Paradoxical embolism. Am J Med 1972;53: 292–8.
4. Hagen PT, Scholz DG, Edwards WD. Incidence and size of patent foramen ovale during the first 10 decades of life: an autopsy study of 965 normal hearts. Mayo Clin Proc 1984;59: 17–20.
5. Strunk BL, Cheitlin MD, Stulbarg MS, Schiller NB. Right-to-left interatrial shunting through a patent foramen ovale despite normal intracardiac pressures. Am J Cardiol 1987;60:413–5.
6. Lynch JJ, Schuchard GH, Gross CM, Wann LS. Prevalence of right-to-left atrial shunting in a healthy population: Detection by Valsalva maneuver contrast echocardiography. Am J Cardiol 1984;53:1478–80.
7. Dubourg O, Bourdarias JP, Farcot JC et al. Contrast echocardiographic visualization of cough-induced right to left shunt through a patent foramen ovale. J Am Coll Cardiol 1984;4:587–94.
8. Webster MWI, Chancellor AM, Smith HJ, Swift DL, Sharpe DN, Bass NM. Patent foramen ovale in young stroke patients. Lancet 1988;2:11–2.
9. Lechat P, Mas JL, Lascault G et al. Prevalence of patent foramen ovale in patients with stroke. N Engl J Med 1988;318:1148–52.
10. Kronik G, Mösslacher H. Positive contrast echocardiography in patients with patent foramen ovale and normal right heart hemodynamics. Am J Cardiol 1982;49:1806–9.
11. Stoddard MF, Keedy DL, Dawkins PR. The cough test is superior to the valsalva maneuver in the delineation of right-to-left shunting through a patent foramen ovale during contrast transesophageal echocardiography. Am Heart J 1993;125:185–9.
12. Gin KG, Huckell VF, Pollick C. Femoral vein delivery of contrast medium enhances transthoracic echocardiographic detection of patent foramen ovale. J Am Coll Cardiol 1993;22: 1994–2000.
13. Langholz D, Louie EK, Konstadt SN, Rao TL, Scanlon PJ. Transesophageal echocardio-

graphic demonstration of distinct mechanisms for right to left shunting across a patent foramen ovale in the absence of pulmonary hypertension. J Am Coll Cardiol 1991;18:1112–7.

14. Cox D, Taylor J, Nanda NC. Refractory hypoxemia in right ventricular infarction from right-to-left shunting via a patent foramen ovale: Efficacy of contrast transesophageal echocardiography. Am J Med 1991;91:653–5.

15. Harpaz D, Motro M, Kaplinsky E, Vered Z. Right-to-left shunt through a patent foramen ovale caused by severe tricuspid regurgitation detected with color doppler echocardiography. J Am Soc Echocardiogr 1992;5:77–80.

16. Thompson RC, Finck SJ, Leventhal JP, Safford RE. Right-to-left shunt across a patent foramen ovale caused by cardiac tamponade: Diagnosis by transesophageal echocardiography. Mayo Clin Proc 1991;66:391–4.

17. Lang I, Steurer G, Weissel M, Burghuber OC. Recurrent paradoxical embolism complicating severe thromboembolic pulmonary hypertension. Eur Heart J 1988;9:678–81.

18. Inoue T, Yamaguchi H, Hayashi T, Morooka S, Takabatake Y. Right-to-left shunting through a patent foramen ovale caused by pulmonary hypertension associated with rheumatoid arthritis and Sjögren's syndrome: a case report. Angiology 1990;92:1082–5.

19. Begin R, Gervais A, Guerin L, Bureau MA. Patent foramen ovale and hypoxemia in chronic obstructive pulmonary disease. Eur J Resp Dis 1981;62:373–6.

20. Kovacs GS, Hill JD, Aberg T, Blesovsky A, Gerbode F. Pathogenesis of arterial hypoxemia in pulmonary embolism. Arch Surg 1966;93:813–5.

21. Moorthy SS, Losasso AM. Patency of the foramen ovale in the critically ill patient. Anesthesiology 1974;41:405–7.

22. Levine BD, Grayburn PA, Voyles WF, Greene R, Roach RC, Hackett PH. Intracardiac shunting across a patent foramen ovale may exacerbate hypoxemia in high-altitude pulmonary edema. Ann Intern Med 1991;114:569–70.

23. Chen WJ, Kuan P, Lien WP, Lin FY. Detection of patent foramen ovale by contrast trans-esophageal echocardiography. Chest 1991;101:1515–20.

24. Valdes-Cruz LM, Pieroni DR, Roland JMA, Varghese PJ. Echocardiographic detection of intracardiac right-to-left shunts following peripheral vein injections. Circulation 1976;54:558–62.

25. Biller J, Johnson MR, Adams HP et al. Further observations on cerebral or retinal ischemia in patients with right-to-left intracardiac shunts. Arch Neurol 1987;44:740–3.

26. DiTullio M, Sacco RL, Gopal A, Mohr JP, Homma S. Patent foramen ovale as a risk factor for cryptogenic stroke. Ann Intern Med 1992;117:461–5.

27. Harvey JR, Teague SM, Anderson JL, Voyles WF, Thadani U. Clinically silent atrial septal defects with evidence for cerebral embolization. Ann Intern Med 1986;105:695–7.

28. Jeanrenaud X, Bogousslavsky J, Payot M, Regli F, Kappenberger L. Foramen ovale permeable et infarctus cerebral du sujet jeune. Schweiz Med Wschr 1990;120:823–9.

29. Siostrzonek P, Zangeneh M, Gössinger H et al. Comparison of transesophageal and trans-thoracic contrast echocardiography for detection of a patent foramen ovale. Am J Cardiol 1991;68:1247–9.

30. Wilmshurst PT, Byrne JC, Webb-Peploe MM. Relation between interatrial shunts and decompression sickness in divers. Lancet 1989;2:1302–6.

31. Guggiari M, Lechat P, Garen-Colonne C, Fusciardi J, Viars P. Early detection of patent foramen ovale by two-dimensional contrast echocardiography for prevention of paradoxical air embolism during sitting position. Anesth Analg 1988;67:192–4.

32. Wu CC, Chen WJ, Chen MF, Liau CS, Chu SH, Lee YT. Left-to-right shunt through patent foramen ovale in adult patients with left-sided cardiac lesions: A transesophageal echocardiographic study. Am Heart J 1993;125:1369–74.

33. Hausmann D, Mügge A, Becht I, Daniel WG. Diagnosis of patent foramen ovale by trans-esophageal echocardiography and association with cerebral and peripheral embolic events. Am J Cardiol 1992;70:668–72.

34. van Camp G, Schulze D, Cosyns B, Vandenbossche JL. Relation between patent foramen ovale and unexplained stroke. Am J Cardiol 1993;71:596–8.
35. Louie EK, Konstadt SN, Rao TL, Scanlon PJ. Transesophageal echocardiographic diagnosis of right to left shunting across the foramen ovale in adults without prior stroke. J Am Coll Cardiol 1993;21:1231–7.
36. de Belder MA, Tourikis L, Leech G, Camm AJ. Risk of patent foramen ovale for thrombo-embolic events in all age groups. Am J Cardiol 1992;69:1316–20.
37. Konstadt SN, Louie EK, Black S, Rao TL, Scanlon P. Intraoperative detection of patent foramen ovale by transesophageal echocardiography. Anesthesiology 1991;74:212–6.
38. Fisher DC, Fisher EA, Budd JH, Rosen SE, Goldman ME. The incidence of patent foramen ovale in 1000 consecutive patients. A contrast transesophageal echocardiography study. Chest 1995;107:1504–9.
39. Stöllberger C, Schneider B, Abzieher F, Wollner T, Meinertz T, Slany J. Diagnosis of patent foramen ovale by transesophageal contrast echocardiography. Am J Cardiol 1993;71:604–6.
40. Mügge A, Daniel WG, Klöpper JW, Lichtlen PR. Visualization of patent foramen ovale by transesophageal color-coded Doppler echocardiography. Am J Cardiol 1988;62:837–9.
41. de Belder MA, Tourikis L, Griffith M, Leech G, Camm AJ. Transesophageal contrast echo-cardiography and color flow mapping: Methods of choice for the detection of shunts at the atrial level? Am Heart J 1992;124:1545–50.
42. Berkompas DC, Sagar KB. Accuracy of color Doppler transesophageal echocardiography for diagnosis of patent foramen ovale. J Am Soc Echocardiogr 1994;7:253–6.
43. Spencer MP, Thomas GI, Nicholls SC, Sauvage LR. Detection of middle cerebral artery emboli during carotid endarterectomy using transcranial doppler ultrasonography. Stroke 1990;21:415–23.
44. Teague SM, Sharma MK. Detection of paradoxical cerebral echo contrast embolization by transcranial doppler ultrasound. Stroke 1991;22:740–5.
45. Karnik R, Stöllberger C, Valentin A, Winkler WB, Slany J. Detection of patent foramen ovale by transcranial doppler ultrasound. Am J Cardiol 1992;69:560–2.
46. Nemec JJ, Marwick TH, Lorig RJ et al. Comparison of transcranial doppler ultrasound and transesophageal contrast echocardiography in the detection of interatrial right-to-left shunts. Am J Cardiol 1991;68:1498–502.
47. Nikutta P, Schneider M, Hausmann D et al. How reliable is transcranial doppler ultrasound in the assessment of patent foramen ovale? J Am Coll Cardiol 1993;21:135A.
48. Nikutta P, Claus G, Hausmann D, Mügge A, Niedermeyer J, Kühn, Daniel WG. Assessment of transpulmonary passage of echo-contrast agents by transesophageal echocardiographic spectral Doppler. J Am Coll Cardiol 1993;21:136A.
49. Klotzsch C, Janssen G, Berlit P. Transesophageal echocardiography and contrast TCD in the detection of a patent foramen ovale: Experience with 111 patients. Neurology 1994;44:1603–6.
50. Jauss M, Kaps M, Keberle M, Haberbosch W, Dorndorf W. A comparison of transesophageal echocardiography and transcranial Doppler sonography with contrast medium for detection of patent foramen ovale. Stroke 1994;25:1265–7.
51. DiTullo M, Sacco RL, Venketasubramanian N, Sherman D, Mohr JP, Homma S. Comparison of diagnostic techniques for the detection of a patent foramen ovale in stroke patients. Stroke 1993;24:1020–4.
52. Cohnheim J. Thrombose und Embolie, Vorlesungen über Allgemeine Pathologie, Vol. 1. Berlin: Hirschwald, 1877:134.
53. Leonard RCF, Neville E, Hall RJC. Paradoxical embolism. A review of cases diagnosed during life. Eur Heart J 1982;3:362–70.
54. Loscalzo J. Paradoxical embolism: Clinical presentation, diagnostic strategies, and therapeutic options. Am Heart J 1986;112:141–5.
55. Robinson FJ. Lodging of an embolus in a patent foramen ovale. Circulation 1950;2:304–5.
56. Sardesi SH, Marshall RJ, Mourant AJ. Paradoxical systemic embolisation through a patent foramen ovale. Lancet 1989;1:732–3.

57. Nelson CW, Snow FR, Barnett M, McRoy L, Wechsler AS, Nixon JV. Impending paradoxical embolism: echocardiographic diagnosis of an intracardiac thrombus crossing a patent foramen ovale. Am Heart J 1991;122:859–62.
58. Gin KG, Thompson CR, Jue J, Ling H. Embolic occlusion of a patent foramen ovale: A cause of false negative contrast echocardiogram. J Am Soc Echocardiogr 1992;5:444–6.
59. Nellessen U, Daniel WG, Matheis G, Oelert H, Depping K, Lichtlen PR. Impending paradoxical embolism from atrial thrombus: Correct diagnosis by transesophageal echocardiography and prevention by surgery. J Am Coll Cardiol 1985;5:1002–4.
60. Speechly-Dick ME, Middleton SJ, Foale RA. Impending paradoxical embolism: a rare but important diagnosis. Br Heart J 1991;65:163–5.
61. Barnard SP, Kulatilake ENP, Azzu AA, Ikram S. Straddle embolus – imminent paradoxical embolus diagnosed by echocardiography and treated surgically. Eur J Cardiothorac Surg 1991;5:105.
62. Ranoux D, Cohen A, Cabanes L, Amarenco P, Bousser MG, Mas JL. Patent foramen ovale: Is stroke due to paradoxical embolism. Stroke 1993;24:31–4.
63. Black S, Cucchiara RF, Nishimura RA, Michenfelder JD. Parameters affecting occurrence of paradoxical air embolism. Anesthesiology 1989;71:235–41.
64. Cucchiara RF, Seward JB, Nishimura RA, Nugent M, Faust RJ. Identification of patent foramen ovale during sitting position craniotomy by transesophageal echocardiography with positive airway pressure. Anesthesiology 1985;63:107–9.
65. Ellis JE, Lichtor JL, Feinstein SB et al. Right heart dysfunction, pulmonary embolism, and paradoxical embolization during liver transplantation. A transesophageal two-dimensional echocardiographic study. Anesth Analg 1989;68:772–6.
66. Moon RE, Camporesi EM, Kisslo JA. Patent foramen ovale and decompression sickness in divers. Lancet 1989;1:513–5.
67. Vik A, Jenssen BM, Brubakk AO. Paradoxical air embolism in pigs with a patent foramen ovale. Undersea Biomed Res 1992;19:361–74.
68. Pell AC, Hughes D, Keating J, Christie J, Busuttil A, Sutherland GR. Brief report: fulminating fat embolism syndrome caused by paradoxical embolism through a patent foramen ovale. N Engl J Med 1993;329:926–9.
69. Vasiljevic JD, Abdulla AK. Coronary embolism by metastatic choriocarcinoma of the uterus: An unusual cause of ischemic heart disease. Gynecol Oncol 1990;38:289–92.
70. Wilson T, Becker SN. Postoperative death from pulmonary and paradoxic emboli containing transitional epithelium. A case report. Obstet Gynecol 1972;39:286–91.
71. Rosenow EC, Osmundson PJ, Brown ML. Pulmonary embolism. Mayo Clin Proc 1981;56:161–78.
72. Stollberger C, Slany J, Schuster I, Leitner H, Winkler WB, Karnik R. The prevalence of deep venous thrombosis in patients with suspected paradoxical embolism. Ann Intern Med 1993;119:461–5.
73. Hausmann D, Mügge A, Daniel WG. Identification of patent foramen ovale permitting paradoxical embolism. J Am Coll Cardiol 1995;26:1030–8.
74. Schneider B, Hofmann T, Justen MH, Meinertz T. Chiari's network: Normal anatomic variant or risk factor for arterial embolic events? J Am Coll Cardiol 1995;26:203–10.
75. Bridges ND, Hellenbrand W, Latson L, Filiano J, Newburger JW, Lock JE. Transcatheter closure of patent foramen ovale after presumed paradoxical embolism. Circulation 1992;86:1902–8.
76. Homma S, DiTullio MR, Sacco RL, Mihalatos D, Li Mandri G, Mohr JP. Characteristics of patent foramen ovale associated with cryptogenic stroke. A biplane transesophageal echocardiographic study. Stroke 1994;25:582–6.
77. Yoshida K, Yoshikawa J, Hoyumi T et al. Detection of left main coronary artery stenosis by transesophageal color Doppler and two-dimensional echocardiography. Circulation 1990;81:1271–6.
78. Yamagishi M, Yasu T, Ohara K, Kuro M, Miyatake K. Detection of coronary blood flow

associated with left main coronary artery stenosis by transesophageal Doppler color flow echocardiography. J Am Coll Cardiol 1991;17:87–93.

79. Redberg RF, Sobol Y, Chou TM et al. Adenosine-induced coronary vasodilatation during transesophageal Doppler echocardiography. Rapid and safe measurement of coronary flow reserve ratio can predict significant left anterior descending coronary stenosis. Circulation 1995;92:190–6.

80. von Bibra H, Stempfle HU, Poll A, Schlief R, Emslander H. Echo contrast agents improve flow display of color Doppler. Echocardiography 1991;5:533–40.

81. Geny B, Mettauer B, Muan B et al. Safety and efficacy of a new transpulmonary echo contrast agent in echocardiographic studies in patients. J Am Coll Cardiol 1993;22:1193–8.

82. Crouse LJ, Cheirif J, Hanly DE et al. Opacification and border delineation improvement in patients with suboptimal endocardial borer definition in routine echocardiography: Results of phase III Albunex multicenter trial. J Am Coll Cardiol 1993;22:1494–500.

83. Porter TR, Xie F, Kricsfeld A, Chiou A, Dabestani A. Improved endocardial border resolution during dobutamine stress echocardiography with intravenous sonicated dextrose albumin. J Am Coll Cardiol 1994;23:1440–3.

84. Schröder K, Agrawal R, Völler H, Schlief R, Schröder R. Improvement of endocardial border delineation in suboptimal stress-echocardiograms using the new left heart contrast agent SHU 508A. Int J Cardiac Imag 1994;10:45–51.

85. von Bibra H, Sutherland G, Becher H, Neudert J, Nihoyannopoulos P. Clinical evaluation of left heart Doppler contrast enhancement by a saccharide-based transpulmonary contrast agent. J Am Coll Cardiol 1995;25:500–8.

86. Iliceto S, Caiati C, Aragona P, Verde R, Schlief R, Rizzon P. Improved Doppler signal intensity in coronary arteries after intravenous peripheral injection of a lung-crossing contrast agent (SHU 508A). J Am Coll Cardiol 1994;23:184–90.

87. Aggarwal KK, Gatewood RP, Nanda NC, Chopra KL. Improved tranesophageal echocardiographic assessment of significant proximal narrowing of the left anterior descending and left circumflex coronary arteries using echo contrast enhancement. Am J Cardiol 1994;73:1131–3.

88. Kozakova M, Palombo C, Zanchi M, Distante A, L'Abbate A. Increased sensitivity of flow detection in the left coronary artery by transesophageal echocardiography after intravenous administration of transpulmonary stable echocontrast agent. J Am Soc Echocardiogr 1994;7:327–36.

89. Dittrich HC, Bales GL, Kuvelas T, Hunt RM, McFerran BA, Greener Y. Myocardial contrast echocardiography in experimental coronary artery occlusion with a new intravenously administered contrast agent. J Am Soc Echocardiogr 1995;8:465–74.

90. Cheirif J, Zoghbi WA, Raizner AE. Assessment of myocardial perfusion in humans by contrast echocardiography, I: Evaluation of regional coronary reserve by peak contrast intensity. J Am Coll Cardiol 1988;11:735–43.

91. Kaul S, Glasheen W, Ruddy TD, Pandian NG, Weyman AE, Okada RD. The importance of defining left ventricular area at risk in vivo during acute myocardial infarction: An experimental evaluation with myocardial contrast two-dimensional echocardiography. Circulation 1987;75:1249–60.

92. Voci P, Bilotta F, Merialdo P, Agati L. Myocardial contrast enhancement after intravenous injection of sonicated albumin microbubbles: A transesophageal echocardiography dipyridamole study. J Am Soc Echocardiogr 1994;7:337–46.

93. Angermann CE, Krüger TM, Junge R et al. Intravenous albunex during transesophageal echocardiography: Quantitative assessment by videodensitometry and integrated backscatter analysis from unprocessed radiofrequency signals. J Am Soc Echocardiogr 1995;8:839–53.

94. Voci P, Bilotta F, Caretta Q, Chiarotti F, Mercanti C, Marino B. Mechanisms of incomplete cardioplegia distribution during coronary artery surgery. An intraoperative transesophageal contrast echocardiography study. Anesthesiology 1993;79:904–12.

95. Allen BS, Winkelman JW, Hanafy H et al. Retrograde cardioplegia does not adequately perfuse the right ventricle. J Thorac Cardiovasc Surg 1995;109:1116–24.

96. Zaroff J, Aronson S, Lee BK, Feinstein SB, Walker R, Wiencek JG. The relationship between immediate outcome after cardiac surgery, homogeneous cardioplegia delivery, and ejection fraction. Chest 1994;106:38–45.
97. Aronson S, Lee BK, Wiencek JG et al. Assessment of myocardial perfusion during CABG surgery with two-dimensional transesophageal contrast echocardiography. Anesthesiology 1991;75:433–40.
98. Asamuna T, Tanabe K, Yoshitomi H, Murakami Y, Sano K, Morioka S. Improved anatomic delineation of left atrium and appendage with myocardial contrast enhancement evaluated by transesophageal echocardiography. Circulation 1995;92(suppl):I-193.
99. Lindert O, Mügge A, Tonduangu DK et al. Transösophageale Echocardiographie mit dem lungengängen Kontrastmittel BY963: Verbesserte Darstellng der linksventrikulären Wandbewegung. Z Kardiol 1995;84(suppl. 3):9.

19. Transesophageal echocardiographic assessment of coronary flow reserve and coronary artery stenosis using echo enhancement

CARLO CAIATI, SABINO ILICETO, NAVIN C. NANDA &
PAOLO RIZZON

INTRODUCTION

Echocardiography has been used to visualize proximal portions of coronary arteries since the advent of the technique [1, 2]. Transesophageal echocardiography (TEE) produces high quality images due to the proximity of the probe to the heart and the use of high frequency transducers, and has reawakened interest in the coronary echocardiographic field. Using transesophageal Doppler echocardiography, evaluation of the left main coronary artery (LMCA) has improved, and this is now a useful tool for detecting left main coronary artery stenosis [3–5]. In addition, transesophageal Doppler echocardiography has been used to record blood flow velocity in proximal left anterior descending coronary artery (LAD) in resting conditions and after pharmacological vasodilatation to explore the maximal amount of flow that the coronary vascular bed can accommodate above normal (coronary flow reserve) [6–8]. The importance of coronary flow reserve assessment lies in the fact that it is considered a better indicator of the functional significance of a coronary stenosis [9] – visual inspection by angiography of the anatomical severity of a coronary stenosis does not accurately reflect its physiological importance [10]. However, several limitations have impeded the widespread clinical application of TEE to coronary artery visualization and, consequently, stenosis detection and coronary blood flow reserve assessment. First, apart from the LMCA, the rest of the proximal left coronary tree is poorly visualized by TEE. In particular, the success rate in imaging an adequate portion of the proximal left anterior descending coronary artery, whose pathology carries important clinical and prognostic implications for the patient [11, 12], is low, with a consequent relatively poor diagnostic impact of TEE on the evaluation of proximal LAD coronary stenosis. On the other hand, TEE coronary flow reserve evaluation has a rather long learning-curve and, even after that, the ability to attain a good quality time–velocity curve (especially the systolic curve) in LMA bifurcation is suboptimal. For these reasons the attention of some research groups has been directed towards a lung-crossing contrast

309

N. C. Nanda, R. Schlief and B. B. Goldberg (eds.), Advances in Echo Imaging Using Contrast Enhancement,
Second Edition, 309–324.
© 1997 Kluwer Academic Publishers.

agent which, when injected i.v., could enhance the intensity of Doppler signal and thus the signal-to-noise ratio in coronary arteries, improving the quality and feasibility of coronary blood flow velocity Doppler recording. In this chapter we will discuss the clinical usefulness of the lung-crossing contrast agent, SHU 508A in improving assessment of coronary flow reserve and detection of proximal left anterior descending coronary artery stenosis during transesophageal Doppler echocardiography.

CONTRAST AGENT (SHU 508A): PHYSICOCHEMICAL PROPERTIES, DOSAGE AND PATIENT TOLERANCE

SHU 508A is a suspension of monosaccharide (galactose) microparticles, with a mean diameter of $2\,\mu m$, in sterile water [13]. The contrast is prepared 2–10 min before its i.v. injection by transferring the amount of sterile water required to obtain the desired concentration into a vial containing the specially manufactured granules of SHU 508 and shaking it vigorously for 5 s. The prepared echo-contrast medium is manually injected (at approximately 2 ml/s) into a vein of the right arm. Contrast injection is immediately followed by a 5 ml bolus of physiological saline solution to flush the cannula and improve the bolus flow. Enhancement of the echo signal is due to microbubbles, physically stabilized in the microparticle suspension, which are stronger reflectors of ultrasound than are red blood cells. These bubbles have a remarkable increase in intravascular stability [14]. Their small dimension and vascular stability allows SHU 508 microbubbles to survive transit through the lung capillaries and, carried by the bloodstream, they reach the left chambers of the heart and the coronaries.

Dosage

In our experience the best dose of SHU 508 for echo enhancement in coronaries was 10 ml of 300 mg/ml concentration. However, with this dose (the highest used), a shadowing effect obscuring coronary imaging and lasting for a few cardiac beats occasionally occurred at the first appearance of the contrast bolus in left atrium, due to the poor penetration of ultrasound through the dense contrast in left atrium. The average duration of echo enhancement is >2 min, although the peak echo enhancement is attained in the first minute.

Patient tolerance

SHU 508 can be considered safe. No serious adverse effects were noticed in our experience or in a multicenter trial on a larger population [15], in which heart rate, ECG and blood pressure did not change. No clinically significant changes were observed in blood chemistry or complete blood count. Reported subjective

Table 1. Pulsed wave Doppler quality in the left anterior descending coronary artery after intravenous injection of SHU 508A.

	Score			
	1	2	3	3 (systolic and diastolic)
Control	10 (29)	17 (49)	8 (23)	2 (11)
1st dose	5 (14)	13 (37)	17 (49)	10 (29)
2nd dose	5 (14)	9 (26)	21 (60)	18 (51)
3rd dose	4 (11)	8 (23)	23 (66)	21 (60%)
4th dose	4 (11)	8 (23)	23 (66)	21 (60%)

Data presented are number (%) of coronary artery segments with a specific score. Score 1 = pulsed wave recording not obtainable; Score 2 = pulsed wave recording obtainable but of suboptimal quality (considering diastolic curve); Score 3 = Doppler signal with complete and well-defined outline of diastolic curve; Score 3 (systolic and diastolic) = Doppler signal with complete and well-defined outline of both diastolic and systolic curves; 1st dose = 5 ml of 200 mg/ml concentration; 2nd dose = 10 ml of 200 mg/ml concentration; 3rd dose = 5 ml of 300 mg/ml concentration; 4th dose = 10 ml of 300 mg/ml concentration. (Modified in part from Iliceto S et al. Improved Doppler signal intensity in coronary arteries after intravenous peripheral injection of a lung-crossing contrast agent (SHU 508A). J Am Coll Cardiol 1994;23:184–90.)

symptoms were a brief burning sensation at the injection site (in 5% of patients), a transient headache (in 1.3% of patients), and a taste sensation (in 0.3% of patients).

TRANSESOPHAGEAL DOPPLER ECHOCARDIOGRAPHIC ASSESSMENT OF CORONARY FLOW RESERVE: ROLE OF ECHO ENHANCEMENT

SHU 508 is a recently developed lung-crossing echo enhancing agent [13, 16, 17] that is a derivative of SHU 454, whose ability to increase Doppler signal intensity as well as signal to noise ratio has been demonstrated in vitro and in vivo studies. Recent studies [13, 14] have shown that i.v. peripheral injection of SHU 508A strongly enhances blood echogenicity not only in the right heart but also in left heart cavities after pulmonary transit.

Our group, for the first time, tested the potential of this contrast agent to increase the intensity of the pulsed wave Doppler signal to the proximal LAD during transesophageal Doppler [18]. Coronary blood flow velocity in the proximal LAD can be evaluated by transesophageal Doppler echocardiography, but an adequate tracing is obtained in a relatively low proportion of patients [6–8]. This could reduce the clinical value of transesophageal Doppler echocardiography in assessing coronary flow reserve. In our study [18] 35 patients underwent transesophageal Doppler echocardiography before and after contrast injection. Color Doppler mapping of coronary flow and pulsed wave Doppler at the proximal LAD was attempted in all patients. A remarkable increase in Doppler signal intensity was evident after contrast injection (Table 1, Figure 1): optimal

312   *Carlo Caiati et al.*

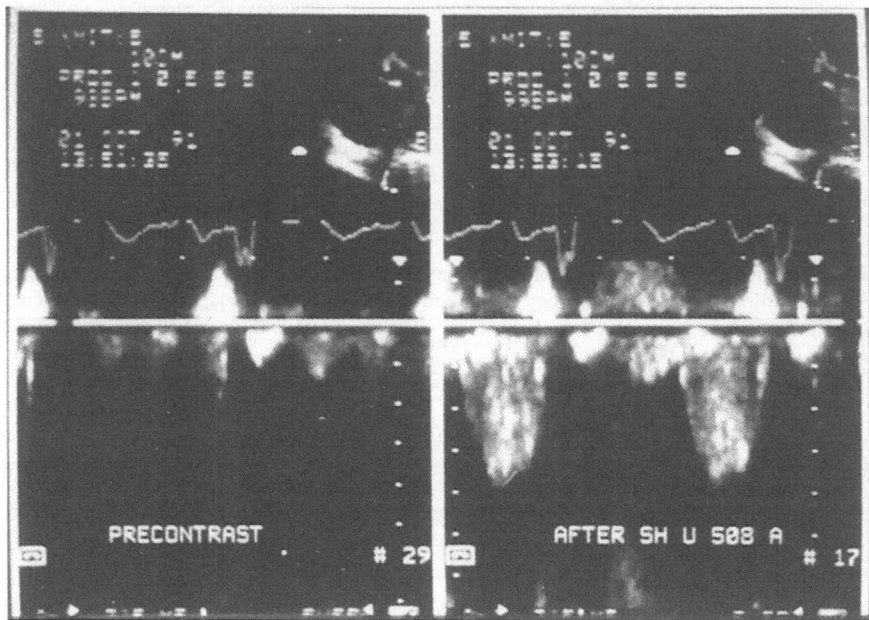

Figure 1. Pulsed wave Doppler recording of coronary blood flow velocity in the left anterior descending coronary artery before (left panel) and after (right panel) i.v. injection of SHU 508A. Before SHU 508A, coronary blood flow velocity is not detected (only spikes and wall artifacts are recorded). After SHU 508A, a typical biphasic coronary blood flow velocity pattern (greater diastolic, small systolic component) is recorded and clearly outlined. (Modified in part from Iliceto S et al. Improved Doppler signal intensity in coronary arteries after intravenous peripheral injection of a lung-crossing contrast agent (SHU 508A). J Am Coll Cardiol 1994;23:184–90.)

PW Doppler signal (Doppler signal with complete and well-defined outline at least of the diastolic wave) was recorded in 23 patients (66%) after contrast enhancement but in only eight patients (22%) before contrast injection ($p < 0.01$; Table 1). This good result was also achieved because color Doppler-guided sample volume positioning was easier owing to the greater available color Doppler flow mapping area in the left anterior descending coronary artery (Figure 2). The width and length of the color Doppler signal in LAD significantly increased after contrast injection (from 5.75 ± 5.32 and 1.51 ± 1.17 to 17.04 ± 8.76 and 4.21 ± 1.78 mm; $p < 0.001$). Our data were confirmed by another similar study in 12 patients [19].

Thus, the evaluation of coronary blood flow velocity and coronary flow reserve by transesophageal Doppler echocardiography can be considerably improved by injecting a small amount of a lung-crossing echocontrast agent (SHU 508A) into a peripheral vein. After SHU 508A injection, the intensity of both color and pulsed

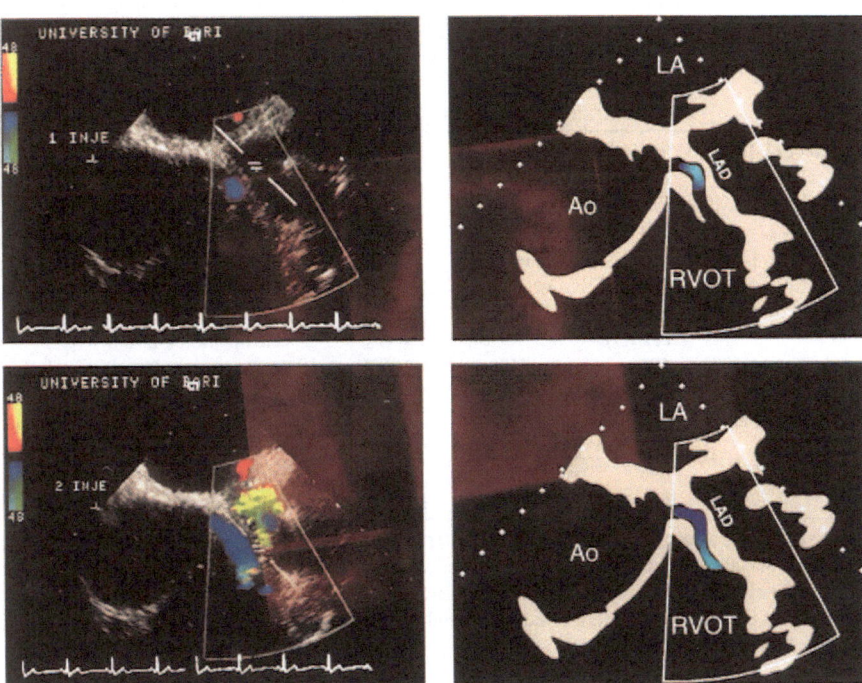

Figure 2. Transesophageal color Doppler flow mapping of blood flow velocity in the left anterior descending coronary artery (LAD) before (upper panels) and after (lower panels) i.v. injection of SHU 508A. The right panels represent the schematics. Baseline color Doppler flow mapping (upper panels) is incomplete (very narrow and short blue signal that does not fill the vessel longitudinal section). After i.v. injection of SHU 508A (lower panels) the color Doppler signal becomes much longer and wider (complete flow mapping). The flow is coded in blue because it recedes from the transducer. Ao=aortic root; LA=left atrium; LAD=left anterior descending coronary artery; RVOT=right ventricle outflow tract. (Modified in part from Iliceto S et al. Improved Doppler signal intensity in coronary arteries after intravenous peripheral injection of a lung-crossing contrast agent (SHU 508A). J Am Coll Cardiol 1994;23:184–90.)

wave Doppler signals increased substantially and the ability to record high quality coronary blood flow reserve tracings improved considerably.

ASSESSMENT OF CORONARY STENOSES BY CONTRAST TRANSESOPHAGEAL DOPPLER ECHOCARDIOGRAPHY

Transesophageal echocardiography has a limited diagnostic impact on the evaluation of proximal left anterior descending coronary artery stenoses [5, 20–25], mainly due to the inadequate B-mode visualization of the more distal part of proximal LAD. The inadequate visualization depends mainly on the poor lateral

314 *Carlo Caiati et al.*

Table 2. Reported studies on the evaluation of LAD by transesophageal echocardiography.

Study	Proximal LAD			Comments
	Visualization	Detection of CAD		
		Sensitivity	Specificity	
Pearce et al. [20]	75% (55/73)	–	–	Intraoperative; no length criteria for visualization
Yamagishi et al. [21]	75% (30/39)	–	–	Postoperative; length criteria for visualization not clearly stated
Iliceto et al. [24]	29% (16/54)	–	–	Awake; no length criteria for visualization
Taams et al. [22]	14% (12/83)	33%	100%	Awake; length criteria for visualization ≥20 cm
Reichert et al. [23]	52% (13/25)	71%	100%	Postoperative; length criteria for visualization ≥10 cm
Samdarshi et al. [25]	93%* (103/111)	79% 48%**	99% 99%**	Intraoperative; no length criteria for visualization
Tardif et al. [5]	69% (31/45)	80%	100%	Intraoperative; length criteria for visualization ≥10 cm

CAD=coronary artery disease; LAD=left anterior descending coronary artery; *Mean length visualized in that study=0.8 cm; **Sensitivity and specificity considering all angiographically identified proximal lesions; the numerator and the denominator of the numbers in parentheses indicate respectively the number of arteries visualized and the total number of patients.

resolution of ultrasound in the far field [2.5 mm for a 5 MHz probe connected to Hewlett-Packard echo-equipment (data from Hewlett-Packard)] relative to a reflector of a few millimeters in diameter (external diameter range of proximal LAD: 2–5 mm [26]). Moreover, coronary arteries can only be visualized intermittently during the cardiac cycle because they move in and out of the tomographic plane, following the heart movement.

In the previous studies where a criterion for LAD visualization was specified (at least 10 mm of the vessel length), the proximal LAD was imaged in a low percentage of patients: i.e., in 14% and in 52% in two studies performed with a single plane probe [22, 23] or in slightly higher percentage (69%) when a multiplane probe was used [5].

In most of the previous studies (Table 2) the examination was performed intra- or postoperatively. It is thus conceivable that in a routine ambulatory study (especially in lightly sedated patients) the visualization rate could be even lower, since insufficient time is available to conduct the study. The reported sensitivity for detecting significant plaque impinging on the lumen ranged from 33% to 80%, but these figures were obtained considering only patients with adequate visualization of the artery (Table 2). The use of color Doppler with B-mode imaging could aid in understanding hemodynamics at the stenosis site so as to improve critical stenosis detection. Significant vessel narrowing (>50%) causes a localized acceleration of blood flow at the stenosis site (convective acceleration

[27]). The convective acceleration in coronaries can be precisely evaluated by Doppler, as shown in a recent animal study [28]. Unfortunately, although the combination of color Doppler and B-mode imaging was better than B-mode imaging alone only for detection of coronary stenosis affecting the left main coronary artery [4, 29], this was not the case for the proximal left anterior descending coronary artery [25]. This results from the fact that the blood flow signal in LAD arises from a vessel of a few millimeters in diameter that is imaged in the far field of the ultrasonographic beam, which is too weak to be resolved. Consequently the signal-to-noise capacity of the instrument is often exceeded. In addition Doppler sampling is technically difficult because of the movement of the heart and the low color Doppler frame rate.

Transesophageal Doppler echocardiographic evaluation of localized disturbed blood flow in a diseased proximal LAD can be improved tremendously by peripheral injection of SHU 508, with a consequent improvement of stenosis detection, as two studies have recently demonstrated [30, 31]. In one recent study [30] we evaluated the ability of transesophageal Doppler echocardiography to improve detection of proximal left anterior descending coronary artery stenosis after echo enhancement by i.v. injection of SHU 508. From our own previous experience [18], we learned that i.v. injection of SHU 508 improves color Doppler and PW spectral signal in the LAD and also increases the length of color Doppler imaging of the vessel regardless of its B-mode visualization quality. Therefore, we hypothesized that an improved blood flow velocity recording in the left anterior descending coronary artery following contrast injection could allow detection of hemodynamically significant left anterior descending coronary stenosis by identifying the localized flow acceleration and disturbance at the site of the stenosis even in the absence of its clear anatomical definition by cross-sectional echocardiography. To verify this hypothesis we evaluated a group of patients undergoing transesophageal Doppler echo imaging of the left anterior descending coronary artery during i.v. injection of SHU 508A and coronary angiography.

Methods

Transesophageal Doppler echocardiography (5 MHz single plane probe connected to either Hewlett-Packard Sonos 1000 or Sonos 500) was performed before and after contrast injection in 31 patients who underwent coronary angiography. Using color Doppler as a guide, pulsed wave Doppler recording of blood flow velocity in the LAD was attempted to detect a localized increase in blood flow velocity. Sampling was performed first where the color signal appeared disturbed and/or aliased if this localized alteration of the color signal was evident; otherwise PW sampling was conducted along the entire course of the color flow imaging of the LAD to detect as high a velocity as possible. Second, sampling was obtained where the color signal appeared normal to obtain a reference value of blood flow velocity in left anterior descending coronary artery. The reference value was obtained from the left main coronary artery bifurcation (in 27 (87%) patients) or in the left main coronary artery (in three (10%) patients) or in a more distal portion of the left anterior descending coronary artery (in one (3%) patient). B-mode evaluation of the vessel was also performed.

Results

Angiography showed a significant proximal left anterior descending coronary artery stenosis in 16 patients (group 1) and no stenosis in 15 patients (group 2). Overall the mean ($\pm$SD) visualized length of the vessel was much greater with contrast color Doppler (18 ± 6 mm) than using either baseline color Doppler (7 ± 5 mm) or B-mode imaging (4 ± 4 mm; overall $p<0.001$; $p<0.05$ length by contrast color Doppler vs. length by both B-mode imaging and baseline color Doppler).

Contrast transesophageal Doppler echocardiography was successful in detecting localized accelerated blood flow velocity at the site of significant vessel narrowing. In 15 of 16 group 1 patients Doppler, after contrast injection, revealed a localized velocity increase of at least 50% of the reference value. Mean ($\pm$SD) percentage increase in velocity was: $150\pm89\%$ (range 0–367%). Figures 3 and 4 show a representative example of left anterior descending coronary artery stenosis detection by Doppler echocardiography after contrast injection. In group 2, echocontrast Doppler revealed a slight localized increase in velocity in four patients and no increase in velocity in the remainder: mean ($\pm$SD) percentage increase in velocity was $5\pm7\%$ (range 0–21%; $p<0.001$ vs. percentage increase in group 1). Considering a percentage velocity increase $\geq50\%$ of the reference value as a positive criterion for detecting a significant stenosis, the sensitivity and specificity were 92% and 100% respectively (Table

---

Figure 3. (Opposite) Color Doppler flow mapping in the proximal left coronary artery before (left upper panel) and after echocontrast enhancement (right upper panel). Lower panels are the corresponding schematics. B-mode visualization of left anterior descending coronary artery (LAD) is restricted to the first few millimeters of the vessel and no significant plaque, impinging on the lumen, is evident. In the precontrast image (left panels) almost no flow is detected in the vessel. After contrast injection (right panels) blood flow velocity in LAD (coded in blue because it moves away from transducer) is clearly recorded not only in the B-mode visualized part but also in a 13 mm long additional segment. In this additional segment of color flow, a localized area of turbulent/aliased signal (indicated by arrowheads) 13 mm distant from coronary bifurcation is clearly depicted. After contrast, blood flow velocity in the left circumflex coronary artery (LCA) is also recorded (it is coded in red because it approaches the transducer). In this patient coronary angiography showed a proximal left anterior descending coronary stenosis (80% narrowing) located 12 mm from the coronary bifurcation. Calibration = 1 cm. Ao = aorta; LA = left atrium; LMA = left main coronary artery; PA = pulmonary artery. (Modified in part from Caiati C et al. Improved Doppler detection of proximal left anterior descending coronary artery stenosis after intravenous injection of a lung contrast agent. A transesophageal Doppler echocardiographic study. J Am Coll Cardiol 1996;27:1413–21.)

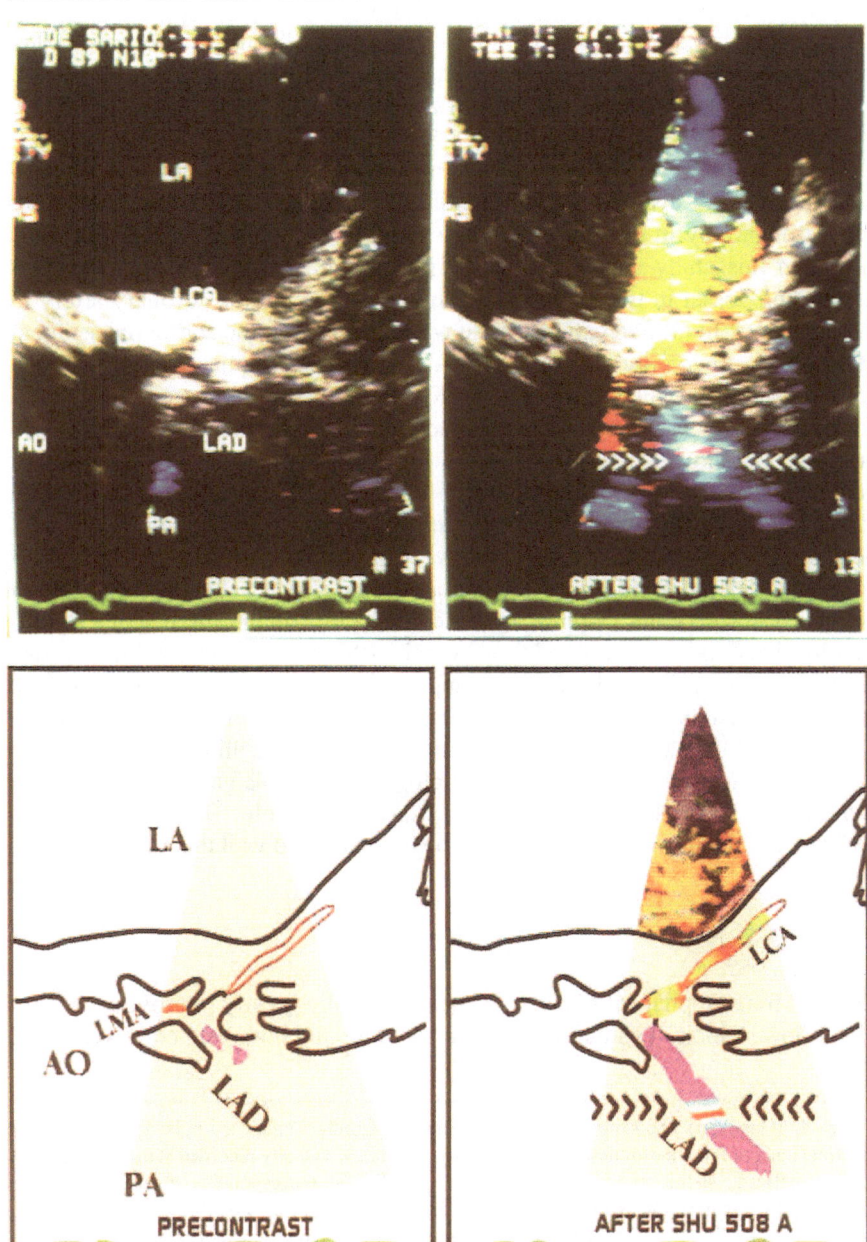

3). The sensitivity of the evaluation before contrast injection or considering B-mode imaging alone was much lower (25% and 19% respectively; $p<0.001$ vs. evaluation after contrast injection). In addition contrast color Doppler located the stenosis along the vessel correctly when compared with angiography (Figure 5).

These results are similar to those of a previous study [31] conducted on 10 patients aimed at evaluating the potential of contrast transesophageal Doppler echocardiography in detecting coronary stenosis in the proximal left coronary artery. In this study, aliased Doppler signals were noted in the left anterior descending vessel in two patients and in the left circumflex coronary artery in three patients. In all five patients these were localized in the additional segment of color flow seen only after SHU 508A. These areas of aliased or accelerated color flow corresponded to significant proximal left anterior descending and left circumflex coronary artery stenosis as judged by coronary angiography. Aliased or accelerated flow signals did not occur in normal arteries or even at the site of non-critical lesions ($\leq50\%$) (Figures 6 and 7). In this study the convective acceleration of blood at the stenosis site was measured only by color Doppler flow mapping. We consider pulsed wave Doppler confirmation of blood flow velocity acceleration (at least 50% increase of reference value) to be important because disturbed color Doppler signals could be seen in the absence of a significant localized increase in blood flow velocity (this occurred in our study in two group 2 patients). The color Doppler appearance indicating turbulence (mosaic aspect), for instance, can be due to parietal irregularities not impinging into the lumen or at left main coronary bifurcation [4]; even the aliasing phenomenon can be present in association with physiological rather than turbulent flow (as a consequence of the low pulse repetition frequency of color Doppler sampling); furthermore, disturbed color signals can be due simply to wall thumps or artifacts.

CONCLUSION

Contrast transesophageal Doppler echocardiography greatly improves coronary blood flow velocity evaluation. A more accurate and complete coronary blood

Figure 4. (Opposite) The same patient as in Figure 3. Contrast-enhanced pulsed wave Doppler spectral signal of left anterior descending coronary blood flow velocity recorded at the first site (left upper panel) and at the second site (right upper panel). The corresponding lower panels show schematically how the color-guided pulsed wave Doppler sampling was performed. At the first site (disturbed color Doppler signal) the velocity is much higher than at the second site (reference velocity), there being a 130% increase of peak diastolic velocity at the first site with respect to that at the second site. The velocity at the second site was obtained after a mild angle correction. Calibration=20 cm/s; other abbreviations as in Figure 3. (Modified in part from Caiati C et al. Improved Doppler detection of proximal left anterior descending coronary artery stenosis after intravenous injection of a lung contrast agent. A transesophageal Doppler echocardiographic study. J Am Coll Cardiol 1996;27:1413–21.)

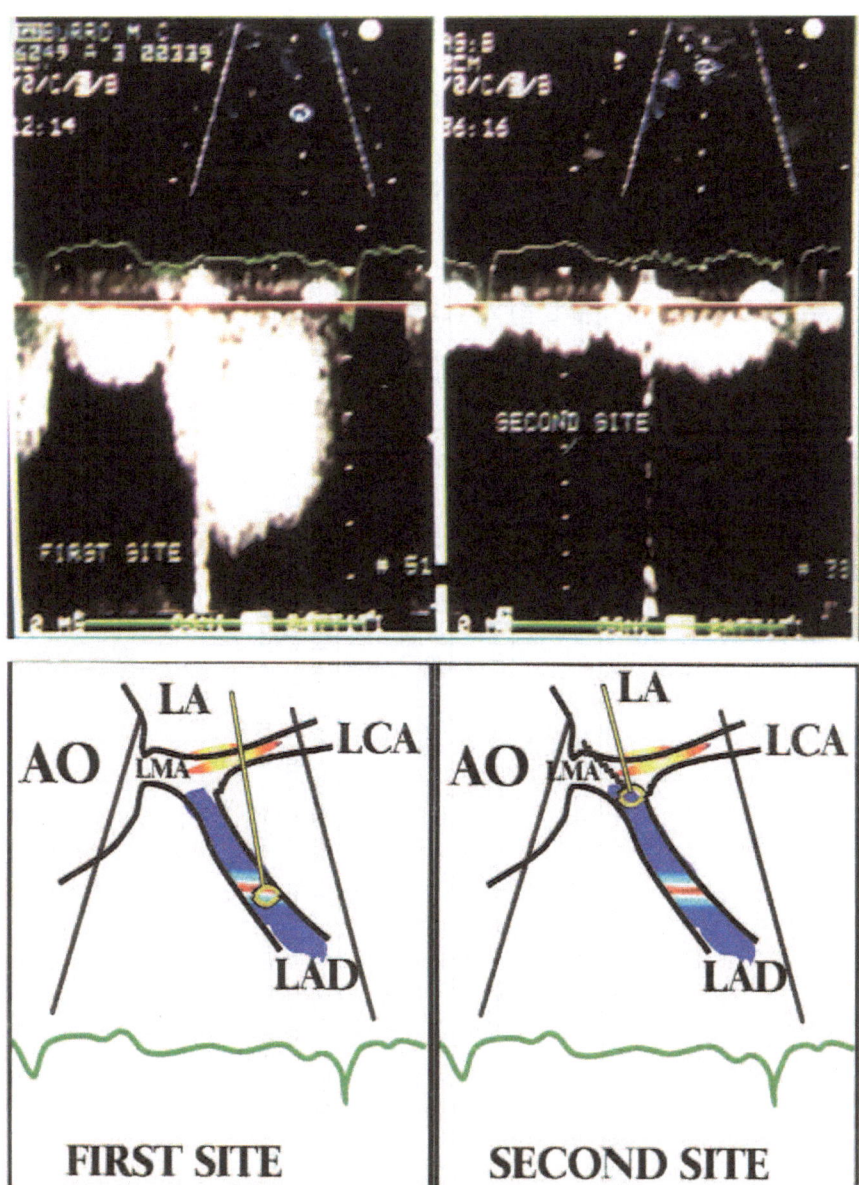

flow velocity Doppler recording renders coronary blood flow reserve assessment more feasible and strongly improves coronary stenosis detection in the proximal LAD. Furthermore, there is preliminary evidence [31] to suggest that this method can also improve coronary stenosis detection in the proximal left circumflex

Table 3. Summary of angiographic and transesophageal results

Angiography	Successful imaging of LAD*			B-mode imaging evaluation		Doppler evaluation before contrast enhancement		Doppler evaluation after contrast enhancement	
	B-mode imaging	Doppler before CE	Doppler after CE	Positive	Negative	Positive	Negative	Positive	Negative
With stenosis (16 patients)	1	2	15	3	13	4	12	15	1
Without stenosis (15 patients)	1	4	15	0	15	0	15	0	15
Sensitivity				19%		25%		93%†	
Specificity				100%		100%		100%	

CE = contrast enhancement; LAD = left anterior descending coronary artery; * = at least 1 cm visualized; † = $p < 0.001$ versus both Doppler before contrast enhancement and B-mode imaging sensitivity. (Modified in part from Caiati C et al. Improved Doppler detection of proximal left anterior descending coronary artery stenosis after intravenous injection of a lung contrast agent. A transesophageal Doppler echocardiographic study. J Am Coll Cardiol 1996;27:1413–21).

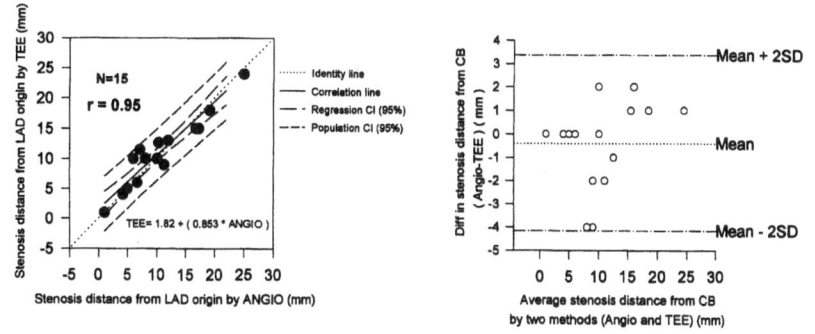

Figure 5. Scattergrams showing comparison of contrast transesophageal Doppler echocardiography (TEE) and angiography (ANGIO) in the localization of proximal left anterior descending coronary (LAD) stenosis in 15 patients. On the left, linear regression of TEE versus ANGIO measurements is graphically represented while, on the right, the difference between angiographic (Angio) and transesophageal echocardiographic contrast-enhanced Doppler measurements (TEE) in locating proximal left anterior descending coronary stenosis (y axis) against their mean (x axis) are plotted (Blandt–Altman method); CB=coronary bifurcation; Diff=difference; Mean=mean difference. (Modified in part from Caiati C et al. Improved Doppler detection of proximal left anterior descending coronary artery stenosis after intravenous injection of a lung contrast agent. A transesophageal Doppler echocardiographic study. J Am Coll Cardiol May 1996;27:1413–18.)

coronary artery (in the study by Aggarwal et al. circumflex coronary stenosis was correctly detected in three patients only by contrast Doppler) and it is conceivable that contrast transesophageal Doppler echocardiography may also refine the evaluation of the left main coronary artery in some doubtful cases. However further studies are needed to confirm the usefulness of contrast transesophageal Doppler echocardiography for left main and circumflex coronary stenosis detection.

In conclusion, transesophageal Doppler echocardiography after contrast enhancement represents an improvement with respect to the traditional ultrasound approaches (namely, B-mode imaging and Doppler echocardiography without contrast injection) for the imaging of coronary arteries.

ACKNOWLEDGEMENT

The authors would like to thank Kate Hunt for her assistance in the preparation of the manuscript.

REFERENCES

1. Weyman A, Feigenbaum H, Dillon J, Johnston K, Eggleton R. Noninvasive visualization of the left main coronary artery by cross-sectional echocardiography. Circulation 1976;54:169–74.

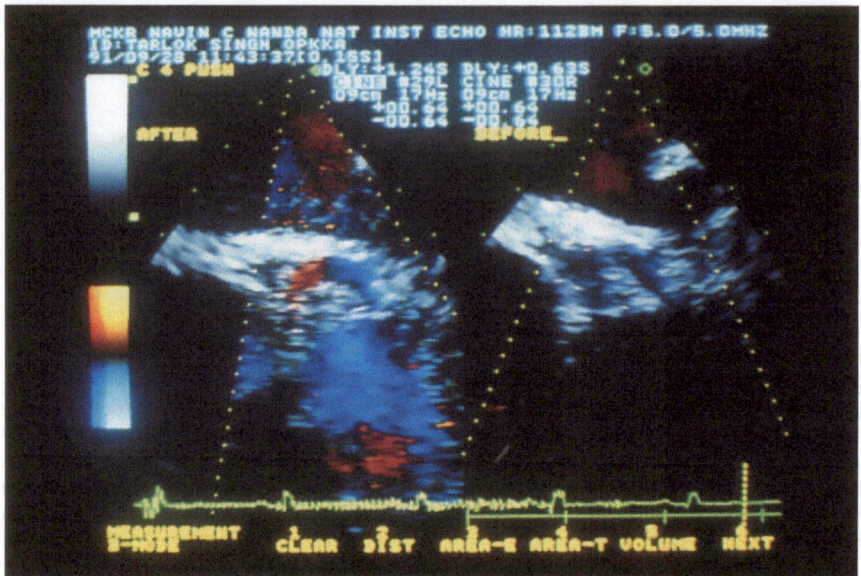

Figure 6. Transesophageal echocardiographic assessment of proximal coronary arteries after echo-contrast enhancement. After contrast injection (left), prominent flow signals fill the lumen of the left main coronary artery and the proximal left anterior descending coronary vessel which show hardly any flow signals at baseline (right). (Reproduced with permission from Aggarwal KK et al. Improved transesophageal echocardiographic assessment of significant proximal narrowing of the left anterior descending and left circumflex coronary arteries using echocontrast enhancement. Am J Cardiol 1994;73:1131–3.)

2. Feigenbaum H. Transthoracic ultrasonic visualization of coronary atherosclerosis. J Am Soc Echocardiogr 1989;2:253–8.
3. Yoshida K, Yoshikawa J, Hozumi T et al. Detection of left main coronary artery stenosis by transesophageal color Doppler and two dimensional echocardiography. Circulation 1990;81: 1271–6.
4. Yamagishi M, Yasu T, Ohara K, Kuro M, Miyatake K. Detection of coronary blood flow velocity associated with left main coronary artery stenosis by transesophageal Doppler color flow echocardiography. J Am Coll Cardiol 1991;17:87–93.
5. Tardif JC, Vannan MA, Taylor K, Schwartz SL, Pandian NG. Delineation of extended lengths of coronary arteries by multiplane transesophageal echocardiography. J Am Coll Cardiol 1994; 24:909–19.
6. Iliceto S, Marangelli V, Memmola C, Rizzon P. Transesophageal Doppler echocardiography evaluation of coronary blood flow velocity in baseline conditions and during dipyridamole-induced coronary vasodilation. Circulation 1991;83:61–9.
7. Stoddard MF, Prince CR, Morris GT. Coronary flow reserve assessment by dobutamine transesophageal echocardiography. J Am Coll Cardiol 1995;25:325–32.
8. Redberg RF, Sobol Y, Chou TM et al. Adenosine-induced coronary vasodilatation during trans-esophageal Doppler echocardiography. Rapid and safe measurement of coronary flow reserve ratio can predict significant left anterior descending coronary stenosis. Circulation 1995;92: 190–6.

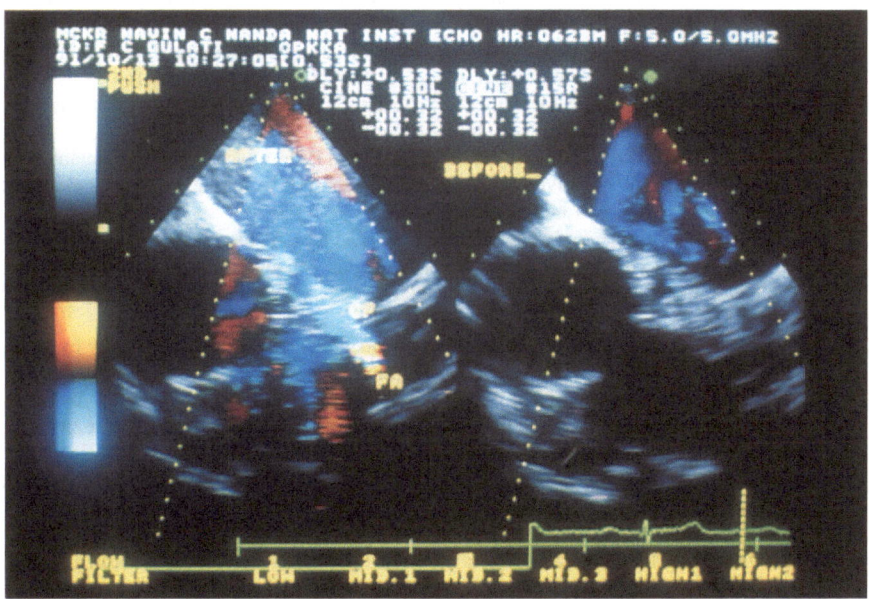

Figure 7. Transesophageal echocardiographic assessment of proximal coronary arteries after echo-contrast enhancement. After baseline (right) virtually no flow signals are noted in the left main coronary artery. After contrast injection (left), prominent flow signals not only completely fill the lumen, but also opacify an additional segment of the proximal left anterior descending coronary vessel that was not visualized at baseline. In addition, flow aliasing (FA) is noted more distally, and its site corresponded to an area of severe stenosis in the left anterior descending artery on the coronary angiogram. CF = the origin of the circumflex vessel. (Reproduced with permission from Aggarwal KK et al. Improved transesophageal echocardiographic assessment of significant proximal narrowing of the left anterior descending and left circumflex coronary arteries using echocontrast enhancement. Am J Cardiol 1994;73:1131–3.)

 9. Hoffman J. Maximal coronary flow and the concept of coronary vascular reserve. Circulation 1984;70:153–9.
10. White C, Wright C, Doty D et al. Does visual interpretation of the coronary angiogram predict the physiologic importance of a coronary stenosis? N Engl J Med 1984;310:819–25.
11. Klein LW, Weintrub WS, Agarwal JB, Seelaus PA, Katz RI, Helfant RH. Prognostic significance of severe narrowing of the proximal portion of the left anterior descending coronary artery. Am J Cardiol 1986;58:42–6.
12. Roberts KB, Califf RM, Harrel FG, Lee KL, Pryor DB, Rosati RA. The prognosis for patients with new-onset angina who have undergone cardiac catheterization. Circulation 1983;68: 970–8.
13. Schlief R, Staks T, Mahaler M, Rufer M, Fritzsch T, Seifert W. Successful opacification of the left heart chambers on echocardiographic examination after intravenous injection of a new saccharide-based contrast agent. Echocardiography 1990;7:1–4.
14. Smith MD, Elion JI, McClure RP et al. Left heart opacification with peripheral venous injection of a new saccharide echo contrast agent in dogs. J Am Coll Cardiol 1989;13:1622.
15. von Bibra H, Sutherland G, Becker H, Neudert J, Nihoyannopoulos N. Clinical evaluation of

left heart Doppler contrast enhancement by a saccharide-based transpulmonary contrast agent. J Am Coll Cardiol 1995;25:500–8.

16. Smith MD, Elion JL, McClure RR. Left heart opacification with peripheral venous injection of a new saccharide echo contrast agent in dogs. J Am Coll Cardiol 1990;25:160–1.

17. Fritzsch T, Hilmann J, Kampfe M, Muller N, Schoobel C, Siegert J. SHU 508 a transpulmonary echo contrast agent: initial experience. Invest Radiol 1990;25:160–1.

18. Iliceto S, Caiati C, Aragona P, Verde R, Schlief R, Rizzon P. Improved Doppler signal intensity in coronary arteries after intravenous peripheral injection of a lung-crossing contrast agent (SHU 508 A). J Am Coll Cardiol 1994;23:184–90.

19. Kozàkovà M, Palombo C, Zanchi M, Distante A, L'Abbate A. Increased sensitivity of flow detection in the left coronary artery by transesophageal echocardiography after intravenous administration of transpulmonary stable echocontrast agent. J Am Soc Echocardiogr 1994;7:327–36.

20. Pearce FB, Sheikh KH, deBruijn NP, Kisslo J. Imaging of the coronary arteries by trans-esophageal echocardiography. J Am Soc Echocardiogr 1989;2:276–83.

21. Yamagishi M, Miyatake K, Beppu S et al. Assessment of coronary blood flow by trans-esophageal two-dimensional pulsed Doppler echocardiography. Am J Cardiol 1988;62:641–4.

22. Taams MA, Gusssenhoven EJ, Cornel JH, Roelandt JRTC, Lancée CT, Brandt MVD. Detection of left coronary artery stenosis by transesophageal echocardiography. Eur Heart J 1988;9:1162–6.

23. Reichert SLA, VIsser CA, Koolen JJ et al. Transesophageal examination of the left coronary artery with a 7.5 MHz annular array two-dimensional color flow Doppler transducer. J Am Soc Echocardiogr 1990;3:118–24.

24. Iliceto S, Memmola C, De Martino G, Piccinni G, Rizzon P. Visualization of the coronary artery using transesophageal echocardiography. In: Erbel R, Khanderia BK, Brennecke R, Meyer J, Seward JB, Tajik AJ, editors. Transesophageal Echocardiography. Berlin: Springer-Verlag, 1989:86–98.

25. Samdarshi TE, Nanda NC, Gatewood RP Jr et al. Usefulness and limitation of transesophageal echocardiography in the assessment of proximal coronary artery stenosis. J Am Coll Cardiol 1992;19:572–80.

26. McAlpine WA. Heart and Coronary Arteries. New York: Springer-Verlag, 1975:180.

27. Yoganathan AP, Cape EG, Sung H-W, Williams F, Jimoh A. Review of hydrodynamic principles for the cardiologist: applications to the study of blood flow and jets by imaging techniques. J Am Coll Cardiol 1988;12:1344–53.

28. Aragam JR, Mai J, Guerrero JL et al. Doppler color flow mapping of epicardial coronary arteries: initial observations. J Am Coll Cardiol 1993;21:478–87.

29. Schrem SS, Tunick PA, Slater J, Kronzon I. Transesophageal echocardiography in the diagnosis of ostial left coronary artery stenosis. J Am Soc Echocardiogr 1990;3:367–73.

30. Caiati C, Aragona PL, Iliceto S, Rizzon P. Improved Doppler detection of proximal left anterior descending coronary artery stenosis after intravenous injection of a lung contrast agent. A transesophageal Doppler echocardiographic study. J Am Coll Cardiol 1996;27:1413–18.

31. Aggarwal KK, Gatewood RP, Nanda NN, Chopra KL. Improved transesophageal echocardio-graphic assessment of significant proximal narrowing of the left anterior descending and left circumflex coronary arteries using echo contrast enhancement. Am J Cardiol 1994;73:1131–3.

20. Left ventricular contrast echocardiography – echoventriculography – stress echocardiography

RAIMUND ERBEL, ROMAN LEISCHIK, CHRISTIAN BRUCH,
RAINER ZOTZ, SUSANNE MOHR-KAHALY, FRANK SCHÖN,
ECKHARD STEINMETZ & RÜDIGER BRENNECKE

INTRODUCTION

Echographic contrast agents were first used for structure identification [1, 2], and pooled data analysis revealed a high level of safety [3]. Special agents have been developed in order to standardize right heart opacification [4] and to enhance echocardiographic Doppler signals [5–9]. The latter effect was not expected [9]. All agents contain between 12 and 35 μl/ml air, as reported by others [9]. The stability lasts for more than 5 min. Injections of up to 20 ml are necessary. The osmolality is between 345 and 1740 mOsm/kg H_2O [9].

   Our interest in contrast echocardiography started when we determined left ventricular volumes by apical two-dimensional echocardiography [10, 11], which produced systematic underestimation of left ventricular volumes compared with cineventriculography. In order to exclude any possible variable such as the effect of contrast agents, differences in contractility, heart rate, respiration and blood pressure, two-dimensional echocardiograms were recorded during cineventriculography [12]. It was noticed that each X-ray contrast injection into the left ventricle resulted in a full opacification of the chamber with echocontrast (Figure 1). This effect was produced by the injection of 40–60 ml of X-ray contrast material and observed in every patient during angiography. We therefore chose small amounts of echocardiographic contrast agents which were used for right heart visualization after injection into the left ventricle.

GELATINE SOLUTION

Left ventricular contrast

At the start of contrast echocardiography, we tested saline, dextrane, polyglycane and other plasma expanders as well as cardiogreen. The contrast effect of saline was produced only following hand agitation and air mixing and the effect was not very reproducible. Cardiogreen is expensive but effective, while dextrane

325

N. C. Nanda, R. Schlief and B. B. Goldberg (eds.), Advances in Echo Imaging Using Contrast Enhancement, Second Edition, 325–345.
© 1997 *Kluwer Academic Publishers.*

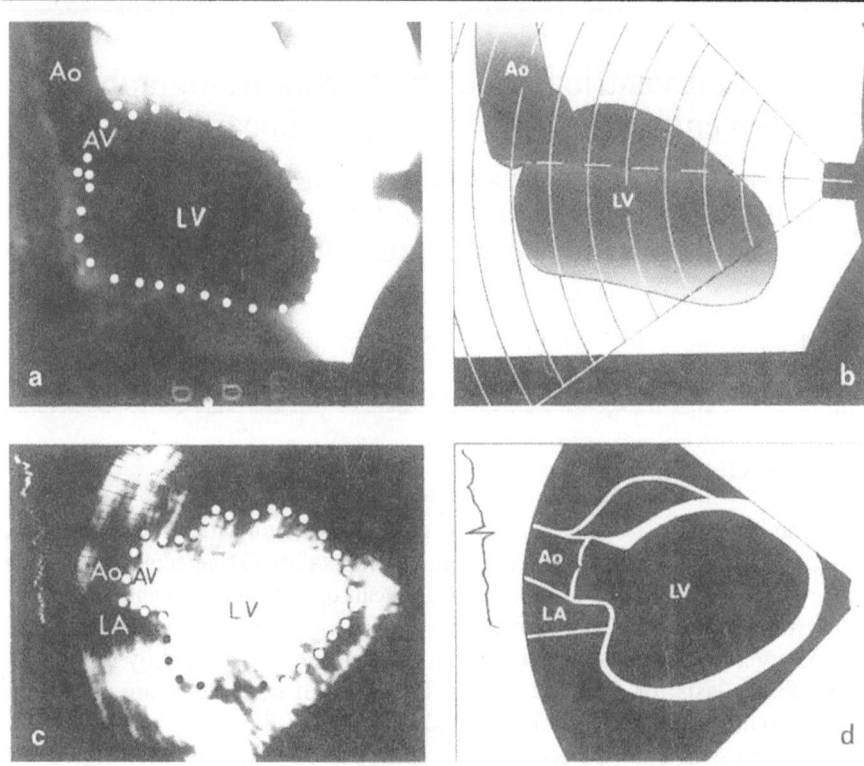

Figure 1. Cineventriculography and simultaneous two-dimensional echocardiogram from the apical approach demonstrating contrast in the left ventricle (LV) and full opacification of the cavity. The position of the echo transducer to receive the apical two-dimensional echocardiogram is visualized.

and polyglycane are not very promising. The plasma expander Gelifundol is a solution used in heart surgery [13] which contains 55 g oxypolygelatine, 5.84 g NaCl, 2.52 g $NaHCO_3$, 0.19 g ethylenediamine tetra-acetic acid and 0.07 g CaCl in 1000 ml. The contrast effect was easily produced and a high reproducibility was noticed. This method has also been used by other groups (e.g. Angelski and Hajkinjak [14]).

For agitation, a stopcock containing 0.1–0.2 ml air is currently used. The gelatine solution in a volume of 10 ml is injected into an empty syringe. The air in the stopcock is enough to produce a milky appearance of the yellow plasma expander which persists for 10–20 s. For right heart contrast, 2–3 ml gelatine solution are injected into a cubital vein. For the left ventricle a rapid injection of 0.5–1 ml (which first fills the pigtail catheter) is used in order to obtain full opacification of the cavity. No parts of the ventricle are empty (Figure 2) and clear separation of cavity and myocardium is produced. Opacification of the left ventricle, however, results in a reduced differentiation of cavity from the

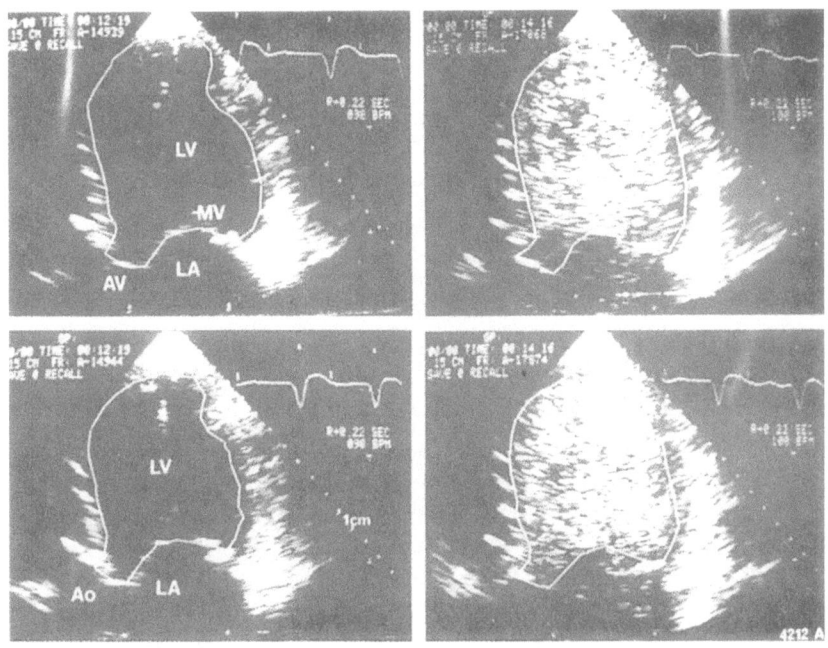

Figure 2. Echoventriculography after injection of 1 ml gelatine solution (Gelifundol) into left ventricle with full opacification and mild mitral regurgitation. LV/LA=left ventricle/atrium, MV=mitral valve, AV=aortic valve, AO=aorta.

myocardium for some seconds. This problem persists during a few heart cycles, before the contrast in the left ventricle decreases and myocardial contrast enhancement appears. During the next 5–10 beats, contrast enhancement of the left ventricular myocardium occurs, and consequently improves the delineation of the endocardium (Figure 3) [13]. This effect is particularly present in patients with left ventricular hypertrophy due to hypertension or aortic disease. Thus, these studies have shown that left ventricular or aortic contrast injection results in an opacification of the myocardium contrast. This method may therefore be used for myocardial perfusion studies.

Safety

Using pigtail catheters with high-fidelity tip manometers, pressure measurements were recorded and the maximal and minimal dP/dt measured before, during and after 1 ml gelatine solution injected into the left ventricle [15].

In 19 patients, 84 injections of 1 ml Gelifundol, 53 into the left ventricle and 31 into the aorta, were performed. During and after the injection, the patients had no sensation and no feeling of the injection. Side-effects such as dizziness,

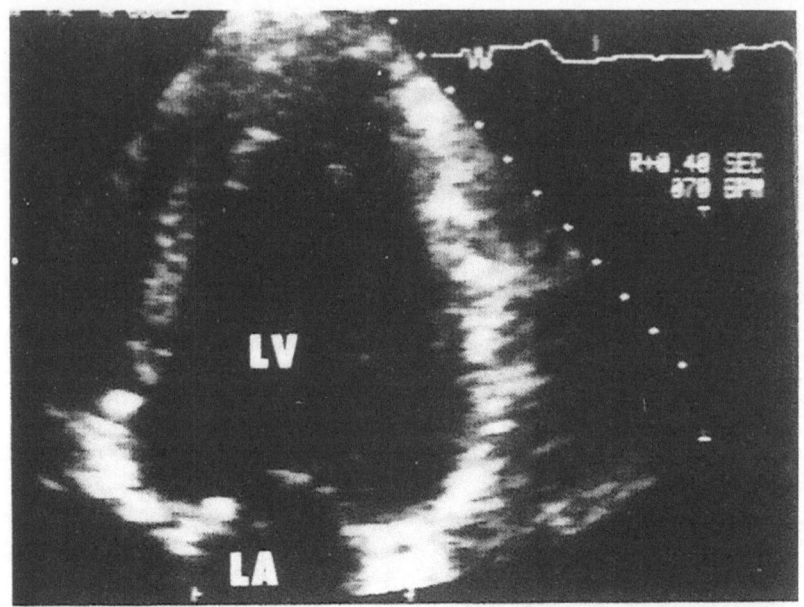

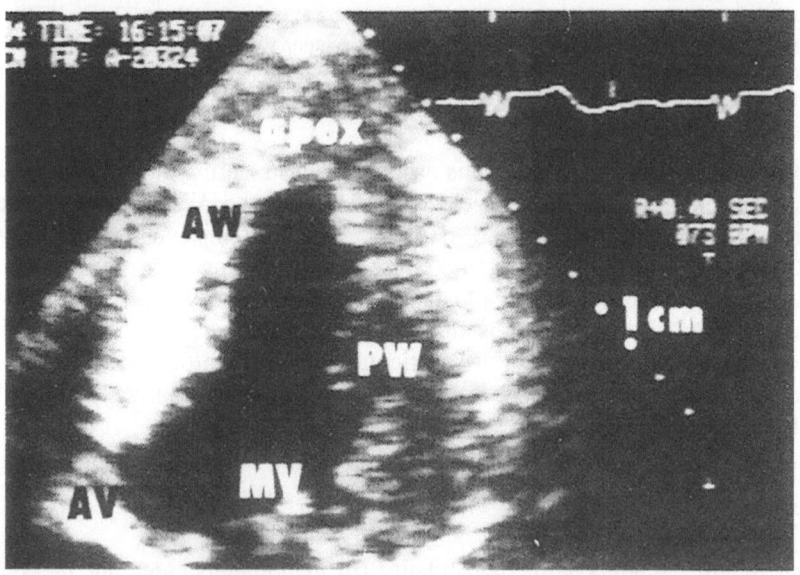

Figure 3. Apical two-dimensional echocardiogram in the RAO equivalent view (a) before and after injection of echocontrast into left ventricle with enhancement of the myocardial contrast, (b) enabling a better outlining of the left ventricular endocardium during the myocardial passage of echocontrast. LA/LV = left atrium/ventricle, AW/PW = anterior/posterior wall, PM = papilllary muscle, AV = aortic valve [13].

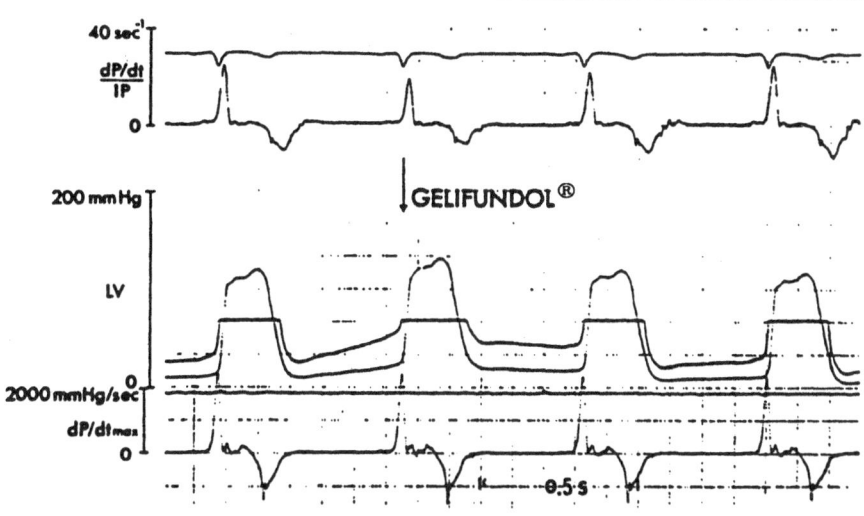

Figure 4. Recording of left ventricular systolic and diastolic pressure using a pigtail high-fidelity tip manometer which demonstrate the dP/dt max and min as well as the parameter dP/dt–IP. Due to a methodological artefact of the cold gelatine solution with the injection, end-diastolic pressure increased [15] without any change of the contractility parameters for 2 beats.

dyspnoea and angina were not observed. All patients were in sinus rhythm. In five patients ventricular extrasystoles occurred. In two patients a transient ST-segment depression was observed. One patient experienced an ST-segment elevation without angina after 21 s. The ECG changes disappeared spontaneously and no heart rate changes were observed [15].

During the X-ray contrast injection for cineventriculography, heart rate increased by 20 beats/min in four of 10 patients; supraventricular/ventricular extraystoles occurred in two patients and multiple extrasystoles in one patient. ST-segment depression was observed in one patient. An increase of end-diastolic pressure resulting from the cold fluid passing the tip manometer was observed for the first beat (Figure 4). For beats 2–10 after the injection no change in relation to the starting level was found. There was no significant change of dP/dt_{max} or dP/dt_{min} (Figures 5, 6).

Injection of gelatine solution into the aortic root caused no change in systolic or diastolic pressure except for the first beat during the injection.

Calculation

Calculation of left ventricular volume in end-diastole and in end-systole showed that the underestimation by echocardiography, compared with cineventriculography, was reduced after contrast injection [15]. Instead of outlining the inner contour of

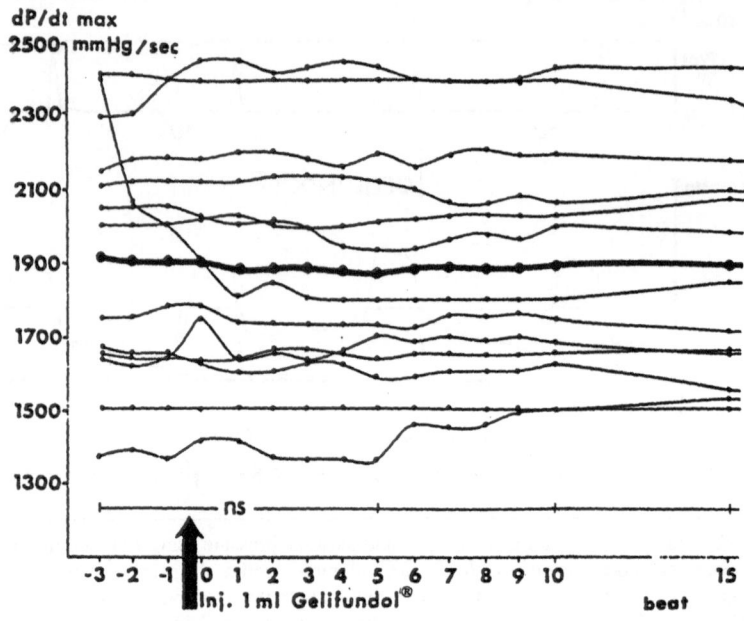

Figure 5. Behavior of dP/dt_{max} before and after injection of 1 ml Gelifundol. No significant change is visible.

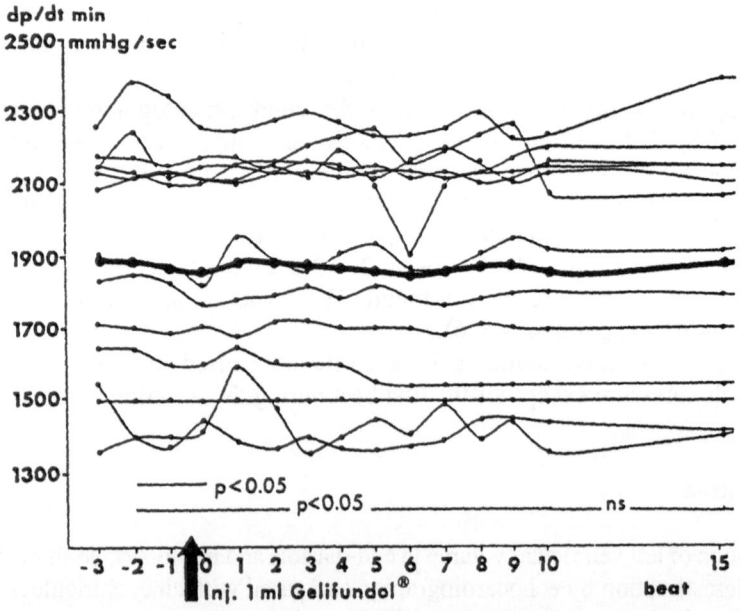

Figure 6. Recording of the dP/dt_{min} before and after injection of 1 ml Gelifundol.

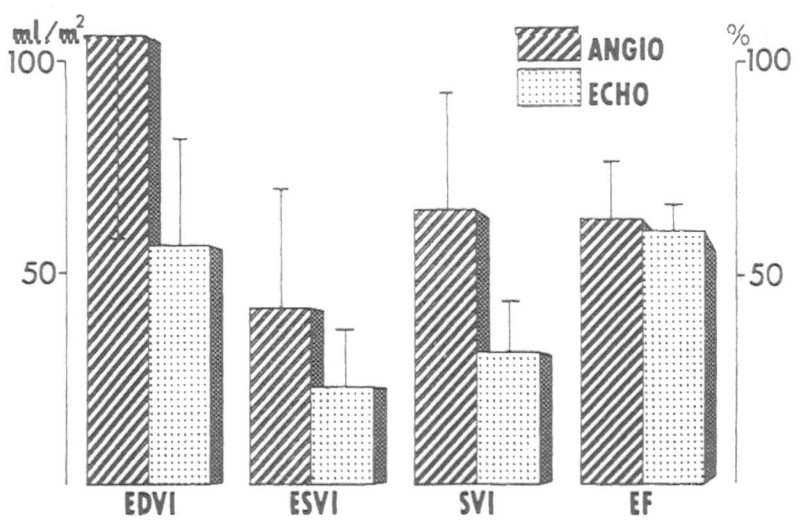

Figure 7. End-diastolic and end-systolic volume (EDVI/ESVI) as well as stroke volume index (SVI) and ejection fraction (EF) determined from angiography and simultaneous recorded apical two-dimensional echocardiograms after contrast injection (*n*=9).

the left ventricular trabeculation, a more distal outlining was seen as the contrast material entered the space between the trabeculations. This was most important for outlining the end-systolic contour. Therefore, the ejection fraction determined by two-dimensional echocardiography was in the same range as that measured by cineventriculography. No systematic differences were present (Figure 7) [13].

Clinical value

Because of the safety and easy use of echoventriculography, we started to use this method in patients with allergic reactions to X-ray contrast agents, or with severe heart failure, renal insufficiency or thyrotoxicosis, in order to determine ventricular volumes and ejection fraction [13, 14, 16, 17]. Mitral and aortic insufficiency can also be calculated [18], which is useful for intraoperative control of mitral valve reconstruction [19, 20].

Between 1984 and 1991 we used this agent in about 100 patients without observing severe side-effects such as emboli, stroke or death. A few patients complained of short periods of transient dizziness.

COLOR SUPERPOSITION

In an attempt to use digital imaging techniques for improvement of left ventricular volumes and ejection fraction, echocontrast subtraction techniques were

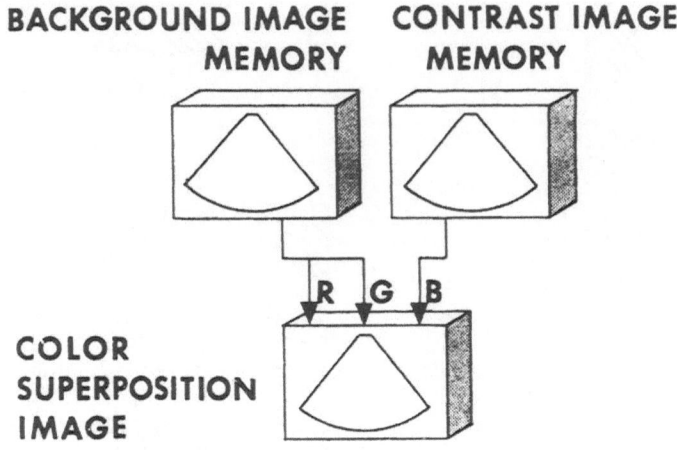

**BACKGROUND IMAGE
MEMORY**

**CONTRAST IMAGE
MEMORY**

R G B

**COLOR
SUPERPOSITION
IMAGE**

Figure 8. Schematic drawing of the color superposition method to enhance echocontrast images. RGB = red-green-blue channel [23].

developed for use in left [16, 21] and right heart [22]. Delineation of the endocardium still remained a problem. The technique worked only in high quality images. After contrast injections, the opacification of the left ventricle reduces the separation of the cavity from the myocardium.

Color superposition was therefore developed [23, 24]. Figure 8 shows the principle of separating the channel: blue (for the contrast image and the image without color) with the red and green resulting in a blue contrast coding. By superposition of both images, the cavity is color-coded. In addition, by using a subtraction technique, the outlining can be enhanced (Figure 9). This technique enhances the ability to outline contrast images. A typical example is shown in Figure 10. It has been shown by our group [25] that left ventricular volumes are larger than those in the negative picture and that ejection fraction determination is no longer significantly underestimated.

This technique is relatively simple and can be used with every echo machine. Analysis of left ventricular function after peripheral contrast injection can therefore be evaluated with higher reproducibility.

STATISTICAL METHODS

Principle

In order to enhance the accuracy of calculation of the left ventricular volume, particularly in patients with reduced image quality, we developed a sophisticated method using statistical principles [26, 27].

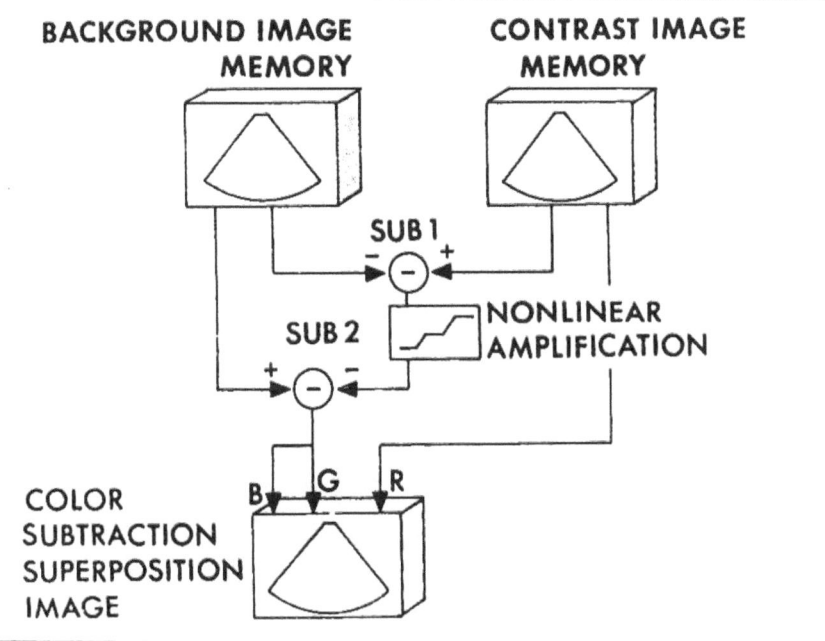

Figure 9. Schematic drawing of the method for subtraction superposition [23].

Before contrast injection, eight heart cycles are ECG-triggered, digitized and stored. For each pixel the ratio of the mean values and the standard errors are calculated and illustrated. Two heart cycles are then sampled following contrast injection and digitized and plotted as illustrated in Figure 11. The plot of all pixels shows, for the first 8 heart cycles, a diffuse scatter at the line of identity. In addition, a dense pixel scatter is found near the X-axis, representing the two contrast heart cycles with different scatters due to the contrast filled cavity. From this scatter, the contrast scatter is selected (Figure 12). The reconstruction of this scatter is shown in Figure 13A–C. The outlining of this contrast image into the native image (Figure 13A) shows that the contour is now within the myocardium at the outer contour of the trabeculation of the myocardium (Figure 13D).

Calculations

Figure 14 demonstrates the correlation of determination of the ejection fraction by apical two-dimensional echocardiography and cineventriculography in 21 consecutive patients studied in the catheterization laboratory. There was no selection of patients, and even patients with reduced image quality were included. Patients were studied on the catheterization table [28]. The slope of the regression

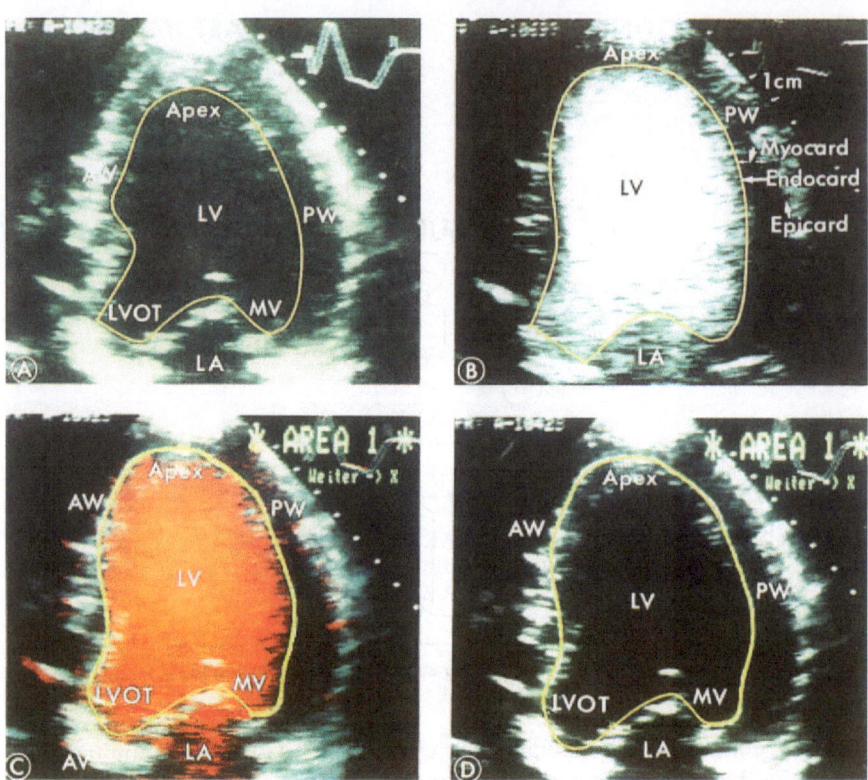

Figure 10. Color superposition. (A) Native apical two-dimensional echocardiogram with outlining of the endocardial contour. (B) Echocontrast injection into the left ventricle with outlining
of the contrast silhouette of the left ventricular cavity. (C) Color superposition of the contrast image
to the native image with outlining of the red area. (D) Superposition of the contrast superposition
outline area to the native image which demonstrates that the contour is within the myocardium
representing the outer contour of the myocardium trabeculation.

line improved and reached the line of identity. The variation decreased and the
standard error of the estimate dropped, when the statistical method (right side)
was used instead of outlining the native picture (left side) [28].

Conclusion

It could thus be demonstrated that digitized echocardiographic images can be
evaluated using statistical methods, yielding important additional information
and enhancing the accuracy of left ventricular volume and ejection fraction determination. This is an additional method which can also be used after peripheral
contrast injection in order to obtain improved volume determination.

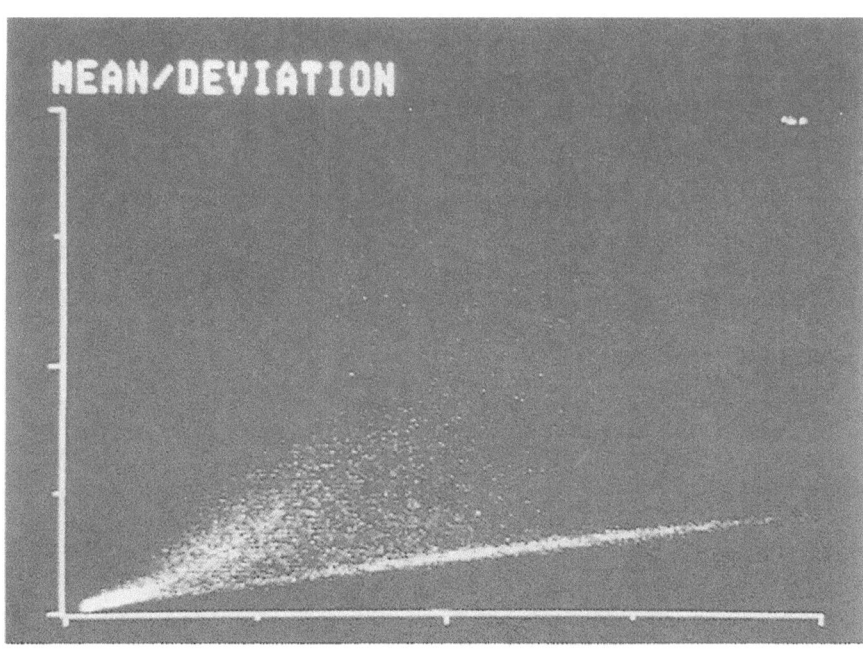

Figure 11. Scattergram of eight native and two contrast images. The scatter of each pixel demonstrates the relation between mean and standard deviation. Close to the X-axis a dense scatter represents the two contrast images.

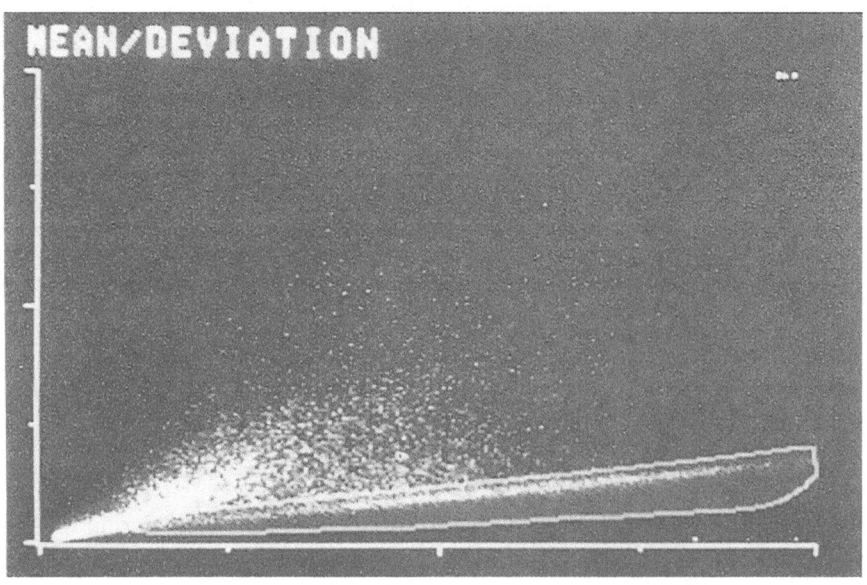

Figure 12. Selection of the two echocontrast images from eight images without contrast. This scattergram is used in construction of the 2D image.

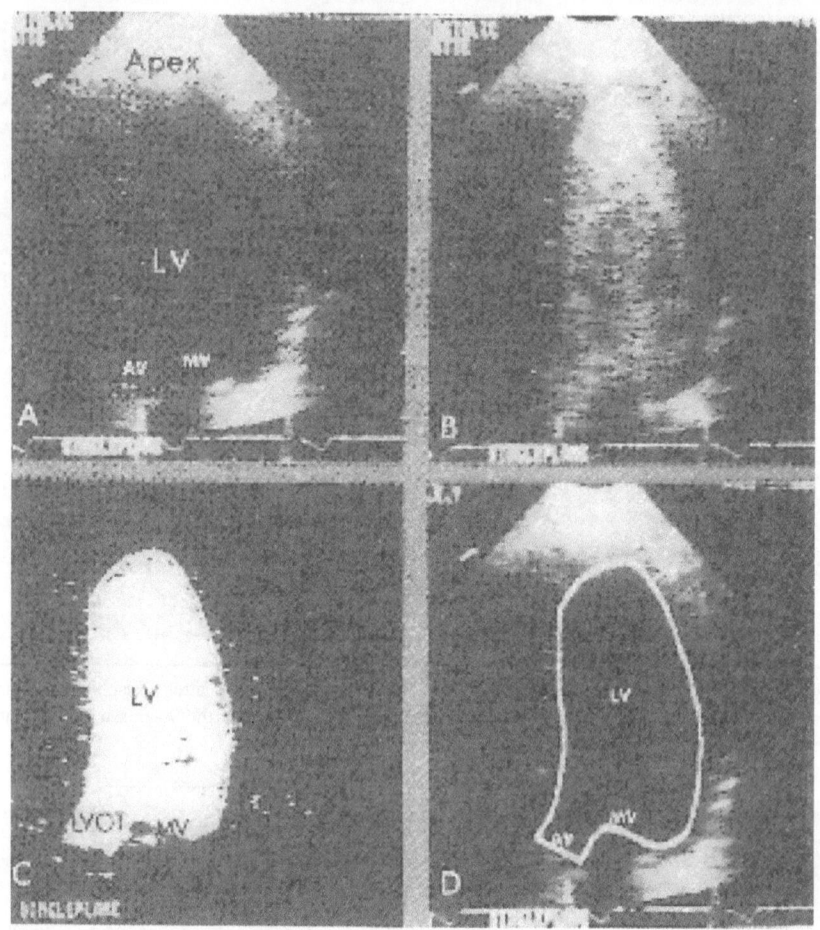

Figure 13. Statistical method for improving the outlining of the left ventricular cavity. A = Native image of lower quality. B = Echocontrast image after injection of 1 ml Gelifundol. C = Statistically reconstructed image from eight native and two contrast images with the selection of the scatter as visualized in Figures 11 and 12. D = Superposition of the outlined image C to the native image A. The outlining demonstrates that the contour is within the myocardium representing the outer part of the trabeculation of the myocardium.

LEFT SIDE CONTRAST AGENTS

Albunex®, SHU 508

Because bubbles dissolve rapidly due to surface tension effects during the traverse of the capillary bed of the lungs, no contrast can be expected in the lungs following i.v. injection of contrast agent [29]. The use of wedge injections for

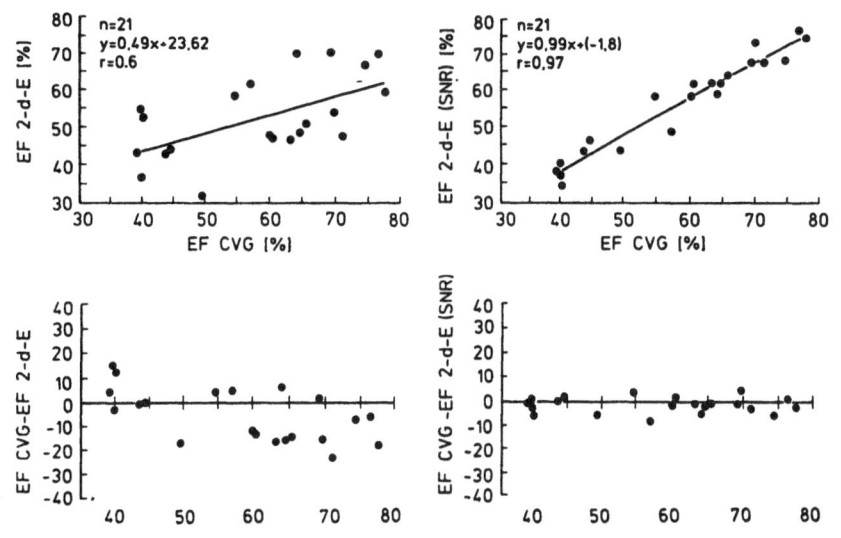

Figure 14. Linear regression between the ejection fraction determined by cineventriculography (CVG) and two-dimensional echocardiography (2d-E) for native images (left side) and statistically analyzed images (right side). The improved regression line near the line of identity with the reduced variability and the smaller scatter is visualized.

left heart echocontrast was attempted, but not used for routine application [17, 30].

The next step was the sonication of echocontrast agents such as radiopaque material and albumin [31]. The bubbles produced were between 6 ± 2 and $8 \pm 4 \, \mu m$, and could thus pass the lungs and opacify the left heart [31, 32]. This method was also used for direct injection of sonicated contrast agents in order to produce myocardial contrast [32–35]. Berwing and Schlepper [36] developed an echocontrast agent consisting of air, lecithin and soya oil fractionate and sodium–iron (III) gluconate complex.

The intensity which is reached in the left ventricle represents roughly 30% of that in the right ventricular cavity (Figure 15). The success rate of opacifying the left ventricle ranges from 70 to 90% using a saccharide (galactose) solution (SHU 508) [38–40]. The success rate seems to be dependent on the pulmonary blood pressure, suggesting that the bubbles are in part destroyed in the pulmonary artery with increasing pressure [41]. Another explanation could be that the ultrasound waves destroy the microcavitations. This can be avoided by use of intermittent imaging.

The apex is not well visualized in consequence of microbubble destruction by ultrasound waves and decreased blood flow in the apical region. In the end-systole the increased pressure in the left ventricle additionally destroys the

338   *Raimund Erbel et al.*

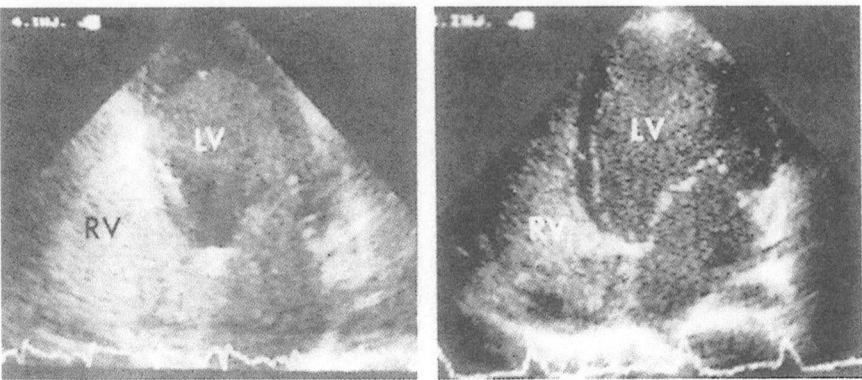

Figure 15. Left heart opacification after intravenous injection of Albunex or SHU 508 i.v. Density in the left ventricle (LV) is approximately 30% of that in the right ventricle (RV).

microcavitations [32]. The contrast effect in the apex can be improved using second harmonic and intermittent imaging. No changes of heart rate or blood pressure were observed and no ECG changes occurred.

We used both drugs to opacify the left ventricle and calculated volumes and ejection fraction for direct contrast injection as described above. The results were comparable to those observed for the direct contrast injection, End-diastolic and end-systolic volumes were larger than those of native echocardiographic images. Determination of ejection fraction was improved and correlated very well to angiography. Reproducibility was improved and side-effects were rare.

HOE 155

In order to be able to produce bubbles with definite size and shape for increasing the reproducibility, a new contrast agent was developed. HOE 155 consists of polymers with encapsulated air (Hoechst, Germany). The first experimental data showed that, after intravenous injection, the right and left heart are opacified [42]. Clinical data are not yet available.

BY 963

BY 963 is a second-generation transpulmonary ultrasound contrast agent comprising an aqueous solution of a phospolipid (of soya origin), a sugar, and additional surface-active ingredients [45]. The essential component of this solution is the surface-active phospholipid, which stabilizes the microbubbles produced during cavitation by covering the microbubbles with a membrane. The

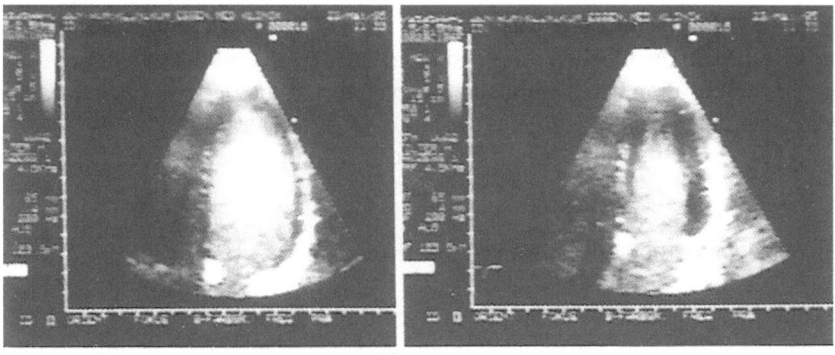

Figure 16. Four-chamber view at exercise: left ventricular cavity after administration of 5 ml of BY 963. (a) End-diastolic frame. (b) End-systolic frame. As described, the enhanced border resolution in the left ventricular cavity improves the reproducibility of calculations of left ventricular volume and ejection fraction at exercise.

mean size of the microbubbles is $3.95\,(\pm 0.23)$mm. BY 963 is supplied as a 5 ml solution which is agitated in a special sterile cavitation chamber containing 0.18 ml air, which is mixed mechanically with a 5 ml standard volume of BY 963. The final solution is prepared by movement of the solution 11 times between two 5 ml syringes through the cavitation chamber [45]. BY 963 produced good opacification of the left ventricular cavity in dogs [45, 46] and in humans [47, 48]. This new echocontrast agent has the potential to increase the myocardial intensity to 20% in dogs [49]. It is also possible to enhance the intensity of the liver [50]. Intravenous administration of BY 963 during exercise echocardiography improves interobserver reproducibility in the quantitative assessment of left ventricular volumes and ejection fraction [48]. The clear delineation of the endocardial border during exercise is shown in Figure 16.

CLINICAL IMPLICATIONS

Echoventriculography – contrast echocardiographic opacification of the left ventricle – permits determination of volume and ejection fraction with a high reproducibility, particularly when the images are outlined by contrast enhancement of the myocardium. Perfusion defects can be determined [43]. Color superposition or statistical methods can be used further to improve delineation of the left ventricular endocardium. Indications for echoventriculography are seen in patients with low image quality. Echoventriculography can be used to avoid cineventriculography in patients with heart failure, reduced renal function, allergic reactions against contrast agents and thyrotoxicosis [44]. In addition, left ventricular contrast injection permits estimation of severity of mitral regurgi-

tation [18] which is used during surgery in order to control mitral valve reconstruction [19, 20]. The widespread use of left heart opacification will reduce the
number of patients in whom contrast cineventriculograms are performed. Thus,
heart catheterization will be limited only to coronary angiography. This may result
in a reduction of costs and an increase in the patient's safety.

Parameters of left and right ventricular function can be obtained [32–36].

Stress echocardiography and left ventricular contrast

One very important field for application of transpulmonary echo-enhancers will
be the use of echocontrast agents in stress echocardiography. Failure to image
the endocardial border during stress echocardiography may have important
implications for clinical decision making in daily routine: the diagnosis of significant coronary disease may be compromised if an ischemic segment is not detected.

Sonicated albumin [51, 52], SHU 508A [53, 54] and BY 963 [48] allow
improved depiction of segmental motion and delineation of the endocardial
border in stress echocardiograms.

Beckmann et al. [53] performed stress echocardiography in 10 patients with
suboptimal image quality without and with injection of SHU 508A. In these
patients, he observed an improvement in the visual detection of the left ventricular border. Interobserver variability was not evaluated.

Schröder et al. [54] evaluated the intraobserver variability in assessing the
delineation score of 10 patients at rest and during pharmacological stress after
i.v. administration of SHU 508A. The variability in the delineation score without
SHU 508A was 13.4%; this improved to 6.4% after i.v. injection of the echocontrast agent. The ejection fraction following i.v. SHU 508A injections could
only be assessed quantitatively in 40% of the patients.

Porter et al. [51] infused sonicated albumin during dobutamine stress
echocardiography. The number of contrast agent-enhanced cardiac cycles in the
left ventricular cavity was significantly greater at peak infusion. Endocardial
border resolution was improved (mainly in lateral segments) in 93% of patients
at low dose and in 95% of patients at peak infusion. There was significantly
better left ventricular peak video intensity at peak dobutamine infusion than
after the same injection at baseline ($p < 0.005$, analysis of variance). This observation seems to be similar to our experience during peak exercise [48]. It is
possible that increased cardiac output improves the transpulmonary passage of
microcavitations and caused the improvement of the contrast effect in the left
ventricular cavity.

Similar to Porter et al. [51], Falcone et al. [52] described a significant increase
in enhancement of the endocardial border with low-dose and high-dose dobutamine after Albunex compared with baseline. Of 179 segments with suboptimal
enhancement at baseline, 137 (77%) became optimal during dobutamine stress
with Albunex. These authors concluded that Albunex improves localization of

LV endocardial contours and that this localization is enhanced during pharmacological stress.

The results of investigation with the contrast agent BY 963 [48] differ from the results described by Schröder et al. [54]. In an exercise echocardiography study with BY 963, good contrast of the left ventricular cavity could be achieved in all volunteers, and the ejection fraction could be estimated in all cases. Schröder et al. [54] described this as being possible in only 40% of patients.

Beckmann et al. [53] found that the end-systolic left ventricular volumes estimated after injection of contrast agent were smaller than those obtained without contrast agent. The improved delineation of the endocardial border after i.v. administration of a contrast agent leads to higher values of ejection fraction. This fact has to be considered for future quantitative studies of left ventricular function during exercise. The 'normal values' of left ventricular ejection fraction estimated during conventional exercise echocardiography have to be re-investigated.

Intravenous administration of BY 963 improves the reproducibility of quantitative analysis of left ventricular function in normal volunteers. The estimation of the end-systolic volume at exercise is improved significantly and the estimated EF values are higher than those obtained from conventional exercise (stress) echocardiograms. The combination of intravenous contrast with left ventricular opacification and stress echocardiography may be an ideal tool to improve interobserver variability for quantitative assessment of left ventricular function, automatic or subjective endocardial border detection and wall motion analysis in patients with suboptimal image quality. In addition, second harmonic technique and intermittent imaging improve and prolong the contrast effect in the left ventricular cavity. The combination of both these techniques permits the use of transpulmonary contrast agents for left ventricular function analysis even in routine patients.

SUMMARY

The opacification of the left ventricle by echocardiographic contrast agents represents an alternative to cineventriculography, as determinations of left ventricular volume and ejection fraction are accurate and highly reproducible even during exercise echocardiography or when methods such as color superposition, statistical imaging techniques or new techniques such as color kinesis are used in order to improve the outlining of the cavity and endocardial border.

Detection of perfusion defects is likely to be possible after i.v. injection of contrast agents [40] in the future, but really useful detection of perfusion defects by i.v. contrast echocardiography is not yet possible. The enhancement of myocardial contrast during the perfusion phase after ventricular opacification improves delineation of the endocardial border. For practical purposes, the direct

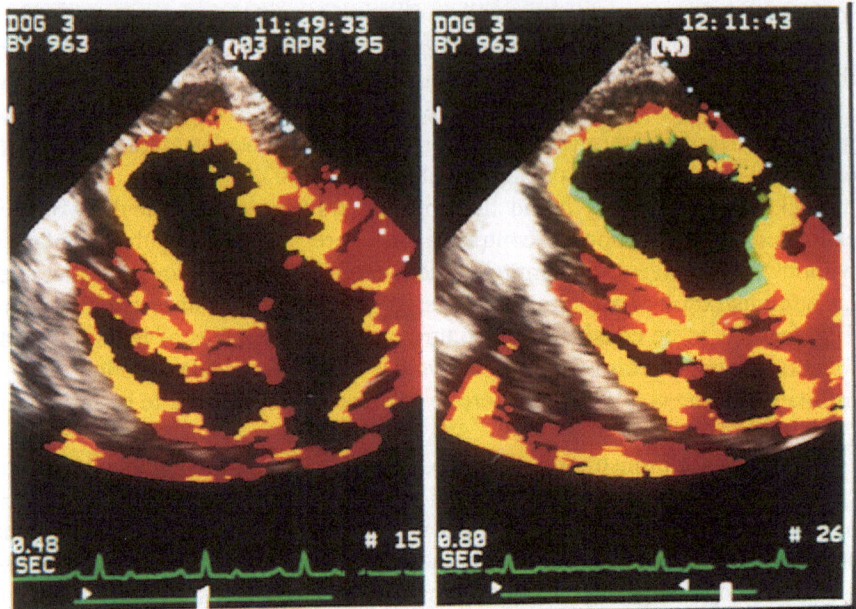

Figure 17. Four-chamber view in the conscious dog. Right: improved imaging of the endocardium using color kinesis after administration of the lung-permeable echocontrast medium (BY 963). Left: original, native image.

injection of echocontrast into the aortic root or by the intracoronary route is inferior to indirect opacification after peripheral venous injection, which can be achieved with sonicated albumin (Albunex, Infoson), SHU 508A (Levovist), BY 963, HOE 155, FSO 69 or Echogen. FDA approval currently exists only for the first generation of echocontrast agents: Albunex and Levovist; the remaining agents are undergoing experimental and clinical trials. Using all these different contrast agents, sophisticated computer techniques have to be built into echocardiographic machines in order to improve the determination of left ventricular volumes, ejection fraction and wall motion abnormalities. The use of new ultrasound techniques such as tissue Doppler imaging [55, 56], color kinesis [57, 58] (Figure 17), second harmonic imaging [59] and intermittent imaging may enhance the feasibility and accuracy of quantitative echocardiography or stress echocardiography. Despite all this potential, the technique may have several limitations (financial implications, unsuccessful left ventricular contrast in some cases). However cineventriculography or radionuclide ventriculography will be rarely performed in the future as echoventriculograms already show left ventricular contraction. In most cases this will result in reduced side-effects and costs.

REFERENCES

1. Gramiak R, Shah PM, Kramer DH. Ultrasound cardiography: Contrast studies in anatomy and function. Radiololgy 1969;939.
2. Kronik G, Hutterer B, Mösslacher H. Diagnose atrialer Links-rects-Shunts mit Hilfe der zweidimensionalen Kontrastechokardiographie. Z Kardiol 1981;70:138–45.
3. Bommer WJ, Shah PM, Allen H, Meltzer R, Kisslo J. The safety of contrast echocardiography: Report of the Committee on Contrast Echocardiography for the American Society of Echocardiography. J Am Coll Cardiol 1984;3:6–13.
4. Grube E, Fritsch T. Verbesserte Reproduzierbarkeit der Kontrastechokardiographie durch SHU 454. Experimentelle Untersuchungen mittels digitaler Subtraktionsechokardiographie. Z Kardiol 1986;75:355–62.
5. Becher H, Schlief R. Improved sensitivity of color Doppler by SHU 454. Am J Cardiol 1989; 64:374–7.
6. Beppu S, Tanabe K, Shimizu T, Ishikura F, Nakatani S, Terasawa A, Matsuda H, Miyatake K. Contrast enhancement of Doppler signals by sonicated albumin for estimating right ventricular systolic pressure. Am J Cardiol 1991;67:1148–50.
7. Zotz R, Duwe L, Erbel R, Frank P, Brennecke R, Meyer J. Rechtsherzechoventrikulographie mittels Gelifundol. Z Kardiol 1989;78:95–100.
8. Von Bibra H, Stempfle HU, Poll A, Schlief R, Emslander H. Echo contrast agents improve flow display of color Doppler. Echocardiography 1991;8:533–40.
9. Irion A. Gegenwärtiger Stand der Entwicklung von Echo-kontrastmitteln. In: Schlepper M, Berwing K, editors. Kontrast-, Doppler-, Farb-Doppler- und transösophageale Echokardiographie, Stuttgart-New York: Schattauer. 1990:9–16.
10. Erbel R, Schweizer P, Meyer J, Grenner H, Krebs W, Effert S. Left ventricular volume and ejection fraction determination by cross-sectional echocardiography in patients with coronary artery disease. Clin Cardiol 1980;3:377–83.
11. Erbel R, Schweizer P, Meyer J, Effert S. Apikale zweidimensionale Echodardiographie. Normalwerte für die monoplane und biplane Bestimmung der Volumina und der Ejektionsfraktion des linken Ventrikels. Dtsch Med Wschr 1982;107:1872–7.
12. Erbel R, Schweizer P, Lambert H et al. Echoventriculography – a simultaneous analysis of two-dimensional echocardiography and cineventriculography. Circulation 1983;67:205–15.
13. Schreiner G, Mohr-Kahaly S, Erbel R, Meyer J. Echoventrikulographie mittels Gelifundol anstelle de cineventrikulographie. In: Erbel R, Meyer J, Brennecke R, editors. Fortschritte der Echokardiographie. Berlin: Springer. 1985:152–7.
14. Angelski R, Hajkinjak J. Contrast echocardiography. Presentation of diagnostic method at patients with atrial septal defect. Zdrav Vestn 1986;55:223–7.
15. Mohr-Kahaly S, Erbel R, Zotz R, Duwe L, Schreiner G, Meyer J. Linksventrikuläre Kontrastechokardiographie mittels Gelifundol – Beurteilung von Hamodynamik, Kontraktilität und EKG-Verlauf. Z Kardiol 1987;76:699–705.
16. Grube E, Lampen M, Becher H. Echokontrastventrikulographie – Bestimmung linksventrikulärer Funktionsparameter unter besonderer Berucksichtigung der digitalen Subtraktionsechocardiographie. Z Kardiol 1986;75:650–8.
17. Roelandt J, Meltzer RS, Serruys PW. Contrast echocardiograpy of the left ventricle. In: Rijsterborgh, editor. Echocardiology. The Hague: Martinus Nijhoff Publishers. 1981:219–32.
18. Reid CL, Kawanishi DT, McKay CR, Elkayam U, Rahimtoola SH, Chandraratna PAN. Accuracy of evaluation of the presence and severity of aortic and mitral regurgitation by contrast 2-dimensional echocardiography. Am J Cardiol 1983;52:519–24.
19. Drexler M, Erbel R, Dahm M, Mohr-Kahaly S, Oelert H, Meyer J. Assessment of successful valve reconstruction by intraoperative transesophageal echocardiography (TEE). Int J Cardiac Imag 1986;2:21–30.
20. Dahm M, Iversen S, Schmidt F, Drexler M, Erbel R, Oelert H. Intraoperative evaluation of

reconstruction of the left ventricular valves by transesophageal echocardiography. Thorac Cardiovasc Surg (special issue) 1987;35:140–2.

21. Monaghan MJ, Quigley PJ, Metcalfe JM, Thomas SD, Jewitt DE. Digital subtraction contrast echocardiography: a new method for the evaluation of regional myocardial perfusion. Br Heart J 1988;59:12–19.

22. Wann LS, Stickels KR, Bamrah VS, Gross CM. Digital processing of contrast echocardiograms: A new technique for measuring right ventricular ejection fraction. Am J Cardiol 1984;53: 1164–8.

23. Clas W, Brennecke R, Zotz R, Erbel R, Jung D, Meyer J. Color superposition: A new modality for contrast echocardiography. Int J Cardiac Imag 1987;2:111–16.

24. Brennecke R, Erbel R. Stand der echokardiographischen Bildgebung. Z Kardiol 1989; 78(Suppl 7):75–80.

25. Zotz R, Brennecke R, Clas W, Erbel R, Meyer J. Farb-superpositionierung linksventrikulärer Kontrastechokardiogramme. In: Grube E, editor. Farb-Doppler- und Kontrast-Echokardiographie. Stuttgart-New York: Georg Thieme 1989:367–77.

26. Steinmetz E, Brennecke R, Schmidtmann I, Kramer H, Erbel R, Meyer J. Quantisierung kardiologischer Ultraschallbilder aufgrund von statistischen Schätzern. Biomed Tech (Erganzungband) 1989;34:136–7.

27. Steinmetz E, Brennecke R, Wittlich N, Schön F, Erbel R, Meyer J. Quantitative comparison of new image processing methods for volumetric analysis of left ventricular contrast echocardiograms. IEEE 1989:371–4.

28. Wittlich N, Schon F, Steinmetz E, Brennecke R, Meyer J. Enhancement of accuracy and reproducibility of echocardiographic left ventricular function determination by image processing. Circulation 1989;80:II-171.

29. Meltzer RS, Tickner EG, Popp RL. Why do the lungs clear ultrasonic contrast? Ultrasound Med Biol 1980;6:263–9.

30. Serruys PW, Meltzer RS, McGhie, Roelandt J. Factors affecting the success for attaining left heart echo contrast after pulmonary wedge injection. In: Meltzer RS, Roelandt J, editors. Contrast Echocardiography. The Hague, Boston, London: Martinus Nijhoff. 1982:120–5.

31. Feinstein SB, Ten Cate FJ, Zwehl W et al. Two-dimensional contrast echocardiography. I. In vitro development and quantitative analysis of echo contrast agents. J Am Coll Cardiol 1984; 3:14–20.

32. Shapiro JR, Reisner SA, Lichtenberg GS, Meltzer RS. Intravenous contrast echocardiography with use of sonicated albumin in humans: systolic disappearance of left ventricular contrast after transpulmonary transmission. J Am Coll Cardiol 1990;16:1603–7.

33. Keller MW, Glasheen W, Teja K, Gear A, Kaul S. Myocardial contrast echocardiography without significant hemodynamic effects or reactive hyperemia: a major advantage in the imaging of regional myocardial perfusion. J Am Coll Cardiol 1988;12:1039–47.

34. Ten Cate FJ, de Jong N, Mittertreiner W, Serruys PW, Roelandt JRTC. Myocardial contrast two-dimensional echocardiography. Int J Cardiac Imag 1989;4:53–6.

35. Vandenberg BF. Myocardial perfusion and contrast echo cardiography: review and new perspectives. Echocardiography 1991;8:65–75.

36. Berwing K,. Schlepper M. Echocardiographic images of the left ventricle by peripheral intravenous injection of echocontrast agent. Am Heart J 1988;115:399–408.

37. Mudra H, Zwehl W, Klauss V, Haufe M, Spec C, Theisen K. Myokardiale Kontrastechokardiographie mit sonikiertem Iopromid (Ultravist 370) vor und nach Koronarangioplastie. Z Kardiol 1991;80:367–72.

38. Schlief R. Ultrasound contrast agents. Radiology 1991;3:198–207.

39. DeMaria AN, Bommer W, Kwan OL, Riggs K, Smith M, Waters J. In vivo correlation of thermodilution cardiac output and videodensitometric indicator-dilution curves obtained from contrast two-dimensional echocardiograms. J Am Coll Cardiol 1984;3:999–1004.

40. Smith MD, Elion JL, McClure RR et al. Left heart opacification with peripheral venous injection of a new saccharide echo contrast agent in dogs. J Am Coll Cardiol 1989;13:1622–8.

41. Zotz RJ, Wagner B. Left heart contrast echocardiography – An independent predictor of pulmonary artery pressure? Circulation 1991;84:II-161.
42. Zotz RJ, Erbel R, Walch A, Krone V. A new biocompatible and biodegradable contrast agent for left heart opacification. Circulation 1991;84:II-161.
43. Schartl M, Fritzsch T, Friedmann W, Lange L. Quantifizierung myokardialer Perfusionsdefekte mittels zweidimensionaler Kontrastechokardiographie. Z Kardiol 1984;73:560–7.
44. Erbel R, Brennecke R, Goerge G et al. Möglichkeiten und Grenzen der zweidimensionalen Echokardiographie in der quantitativen Bildanalyse. Z Kardiol 1989;78(Suppl 7):131–42.
45. Leischik R, Beller K-D, Erbel R. Comparison of new intravenous echo contrast agent (BY 963) with Albunex. Basic Res Cardiol 1996;91:101–9.
46. Leischik R, Beller K-D, Erbel R. The new transpulmonary echo contrast agent BY 963 and its evaluation in non-anaesthetized dogs. Eur Heart J 1995;16:576.
47. Belz GG, Breithaupt K, Butzer R, Bliesath H, De May C. Image quality and safety following intravenous BY 963, a new transpulmonary echo contrast medium in man. J Am Coll Cardiol. 1994;(Suppl 24A).
48. Leischik R, Kuhlmann C, Haude M, Ge J, Liu F, Erbel R. Improved reproducibility of quantitative analysis of left ventricular function during exercise echocardiography after intravenous injection of BY 963. J Am Soc Echocardiogr 1996;9:201.
49. Leischik R, Beller K-D, Erbel R. Successful myocardial opacification from intravenous injection of a new contrast agent. Proc World Ultrasonic Congr 1995;2:1029–32.
50. Leischik R, Beller K-D, Erbel R. Successful opacification of the liver after intravenous injection of a new ultrasound contrast agent (BY 963). Proc World Ultrasonic Congr 1995;2:1039–42.
51. Porter TR, Xie F, Kricsfeld A, Chiou A, Dabestani A. Improved endocardial border resolution during dobutamine stress echocardiography with intravenous sonicated dextrose albumin. J Am Coll Cardiol 1994;23:1440–3.
52. Falcone RA, Marcovitz PA, Perez JE, Dittrich HC, Hopkins WE, Armstrong WF. Intravenous albunex during dobutamine stress echocardiography: Enhanced localization of left ventricular endocardial borders. Am Heart J 1995;130:254–8.
53. Beckmann S, Schartl M, Bocksch W, Paeprer H. Stressechokardiography: Assessment of left ventricular function after administration of the transpulmonary contrast agent SHU 508 A. Z Kardiol 1993;82:317–23.
54. Schröder K, Agrawal R, Völler H, Schlief R, Schröder R. Improvement of endocardial border delineation in suboptimal stress-echocardiograms using the new left heart contrast agent SH U 508 A. Int J Cardiac Imaging 1994;10:45–51.
55. Erbel R, Nesser HJ, Drozdz J, editors. Atlas of Tissue Doppler Echocardiography (TDE) 1995. Steinkopfverlag, Germany.
56. Sutherland GR, Stewart MJ, Groundstroem KWE et al. Color Doppler myocardial imaging: a new technique for the assessment of myocardial function. J Am Soc Echocardiogr 1994;7:441–58.
57. Leischik R. Can wall motion be analyzed using automated color-coded endocardial detection? Color kinesis, a new on-line stress echo technique? Cardiogram 1995;3:15–17.
58. Lang RM, Vignon P, Weinert L et al. Echocardiographic quantification of regional left ventricular wall motion with color kinesis. Circulation 1996;93:1877–85.
59. Burns P. Potential advantages of harmonic imaging. Adv Echo-contrast 1994;4:17–19.

21. Contrast color flow Doppler evaluation of the left ventricle

JOEL S. RAICHLEN & NAVIN C. NANDA

INTRODUCTION

The development of two-dimensional echocardiography represented a major advance in non-invasive imaging of the heart. In regard to the evaluation of the left ventricle, this technique enabled one to determine the size of the left ventricular cavity at end-diastole and end-systole which provided a measure of stroke volume and permitted estimation of left ventricular ejection fraction. Two-dimensional echocardiography also enabled assessment of the thickness of the walls of the heart and the change in wall thickness from diastole to systole. Assessment of regional wall thickening and segmental wall motion made echocardiography the non-invasive study of choice for the evaluation of patients with coronary artery disease. A limitation of echocardiography results from the fact that physical characteristics of many patients preclude generation of images that are of sufficiently high quality to permit complete assessment of regional and global left ventricular function. This is most important in the assessment of ischemic heart disease, wherein incomplete echocardiograms are observed in up to 36% of patients [1]. This restriction has prompted the development of echocardiographic contrast agents as a means of enhancing the diagnostic capabilities of clinical two-dimensional echocardiograms.

The major objective of intravenous contrast agents has been to generate high intensity ultrasound reflections within the blood pool. The goal in the left heart is to opacify the left ventricular cavity so as to define the endocardial borders in regions that could not be visualized in a conventional (unenhanced) clinical echocardiogram. As a consequence investigators have sought to develop agents that are small enough to cross the pulmonary circuit and enter the left heart following an intravenous injection.

N. C. Nanda, R. Schlief and B. B. Goldberg (eds.), Advances in Echo Imaging Using Contrast Enhancement, Second Edition, 347–360.

APPROACH TO ENHANCING ULTRASOUND REFLECTIONS

It has been known for some time that bubbles provide highly reflective surfaces that enhance the echogenicity of blood when injected intravenously. Following agitation, fluids such as saline and renografin provide high intensity contrast effects in the right heart, but are ineffective in achieving left heart opacification in the absence of intrapulmonary shunts. This limitation occurs because bubbles created by agitating fluids that persist long enough to reach the left heart are too large to permit transpulmonary passage. Sonication of fluids produces micro-bubbles that are $<10 \mu m$ in diameter, and are therefore able to cross the pulmonary capillary circuit and enter the left heart. The success and limitations of sonicated fluids led to the development of the first generation of contrast agents, all of which utilize microbubbles of air. Albunex and Levovist (SHU 508A) were the first stable agents to provide microbubbles capable of transpulmonary passage, enabling opacification of the left ventricular cavity after an intravenous injection [2, 3]. In patients with suboptimal echocardiograms, complete opacification of the left ventricular cavity at end-diastole provides a means of determining left ventricular end-diastolic volume and, by enhancing delineation of endocardial borders, enables assessment of left ventricular wall thickness.

LIMITATIONS OF THE FIRST-GENERATION CONTRAST AGENTS

The success of contrast agents that utilize microbubbles of air has been limited in several regards. First, they are not uniformly successful in achieving a high level of left ventricular cavity opacification in most patients. The second problem revolves around the fact that the intensity of contrast effects is markedly reduced during systole compared with that achieved in diastole, particularly for sonicated albumin solutions [4]. A third problem results from the short duration of left-sided contrast effects which precludes evaluation of multiple image planes following a single contrast injection. These limitations have fostered a search for means of enhancing the efficacy of the microbubble contrast agents.

APPROACHES TO ENHANCING CONVENTIONAL CONTRAST EFFECTS

A number of techniques have been utilized to improve the echogenicity of two-dimensional contrast imaging. Methods that utilize standard equipment include varying the power and gain settings and increasing the dynamic range of transducers to increase the persistence of microbubbles, and imaging the contrast-filled left ventricular cavity with higher frequency transducers to improve the detection of microbubbles. Manufacturers have attempted to alter ultrasound equipment in an effort to improve contrast resolution. Transducers have been modified to enable excitation at a lower frequency while imaging at twice that frequency. This technique (called harmonic imaging) takes advantage of the fact that microbubbles resonate and produce acoustic reflections at multiples of the excitation frequency to which they are exposed. Echocardiographic machines have been modified to permit intermittent imaging during which images are made at a frequency less than one image per cardiac cycle. This approach is used to eliminate the destructive effects of the ultrasound beam on the microbubbles themselves. Overall, these approaches have not been uniformly successful.

CONTRAST DOPPLER ECHOCARDIOGRAPHY OF THE RIGHT HEART

Another approach to enhancing the contrast effects of the first-generation agents involves Doppler echocardiography. It has long been recognized that the amplitude of reflection necessary to generate Doppler signals is much less than that required for two-dimensional imaging. This is readily apparent when one considers that Doppler signals are reflected from red blood cells, which generate the lowest amplitude signal within a two-dimensional echocardiogram. Early contrast agents were found to enhance continuous wave Doppler echocardiograms, improving assessment of right-sided valve lesions such as pulmonic stenosis [5]. Today agitated saline is commonly applied to enhance right sided Doppler signals for measuring RV–RA gradients by continuous wave Doppler [6].

The first commercial echocardiographic contrast agents which utilized relatively large microbubbles (too large to cross the pulmonary circuit) enhanced the signal-to-noise ratio of conventional and color flow Doppler images. The higher sensitivity of contrast color flow images improved the identification of blood flowing through the right heart [7, 8], the area of tricuspid regurgitation and the detection of left to right shunts in atrial septal defects [9].

CONTRAST DOPPLER ECHOCARDIOGRAPHY OF THE LEFT HEART

Microbubble-containing agents that are able to cross the pulmonary circuit have been valuable for enhancing Doppler signals in the left heart. Initial investigations focused on enhancing conventional Doppler signals. Contrast was found to facilitate the estimation of aortic valve gradients by continuous wave Doppler when weak signals were demonstrated in suspected aortic stenosis [10, 11]. Echocardiographic contrast agents have also been shown to improve pulsed Doppler tracings of pulmonary venous flow and to enhance the spectral and color Doppler displays of regurgitant jets in the left heart [11]. Terasawa and colleagues found that sonicated albumin increased the mitral regurgitant signal area by an average of 59%, presumably by enhancing Doppler flow signal-to-noise ratios [12]. They also found that the pulsed signal enhancement persisted significantly longer than two-dimensional contrast enhancement. von Bibra and associates found a more marked effect on mitral regurgitant signal areas using a galactose based contrast agent (SHU 508A). They showed ≥170% increase in regurgitant jet area and ≥200% increase in normal antegrade flow area in both mitral regurgitation patients and controls [13]. Miyatake and co-workers observed that galactose-based SH/TA-508 enhanced color Doppler signals in 80 to 93% of patients in a phase II trial involving 18 institutions in Japan [14].

CONTRAST COLOR DOPPLER OF THE LEFT VENTRICLE

The poor contrast effect observed in many patients receiving the first-generation contrast agents, combined with the ability of these agents to enhance the detection of blood flow using color Doppler imaging, has prompted investigators to use contrast color Doppler to improve visualization of the endocardial borders of the left ventricular cavity. This approach has great potential in the patients with poor global or regional left ventricular function, who are least likely to have complete filling of the left ventricle following intravenous contrast administration. Enhancement of two-dimensional images requires a sufficient quantity of contrast to be delivered throughout the left ventricular cavity. As a consequence, enhancement would be least in akinetic regions, particularly when asynergy involves the distal ventricle (Figures 1A, B). The small amount of contrast that enters akinetic regions is probably trapped there in a slowly swirling pattern of flow similar to that observed when spontaneous contrast is found in a dilated left or right atrium. These trapped microbubbles produce a prominent mosaic pattern on color Doppler images (Figures 1C, D). The sensitivity of color Doppler contrast in general, and its potential for enhancing asynergic regions, makes this technique attractive as a means of enhancing opacification of the left ventricular apex. This promise was realized in a study by Firschke and colleagues that assessed the efficacy of SHU 508A in a series of 32 patients, 25 of whom had inadequate visualization of the left ventricular apex [15]. They found that contrast-enhanced i, Doppler images provided more complete delineation of the borders of the left ventricular apex than two-dimensional contrast images, both in systole and diastole.

Two-dimensional echocardiograms with incomplete endocardial border delineation cannot be digitized to estimate left ventricular volume or ejection fraction. If contrast administration fully opacifies the left ventricular cavity it will delineate the undefined endocardium, but it may also obscure previously demarcated borders that have similar echo-intensity to that of the contrast within the cavity (Figure 2). Agrawal and associates extended the application of contrast color Doppler imaging to assess this problem [16]. They used Levovist-enhanced color Doppler images to improve assessment of the entire left ventricular chamber and permit assessment of left ventricular cavity size. In an examination of 14 patients, they found that conventional color Doppler images

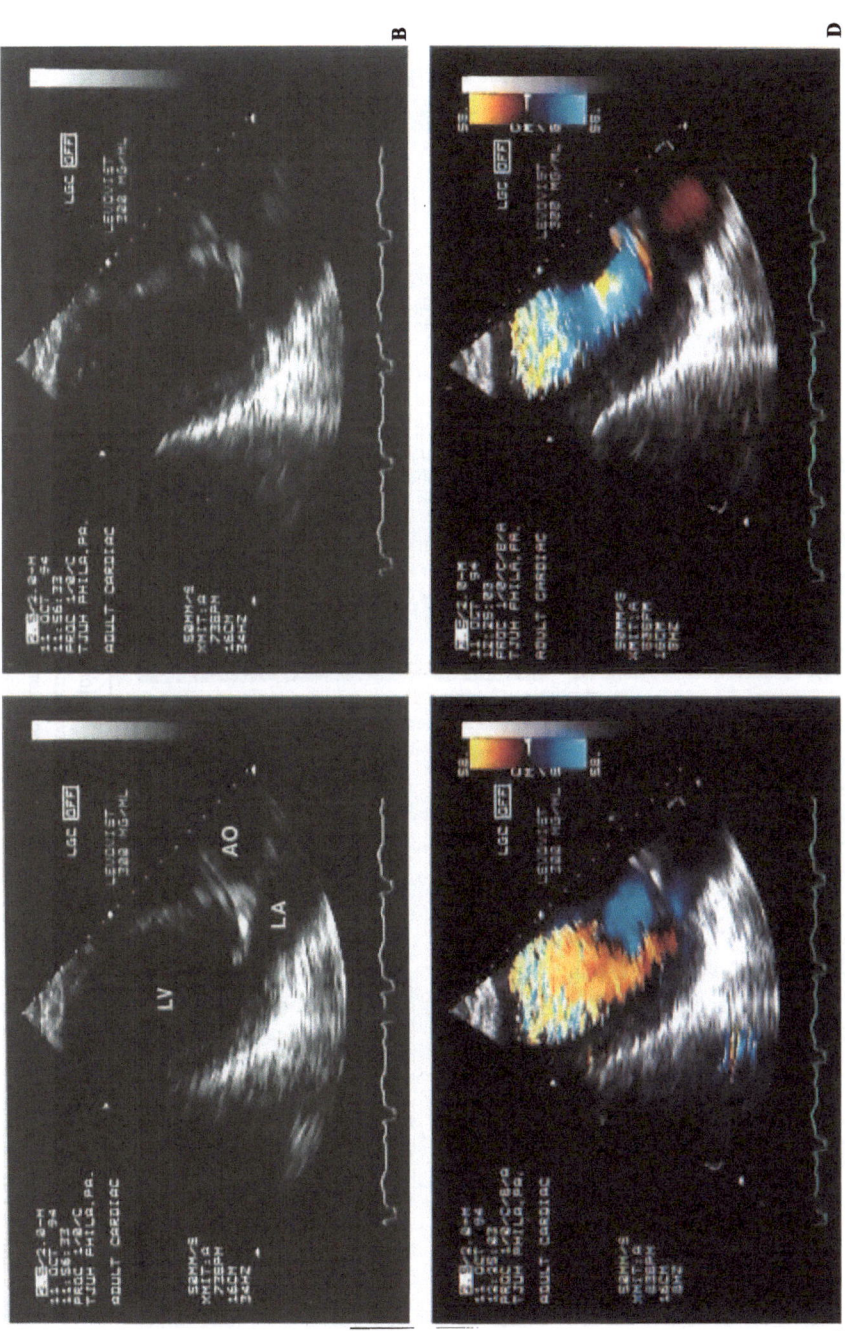

Figure 1. Long-axis view of the left ventricle (LV) (A) at end-diastole and (B) at end-systole showing poor definition of the anteroseptal wall and no delineation of the endocardial border of the distal posterior wall and apex. Two-dimensional images obtained following injection of contrast showed no opacification of the left ventricular cavity. Levovist-enhanced color flow images (C) at end-diastole and (D) at end-systole show normal contraction of the proximal walls and akinesis of the distal anteroseptal and periapical walls. The mosaic pattern in the distal ventricle is consistent with a swirling motion of the blood in that akinetic region. AO = aorta, LA = left atrium.

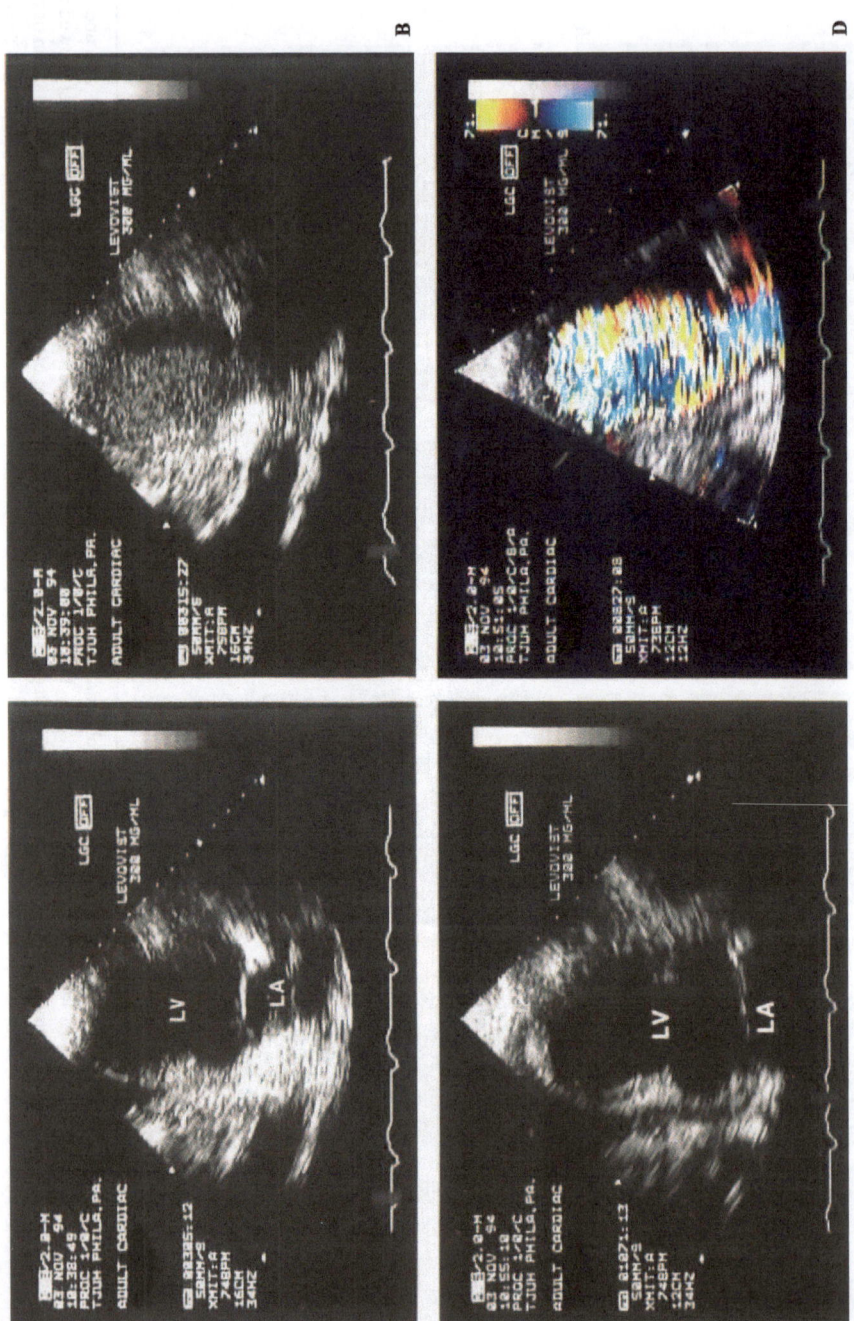

Figure 2. (A) Apical two chamber view of the left ventricle (LV) showing inadequate delineation of the endocardium of the anterior wall at end-systole. (B) Levovist contrast echocardiogram showing well delineated anterior wall endocardium at end-systole. There is poor visualization of the inferior wall and apex due to contrast opacification of the cavity making it of similar echo-intensity as the myocardium. (C) Apical two-chamber echocardiogram showing inadequate definition of the endocardial border of the anterior wall at end-systole. (D) Levovist-enhanced color Doppler image showing complete opacification of the left ventricle at end-systole resulting in good delineation of all endocardial borders. LA=left atrium.

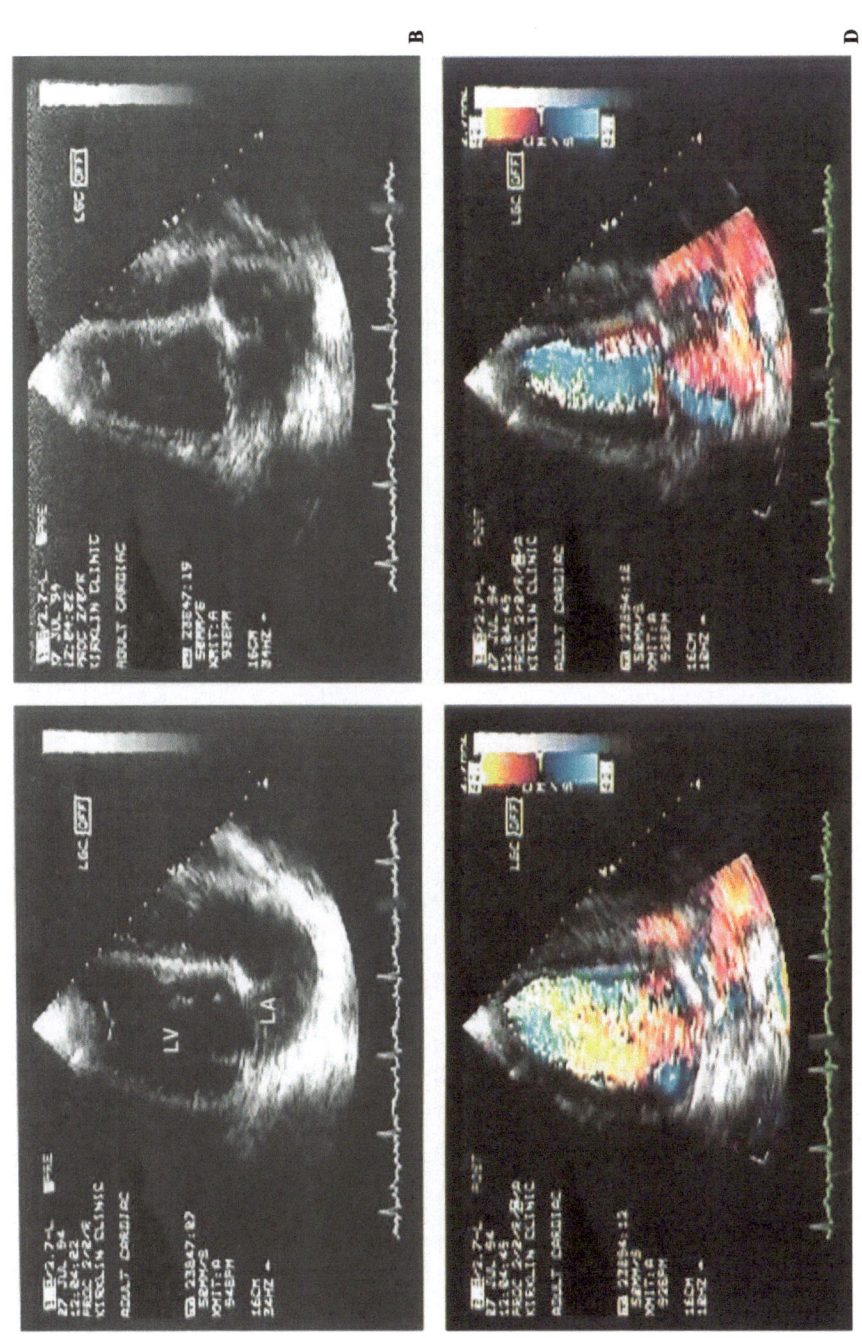

Figure 3. Two-dimensional echocardiograms (A) at end-diastole and (B) at end-systole showing fair definition of the endocardial borders of the left ventricle (LV). Levovist contrast-enhanced color flow Doppler images (C) at end-diastole and (D) at end-systole demonstrate complete filling of the LV cavity in both portions of the cardiac cycle. LA = left atrium, RA = right atrium, RV = right ventricle.

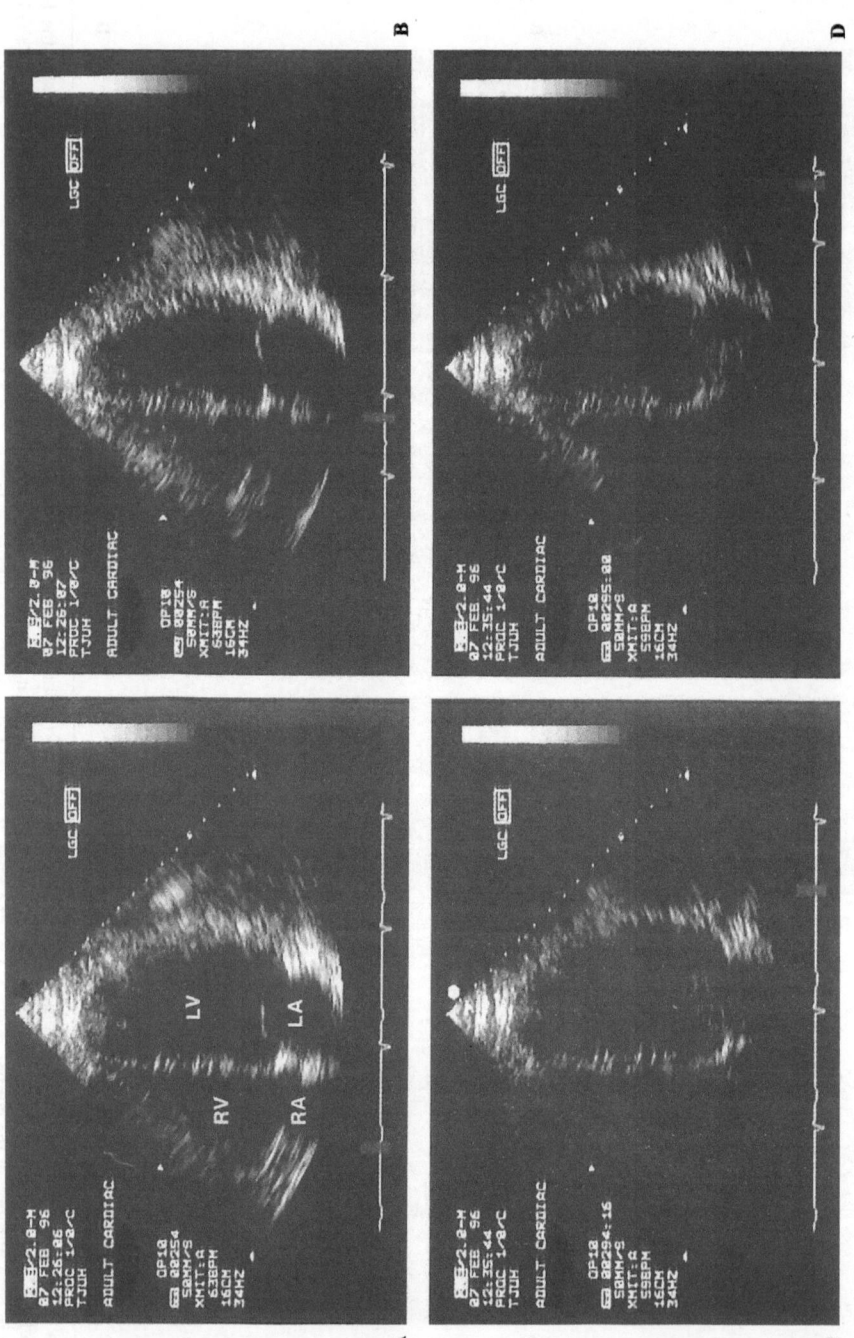

Figure 4. Apical four-chamber view of the left ventricle (LV) showing inadequate delineation of the endocardial border of the lateral wall (A) at end-diastole and (B) at end-systole. Two-dimensional images (C) at end-diastole and (D) at end-systole following injection of Albunex show loss of definition of the right heart and very faint contrast effects within the LV cavity. Color flow images (E) at end-diastole and (F) at end-systole following Albunex administration show good definition of the endocardial borders of the LV. LA=left atrium, RA=right atrium, RV=right ventricle.

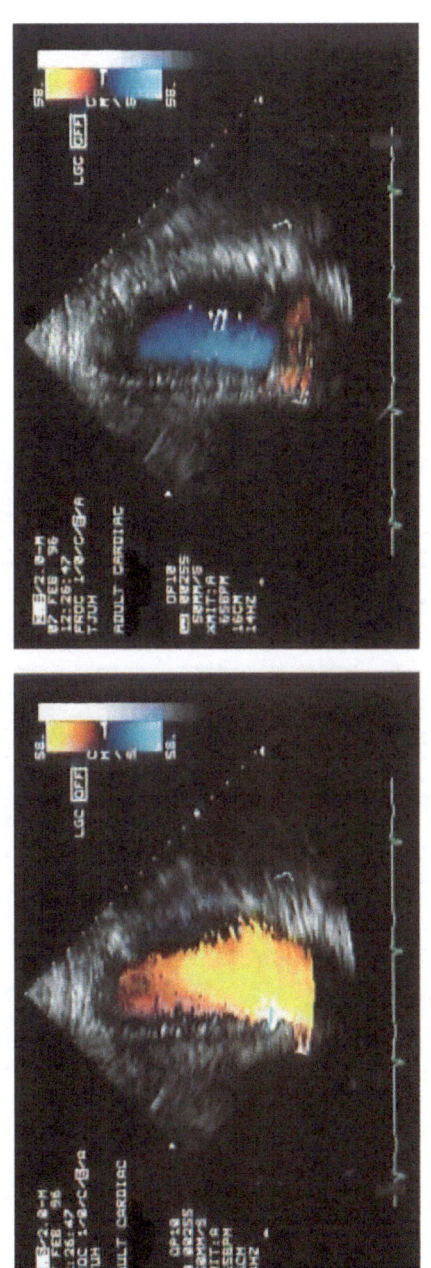

Figure 4. *Continued.*

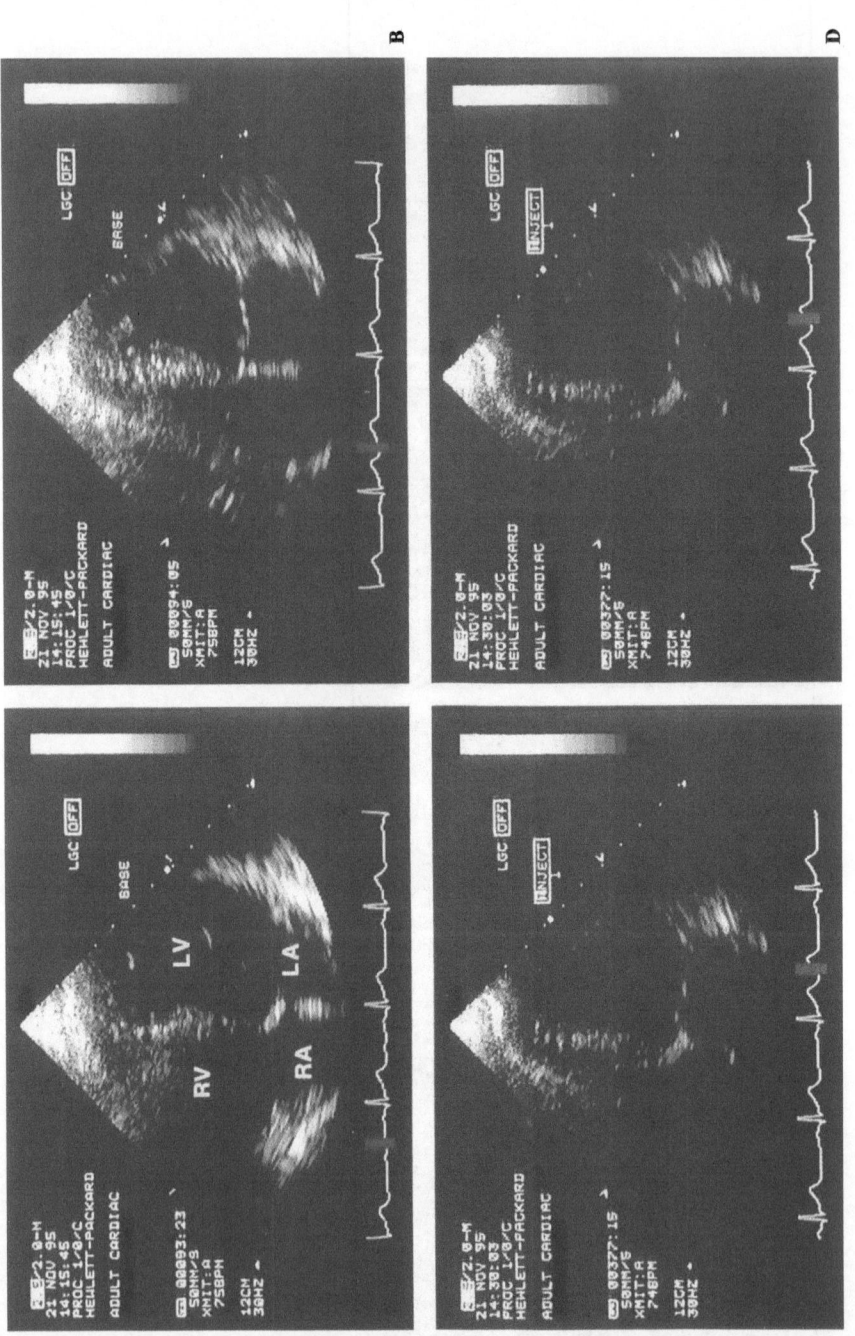

Figure 5. Apical four-chamber views of the left ventricle (LV) (A) at end-diastole and (B) at end-systole showing inadequate visualization of the lateral wall. (C) Two-dimensional image at end-diastole following the injection of Albunex showing loss of definition of the right heart and faint contrast effects in the mid- and distal portions of the LV cavity. (D) Corresponding end-systolic image showing no contrast effect within the LV. Color flow Doppler images (E) at end-diastole and (F) at end-systole following Albunex administration show clear delineation of the endocardial border of the lateral wall. LA = left atrium, RA = right atrium, RV = right ventricle.

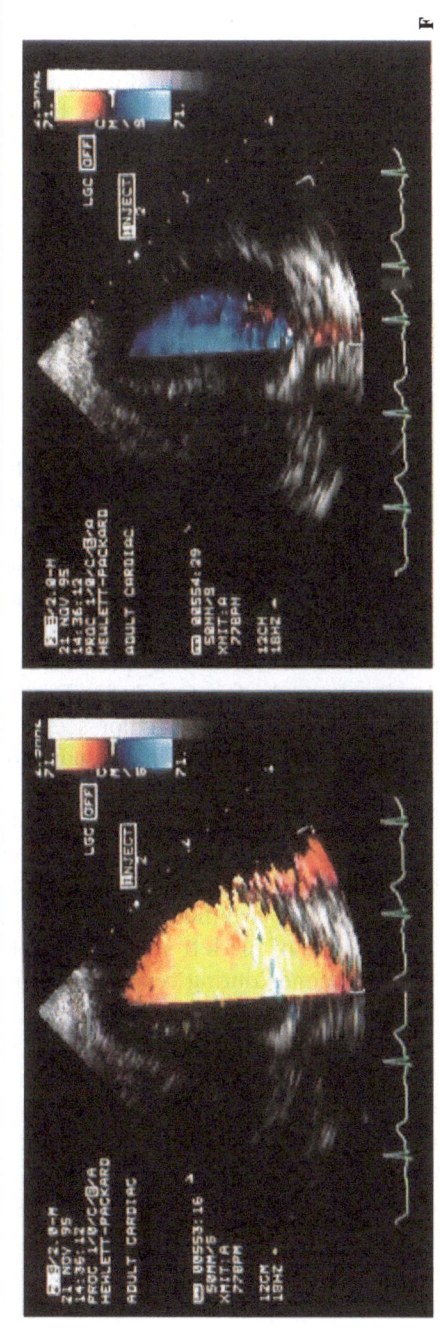

Figure 5. *Continued.*

and contrast two-dimensional echocardiograms only partially filled the left ventricular chamber in systole and diastole. Contrast color Doppler imaging provided a significant improvement over these imaging techniques, complete filling the left ventricular chamber at both end-diastole and end-systole in all patients [16] (Figure 3). Complete filling of the left ventricular cavity at end-diastole and end-systole should permit determination of end-diastolic and end-systolic volumes which can be used to estimate stroke volume and ejection fraction.

Assessment of regional ventricular function requires visualization of endocardial borders throughout systole. In patients with poorly visualized endocardium, echocardiograms enhanced by first-generation contrast agents frequently have effects that dominate in diastole. In a study of nine volunteers, Shapiro and co-workers found that left ventricular cavity opacification by sonicated albumin rapidly diminished beginning early in systole and became almost absent by end-systole [4]. Raichlen and associates similarly found poorer delineation of endocardial borders in systole than diastole following Albunex administration in 27 patients with suboptimal echocardiograms. They showed that the addition of color Doppler imaging during contrast administration significantly increased the ability to visualize systolic wall motion over that of contrast two-dimensional images (Figure 4), despite the fact that the duration of color Doppler enhancement was not different than that of two-dimensional image enhancement. Color Doppler images that encompass the entire left ventricle are associated with very slow frame rates, and this frequently precludes examination of wall motion. Assessment of regional systolic function during Doppler imaging was facilitated by placement of a relatively narrow color sector around the region of interest to generate an adequate frame rate for visualization of wall motion (Figure 5).

Raichlen and colleagues also evaluated the improvement of contrast effects with color Doppler imaging using Levovist in a series of patients with suboptimal two-dimensional echocardiograms [17]. As in the Albunex study, each patient received separate injections of contrast during two-dimensional imaging and color Doppler imaging of the undefined region of the left ventricle. Contrast effects were observed for an average of 27 seconds during two-dimensional imaging, versus 222 seconds during color Doppler imaging (Figures 1 and 2). The ability to determine wall thickness in end-diastolic images was improved with the addition of color Doppler imaging. The regions that could not be delineated in the conventional images became well defined in only two of the 10 patients during two-dimensional contrast imaging. During color Doppler contrast imaging, nine of the 10 patients had adequate or well-defined endocardial borders. In the evaluation of systolic wall motion, contrast-enhanced images provided significantly higher scores during color flow than conventional two-dimensional imaging [17]. The markedly prolonged duration of effects observed during Levovist color Doppler interrogation should allow assessment of multiple imaging planes following a single contrast injection.

These studies indicate that the color Doppler contrast imaging technique has great promise as a means of enhancing the assessment of patients with inadequate images on conventional echocardiograms. Prolonged contrast effects and improved delineation of the endocardial borders of the left ventricle in systole and diastole should enhance the ability to assess wall thickness and regional wall motion. The ability to perform these studies using conventional echocardiographic equipment and commercially available first-generation contrast agents, makes these techniques readily available for application in all patients with suboptimal echocardiograms.

REFERENCES

1. Broderick TM, Bourdillon PDV, Ryan T, Feigenbaum H, Dillon JC, Armstrong WF. Comparison of regional and global left ventricular function by serial echocardiograms after reperfusion in acute myocardial infarction. J Am Soc Echocardiogr 1989;2:315–23.
2. Crouse LJ, Cheirif J, Hanly DE et al. Opacification and border delineation improvement in patients with suboptimal endocardial border definition in routine echocardiography: results of the phase III Albunex multicenter trial. J Am Coll Cardiol 1993;22:1494–500.
3. Schlief R, Staks T, Mahler M, Rufer M, Fritzsch T, Seifert W. Successful opacification of the left heart chambers on echocardiographic examination after intravenous injection of a new saccharide based contrast agent. Echocardiography 1990;7:61–4.
4. Shapiro JR, Reisner SA, Lichtenberg GS, Meltzer RS. Intravenous contrast echocardiography with use of sonicated albumin in humans: systolic disappearance of left ventricular contrast after transpulmonary transmission. J Am Coll Cardiol 1990;16:1603–7.
5. Hagler DJ, Currie PJ, Seward JB, Tajik AJ, Mair DD, Ritter DG. Echocardiographic contrast enhancement of poor or weak continuous-wave Doppler signals. Echocardiography 1987;4: 63–7.
6. Beard JT, Byrd BF. Saline contrast enhancement of trivial Doppler tricuspid regurgitation signals for estimating pulmonary artery pressure. Am J Cardiol 1988;62:486–8.
7. Becher H, Schlief R. Improved sensitivity of color Doppler by SHU 454. Am J Cardiol 1989; 64:374–7.
8. Becher H, Zahler K, Grube E, Schlief R, Luderitz B. Improvement of color flow mapping Doppler echocardiography of right heart chambers by intravenous injection of SHU 454. Z Kardiol 1988;77:227–32.
9. von Bibra H, Hartmann F, Petrik M, Schlief R, Renner U, Blomer H. Contrast color-coded Doppler flow imaging – improved diagnosis of right heart disease after intravenous injection of Echovist. Z Kardiol 1989;78:101–8.
10. Nakatani A, Imanishi T, Terasawa A, Beppu H, Nagata S, Miyatake K. Clinical application of transpulmonary contrast-enhanced Doppler technique in the assessment of severity of aortic stenosis. J Am Coll Cardiol 1992;20:973–8.
11. von Bibra H, Sutherland G, Becher H, Neudert J, Nihoyannopoulos P. Clinical evaluation of left heart Doppler contrast enhancement by a saccharide-based transpulmonary contrast agent. J Am Coll Cardiol 1995;25:500–8.
12. Terasawa A, Miyatake K, Nakatani S, Yamagishi M, Matsuda H, Beppu S. Enhancement of Doppler flow signals in the left heart chambers by intravenous injection of sonicated albumin. J Am Coll Cardiol 1993;21:737–42.
13. von Bibra H, Becher H, Firschke C, Schlief R, Emslander HP, Schomig A. Enhancement of mitral regurgitation and normal left atrial color Doppler flow signals with peripheral venous injection of a saccharide-based contrast agent. J Am Coll Cardiol 1993;22:521–8.
14. Miyatake K, Yamaguchi T, Takatsuji H et al. SH/TA-508 clinical phase II study: dose evaluation of SH/TA-508 in echocardiography. J Cardiol 1995;26:111–33.

15. Firschke C, Tuchnitz A, von Bibra H, Klinik IM. Improved visualization of left ventricular apex by intravenous SHU 508 A: contrast 2-dimensional echocardiography versus contrast enhanced color Doppler flow imaging. Circulation 1994;90:1–68.
16. Agrawal G, Cape EG, Tirtaman C, Lee FT, Fan P, Nanda NC. Color Doppler/contrast combination provides complete delineation of left ventricular cavity. J Am Coll Cardiol 1996;27: 77A.
17. Raichlen JS, Sheth VR, Kutner HJ, Maniet AR. Contrast enhancement of left ventricular imaging with SHU 508A: two-dimensional versus color flow Doppler echocardiography. J Am Soc Echocardiogr 1995;8:364.

22. Usefulness of echo enhancement in stress echocardiography (USA experience)

JACQUELINE WIEWALL WINKELMANN, ROMY JILL BLOCK &
STEVEN B. FEINSTEIN

CONTRAST ECHOCARDIOGRAPHY

Background

Attempts to improve the diagnostic evaluation of cardiac structures, function and myocardial perfusion have motivated many past and present developmental efforts. Recent advances in instrumentation and sophisticated computer techniques have led to a much more refined and validated field. Contrast echocardiography (CE) uses intravascular injections of microbubbles in order to provide a safe, non-invasive and relatively inexpensive means of assessing myocardial perfusion. The development of CE can be traced back to Gramiak and Shah in 1968, who observed a contrast effect after injecting agitated indocyanine green during coronary arteriography [1]. Early applications of CE employed relatively large bubbles, the use of which was limited to the identification of cardiac structures in the right ventricle, including detection of atrial septal defects and enhancement of the tricuspid regurgitant jet signal [2]. Since then, the field has been undergoing constant developments in attempts to perfect the ultrasound technology, the quantitation methods and the microbubbles themselves. With the advent of microbubbles capable of transpulmonary passage, these efforts have expanded the uses of CE to include the assessment of myocardial perfusion, assessment of risk area during coronary occlusion and following reperfusion, assessment of collateral flow and myocardial viability, and enhancement of endocardial border detection during stress echocardiography [3].

Successful developments in the field of contrast echocardiography allow its role in different facets of coronary artery disease to continually evolve. In this new era of contrast echo, physicians have been able to utilize the field not only for diagnostic purposes, but more importantly, for clinical decision making and risk management. These applications include risk stratification prior to non-cardiac surgery, the establishment of the no-reflow phenomenon after reperfusion of an occluded coronary artery, and the improvement of segmental wall motion

361

N. C. Nanda, R. Schlief and B. B. Goldberg (eds.), Advances in Echo Imaging Using Contrast Enhancement, Second Edition, 361–369.
© *1997 Kluwer Academic Publishers.*

depiction and endocardial border delineation during stress echocardiography [4].

Contrast agents

Early contrast agents were produced by manually agitating solutions of saline, hydrogen peroxide, or even blood [5]. These hand-agitated bubbles had two major limitations: relatively short and variable half-lives, and relatively large sizes and variable diameters. More recently sonication has emerged as the technique of choice for producing smaller, more uniformly distributed microbubbles with more stable half-lives capable of transpulmonary passage [6]. Since the emergence of sonication, several substances have been used for creating stable microbubbles, including dextrose, Renograffin and albumin. In addition, a variety of biological compounds have been developed for use as outer shells and non-diffusible gases in order to create bubbles capable of transpulmonary passage and myocardial perfusion. New developments in this area have allowed investigators to use microbubbles capable of transpulmonary passage for in-depth studies of the left ventricle, including its wall motion, endocardial borders and myocardial perfusion.

STRESS ECHOCARDIOGRAPHY

Stress echocardiography originated from the idea that abnormal cardiac function is often manifested only when increased oxygen consumption results in increased cardiac demands that cannot be met by usual compensatory changes. Increased cardiac demands can be induced by exercise testing or with appropriate pharmacological interventions. Exercise stress echocardiography can be performed by recording images of the left ventricle immediately preceding and following treadmill exercise testing or by recording images during supine or bicycle exercise. Parameters measured by exercise testing include global ventricular function, ventricular volumes and ejection fraction. Some of the most common indications for exercise testing include suspected coronary artery disease, investigation of the extent of vessel involvement, and evaluation of cardiac function following revascularization [7]. Echocardiographic images recorded during ischemia show wall motion abnormalities, allowing detection of coronary artery disease. However, it is not always feasible to visualize clearly all segments of a two-dimensional echocardiographic image (Table 1).

Pharmacological stress echocardiography is the non-invasive imaging modality of choice when the patient is unable to exercise or there are other limiting factors. Agents used for pharmacological stress testing include dobutamine, a potent β-agonist, which increases heart rate, myocardial contractility, myocardial oxygen demand and peripheral vasodilation. Testing can also be performed using

Table 1. Stress echocardiography: advantages and disadvantages.

Type of stress echocardiography	Advantages	Disadvantages
Exercise		
Treadmill	Widely available	Imaging post-test only
Bicycle	Imaging during exercise	Imaging technically difficult
Pharmacological		
Dobutamine	1. Continuous imaging	1. Potential adverse effects of dobutamine
	2. Does not require physical activity of the patient	2. Optimal imaging technically difficult
Dipyridamole	1. Continuous imaging	1. Potential adverse effects of dipyridamole
	2. Does not require physical activity of the patient	2. Relative flow inequality rather than ischemia *per se*

adenosine, which vasodilates the vessels to the point of 'stealing' blood from stenosed vessels, resulting in ischemia. Dipyridamole has also been proposed by some investigators due to its distinct pattern of distribution of blood flow among the coronary arteries. After infusion of dipyridamole, an increase in blood flow can be appreciated in the normal arteries, with a relative decrease in flow in diseased vessels. So far, this modality has proved to be more effective than perfusion imaging with thallium.

Regardless of the method used to achieve ischemia, the diagnosis of coronary artery disease is based on a comparison of resting and stress echocardiographic images. However, good-quality studies need to be obtained in order for wall motion abnormalities to be detected with relative confidence. Good quality studies imply that the true endocardial border can be accurately discerned in both rest and stress images.

Clinical use of echocontrast for enhancement of stress testing

As previously noted, stress echocardiography has emerged as a safe and accurate non-invasive method for the evaluation of known or suspected coronary artery disease. Two-dimensional echocardiography provides multiple tomographic images of the left ventricle during both rest and peak stress, allowing a detailed assessment of regional wall motion and endocardial thickening by which to evaluate the existence of ischemia. However, detailed imaging of the ventricular myocardium and endocardium is not possible in about 10–20% of the patients, due to acoustic impedance caused by body habitus, lung disease or other technical factors [8]. This results in technically limited echocardiograms not suitable for adequate assessment of coronary artery disease. Specifically, identification of the endocardial surfaces of the apex and lateral walls is difficult in the apical views. The i.v. administration of a contrast agent has been shown to enhance endocardial border delineation by opacifying the left ventricle, thus improving interpretation of regional wall motion, cardiac output, ejection fractions and left ventriculograms during routine stress echocardiography [9–15]. This allows physicians to make accurate clinical decisions as well as to assess risk management.

Early studies performed by Feinstein and associates evaluated the safety and efficacy of the only FDA-approved contrast agent available in the USA, Albunex, a 5% solution of sonicated human serum albumin. Low doses of i.v. administered Albunex in patients were shown to achieve transpulmonary opacification of the left ventricle in 151 (63%) of 240 injections [9]. The study suggested that a transpulmonary contrast agent had the ability to enhance the detection of wall motion abnormalities. Crouse and colleagues assessed the specificity of improved endocardial border enhancement in patients with inadequate border delineation at rest [10,11]. Of the 175 patients studied, 145 (83%) showed an improvement in endocardial border delineation after one dose of i.v. Albunex [11]. The use of contrast echocardiography also improved the graders' confidence in assessing regional wall motion in 67 (77%) of the 87 patients in whom investigators were not confident in assessing the baseline images [11].

Dobutamine stress echocardiography (DSE)

Recent studies have shown that the administration of contrast agents during DSE improves the endocardial border detection by highlighting wall motion abnormalities that may not have otherwise been detected. Porter and associates showed a significant improvement in left ventricular opacification (LVO) at peak dobutamine infusion when compared with rest following i.v. administration of albumin sonicated with dextrose [12]. Of the 50 patients analyzed using acoustic densitometry software (Hewlett Packard), 67% achieved LVO at rest and 92% at peak dobutamine. Among the 37 patients in whom at least one suboptimal endocardial segment was visualized, improved border resolution was seen in 68% with contrast enhancement at the baseline, and in 95% at peak dobutamine infusion [12]. Other studies have described improvement in localization of the left ventricular endocardial borders [13]. The greatest improvement with contrast enhancement has been noted following peak dobutamine in images with inadequate baseline border delineation. Figure 1 depicts a four-chamber image of an inadequate baseline precontrast enhancement and the same patient following i.v. injection of 5 cm³ of Albunex during peak DSE. Note the improved endocardial border resolution in the left ventricle.

Current studies under evaluation by Feinstein and colleagues have shown improvement in the assessment of lateral and apical endocardial border resolution, ventricular function, regional wall motion, speed of interpretation and improved confidence of reading in patients undergoing DSE with contrast enhancement [14]. The use of Albunex decreased the intraobserver variability: no significant difference was found among the five graders [14]. Overall, the use of contrast

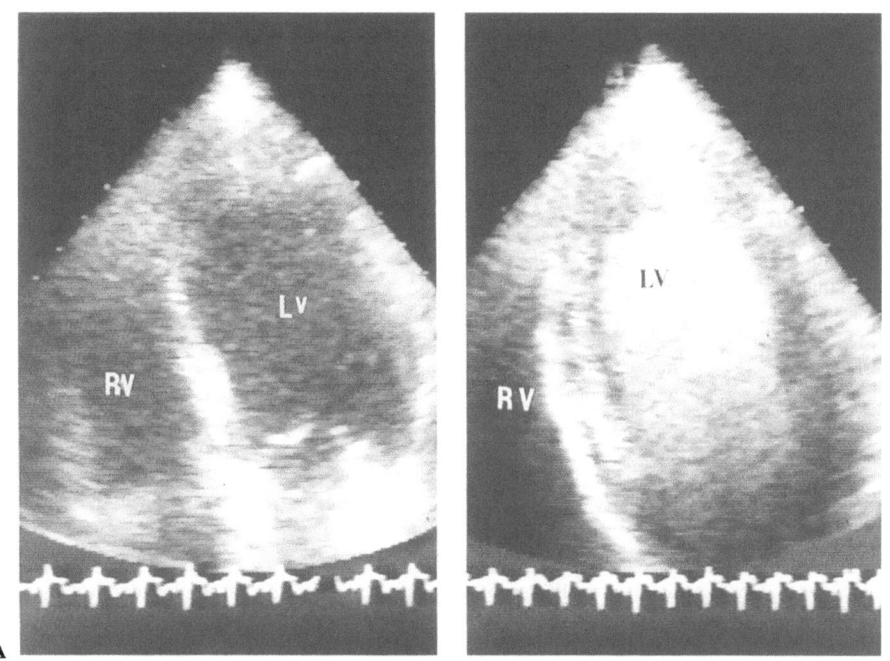

Figure 1. (A) Four-chambered view of a patient with inadequate endocardial border delineation at peak dobutamine during a routine stress echocardiogram. (B) Same patient at peak dobutamine following the i.v. administration of $5\,cm^3$ of the contrast agent Albunex. Notice the full left ventricular opacification as well as the improvement in endocardial border resolution. This patient's baseline study was uninterpretable while the contrast study allowed a more accurate assessment of a stress echocardiogram.

echocardiography during routine DSE has proved to be an effective method for improving assessment of stress induced ischemia in patients with suspected coronary artery disease.

Exercise echocardiography

Limited data have been reported on the use of contrast echocardiography during exercise echocardiography (EE). Chan and associates have recently described a significant increase in the visualization of the endocardial borders in patients undergoing EE following contrast enhancement. The use of i.v. Albunex increased the overall percentage of endocardial depiction from 86% to 91% [15]. The safety and feasibility of this method have indicated its potential for improving inadequate images during routine EE.

Table 2. Current transpulmonary ultrasound contrast agents (Europe and USA).

Company	Agent	Characteristics	Stage of testing
Mallinckrodt/MBI	Albunex*	Sonicated human serum albumin	FDA approved 1994
Nycomed	Infoson*	Sonicated human serum albumin	Approved
Mallinckrodt/MBI	FS069	Perfluorocarbon	Phase III
Schering AG	Levovist SHU 508A	Saccharide based	Phase III
Schering AG	SHU 454	N/A	Approved Europe 1991
Schering AG	SHU 563A	N/A	Pre-clinical
Bracco	BR-1	Non-diffusible gas	Phase I
IMARx	Aerosomes MRX-115	Liposomes	Phase I
Alliance	Imagent US/AFO 145	Non-diffusible gas	Pre-clinical
Sonus	Echogen	Phase-shift technology	Phase III
Nycomed	NUS	N/A	Pre-clinical
Andarias (Delta)	Quantison	Albumin based	Phase I
Acusphere	Polyphazenes	N/A	Pre-clinical
Byk Gulden	BY 963	Phospholipid saccharide	Phase I

*Albunex and Infoson are the same FDA approved sonicated human serum albumin contrast agent. Infoson is licensed and manufactured in Europe by Nycomed, while Albunex is produced and distributed by Mallinckrodt and Molecular Biosystems Inc. in the USA.

NEW CONTRAST AGENTS

Albunex is the only contrast agent currently available commercially for routine use in the echocardiographic laboratory in the USA, although several contrast agents are currently under development. Phase III trials are now assessing the safety and feasibility of a second-generation perflourocarbon agent, FS069 (Molecular Biosystems Inc.), in patients. The efficacy of the transpulmonary opacification of FS069 is being compared with that of Albunex at rest. Other new agents such as QW3600 (Echogen), Aerosome and AF0146 (Imagent) also have the potential to improve endocardial border delineation and are undergoing clinical trials during stress echocardiography (Table 2). These agents tend to persist longer in the cavity due to use of relatively non-diffusible gases and the inclusion of biological materials such as phospholipids and saccharides in the outer shells of the bubbles [16]. The newer agents may also opacify the left ventricle more densely due to a decrease in acoustic destruction of microbubbles.

EUROPEAN PERSPECTIVE

Similar experiences with contrast enhancement during stress echocardiography have been occurring in Europe with the agent SHU 508A(Levovist; Schering) and Infoson (the European tradename of Albunex manufactured by Nycomed).

Studies have shown significant improvement in endocardial border resolution after contrast infusion during exercise and dobutamine stress echocardiography [17, 18]. Wall motion analysis was found to be significantly improved, with agreement with the gold standard angiography increasing from 64.5% to 90.3% following i.v. administration of SHU 508A in the 10 patients assessed with an endocardial dropout [18]. Other agents under development such as Quantison, BY 963 and BR-1 may prove to have similar capabilities (Table 1).

COST EFFECTIVENESS

The use of ultrasound contrast agents in combination with stress echocardiography has been shown to be cost-effective in a recent retrospective review of 906 patients [19]. The official echo reports were used to define efficiency (i.e. a limited conclusion was considered inefficient). The study compared the efficiency of supplying a definitive diagnosis in patients for whom Albunex was not available (pre-Albunex era) to those exercise studies in which Albunex was available (post-Albunex era). The use of Albunex was based on the assessment of endocardial wall identification in patients during the rest imaging. Overall results revealed an efficiency of 91.5% in 284 patients studied in the pre-Albunex era, compared with 93.4% in the post-Albunex era. In addition, the number of redundant tests (i.e. thallium studies and cardiac catheterization studies) was reduced by 47% in the post-Albunex era. Based on the assumption that the additional costs associated with Albunex were not reimbursed (assumed a 100% capitated cost system) and a presumptive actual cost of performing echocardiograms, thallium tests and coronary angiography, the authors concluded that the use of Albunex in 15% of patients undergoing stress echo studies resulted in an increased efficiency of providing definitive diagnosis, reduction in redundant testing and potentially lower overall costs to the health care-provider system. This study did not address the cost reductions associated with a reduced false-positive or false-negative diagnosis or the benefits to the patients associated with reduced testing time and the risks associated with redundant invasive studies.

The limitations of the study included the retrospective analysis of data, the lack of defined indications for the use of contrast agents, the lack of uniform grading systems by the echocardiographers and in-house biases. However, a prospective study of the cost-effectiveness of contrast echocardiography associated with stress echocardiography is anticipated. This approach will be useful if new technologies (including ultrasound contrast materials) are to be introduced into existing clinical markets and diagnostic treatment algorithms.

CONCLUSIONS

The use of contrast enhancement during stress echocardiography has been established as a safe and efficient non-invasive tool in routine echocardiograms in order to detect the early stages of coronary artery disease. Contrast enhancement provides a significant improvement in endocardial border delineation allowing the physician to accurately assess regional wall motion, cardiac output, ventricular function and ejection fraction. The use of contrast is invaluable in analyzing the degree of suspected heart disease in the 10–20% of patients with inadequately visualized ventricular borders. By increasing the sensitivity and the accuracy of DSE in detecting possible wall motion abnormalities or ischemic myocardium, CE improves the reader confidence as well as the speed of interpretation.

REFERENCES:

1. Gramiak R, Shah PM. Echocardiography of the aortic root. Invest Radiol 1968;3:356–66.
2. Otto CM, Pearlman AS. Textbook of clinical echocardiography. Philadelphia, PA: WB Saunders Company, 1995.
3. Dagianti A, Feigenbaum H, eds. Echocardiography 1993. Amsterdam: Elsevier Science Publishers, 1993.
4. Crouse LJ. Sonicated serum albumin in contrast echocardiography: improved segmental wall motion depiction and implications for stress echocardiography. Am J Cardiol 1992;69: 42H–45H.
5. Feigenbaum H, Stone JM, Lee DA. Identification of ultrasound echoes from the LV by the use of indocyanine green. Circulation 1970;41:613–20.
6. De Jong N, Ten Cate FJ, Lancee CT, Roelandt TC, Bom N. Principles and recent developments in ultrasound contrast agents. Ultrasonics 1991;29:324–30.
7. Popp RL. Echocardiography. N Engl J Med 1990;323:101–9.
8. Freeman AP, Giles RW, Walsh WF, Fisher R, Murray IPC, Wilcken DEL. Regional left ventricular wall motion assessment: Comparison of two-dimensional echocardiography and radionuclide angiography with contrast angiography in healed myocardial infarction. Am J Cardiol 1985;56:8–12.
9. Feinstein SB, Cheirif J, Ten Cate FJ et al. Safety and efficacy of a new transpulmonary ultrasound contrast agent: Initial multicenter clinical results. J Am Coll Cardiol 1990;16:316–24.
10. Crouse L. Sonicated serum albumin in contrast echocardiography: Improved segmental wall motion depiction and implications for stress echocardiography. Am J Cardiol 1992;69: 42H–45H.
11. Crouse L, Cheirif J, Hanley D et al. Opacification and border delineation improvement in patients with suboptimal endocardial border definition in routine echocardiography: Results of the phase III Albunex multicenter trial. J Am Coll Cardiol 1993;22:1494–500.
12. Porter T, Feng X, Kricsfeld A, Chiou A, Dabestani A. Improved endocardial border resolution during dobutamine stress echocardiography with intravenous sonicated dextrose albumin. J Am Coll Cardiol 1994;23:1440–3.
13. Falcone R, Marcovitz P, Perez J, Dittrich H, Hopkins W, Armstrong W. Intravenous Albunex during dobutamine stress echocardiography: Enhanced localization of left ventricular endocardial borders. Am Heart J 1995;130:254–8.
14. Emani V, Block R, Devries S et al. Albunex in dobutamine stress echocardiography: observations of endocardial enhancement, ventricular function, and wall motion. J Am Soc Echocardiogr (submitted).
15. Yvorchuk K, Sochowski R, Chan K. Sonicated albumin in exercise echocardiography: Technique and feasibility of a novel approach to enhance endocardial border visualization. J Am Soc Echocardiogr 1996;9:462–9.

16. Ten Cate FJ, Aiazian A, de Jong N. Advances in clinical contrast echocardiography: The Thorax-center Experience. Thoraxcenter J 1995;7:13–14.
17. Beckmann S, Schartl M, Bocksch W, Paeprer H. Stress echocardiography: Evaluation of left ventricular function after administration of the transpulmonary echo contrast medium SHU 508 A. Z Kardiol 1993;82:317–23.
18. Schroder K, Agrawal R, Boller H, Schlief R, Schroder R. Improvement of endocardial border delineation in suboptimal stress echocardiograms using the new left heart contrast agent SHU 508 A. Int J Card Imaging 1994;10:45–51.
19. Tejpal Y, Block R, Sagireddy B, Emani V, Devries S, Feinstein S. Cost-effectiveness in patients undergoing dobutamine stress echocardiography with and without ultrasound contrast. J Am Soc Echocardiogr 1996;9:40Z–A801F.

23. Assessment of myocardial perfusion with various intravenous echo-enhancing agents

DANIELE ROVAI, ANTHONY N. DeMARIA, MASSIMO LOMBARDI,
CECILIA MARINI, JAN SCHNEIDER & ANTONIO L'ABBATE

BACKGROUND

The injection of echocontrast agents directly into the coronary circulation enhances the ultrasound signal backscattered from the myocardium. This increase in myocardial reflectivity provides accurate information about the spatial distribution of myocardial perfusion [1–9] and allows the assessment of coronary blood flow [10–11]. However, this approach to myocardial contrast echocardiography requires cardiac catheterization, which has so far limited the widespread application of the technique. To overcome this problem, investigators as well as pharmaceutical companies have sought to develop contrast agents which could achieve myocardial contrast enhancement following peripheral venous injection. Attainment of this goal could remarkably expand the application of myocardial contrast echocardiography and even conveys the potential to modify the clinical approach to patients with known or suspected coronary artery disease.

In order to study myocardial perfusion by intravenous administration of echo-enhancing agents, some basic principles of coronary anatomy and physiology should be recalled. Only a small fraction of cardiac output (5–10%) contributes to coronary circulation under physiological conditions. Since cardiac output at rest averages about 5 l/min, resting coronary blood flow ranges from 250 to 500 ml/min. With regard to anatomy, the coronary circulation consists of two major coronary arteries which branch into smaller vessels and finally end in the capillary network. The capillaries are vessels with a diameter of only approximately 5 μm, through which the red blood cells flow in a single line. The capillary network is very dense, with an average intercapillary distance of 20 μm [12].

In view of these physiological principles, two problems capable of limiting the study of myocardial perfusion by intravenous contrast administration are apparent. First, the number of microbubbles reaching the myocardium is much smaller than the dose injected into the vein. Assuming that microbubbles can

N. C. Nanda, R. Schlief and B. B. Goldberg (eds.), Advances in Echo Imaging Using Contrast Enhancement,
Second Edition, 371–385.

flow freely throughout the microcirculation and that they are sufficiently stable so as not to be destroyed in the different areas of the circulation, only 5–10 of every 100 microbubbles which reach the aortic root will effectively traverse the coronary circulation. Second, the number of microbubbles present in a defined volume of myocardium is much smaller than the number in the same volume of blood. If the concentration of microbubbles in the blood flowing through the aortic root is assumed to be the same as in the blood flowing in the coronary circulation, the concentration of microbubbles in the myocardium (which consists of both muscle and intramyocardial blood) will be lower than that of pure blood (according to the ratio between intramyocardial blood content and tissue volume). This ratio varies with perfusion pressure, extravascular compressive forces and the tone of coronary microvasculature and has been reported to be about 6–7% under resting conditions [13]. Thus, the concentration of microbubbles in the myocardium should be equal to 6–7% of their concentration in blood.

The small fraction of the total dose which perfuses the myocardium and the low tissue concentration of microbubbles are both responsible for the reduced intensity of myocardial contrast echo signals as compared to that of intracavitary blood. The study of myocardial perfusion following intravenous administration of echo-enhancing agents is based on the analysis of this relatively low intensity signal. Therefore, to make such analysis feasible and useful, different approaches to overcome these limitations have been proposed over the past few years.

PULMONARY WEDGE INJECTIONS OF HAND-AGITATED AGENTS

The echo-enhancing properties of conventional contrast agents such as hand-agitated saline, indocyanine green or CO_2 are lost during their first passage through the pulmonary circulation. This is due to the fact that the microbubbles with a diameter greater than the pulmonary capillary lumen are mechanically blocked while those with a smaller diameter collapse very quickly due to their high surface tension [14]. In an attempt to visualize the left heart a technique using venous catheterization was developed by Reale et al. [15] and by Meltzer et al. [16]. A balloon-tipped catheter was advanced into a pulmonary artery branch, the balloon was inflated and then, after injection of an echo-enhancing agent, was rapidly deflated. M-mode or two-dimensional echocardiography was recorded. Echoes appeared in the cavities of the left heart following such injections, and complications were absent. However, the echocontrast enhancement was limited to the cavities and did not involve the myocardium. A significant improvement toward the non-invasive visualization of the left heart has subsequently been achieved by standardized, lung-crossing contrast agents, produced by pharmaceutical companies.

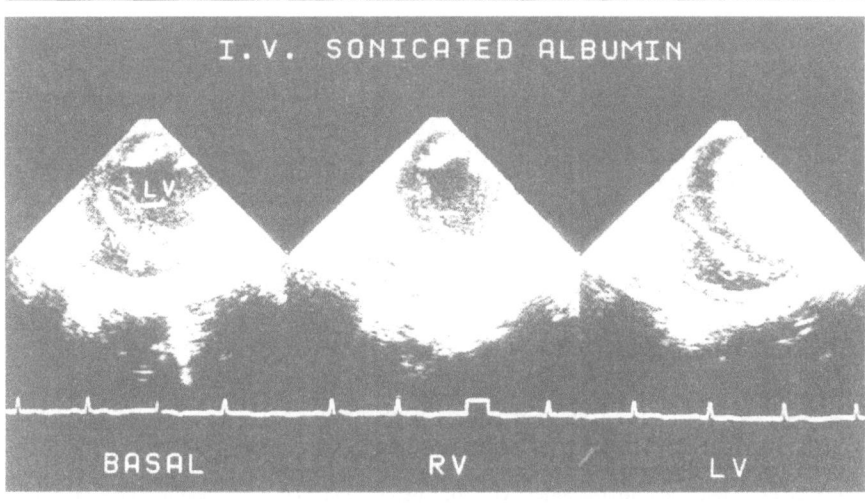

Figure 1. Epicardial echo of a dog before and after intravenous injection of Albunex. Although an evident contrast enhancement can be observed in both right and left ventricular cavities, no clear myocardial contrast enhancement can be seen.

FIRST-GENERATION COMMERCIAL LUNG-CROSSING AGENTS

Conventional signal processing

The visualization of the left heart after intravenous contrast administration was made possible by two contrast agents: Albunex (MBI, California) [17, 18] and Levovist (Schering AG, Germany) [19, 20]. Albunex comprises microcapsules with a median diameter of 4 μm, which are surrounded by an albumin coat. They are produced by sonication of a solution of human serum albumin. Levovist contains microbubbles with an average diameter ranging from 2–8 μm, which are produced by combining galactose microparticles with sterile water, subjecting the mixture to hand agitation for 5 s and allowing equilibration for 2 min. Levovist microbubbles are made stable by adding a small amount of a surfactant to the formulation. Both agents contain atmospheric air inside.

Although these agents are able to opacify the left ventricular cavity after venous injection in the majority of patients, neither is able to consistently opacify the myocardium. Figure 1 shows the epicardial echo of a dog before and after intravenous injection of Albunex. Although contrast opacification of both the right and left cavity is evident, no clear myocardial contrast enhancement can be seen. Figure 2 depicts the apical four-chamber view of a patient before and after intravenous injection of Levovist. Again, despite the evident contrast enhancement of cardiac cavities, no myocardial contrast effect can be detected. As mentioned above, both the small fraction of the dose perfusing the myocardium and

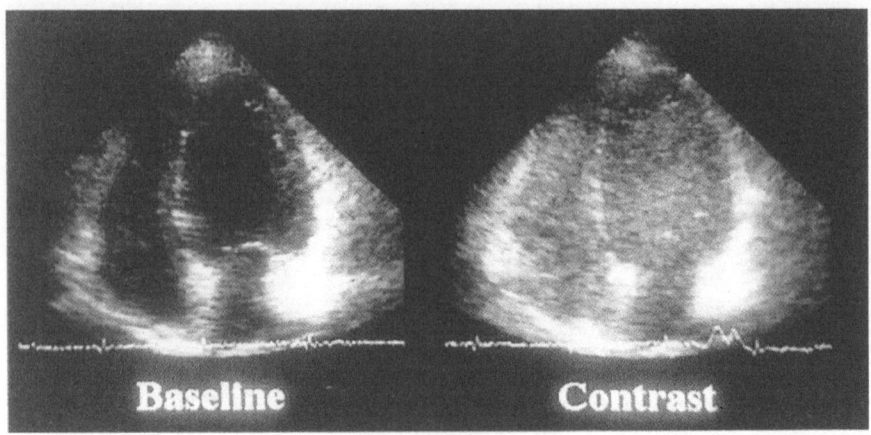

Figure 2. Left ventricular four-chamber view of a patient under baseline condition and after intravenous injection of Levovist. Despite the evident contrast enhancement of the cardiac cavities, no substantial myocardial contrast effect can be observed.

the low concentration of the contrast agent in the myocardial tissue limit the study of myocardial perfusion after intravenous injection. However, the ability of these two agents to enhance contrast in the myocardium may be hampered by two additional factors: possible fragility (perhaps by pressure or ultrasonic energy) of these agents, resulting in destruction of microbubbles as they flow through the cardiac cavities and the coronary circulation, and the limited sensitivity of conventional scanners in detecting low concentrations of microbubbles.

Improving protocol and display

In an attempt to visualize myocardial perfusion by Albunex administration, Villanueva et al. modified the technique of contrast injection and the display of echocontrast images in an animal model [21]. These authors increased the dose of Albunex 10 times over the conventional diagnostic amount. Furthermore, they selected microbubbles with a diameter slightly larger than those used for conventional imaging (thus increasing contrast backscatter) and injected the contrast agent directly into the right atrial cavity. In addition, the injections were performed after premedication of the animals with dipyridamole in order to increase the ratio between coronary blood flow and cardiac output and thus the fraction of the dose perfusing the myocardium. Finally, to improve the detection of the contrast effect, the cross-sectional images were digitally subtracted and color-coded. This approach allowed the visualization of myocardial perfusion, including the identification of myocardial perfusion defects located in the anterior wall [22]. Recently myocardial opacification after intravenous injection of

Electronic Signal Processing

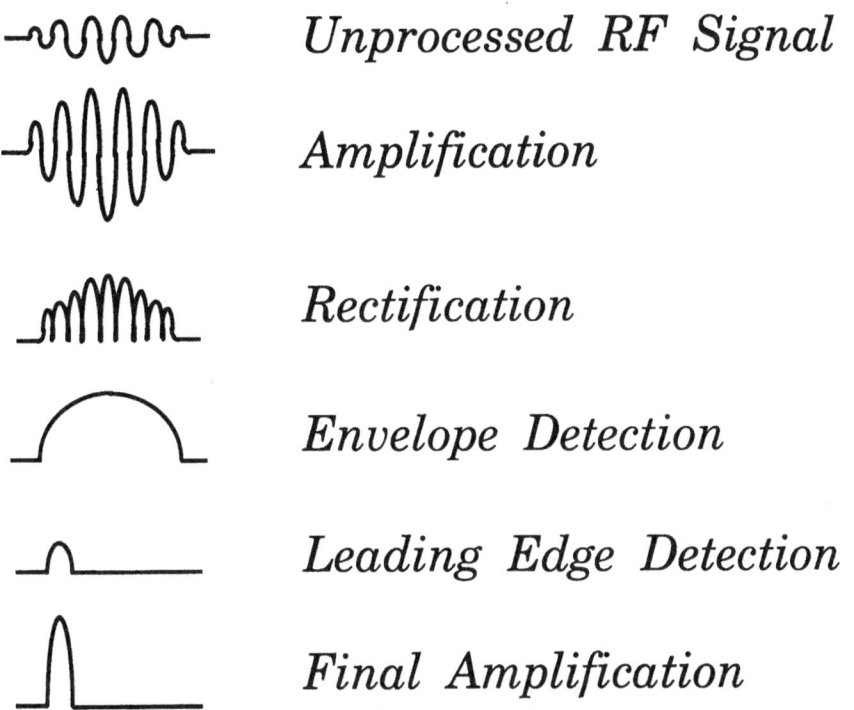

Figure 3. Schematic diagram of electronic signal processing in conventional scanners.

sonicated albumin was observed in 11 patients by Voci et al. [23] who combined transesophageal echocardiography with dipyridamole vasodilatation and intermittent imaging.

Radio-frequency analysis

The processing of an echocardiographic image usually results in considerable compression and distortion of the original backscattered signal. As shown schematically in Figure 3, the return signal to the transducer is initially amplified and rectified (i.e. the negative waves are transformed into the corresponding positive waves). The envelope and the leading edge of this signal are then detected and finally the signal is once more amplified, usually in a non-linear fashion. The signal is further processed in the scan converter and additional loss of data integrity occurs during the recording of the images on videotape. The main goal of this manipulation is to produce images pleasant to the eye; however this heavy processing may impair the ability to detect small ultrasound targets such as microbubbles.

In order to improve the detection of low concentrations of contrast agent, the analysis of the native radio-frequency signal has been proposed by Monaghan et al. [24]. These authors carried out this approach from a region of interest located in the left ventricular myocardium of 30 patients. Analysis of mean integrated backscatter (a measure of the total ultrasound energy) was performed before, during and after Albunex injection. The data were also analyzed for a possible shift in the frequency spectrum, which could result from resonance of Albunex contrast microbubbles. Digital radio-frequency data were obtained from 76% of patients and showed a significant increase in mean integrated back-scatter as well as a shift in frequency, which enabled the construction of time – intensity curves, representing wash-in and wash-out of contrast from the myocardium.

The main limitations of the radio-frequency approach are the huge amount of computer memory required to store the digital images, the long duration required for data analysis, and the unsolved problem of attenuation artifacts. Angermann et al. also analyzed the native radio-frequency signal to detect myocardial per-fusion [25]. This group combined radio-frequency analysis with transesophageal echocardiography before and after intravenous administration of Albunex in man. The incidence of myocardial contrast enhancement observed was lower than that reported by Monaghan. This variance could be attributed to different transducer frequencies: Monaghan employed a transducer with a 2.5 mHz center frequency, whereas Angermann utilized a 5 MHz probe. It has recently been demonstrated that the intensity of signal backscatter from a contrast agent increases if the agent is stimulated at the frequency at which the microbubbles start to resonate (resonance frequency) [26]. Thus, the greater sensitivity found by Monaghan in detecting low myocardial concentrations of Albunex may be attributed to a transducer operating closer to the resonance frequency of Albunex, which is about 2 MHz [27]. In addition, recent studies have shown that the energy of the ultrasound beam may destroy contrast microbubbles [28, 29]. For this reason, the intensity of myocardial signal after contrast administration increases if the myocardium is stimulated intermittently, as performed by Monaghan, instead of in continuous mode, as performed by Angermann.

Harmonic imaging

Contrast microbubbles can undergo periodic oscillations (expansion and con-traction) due to their interaction with ultrasound [26]. These oscillations are maximal when the microbubbles are stimulated at their resonance frequency. This phenomenon depends on the size and elastic properties of the microbubbles, the

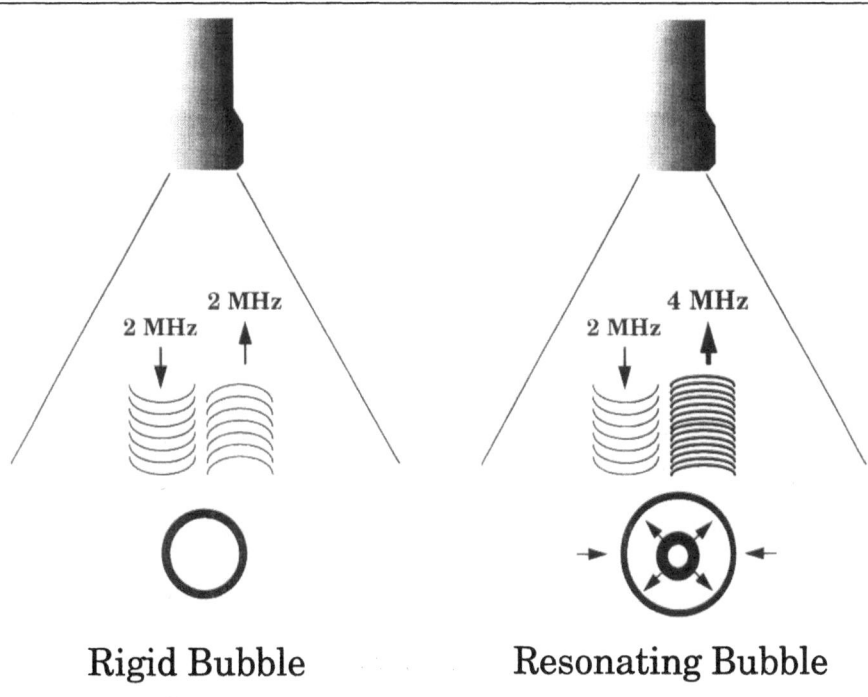

Rigid Bubble   **Resonating Bubble**

Figure 4. Two different behaviors of contrast microbubbles are schematically shown. If a micro-bubble does not resonate (left panel), the backscattered signal has the same frequency as the stimulating one. When a microbubble resonates (right panel), the backscattered signal has a frequency which is a multiple of the stimulating one and has a greater intensity.

pressure of the fluid in which they are suspended, and probably on their proximity to solid structures. Resonance frequency varies among the different contrast agents and probably with the different sizes of microbubbles within the same agent. When contrast microbubbles are stimulated at their resonance frequency two interesting phenomena occur. First, the amplitude of the backscattered signal increases remarkably; second, the frequency of the backscattered signal becomes a multiple of the stimulating frequency (Figure 4), e.g. if a signal is transmitted at 2 MHz and a contrast agent resonates at a frequency of 2 MHz, the backscattered signals will have frequencies of 4, 6 or 8 MHz. Prototype scanners able to generate ultrasound with a given frequency and to receive at double that frequency (second harmonic) are now being produced by different companies. The advantage of harmonic imaging over conventional imaging is that biologic tissues do not resonate. This technique may therefore improve the detection of low concentrations of microbubbles. Keramati demonstrated the ability of second harmonic imaging to amplify the intensity of cavity contrast signals relative to myocardium in patients following administration of Levovist and Albunex respectively [30].

With these considerations in mind, Mulvagh et al. injected Levovist i.v. in dogs and performed echocardiography of the left ventricular short-axis using harmonic imaging [31]. While Levovist was not able to opacify the myocardium using conventional imaging, intramural myocardial vessels were easily visualized using harmonic imaging. It was also possible to estimate coronary flow velocity within the intramural vessels by means of pulsed wave Doppler. These coronary velocity signals closely resembled those conventionally obtained by electro-magnetic flowmeters or intracoronary Doppler catheters. A series of further studies is currently underway in several centers.

HIGH PERSISTENCE AGENTS

One of the main characteristics of the first-generation of commercial contrast agents is their ability freely to cross the microcirculation. Both Albunex and Levovist are well tolerated after i.v. injection in humans [18, 20]. Albunex has been injected into the coronary arteries without noticeable side effects [32]. The behavior of these agents resembles that of 'free-flowing' vascular tracers such as radioactive albumin or labeled red blood cells [33], except for the sensitivity to pressure and the limited stability of sonicated agents [34–36]. Recently, an innovative ultrasound contrast agent has been produced (QW3600, EchoGen, Sonus Pharmaceutical), whose prolonged persistence in the myocardium resembles that of tracers such as thallium [37, 38]. The echocontrast effect in the myocardium obtained after i.v. administration of this agent may persist for several minutes, depending upon dose and mode of administration.

The contrast agent EchoGen consists of an emulsion of dodecafluoropentane, which has a higher density than air and a low solubility in blood, resulting in a limited tendency to diffuse out of the microbubbles. Dodecafluoropentane displays a phase shift at 28.5°C: it is a fluid at room temperature and shifts to a gaseous phase at body temperature. Soon after venous administration of this agent ultrasonic reflectances are apparent in both cardiac cavities and the myocardium. A few minutes later the contrast in the cardiac cavities is com-pletely washed out whereas the myocardium still shows pronounced signal enhancement. In a recent study Grayburn et al. [39] administered EchoGen intra-venously in a canine model of acute coronary occlusion. Myocardial contrast echocardiography accurately defined the extent of the area at risk and the infarct size, which were measured independently in gross pathological specimen stained with monostral blue and triphenyltetrazolium, respectively. Transient hemodynamic changes may occur following injection of this agent. Pulmonary artery pressure and pulmonary vascular resistance may increase at high doses of contrast, while blood pressure may decrease and heart rate increase.

The persistence of EchoGen in the myocardium as well as the cause of the hemodynamic deterioration are still under investigation. Recently, Beppu et al. [40] studied this agent in the microcirculation by microscopy. Once the microbubbles reached the microcirculation their diameter increased up to twice their initial size. This enlargement of the microbubbles within the tissue might explain both the persistence of myocardial contrast and the embolic side effects of this agent. The efficacy and safety of EchoGen in man are presently being investigated [41]. A new formulation of dodecafluoropentane has recently been developed which is capable of dense myocardial opacification without producing hemodynamic effects.

NEW LONG PERSISTENCE CONTRAST AGENTS

A new group of contrast agents has recently been introduced. These compounds are characterized by a long persistence and the ability to freely cross the microcirculation in the manner of flow tracers. Compared with Albunex and Levovist, the behavior of these agents is closer to that of an intravascular 'free-flowing' tracer such as labeled red blood cells or albumin. These agents include perfluorocarbon-exposed sonicated dextrose albumin (PESDA) [29], FS069 (Molecular Biosystem Inc., California) [42], BR1 (Bracco Research, Switzerland) [43], AFO150 (Imagent, US Alliance Pharmaceutical Corp., California) [44], aerosomes (MRX, Arizona) [45] and others under investigation. The stability of these agents is largely dependent upon the constituent gases, all of which are fluorocarbons. In smaller microbubbles, a higher surface tension is associated with a higher internal pressure, leading to gradients with respect to the surrounding medium. Due to these gradients, the gases tend to rapidly diffuse out of the microbubbles, causing their shrinkage and progressive loss of ultrasound reflectivity. The solubility of the gases in the blood and their ability to diffuse across the microbubble shell, if present, play an additional role in determining persistence. To improve the stability of contrast microbubbles, gases with low solubility in blood (such as sulphur hexafluoride) have been used in these new agents [43, 46]. The composition of the shell varies among the different agents from proteins to phospholipids and others.

The intravenous injection of these agents results in an enhancement of the ultrasound signal in the ventricular cavities, which lasts from a few seconds (at the lowest dose) to several minutes. During peak contrast enhancement of the left ventricular cavity myocardial opacification has been observed with all agents. This increase in myocardial reflectivity is visible using conventional scanners and reflects myocardial tissue perfusion. In fact, the lack of contrast enhancement in an area of myocardium after venous injection of these agents can be used to identify the area of myocardium at risk due to acute coronary occlusion [29, 39]. The extent of myocardial perfusion deficits as assessed by contrast echocardiography has been compared with the extent of the deficits detected by reference techniques. Ultrasound contrast appeared to be a reliable tool to assess the location and extent of myocardial perfusion defects.

UNSOLVED ISSUES

Anisotropy

In conventional echocardiographic images the reflectivity of the myocardium varies widely among different ventricular walls. In the parasternal short axis view, the reflectivity of the interventricular septum is greater than that of the posterior wall which, on the other hand, is higher than that of the lateral wall. This heterogeneity is due to several factors, including the attenuation of the ultrasound beam in the myocardium, the different orientation of the myocardial fibers in the various myocardial layers and the angle of incidence between myocardial fibers and the ultrasound beam, a property termed anisotropy [47–49]. The presence of echocontrast agents in the coronary circulation further enhances this gray level variability due to contrast-induced attenuation phenomena. Myocardial reflectivity also varies during the cardiac cycle [50], a factor which is also influenced by the presence of contrast. To quantify myocardial perfusion in individual ventricular segments it would be tempting to merely compare their reflectivity, and assume that the higher the intramyocardial blood content, the greater will be the number of microbubbles and thus myocardial reflectivity. Unfortunately, due to the above mentioned regional gray level variability, this approach is not feasible. Algorithms to correct for the phenomenon of anisotropy are presently being developed and might be extremely useful in the assessment of myocardial perfusion by contrast echocardiography.

Contrast-induced attenuation

An unsolved issue of considerable importance in the study of myocardial perfusion by contrast echocardiography is signal attenuation induced by the contrast agent itself. In recent studies using an in vitro flow model the intensity of the backscattered signal was measured at increasing concentrations of contrast agent [51, 52]. At low doses the intensity of the backscattered signal increases quite linearly with concentration. At higher concentrations, however, the relationship becomes curvilinear and, at even higher doses, the curve plateaus and finally declines due to contrast-induced attenuation. The echocardiographic image may be totally eliminated at very high concentrations of microbubbles (shadowing), with the microbubbles acting like an ultrasonic shield. From the practical point of view this phenomenon may limit the assessment of myocardial perfusion following intravenous contrast administration in segments which are shadowed during the period of contrast enhancement. Ventricular cavities filled with contrast can attenuate the ventricular walls located posteriorly thus preventing comparative measurements.

Thresholding

In standard gray-scale echocardiograms the reflectivity of the myocardium is substantially higher than that of blood. Due to this difference in background intensity, a low concentration of microbubbles can be more easily identified in the 'dark' left ventricular cavity than in the 'gray' myocardium. In other words, a critical concentration of microbubbles has to be reached in the myocardium before it can be visualized by conventional scanners. For this reason, the time–intensity curves recorded in the myocardium can be surprisingly short compared with those recorded in the left ventricular cavity after intravenous injection. According to both coronary physiology and indicator dilution theory this observation is unrealistic and suggests the occurrence of a thresholding artifact which eliminates the lower portion of the time–intensity curves.

This threshold effect has been demonstrated in vitro by collecting radio-frequency data from a vessel after bolus injection of ultrasound contrast [53]. Data were collected both before and after the interposition of biological tissue between the vessel and the transducer. Because of the threshold effect the mean transit time of microbubbles was shorter in the presence of the tissue than without. In this study an attempt was also made to correct for signal attenuation due to the intervening tissue by increasing the amplification of the reflected ultrasound signal.

Dispersed input function

After intravenous contrast administration the microbubble blood concentration progressively decreases from the peripheral vein to the aortic root since the microbubbles are diluted in the cardiac and lung blood volume. The concentration of microbubbles in a cross-sectional area of myocardium is, in addition, lower than the concentration in blood. To solve this problem a mathematical approach has been proposed by Feinstein et al. [54] based on indicator dilution theory [55]. According to this approach an input function can be obtained from a region of interest located in the aortic root and a residue function by a region of interest located in the myocardium. The deconvolution of the input function with the residue function enables the derivation of a myocardial transfer function. The latter should represent the function which would be recorded in the myocardium had the contrast injection been proximal (coronary ostia) and instantaneous instead of remote (vein). However, deconvolution is in practice a procedure which produces too much noise to be clinically useful, thus alternative methods must be found.

COMPARISON WITH OTHER IMAGING TECHNIQUES

Myocardial perfusion can currently be assessed by various methods in the clinical setting, most commonly myocardial scintigraphy with either thallium-

201 [56, 57] or technetium-99m isonitrile [58]. Both of these isotopes are used to obtain information regarding myocardial perfusion; thallium also allows assessment of cellular integrity. Positron emission tomography is considered the gold standard for both the non-invasive measurement of coronary blood flow and the evaluation of regional metabolism [59, 60]. At present, positron emission tomography is the only technique which allows non-invasive quantification of myocardial blood flow per unit of tissue. Compared with these techniques, myocardial contrast echocardiography does not require the use of radioactive materials, has a much better spatial and temporal resolution, a lower cost and a wider clinical availability.

Magnetic resonance imaging is a non-invasive, non-ionizing imaging technique which is attracting a growing interest for the evaluation of myocardial perfusion. A number of studies have shown the possibility of using ultrafast sequences to obtain one or more slices of the heart within a single heart beat and to follow the first pass of a bolus of contrast agent through the myocardium. This method permits relative quantification of regional myocardial blood flow in animals [61] and evaluation of the distribution of myocardial perfusion in man [62]. Recently, ultrafast CT has also been successfully applied to evaluate myocardial perfusion [63, 64]. The advantages of contrast echocardiography over both these methods include, in addition to the lower costs and the wider availability, the user-friendly nature of echocardiography.

Tremendous advances have been achieved in the research of myocardial contrast echocardiography over the past few years. The ability to study myocardial perfusion by intravenous injection of echo-enhancing agents has been impressively demonstrated in several experimental studies using different contrast agents and scanners. Due to this continuous progress it is reasonable to hypothesize that myocardial echocontrast will become a widely used clinical reality in the next few years.

REFERENCES

1. Kaul S, Pandian NG, Okada RD et al. Contrast echocardiography in acute myocardial ischemia. I. In vivo determination of total left ventricular 'area at risk'. J Am Coll Cardiol 1984;4: 1272–82.
2. Kaul S, Glasheen W, Ruddy TD et al. The importance of defining left ventricular area at risk in vivo during acute myocardial infarction: an experimental evaluation with myocardial contrast two-dimensional echocardiography. Circulation 1987;75:1249–60.
3. Kemper AJ, O'Boyle JE, Cohen CA et al. Hydrogen peroxide contrast echocardiography: quantification in vivo of myocardial risk area during coronary occlusion and of the necrotic area remaining after myocardial reperfusion. Circulation 1984;70:309–17.
4. Kemper AJ, Force T, Perkins L et al. In vivo prediction of the transmural extent of experimental acute myocardial infarction using contrast echocardiography. J Am Coll Cardiol 1986;8:143–9.
5. Villaneuva FS, Glasheen WP, Sklenar J, Kaul S. Characterization of spatial patterns of flow within the reperfused myocardium by myocardial contrast echocardiography. Implications in determining extent of myocardial salvage. Circulation 1993;88:2596–606.

6. Ito H, Tomooka T, Sakai N et al. Lack of myocardial perfusion immediately after successful thrombolysis. A predictor of poor recovery of left ventricular function in anterior myocardial infarction. Circulation 1992;85:1699–705.
7. Lim YJ, Nanto S, Masuyama T et al. Myocardial salvage: its assessment and prediction by the analysis of serial myocardial contrast echocardiograms in patients with acute myocardial infarction. Am Heart J 1994;128:649–56.
8. Sabia PJ, Powers ER, Jayaweera AR et al. Functional significance of collateral blood flow in patients with recent acute myocardial infarction. A study using myocardial contrast echocardiography. Circulation 1992;85:2080–9.
9. Sabia PJ, Powers ER, Ragosta M et al. An association between collateral blood flow and myocardial viability in patients with recent myocardial infarction. N Engl J Med 1992;327: 1825–31.
10. Kaul S, Kelly P, Oliner JD et al. Assessment of regional myocardial blood flow with myocardial contrast two-dimensional echocardiography. J Am Coll Cardiol 1989;13:468–82.
11. Rovai D, Ghelardini G, Lombardi M et al. Myocardial washout of sonicated iopamidol reflects coronary blood flow in the absence of autoregulation. J Am Coll Cardiol 1992;20:1417–24.
12. Spaan JAE. Basic coronary physiology. Heart function and coronary flow. In: Spaan JAE, editor. Coronary blood flow. Mechanics, distribution, and control. Dordrecht: Kluwer Academic Publishers, 1991:1–4.
13. Spaan JAE. Structure and perfusion of the capillary bed. In: Spaan JAE, editor. Coronary blood flow. Mechanics, distribution, and control. Dordrecht: Kluwer Academic Publishers, 1991: 69–86.
14. Meltzer RS, Tickner EG, Popp RL. Why do the lungs clear ultrasonic contrast? Ultrasound Med Biol 1980;6:263–9.
15. Reale A, Pizzuto F, Gioffrè PA et al. Contrast echocardiography transmission of echoes to the left heart across the pulmonary vascular bed. Eur Heart J 1980;1:101–6.
16. Meltzer RS, Serruys PW, McGhie J et al. Pulmonary wedge injections yielding left-sided echocardiographic contrast. Br Heart J 1980;44:390–4.
17. Keller MW, Glasheen W, Kaul S. Albunex. A safe and effective commercially produced agent for myocardial contrast echocardiography. J Am Soc Echocardiogr 1989;2:48–52.
18. Feinstein SB, Cheirif J, Ten Cate FJ et al. Safety and efficacy of a new transpulmonary contrast agent: initial multicenter clinical results. J Am Coll Cardiol 1990;16:316–24.
19. Smith MD, Elion JL, McClure RR et al. Left heart opacification with peripheral venous injection of a new saccharide echo contrast agent in dogs. J Am Coll Cardiol 1989;13:1622–8.
20. Schlief R, Staks T, Mahler M et al. Successful opacification of the left heart chambers on echocardiographic examination after intravenous injection of a new saccharide based contrast agent. Echocardiography 1990;7:61–4.
21. Villaneuva FS, Glasheen WP, Sklenar J et al. Successful and reproducible myocardial opacification during two-dimensional echocardiography from right atrial injection of contrast. Circulation 1992;85:1557–64.
22. Villaneuva FS, Glasheen WP, Sklenar J, Kaul S. Assessment of risk area during coronary occlusion and infarct size after reperfusion with myocardial contrast echocardiography using left and right atrial injections of contrast. Circulation 1993;88:596–604.
23. Voci P, Bilotta F, Merialdo P et al. Myocardial contrast enhancement after intravenous injection of sonicated albumin microbubbles: a transesophageal echocardiography dipyridamole study. J Am Soc Echocardiogr 1994;7:337–46.
24. Monaghan MJ, Metcalfe JM, Odunlami S et al. Digital radiofrequency echocardiography in the detection of myocardial contrast following intravenous administration of Albunex. Eur Heart J 1993;14:1200–9.
25. Angermann CE, Kruger TM, Junge R et al. Intravenous Albunex during transesophageal echocardiography: quantitative assessment by videodensitometry and integrated backscatter analysis from unprocessed radiofrequency signals. J Am Soc Echocardiogr 1995;8:839–53.

26. Schrope BA, Newhouse VL. Second harmonic ultrasound blood perfusion measurement. Ultrasound Med Biol 1993;19:567–79.
27. De Jong N. Basic principles of ultrasound contrast agents. In: De Jong N, editor. Acoustic Properties of Ultrasound Contrast Agents. Woerden: Zuidam and Zonen bv., 1993:17–36.
28. Wray RA, Zoghbi WA, Quinones MA, Cheirif J. Contrast echocardiography: relation of acoustic power and time gain compensation to contrast intensity duration. J Am Soc Echocardiogr 1992;4:286.
29. Porter TR, Xie F. Transient myocardial contrast after initial exposure to diagnostic ultrasound pressures with minute doses of intravenously injected microbubbles. Demonstration and potential mechanisms. Circulation 1995;92:2391–5.
30. Keramati S, Cotter B, Kwan OL, Calisi C, DeMaria AN. Effect of second harmonic imaging upon left ventricular contrast enhancement produced by Albunex®. J Am Coll Cardiol 1996; 27(Suppl A):127A.
31. Mulvagh SL, Foley DA, Aeschbaker BC, Klarich KK. Second harmonic contrast echocardiography enables non-invasive evaluation of coronary blood flow. Eur Heart J 1995;16(Suppl): 393.
32. Ten Cate FJ, Widimsky P, Cornel JH et al. Intracoronary albunex. Its effects on left ventricular hemodynamics, function, and coronary sinus flow in humans. Circulation 1993;88:2123–7.
33. Rovai D, Lombardi M, Distante A, L'Abbate A. Myocardial perfusion by contrast echocardiography. From off-line processing to radio frequency analysis. Circulation 1991;83(Suppl III):III-97–III-103.
34. Shapiro JR, Reisner SA, Lichtenberg GS, Meltzer RS. Intravenous contrast echocardiography with use of sonicated albumin in humans. Systolic disappearance of left ventricular contrast after transpulmonary transmission. J Am Coll Cardiol 1990;16:1603–7.
35. Schlief R, Schurmann R, Balzer T, Zomack M, Niendorf HP. Saccharide based contrast agents. In: Nanda NC, Schlief R, editors. Advances in Echo Imaging Using Contrast Enhancement. Dordrecht: Kluwer Academic Publishers, 1993;71–96.
36. Rovai D, Ghelardini G, Trivella MG et al. Intracoronary air-filled albumin microspheres for myocardial blood flow measurement. J Am Coll Cardiol 1993;22:2014–21.
37. DeMaria AN, Dittrich H, Kwuan OI, Kimura B. Myocardial opacification produced by peripheral venous injection of a new ultrasonic contrast agent. Circulation 1993;88(Suppl):I-401.
38. Beppu S, Matsuda H, Shishido T, Miyatake K. Success of myocardial contrast echocardiography by peripheral venous injection method: visualisation of area at risk. Circulation 88; 4:I-401.
39. Grayburn PA, Erickson JM, Escobar J et al. Peripheral intravenous myocardial contrast echocardiography using a 2% dodecafluoropentane emulsion: identification of myocardial risk area and infarct size in the canine model of ischemia. J Am Coll Cardiol 1995;26:1340–7.
40. Beppu S. Newer contrast agents. Heart Vessel 1995(Suppl);11:14.
41. Cotter B, Keramati S, Kwan OL et al. Myocardial opacification with lose dose activated EchoGen in patients: initial experience. Circulation 1995;92(Suppl):I-192.
42. Dittrich HC, Bales GL, Kuvelas T et al. Myocardial contrast echocardiography in experimental coronary artery occlusion with a new intravenously administered contrast agent. J Am Soc Echocardiogr 1995;8:465–74.
43. Rovai D, Lubrano V, Vassalle C et al. Detection of myocardial perfusion deficits by intravenous administration of the ultrasonic contrast agent BR1. Circulation 1995;92(Suppl I):I-659.
44. Mulvagh SL, Foley DA, Aeschbacher BC et al. A new intravenous perfluorochemical echocardiographic contrast agent, Imagent, US: imaging characteristics and hemodynamic profile. J Am Soc Echocardiogr 1995;8:345.
45. Sutherland GR, Grauer SE, Moran C et al. Aerosomes MRX 115 echo contrast agent demonstrates myocardial opacification after intravenous injection in humans, without significant side effects, in a phase I clinical trial. Circulation 1995;92(Suppl):2213.
46. Porter TR, Xie F. Visually discernible myocardial echocardiographic contrast after intravenous

injection of sonicated dextrose albumin microbubbles containing high molecular weight soluble gases. J Am Coll Cardiol 1995;25:509–15.

47. Olshansky B, Collins SM, Skorton DJ et al. Variation of left ventricular myocardial grey level on two-dimensional echocardiograms as a result of cardiac contraction. Circulation 1984;70: 972–7.

48. Mottley JG, Miller JG. Anisotropy of the ultrasonic backscatter of myocardial tissue. I. Theory and measurements in vitro. J Acoust Soc Am 1988;83:755–61.

49. Madaras EI, Perez JE, Sobel BE et al. Anisotropy of the ultrasonic backscatter of myocardial tissue. II. Theory and measurements in vivo. J Acoust Soc Am 1988;83:762–9.

50. Madaras EI, Barzilai B, Perez JE et al. Changes in myocardial backscatter throughout the cardiac cycle. Ultrason Imag 1983;5:229–32.

51. Powsner SM, Keller MW, Saniie J, Feinstein SB. Quantitation of echocontrast effects. Am J Physiol Imaging 1986;1:124–8.

52. De Pieri G, Rovai D, Mazzarisi A et al. Computerized radio-frequency analysis of concentration and 'decay' of echo contrast agents. In: IEEE Comp. Society, editors. Computers in Cardiology. Washington DC: The Computers Society Press, 1988;211–4.

53. Rovai D, Lombardi M, Mazzarisi A et al. Flow quantitation by contrast echocardiography. Effects of intervening tissue and the angle of incidence between flow and ultrasonic beam. Int J Cardiac Imaging 1993;9:21–7.

54. Feinstein SB, Voci P, Segil LJ, Harper PV. Contrast echocardiography. In: Marcus ML, Schelbert HR, Skorton DJ, Wolf GL, eds. Cardiac Imaging. Philadelphia: W.B. Saunders Company, 1991.

55. Zierler KL. Equations for measuring blood flow by external measurement of radioisotopes. Circ Res 1965;14:309.

56. Kiat H, Berman DS, Maddahi J. Comparison of planar and tomographic thallium-201 imaging methods for the evaluation of coronary artery disease. J Am Coll Cardiol 1989;13:613–6.

57. Leppo JA, Boucher CA, Okada RD et al. Serial thallium-201 myocardial imaging after dypiridamole infusion: diagnostic utility in detecting coronary stenoses and relationship to regional wall motion. Circulation 1982;66:649–57.

58. Marzullo P, Sambuceti G, Parodi O. The role of sestamibi scintigraphy in the radioisotopic assessment of myocardial viability. J Nucl Med 1992;33:1925–30.

59. Shelbert HR, Phelps ME, Huang SC et al. N-13 ammonia as an indicator of myocardial blood flow. Circulation 1981;63:1259–72.

60. Ratib O, Phelps ME, Huang SC et al. Positron tomography with deoxyglucose for estimating local myocardial glucose metabolism. J Nucl Med 1982;23:577–86.

61. Wilke N, Simm C, Zhang J. Contrast-enhanced first pass myocardial perfusion imaging: correlation between myocardial blood flow in dogs at rest and during hyperaemia. Magn Res Med 1993;29:485–97.

62. Lombardi M, Kvaerness J, Soma J et al. Relationship between function, perfusion and contractile reserve early after myocardial infarction: a dynamic magnetic resonance and stress echocardiographic study in man. J Am Coll Cardiol 1995;2(Suppl):12A.

63. Weiss RM, Otoadese EA, Noel MP et al. Quantitation of absolute regional myocardial perfusion using cine computed tomography. J Am Coll Cardiol 1994;23:1186–93.

64. Brundage BH. Beyond perfusion with ultrafast computed tomography. Am J Cardiol 1995;75: 69D–73D.

24. Myocardial contrast echocardiography in ischemic heart disease: Osaka experience

TOHRU MASUYAMA, HIROSHI ITO, YOUNG-JAE LIM & AKIRA KITABATAKE

INTRODUCTION

Coronary angiography is a popular method for analyzing coronary circulation in humans. This method is certainly useful for assessing the epicardial coronary artery; however, it is of little use for assessing coronary microvasculature. Myocardial contrast echocardiography (MCE) is a relatively new method of visualizing the territory of the coronary arteries through the intracoronary injection of microbubbles. We recently showed that MCE provides information about the territory of the coronary circulation but also the degree of damage of coronary microvasculature. We have studied coronary microvasculature in patients with ischemic heart disease using MCE, and summarize our MCE data here.

MYOCARDIAL CONTRAST ECHOCARDIOGRAPHY (MCE) IN HUMANS

Functional border zone

The size of the damaged or ischemic area can be estimated by measuring the abnormal contractile area using two-dimensional echocardiography or left ventriculography, using the assumption that the dyskinetic area is equivalent to the malperfused area. However, several animal studies have demonstrated the presence of non-ischemic regional dysfunction at the lateral borders of the damaged or ischemic myocardium [1–3]. The concept of a functional border zone (FBZ) was recently introduced [2, 3] to identify such myocardium. The existence of an FBZ would explain the disparity between the regional wall motion and the pathological infarct size in animal and clinical studies [4]. If the FBZ does exist, the malperfused areas could not be precisely assessed from the extent of the abnormal contractile segment. Although the presence of an FBZ, defined as the non-ischemic but asynergic myocardium adjacent to the ischemic

387

N. C. Nanda, R. Schlief and B. B. Goldberg (eds.), Advances in Echo Imaging Using Contrast Enhancement, Second Edition, 387–399.
© 1997 *Kluwer Academic Publishers.*

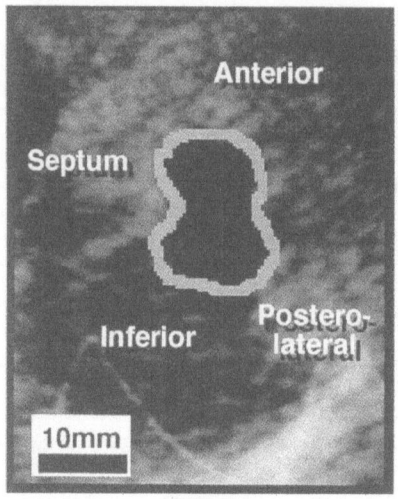 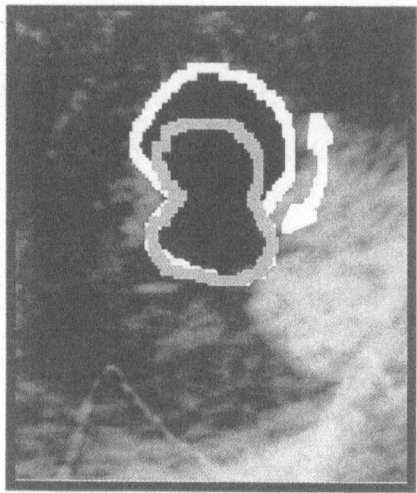

Control   **LAD occlusion (PTCA)**

Figure 1. Representative myocardial contrast echocardiographic images of the short-axis view obtained at control (left panel) and during left anterior descending (LAD) coronary arterial occlusion (right panel) in a patient undergoing coronary angioplasty (PTCA). Thick gray and white lines represent endocardial edges determined in the images at control and during coronary occlusion, respectively. The gray line in the right image is the superimposed line that represents the endocardial edge determined in the control study. The arrow in the right panel indicates the ends of the functional border zone. [Reproduced with permission from the American Heart Association (Circulation 1993;88:447–53).]

area, has been demonstrated in animals hearts, it is not known whether this zone exists in humans.

MCE was performed before and during balloon inflation in the area of coronary stenosis by injecting contrast medium through the guiding catheter in 13 patients with effort angina who underwent successful coronary angioplasty [5]. The area showing MCE defect during balloon inflation was determined with reference to the preangioplasty MCE and was regarded as an ischemic area. Both the anteroseptal and posterolateral regions were well opacified in the control MCE study, and no contrast defect was observed. In the MCE observed during balloon inflation, the anteroseptal region was not opacified while good opacification was obtained in the posterolateral region (Figure 1). The size of the FBZ was assessed by measuring the length of the endocardium that showed asynergy in the echo-enhanced (non-ischemic) area. The FBZ measured was 13 ± 4 mm in the short-axis view ($n=5$), and 16 ± 9 mm in the long-axis view ($n=8$). The mechanism of FBZ is not fully clarified, but it has been considered as a mechanical phenomenon that is dictated by the geometry of the left ventricle and the relative levels of stiffness in ischemic and non-ischemic areas, but which is unrelated to flow restriction or relative ischemia [6]. This MCE study clearly showed that non-ischemic contractile dysfunction exists even in human hearts. The presence of an FBZ may limit the use of wall motion analysis in assessing the risk or ischemic area in patients with myocardial infarction. MCE appears to be a unique technique for assessing the risk or ischemic area.

Changes in the transmural flow distribution during pacing-induced myocardial ischemia

Myocardial ischemia is clinically anticipated from ST segment depression in the electrocardiogram, because reduction in coronary flow causes a shift in transmural myocardial blood flow distribution towards epicardial site, producing myocardial ischemia predominantly in the subendocardial layer [7]. The changes in transmural myocardial blood flow distribution during myocardial ischemia are well demonstrated and visualized in experimental studies using radionuclide microsphere methods [8]. However, no imaging method has been available to visualize transmural myocardial blood flow distribution in humans. We used MCE to visualize changes in transmural myocardial blood flow distribution provoked by rapid atrial pacing in patients with coronary artery disease.

MCE was performed in patients with coronary artery disease by injection of the contrast agent into the left and right coronary arteries before and just after rapid atrial pacing for 3 min at a rate of 150 bpm [9]. Representative two-dimensional images of MCE before and after rapid pacing are shown in Figure 2. Injection of contrast agents into the stenotic coronary artery stained both endocardial and epicardial halves of the supplied myocardium before rapid pacing. However, after rapid pacing, the injection into the stenotic coronary artery generally did not stain the endocardial half of the supplied myocardium as much as it did the epicardial half. MCE images of pacing-induced myocardial ischemia were obtained in 11 patients with effort angina due to single vessel coronary artery disease and quantitated with an off-line image analyzer. Gray levels were measured for epicardial and endocardial halves of the anteroseptal, posterolateral and inferior segments of the left ventricle in each patient. The gray level before injection was subtracted from the gray level after injection, and the subtracted gray level was used to calculate the endo:epi gray level ratio. The endo:epi gray level ratios before rapid pacing for the segments supplied by the non-stenotic coronary artery and for the segments supplied by the stenotic coronary artery were 0.99 ± 0.23 and 0.93 ± 0.18, respectively (no significant difference). In the segments supplied by the non-stenotic coronary artery, the endo:epi gray level ratio decreased slightly, but not significantly, from 0.99 ± 0.23 to 0.88 ± 0.20 after rapid pacing. In the segments supplied by the stenotic coronary artery, it decreased significantly from 0.93 ± 0.18 to 0.40 ± 0.21 after rapid pacing ($p < 0.01$). The endo:epi gray level ratios were nearly equal to 1 before rapid pacing in both the segments supplied by the non-stenotic coronary artery and in those supplied by the stenotic coronary artery.

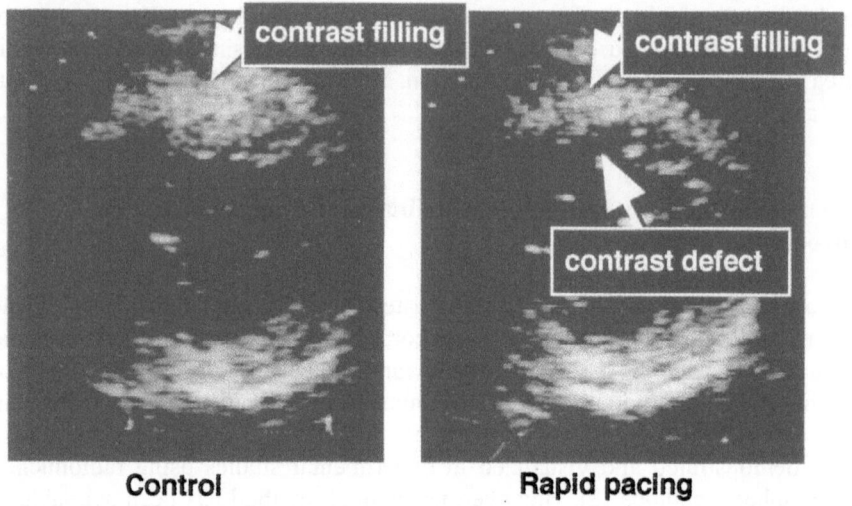

Figure 2. End-diastolic myocardial contrast echocardiograms at the left coronary artery contrast injection before (left) and after rapid pacing (right) in a patient with severe coronary stenosis in the proximal portion of the left anterior descending artery. The enhancement of the gray levels is observed in endocardial and epicardial halves of the anteroseptal and posterolateral segments before rapid pacing (control). However, just after rapid pacing, the enhancement is observed only in the epicardial half.

These data indicate that there was no maldistribution of transmural blood flow before rapid pacing not only in the segment with a non-stenotic coronary artery but also in the segment with a stenotic coronary artery. This is comparable to the results of an experimental study [10] which showed that reduction of the coronary flow by 20% caused no myocardial maldistribution of radioactive microsphere at rest. The endo:epi gray level ratio significantly decreased after rapid pacing only in the segment supplied by a stenotic coronary artery. This result is also comparable to that of an experimental study [11] in which MCE was demonstrated to be useful in assessing changes in regional endo:epi flow ratio induced by intravenous dipyridamole in dogs subjected to a critical coronary stenosis that did not alter resting coronary flow. The abnormal transmural blood flow distribution at rapid pacing disappeared after successful coronary angioplasty [12].

Dynamic collateral function and myocardial ischemia at stress

In patients with old myocardial infarction, development of coronary collaterals in the acute phase often results in the co-existence of viable and fibrotic myocardium. In such cases, even though collateral function is enough to perfuse the

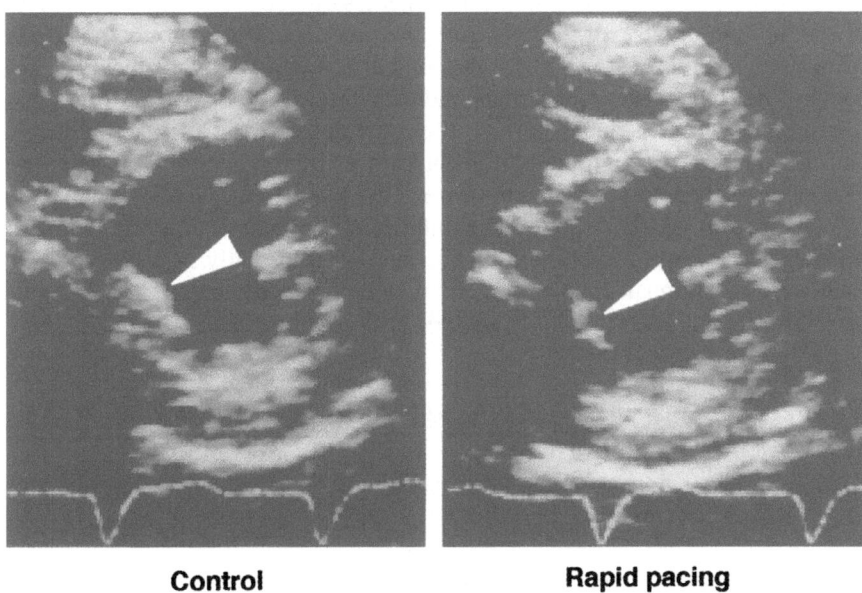

Control   **Rapid pacing**

Figure 3. Myocardial contrast echocardiograms in a patient with well developed angiographical collaterals from the left to right coronary artery. The inferior area of the left ventricle showed good myo-cardial contrast enhancement before rapid atrial pacing (indicated by the arrow in the left panel) but showed poor myocardial contrast enhancement immediately after rapid atrial pacing (indicated by the arrow in the right panel) with the contrast injection into the donor coronary artery (left coronary artery).

myocardium at rest, it may be insufficient to prevent stress-induced myocardial ischemia, which may thus develop in the infarct regions during rapid atrial pacing. Although resting collateral function can be assessed by coronary angiography, its role in preventing stress-induced myocardial ischemia is not predictable from the resting data. We performed MCE before and immediately after rapid atrial pacing in patients with a totally occluded coronary artery and various degrees of collateral development [13]. In all patients, contrast enhancement was observed in the infarct region following injection of contrast medium into the donor coronary artery at rest; however, contrast enhancement was not always observed after rapid atrial pacing. Stress-induced wall motion abnormality was significant only in patients with poor contrast enhancement at pacing (Figure 3), suggesting that MCE findings reflect dynamic collateral function.

Value of MCE

The data discussed so far indicate that MCE provides interesting information about myocardial perfusion that is not obtainable with any conventional methods.

Myocardial perfusion abnormality cannot necessarily be assessed by the analysis of wall motion because of the existence of a functional border zone. In addition MCE provides information about transmural distribution of myocardial perfusion which is not detectable with radionuclide technique. Finally, we showed a disparity between MCE and coronary arteriography. Even if coronary filling is detected in coronary angiogram, it is not known whether the filling is adequate. This can be only assessed by the measurement of flow at the level of microvessel or myocardium. MCE thus provides interesting and useful information about myocardial perfusion that cannot be obtained with currently available methods such as conventional echocardiography, left ventriculography, electrocardiography, radionuclide techniques and coronary arteriography.

'NO REFLOW' PHENOMENON IN PATIENTS WITH ACUTE MYOCARDIAL INFARCTION

'No reflow' phenomenon

In recent years, an interest in preserving acutely ischemic myocardium has led to the development of interventional techniques for restoring coronary perfusion to the jeopardized myocardium [14, 15]. Coronary arteriography has been used to evaluate the success of these techniques, but it does not provide an assessment of myocardial perfusion. MCE defects obtained before coronary reflow denote the size of the risk area, and those obtained after successful coronary reflow denote regions of no reflow. MCE was performed before and immediately after successful reflow with intracoronary injection of sonicated Ioxaglate in 39 patients with acute anterior myocardial infarction [16]. The average segmental score by two-dimensional echocardiography (graded 0, normal, to 3, akinetic/dyskinetic) and global ejection fraction was measured at 1 day and at 4 weeks after reflow. Hypokinesis in the infarct region was assessed by the center line method and expressed in terms of standard deviations (regional wall motion [RWM]: SD/chord) of normal. Immediately after reflow 30 of 39 patients showed significant contrast enhancement within the risk area (MCE reflow; Figure 4). The other nine patients (23%) showed a residual contrast defect in the risk area (MCE no reflow; Figure 5). Although the no reflow phenomenon has been recognized during thrombolysis after prolonged coronary occlusion in animal studies [17, 18], the mechanism of this phenomenon is still unclear. This phenomenon has been associated with extensive myocardial necrosis and microcirculatory damage in animals [19, 20]. The animal studies [17, 18] suggested that the no reflow phenomenon is related to paucity of microcirculation following its destruction by the infarction process.

There were no significant differences in the elapsed time, angiographic collateral grade, and degree of residual stenosis between patients with and without MCE-demonstrable reflow [16]. Improvement in global and regional left ventricular

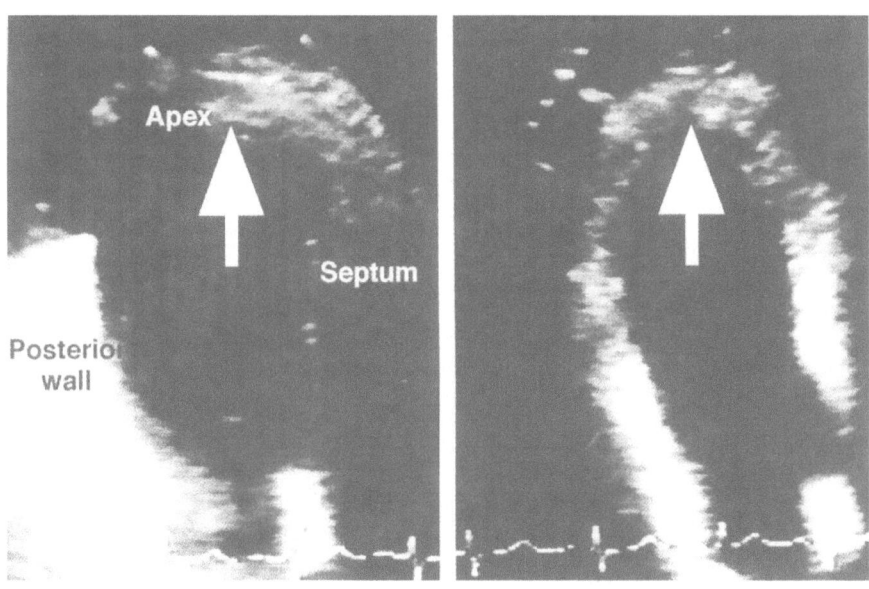

Pre reflow　　　　　**Post reflow**

Figure 4. Myocardial contrast echocardiograms of apical long-axis view in a patient undergoing emergent coronary angioplasty within 4 h of symptom onset. Initial coronary arteriogram showed total occlusion at the proximal left anterior descending coronary artery. The interventricular septum and cardiac apex exhibit an area of contrast defect after contrast injection into the left coronary artery and is defined as the risk area (arrow in the left panel). Immediately after angioplasty, injection of sonicated microbubbles into the left coronary artery produces significant contrast enhancement within the risk area, indicating recovery of myocardial flow (arrow in the right panel). [Reproduced with permission from the American Heart Association (Circulation 1992;85:1699–705).]

function was greater in patients with reflow than in patients with no reflow on MCE (average segmental score 0.44 ± 0.41 vs. 0.97 ± 0.36, $p < 0.01$; LV ejection fraction 56 ± 13 vs. 43 ± 9, $p < 0.05$; RWM -1.87 ± 0.85 vs. -3.18 ± 0.52, $p < 0.005$). Thus, MCE demonstrates that angiographically successful reflow cannot be used as an indicator of successful myocardial reperfusion in patients with acute myocardial infarction. The residual contrast defect in the risk area demonstrated immediately after reflow (MCE no reflow) is a predictor of poor functional recovery of the postischemic myocardium.

Serial changes in MCE reflow and no reflow

Although myocardial perfusion has been clarified shortly after coronary reflow, little is known about its changes in the convalescent stage. Experimental studies

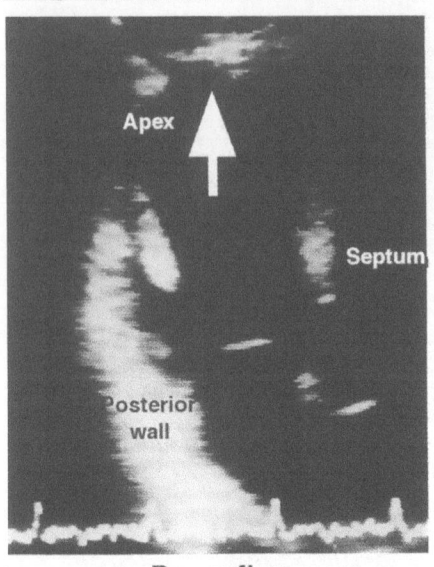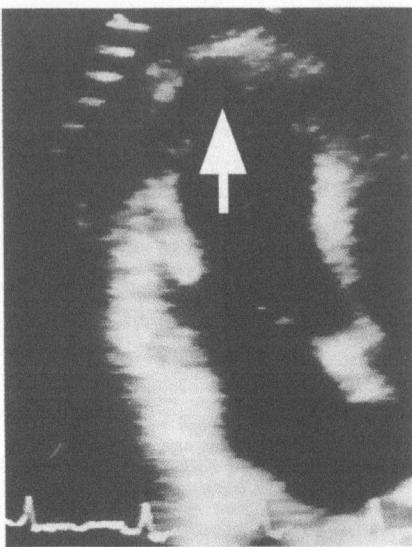

Figure 5. Myocardial contrast echocardiograms of apical long-axis view before (left panel) and after (right panel) thrombolysis in a patient with acute myocardial infarction. This patient had total occlusion just distal to the first septal branch in the left anterior descending coronary artery and received reflow 4 h after symptom onset by coronary angioplasty. After reflow, the distal portion of interventricular septum and cardiac apex exhibit residual contrast defect, indicating the no reflow phenomenon (arrow in the right panel). [Reproduced with permission from the American Heart Association (Circulation 1992;85:1699−705).]

have shown that flow to areas that initially received adequate perfusion progressively reduces for several hours after coronary reflow, suggesting that the microvascular damage may progress even after coronary reflow [21, 22]. In contrast, Henes et al. have demonstrated in patients with reperfused myocardial infarction that myocardial perfusion may return rapidly (18±6 h) to normal values after reperfusion [23]. It is still not known whether the microvascular damage caused by ischemia is reversible or progressive in the convalescent stage in patients with reperfused acute myocardial infarction. This was clarified by performing myocardial contrast echocardiography serially before reperfusion, immediately after reperfusion and in the chronic stage following injection of sonicated Ioxaglate into the right and left coronary artery in 28 patients with acute myocardial infarction [24]. Contrast defect was observed even after reperfusion in eight (29%) patients, while contrast defect disappeared after reperfusion in the other 20 patients. In four (20%) of the 20 patients, contrast defect reappeared in the chronic stage despite patency of the infarct-related vessel (Figure 6). The exact mechanism for the late reappearance of contrast defect cannot be explained from the data of the current study, particularly because of the lack of histological

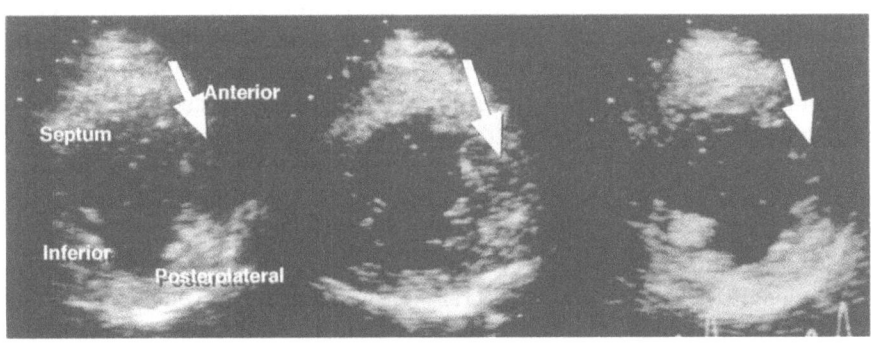

Before reperfusion   Just after reperfusion   Chronic phase
(Acute phase)

Figure 6. Myocardial contrast echocardiograms in a patient with anterior myocardial infarction before reperfusion (left panel), immediately after reperfusion (middle panel) and in the chronic stage (right panel). Anterolateral region showed good myocardial opacification immediately after reperfusion but poor myocardial opacification in the chronic stage (arrows). [Reproduced with permission from Mosby-Year Book, Inc. (Am Heart J 1994;128:649–56).]

data. We can only speculate that reperfusion injury and the time delay between cell death and destruction of the capillary network may be possible causes for the late reappearance of contrast defect [25, 26]. Left ventricular function recovery of the risk area in the chronic stage as assessed with regional wall motion and wall thickness was better in the patients with MCE reflow after reperfusion than in those with persistent or recurrent MCE no reflow. This study clearly showed that MCE reflow immediately after reperfusion does not necessarily imply myocardial salvage in the chronic stage.

More recently, MCE was analyzed quantitatively to assess the temporal changes in myocardial perfusion after reflow both in patients with and without MCE no reflow and to investigate the relationship between MCE findings and myocardial viability [27]. In this study, microvascular function was assessed with MCE on the day of infarction and 4 weeks later in patients with reperfused anterior wall myocardial infarction. Corrected myocardial video intensity and the extent of no or extremely low reflow area were compared between acute and chronic stage MCE to elucidate whether ischemic microvascular damage is reversible.

In this study, MCE was performed with the intracoronary injection of sonicated microbubbles before and shortly after coronary reflow and a month later in 45 patients with anterior wall acute myocardial infarction. MCE before reflow was analyzed to determine the risk area as an area of contrast defect in the apical long-axis view. MCE images after reperfusion were analyzed to determine peak contrast intensity, which should be in proportion to the concentration of microbubbles within microvasculature in the infarct and normal myocardium, and the ratio of these (PI ratio) was used to assess microvascular integrity. Areas of residual contrast defect were expressed as a ratio to those of left ventricular myocardium (RCD ratio) in order to assess the spatial extent of the MCE no reflow. Regional wall motion (RWM: SD/chord) in the territory of the left anterior descending coronary artery was determined by the use of the center line method in both the acute and late stages. Although the PI ratio was extremely low shortly after coronary reflow it increased in the late stage of the acute myocardial infarction, with improvement in regional contractile function (RWM: -3.2 ± 0.5 vs. -2.6 ± 1.0, $p < 0.01$, PI ratio: 0.44 ± 0.25 vs. 0.60 ± 0.29, $p < 0.01$). Reduction in the RCD ratio was observed even in 15 patients with MCE no reflow in the acute stage (0.33 ± 0.09 vs. 0.16 ± 0.11, $p < 0.01$). Thus, it was concluded that the recovery from ischemic microvascular damage is generally observed in the late stage of acute myocardial infarction in association with improvement in myocardial contractile function, and that the degree of the improvement in contractile function and microvascular integrity varies among patients.

Myocardial viability

There is little information on the relationship between the microvascular integrity and myocardial viability in the clinical setting. Since there should be a low density of microvascular beds in the regions of necrosis or scar, it may be hypothesized that the density of microvessels should be in proportion to the amount of residual myocardium, and a major determinant of myocardial viability. Thus, the relationship between corrected myocardial video intensity and residual contractile function was further studied to clarify whether myocardial viability can be assessed with MCE shortly after coronary reflow or in the chronic stage [27]. When the relationship between residual contractile function and the microvascular integrity was investigated, a significant correlation was found between the PI ratio and RWM ($r = 0.73$, $p < 0.001$) in the late stage of the acute myocardial infarction. The contrast peak intensity in the late stage of infarction may provide a useful estimate of the myocardial viability.

Clinical implications of 'no-reflow' phenomenon

The prognostic value of MCE no reflow was examined by comparing complications, left ventricular morphology and in-hospital survival following acute myocardial infarction in 126 patients with and without MCE no reflow following a first anterior myocardial infarction [28]. All patients received successful coronary reflow within 24 h of the onset and underwent MCE before and shortly after coronary reflow with the intracoronary injection of sonicated microbubbles. From contrast reperfusion patterns, patients were divided into two subsets, MCE reflow and MCE no reflow. Frequencies of the following complications were compared between the two subsets: intractable arrhythmias, pericardial effusion, early congestive heart failure (within 3 days after the onset), late congestive heart failure (4 days or more after the onset), coronary events and in-hospital death. Forty-seven patients (37%) manifested MCE no reflow and 79 showed MCE reflow. There was no difference in the frequency of arrhythmias or coronary events between the two subsets. Pericardial effusion and congestive heart failure were observed more frequently in patients with MCE no reflow than in those with MCE reflow (26 vs. 4%, $p < 0.05$; 45 vs. 15%, $p < 0.05$, respectively). Congestive heart failure tended to be prolonged in those with MCE no reflow and three patients (7%) of this subset died of pump failure. Left ventricular end-diastolic volume progressively increased in the convalescent stage in patients with MCE no reflow (acute vs. late, 145 ± 43 vs. 169 ± 60 ml, $p < 0.001$), whereas it decreased in those with MCE reflow (154 ± 42 vs. 144 ± 44 ml, $p < 0.01$). The substantial MCE no reflow phenomenon at reperfusion therefore conveys useful information about an outcome of coronary intervention and left ventricular remodeling in individual patients with anterior wall acute myocardial infarction, although these are suggestive results in a limited number of patients.

FUTURE DIRECTION

MCE is inexpensive and can be performed in the catheterization laboratory. It may also provide clinically useful information because MCE can clearly delineate the borders of the damaged or ischemic myocardium, to assess the transmural flow distribution, to evaluate the condition of coronary microvasculature in the early and convalescent stages of myocardial infarction in patients and to estimate myocardial viability. Sonicated albumin solution has the potential to produce contrast enhancement of the left-sided chambers as well as the myocardium through the contrast injection into the right atrium or peripheral veins [29]. Such new contrast media should expand the clinical application of MCE. With advent of such contrast media, MCE may be performed to serially assess the myocardial flow and viability after coronary reperfusion at the bedside in the intensive or coronary care unit.

REFERENCES

1. Kerber RE, Marcus ML, Ehrhardt J, Wilson R, Abbund FM. Correlation between echocardiographically demonstrated segmental dyskinesis and regional myocardial perfusion. Circulation 1975;52:1097–104.
2. Gallagher KP, Gerren RA, Stiring MC et al. The distribution of functional impairment across the lateral border of acutely ischemic myocardium. Circ Res 1986;58:570–83.
3. Buda AJ, Zotz RJ, Gallagher KP. Characterization of the functional border zone around regionally ischemic myocardium using circumferential flow-function maps. J Am Coll Cardiol 1986;8:150–8.
4. Shen WK, Khandheria BK, Edwards WD et al. Value and limitations of two-dimensional echocardiography in predicting myocardial infarct size. Am J Cardiol 1991;68:1143–9.

5. Nanto S, Masuyama T, Lim YJ, Hori M, Kodama K, Kamada T. Demonstration of functional border zone with myocardial contrast echocardiography in human hearts. Simultaneous analysis of myocardial perfusion and wall motion abnormalities. Circulation 1993;88:447–53.
6. Bogen DK, Rabinowitz SA, Needleman A, McMahon TA, Abelmann WH. An analysis of the mechanical disadvantage of myocardial infarction in the canine left ventricle. Circ Res 1980; 47:728–41.
7. Guyton RA, McClenathan JH, Newman GE, Michaelis LL. Significance of subendocardial ST segment elevation caused by coronary stenosis in the dog. Epicardial ST segment depression, local ischemia and subsequent necrosis. Am J Cardiol 1977;40:373–80.
8. Neill WA, Oxendine J, Phelps N, Anderson RP. Subendocardial ischemia provoked by tachycardia in conscious dogs with coronary stenosis. Am J Cardiol 1975;35:30–6.
9. Lim YJ, Nanto S, Masuyama T et al. Visualization of subendocardial myocardial ischemia with myocardial contrast echocardiography in humans. Circulation 1989;79:233–44.
10. Neil WA, Oxendine J, Phelps N, Anderson RP. Subendocardial ischemia provoked by tachycardia in conscious dogs with coronary stenosis. Am J Cardiol 1975;35:30–6.
11. Cheirif J, Zoghibi WA, Bolli R, O'Neill PG, Hoyt B, Quinones MA. Assessment of regional myocardial perfusion by contrast echocardiography. 2. Detection of changes in transmural and subendocardial perfusion during dipyridamole-induced hyperemia in a model of critical coronary stenosis. J Am Coll Cardiol 1989;14:1555–65.
12. Lim YJ, Nanto S, Masuyama T, Hori M. Subendocardial myocardial ischemia as assessed with myocardial contrast echocardiography in patients with ischemic heart diseases. Biorheology 1993;30:349–58.
13. Lim YJ, Masuyama T, Kohama A, Nanto S. Stress myocardial contrast echocardiography in the assessment of dynamic collateral function. Circulation 1992;86(Suppl I):I-523.
14. Braunwald E. The aggressive treatment of acute myocardial infarction. Circulation 1985; 71:1087–92.
15. Przyklenk K, Vivalidi MT, Scoen FJ, Arnold JMO, Kloner RA. Salvage of ischemic myocardium by reperfusion: importance of collateral blood flow and myocardial oxygen demand during occlusion. Cardiovasc Res 1986;20:403–14.
16. Ito H, Tomooka T, Sakai N et al. Lack of myocardial perfusion immediately after successful thrombolysis: a predictor of poor recovery of left ventricular function in anterior myocardial infarction. Circulation 1992;85:1699–705.
17. Kloner RA, Ganote CE, Jennings RB. The 'no reflow' phenomenon after temporary coronary occlusion in the dog. J Clin Invest 1974;54:1496–508.
18. Kloner RA, Rude RE, Carlson N, Maroko PR, De Boer LWV, Braunwald E. Ultrastructural evidence of microvascular damage and myocardial cell injury after coronary artery occlusion: which comes first? Circulation 1980;62:945–52.
19. Przyklenk K, Kloner RA. 'Reperfusion injury' by oxygen-derived free radicals? Effect of superoxide dismutase plus catalase, given at the time of reperfusion, on myocardial infarct size, contractile function, coronary microvasculature, and regional myocardial blood flow. Circ Res 1989;64:86–96.
20. Rosmon FJL, Hook BS, Kunkel SL, Abrams GD, Schork A, Lucchesi BR. Reduction of the extent of ischemic myocardial injury by neutrophil depletion in the dog. Circulation 1983;67: 1016–23.
21. Ambrosio G, Weisman HF, Mnnisi JA, Becker LC. Progressive impairment of regional myocardial perfusion after initial restoration of postischemic blood flow. Circulation 1989;80: 1841–61.
22. Jeremy RW, Links JM, Becker CC. Progressive failure of coronary flow during reperfusion of myocardial infarction: documentation of the no reflow phenomenon with positron emission tomography. J Am Coll Cardiol 1990;16:695–704.
23. Hanes CG, Bergman SR, Perez JE, Sobel BE, Geltman EM. The time course of restoration of nutritive perfusion, myocardial oxygen consumption, and regional function after coronary thrombolysis. Coronary Artery Dis 1990;1:687–96.

24. Lim YJ, Nanto S, Masuyama T, Kohama A, Hori M, Kamada T. Myocardial salvage: its assessment and prediction by the analysis of serial myocardial contrast echocardiograms in patients with acute myocardial infarction. Am Heart J 1994;128:649–56.

25. Olafsson B, Forman MB, Puett DW et al. Reduction of reperfusion injury in the canine preparation by intracoronary adenosine: importance of the endothelium and the no-reflow phenomenon. Circulation 1987;76:1135–45.

26. Kloner RA, Rude RE, Carlson N, Maroko PR, De Boer LWV, Braunwald E. Ultrastructural evidence of microvascular damage and myocardial cell injury after coronary artery occlusion: which comes first? Circulation 1980;62:945–52.

27. Ito H, Iwakura K, Oh H et al. Temporal changes in myocardial perfusion patterns in patients with reperfused anterior wall myocardial infarction: their relation to myocardial viability. Circulation 1995;91:656–62.

28. Ito H, Maruyama A, Iwakura K et al. Clinical implications of 'no reflow' phenomenon: a predictor of complications and left ventricular remodeling in reperfused anterior myocardial infarction. Circulation 1996;93:223–8.

29. Villanueva FS, Glasheen WP, Sklenar J, Jayaweera AR, Kaul S. Successful and reproducible myocardial opacification during two-dimensional echocardiography from right heart injection of contrast. Circulation 1992;85:1557–64.

25. Assessment of myocardial viability post-infarction with myocardial contrast echocardiography

CHRISTIAN FIRSCHKE & SANJIV KAUL

BACKGROUND

In a patient with acute myocardial infarction, clinical, electrocardiographic and angiographic information may be inadequate for determining the presence of viable tissue. If wall thickening is present, the myocardium is viable; if wall thickening is absent, the myocardium may or may not be viable [1]. The basis for the underlying myocardial dysfunction may be multifactorial in a single patient or even in a single myocardial segment. Myocardial contrast echocardiography (MCE) can define regions of viability within the infarct zone based on its ability to demonstrate the spatial distribution of microvascular perfusion [2]. The magnitude and spatial extent of microvascular perfusion within the infarct zone is the predominant determinant of the extent of myocellular viability.

In this regard, it is important to first define the term 'viability'. According to the *Oxford English Dictionary*, 'viable' means 'capable of living' [3]. Unfortunately, the terms living and capable of contracting in the presence of adequate blood flow have been used interchangeably for the myocardium. It has been suggested that viable myocardium is that which demonstrates thickening after blood flow has been restored to it [4]. Most clinical studies of viability are based on this definition. This definition, however, ignores a fundamental physiological principle, which is that at rest most left ventricular wall thickening occurs as a result of endocardial thickening (Figure 1); the middle layer of the myocardium contributes only a modest amount to wall thickening and the contribution of the epicardium is negligible [5, 6]. Thus, if the endocardium is necrosed, wall thickening will be significantly diminished at rest, even if blood flow is restored to the rest of the living myocardium within the ventricular wall. In experiments performed in dogs where the coronary artery was occluded and regional function was assessed 2 days later, complete loss of wall thickening was noted when the infarction involved ≥20% of the wall thickness. When infarction involved <20% of the wall thickness, contractile function was present, albeit reduced (Figure 2) [1].

In a post-infarct patient, there are two questions of major clinical relevance:

401

N. C. Nanda, R. Schlief and B. B. Goldberg (eds.), Advances in Echo Imaging Using Contrast Enhancement,
Second Edition, 401–414.
© 1997 Kluwer Academic Publishers.

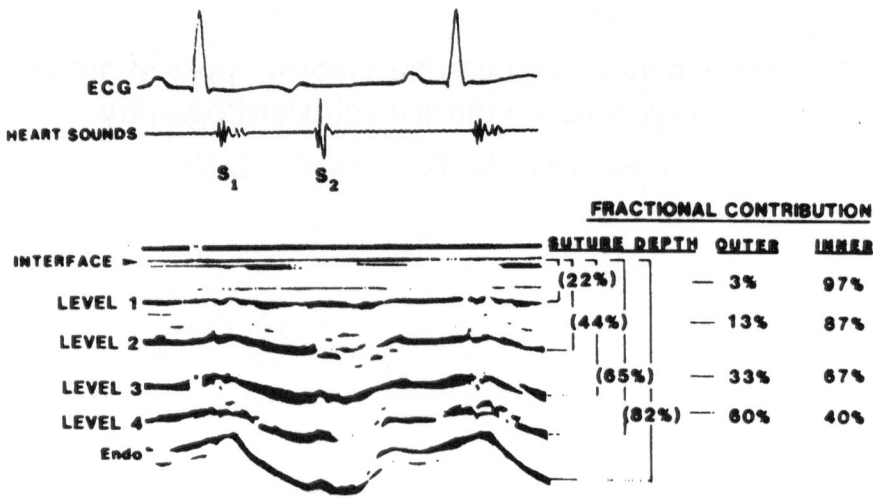

Figure 1. Example of an M-mode echocardiographic tracing where sutures have been placed within the myocardium at various depths. The excursion of each suture during systole indicates the amount of thickening between that suture and the endocardium or between that suture and the next. Contribution of the different myocardial layers to total wall thickening decreases from the endocardium to the epicardium. [Reproduced from Myers JH et al. Circulation 1986;74:164, with permission of the American Heart Association.]

Is there any viable myocardium? Is that myocardium susceptible to ischemia? Since tissue has to be viable in order to become ischemic, any imaging method that can detect ischemia will answer the second question. A patient with a moderate to large area of myocardium, susceptible to ischemia, is a candidate for a revascularization procedure.

The value of assessing viability in post-infarction patients who do not have ischemia is unknown. If most of the myocardium within an infarct region is viable and is not susceptible to ischemia, then recovery in function of that region will automatically occur within weeks [7, 8]. Figure 3 illustrates the natural history of wall motion recovery in patients who receive thrombolytic therapy within the first 4 h of chest pain: about one-third (which constitutes approximately one-half of those with an open infarct-related artery) show recovery in function within the first 2 weeks after thrombolytic therapy. If thrombolytic therapy or angioplasty is provided very early (within the first hour), then recovery of function is seen in almost all patients with an open infarct-related artery [9, 10]. If the endocardium is necrosed, however, spontaneous recovery of function will not occur. Although not proven, knowing whether or not there is substantial viable myocardium in the middle and outer layers of the myocardium, which is not susceptible to ischemia, may provide prognostic information.

Although the middle and outer layers of the myocardium contribute little to

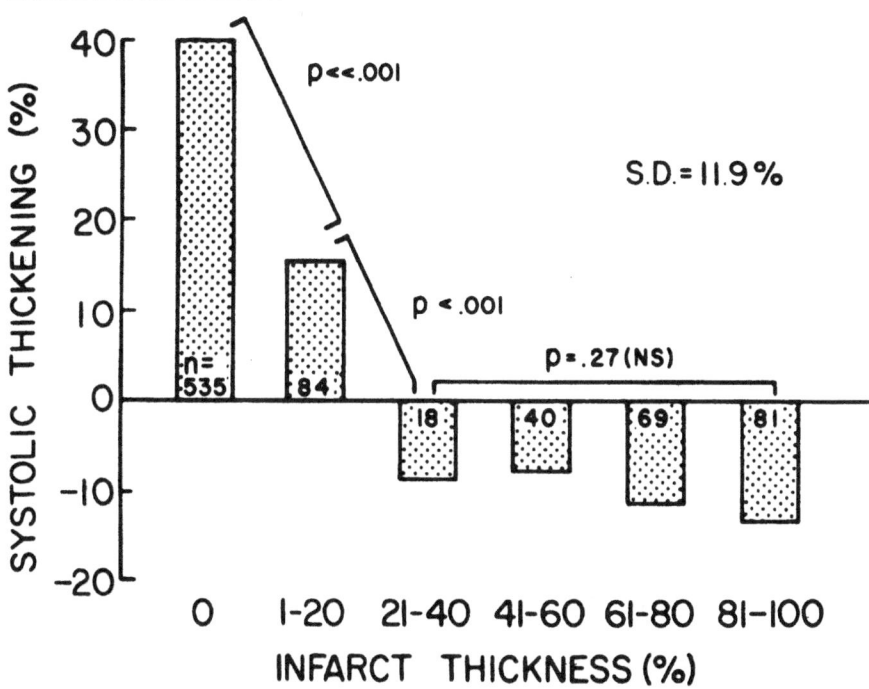

Figure 2. Relation between the transmural extent of infarction (x-axis) and regional wall thickening (y-axis) in dogs with 48 h of coronary occlusion. Some thickening is maintained as long as ≤20% of the transmural thickness is infarcted. Infarctions involving >20% thickness of the myocardium results in wall thinning. There is no relation between percent wall thinning and the transmural extent of infarction. [Reproduced from Lieberman AN et al. Circulation. 1981;63:739, with permission of the American Heart Association.]

overall thickening at rest, their thickening increases with catecholamine stimulation [11] which may contribute to overall wall thickening and increase in global left ventricular systolic performance during exercise and other periods of stress. The presence of viable myocytes in the outer layers of the ventricular wall may also contribute towards maintaining left ventricular shape and size by preventing infarct expansion, and consequent heart failure and late mortality after acute myocardial infarction [7, 12].

Myocellular viability can be demonstrated by different principles: presence of metabolic activity on positron emission tomography or magnetic resonance spectroscopy, identification of intact cell membrane on ^{201}Tl activity, assessment of mitochondrial function by ^{99m}Tc agents, and presence of contractile reserve by dobutamine echocardiography or magnetic resonance imaging. MCE, in contradistinction, defines the topography of microvascular perfusion, which is an important determinant of myocellular integrity during coronary occlusion [13, 14].

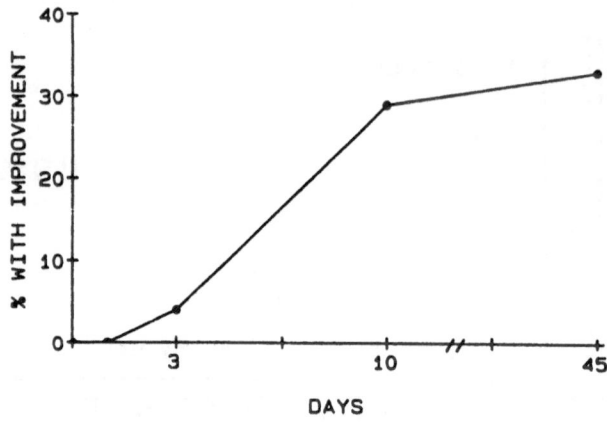

Figure 3. Percentage of patients (y-axis) demonstrating improvement in regional function over time (x-axis) after receiving intravenous streptokinase within 4 h of chest pain. The majority who showed recovery in function (approximately 40% of all patients; all with open infarct-related arteries) did so between 3 and 10 days. There were an equal number of patients with open infarct-related arteries who did not demonstrate spontaneous recovery in regional function. [Reproduced from Touchstone DA et al. J Am Coll Cardiol 1989;13:1506, with permission of the American College of Cardiology.]

During coronary occlusion, if residual flow to the ischemic myocardium is very low (<0.15 ml/min/g), myocytes within that region will start undergoing necrosis within 45–60 h after occlusion [13, 14]. On the other hand, if residual flow is higher, myocytes may remain viable for extended periods of time, although because of low flow they will not contrast normally. Defining regions with adequate residual flow, therefore, defines regions with myocellular viability. When the infarct-related artery is occluded, the only way to demonstrate adequate residual flow is through collaterals [13, 14].

After reflow, the zones that had adequate residual flow during occlusion continue to be viable, while areas that did not receive adequate flow during occlusion become necrotic. These areas of necrosis either have no flow (no-reflow phenomenon [15]), low flow (low-reflow phenomenon [16]) or even normal or high flows [17] due to reactive hyperemia seen for the first few hours after reflow.

The zones of no reflow or low reflow are identifiable on MCE as areas with very little contrast enhancement. Areas within the infarction not showing abnormal flow at rest, nevertheless, have microvascular injury that manifests as abnormal flow reserve, which can be unmasked by use of coronary vasodilators. The entire zone of microvascular injury (no flow, low flow, or abnormal flow reserve) is located within the borders of irreversibly injured myocardium. Thus, by defining abnormal microvascular function after reflow, MCE can be used to define infarct size, and hence, the extent of myocellular viability.

DEMONSTRATION OF COLLATERAL FLOW DURING CORONARY OCCLUSION

As stated earlier, since viability can only be present in tissue with some residual flow (>0.15 ml/min/g) [13, 14], it follows that detection of residual flow to the infarct zone via collaterals can indicate regions that are likely to be viable. Conversely, regions not receiving adequate tissue perfusion via collaterals are not likely to survive prolonged periods of coronary occlusion. Because of the low number of bubbles entering the microvasculature, no contrast effect is seen at flows <0.15 ml/min/g [18]. The presence of contrast within the infarct zone thus usually defines regions with residual blood flow at levels capable of maintaining myocellular viability [13, 14]. The regions within an infarct zone that opacify during coronary occlusion are the same regions that opacify after reflow [19], which implies that myocardium that receives adequate collateral flow during occlusion tends to have preserved microvasculature and thus intact myocellular viability after reperfusion.

Figure 4 illustrates MCE images after right and left main coronary injections of contrast in a patient with a recent anterior infarction and akinesia within the infarct zone [19]. Right coronary artery injection of contrast demonstrates opacification not only in the bed supplied by that artery but also in the medial half of the occluded left anterior descending arterial bed. Similarly, injection of contrast into the left main artery demonstrates not only opacification of the left circumflex artery which is patent, but also of the lateral half of the occluded left anterior descending arterial bed. In this patient, therefore, MCE demonstrates the presence of both right-to-left and left-to-left collaterals [19]. The demonstration of collateral perfusion is more accurate with MCE than with coronary angiography [19] since the latter can only define vessels >100 μm in size [20], while most collateral channels are much smaller [21]. The spatial extent of collateral perfusion within an infarct zone by MCE, therefore, correlates poorly with angiographic collaterals [19].

Regions that demonstrate adequate collateral perfusion during coronary occlusion on MCE show improved function after anterograde flow is restored [22]. Figure 5 illustrates the baseline and 1 month wall motion scores in patients who have undergone revascularization (by angioplasty) of the occluded infarct-related artery. Those with good collateral perfusion, defined as supplying >50% of the infarct bed, showed improvement in function 1 month later, while those with poor collateral flow did not show such an improvement. Interestingly, as long as there was adequate residual collateral perfusion within the infarct zone, function improved even when anterograde flow was established days to weeks after acute myocardial infarction [22]. These data indicate that in addition to the duration of coronary occlusion, the extent of residual collateral perfusion as assessed by MCE, is a major determinant of infarct size. Thus, myocardial regions that suffer an infarction and do not undergo immediate reperfusion are still likely to be viable if they have adequate residual collateral perfusion.

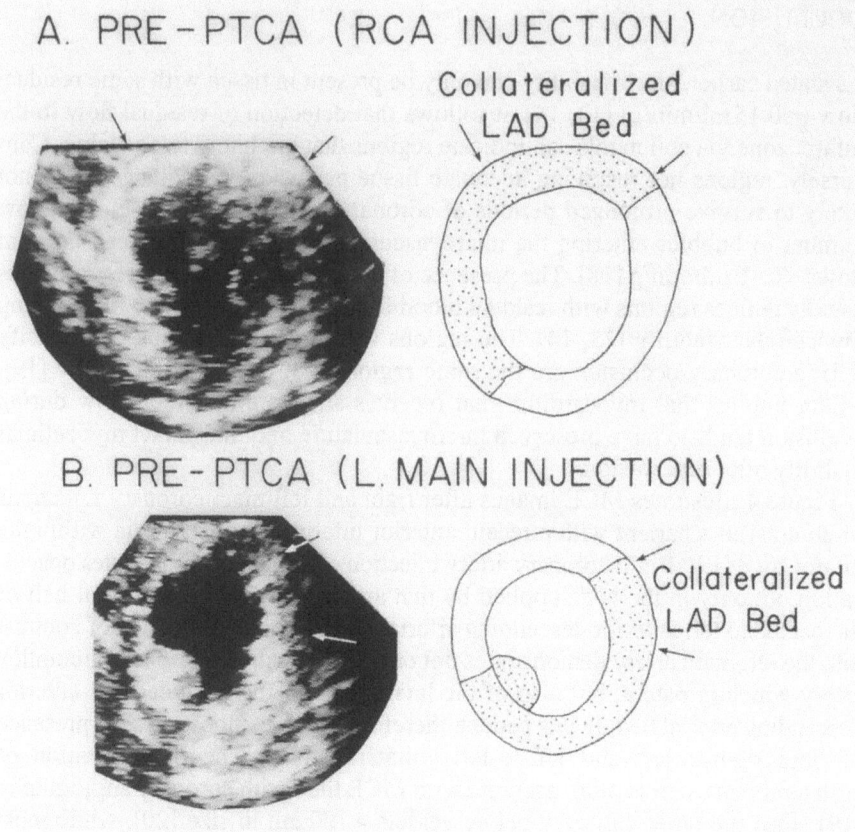

Figure 4. (A) Right-to-left collateral perfusion when microbubbles are injected into the right coronary artery and (B) left-to-left collateral perfusion when microbubbles are injected into the left main artery in a patient with a recent anteroseptal myocardial infarction and an occluded left anterior descending artery. See text for details. [Reproduced from Sabia et al. Circulation 1992;85:2080, with permission of the American Heart Association.]

ASSESSMENT OF INFARCT SIZE AFTER REPERFUSION

As stated earlier, after reperfusion, in addition to cellular necrosis, there is microvascular damage in regions with myocardial infarction. The microvascular damage can take the form of capillary disruption, vascular plugging with white cells and other debris, or vascular obliteration resulting from interstitial edema [15–17]. The areas of vascular damage are located within the borders of the infarct and approximate the infarct size. Regions within the risk areas that do not undergo necrosis during coronary occlusion, because they have at least an intermediate level of flow, do not suffer microvascular damage [15–17]. Thus, the

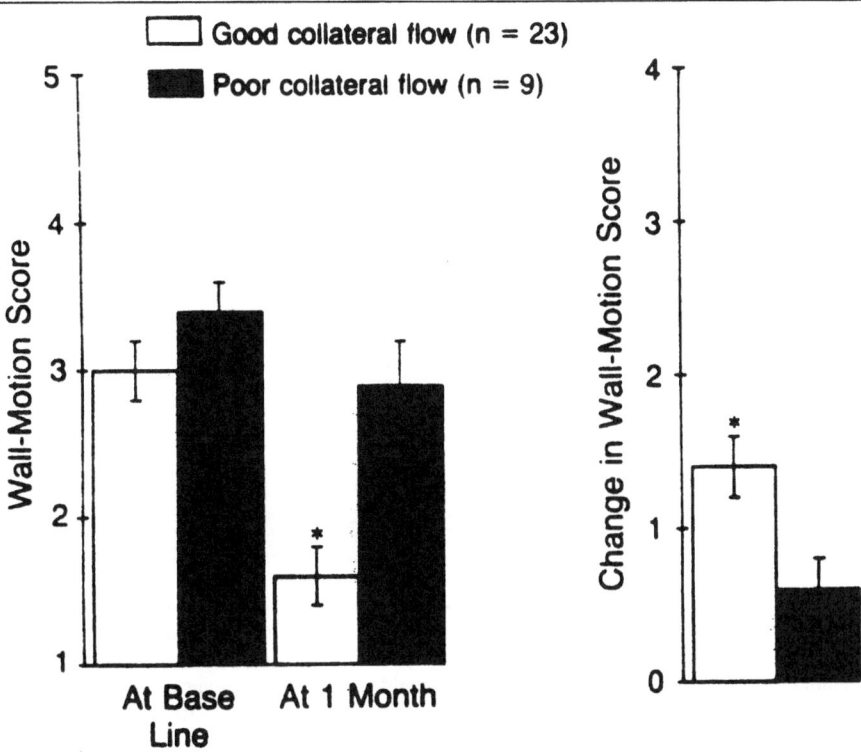

Figure 5. Wall motion scores at baseline and 1 month after successful angioplasty (where 1=normal and 5=dyskinesia) and percentage of the infarct bed supplied by collateral blood flow as defined by MCE before angioplasty (left panel). Change in wall motion score between baseline and one month post-angioplasty based on the extent of collateral perfusion on MCE (right panel). See text for details. [Reproduced from Sabia et al. N Engl J Med 1992;372:1825, with permission of the Massachusetts Medical Society.]

spatial extent of microvascular damage reflects the extent of cellular damage and regions without microvascular damage indicate the presence of viable myocardium.

This principle has been used to define infarct size and the extent of myocardial viability after reperfusion using MCE [23–28]. Regions with low reflow or no reflow with this technique have been shown not to improve in thickening despite a patent infarct-related artery [23–28]. In comparison, regions showing adequate flow after reperfusion do show improved thickening if the infarct-related artery remains patent [23–28]. Figure 6 depicts a patient with an anteroapical infarction who received thrombolytic therapy [24]. Six days later, angiography revealed good flow in the left anterior descending artery. Injection of microbubbles into the left main artery, however, showed lack of tissue perfusion in the apex and low tissue perfusion in the mid- and lower interventricular septum, while the

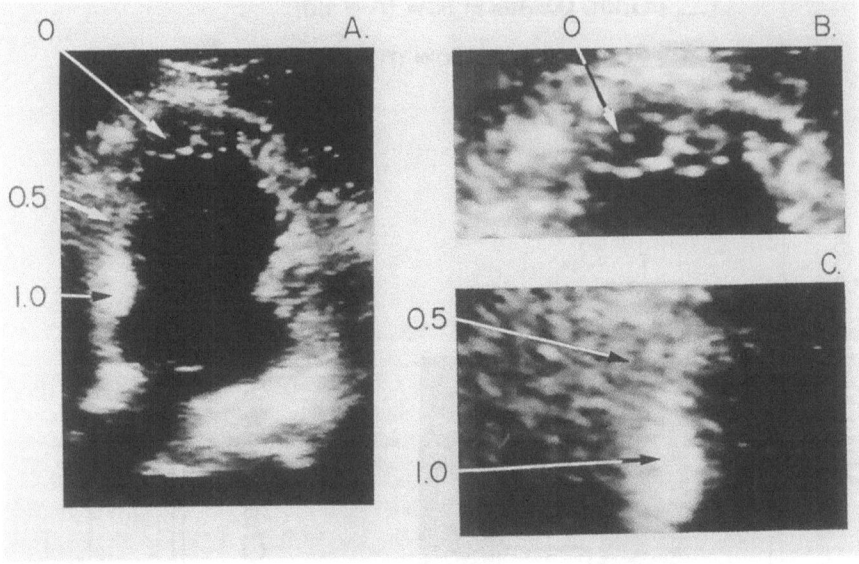

Figure 6. Three different perfusion patterns in the same bed in a patient with a recent antero-septal infarction and an open left anterior descending artery after thrombolysis. The upper interventricular septum shows normal homogeneous perfusion, the middle to lower interventricular septum depicts patchy perfusion, while no perfusion is noted in the apex. [Reproduced from Ragosta et al. Circulation 1994;89:2562, with permission of the American Heart Association.]

upper interventricular septum demonstrated adequate perfusion. One month later, the upper interventricular septum showed marked improvement in function, the mid- and lower interventricular septum demonstrated some improvement, while the apex showed no improvement [24].

Figure 7 illustrates the extent of microvascular perfusion post-infarction on the x-axis, while the y-axis depicts one month post-revascularization wall motion score, baseline and 1 month post-revascularization wall motion scores, and improvement in wall motion score 1 month after revascularization. The higher the wall motion score, the worse the function within the infarct bed. Because of myocardial stunning and/or hibernation, although the baseline wall motion score is similar for all degrees of microvascular perfusion within the infarct, the 1 month wall motion score is inversely related to the perfusion score [24]. The results of these studies indicate that the spatial extent of microvascular perfusion as assessed with MCE reflects the extent of myocellular viability after reperfusion in patients with acute myocardial infarction.

The data discussed above [24–28] are from patients studied 1 day to 4 weeks after their myocardial infarction, where hyperemia in those with reflow had abated. The information obtained from MCE immediately after reflow is established is more complex because of reactive hyperemia. It takes several hours or even a

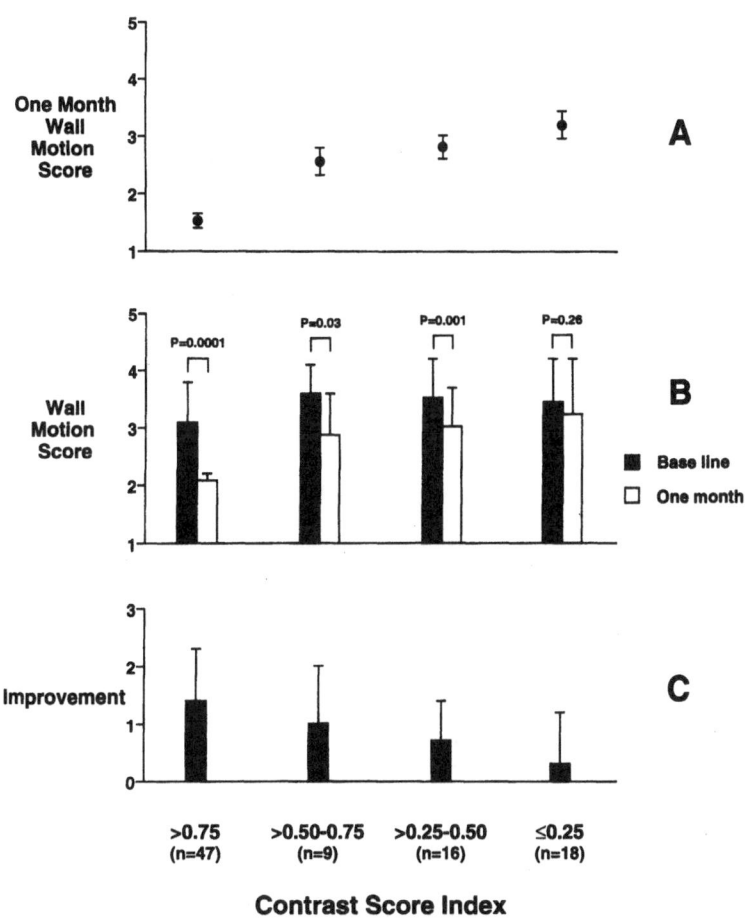

Figure 7. Relation between wall motion score and the extent of post-infarction microvascular perfusion. On the y-axis is depicted one month post-revascularization wall motion score (A), baseline and 1 month post-revascularization wall motion scores (B), and improvement in wall motion score 1 month after revascularization (C) based on the extent of microvascular perfusion (x-axis). The higher the wall motion score, the worse the function within the infarct bed. The higher the perfusion score, the greater the extent of microvascular integrity. [Adapted from Ragosta et al. Circulation 1994;89:2562, with permission of the American Heart Association.]

day after reflow has been established, for reactive hyperemia to abate. The degree of hyperemia is inversely related to the amount of microvascular damage. Thus, if there is severe necrosis and microvascular damage, the ability to mount a hyperemic response (which is modulated by the intact microvasculature) is greatly diminished and MCE may more closely reflect the degree of myocardial damage. On the other hand, if microvascular damage is only moderate, significant reactive

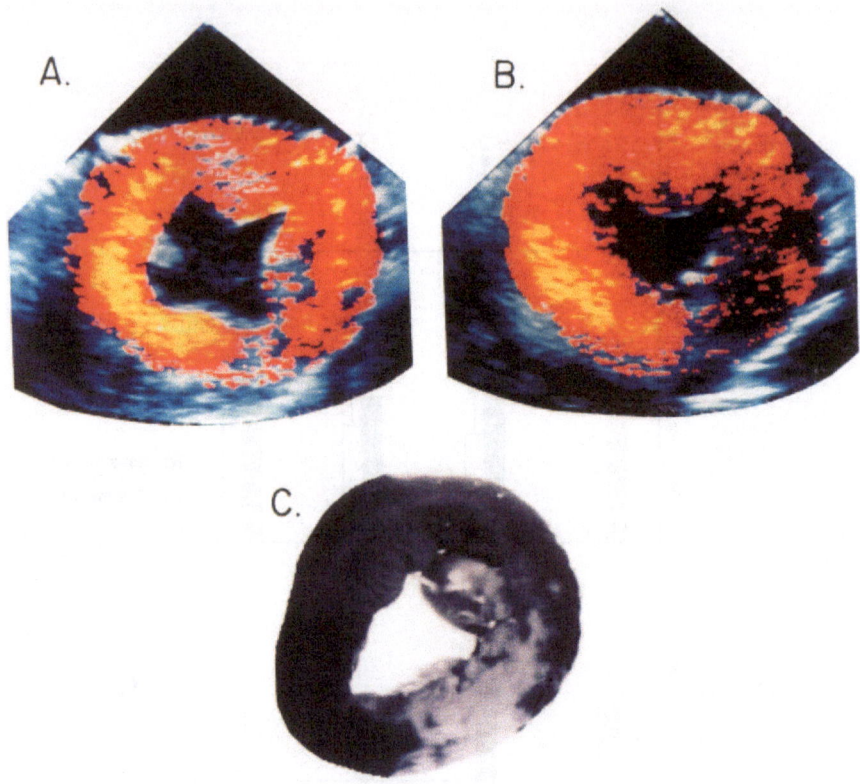

Figure 8. Color-coded MCE perfusion defects after aortic root injection of microbubbles 15 min after reflow is achieved in the left circumflex artery of a dog which had been occluded for 5 h: (A) before and (B) during infusion of a coronary vasodilator. It is clear that without a coronary vasodilator, the infarct size on triphenyl tetrazolium chloride-staining of the heart (C) is underestimated on CCE, while in the presence of the vasodilator, its estimation is accurate (see text for details).

hyperemia may occur and MCE may underestimate the degree of myocardial necrosis. The issue is further complicated by the degree of residual stenosis present in the infarct-related artery. If the stenosis is critical and completely attenuates the hyperemic response, MCE will accurately define the region of necrosis. Conversely, if the stenosis is not critical, the degree of hyperemic attenuation may be variable and MCE may underestimate tissue necrosis by a variable amount.

Because of microvascular damage within the infarct zone, the reactive hyperemia is less than could be mounted within the normal myocardium [29, 30]. This principle can be exploited to accurately define infarct size in the few hours after reflow despite the presence of hyperemia within the infarct zone [31]. When a coronary vasodilator is infused intravenously, flow in the normal bed (anteroseptal myocardium in Figure 8) increases maximally (approximately

4- to 5-fold), while the flow within the infarct bed (lateral myocardium in Figure 8) cannot increase any further because maximal vasodilatation has already occurred in this bed [31]. Thus, relative to the normal bed, the infarct bed is seen to have reduced flow during infusion compared with prior to infusion of the vasodilator. This region of reduced relative flow is almost identical to the region of infarction as determined by triphenyl tetrazolium chloride staining of the myocardium. The color-coding algorithm used ascribes the color white to the pixel with the maximal intensity, and all other pixels are assigned colors which indicate lower intensities. In this manner, all intensities are relative to the brightest pixel within the myocardium. For this reason, the infarct zone is seen as an area with relatively low flow. These data indicate that even in the presence of reactive hyperemia, MCE can be used to define infarct size and hence myocardial viability after reflow. Although these results are obtained from canine experiments, they have major application in the clinical setting.

The data discussed above are based on intracoronary or intra-aortic injections of contrast applied as an adjunct to coronary angiography in the cardiac catheterization laboratory. The information on the status of the coronary micro-vasculature gained using this approach can be useful in determining revascular-ization strategies. The additional time required to obtain these data is 5–10 min, although at present an echocardiographer is required to assist the angiographer in making judgments based on the data obtained. With the advent of intracardiac ultrasound, which yields images of excellent quality and data which can be interpreted by the angiographer, this technique may find increasing application in the cardiac catheterization laboratory.

With the advent of newer contrast agents [32] and better forms of echocardi-ographic data acquisition such as intermittent harmonic imaging [33], similar results are now available using venous injection of contrast. Figure 9A is an example of a short-axis view using a venous injection of 1 ml of a new second-generation contrast agent at baseline prior to coronary occlusion. It shows hom-ogeneous contrast effect except for some posterior wall attenuation from the presence of contrast in the left ventricular cavity. During coronary occlusion, the risk area (area at risk for necrosis if occlusion remains indefinitely) is clearly defined (Figure 9B). Images acquired after reflow in the presence of a coronary vasodilator demonstrate a perfusion defect (Figure 9C) that corresponds in size and topology to the infarction defined on tissue staining (Figure 9D). These results set the stage for the use of MCE for defining myocardial viability in patients with infarction using venous contrast agents.

SUMMARY

The pathophysiological principles and experimental and clinical data described in this chapter have laid a strong foundation for the use of MCE for the assessment of myocardial viability. Advances in microbubble and ultrasound engineering

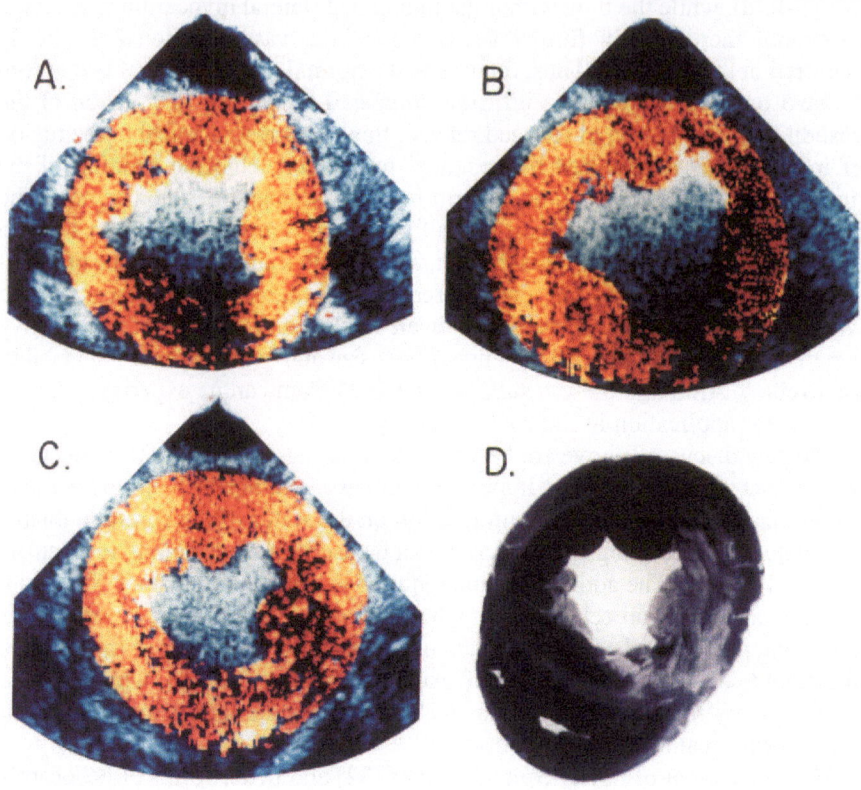

Figure 9. Examples of a color-coded MCE after a venous injection of 1 ml of a new second-generation contrast agent. (A) at baseline showing homogeneous contrast effect except for some posterior wall attenuation from the presence of contrast in the left ventricular cavity. (B) During coronary occlusion, where the risk area is clearly defined. (C) After reflow (in the presence of a coronary vasodilator) where a perfusion defect corresponding in size and topology to the infarction defined on tissue staining (D).

will add fillip to the clinical assessment of myocardial viability using this technique. MCE is likely to become one of the most utilized techniques for the assessment of post-infarction myocardial viability in the future.

ACKNOWLEDGMENTS

Supported in part by a grant (R01-HL48890) from the National Institutes of Health, Bethesda, Maryland, USA, to Dr Kaul. Dr Firschke is a recipient of grants from Deutsche Gesellschaft für Ultraschall in der Medizin, Stuttgart, Germany, and the Fulbright Commission, Bonn, Germany and Dr Kaul is an Established Investigator of the American Heart Association, Dallas, Texas, USA.

REFERENCES

1. Lieberman AN, Weiss JL, Jugdutt BI et al. Two-dimensional echocardiography and infarct size: Relationship of regional wall motion and thinning to the extent of myocardial infarction in the dog. Circulation 1981;63:739–46.
2. Kaul S, Force T. Assessment of myocardial perfusion with contrast two-dimensional echocardiography. In: Principles and Practice of Echocardiography. 2nd edn. Weyman AE, editor. Philadelphia: Lea and Febiger, 1993;687–720.
3. Oxford English Dictionary. Oxford: Oxford University Press, 1971.
4. Gropler RJ, Bergmann SR. Myocardial viability – what is the definition? J Nucl Med 1991; 32:10–2.
5. Myers JH, Stirling MC, Choy M, Buda AJ, Gallagher KP. Direct measurement of inner and outer wall thickening dynamics with epicardial echocardiography. Circulation 1986;74:164–72.
6. Weintraub WS, Hattori S, Aggarwal JB, Bodenheimer MM, Banka V, Halfant RH. The relationship between myocardial blood flow and contraction by myocardial layer in the canine left ventricle during ischemia. Circ Res 1981;48:430–8.
7. Touchstone DA, Beller GA, Nygaard TW, Tedesco C, Kaul S. Effects of successful intravenous reperfusion therapy on regional myocardial function and geometry in man: A tomographic assessment using two-dimensional echocardiography. J Am Coll Cardiol 1989;13:1506–13.
8. Penco M, Romano S, Agati L et al. Influence of reperfusion induced by thrombolytic treatment on natural history of left ventricular regional motion abnormality in acute myocardial infarction. Am J Cardiol 1993;71:1015–20.
9. Widemsky, Cervenka V, Visek V, Sladkova T, Dvorak J, Drdlicka S. First month course of left ventricular asynergy after intracoronary thrombolysis in acute myocardial infarction: A longitudinal echocardiographic study. Eur Heart J 1985;6:759–65.
10. Charuzi Y, Beeder C, Marshall LA et al. Improvement in regional and global left ventricular function after intracoronary thrombolysis: assessment with two-dimensional echocardiography. Am J Cardiol 1984;53:662–5.
11. Sklenar J, Villanueva FS, Glasheen WP, Ismail S, Goodman NC, Kaul S. Dobutamine echocardiography for determining the extent of myocardial salvage after reperfusion: an experimental evaluation. Circulation 1994;90:1503–12.
12. Marino P, Zanolla L, Zardini P. Effect of streptokinase on left ventricular modeling and function after myocardial infarction: The GISSI (Gruppo Italiano per Studio della Streptochinasi nell'Infarto Miocardico) Trial. J Am Coll Cardiol 1989;14:1149–58.
13. Reimer KA, Jennings RB. The 'wavefront phenomenon' of myocardial ischemic cell death. II. Transmural progression of necrosis within the framework of ischemic bed size (myocardium at risk) and collateral flow. Lab Invest 1979;40:633–44.
14. Jugdutt BI, Hutchins GM, Bulkley BM, Becker LC. Myocardial infarction in the conscious dog: Three-dimensional mapping of infarct, collateral flow and region at risk. Circulation 1979;60:1141–50.
15. Kloner RA, Ganote CE, Jennings RB. The 'no-reflow phenomenon' after temporary coronary occlusion in the dog. J Clin Invest 1974;54:1496–508.
16. White FC, Sanders M, Bloor CM. Regional redistribution of myocardial blood flow after coronary occlusion and reperfusion in the conscious dog. Am J Cardiol 1978;42:234–43.
17. West PN, Connors JP, Clark RE, Weldon CS. Compromised microvascular integrity in ischemic myocardium. Lab Invest 1978;38:677–84.
18. Kaul S, Kelly P, Oliner JD, Glasheen WP, Keller MW, Watson DD. Assessment of regional

myocardial blood flow with myocardial contrast two-dimensional echocardiography. J Am Coll Cardiol 1989;13:468–82.

19. Sabia PJ, Powers ER, Jayaweera AR, Ragosta M, Kaul S. Functional significance of collateral blood flow in patients with recent acute myocardial infarction. A study using myocardial contrast echocardiography. Circulation 1992;85:2080–9.
20. Gensini GG, daCosta BCB. The coronary collateral circulation in living man. Am J Cardiol 1969;24:393–400.
21. Cohen MV. Morphological considerations of the coronary collateral circulation in man. In: Coronary Collaterals. New York: Futura Publishing Co. 1985:1–91.
22. Sabia PJ, Powers ER, Ragosta M, Sarembock IJ, Burwell L, Kaul S. An association between collateral blood flow and myocardial viability in patients with recent myocardial infarction. N Engl J Med 1992;327:1825–31.
23. Ito H, Tomooka T, Sakai N, Yu H et al. Lack of myocardial perfusion immediately after successful thrombolysis. A predictor of poor recovery of left ventricular function in anterior myocardial infarction. Circulation 1992;85:1699–705.
24. Ragosta M, Camarano GP, Kaul S, Powers E, Gimple LW. Microvascular integrity indicates myocellular viability in patients with recent myocardial infarction: new insights using myocardial contrast echocardiography. Circulation 1994;89:2562–9.
25. Lim Y-J, Nanto S, Masuyama T et al. Myocardial salvage: Its assessment and prediction by the analysis of serial myocardial contrast echocardiograms in patients with acute myocardial infarction. Am Heart J 1994;128:649–56.
26. Agati L, Voci P, Bilotta F et al. Influence of residual perfusion within the infarct zone on the natural history of left ventricular dysfunction after acute myocardial infarction: a myocardial contrast echocardiographic study. J Am Coll Cardiol 1994;24:336–42.
27. Camarano G, Ragosta M, Gimple LW, Powers ER, Kaul S. Identification of viable myocardium with contrast echocardiography in patients with poor left ventricular systolic function caused by recent or remote myocardial infarction. Am J Cardiol 1995;75:215–9.
28. Villanueva FS, Glasheen WD, Sklenar J, Kaul S. Assessment of risk area during coronary occlusion and infarct size after reperfusion with myocardial contrast echocardiography using left and right atrial injections of contrast. Circulation 1993;88:596–604.
29. Johnson WB, Malone SA, Pantely G et al. No reflow and extent of infarction during maximal vasodilatation in the porcine heart. Circulation 1988;78:462–72.
30. Vanhaecke J, Flameng W, Borgers M et al. Evidence for decreased coronary flow reserve in viable postischemic myocardium. Circ Res 1990;67:1201–10.
31. Villanueva FS, Glasheen WP, Sklenar J, Kaul S. Characterization of spatial patterns of flow within the reperfused myocardium using myocardial contrast echocardiography: Implications for determining extent of myocardial salvage. Circulation 1993;88:2596–606.
32. Skyba DM, Camarano G, Goodman NC, Price RJ, Skalak TC, Kaul S. Hemodynamic characteristics, myocardial kinetics, and microvascular rheology of FS-069, a second-generation echocardiographic contrast agent capable of producing myocardial opacification from a venous injection. J Am Coll Cardiol 1996;28:1292–300.
33. Porter TR, Xie F. Transient myocardial contrast after initial exposure to diagnostic ultrasound pressures with minute doses of intravenously injected microbubbles. Demonstration of potential mechanisms. Circulation 1995;92:2391–5.

26. Can myocardial echocontrast replace thallium?

SHINTARU BEPPU

INTRODUCTION

Myocardial contrast echocardiography (MCE) is a method for visualizing the extent of the perfused area of the heart by injecting microbubbles into the coronary artery. Microbubbles are distributed via coronary blood flow and opacify the perfused area. The opacified area corresponds closely to the territory of the coronary artery, and animal experiments have demonstrated that the non-perfused area is not opacified by MCE. The equivalence of the non-opacified and non-perfused area is evidenced by the Evans Blue stain method and TTC stain [1]. As expected, non-perfused wall segments show abnormal motion.

In human subjects, abnormalities on MCE and wall motion abnormality correspond closely. In a case of anterior myocardial infarction, for instance, the upper septum and the anterior wall are akinetic and these wall segments are not opacified on MCE (Figure 1). This chapter draws on our experience with MCE in patients with ischemic heart disease. The short-axis view of the left ventricle at the papillary muscle level was obtained to compare wall motion abnormalities with MCE. Myocardial opacification by MCE was graded into three levels (good, poor and none) and wall motion abnormality was also graded into three levels (normal motion, hypokinetic motion, and akinetic or dyskinetic motion) in the anterior, inferior and lateral wall segments of the left ventricle. The wall motion abnormalities coincided with the results of MCE in 92% of the segments examined (Table 1). This finding indicates that wall motion abnormality corresponds to myocardial perfusion abnormality.

The severity of coronary stenosis, pathological changes in the myocardium, and myocardial ischemia vary significantly among patients. MCE results may be complex due to these variable factors, although it has been recognized that MCE demonstrates the extent and magnitude of myocardial perfusion. Conventionally, thallium-201 scintigraphy is the best method for assessing myocardial perfusion. In this section, MCE is compared with thallium examination.

415

N. C. Nanda, R. Schlief and B. B. Goldberg (eds.), Advances in Echo Imaging Using Contrast Enhancement, Second Edition, 415–426.
© 1997 *Kluwer Academic Publishers.*

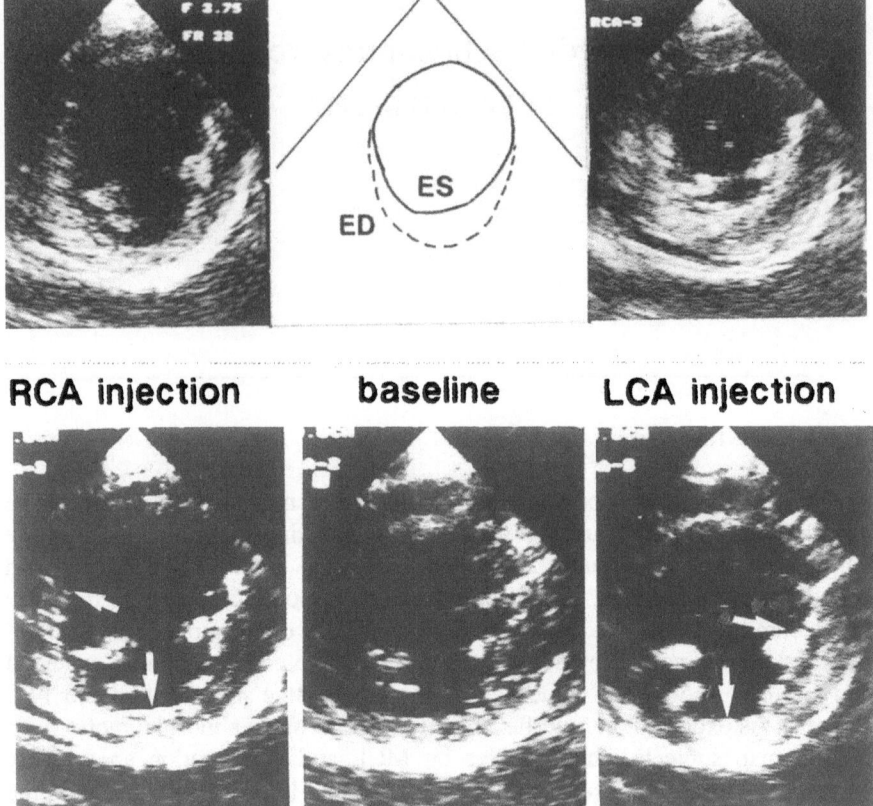

Figure 1. MCE concordant with wall motion abnormality (in a case of anterior myocardial infarction). The upper septum and the anterior wall are kinetic and not opacified by contrast echoes, while the lateral and inferior walls are opacified by left and right coronary injections, respectively. The akinetic area corresponds to the non-opacified area.

Table 1. Relationship between wall motion abnormalities and myocardial opacification (132 patients).

	Wall motion		
	Normal	Hypo	None
M Good	261	11	10
C Poor	2	12	2
E Absent	0	0	47

DIAGNOSIS OF MYOCARDIAL ISCHEMIA

Severity of stenosis of the coronary artery is one of the major underlying factors for myocardial ischemia. In canine experiments using MCE, the washout rate derived from the time–intensity curve of the MCE was prolonged according to the severity of the coronary stenosis [2]. Myocardial ischemia is diagnosed by exercise or a pharmacological stress test. A thallium-201 examination following a stress test is a reliable diagnostic tool for myocardial ischemia. With MCE, on the other hand, an exercise stress test is not usually performed, because the contrast agent needs to be injected directly into the coronary arteries or the root of the aorta via a catheter. Instead, pharmacological stress or atrial pacing are generally used to induce stress with MCE. In dogs, Cheirif et al. observed a hypoperfused area during dipyridamole-induced hyperemia [3]. In mild coronary stenosis, no contrast defect was evident, even during dipyridamole administration. Thallium-201 SPECT demonstrated a perfusion defect in all subjects using dipyridamole stress. They concluded that MCE was useful in visualizing the area at risk due to moderate to severe stenosis, but that the sensitivity of MCE was inferior to that of thallium scintigraphy.

However, MCE has methodological advantages, since the cross-section of the myocardium is more clearly demonstrated than is the case with thallium scintigraphy. Therefore, MCE can indicate abnormal blood distribution across the myocardium from the endocardium to the epicardium. In human subjects, pacing-induced endocardial ischemia was nicely demonstrated by MCE [4]. Results indicated a low peak intensity at the endocardial half of the left ventricular wall, while the epicardial half showed a high peak intensity during stress. In canine experiments, however, MCE failed to indicate the abnormal distribution of the blood induced by papaverine across the myocardium from the endocardium to the epicardium [5]. It is true that MCE has some methodological limitations, such as artifacts induced by ultrasound. The attenuation effect of contrast echoes may obscure the image behind it. Although these are opposite findings, it should be noted that MCE can be used to visualize small changes in perfusion of small areas. In this way, it is superior to the thallium scintigraphy. At present, MCE is not widely used for diagnosing myocardial ischemia, because it is limited to a catheterization room. If MCE were non-invasive and easy to perform, for example, by the intravenous route, it could replace thallium scintigraphy.

ASSESSMENT OF MYOCARDIAL VIABILITY

In general, regional wall motion abnormality due to myocardial infarction is stable and irreversible. In some patients, however, normal motion can be recovered at the remote phase of myocardial infarction or after relief of chronic ischemia. In the clinical setting, it is very important to diagnose whether the asynergic wall segment is viable.

418   *Shintaro Beppu*

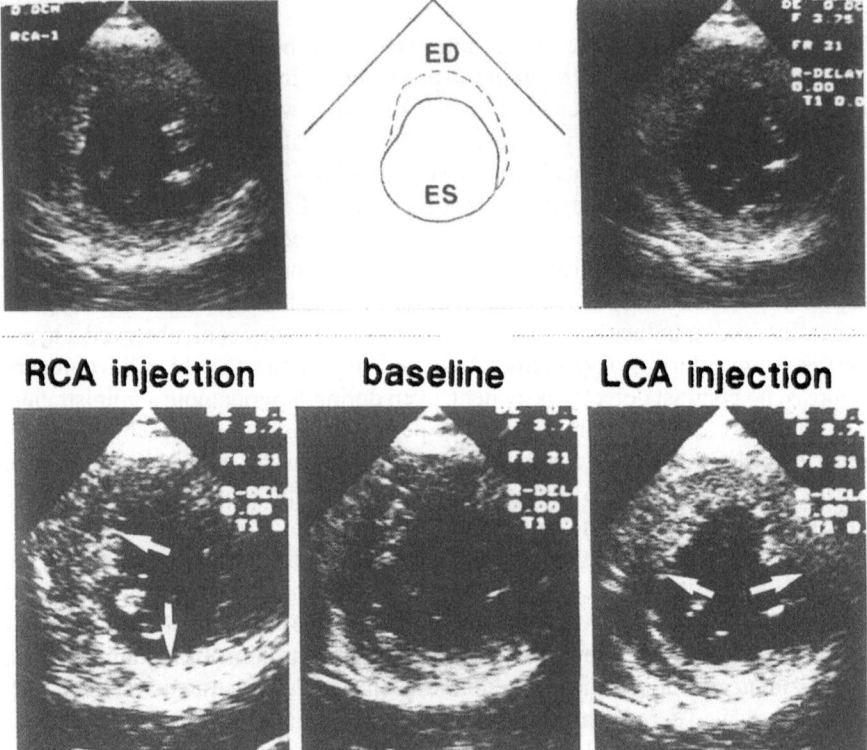

RCA injection   baseline   LCA injection

Figure 2. MCE discordant with wall motion abnormality in a case of inferior myocardial infarction. The inferior wall is akinetic, but opacified by contrast echoes.

Wall motion abnormality and MCE

As described earlier, the intensity of myocardial opacification corresponds to the severity of wall motion abnormality in 92% of segments. No case of normal wall motion without myocardial opacification has been reported, but in 8% of the segments the myocardium is opacified well despite abnormal wall motion. In the case of inferior infarction, for instance, the akinetic inferior wall was opacified well by MCE (Figure 2). Akinetic motion in the inferior wall improved to hypokinesis at the remote phase. Among the patients who could be evaluated at the remote phase, wall motion became normal in some segments with good opacification, while it did not recover at all in segments lacking good opacification (Figure 3). A ventricular wall segment with positive MCE does not become thin in the remote phase. Other investigators also found that akinetic segments with good perfusion recover their motion at the remote phase of myocardial infarction [6, 7].

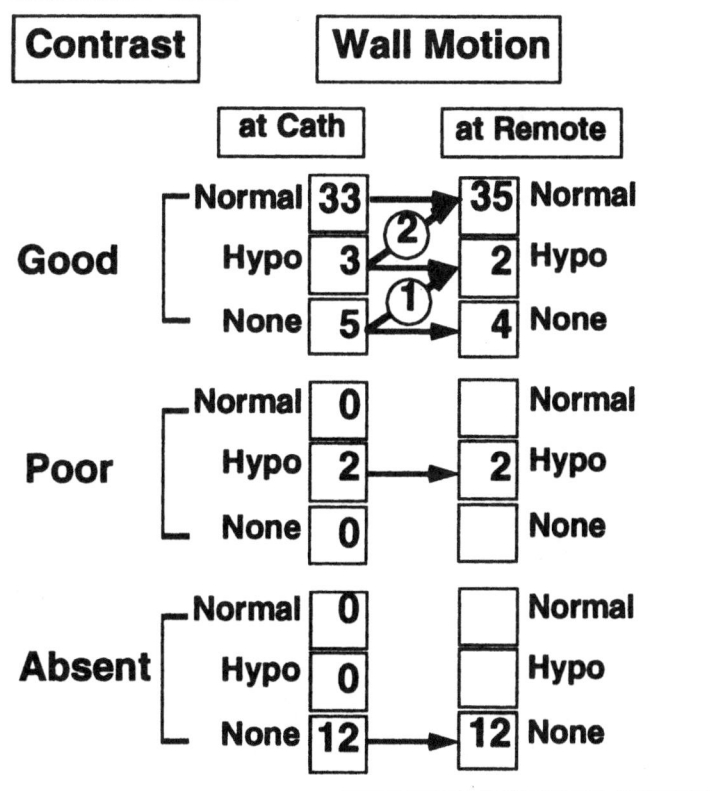

Figure 3. Time course of wall motion abnormalities. Regional wall motion at the time of MCE and about 2 years later are shown, and are classified into three groups based on MCE findings. The akinetic and non-opacified segments did not move, even in the remote phase. Some of the asynergic segments recovered their motion in areas that were well-opacified.

LOW-DOSE DOBUTAMINE STRESS ECHO AND MCE

Low-dose dobutamine encourages motion in the akinetic but viable myocardium. Low-dose dobutamine stress echocardiography can diagnose a stunned or hibernating myocardium [8, 9]. A few studies have compared the diagnostic accuracy of low-dose dobutamine with that of thallium-201 scintigraphy in diagnosing myocardial viability. Charney found a sensitivity of 78% and a specificity of 86% using 5 µg/kg/min dobutamine, while sensitivity was 95% and specificity was 85% using resting thallium scintigraphy [10]. Thus, low-dose dobutamine echocardiography is a useful method for assessing a viable myocardium. In patients with inferior myocardial infarction, in whom the inferior wall was akinetic almost normal motion of the inferior wall was seen after dobutamine infusion (Figure 4). MCE showed that the infarcted area was perfused well from the left coronary artery. In this study, a combination of isosorbide dinitrate and

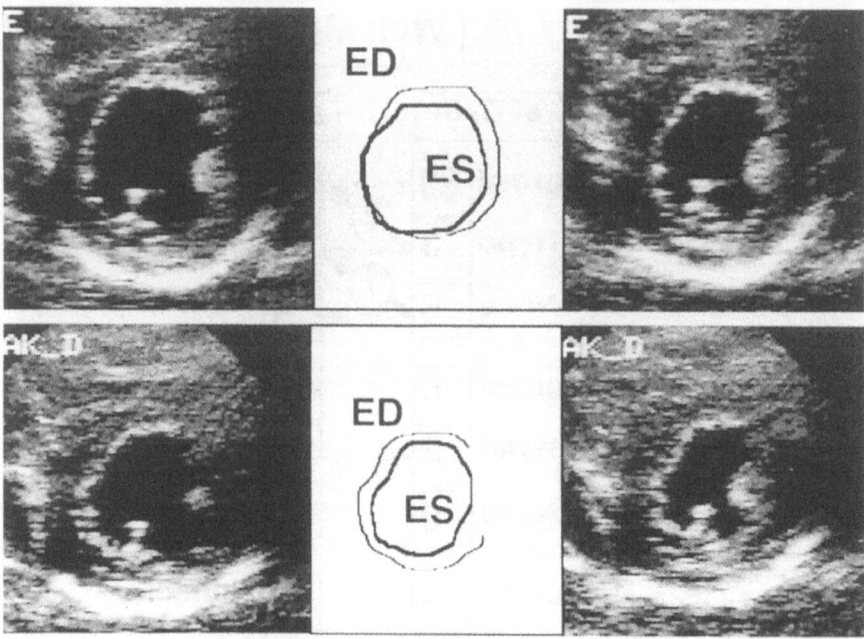

Figure 4. Low-dose dobutamine echo. The inferior wall segment moves almost normally during low-dose dobutamine administration (lower panel), while this segment is akinetic at baseline (upper panel). The left is end-diastolic image and the right is end-systolic image.

low-dose dobutamine (5 μg/kg/min) was used. The anterior, inferior and lateral wall segments in the short-axis view at the level of the papillary muscle were selected to assess wall motion abnormalities (normal, hypokinetic, and akinetic or dyskinetic) and the opacification of the myocardium (good, poor, and absent) in a study of multiple patients (Figure 5). Following infusion of dobutamine all hypokinetic segments moved normally, and some of the akinetic segments moved normally or hypokinetically, although most remained akinetic. None of the akinetic segments lacking contrast opacification responded to dobutamine, but almost all akinetic segments with contrast opacification did respond to dobutamine. This finding indicates that the presence of microvasculature is essential in retaining viability of the myocardium, irrespective of the absence or presence of wall motion abnormality. However, not all myocardial cells hibernate homogeneously in an ischemic region. The proportion of infarcted, viable and normal myocardial cells varies among patients. When all myocardial cells in the akinetic area are hibernating, or normal myocardial cells are scattered among the infarcted myocardial cells, the wall segment will move during dobutamine administration. Pierard et al. revealed that the asynergic wall segments with high glucose/perfusion ratio in FDG imaging using positron emission tomography

Baseline   DOB

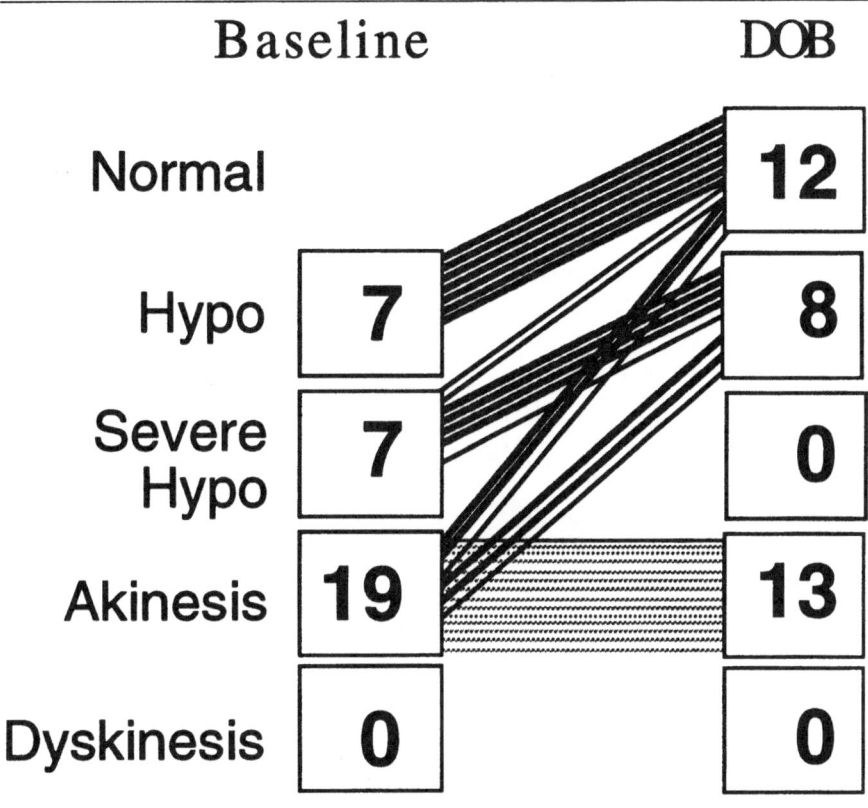

Figure 5. Relationship between results of dobutamine and MCE. Assessment of wall motion at baseline and during dobutamine are connected by a line according to MCE results: thick lines: good, thin lines: poor, and dotted lines: no opacification. Note that the wall segments which recover their motion by dobutamine is opacified in MCE.

will not recover their motion at the remote phase, even if the wall thickening increases during dobutamine administration [8].

The subjects of this study were the patients with chronic myocardial infarction, but an almost identical relationship between findings of wall motion abnormalities and MCE is present in those with chronic infarction.

COMPARISON WITH THALLIUM SCINTIGRAPHY AND MCE

Thallium scintigraphy is the standard method used to assess myocardial perfusion and is usually performed during and after exercise stress to detect both myocardial ischemia and viability. However, late redistribution imaging 24 h after exercise may not detect viable myocardium and a reinjection technique is recommended

Table 2. Thallium uptake and MCE opacification.

	Thallium SPECT		
	Normal	Hypo	Defect
M Good	55	4	2
C Poor	0	5	0
E Absent	0	0	13

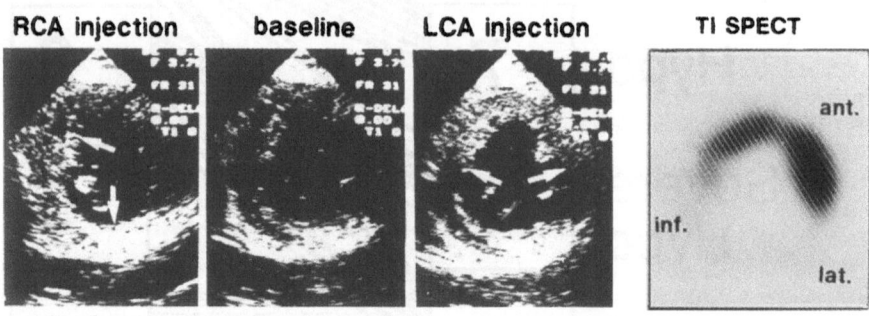

Figure 6. MCE and thallium SPECT in a case of inferior and lateral infarction. In the lateral wall segment, both MCE and SPECT show negative findings. However, the inferior wall segment is opacified well in MCE imaging, while this wall segment does not take up thallium.

for assessing myocardial viability [11]. The scintigraphic image obtained on reinjection of thallium after exercise is equivalent to the resting image before exercise. In the following study, resting thallium SPECT was administered to assess myocardial viability. The short-axis section at the papillary muscle level was selected to compare the image obtained with MCE and thallium SPECT. The left ventricular wall was divided into anterior, inferior and lateral wall segments. Myocardial opacification by MCE was classified into three levels: good, poor and none. Thallium uptake was also classified into three levels: good, poor, and none. In most segments (92% of all segments), the results from the two methods were concordant (Table 2). However, there were some segments with positive opacification on MCE but negative thallium uptake. In a case of inferior myocardial infarction, for instance, the inferior wall was opacified well by MCE but the segment did not take up thallium (Figure 6). Routine thallium scintigraphy would report this segment as non-viable. However, motion of the inferior wall segment recovered at the remote phase in this patient. Both MCE and thallium scintigraphy allow visualization of myocardial perfusion: what is the reason for such discrepancies among their results? Myocardial circulation can be compared to the branches and leaves of a tree (Figure 7). Although coronary angiography indicates a spread of branches or a broken branch, it does not offer information on microcirculation. The uptake of thallium corresponds to healthy leaves of the tree, colored green. If active transport of potassium is absent and

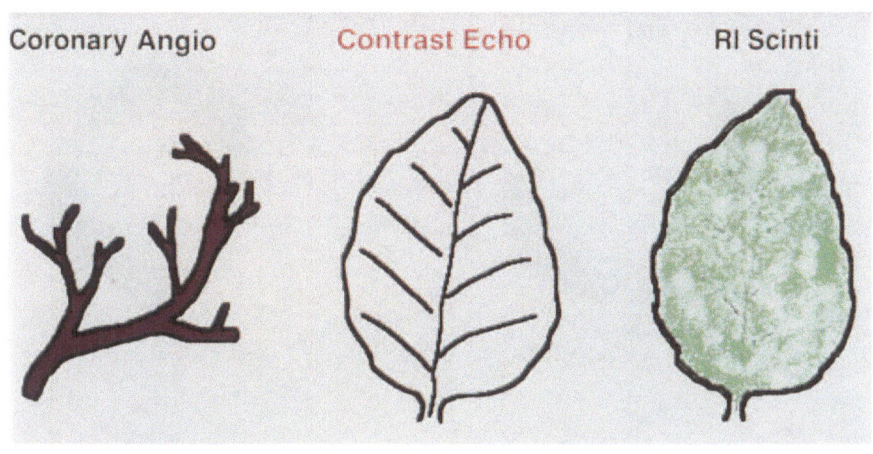

Figure 7. Schema of coronary circulation visualized by coronary angiography, MCE and thallium scintigraphy. Coronary angiography represents a stem and branches of a tree. MCE indicates veins of leaves, while thallium shows the result of photosynthesis.

thallium is not absorbed, the 'leaves' will not be green. In sarcoidosis, for instance, thallium scintigraphy shows a dilated heart with many perfusion defects before treatment; these perfusion defects decrease after proper treatment (Figure 8). In this case, thallium scintigraphy may not detect the viable myocardium. On the other hand, MCE examines the presence of the veins of a leaf. Veins correspond to the microchannels which distribute blood to myocardial cells. Just as veins deliver nutrients and produce healthy leaves, the presence of microchannels indicates healthy myocardial cells.

Myocardial contrast echocardiography is a unique method for examining the presence of veins of a leaf. The akinetic segment should be viable if the area is opacified by the contrast echo.

REPERFUSION AND MCE

Reperfusion is an indispensable therapy in acute myocardial infarction. Currently, however, the prognosis of the reperfused area cannot be determined immediately after reperfusion. Ito et al. indicated that MCE imaging revealed relatively many cases of 'no-reflow' of microcirculation, despite the reopening of the obstructed coronary artery [12], indicating that recanalization does not always lead to reperfusion. Ito et al. also indicated that the outcome of recanalization would be different between non-reperfused and reperfused patients. The area at risk will

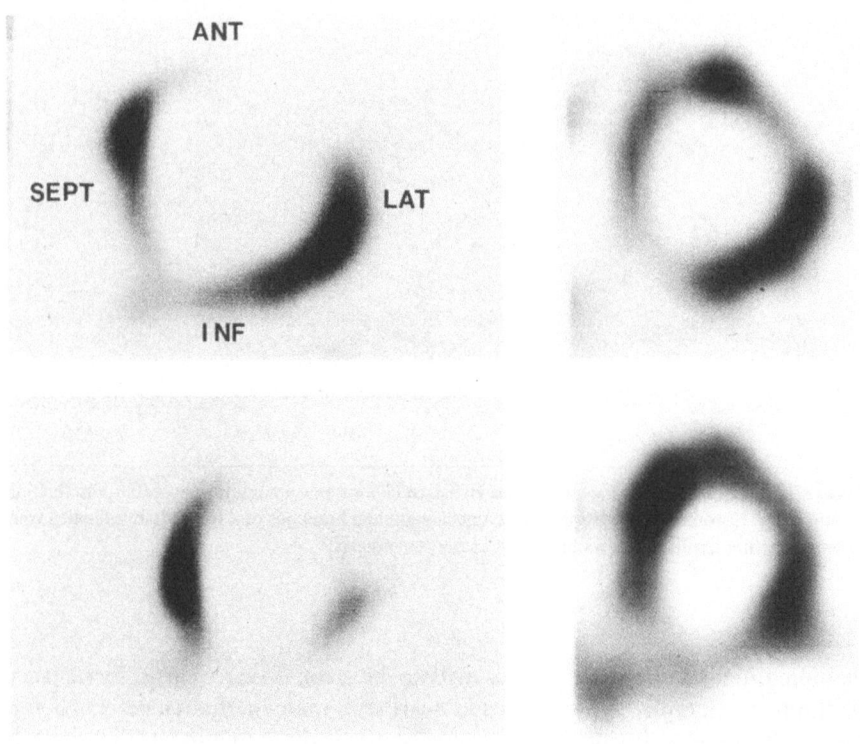

Figure 8. Thallium scintigraphy in a case of sarcoidosis before (left column) and after treatment (right column).

recover normal motion if the area is opacified by contrast soon after recanalization, but will not recover if the area is not opacified. They declared that recanalization of microvasculature is necessary to salvage wall motion.

In the acute phase soon after onset of myocardial infarction, MCE is a fairly reliable technique when compared with isotopic techniques.

INTRAVENOUS MCE

Currently, the contrast agent has to be injected directly into the coronary artery to opacify the myocardium. However, technologies which increase applications for MCE are being developed. Using the second harmonic of ultrasound waves is a promising technique for demonstrating myocardial perfusion. Long-life bubbles composed of fluorocarbon makes intravenous MCE possible. Such technologies will enable the myocardial perfusion bed to be clearly demonstrated even by peripheral venous injection.

CAN MCE REPLACE THALLIUM?

MCE is inferior to thallium scintigraphy in many ways. For example, image quality depends on the examiner's technique completely. The lungs and bones disturb data acquisition. Also, plane setting of cross-sections can be ambiguous, whereas any slice planes of SPECT are exactly parallel.

However, MCE findings are unique and useful for understanding coronary microcirculation, as described above. Moreover, additional advantages will exist when intravenous MCE is available. Sequential changes in the extent and degree of myocardial perfusion after recanalization can be evaluated precisely, while thallium cannot be administered frequently enough to examine changes over time after reperfusion. Myocardial ischemia can be diagnosed quickly, correctly and completely non-invasively. MCE allows assessment of regional wall motion abnormality at the time of examination. This technique will have important applications with exercise and pharmacological stress tests and in emergency care units, because it can differentiate myocardial ischemia from other causes of chest pain. The cost of MCE examination is lower than a thallium examination, with respect to examination space, equipment, and agents used for imaging.

Thus, we conclude that MCE will become a more popular method for assessing myocardial perfusion in the near future.

REFERENCES

1. Sakamaki T, Tei C, Meerbaum S et al. Verification of myocardial contrast two-dimensional echocardiographic assessment of perfusion defects in ischemic myocardium. J Am Coll Cardiol 1984;3:34–8.
2. Tei C, Kondo S, Meerbaum S et al. Correlation of myocardial echo contrast disappearance rate ('washout') and severity of experimental coronary stenosis. J Am Coll Cardiol 1984;3:39–46.
3. Cheirif J, Desir RM, Bolli R et al. Relation of perfusion defects observed with myocardial contrast echocardiography to the severity of coronary stenosis: Correlation with thallium-201 single-photon emission tomography. J Am Coll Cardiol 1992;19:1343–9.
4. Lim YJ, Nanto S, Masuyama T et al. Visualization of subendocardial myocardial ischemia with myocardial contrast echocardiography in humans. Circulation 1989;79:233–44.
5. Kaul S, Jayaweera AR, Glasheen WP et al. Myocardial contrast echocardiography and the transmural distribution of flow: A critical appraisal during myocardial ischemia not associated with infarction. J Am Coll Cardiol 1992;20:1005–16.
6. Ragosta M, Camarano G, Kaul S et al. Microvascular integrity indicates myocellular viability in patients with recent myocardial infarction: New insights using myocardial contrast echocardiography. Circulation 1994;89:2562–9.
7. Camarano G, Ragosta M, Gimple LW, Powers ER, Kaul S. Identification of viable myocardium with contrast echocardiography in patients with poor left ventricular systolic function caused by recent or remote myocardial infarction. Am J Cardiol 1995;75:2145–9.
8. Pierard LA, De LCM, Berthe C, Rigo P, Kulbertus HE. Identification of viable myocardium by echocardiography during dobutamine infusion in patients with myocardial infarction after thrombolytic therapy: comparison with positron emission tomography. J Am Coll Cardiol 1990; 15:1021–31.
9. Baer FM, Voth E, Deutsch HJ et al. Assessment of viable myocardium by dobutamine trans-

esophageal echocardiography and comparison with fluorine-18 fluorodeoxyglucose positron emission tomography. J Am Coll Cardiol 1994;24:343–53.
10. Charney R, Schwinger ME, Chun J et al. Dobutamine echocardiography and resting-redistribution thallium-201 scintigraphy predicts recovery of hibernating myocardium after coronary revascularization. Am Heart J 1994;128:864–9.
11. Kayden DS, Sigal S, Soufer R et al. Thallium-201 for assessment of myocardial viability: Quantitative comparison of 24-hour redistribution imaging with imaging after reinjection at rest. J Am Coll Cardiol 1991;18:1480–6.
12. Ito H, Tomooka T, Sakai N et al. Lack of myocardial perfusion immediately after successful thrombolysis. A predictor of poor recovery of left ventricular function in anterior myocardial infarction. Circulation 1992;85:1699–705.

27. Intraoperative applications of contrast echocardiography

JANINE R. SHAPIRO

INTRODUCTION

Myocardial perfusion imaging using contrast echocardiography has been reported in animal studies since 1982 [1, 2]. It has been shown to adequately define the area of perfusion deficits produced by experimental coronary occlusion [3] and to correlate closely with the area at risk of infarction [4–6]. The technique has been applied, in conjunction with coronary arteriography and coronary angioplasty in humans to the study of regional myocardial perfusion patterns [7–10]. Recently, contrast echocardiography has been used to evaluate myocardial perfusion in the setting of cardiac surgery and revascularization. This chapter will summarize the intraoperative applications of contrast echocardiography.

INTRAOPERATIVE ASSESSMENT OF REGIONAL MYOCARDIAL PERFUSION AND GRAFT PATENCY FOLLOWING REVASCULARIZATION

Intraoperative myocardial perfusion imaging using contrast echocardiography has been used to delineate myocardial perfusion patterns before, during and after coronary artery bypass surgery. It provides a unique opportunity to measure the outcome of an anatomical intervention. Feinstein and Aronson have described their technique for intraoperative studies in recent reviews [11, 12]. After initiation of cardiopulmonary bypass and following the application of the aortic cross-clamp, the contrast agent is delivered directly into the aortic root via the antegrade cardioplegia catheter. The injection is synchronized to the delivery of cardioplegia in order to minimize aortic regurgitation of contrast into the ventricular cavity. The procedure is repeated at the completion of coronary bypass grafting, when warm blood or cardioplegia is delivered to the revascularized heart via the antegrade aortic root catheter. As the myocardial perfusion region supplied by coronary arteries has been shown to be dependent on the magnitude

427

N. C. Nanda, R. Schlief and B. B. Goldberg (eds.), Advances in Echo Imaging Using Contrast Enhancement, Second Edition, 427–432.
© *1997 Kluwer Academic Publishers.*

of blood flow to that region [13], constant flow of the carrier cardioplegia solution must be used to deliver the contrast agent. Injections can also be made directly into the proximal end of a saphenous vein coronary graft anastomosis. The latter can provide useful information about the perfusion distribution of the bypass graft.

Real-time visual assessment of the regional perfusion region as noted on the video screen can be made on-line and can have potential great importance in intraoperative decision making. Semi-quantitative analyses of regional myocardial blood flow can be performed off-line using videodensitometry.

Smith et al. [14] first described the use of TEE and contrast echocardiography to detect the regional perfusion during coronary bypass surgery. Kabas and co-workers [15] evaluated the safety and efficacy of intraoperative perfusion contrast echocardiography in assessing the results of coronary artery bypass grafting in 20 patients. Direct injections of sonicated Renograffin-76 microbubbles into completed saphenous vein grafts resulted in contrast enhancement in the areas of myocardium subserved by each graft. Subsequently, Aronson et al. [16], using transesophageal echocardiography, showed myocardial enhancement with injections of sonicated Renograffin-76 into the aortic root before and after revascularization as well as with injections directly into saphenous vein grafts. Areas with perfusion defects after revascularization correlated with regional wall motion abnormalities after separation from cardiopulmonary bypass.

Villanueva et al. [17] evaluated the feasibility of on-line quantitative myocardial contrast echocardiography in 21 patients to assess objectively the success of coronary artery bypass graft operations. The technique was successful two-thirds of the time. Three patterns of contrast enhancement were noted: (1) reduced contrast effect before with improvement after coronary artery bypass graft operations; (2) adequate contrast effect before and after revascularization; (3) no contrast effect either before or after in one patient with previous infarction. Other investigators have also successfully used myocardial contrast echocardiography for the intraoperative assessment of regional bypass perfusion [18, 19].

ASSESSMENT OF THE DISTRIBUTION OF CARDIOPLEGIA SOLUTION

Goldman and Mindich [20] first observed the spontaneous contrast effect of routine aortic root cardioplegic solution infusion during coronary artery bypass grafting and used this information to estimate hypoperfused areas of myocardium and direct the sequence of bypass grafting. Recently, myocardial contrast echocardiography has been used to study the distribution of cardioplegia solution into the heart. The contrast agent is injected into the antegrade cardioplegia line or into the cardioplegia line placed into the coronary sinus. Contrast enhancement of antegrade or retrograde-delivered cardioplegia can then be appreciated.

The determination of cardioplegic perfusion can be useful in assessing the adequacy of myocardial protection by the cardioplegic solution. Recent studies have focused on evaluating the efficacy of antegrade cardioplegia delivery through the aortic root versus the retrograde delivery through the coronary sinus. Studies have shown the benefit of retrograde cardioplegia in coronary artery bypass surgery in patients with significant coronary artery stenosis and in aortic valve surgery [21, 22]. However, the distribution of retrograde cardioplegia is not well known. Aronson et al. [23] studied 19 patients undergoing cardiac surgery and showed that retrograde cardioplegia resulted in contrast enhancement of all segments of the left ventricle and the interventricular septum. In four patients the right ventricle was partially enhanced, while in the remaining 15 patients the right ventricle was not imaged adequately to allow assessment of cardioplegic distribution.

ASSESSMENT OF CORONARY COLLATERAL FLOW

Myocardial contrast echocardiography has been shown to be a valuable technique for the study of the coronary collateral circulation and to be superior to coronary angiography for the demonstration of coronary blood vessels. Spotnitz and co-workers [24] demonstrated intraoperatively significant collateral flow in three of six patients undergoing coronary artery bypass surgery in whom coronary collaterals were not seen with coronary angiography. The identification of collateral flow can provide useful information on collateral reserve and myocardial viability. Further, it can allow direct real-time evaluation of the extent and distribution of the perfusion area of the native coronary vessel selected for bypass grafting and help in surgical decision making [25].

IDENTIFICATION OF STUNNED MYOCARDIUM

An important potential application of intraoperative myocardial contrast echocardiography is in the evaluation of left ventricular systolic dysfunction after discontinuation of cardiopulmonary bypass. This technique can be used to differentiate between reversibly stunned (or hibernating) myocardium from irreversibly damaged (infarcted) myocardium [26]. In both cases, segmental function would be impaired, but stunned myocardium would be characterized by normal perfusion. This distinction can have important therapeutic applications in identifying those patients who could benefit from prolonged circulatory support.

OTHER INTRAOPERATIVE CARDIAC APPLICATIONS OF CONTRAST ECHOCARDIOGRAPHY

Contrast echocardiography has been used intraoperatively to evaluate valvular competence and the success of valve repair [27, 28], assess ventricular septal defect closure [29], guide percutaneous pericardiocentesis [30], and to detect right-to-left shunt in patients undergoing neurosurgical procedures in the sitting position [31].

NON-CARDIAC INTRAOPERATIVE APPLICATIONS OF CONTRAST ECHOCARDIOGRAPHY

Contrast echocardiography is currently being used in clinical trials to assess renal perfusion patterns intraoperatively in kidney transplant patients. This technique may also have an important application in the evaluation of patients with perioperative acute renal failure. Recently, Scott and co-workers [32] used the flush solution in the portal cannula to study the presence and degree of portapulmonary shunts in patients undergoing orthotopic liver transplantation, and their contribution to hypoxemia. Several experimental techniques are also currently being evaluated for the study of skeletal muscle and tissue perfusion.

CONCLUSION

Contrast echocardiography has important potential intraoperative applications including the assessment of regional myocardial perfusion, coronary artery bypass graft patency, cardioplegia and coronary artery collateral vessel distribution, as well as in the assessment of renal and tissue perfusion. With the continued development of contrast agents and improved sensitivity of commercial ultrasound imaging devices and computer-aided software analysis programs, the future of intraoperative myocardial contrast perfusion imaging looks promising.

REFERENCES

1. Meltzer RS, Vermeulen HWJ, Valk NK, Verdouw PD, Lancee CT, Roelandt J. New echocardiographic contrast agents: transmission through the lungs and myocardial perfusion imaging. J Cardiovasc Ultrason 1982;1:277–82.
2. Armstrong WF, Meuller TM, Kinney EL, Tickner EG, Dillon JC, Feigenbaum H. Assessment of myocardial perfusion abnormalities with contrast-enhanced two-dimensional echocardiography. Circulation 1982;66:166–73.
3. Tei C, Sakamaki T, Shah PM et al. Myocardial contrast echocardiography: A reproducible technique of myocardial opacification for identifying regional perfusion deficits. Circulation 1983;67:585–93.
4. Kaul S, Pandian NG, Okada RD, Pohost GM, Weyman AE. Contrast echocardiography in acute myocardial ischemia, I: In vivo determination of total left ventricular 'area at risk'. J Am Coll Cardiol 1984;4:1272–82.
5. Kaul S, Gillam LD, Weyman AE. Contrast echocardiography in acute myocardial ischemia, II: The effect of site of injection of contrast agents on the estimation of area at risk for necrosis after coronary occlusion. J Am Coll Cardiol 1985;6:825–30.
6. Kemper AJ, Force T, Perkins L, Gilfoil M, Parisi AF. In vivo prediction of the transmural

extent of experimental acute myocardial infarction using contrast echocardiography. J Am Coll Cardiol 1986;8:143–9.

7. Feinstein SB, Lang RM, Dick C et al. Contrast echocardiography during coronary arteriography in humans: Perfusion and anatomic studies. J Am Coll Cardiol 1988;11:59–65.

8. Lang RM, Feinstein SB, Feldman T, Neumann A, Chua KG, Borow KM. Contrast echocardiography for evaluation of myocardial perfusion: Effects of coronary angioplasty. J Am Coll Cardiol 1986;8:232–5.

9. Griffin B, Timmis AD, Henderson RA, Sowton E. Contrast perfusion echocardiography: Identification of area at risk of dyskinesis during percutaneous transluminal coronary angioplasty. Am Heart J 1987;114:497–502.

10. Reisner SA, Ong LS, Lichtenberg GS et al. Quantitative assessment of the immediate results of coronary angioplasty by myocardial contrast echocardiography. J Am Coll Cardiol 1989;13: 852–6.

11. Feinstein SB, Aronson S. Advances in contrast echocardiography: Intraoperative perfusion assessment. Am J Physiol Imag 1992;314:155–9.

12. Aronson S, Wiencek JG. Intraoperative perfusion echocardiography. J Cardiothorac Vasc Anesth 1994;8:97–107.

13. Vandenberg BF, Feinstein SB, Kieso RA, Hunt M, Kerber RE. Myocardial risk area and peak gray level measurement by contrast echocardiography: Effects of microbubble size and concentration injection rate on coronary vasodilatation. Am Heart J 1988;115:733–9.

14. Smith JS, Feinstein SB, Kapelanski D et al. Intraoperative determination of myocardial perfusion using contrast echocardiography. Anesthesiology 1986;65:A27.

15. Kabas JS, Kisslo J, Flick CL et al. Intraoperative perfusion contrast echocardiography: Initial experience during coronary artery bypass grafting. J Thorac Cardiovasc Surg 1990;99:536–42.

16. Aronson S, Lee BK, Wiencek JG et al. Assessment of myocardial perfusion during CABG surgery with two-dimensional transesophageal contrast echocardiography. Anesthesiology 1991; 75:433–40.

17. Villanueva FS, Spotnitz WD, Jayaweera AR, Dent J, Gimple LW, Kaul S. On-line intraoperative quantitation of regional myocardial perfusion during coronary artery bypass graft operations with myocardial contrast two-dimensional echocardiography. J Thorac Cardiovasc Surg 1992;104:1524–31.

18. Mudra H, Zwehl W, Klauss V et al. Intraoperative myocardial contrast echocardiography for assessment of regional bypass perfusion. Am J Cardiol 1990;66:1077–81.

19. Spotnitz WD, Kaul S. Intraoperative assessment of myocardial perfusion using contrast echocardiography. Echocardiography 1990;65:209–27.

20. Goldman ME, Mindich BP. Intraoperative cardioplegic contrast echocardiography for assessing myocardial perfusion during open heart surgery. J Am Coll Cardiol 1984;4:1029–34.

21. Gundy SR, Kirsh MM. A comparison of retrograde cardioplegia versus antegrade cardioplegia in the presence of coronary artery obstruction. Ann Thorac Surg 1986;38:124–7.

22. Menasché P, Subayi J-B, Piwnica A. Retrograde coronary sinus cardioplegia for aortic valve operations: A clinical report on 500 patients. Ann Thorac Surg 1990;49:556–64.

23. Aronson S, Zaroff JG, Lee BK. Myocardial distribution of retrograde delivered cardioplegia in patients undergoing cardiac surgery. J Cardiovasc Surg 1993;105:214–21.

24. Spotnitz WD, Matthew TL, Keller MW, Powers ER, Kaul S. Intraoperative demonstration of coronary collateral flow using myocardial contrast two-dimensional echocardiography. Am J Cardiol 1990;65:1259–61.

25. Hirata N, Nakano S, Sakai K, Sakaki S, Matsuda H. Extent and distribution of the perfusion areas of the coronary artery selected for bypass grafting: Assessment by intraoperative myocardial contrast echocardiography. J Thorac Cardiovasc Surg 1994;107:323–5.

26. Aronson S. Identifying stunned myocardium during cardiac surgery: The role of myocardial contrast echocardiography. J Card Surg 1993;8(Suppl):224–7.

27. Goldman ME, Fuster V, Guarino T, Mindich BP. Intraoperative echocardiography for the

evaluation of valvular regurgitation: Experience in 263 patients. Circulation 1986;74:1143–9.

28. Goldman ME, Guarino T, Fuster V, Mindich B. The necessity for tricuspid valve repair can be determined intraoperatively by two-dimensional echocardiography. J Thorac Cardiovasc Surg 1987;94:542–50.

29. Stümper O, Fraser AG, Elzenga N et al. Assessment of ventricular septal defect closure by intraoperative epicardial ultrasound. J Am Coll Cardiol 1990;16:1672–9.

30. Susini G, Pepi M, Sisillo E et al. Percutaneous pericardiocentesis versus subxiphoid pericardiotomy in cardiac tamponade due to postoperative pericardial effusion. J Cardiothorac Vasc Anesth 1993;7:178–83.

31. Black S, Muzzi DA, Nishimura RA, Cucchiara RF. Preoperative and intraoperative echocardiography to detect right-to-left shunt in patients undergoing neurosurgical procedures in the sitting position. Anesthesiology 1990;72:436–8.

32. Scott VL, De Wolf AM, Daniel ED, Keen FK, Kang Y. Porta-pulmonary shunt identified intraoperatively by transesophageal echocardiography. Anesthesiology 1994;81:3A.

28. Harmonic imaging during contrast echocardiography: basic principles and potential clinical value

HECTOR R. VILLARRAGA, DAVID A. FOLEY,
SANG MAN CHUNG, NAVIN C. NANDA & SHARON L. MULVAGH

INTRODUCTION

Contrast echocardiography has evolved rapidly, due to major recent advances in both contrast agent development and ultrasound equipment technology. Transpulmonary passage of intravenously administered contrast agents and resultant left ventricular opacification has been demonstrated in humans [1, 2]. Newer agents, containing various stabilizing gases, produce both left ventricular and myocardial opacification after intravenous injection in animal models during normal and altered perfusion states [3–9]. Experimental human studies indicate similar findings [10, 11], suggesting enormous potential for the clinical application of these new agents. However, the myocardial contrast effect is variably detected when imaged with standard commercially available ultrasound equipment, and if higher doses are utilized to improve detectability, attenuation from within the ventricles frequently interferes with complete visualization of the myocardium.

Harmonic ultrasound imaging is a method of enhancing the detection of ultrasound contrast agents within blood-containing cavities and vascularized tissue [12, 13]. This novel detection method is enabled by the non-linear emission of harmonics by resonant microbubbles pulsating in an ultrasonic field [14, 15]. Prototype ultrasound systems currently under investigation employing this principle utilize specialized transducers and software to transmit ultrasound at the fundamental frequency (f) and receive ultrasound at a multiple of this frequency (i.e. $2f$, $3f$, etc.). For example, to detect the second harmonic of ultrasound transmitted at 2.5 MHz, the system would receive signals at 5 MHz. Detection of subharmonics in addition to supraharmonics is theoretically possible, although preclinical and clinical evaluations to date have focused upon detection of the second harmonic ($2f$). Contrast-containing structures are enhanced relative to the surrounding non-contrast-filled structures, thereby increasing the strength of the 'signal' relative to the background noise. This has significant implications for clinical detection of myocardial perfusion by contrast echocardiography. The basic principles of harmonic imaging will be reviewed, followed by preclinical

433

N. C. Nanda, R. Schlief and B. B. Goldberg (eds.), Advances in Echo Imaging Using Contrast Enhancement,
Second Edition, 433–450.
© 1997 *Kluwer Academic Publishers.*

and clinical experience with various gas-filled microbubbles demonstrating promising potential applications.

BASIC PRINCIPLES

Ultrasound physics

Ultrasound is a form of mechanical energy that propagates as waves through a deformable medium of gas, liquid or solid. The interaction of the transmitted ultrasound with the propagating medium is determined by the composition of the medium and the frequency of the ultrasound. In clinical imaging, reflection of ultrasound by the targeted body tissues is the property used to produce images. Reflection occurs at interfaces between tissues with differing acoustic impedances and is characterized by two types of signals: specular and non-specular. Specular reflection occurs when the interface is longer than the wavelength of the ultrasound beam. The amount of ultrasound reflected is directly related to the insonifying power of the signal. In specular reflection, the angle of reflection equals the angle of incidence. Non-specular reflection, or scattering, occurs when the interfaces are smaller than the ultrasound wavelength, and results in a signal which is emitted in all directions. The strength of the scattered signal is related to a power of the frequency, generally $f^2 - f^4$. Signals reflected back to the transducer are termed backscatter.

Traditionally, the most important property of ultrasonic contrast agents has been their capacity to enhance the backscatter signal by creating numerous small reflective surfaces and produce 2-dimensional images of greater definition. Free or encapsulated gas bubbles have been considered superior to colloidal suspensions, emulsions or aqueous solutions in their ability to enhance backscatter [16]. However, bubbles also exhibit properties more complex than other scatterers. When placed in an ultrasound field, the bubbles oscillate in a complicated manner dependent upon the size and composition of the bubble, the nature of the surrounding medium, and the ultrasound frequency [14]. This oscillation of bubbles results in the emission of sound waves with a wide range of frequencies, a feature which is termed non-linear behavior. The emission of ultrasound is greatest when the bubbles are insonified at their optimal resonant first harmonic, or fundamental frequency. In this case, higher harmonic ($2f$, $3f$, $4f$, etc.) signals will be emitted very efficiently and their amplitude may be greater than the signals at the insonating frequency. Imaging of these harmonic ultrasound signals is undergoing intensive investigation and development, and is the topic of this discussion.

Historical perspective and overview

The first theoretical description of the behavior of bubbles exposed to an external pressure field was established by Lord Rayleigh in 1917 through simple observations of a boiling tea kettle [17], and this served as the basis for extensive complex numerical and in vitro investigations exploring the non-linear behavior of small gas bubbles in an ultrasonic field [18–21]. Tucker and Welsby proposed that the detection of second harmonic ultrasound signals was a very sensitive indicator for the presence of intravascular bubbles in subjects being monitored for decompression sickness [22]. The non-linear second harmonic emissions from resonant bubbles were characterized in vitro by Miller [15] and shown to be related to bubble size, with a smaller bubble size (4.2 μm) producing a second harmonic response of a 40-fold greater magnitude than larger bubbles (500 μm). DeJong extended these observations and determined that for a bubble with a radius of 2 μm in an aqueous medium under normal atmospheric conditions, when insonated at its resonant frequency of 1.64 MHz, the second harmonic generated at 3.28 MHz had a scattering cross-section nearly equal to that of the fundamental frequency, yet there was virtually no energy scattered for larger bubbles exposed to the same insonating frequency [14].

Experimental in vitro studies have suggested that the detection of non-linear ultrasound contrast agents within blood-containing cavities and vascularized tissue could provide the ability to allow real-time imaging of flowing blood and, ultimately, organ perfusion with ultrasound [12, 13, 23]. Simulated capillary blood flow measurements using degassed vs. stabilized microbubbles within suspensions of saccharide microparticles (Levovist, Schering AG) have clearly demonstrated the non-linear (harmonic) properties of microbubble solutions [12] (Figure 1). Myocardial perfusion has been historically difficult to quantify because of the characteristically slow movement of red blood cells through the capillaries, and the small amount of blood contained within a perfused volume. The velocity of red blood cells as they travel through the capillary network approximates 1 mm/s and the volume of blood contained within myocardial tissue approximates 5–10% [24]. Thus the overwhelming tissue signal obscures detection of the blood signal. However, tissue artifact and inadequate blood signal intensity can be overcome by studying the second harmonic component of the backscattered echo, which is much greater in magnitude for the contrast agent than for the tissue. This enables detection of slow, small volume blood flow in the presence of the more prevalent tissue, whose reflection at the fundamental frequency overwhelms the signal from the blood. Using a continuous wave Doppler system with transmit and receive transducers of 2.25 MHz and 5.0 MHz, respectively, Schrope and Newhouse elegantly demonstrated that analysis of contrast agent-enhanced backscattered ultrasonic signal at the second harmonic frequency unambiguously defined blood perfusion as indicated by small vessel flow in rabbit ear capillaries [13]. They recognized that if these results are to be extrapolated to deeper organ tissues (such as the heart or kidney), such investigations would be performed not on isolated capillary beds but on the collection of vessels that perfuse the tissues, including arterioles, venules and, perhaps, larger conduit vessels which traverse the sample volume. Harmonic imaging could, therefore, potentially allow measurement of relative changes in

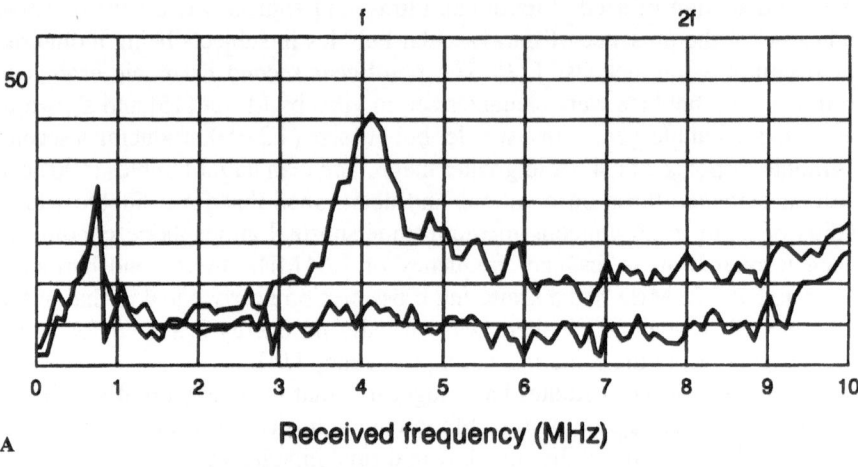

A

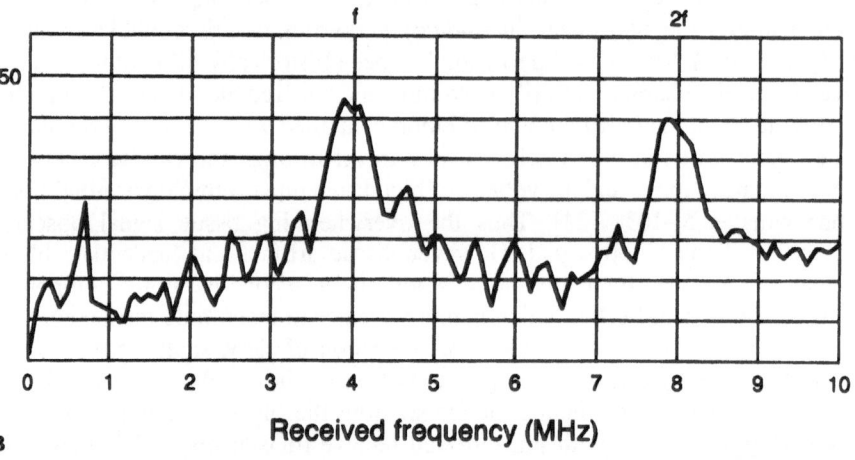

B

Figure 1. Frequency domain spectra for (A) lower curve reference material (deionized water, plasma, blood), upper curve: linear contrast medium (particles); (B) non-linear contrast medium (Levovist, Schering AG), low concentration (5×10^{-6}), received frequency = 4 MHz; second harmonic visible. Peak at 0.67 MHz in (A) and (B) is background radio-frequency pickup; vertical scale is 5 dB/division. [Reproduced from Schrope BA et al. Ultrasonic Imaging 1992;14:134–58.]

blood perfusion and echocardiographic assessment of myocardial perfusion to determine regions of inadequate or absent blood supply could be envisioned. Indeed, we have been able to demonstrate the detection and measurement of coronary blood flow during transthoracic harmonic imaging of non-linear contrast agents in the canine model [25]. Similarly, detection of myocardial tissue contrast in preclinical and early clinical studies during harmonic imaging confirm the dramatic improvement in signal-to-noise ratio when using this technique, although best results have been obtained when transmission of ultrasound is intermittently suspended (transient-response imaging or TRI) [26, 27].

Contrast agents

Recent advances in contrast agent development have resulted in the development of free or coated microbubbles (i.e. albumin, galactose, phospholipids, surfactants) containing inert, relatively heavy, poorly soluble gases (i.e. perfluorocarbons, sulphur hexafluoride). These microbubbles range in size between 1 and 10 μm and persist in the circulation for up to several minutes.

The myocardial contrast effect of these new agents is variably detectable depending upon the agent and imaging technique. If higher doses are utilized to increase the myocardial microbubble concentration, attenuation from within the right and left ventricles frequently interferes with complete visualization of the myocardium. Harmonic imaging can enhance microbubble detection, enabling the utilization of lower doses of contrast agent, thereby reducing the problem of attenuation.

The resonance phenomenon that results in the generation of harmonic signals is dependent not only upon the bubble radius, which determines the resonant frequency, but also on the inter-related factors of acoustic pressure and velocity of the wavefront, the characteristics of the surrounding medium, and the composition and viscoelastic properties of the microbubble shell [14]. The second harmonic potential of many of these new contrast agents has been demonstrated [28] (Figure 2) and a multitude of clinical applications are evolving (see below). Fortunately, the resonant frequency of the microbubbles utilized in contrast echocardiography (median size <6 μm), coincides with the range of ultrasound routinely employed during transthoracic echocardiography (2–4 MHz). Transmission of ultrasound at or near the resonant frequency of the contrast microbubbles enhances the harmonic signals and improves the detection of contrast agent by second harmonic imaging during clinical cardiac ultrasound examinations. Most contrast agents are composed of microbubbles with a range of sizes rather than a single uniform dimension. This feature may be an additional clinical advantage for second harmonic imaging, as the ultrasound transducers may not need to be precisely tuned to the bubbles, but can employ a broad bandwidth and still reliably produce harmonic effects.

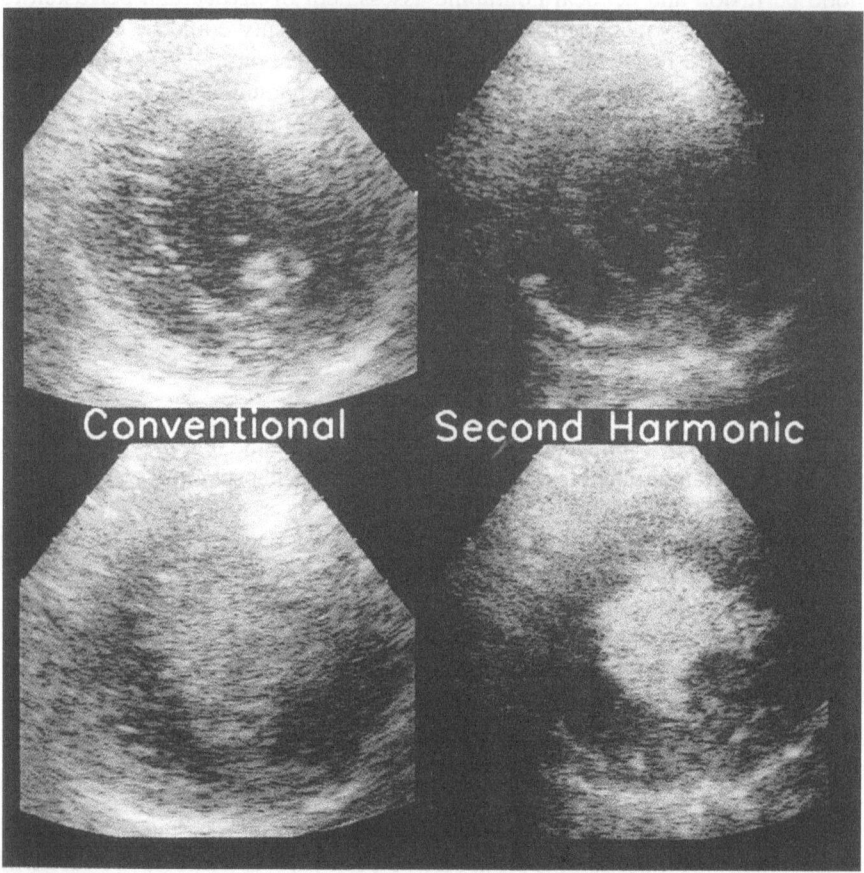

Figure 2. Transthoracic short-axis mid-papillary muscle level images in closed chest dogs before (upper panels) and after (lower panels) femoral vein injection of MRX-115 (Aerosomes) during fundamental (conventional) imaging (left panels) and second harmonic imaging (right panels) (2.5 MHz transmit, 5.0 MHz receive; Acuson prototype).

Ultrasound equipment

In addition to having a contrast agent capable of harmonic resonance, second harmonic imaging requires echocardiographic equipment capable of exploiting this phenomenon. Several ultrasound manufacturers are now developing or modifying existing transducers and imaging systems to permit harmonic imaging. Although laboratory measurements of harmonic signals have employed separate transducers for transmission and reception, this is not practical for clinical imaging. Instead, clinical systems utilize broad band transducers which are driven at a low frequency (usually 1.8–2.5 MHz) and receive across a wide range of frequencies (usually 3.6–5.0 MHz), permitting detection of the second harmonic. Electronic signal processing is then employed to process and display the harmonic frequencies. Because there is overlap between the actual transmitted frequency range and the second harmonic signals which are processed, the images usually contain some higher frequency fundamental image data. This results in the display of the background tissues, which theoretically do not emit significant harmonic signals. The background tissues are displayed relatively weakly compared with signal from the contrast agent so the theoretical advantage of 'improved signal-to-background noise' of second harmonic imaging is maintained, and the fundamental tissue image provides an orienting reference when contrast is not present, facilitating clinical imaging. Second harmonic principles have been implemented not only for gray scale imaging, but also for color flow, pulsed wave and continuous wave Doppler. While these other modalities have been less extensively evaluated than gray scale imaging, their sensitivity and specificity for contrast agents appear to be improved with in vivo gains of more than 30 dB demonstrated [29].

HARMONIC IMAGING OF THE HEART

Experimental observations

Very recent and ongoing experimental studies have demonstrated that the harmonic signals generated by oscillating microbubbles improve the identification of contrast agents within the left ventricular blood pool, myocardial tissue and coronary vasculature.

Left ventricular blood pool cavity enhancement by second harmonic contrast echocardiography

Left ventricular opacification can be utilized in general echocardiography as well as in special imaging situations such as stress echocardiography. In dogs given various intravenously injected contrast agents, second harmonic ultrasound imaging resulted in a 15–70% increase in left ventricular cavity contrast intensity, as well as increased duration of detection [30]. Qualitative and quantitative assessment of left ventricular opacification during second harmonic imaging of peripherally injected contrast agents has recently been undertaken in humans, and has demonstrated improved left ventricular cavity enhancement compared with fundamental contrast enhanced imaging. DeMaria and co-workers reported a 7-fold increase in the ratio of left ventricular cavity to myocardial intensity in five patients undergoing second harmonic imaging compared with standard trans-thoracic echocardiography during injections of Levovist (Schering AG, Berlex USA) [31]. Similarly, we found a 35–105% increase in the background-subtracted pixel intensity within the left ventricular cavity in 10 patients undergoing stress

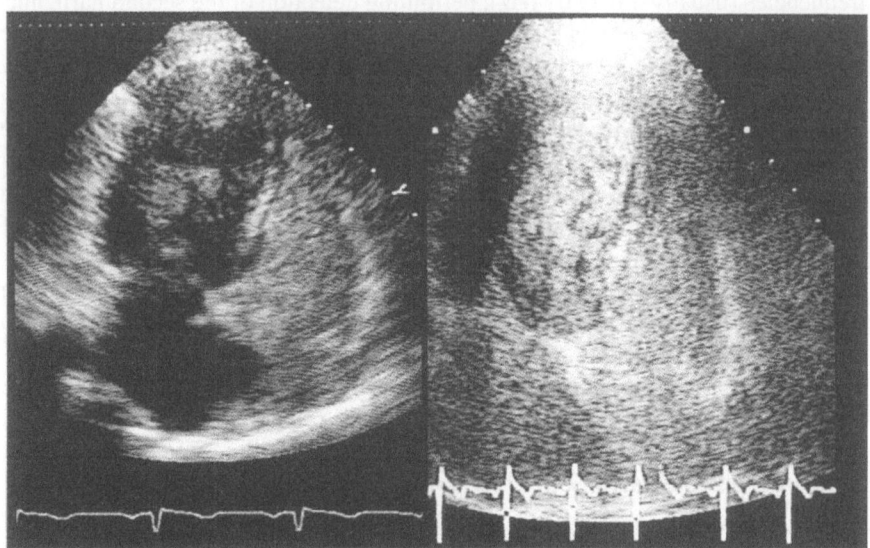

Figure 3. Transthoracic apical four-chamber (left ventricule on left; apex up) images obtained after intravenous injection of Levovist (Berlex, USA) 3.2 g in a 63-year-old female. Left: fundamental imaging (Hewlett-Packard (HP) Sonos 1500). Right: second harmonic imaging (2.5 MHz transmit, 5.0 MHz receive; Acuson prototype).

echocardiography and peripheral injections of Levovist [32] (Figure 3). Similar objective increases in pixel intensity during second harmonic imaging have been observed with Albunex (Mallinckrodt Medical, USA) in patients undergoing pharmacological stress echocardiography [33] (Figure 4), confirming the visual impression of improved left ventricular cavity delineation with second harmonic imaging compared with fundamental contrast echocardiography. We have currently studied 14 patients referred for diagnostic echocardiography with MRX-115 (Aerosomes, ImaRx Pharmaceutical Corp., USA), an intravenously injected perfluorocarbon containing contrast agent composed of microbubbles contained within a liposomal membrane, and have noted markedly persistent and intense detection of contrast effect within the left ventricular cavity during second harmonic imaging, long after the contrast is no longer detected using fundamental imaging (Figure 5). Importantly, no adverse hemodynamic effects or symptoms have been observed when using this agent.

Myocardial perfusion detection during second harmonic contrast echocardiography

Newer contrast agents containing stabilizing gases, primarily composed of perfluorocarbons, have been demonstrated to produce both left ventricular and

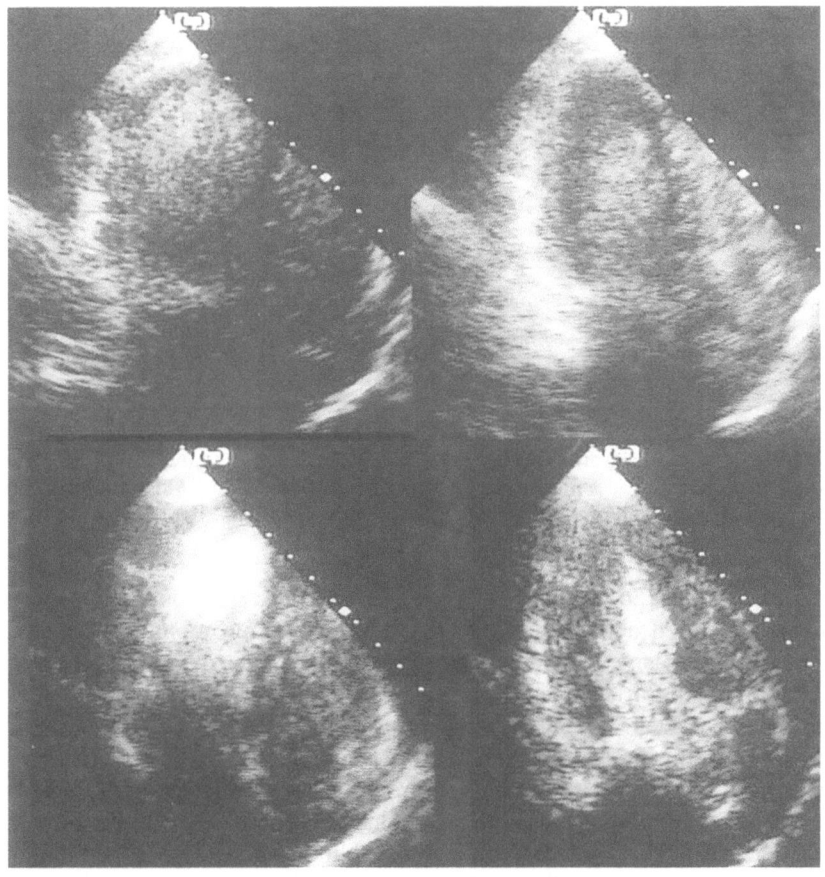

Figure 4. End-diastolic (left panels) and peak systolic (right panels) transthoracic apical long axis (left ventricle on left; apex up) images obtained during dobutamine infusion (30 μg/kg/min) and 10 ml intravenous Albunex in a 65-year-old female. Upper panels: fundamental imaging (HP Sonos 2500, 2.5 MHz). Lower panels: second harmonic imaging (HP Sonos 2500, 1.8 MHz transmit, 3.6 MHz receive).

myocardial opacification after intravenous injection during normal and altered perfusion states [3–11]. Our experience with second harmonic imaging of these agents during continuous delivery of ultrasound has revealed the presence of an inhomogeneous distribution of myocardial contrast consistent with delineation of intramyocardial coronary vessels [25] (see below).

Very recent preclinical and preliminary clinical studies suggest that the combination of second harmonic imaging, and intermittent triggered acquisition, or 'transient-response imaging (TRI)', provides enhanced detection of myocardial perfusion using intravenously injected perfluorocarbon-exposed sonicated

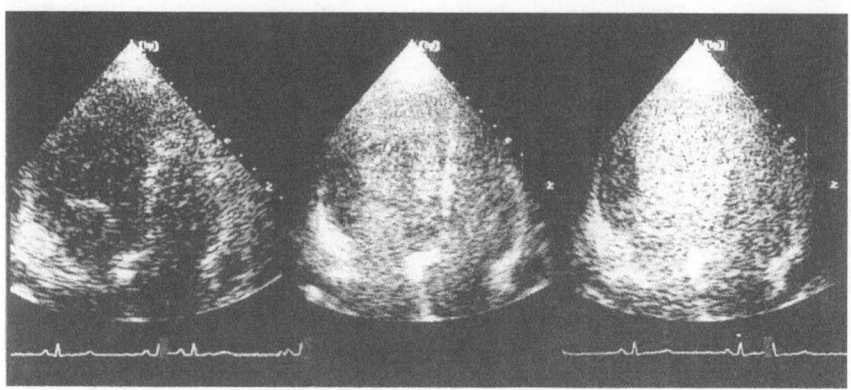

Figure 5. Transthoracic apical four-chamber (left ventricle on left; apex up) images before (left), 30 s after (center), and 5 min after (right) intravenous administration of Aerosomes in a 62-year-old female. Left and center images with fundamental imaging (HP Sonos 1500, 2.5 MHz) and right image with second harmonic imaging (HP Sonos 2500, 1.8 MHz transmit, 3.6 MHz receive).

dextrose albumin [26, 27]. We have made similar observations in the cohort of patients described above who received intravenous injections of Aerosomes (Figure 6A). Intensive clinical investigation in this area is ongoing as the newer generation of contrast agents is progressing through developmental studies. Preliminary data indicate that myocardial perfusion detection using second harmonic imaging of intravenously injected contrast agents may be reliably achieved in humans, especially when the physical properties of the transmitted ultrasound are optimized to reduce destruction and enhance detection of the microbubbles. Post-processing techniques applying color-coding to improve differentiation of gray levels, as utilized in the nuclear cardiology laboratory, may enhance the detection and evaluation of regional myocardial blood flow abnormalities (Figure 6B).

Coronary visualization and blood flow measurements during second harmonic contrast echocardiography

We have non-invasively demonstrated [25] myocardial blood flow in intramyocardial coronary arteries during second harmonic ultrasound imaging in the

Figure 6. (Opposite) (A) Second harmonic transthoracic images (HP Sonos 2500, 1.8 MHz transmit, 3.6 MHz receive) 6 min after administration of Aerosomes (same patient and orientation as Figure 5). Left: single end-diastolic frame during continuous ultrasound transmission. Right: single end-diastolic frame after five cardiac cycles of intermittent (R-wave only) ultrasound transmission. Note the myocardial opacification. (B) Application of color-coding to right image of (A).

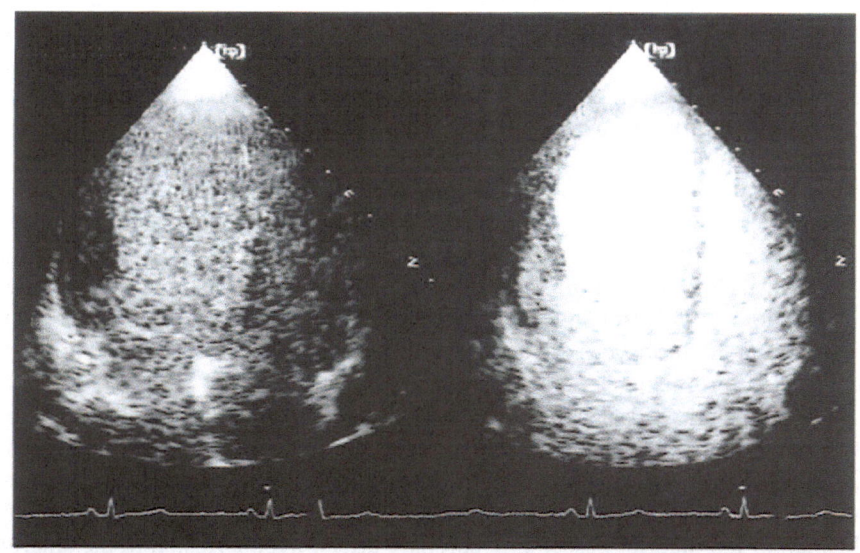

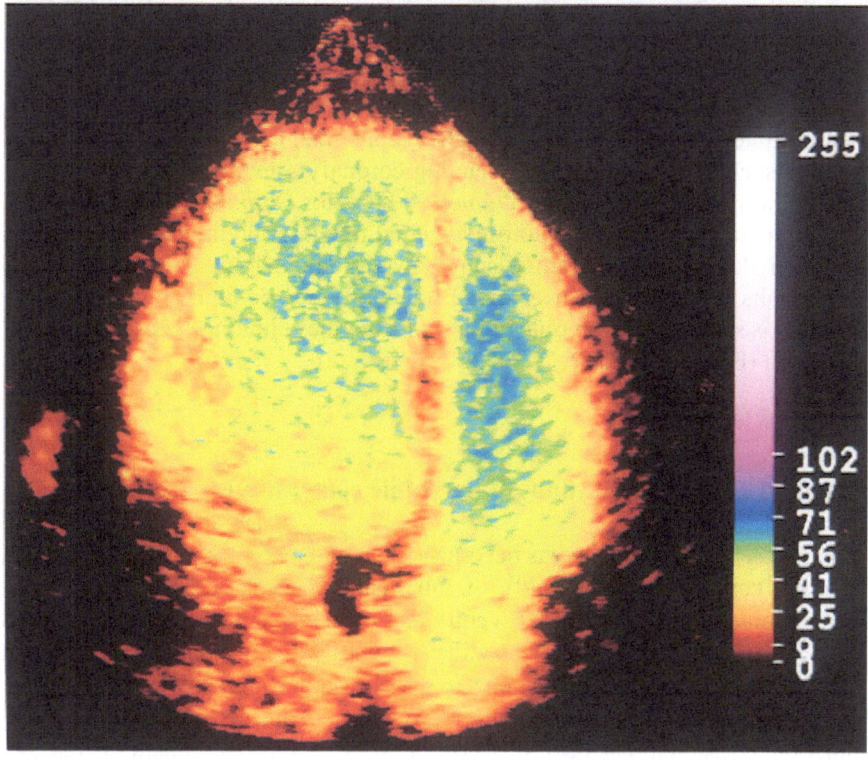

closed chest dog during intravenous injections of the perfluorocarbon contrast agent AF0145 (Imagent US, Alliance Pharmaceutical Corp., USA). The novel appearance of heterogeneous, linear branching opacified structures within the myocardium (Figure 7A) was confirmed to represent coronary arterial vascular arcades by comparison with simultaneous coronary Doppler flow signals measured using an intracoronary wire (Figure 7B). The Doppler signals were measured at baseline and during adenosine administration; the coronary vaso-dilator reserve ratio determined using the non-invasive harmonic contrast Doppler and invasive Doppler flow wire methodologies were similar, suggesting that the non-invasive assessment of coronary vasculature and measurement of coronary vasodilator reserve is possible using second harmonic contrast echocardiography. These observations have yet to be extended into the clinical arena, but preliminary data appear promising.

Experimental work performed at the University of Alabama at Birmingham has shown the superiority of harmonic imaging over fundamental imaging in both left ventricular cavity and myocardial contrast enhancement in canine as well as pig models. The contrast agent used was Levovist. Following temporary ligation of left anterior descending coronary artery, hypoperfusion in the anterior ventricular septum was clearly delineated during harmonic imaging; following release of ligation, contrast enhancement re-developed in the anterior septum to practically the same level as at baseline. Thus, the usefulness of harmonic imaging in not only detecting myocardial ischemia but also the effects of reperfusion was demonstrated. In addition, flow signals in epicardial as well as intramyocardial arteries such as the posterior descending coronary artery and septal perforator arteries were also more clearly delineated by color Doppler when examined in harmonic mode as compared to fundamental imaging. Preliminary studies have also been done in our laboratory using power mode imaging in harmonic mode. The results suggest that power mode imaging when combined with harmonic imaging is superior to harmonic mode alone but these findings need to be further validated (Figures 8 and 9).

Potential clinical applications of harmonic contrast echocardiography

Contrast echocardiography has become a clinical reality with the immediate and pending commercial availability of intravenously injected agents capable of transpulmonary passage and left ventricular cavity opacification with improved endocardial border detection. Preliminary studies indicate a potential application of echocardiographic contrast agents in stress echocardiography, where enhancement of regional wall motion abnormality analysis and left ventricular systolic function assessment has been observed. Early experiences indicate that second harmonic imaging further improves the left ventricular cavity contrast effect during pharmacologic stress echocardiography, enabling more confident assessments

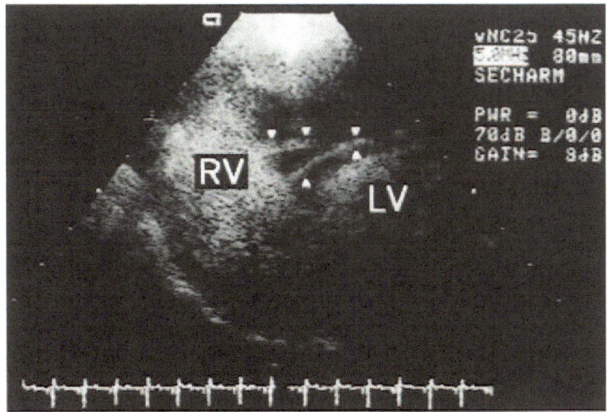

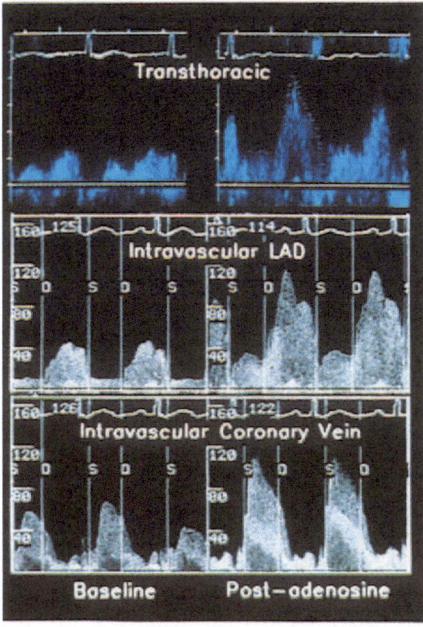

Figure 7. (A) Modified basal short-axis view during second harmonic imaging of closed chest dog after 30 mg i.v. AF0145 (Imagent®US) demonstrating proximal septal intramyocardial vessels (arrows). (B) Second harmonic pulsed wave Doppler after AF0145 (40 mg i.v.) (top panel) and comparison to invasively measured coronary arterial (left anterior descending artery; middle panel) and venous (coronary sinus; bottom panel) Doppler flows at baseline (left panels) and at peak adenosine (140 μg intracoronary) effect (right panels). [Reproduced from Mulvagh SL et al. J Am Coll Cardiol 1996;27:x–x.]

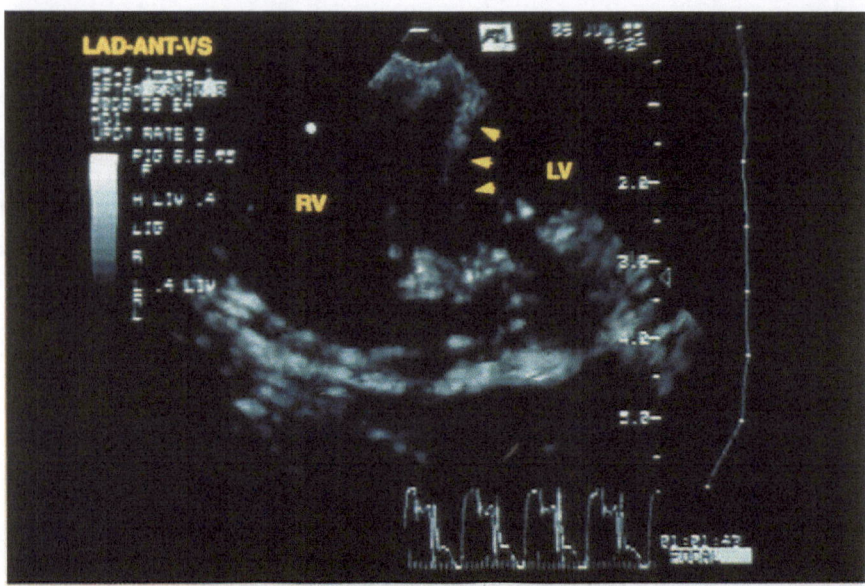

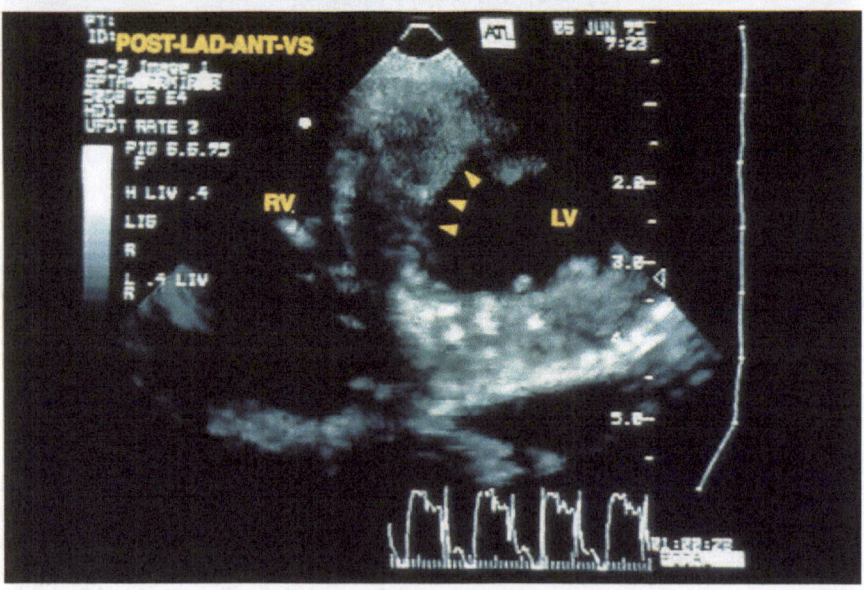

Figure 8. Transthoracic contrast echocardiography during harmonic imaging in an open-chest pig model. (A) Short-axis view showing absence of contrast echoes in the anterior ventricular septum (arrows) following proximal left anterior descending coronary artery ligation and injection of Levovist, 0.4 cm³/kg. (B) Following release of ligation, contrast echoes opacify at anterior septum (arrows) indicative of reperfusion. LV = left ventricle; RV = right ventricle.

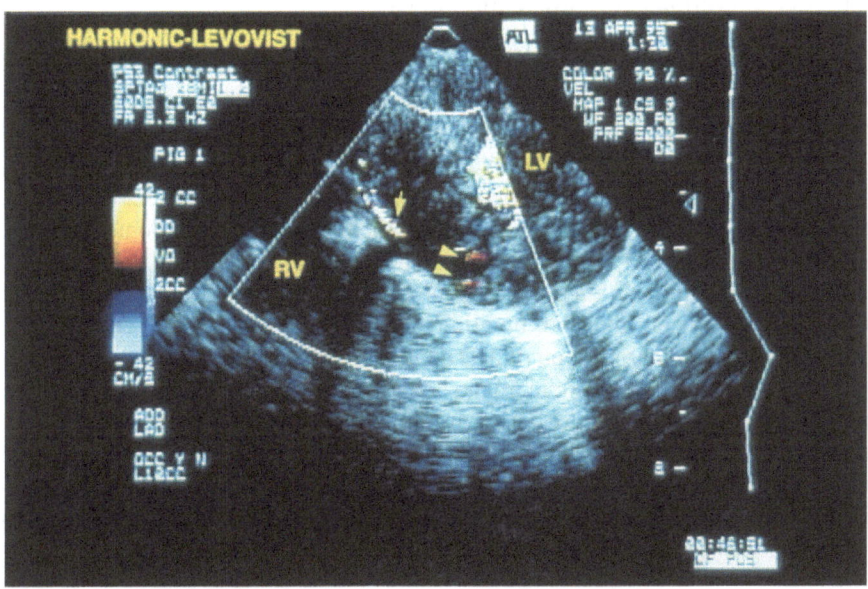

Figure 9. Transthoracic color Doppler contrast echocardiography during harmonic imaging in an open-chest pig model. Color Doppler examination during harmonic imaging following injection of Levovist demonstrates prominent flow signals in the posterior descending coronary artery (arrow) and in the intramyocardial coronary arteries (arrowheads). LV=left ventricle; RV=right ventricle.

of regional wall motion abnormalities when compared with fundamental imaging. Newer perfluorocarbon-containing echocardiographic contrast agents appear to possess the potential to provide myocardial perfusion assessment in patients with cardiac disease, particularly when second harmonic and intermittent, rather than continuous, ultrasound delivery are combined. The application of three-dimensional echocardiographic epicardial acquisition techniques utilizing these principles has demonstrated enhanced delineation of myocardial perfusion defects in open-chest dogs [34].

These new agents and technologies are currently undergoing intensive preclinical and clinical investigation [35]. Early preliminary results appear promising, especially for the detection of coronary artery disease and the evaluation of acute myocardial infarction both before and after attempted reperfusion, but have yet to be rigorously compared with current methods of myocardial perfusion assessment. Clinical implications of such a completely non-invasive, non-ionizing, portable, relatively inexpensive technique yielding immediate results and capable of serial investigations would have major impact upon the diagnosis, evaluation and treatment of both acute and chronic coronary ischemic syndromes.

448   *Hector R. Villarraga et al.*

Current limitations and future directions

At present, there is no commercially available, clinically approved myocardial contrast agent which can be injected intravenously with successful passage through the pulmonary microvasculature to result in myocardial perfusion and coronary vascular imaging. However, multiple agents are in various phases of preclinical and clinical development and appear to have the potential to play a key role in the future non-invasive management of acute and chronic ischemic syndromes. As of yet, there is no commercially available second harmonic ultrasound machine, but their development is an active area of investigation and these instruments are likely to be clinically available in the near future. In the very near future, incorporation of 3-D acquisition and reconstruction techniques may facilitate second harmonic image acquisition, either by the transthoracic or transesophageal approach, resulting in enhanced interpretation and display of contrast echocardiographic data. With these rapid technological advances, we are fast approaching a new definition of the 'complete' echocardiographic examination which will include simultaneous assessment of ventricular function, perfusion and coronary pathophysiology.

REFERENCES

1. Feinstein SB, Cheirif J, Ten Cate FJ et al. Safety and efficacy of a new transpulmonary ultrasound contrast agent: initial multicenter clinical results. J Am Coll Cardiol 1990;16:316–34.
2. Schlief R, Schurmann R, Niendorf HP. Blood-pool enhancement with SHU 508 A. Results of phase II clinical trials. Invest Radiol 1991;(Suppl 1):S188–9.
3. Porter TR, Xie F. Visually discernible myocardial echocardiographic contrast after intravenous injection of sonicated dextrose albumin microbubbles containing high molecular weight, less soluble gases. J Am Coll Cardiol 1995;25:509–15.
4. Grayburn PA, Erikson JM, Escobar J, Womack L, Velasco CE. Peripheral intravenous myocardial contrast echocardiography using a 2% dodecafluoropentane emulsion: identification of myocardial risk area and infarct size in the canine model of ischemia. J Am Coll Cardiol 1995;26:1340–7.
5. Dittrich HC, Bales GL, Kuvelas T, Hunt RM, McFerran BA, Greener Y. Myocardial contrast echocardiography in experimental coronary artery occlusion with a new intravenously administered contrast agent. J Am Soc Echocardiogr 1995;8:465–74.
6. Porter TR, Feng X, Kricsfeld A, Kilzer K. Noninvasive identification of acute myocardial ischemia and reperfusion with contrast ultrasound using intravenous perfluoropropane-exposed sonicated dextrose albumin. J Am Coll Cardiol 1995;26:33–40.
7. Ismail S, Jayaweera AR, Goodman NC, Camarano GP, Skyba DM, Kaul S. Detection of coronary stenoses and quantification of the degree and spatial extent of blood flow mismatch during coronary hyperemia with myocardial contrast echocardiography. Circulation 1995;91:921–30.
8. Grauer S, Pantely GA, Xu J et al. Aerosomes MRX-115: echocardiographic and hemodynamic characteristics of new echo contrast agent that produces myocardial opacification after intravenous injection in pigs. Circulation 1994;90:I-556.
9. Mulvagh SL, Foley DA, Aeschbacher BC, Klarich KW, Seward JB. A new intravenous perfluorochemical echocardiographic contrast agent, Imagent® US: imaging characteristics and hemodynamic profile [abstr]. J Am Soc Echocardiogr 1995;8:345.

10. Sutherland GR, Grauer SE, Moran C, Ishii M, Sahn DJ. Aerosomes MRX-115 echo contrast agent demonstrates myocardial opacification after intravenous injection in humans, without significant side effects, in a phase I clinical trial. Circulation 1995;92:I-463.

11. Dittrich HC, Kuvelas T, Dadd K et al. Safety and efficacy of the ultrasound contrast agent FS069 in normal humans: results of a phase I trial. Circulation 1995;92:I-464.

12. Schrope BA, Newhouse VL, Uhlendorf V. Simulated capillary blood flow measurement using a nonlinear ultrasonic contrast agent. Ultrasonic Imaging 1992;14:134–58.

13. Schrope B, Newhouse VL. Second harmonic ultrasound blood perfusion measurement. Ultrasound Med Biol 1993;19:567–79.

14. de Jong N, Ten Cate FJ, Lancee CT, Roelandt JR, Bom N. Principles and recent developments in ultrasound contrast agents. Ultrasonics 1991;29:324–30.

15. Miller DL. Ultrasonic detection of resonant cavitation bubbles in a flow tube by their second-harmonic emissions. Ultrasonics 1981;19:217–24.

16. Ophir J, Parker KJ. Contrast agents in diagnostic ultrasound. Ultrasound Med Biol 1989;15: 319–33.

17. Lord Rayleigh. On the pressure developed by a liquid during the collapse of a spherical cavity. Philos Mag 1917;34:94–8.

18. Eatock BC, Nishi RY. Numerical studies of the spectrum of low-intensity ultrasound scattered by bubbles. J Acoust Soc Am 1985;77:1692–701.

19. Lauterborn W. Numerical investigations of nonlinear oscillations of gas bubbles in liquids. J Acoust Soc Am 1976;59:283–93.

20. Prosperetti A. Nonlinear oscillations of gas bubbles in liquids: steady-state solutions and the connection between subharmonic signal and cavitation. J Acoust Soc Am 1974;56:878–85.

21. de Jong N, Cornet R, Lancée CT. Higher harmonics of vibrating gas-filled microspheres. Part One: simulations. Part two: measurements. Ultrasonics 1994;32:447–59.

22. Tucker DG, Welsby VG. Ultrasonic monitoring of decompression. Lancet 1968;1:1253.

23. Burns PN. Ultrasound contrast agents in radiological diagnosis. Radiol Med 1987;1–5:71–82.

24. Wu X, Ewert DL, Liu YH, Ritman EL. In vivo relation of intramyocardial blood volume to myocardial perfusion. Circulation 1992;85:730–7.

25. Mulvagh SL, Foley DA, Aeschbacher BC, Klarich KW, Seward JB. Second harmonic imaging of intravenously administered echocardiographic contrast: visualization of coronary arteries and measurement of coronary blood flow. J Am Coll Cardiol 1996;27:1519–25.

26. Porter TR, Xie F. Transient myocardial contrast after initial exposure to diagnostic ultrasound pressures with minute doses of intravenously injected microbubbles. Circulation 1995;92: 2391–5.

27. Porter TR, Armbruster R, Holdeman K, Xie F. Second harmonic transient response imaging with intravenous perfluorocarbon-exposed sonicated dextrose albumin in patients with previous myocardial infarction: initial clinical experience. J Am Coll Cardiol 1996;27:76A.

28. Foley DA, Mulvagh SL, Aeschbacher BC, Klarich KW, Seward JB. Second harmonic imaging of echocardiographic contrast agents: unique features and promising characteristics. J Am Coll Cardiol 1995:16A.

29. Burns PN, Powers JE, Hope-Simpson D et al. Harmonic power mode Doppler using microbubble contrast agents: an improved method for small vessel flow imaging. Inst Electronics Eng 1994;3:1547–50.

30. Villarraga HR, Foley DA, Pellikka PA et al. Second harmonic imaging of myocardial contrast agents: qualitative and quantitative observations from animal and human studies. Workshop on: Second Harmonic Technology in Cardiac Ultrasound. Costa Mesa, CA, November 11th, 1995.

31. Mahmud E, Cotter B, Kimura B et al. Second harmonic imaging enhances contrast echocardiography in patients with cardiac disease: demonstration of feasibility. J Am Coll Cardiol 1995; 25:39A.

32. Allen MR, Pellikka PA, Villarraga HR, Foley DA, Pumper GM, Mulvagh SL. Second harmonic enhancement of echocardiographic contrast intensity and duration in patients. Eur Heart J 1996;17:169.

33. Villarraga HR, Pumper GM, Foley DA, Pellikka PA, Mulvagh SL. Enhancement of left ventricular opacification by Albunex® using second harmonic imaging during dobutamine stress echocardiography [abstr]. J Am Soc Echocardiogr 1996: in press.
34. Cao QL, Masani N, Delabays A et al. Harmonic imaging and single frame 'triggered mode' data acquisition enhance delineation of myocardial perfusion defects by volume-rendered 3-dimensional echocardiography. J Am Coll Cardiol 1996;27:21A.
35. Villarraga HR, Foley DA, Mulvagh SL. Contrast Echocardiography 1996: A Review. Texas Heart J 1996;23:90–7.

29. Transient response imaging during contrast echocardiography

THOMAS R. PORTER

INTRODUCTION

Although newer generation ultrasound contrast agents appear to improve the amount of myocardial contrast produced from an intravenous injection of contrast medium, large doses are required to produce this effect, and these cause acoustic shadowing of myocardium in the far field of insonation [1–3]. Two advances in ultrasound imaging techniques permit the use of lower doses of i.v. microbubbles while producing greater myocardial contrast. One of these advances is second harmonic imaging, which will be discussed in other chapters. Harmonic imaging significantly improves the signal-to-noise ratio by taking advantage of the non-linear reflective characteristics of microbubbles [4, 5]. The second advance involves a technique referred to as transient response imaging or intermittent imaging. The purpose of this chapter is to define this new imaging modality, and present data regarding its clinical efficacy in defining myocardial perfusion.

TRANSIENT RESPONSE IMAGING: BACKGROUND

The peak negative rarefraction pressures released from diagnostic ultrasound transducers, which range from 0.5 to 2.2 MPa [6] are capable of destroying microbubbles filled with room air [7–9]. This destruction results in a decrease in the videointensity produced by these contrast agents and occurs at a rate directly related to the peak negative pressure output from the transducer (Figure 1). The mechanism for destruction has been attributed to cavitation. Less frequent exposure to these pressures would, therefore, be expected to destroy fewer microbubbles. This would translate in vivo to greater ultrasound contrast following i.v. injection of microbubbles.

A second phenomenon regarding diagnostic ultrasound interaction with bubbles is also observed with room air-containing microbubbles. Upon initial exposure to diagnostic ultrasound pressures, isolated microbubbles transiently

451

N. C. Nanda, R. Schlief and B. B. Goldberg (eds.), Advances in Echo Imaging Using Contrast Enhancement, Second Edition, 451–464.
© 1997 *Kluwer Academic Publishers.*

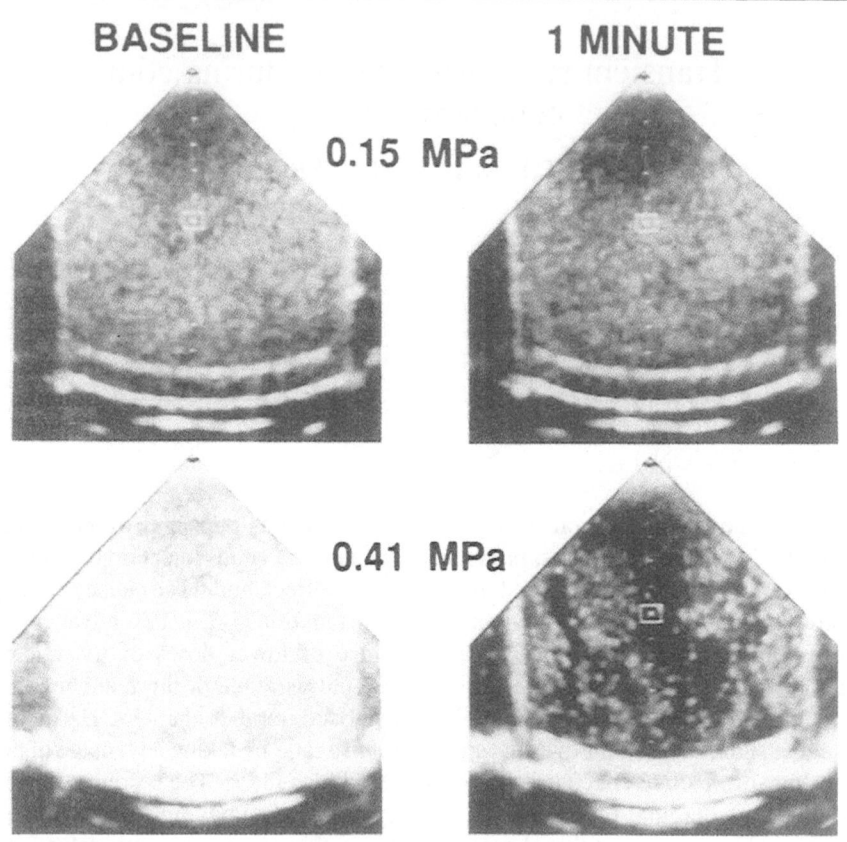

Figure 1. The decrease in relative integrated backscatter from microspheres within the scanning chamber after 1 mm exposure to different acoustic pressures. The 0.41 MPa peak negative pressure produced a significantly greater decrease in backscatter than 0.15 MPa insonation. [Reprinted with permission from Vandenberg BF, Melton HE. Acoustic liability of albumin microspheres. J Am Soc Echocardiogr 1994;7:582–9.]

increase in size, sometimes achieving several times their steady-state radius [10, 11]. The magnitude of this change in bubble size is indirectly related to the insonating frequency (Figure 2). This phenomenon has been observed with sonicated albumin-coated microbubbles [12], but the duration of this transient increase in size (referred to as transient cavitation) is very brief (Figure 3). The magnitude of the increase in microbubble size is also dependent upon the initial microbubble radius: smaller (<2.0 mm) microbubbles show the greatest increase in size (Figure 2). Interestingly, this transient increase in bubble radius upon initial exposure to diagnostic ultrasound pressures also exhibits harmonic features [13].

Both diagnostic ultrasound-induced microbubble destruction and transient cavitation may play a role in the phenomenon of transient response imaging [14].

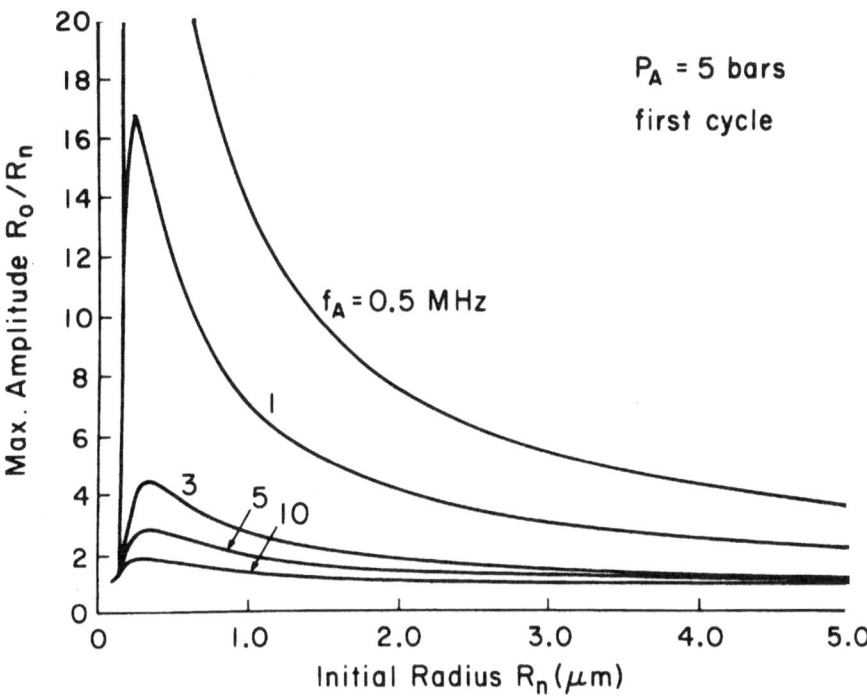

Figure 2. Demonstration of the change in microbubble radius that would be expected upon initial exposure to ultrasound during the first pressure cycle as a function of initial bubble radius and insonating frequency using a peak pressure of 5 bars. Note that the lower transducer frequencies and smaller microbubbles would be expected to exhibit the greatest transient increase in microbubble radius. R_o = Maximal microbubble radius, R_n = Initial radius. [Reprinted with permission from Flynn HG, Church CC. Erratum: transient pulsations of small gas bubbles in water. (J Acoust Soc Am 1988;84:985–98) J Acoust Soc Am 1988;84:1863–1876 (Abstract).]

This phenomenon was first observed when diagnostic ultrasound was suspended for a period of time following i.v. injection of perfluorocarbon-exposed sonicated dextrose albumin in open-chest dogs. After freezing transducer output for over 40 s following injection, a brief, but dramatic, increase in both myocardial and left ventricular cavity contrast was evident when ultrasound transmission was resumed (Figure 4). This phenomenon was reproducible following each injection, and demonstrable for over 1 min after i.v. injection of perfluorocarbon-exposed sonicated dextrose albumin (PESDA) microbubbles [15]. Because of the possibility that the increased contrast was due to transient cavitation of bubbles exposed to intermittent pulses of ultrasound, the imaging technique was altered by triggering just one frame rate every cardiac cycle throughout the i.v. injection of MB. This also produced bright myocardial contrast that was much greater than that produced with conventional 30 Hz frame rate imaging. This phenomenon could be

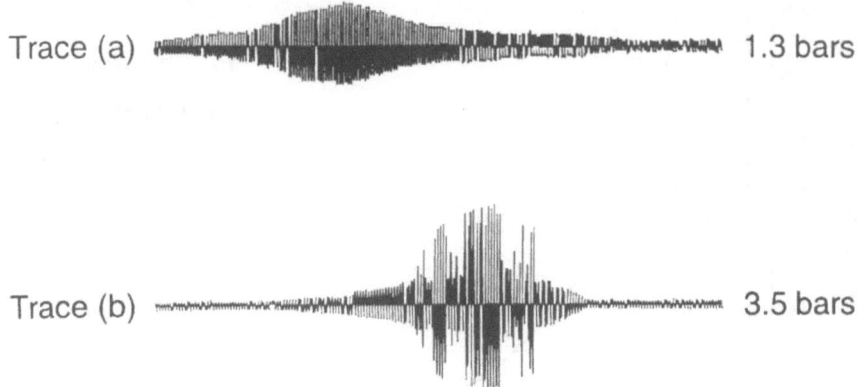

Trace (a)   1.3 bars

Trace (b)   3.5 bars

Figure 3. A digital oscilloscope recording from a cavitation detector which is measuring the transient cavitation (y axis) that occurred when sonicated albumin microbubbles were initially exposed to diagnostic ultrasound pressures of 1.3 and 3.5 bars (1 MHz). Time is along the x axis; the total time elapsed for this transient cavitation was <200 m. [Reprinted with permission from Roy RA, Church CC, Calabrese A. Cavitation produced by short pulses of ultrasound. In: Hamilton AF, Blackstock DT, editors. Frontiers of Nonlinear Acoustics: Proceedings of 12th International Symposium of Nonlinear Acoustics. Elsevier, London. 1990;476–81.]

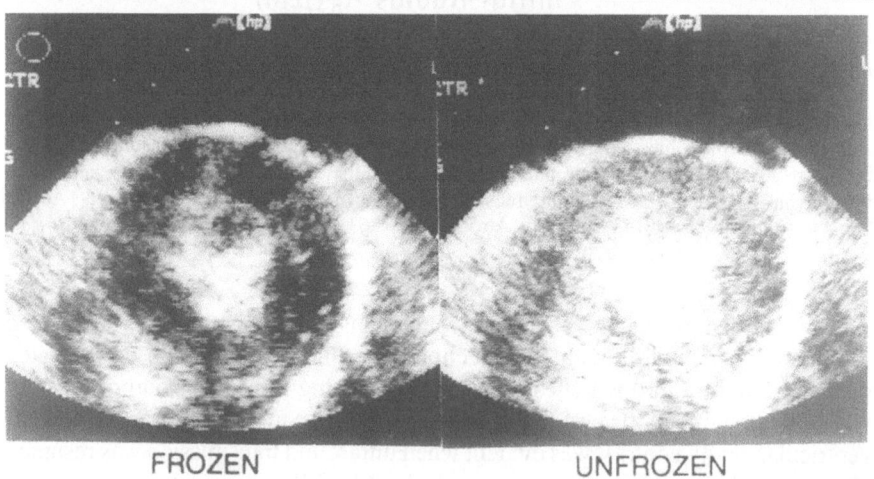

FROZEN   UNFROZEN

Figure 4. An epicardial short-axis view obtained with a 3.5 MHz transducer over 60 s after an i.v. injection of perfluorocarbon-exposed sonicated dextrose albumin. The transducer was frozen (i.e., no ultrasound transmission) for a period of >30 s (left panel) after i.v. injection of contrast. On the right panel is demonstrated one of the first frames from the short axis after the freeze button on the ultrasound system was turned off (unfrozen). There is a dramatic, but transient, increase in myocardial contrast that lasted slightly more than one cardiac cycle when using conventional 30 Hz frame rate imaging.

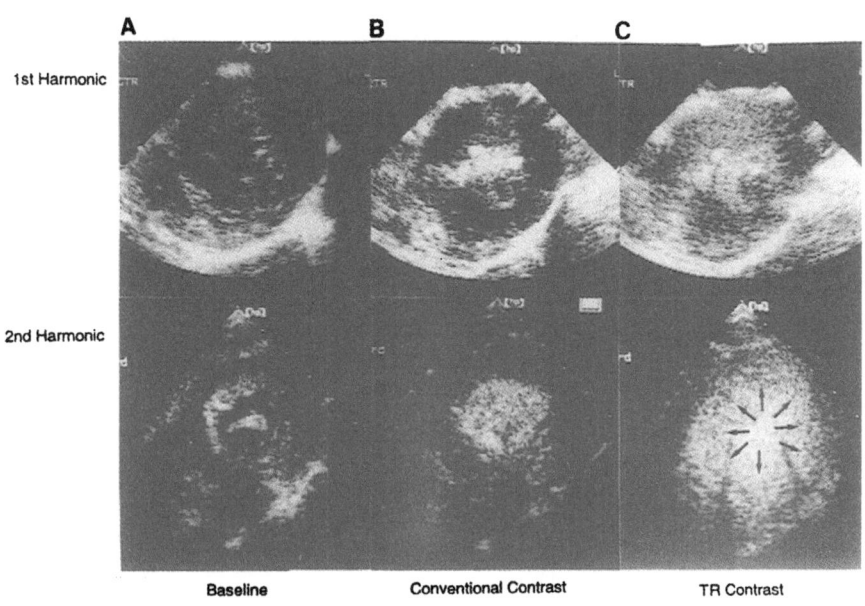

Figure 5. Panels taken from the same dog before and after the same i.v. dose of perfluorocarbon-exposed sonicated dextrose albumin. The top panels are with first harmonic imaging (2.0–2.5 MHz transmit and receive) using either conventional 30 Hz frame rates (B), or with transient response (TR) imaging (C). The lower panels are with a second harmonic receiving frequency (4.0–5.0 MHz received frequency). With both first and second harmonic imaging, there is a dramatic increase in myocardial contrast with transient response imaging. [Reproduced from ref. 15 with permission.]

visualized to an even greater degree with harmonic imaging (Figure 5). The increase in myocardial contrast with transient response imaging was so dramatic with second harmonic imaging that contrast doses in the dog could be reduced to 0.0025 ml/kg i.v. PESDA, whereas 0.03 ml/kg (12 times the dose) was required to produce any myocardial contrast with conventional imaging (30 Hz frame rates).

On the basis of these initial observations, in vitro and further in vivo studies have been performed to discern the mechanism for this phenomenon.

IN VITRO STUDIES WITH TRANSIENT RESPONSE IMAGING

As mentioned previously, Vandenberg et al. have demonstrated that albumin microbubbles are destroyed by diagnostic ultrasound pressures [7]. These observations, combined with the demonstration of markedly greater contrast with reduced ultrasound frame rates, led to an in vitro study of the effects of 30 Hz vs. triggered (0.5–1.0 Hz) frame rates on PESDA microbubble concentration. An in vitro flow chamber was designed in which PESDA was insonated at a fre-

Table 1. Comparison of effluent microbubbles concentration after exposure to conventional vs. triggered ultrasound as well as difference in videointensity. [Reproduced from ref. 14 with permission.]

	No ultrasound	0.5–1.0 Hz	30 Hz
Microbubble concentration			
Hemocytometry (number/field[a])	$41 \pm 19^{*}$	$40 \pm 14^{*}$	21 ± 9
Coulter counter ($\times 10^4$/ml)	$6.1 \pm 1.2^{*}$	$6.6 \pm 1.2^{*}$	3.2 ± 1.4
Videointensity[b]			
20 ml/min		$131 \pm 28^{**}$	108 ± 31
40 ml/min		$146 \pm 47^{**}$	138 ± 46
100 ml/min		$117 \pm 43^{**}$	111 ± 44

$^{*}p < 0.05$ compared with 30 Hz; $^{**}p < 0.001$ compared with 30 Hz.
[a]Field = 36 mm^2.
[b]Using 2.0 MHz frequency (first and second harmonic).

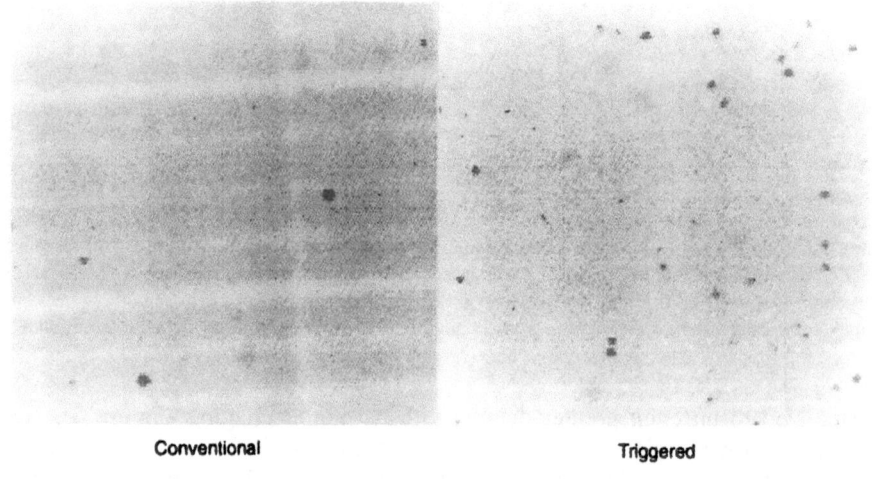

Conventional   Triggered

Figure 6. Hemocytometer slides demonstrating the difference in microbubble concentration that occur when the microbubble has been exposed to ultrasound transmission rates of 30 Hz (conventional) vs. triggered (1 Hz) frame rates. The triggered frame rate imaging resulted in significantly larger numbers of microbubbles remaining in the effluent. [Reproduced from ref. 14 with permission.]

quency of 2.0 MHz (1.1 MPa peak negative pressure) following injection of the microbubbles proximal to the chamber. After insonation, a portion of the effluent solution exiting the flow chamber was then sampled for microbubble concentration. The effluent PESDA microbubble concentration measured following 30 Hz frame rates was significantly lower than when ultrasound was either delivered at 0.5–1.0 Hz or when ultrasound was frozen when injections were given [14] (Table 1). Figure 6 shows an example from one hemocytometer slide, demonstrating the greater microbubble destruction that occurs when using 30 Hz frame rates compared with 0.5–1.0 Hz.

In this same experiment, the flow system was closed following the sampling of a portion of the effluent microbubbles, and the remaining microbubbles were repeatedly circulated through the flow chamber (scanning chamber). This constant number of PESDA microbubbles was then insonated at different frequencies (2.0, 2.5, and 3.5 Mz) as well with one harmonic transducer (2.0 Mz transmit, 4.0 Mz receiving). The videointensity in the flow cell was measured at both the 30 Hz frame rate and the 0.5–1.0 Hz frame rate. Videointensity with these different frame rates and transmit frequencies was compared at flow rates of 0, 20, 40, and 100 ml/min. Despite the presence of a constant number of microbubbles, the 0.5–1.0 Hz frame rates still produced greater videointensity than the 30 Hz frame rate (Table 2), especially when using the lower 2.0 Mz, transducer, and 0 and 20 ml/min flow rates (Figure 7). These observations illustrate two important phenomena concerning transient response imaging. First, diagnostic ultrasound destroys significantly more perfluorocarbon-containing microbubbles when using conventional 30 Hz frame rates than at one frame/s. Second, this lower frame rate produces greater videointensity from a constant concentration of PESDA microbubbles than 30 Hz frame rates, especially at lower flow rates and lower transducer frequencies.

These observations indicate that the significantly better contrast obtained with transient response imaging may be due to both less microbubble destruction as well as enhanced contrast from the same number of microbubbles. Further studies are being performed to define better the mechanisms responsible for this new imaging technique. However, the dramatically improved contrast obtained with transient response imaging has led to several in vivo studies in animals and humans to study the potential clinical applicability of transient myocardial contrast.

TRANSIENT RESPONSE IMAGING IN ANIMAL STUDIES

As stated previously, when ultrasound transmission was triggered (1 frame per cardiac cycle) in both closed and open-chest dogs, the doses of PESDA required to produce myocardial contrast were several fold less than that required to produce contrast with conventional 30 Hz frame rates. This had three advantages. First, it significantly reduced the attenuation of more distal structures due to left ventricular cavity contrast. Second, even though the higher doses had minimal hemodynamic effects, the lower doses further reduced any potential safety risk with i.v. perfluorocarbon-containing microbubbles. Finally, the lower doses required to produce myocardial contrast significantly reduced the cost of contrast agent production.

The dramatic myocardial contrast produced with transient response imaging also allowed one more easily to visualize the area at risk during acute ischemia (Figure 8). We demonstrated a very close correlation between planimetered area at risk with transient response imaging following low intravenous doses of PESDA and actual risk area measured with Monastral blue (r=0.99).

Table 2. Contrast videointensity produced with triggered imaging over conventional imaging at different transducer frequencies and flow rates. [Reproduced from ref. 14 with permission.]

Transducer frequency (MHz)	Maximum rarefaction pressure (MPa)	Flow rate (ml/min)							
		0		20		40		100	
		CON	TRI	CON	TRI	CON	TRI	CON	TRI
2.0 (1st harmonic)	1.3	155±16	172±35	131±30	149±20	151±43	161±49	121±39	128±38
2.5 (1st harmonic)	1.6	140±25	149±32	104±11	120±16	151±24	164±23	103±13	111±11
3.5 (1st harmonic)	1.9	106±35	117±34	92±21	96±17	112±14	122±14	105±31	109±32
4.0 (2nd harmonic)	0.9	65±28	84±29	80±11	108±13*	102±9	109±11	86±11	90±11

*$p<0.05$, compared with the increase in other categories except for 2.0 (1st) at flow rates of 0 and 20 ml/min and 4.0 (2nd) at 0 flow rate.

CON = 30 Hz frame rate imaging.

TRI = triggered (0.5–1.0 Hz) frame rates.

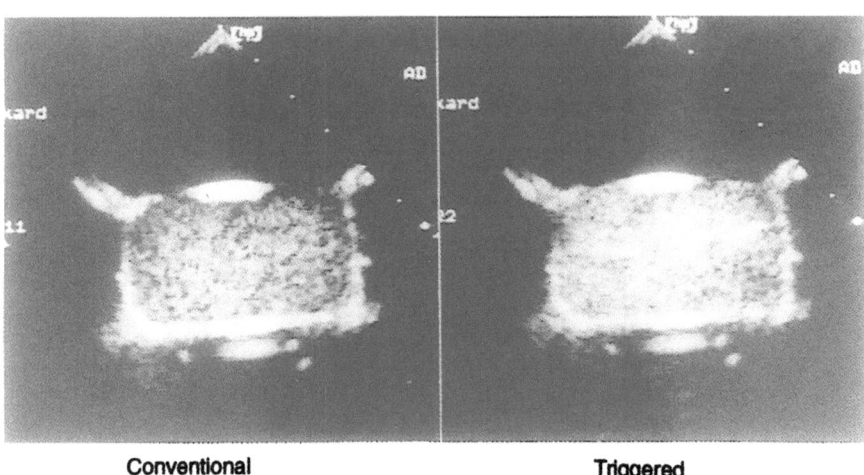

Conventional   Triggered

Figure 7. Videointensity from the scanning chamber with a constant number of PESDA micro-bubbles (zero flow rate) during two different ultrasound frame rates (conventional and triggered). Note the visually evident increase in videointensity during triggered frame rate imaging. [Reproduced from ref. 14 with permission.]

We also discovered that visually evident perfusion defects could be identified during dobutamine stress in the perfusion bed of dogs who had (by quantitative angiography) a >50% diameter stenosis [16] with doses of PESDA between 0.005 and 0.01 ml/kg i.v. Visually evident myocardial contrast defects were detectable with transient response imaging in nine of 10 dogs at peak dobutamine stress, but in only one dog using conventional imaging. Furthermore, the peak myocardial videointensity ratio in the stenosed perfusion bed divided by the normally perfused bed when using transient response imaging decreased in all dogs during dobutamine stress. The magnitude of this decrease in the peak myocardial videointensity ratio during dobutamine was directly correlated to the minimal coronary lumen diameter at the lesion site ($r=0.70, p=0.02$). These animal findings indicate that transient response imaging can be used to identify rest and stress-induced myocardial perfusion abnormalities. Further work is needed in this model to determine the ability of transient response imaging to detect myocardial viability in the setting of reperfusion following acute myocardial infarction. Nonetheless, these preliminary observations in animals led to the application of transient response imaging in humans.

TRANSIENT RESPONSE IMAGING IN HUMANS: VALIDATION AND CLINICAL APPLICATIONS

In humans, we first studied the ability of transient response imaging to produce myocardial contrast using transthoracic harmonic imaging following an i.v.

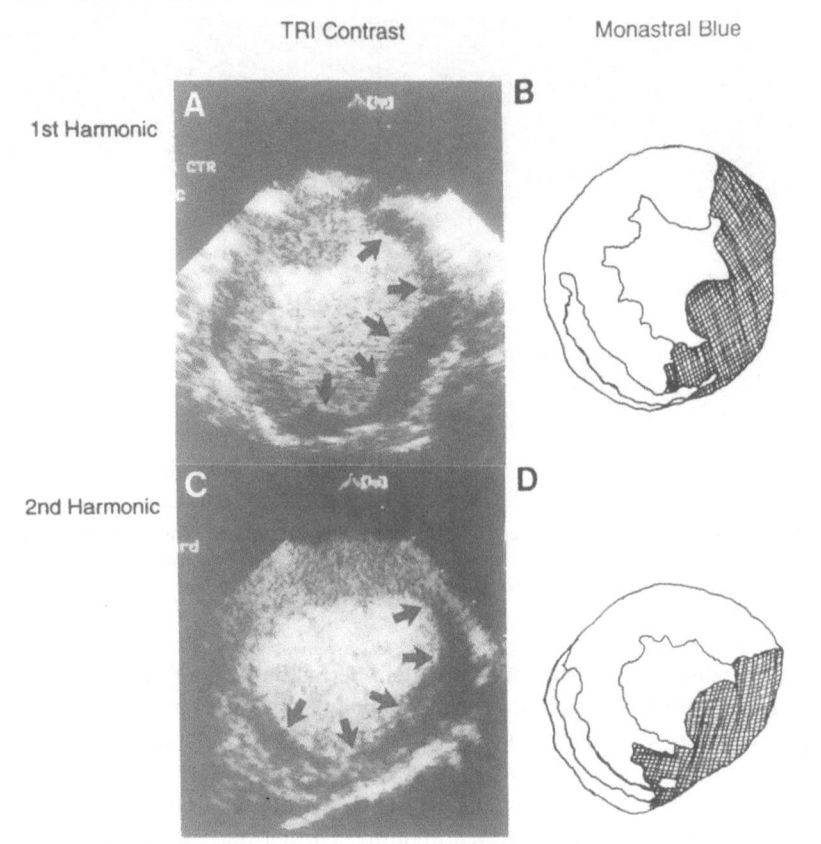

Figure 8. Demonstration in two different dogs of how the myocardial contrast produced with transient response imaging (TRI) allows the accurate delineation of area at risk during acute ischemia. Whether first or second harmonic, the contrast defect is easily visualized (arrows), and correlates well with post mortem Monastral blue staining (ischemic area is demonstrated in the cross-hatched region). [Reproduced from ref. 15 with permission.]

injection of contrast [17]. In 15 patients with normal resting wall motion, 0.0025–0.01 ml/kg PESDA was administered i.v. and the visual myocardial contrast produced with transient response imaging was compared with that following conventional 30 Hz frame rates (2.0 MHz transit, 4.0 MHz received frequency). None of the patients in this study had any significant side effects from these injections. Visually evident anterior or apical myocardial contrast was observed in 14 of the 15 patients when using transient response imaging, but in only seven patients with conventional imaging (Figure 9). Posterior or basal myocardial contrast was evident in 10 of the patients with transient response imaging, but only one patient with conventional harmonic imaging.

One important modification of the technique of transient response imaging

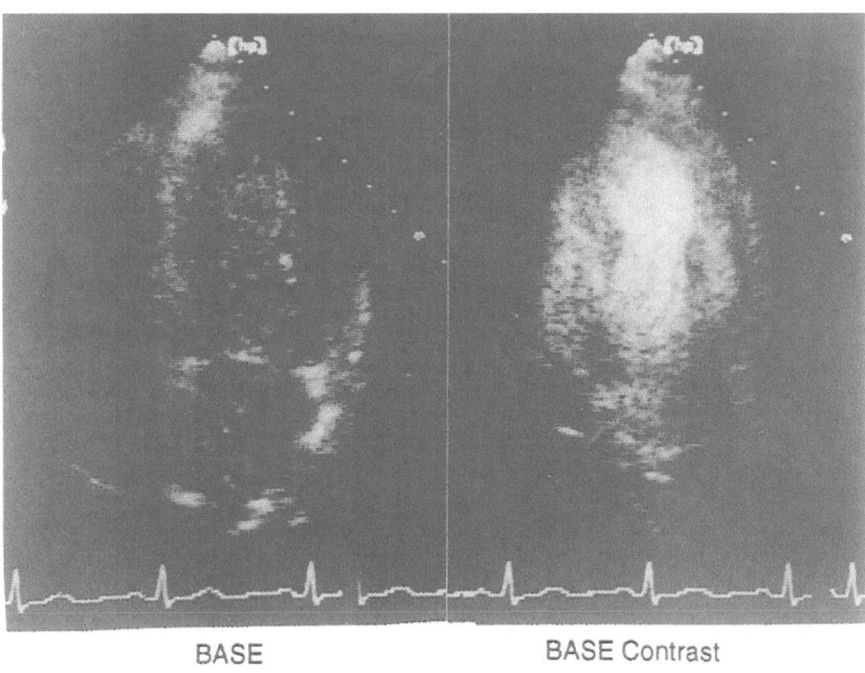

BASE   BASE Contrast

Figure 9. This apical four chamber view before (base) and after i.v. PESDA in one patient using transient response imaging (base contrast) demonstrates how myocardial contrast enhancement can occur with this technique in this normal human. [Reproduced from ref. 17 with permission.]

has further improved our ability to detect myocardial perfusion in multiple views following a single i.v. injection of contrast. Using a second harmonic transducer (1.8–2.0 MHz transmit frequency, 3.6–4.0 MHz receiving frequency), we demonstrated that the myocardial contrast produced by triggering frame rates to just once every 5–10 cardiac cycles was significantly better than once every cardiac cycle [18]. Using longer intervals between frame rates allowed the detection of visual myocardial contrast up to 120 s after i.v. injection of PESDA. Using this extended time interval between ultrasound pulses, visual myocardial contrast following intravenous PESDA could be observed in at least two parasternal or apical windows in 23 of 27 patients. In addition, a contrast defect could be visualized in nine of the patients (33%); six of these abnormalities were confirmed by simultaneous rest thallium images, and in two the contrast defect corresponded to the site of myocardial infarction on the resting 12 lead electrocardiogram. In one of these patients (Figure 10), i.v. PESDA was given following thrombolytic therapy for an acute anteroseptal myocardial infarction. Despite persistent anteroseptal and apical akinesis, it was evident that the apical myocardium filled with contrast following i.v. PESDA only when ultrasound

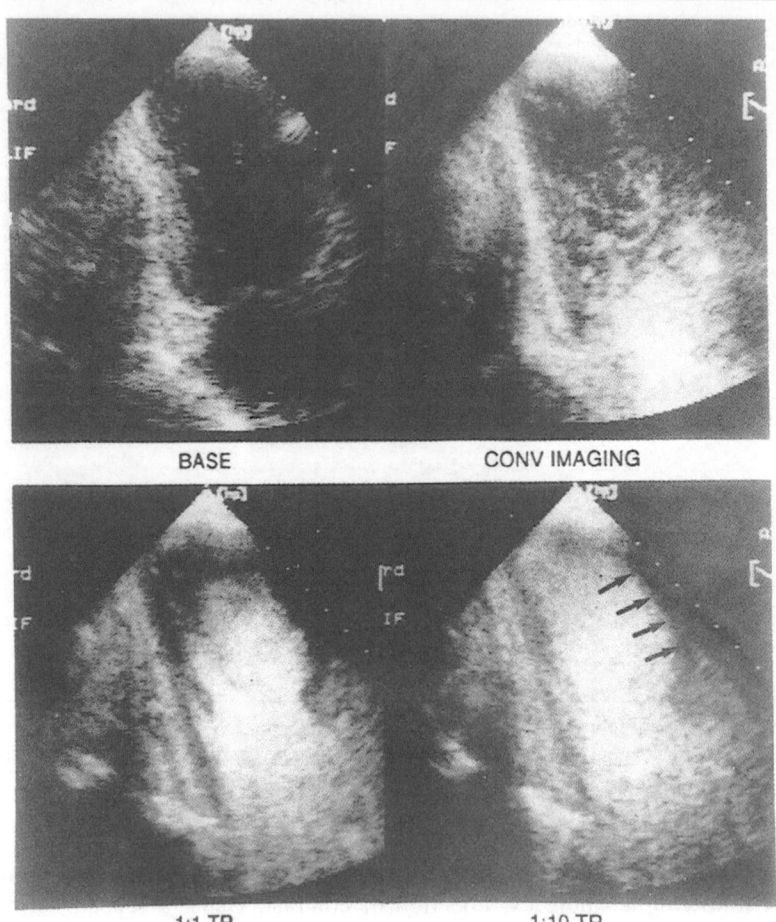

Figure 10. Echocardiographic images from the apical three-chamber views before i.v. injection of PESDA (base) in a patient with a recent anteroseptal infarction. The images were obtained during conventional harmonic imaging at 30 Hz (conv imaging) and subsequent transient response imaging where ultrasound transmission occurred once every cardiac cycle (1:1 TR) vs. once every 10 cardiac cycles (1:10 TR). Only with 1:10 TR could the residual contrast defect in the anteroseptal myocardium (arrows) be visualized, and essentially normal perfusion to the akinetic apex be observed.

frame rates triggered to once every 10 cardiac cycles, thus indicating reperfusion to this region.

We have subsequently applied transient response imaging to detect myocardial contrast defects in patients undergoing dipyridamole stress testing, and correlated the contrast appearance with the results obtained by dual isotope nuclear imaging performed the same day [19]. The visual myocardial contrast following administration of PESDA in doses of 0.00125–0.01 ml/kg i.v. using transient response imaging was compared with resting thallium uptake and stress Tc-99 sestamibi uptake. The myocardium was divided into six regions consisting of the basal, mid and apical segments of the anterior, anteroseptal, septal, inferior, posterior and a lateral walls for each patient. Myocardial contrast was judged by two independent reviewers to be either 0=no uptake (abnormal) or 1 to 2+ (normal contrast). Among the 296 regions compared there was agreement between nuclear uptake (normal vs. abnormal) and contrast appearance in 254 of these regions (86%). Furthermore, in the eight patients with contrast defects at rest or during dipyridamole stress, five had a nuclear defect in a similar region. In only four of the patients did ultrasound attenuation in the inferior, posterior and lateral regions prevent evaluation of contrast appearance.

TRANSIENT RESPONSE IMAGING: FUTURE DIRECTIONS

These preliminary results in humans emphasize the significance of the development of transient response imaging in making contrast echocardiography a non-invasive method of assessing myocardial perfusion. At present, we are limited by both the clinical lack of availability of the newer generation ultrasound contrast agents that will consistently produce myocardial contrast with transient response imaging, and the absence of commercially available harmonic imaging transducers. Based on our initial findings in humans, it appears that non-invasive assessment of myocardial perfusion with i.v. PESDA in a wide variety of patients is possible when using harmonic transient response imaging. It is imperative, therefore, that the manufacturers of both newer generation fluorocarbon-containing microbubbles and ultrasound scanners proceed as rapidly as possible to make these contrast agents and harmonic scanners available. This will permit the evaluation of transient response imaging in a larger patient population, and validate this technique as a non-invasive method of rapidly determining myocardial perfusion without the need of ionizing radiation.

REFERENCES

1. Carstensen EL, Duck FA, Meltzer RS, Schwarz KQ, Keller B. Bioeffects in echocardiography. Echocardiography 1992;9:605–23.
2. Porter TR, Xie F, Kricsfeld A. Myocardial ultrasound contrast with intravenous perfluoro-propane-enhanced sonicated dextrose albumin: initial clinical experience in humans. J Am Coll Cardiol 1995;39A.
3. Villanueva F, Glasheen WP, Sklenar J, Jayaweera AR, Kaul S. Successful and reproducible myocardial opacification during two-dimensional echocardiography from right heart injection of contrast. Circulation 1992;85:1557–64.
4. Villaneuva F, Glasheen WP, Sklenar J, Kaul S. Assessment of risk area coronary occlusion and infarct size after reperfusion with myocardial contrast echocardiography using left and right atrial injections of contrast. Circulation 1993;88:596–604.

5. Schrope BA, Newhouse VL. Second harmonic ultrasound blood perfusion measurement. Ultrasound Med Biol 1993;19:567–79.
6. De Jong N, Ten Cate FJ, Lancee CT, Roelandt JR, Bom N. Principles and recent developments in ultrasound contrast agents. Ultrasonics 1991;29:324–30.
7. Mor-Avi V, Robinson K, Shroff S, Lang RM. Stability of albunex microspheres under ultrasonic irradiation: an in vitro study. J Am Soc Echocardiogr 1994;7:S29.
8. Vandenberg BF, Melton HE. Acoustic liability of albumin microspheres. J Am Soc Echocardiogr 1994;7:582–9.
9. Wray RA, Zoghbi WA, Quinones MA, Cheirif J. Contrast echocardiography: relation of acoustic power and time gain compensation to contrast intensity duration. J Am Soc Echocardiogr 1992;4:286.
10. Holt RG, Crum LA. Acoustically forced oscillations of air bubbles in water: experimental results. J Acoust Soc Am 1992;91:1924–32.
11. Flynn HG, Church CC. Erratum: transient pulsations of small gas bubbles in water. [J Acoust Soc Am 1988;84:985–98] J Acoust Soc Am 1988;84:1863–76.
12. Crum LA, Roy RA, Dinno MA, Church CC. Acoustic cavitation produced by microsecond pulses of ultrasound: a discussion of some selected results. J Acoust Soc Am 1992;91:1113–9.
13. Roy RA, Church CC, Calabrese A. Cavitation produced by short pulses of ultrasound. In: Hamilton AF, Blackstock DT, editors. Frontiers of Nonlinear Acoustics: Proceedings of 12th International Symposium of Nonlinear Acoustics. Elsevier, London. 1990;476–81.
14. Porter TR, Xie F, Li S, D'Sa A, Rafter P. Increased ultrasound contrast and decreased microbubble destruction rates using triggered ultrasound imaging. J Am Soc Echocardiogr 1996;9: 599–605.
15. Porter T, Xie F. Transient myocardial contrast after initial exposure to diagnostic ultrasound pressures with minute doses of intravenously injected microbubbles: Demonstration and potential mechanisms. Circulation 1995;92:2391–5.
16. Porter T, Xie F, Kilzer K, Deligonul U. Transient response imaging with intravenous perfluorocarbon-exposed sonicated dextrose albumin detects the spatial extent of ischemia during dobutamine stress echocardiography. J Am Coll Cardiol 1996 (Abstract).
17. Porter TR, Xie F, Kricsfeld D, Armbruster RW. Improved myocardial contrast with second harmonic transient ultrasound response imaging in humans using intravenous perfluorocarbon-exposed sonicated dextrose albumin. J Am Coll Cardiol 1996;27:1497–501.
18. Porter T, Xie F. Enhanced myocardial contrast over one minute after intravenous injection of perfluorocarbon exposed sonicated dextrose albumin using prolonged suspension of ultrasound transmission. J Am Coll Cardiol 1996 (Abstract).
19. Porter T, Li S, Kricsfeld D, Armbruster R. Detection of myocardial perfusion in multiple echocardiographic windows with one intravenous injection of microbubbles using transient response second harmonic imaging. J Am Coll Cardiol 1997 (in press).

30. Three-dimensional echocardiographic assessment of myocardial perfusion abnormalities

JIEFEN YAO, NAVROZ D. MASANI & NATESA G. PANDIAN

INTRODUCTION

During the last decade a number of major advances in ultrasound technology and contrast agent manufacture have brought the technique of contrast echocardiography to the threshold of clinical application. Intravenous injections of currently available contrast agents aid in the diagnosis of intracardiac shunts, enhance Doppler signals and opacify the left ventricle. Work is in progress to refine the technique so that myocardial perfusion can be assessed by i.v. contrast administration. A considerable body of experimental work in animal models has shown that contrast agents act as tracers of blood flow and enhance myocardial ultrasound signals and that the presence or absence of perfusion abnormalities can be identified [1–9]. It has been shown that in the presence of coronary occlusion, the area supplied by the occluded coronary artery exhibits lack of contrast enhancement and the 'area at risk' of infarction can be delineated and quantified [10–13]. Contrast echocardiography has also been used to identify the collateral blood supply to ischemic myocardium [14–16]. In the setting of coronary stenosis hypoperfused areas can be identified with contrast administration at rest or in conjunction with a vasodilator [17]. Most of these investigations have been performed in animal models with contrast administration into the left-sided chambers or within the coronary arteries [9, 18]. In humans, contrast echocardiography has been undertaken primarily in invasive laboratories using left heart injections. Studies utilizing recent advances in ultrasound instrumentation and microbubble technology indicate that newer generation contrast agents, when combined with innovative imaging modalities such as transient response imaging, harmonic imaging and power Doppler, can opacify perfused myocardium and can detect hypoperfusion following i.v. injections [19–20] (Figures 1, 2). As contrast echocardiography to assess perfusion abnormalities approaches clinical use, it is also important to explore more versatile and comprehensive means of displaying and quantifying perfused and non-perfused myocardial regions. Current developments in three-dimensional echocardiography suggest that it is

465

N. C. Nanda, R. Schlief and B. B. Goldberg (eds.), Advances in Echo Imaging Using Contrast Enhancement, Second Edition, 465–477.
© 1997 *Kluwer Academic Publishers.*

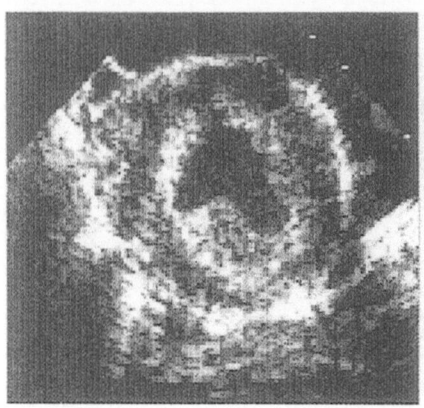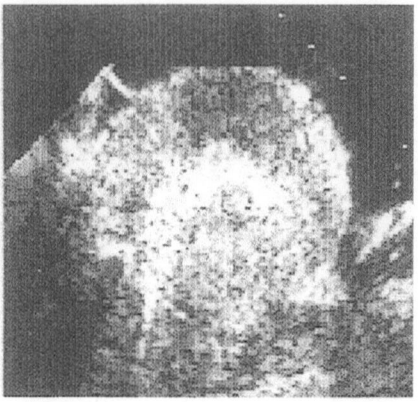

Figure 1. Experimental myocardial contrast echocardiographic study in a open-chest canine model. Left: a two-dimensional short-axis view of the left ventricle at baseline. Right: after i.v. injection of a contrast agent the LV cavity is filled with contrast and the myocardium is opacified homogeneously.

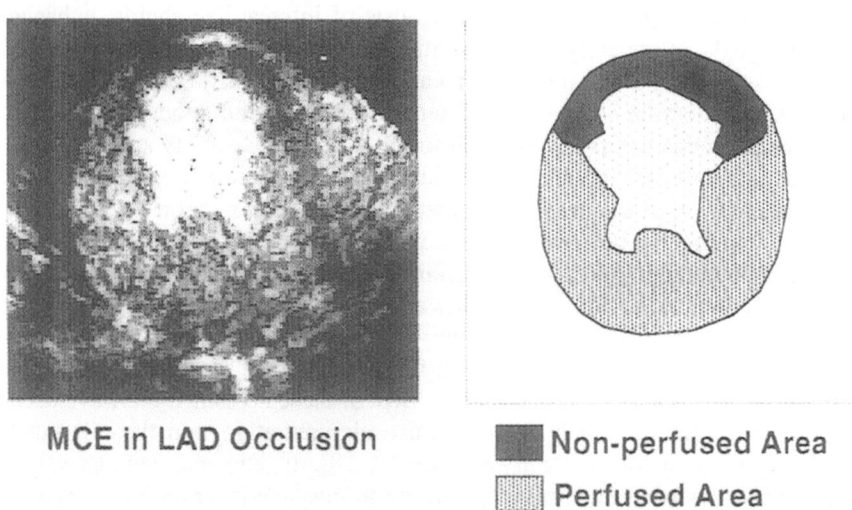

MCE in LAD Occlusion

Non-perfused Area
Perfused Area

Figure 2. Myocardial contrast echocardiography in the same experimental study as Figure 1 after occlusion of the left anterior descending coronary. An anterior perfusion defect can be seen in the left image, as shown in the schematic on the right.

possible to delineate regions with perfusion abnormalities in a more anatomically appropriate three-dimensional format.

CURRENT APPROACHES TO THE DELINEATION OF ISCHEMIC MYOCARDIAL REGIONS

Previous studies in animals and humans employed conventional two-dimensional echocardiography to image the left ventricle and assess regions with perfusion defects. In almost all cases, the left ventricle was imaged in short-axis planes [9, 15, 18]. While short-axis views can provide up to five transverse images of the ventricle, a major portion of the myocardium, particularly the apex, is not interrogated. Commonly, the apex of the left ventricle is affected by perfusion abnormalities related to disease in the left anterior descending coronary artery and, not infrequently, the posterior descending artery. Thus, contrast echocardiography limited to short-axis imaging does not provide a comprehensive assessment of all coronary territories and may result in an inadequate examination of myocardial perfusion. In studies that employed imaging from the apex, only one or two views were generally obtained [17, 21]. Since myocardial perfusion territories have a three-dimensional extent, visualizing the myocardium in selected two-dimensional views could hinder optimal visualization, delineation and quantification of perfusion defects. Three-dimensional echocardiography could serve to overcome these limitations associated with two-dimensional echocardiography.

POTENTIAL OF THREE-DIMENSIONAL ECHOCARDIOGRAPHY

Volume-rendered three-dimensional echocardiographic reconstruction of the heart has now become a clinical reality. From sequentially acquired two-dimensional images, gated to ECG and respiration, dynamic three-dimensional reconstruction can be used to display intracardiac anatomy clearly, in projections hitherto unavailable. Three-dimensional echocardiography, using both transthoracic and transesophageal windows, has allowed visualization of intracardiac structures and great vessels and delineation of their pathology in a comprehensive, realistic manner [22]. Abnormalities of cardiac chamber size, geometry and function, lesions in valves, intracardiac septa and aortic disorders can all be visualized in orientations that provide information of diagnostic and therapeutic value [23–26]. The accuracy of three-dimensional echocardiography in the quantification of left and right ventricular chamber volumes and lesions such as atrial and ventricular septal defects and cardiac masses has been demonstrated [27, 28]. Emerging clinical experience indicates the strong potential of three-dimensional echocardiography in qualitative and quantitative diagnostic appraisal of various cardiac problems. Combined with ultrasound contrast agents, three-dimensional echocardiography could be valuable in the assessment of myocardial perfusion

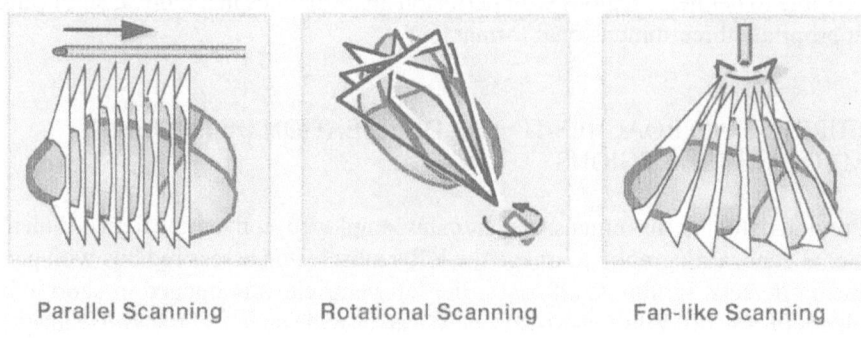

Parallel Scanning   Rotational Scanning   Fan-like Scanning

Figure 3. Schematics of three modes of data acquisition for volume-rendered three-dimensional echocardiography.

abnormalities. Preliminary experience in animal models indicates its potential to improve the display and quantitative measurement of myocardial perfusion defects [29–31]. The methodology of three-dimensional echocardiographic data acquisition and processing, display and quantification of contrast-enhanced and contrast-devoid myocardial regions will be discussed in the following section.

IMAGE ACQUISITION FOR THREE-DIMENSIONAL ECHOCARDIOGRAPHY

Volume-rendered three-dimensional echocardiography employs sequential data acquisition obtained from computer-oriented transducer movements in a parallel, rotational or fan-like manner (Figure 3). In parallel scanning a series of equidistant short axis slices of the left ventricle are obtained with the transducer moved linearly along its longitudinal axis [32]. In clinical practice, as well as in experimental studies, rotational and fan-like scanning are more frequently used [33, 34]. The left ventricle can be imaged from the apical window with the transducer rotated from 0° through 180° (rotational scanning) in predefined steps (usually 1–3°) controlled by a computer steering logic system or from the parasternal window with the transducer moved in a 90° fan-like arc 'sweeping' through the left ventricle (fan-like imaging). During data acquisition, the subjects or patients being studied are required to stay still. A carriage device for the transducer is held immobile either by the examiner or using a mechanical arm. ECG and respiration gating are needed in in vivo studies for registration of the acquired images according to their spatial and temporal locations and for minimizing artifacts caused by respiratory movement. The two-dimensional images acquired are stored in a three-dimensional image processing computer for post-processing and data analysis. The acquisition process takes about 1–3 min.

DATA PROCESSING FOR THREE-DIMENSIONAL RECONSTRUCTION

Following data acquisition, the collected two-dimensional images are digitized and reformatted into pixels. All the acquired images are realigned according to their spatial and temporal relationships. Spaces between adjacent slices are interpolated by the computer and thus pixel-based two-dimensional images are turned into a voxel-based three-dimensional data set arranged along Cartesian coordinates.

VOLUME-RENDERED THREE-DIMENSIONAL DISPLAY PROJECTIONS

The volume-rendered three-dimensional data set can be displayed in various ways. Secondary two-dimensional images can be derived from the three-dimensional data set using anyplane (free-orientation of the image plane) or paraplane (with image planes parallel to each other) cuts, including novel two-dimensional images which cannot be obtained using conventional echocardiography [35]. Three-dimensional reconstruction adds the perception of depth to the display resulting in a more realistic, anatomically correct rendering. Dynamic three-dimensional reconstructions of cardiac structures can be displayed in projections simulating the surgeon's view, or from new anatomical projections, as if cardiac walls have been 'electronically dissected' [36–39]. In this way, conventional cut planes such as the apical two-chamber view can be used to view the interventricular septum, short-axis imaging of the mitral or tricuspid valve can be performed from above or below, and novel cut planes can be displayed for better evaluation of, for example, the atrial septum. Shape, size and location of cardiac abnormalities such as congenital defects, intracardiac masses and abnormal blood flows (intracardiac shunts or regurgitant jets) can be appreciated from multiple views looking at the abnormalities either en face or with different angles.

QUANTIFICATION OF PERFUSION DEFECTS

Volume-rendered three-dimensional echocardiography has been well validated for quantification of ventricular myocardial mass or chamber volume. The mass of specific regions of the myocardium such as perfusion defect mass or infarct mass can be quantified in a similar way. For measurement of hypoperfused myocardium, the three-dimensional data set of the left ventricle, acquired during contrast injection and myocardial opacification, is viewed in parallel equidistant short-axis slices. By visual or videodensitometric evaluation, regions of hypoperfused myocardium, which show less contrast opacification compared with normally perfused regions, are delineated. These regions are contoured in each slice and their area computed by the computer. The product of this area and the

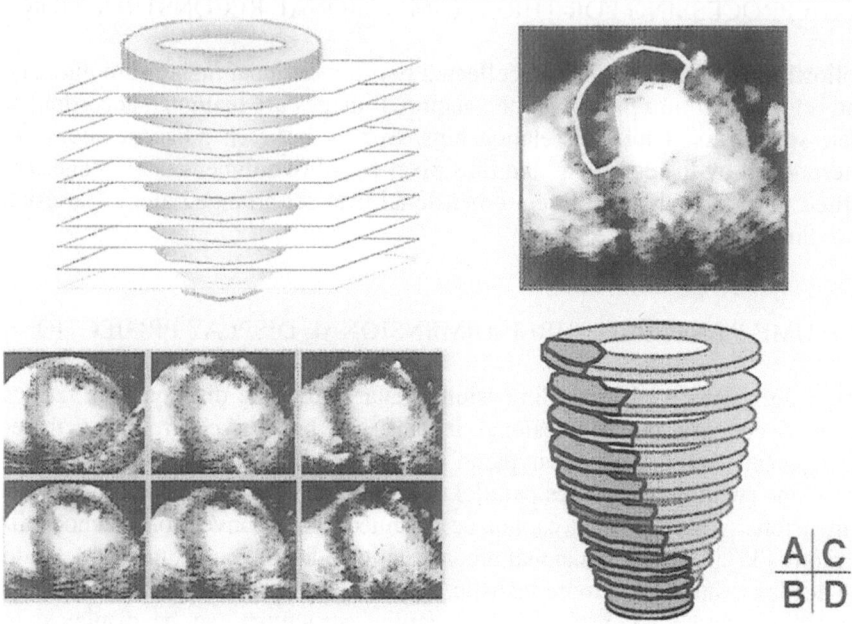

Figure 4. Schematics of three-dimensional echocardiographic quantification of regional myocardial perfusion defect. (A) The three-dimensional data set of the left ventricle is transversely dissected into parallel equidistant slices. (B) Visualization of the myocardial perfusion on each short axis slice. (C) Contouring the perfusion defect area on each slice. (D) Volume of the contoured three-dimensional perfusion defect myocardial region calculated by the computer.

known slice thickness yields the volume of the perfusion defect of each slice and their summation gives its total volume (Figure 4). The mass of this region can be calculated by multiplying its volume by the specific density of myocardium (1.04 g/ml). The total mass of the left ventricle is measured in a similar manner by contouring the whole myocardium in each short-axis slice and the perfusion defect can be expressed as a percentage of the left ventricle.

CURRENT EXPERIENCE WITH THREE-DIMENSIONAL CONTRAST ECHOCARDIOGRAPHY

Our experience with experimental canine models has demonstrated the feasibility of three-dimensional reconstruction, delineation and quantification of ischemic and infarcted myocardium associated with transient and permanent coronary occlusion [29, 30]. Two-dimensional cut-sections in anyplane and paraplane orientations clearly depicted normally perfused contrast enhanced areas and non-perfused regions with lack of contrast enhancement (Figure 5). Following

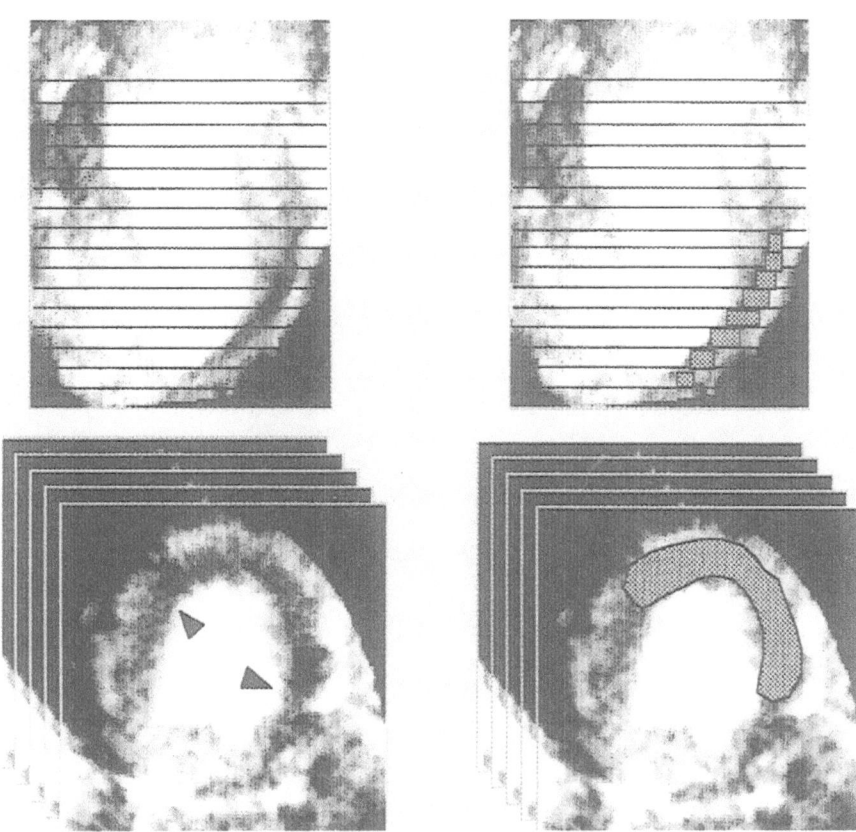

Figure 5. Three-dimensional contrast echocardiographic quantification of myocardial perfusion deficit region. The upper left panel shows the reference image in a longitudinal view. The longitudinal extent of the perfusion defect is shown clearly. Lower left: the left ventricle is electronically segmented into multiple equidistant transverse slices. Regions of perfused and non-perfused myocardium are well delineated in short-axis images (arrows). Lower right: the left ventricular myocardial region with perfusion defect is contoured and masked on each transverse slice. The area of this region is calculated automatically by the computer. Upper right: the longitudinal extent of the mask is displayed in the reference image. With the area of non-perfused region on each slice and the known distance between slices, the volume of this region can be derived. The absolute mass of non-perfused region can be calculated from the volume and the specific density of the myocardium.

contouring of these regions in paraplane slices, they could be extracted and displayed in three-dimensions and viewed from various angles. Such visualization helped provide better perception of the size, topography and geometry of ischemic and infarcted territories (Figures 6, 7). They corresponded well with the sites of coronary occlusion and, as expected, distal coronary occlusions resulted in smaller zones of ischemia and proximal occlusion in larger territories. We

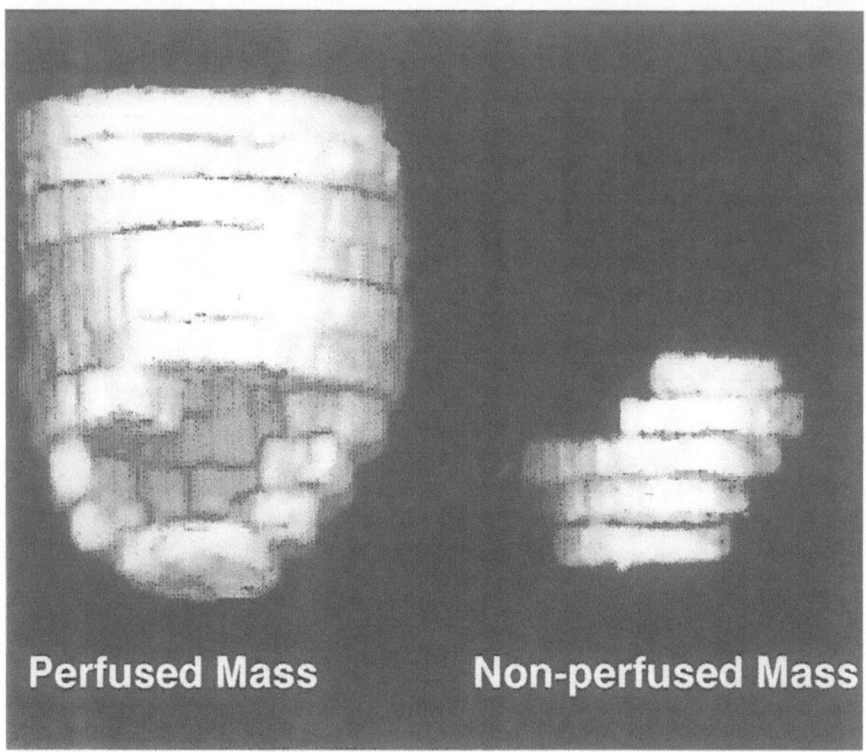

Figure 6. Posterior views of extracted three-dimensional perfused (left) and non-perfused (right) left ventricular myocardial masses in an experimental study of mid-posterior descending coronary artery occlusion. The spatial extent and location of the defect can be appreciated readily.

were able to quantify the mass of the non-opacified myocardium and compute the proportion of infarcted left ventricle. The mass of perfusion defect regions showed an excellent correlation with infarct size determined by triphenyl tetrazolium chloride staining ($r = 0.93$). The extent of myocardium devoid of contrast enhancement, expressed as percentage of left ventricle involved, also corresponded well with similar anatomical measurements ($r = 0.90$) [30]. Volume-rendered three-dimensional projections, as well as anyplane and paraplane sections derived from the three-dimensional data set, depicted myocardial segments that exhibited contractile dysfunction. The dysfunctional regions of the myocardium were analyzed in a manner similar to the extraction and quantification of perfusion deficit regions. The mass of dysfunctional myocardium, measured in grams, also corresponded well with anatomical infarct mass ($r = 0.92$) and the percentage of left ventricle involved in dysfunction correlated well with percentage of left ventricle involved in infarction ($r = 0.93$) [40]. Our observations suggest that when an effective transpulmonary contrast agent becomes available

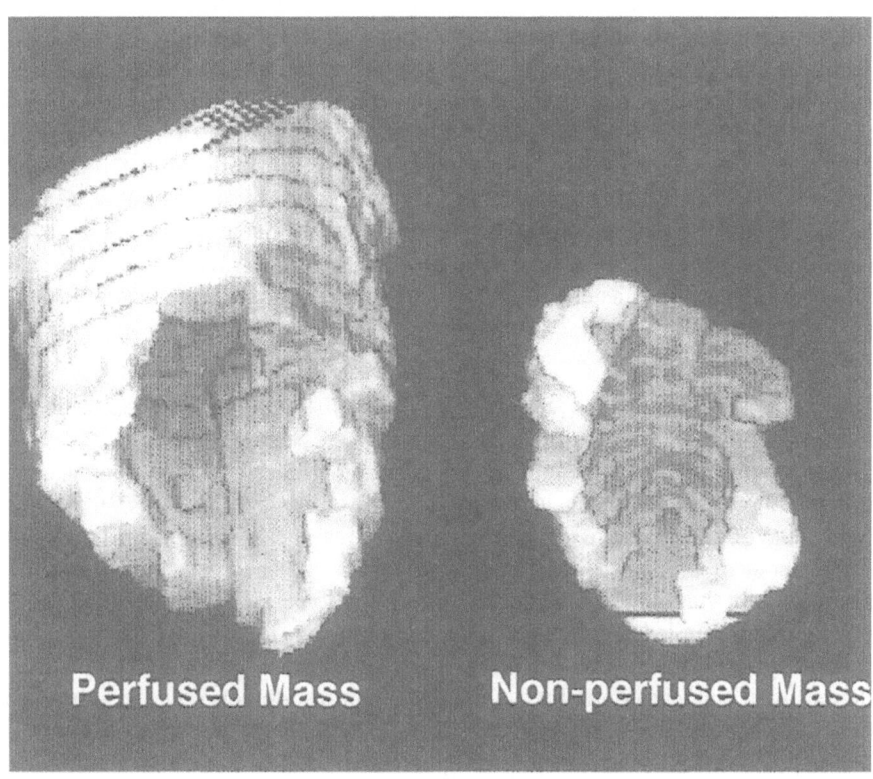

Figure 7. Extracted regions of perfused (left) and non-perfused (right) left ventricular myocardium are displayed in a three-dimensional format (anterior projection) from an experimental contrast study in a dog with proximal left anterior descending coronary occlusion. The defect is larger than that in Figure 6 and involves more of the apex.

for use in humans, it should be possible not only to visualize perfusion abnormalities but also to display them in three-dimensions and quantify the mass of ischemic zones.

CURRENT PROBLEMS OF THREE-DIMENSIONAL MYOCARDIAL CONTRAST ECHOCARDIOGRAPHY

Contrast echocardiography with three-dimensional reconstruction is subject to the same limitations as two-dimensional contrast echocardiography. Intense opacification of the left ventricular cavity results in attenuation of signals from posterior cardiac structures. Myocardial opacification is affected by non-linear signal processing and unfavorable signal-to-noise characteristics of conventional ultrasound systems. Improvements in these aspects of two-dimensional imaging will be directly applicable to three-dimensional echocardiography.

Three-dimensional echocardiography also presents some practical difficulties. In adults with suboptimal acoustic windows, it may be difficult to encompass the complete left ventricle and dropout may occur at the apex or anterior wall. Data acquisition takes 1–3 min and a steady state contrast effect must, therefore, be achieved; we have overcome this in animal studies by using multiple bolus doses of transpulmonary contrast agents. At present, quantitation of perfusion defects is laborious; the process of data acquisition, processing and analysis lasts 20–60 min, although faster processors and semi-automated analysis programs are expected to address these problems.

CLINICAL POTENTIAL AND FUTURE PERSPECTIVES

The accurate depiction and quantitation of perfusion defects by three-dimensional echocardiography could be valuable in a number of clinical situations. In patients with acute coronary occlusion, the size of the area at risk is the most important determinant of the subsequent amount of myocardial necrosis which, in turn, has important therapeutic and prognostic implications. Three-dimensional echocardiography coupled to contrast echocardiography has great potential to quantify accurately the area at risk and infarct size. The precision with which three-dimensional echocardiography can delineate contrast defects could also be used to determine the success of reperfusion and extent of salvaged myocardium.

The potential for contrast echocardiography to distinguish viable from infarcted myocardium could be refined by three-dimensional reconstruction: decisions regarding revascularization are based upon the presence, location and extent of viable myocardium and, in the future, three-dimensional echocardiography should be able to provide this information in more detail than two-dimensional echocardiography.

We are familiar with the ability of three-dimensional echocardiography to calculate left ventricular volumes; semi-automated processing incorporating border detection is an evolution of this technique. Combining this with left ventricular cavity opacification using contrast agents promises to facilitate the precise measurement of left ventricular volumes, even in patients with suboptimal acoustic windows.

In the future, we may be able to perform densitometric processing of the entire digital three-dimensional data set so that regional myocardial blood volume and flow can be assessed throughout the left ventricle.

CONCLUSION

Three-dimensional echocardiography, in combination with i.v. contrast agents, has great potential in delineating and quantifying regional myocardial perfusion defects. For the clinical assessment of myocardial 'area at risk' or infarct size and for the evaluation of success of reperfusion or revascularization, three-dimensional contrast echocardiography promises to enhance our ability to display and measure accurately perfusion abnormalities.

REFERENCES

1. Tei C, Sakamaki T, Shah PM et al. Myocardial contrast echocardiography: a reproductive technique of myocardial opacification for identifying regional perfusion defects. Circulation 1983; 67:585–93.
2. Kaul S, Pandian NG, Okada RD, Pohost GM, Weyman AE. Contrast echocardiography in acute myocardial ischemia: I. In-vivo determination of total left ventricular 'area at risk'. J Am Coll Cardiol 1984;4:1272–82.
3. Armstrong WF, Kinney EL, Mueller TM, Tickner EG, Dillon JC, Feigenbaum H. Assessment of myocardial perfusion abnormalities with contrast enhanced two-dimensional echocardiography. Circulation 1982;66:166–73.
4. Feinstein SB, TenCate FJ, Zwehl W et al. Two-dimensional echocardiography. I. In vitro development and quantitative analysis of echo contrast agents. J Am Coll Cardiol 1984;3: 14–20.
5. Kaul S, Kelly P, Oliner JD, Glasheen WP, Keller MW, Watson DD. Assessment of regional myocardial blood flow in-vivo with myocardial contrast two-dimensional echocardiography. J Am Coll Cardiol 1989;13:468–82.
6. Villanueva FS, Glasheen WP, Sklenar J, Jayaweera AR, Kaul S. Successful and reproducible myocardial opacification during two-dimensional echocardiography from right heart injection of contrast. Circulation 1992;85:1557–64.
7. Kaul S, Pandian NG, Gillam LD, Newell JB, Okada RD, Weyman AE. Contrast echocardiography in acute myocardial ischemia: III. An in-vivo comparison of the 'area at risk' for necrosis. J Am Coll Cardiol 1986;7:383–92.
8. Kaul S, Glasheen W, Ruddy TD, Pandian NG, Weyman AE, Okada RD. The importance of defining left ventricular area at risk in-vivo during acute myocardial infarction: an experimental evaluation with myocardial contrast two-dimensional echocardiography. Circulation 1987;75:1249–60.
9. Sakamaki T, Tei C, Meerbaum S et al. Verification of myocardial contrast two-dimensional echocardiographic assessment of perfusion defects in ischemic myocardium. J Am Coll Cardiol 1984;3:34–8.
10. Desir RM, Cheirif J, Bolli R, Zoghbi WA, Hoyt BD, Quiñones MA. Assessment of regional myocardial perfusion with myocardial contrast echocardiography in a canine model of varying degrees of coronary stenosis. Am Heart J 1994;127:56–63.
11. Jayaweera AR, Matthew TL, Sklenar J, Spotnitz WD, Watson DD, Kaul S. Method for the quantitation of myocardial perfusion during myocardial contrast two-dimensional echocardiography. J Am Soc Echocardiogr 1990;3:91–8.
12. Villanueva FS, Glasheen WP, Sklenar J, Kaul S. Assessment of risk area during coronary occlusion and infarct size after reperfusion with myocardial contrast echocardiography using left and right atrial injections of contrast. Circulation 1993;88:596–604.
13. Shapiro JR, Reisner SA, Amico AF, Kelly PF, Meltzer RS. Reproducibility of quantitative myocardial contrast echocardiography. J Am Coll Cardiol 1990;15:602–9.
14. Grill HP, Brinker JA, Taube JC et al. Contrast echocardiography mapping of collateralized myocardium in humans before and after coronary angioplasty. J Am Coll Cardiol 1990;16: 1594–600.
15. Sabia PJ, Powers ER, Jayaweera AR, Ragosta M, Kaul S. Functional significance of collateral

blood flow in patients with recent acute myocardial infarction. A study using myocardial contrast echocardiography. Circulation 1992;85:2080–9.

16. Cheirif J, Narkiewicz-Jodko JB, Hawkins HK, Bravenec JS, Quiñones MA, Mickelson JK. Myocardial contrast echocardiography: relation of collateral perfusion to extent of injury and severity of contractile dysfunction in a canine model of coronary thrombosis and reperfusion. J Am Coll Cardiol 1995;26:537–46.

17. Agati L, Voci P, Bilotta F et al. Dipyridamole myocardial contrast echocardiography in patients with single-vessel coronary artery disease: perfusion, anatomic, and functional correlates. Am Heart J 1994;128:28–35.

18. Skyba DM, Jayaweera AR, Goodman NC, Ismail S, Camarano G, Kaul S. Quantification of myocardial perfusion with myocardial contrast echocardiography during left atrial injection of contrast. Implications for venous injection. Circulation 1994;90:1513–21.

19. Schrope BA, Newhouse VL. Second harmonic ultrasonic blood perfusion measurement. Ultrasound Med Biol 1993;19:567–79.

20. Porter TR, Xie F. Transient myocardial contrast after initial exposure to diagnostic ultrasound pressures with minute doses of intravenously injected microbubbles. Demonstration and potential mechanisms. Circulation 1995;92:2391–5.

21. Schlief R, Staks T, Mahler M, Rufer M, Fritzsch T, Seifert W. Successful opacification on the left heart chambers on echocardiographic examination after intravenous injection of a new saccharide based contrast agent. Echocardiography 1990;7:61–4.

22. Pandian NG, Roelandt J, Nanda NC et al. Dynamic three-dimensional echocardiography: methods and clinical potential. Echocardiography 1994;11:237–59.

23. Pandian NG, Nanda NC, Schwartz SL et al. Three-dimensional and four-dimensional trans-esophageal echocardiographic imaging of the heart and aorta in humans using a computed tomographic imaging probe. Echocardiography 1992;9:677–87.

24. Roelandt JRTC, Cate FJ, Vletter WB, Taams MA. Ultrasonic dynamic three-dimensional visualization of the heart with a multiplane transesophageal imaging transducer. J Am Soc Echocardiogr 1994;7:217–29.

25. Vogel M, Losch S. Dynamic three-dimensional echocardiography with a computed tomography imaging probe: initial clinical experience with transthoracic application in infants and children with congenital heart defects. Br Heart J 1994;71:462–7.

26. Marx G, Fulton DR, Pandian NG et al. Delineation of site, relative size and dynamic geometry of atrial septal defects by real-time three-dimensional echocardiography. J Am Coll Cardiol 1995;25:482–90.

27. Jiang L, Siu SC, Handschumacher MD et al. Three-dimensional echocardiography: in vivo validation for right ventricular volume and function. Circulation 1994;89:2342–50.

28. Siu SC, Rivera JM, Guerrero JL et al. Three-dimensional echocardiography: in vivo validation for left ventricular volume and function. Circulation 1993;88:17–23.

29. Delabays A, Cao QL, Sugeng L et al. Is there a critical amount of ischemic myocardium required to produce a detectable wall motion abnormality? A quantitative 3-dimensional echocardiographic contrast study. Circulation 1995;92:I-263.

30. Delabays A, Cao QL, Yao J et al. Contrast three-dimensional echocardiography in acute myocardial infarction: 3-D reconstruction of perfusion defects yields accurate estimate of infarct mass and extent (abstract). J Am Coll Cardiol 1996;27(Suppl A):63A.

31. Kasprzak JD, Ten Cate FJ, Sutton A, Vletter WB, Meegen JV. Dynamic three dimensional myocardial perfusion echocardiography using a deposit contrast agent. Circulation 1995;92: I-261.

32. Fulton DR, Marx GR, Pandian NG et al. Dynamic three-dimensional echocardiographic imaging of congenital heart defects in infants and children by computer-controlled tomographic parallel slicing using a single integrated ultrasound instrument. Echocardiography 1994;11:155–64.

33. Delabays A, Pandian NG, Cao QL et al. Transthoracic real-time three-dimensional echocardiography using a fan-like scanning approach for data acquisition: methods, strength, problems, and initial clinical experience. Echocardiography 1995;12:49–59.

34. Ludomirsky A, Vermilion R, Nesser J et al. Transthoracic real-time three-dimensional echo-cardiography using the rotational scanning approach for data acquisition. Echocardiography 1994;11:599–606.
35. Roelandt J, Salustri A, Vletter W, Nosir Y, Bruining N. Precordial multiplane echocardiography for dynamic anyplane, paraplane and three-dimensional imaging of the heart. Thoraxcentre J 1994;6:4–13.
36. Schwartz SL, Cao QL, Azevedo J, Pandian NG. Simulation of intraoperative visualization of cardiac structures and study of dynamic surgical anatomy with real-time three-dimensional echocardiography. Am J Cardiol 1994;73:501–7.
37. Salustri A, Spitaels S, McGhie J, Vletter W, Roelandt JRTC. Transthoracic three-dimensional echocardiography in adult patients with congenital heart disease. J Am Coll Cardiol 1995;26: 759–67.
38. Nanda NC, Roychoudhury D, Chung SM, Kim KS, Ostlund V, Klas B. Quantitative assessment of normal and stenotic aortic valve using transesophageal three-dimensional echocardiography. Echocardiography 1994;11:617–25.
39. Cao QL, Pandian NG, Azevedo J et al. Enhanced comprehension of dynamic cardiovascular anatomy by three-dimensional echocardiography with the use of mixed shading techniques. Echocardiography 1994;11:627–33.
40. Yao J, Cao QL, Delabays A et al. How well does 3-dimensional echocardiographic quantification of dysfunctional left ventricular mass reflect actual anatomic infarct mass? Experimental studies. J Am Coll Cardiol 1996;27(Suppl A):49A.

31. Three-dimensional visualization of myocardial perfusion using contrast echocardiography

ARIC A. AIAZIAN, FOLKERT J. TEN CATE &
JOS R. T. C. ROELANDT

INTRODUCTION

The need for an imaging technique that allows visualization of myocardial per-fusion during diagnostic and interventional procedures has been increasing in proportion to the number of these procedures performed every year [1, 2]. Myo-cardial contrast echocardiography (MCE) is a promising technique for assess-ment of myocardial perfusion, especially in patients undergoing intracoronary catheterization [3–5] or bypass surgery [6]. The data provided by MCE are mainly limited by the acquisition of only one cross-sectional view of the left ventricle and the inability to quantitate myocardial blood flow with the presently available echocontrast agents. These limitations do not allow the evaluation of the whole left ventricle and the reconstruction of a three-dimensional image of the myocardial perfusion similar to those obtained with PET and SPECT. Three-dimensional reconstruction using echocardiographic images obtained by a rotating ultrasound transducer has been used to visualize and quantitate the myocardial area at risk during MCE [7]. However, as a result of the relatively long image acquisition time this approach would require injection of a deposit echocontrast agent which stays in the myocardium for a long time. Currently, such an agent is not yet available for clinical use.

We developed a modified technique of MCE using sonicated contrast agents with subsequent reconstruction of a three-dimensional (3D) myocardial perfusion map [8]. The technique has practical advantages and can be readily used in the clinical setting.

TECHNIQUE OF IMAGE ACQUISITION

We recently introduced a technique using multiplane MCE following several injections of an echocontrast agent into the right coronary artery (RCA) and the left coronary artery (LCA) during cardiac catheterization. Digital analysis of the

479

N. C. Nanda, R. Schlief and B. B. Goldberg (eds.), Advances in Echo Imaging Using Contrast Enhancement,
Second Edition, 479–489.
© 1997 *Kluwer Academic Publishers.*

images allows reconstruction of a three-dimensional perfusion map containing 16 standard myocardial segments. A similar procedure has been used for the echo evaluation [8, 9]. The images obtained during MCE are recorded with the ultrasound transducer in the apical position and the patient in a steady position. In our experience about 5% of the patients are not suitable candidates for multi-plane MCE and 3D reconstruction of the perfusion map due to a poor ultrasound window. It is important to use constant machine parameters throughout the study. Therefore it is important to check the quality of the ultrasound image quality in a supine position prior to cardiac catheterization.

The acquisition of the images is started before echocontrast injection (baseline recording). Three apical views of the heart (four-chamber, two-chamber and long-axis views) are recorded on videotape or on hard disk of the image analysis system in digital format (Axle, the Netherlands), using the ECG triggered mode over a period of 1 min. The same views are then recorded after injection of echocontrast into the RCA and LCA. For each view the recording is continued for approximately 1 min to allow complete wash-out of the contrast agent. A total of six injections of echocontrast is, therefore, required. The whole procedure takes approximately 6 min.

We found that three standard apical views are sufficient for 3D reconstruction of a perfusion map containing 16 myocardial segments of the left ventricle. If the images are recorded on videotape, the end-diastolic frames of each cardiac cycle are digitized off-line for quantitative analysis and 3D reconstruction of the perfusion map.

Digital image analysis and three-dimensional reconstruction of perfusion map

Since quantitation of myocardial perfusion is impossible with the presently available echocontrast agents and ultrasound equipment, a semi-quantitative approach similar to that used in nuclear cardiology has been developed.

The end-diastolic frames are semi-automatically analyzed by a dedicated system for contrast echocardiography (Axle, The Netherlands). The left ventricular myocardium is divided into segments [12] (Figure 1) and in each segment the time-course of the contrast intensity is automatically calculated and plotted as a time–intensity curve after subtraction of the baseline intensity. Peak intensities measured automatically by the system show an excellent correlation with manual measurements of peak intensities from the same time–intensity curves for both sonicated contrast agent (Figure 2A) and Albunex (Figure 2B). The quantitative data are stored as a dataset with quantitative parameters of each myocardial segment. For rapid comparison of peak contrast, intensity of different segments is either encoded in shades of gray or color-coded. Using this dataset a 3D perfusion map of the left ventricle is constructed and displayed (Figures 3C, D). The color scheme (or gray scale) for each patient is the same, but the

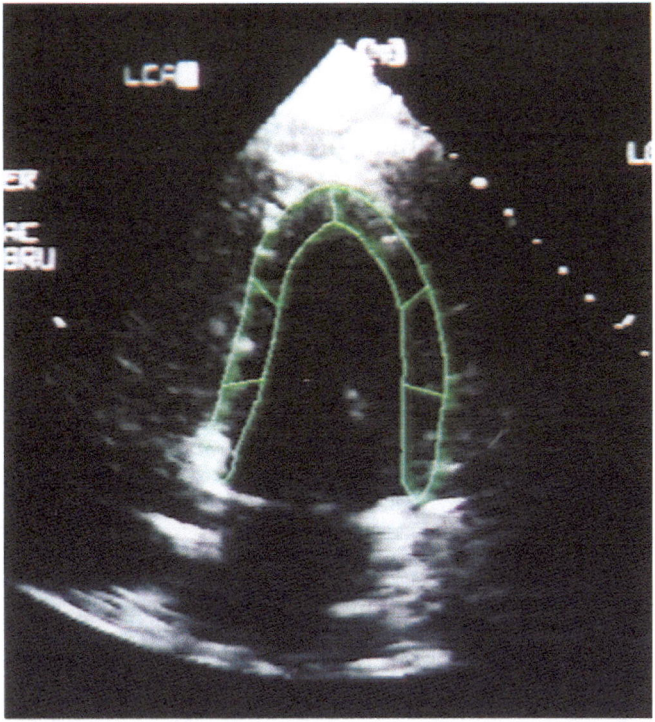

Figure 1. In every frame the ultrasound image of the myocardium is automatically analyzed by the system according to the total number of the segments indicated by the operator. The outline of the segments can be seen over the original two-dimensional ultrasound image.

absolute numbers are different. The minimal intensity increase in all instances is displayed as dark blue (or black) but the contrast opacification in intensity units is different in each patient, as shown by the numbers along the bars (Figure 4).

These relative intensity units allow comparison of different regions with different intensities after interventional procedures. Values <5 intensity units are not considered to be caused by the contrast injections but reflect small fluctuations in the background intensity.

INITIAL CLINICAL EXPERIENCE

A dataset is obtained for each patient following LCA and RCA injections. The results are displayed as a pair of images representing the 3D perfusion map after LCA and after RCA contrast injection.

The separate reconstruction of LCA and RCA perfusion maps allows estimation

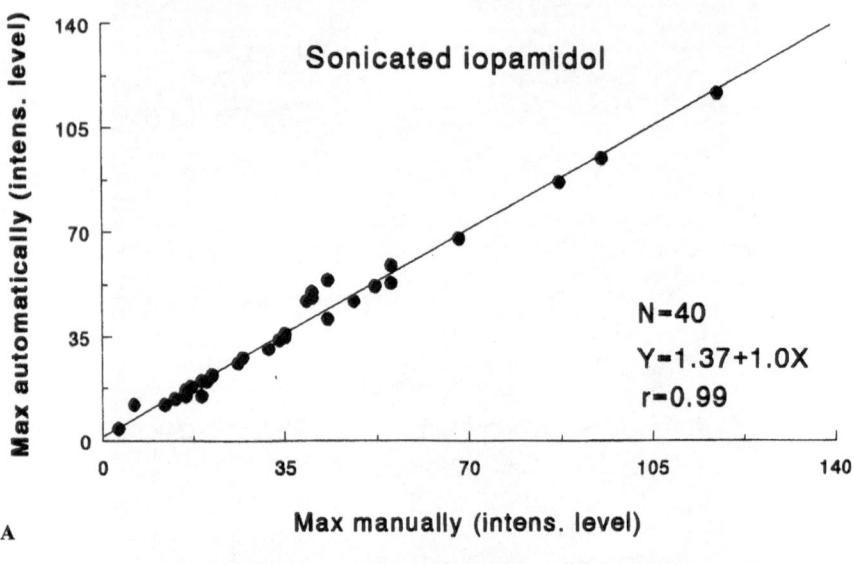

A

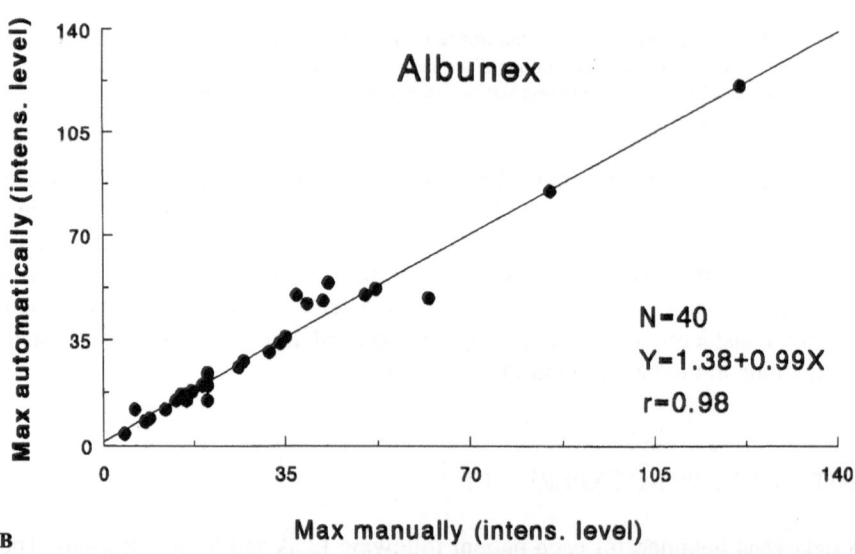

B

Figure 2. Correlation between the measurements of the peak contrast intensity increase (in relative intensity units, IU) obtained automatically and manually (by manually driven cursor) from the same time–intensity curves. The correlation was high for the measurements of the myocardial time–intensity curves obtained during the intracoronary injections of sonicated Iopamidol (A) and Albunex (B).

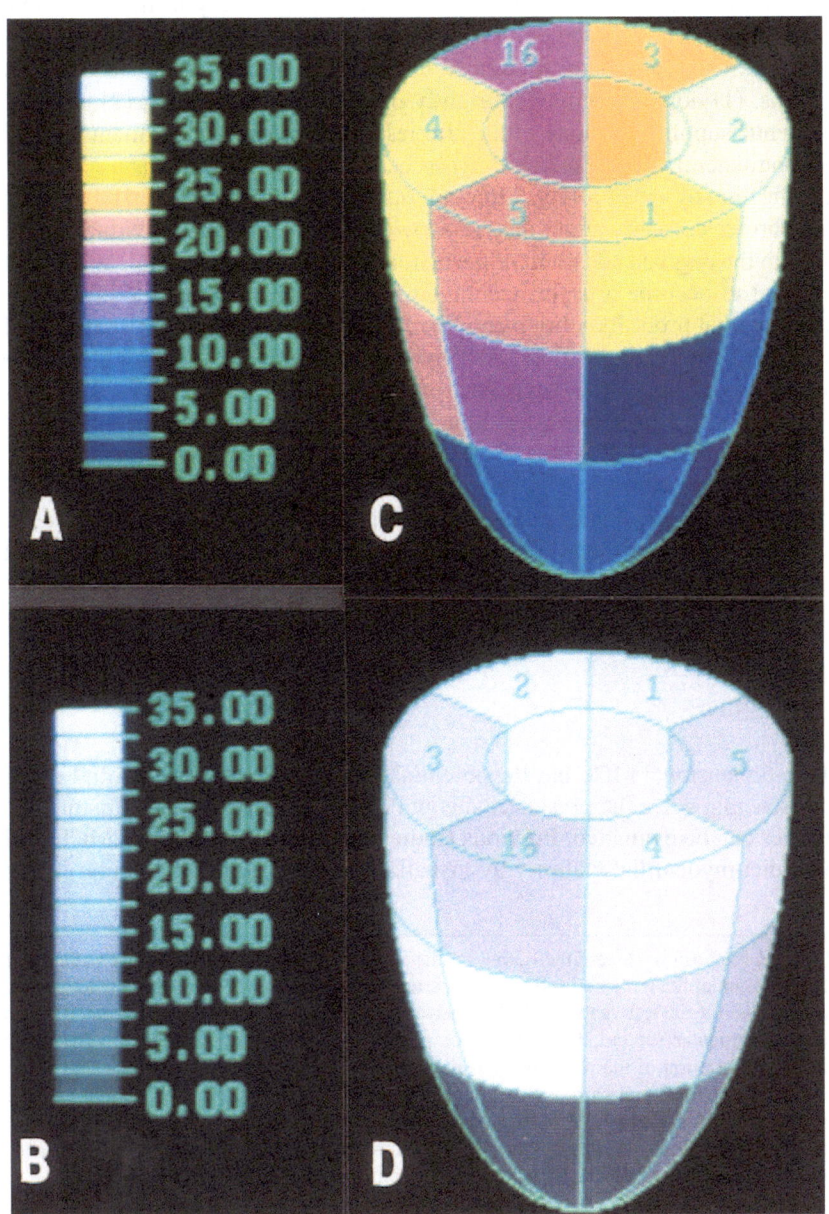

Figure 3. (A) The color bar used for color-coding the intensity increase and other parameters of myocardial opacification. The color-coding of the myocardium is performed from low to high parameter value (intensity increase here) as blue, violet, red, yellow and white. (B) If preferred, the same information can be encoded in shades of gray. (C) An example of a 3D image of left ventricular peak intensity increase following an echocontrast injection into the right coronary artery in a patient with multiple lesions in the left coronary artery. (D) The same data as in (C) displayed in the shades of gray.

484   Aric A. Aiazian et al.

of the relative contribution of each major coronary artery to the perfusion of individual myocardial segments (Figures 4A, B) and classification of the segments into different patterns. A practical clinical classification is used for these patterns: (1) normal segments, brightly opacified by one artery; (2) non-viable segments, supplied by none of the arteries; (3) collateralized segments, supplied by contralateral or both arteries; (4) ischemic segments, which show decreased opacification by either artery. Although our initial clinical experience in patients with chronic coronary artery disease showed significant intra-individual variation for both the degree and extent of perfusion, we feel that multiplane MCE has the potential to become a useful technique for the assessment of the effects of an interventional procedure. For example in 12 patients with chronic total occlusion of one major coronary artery, 3D myocardial contrast echocardiography performed before and after successful opening of the occluded vessel allowed visualization of the myocardial blood flow redistribution following the intervention (Figures 4, 5).

Using this method, we observed that, following successful opening of an occluded artery, previously collateralized segments sometimes remained perfused by both coronary arteries. Pathophysiologically it would have been expected that the collateral circulation would cease to function because of the change in perfusion pressure.

DISCUSSION

Three-dimensional MCE has the potential for assessment of myocardial viability and ischemia since the presence, site and extent of myocardial perfusion abnormalities can be evaluated. Previous studies have demonstrated that MCE is able to predict myocardial viability by visualizing preserved microvascular integrity

---

Figure 4. (Opposite) Three-dimensional perfusion maps of the left ventricle in a patient with a total occlusion of the left anterior descending artery (LAD). AW, anterior wall of the left ventricle; FW, free wall of the left ventricle; PW, posterior wall of the left ventricle; IVSa, anterior part of the interventricular septum; IVSp, posterior part of the interventricular septum. The numbers indicate the location of the respective standard myocardial segments. The anterior and posterior views of the left ventricle are displayed following the same echocontrast injection. (A) Three-dimensional perfusion map after a left coronary artery injection. A perfusion defect is observed in the anterior wall, anterior part of interventricular septum and apex. The inferior wall and the posterior part of interventricular septum are opacified by the left coronary artery. (B) A right coronary artery injection in the same patient shows that part of the anterior wall, free wall and the anterior part of interventricular septum are supplied by collaterals from the right coronary artery. Side by side assessment of the left (a) and right (b) coronary artery injection allowed to observe absence of opacification by either artery in the apex. (C) Following the successful opening of the LAD. The left coronary artery supplies the anterior wall, free wall and the anterior part of the interventricular septum. (D) Three-dimensional image of a right coronary artery injection after intervention shows disappearance of collateral perfusion from the right coronary artery to the anterior wall and the anterior part of interventricular septum.

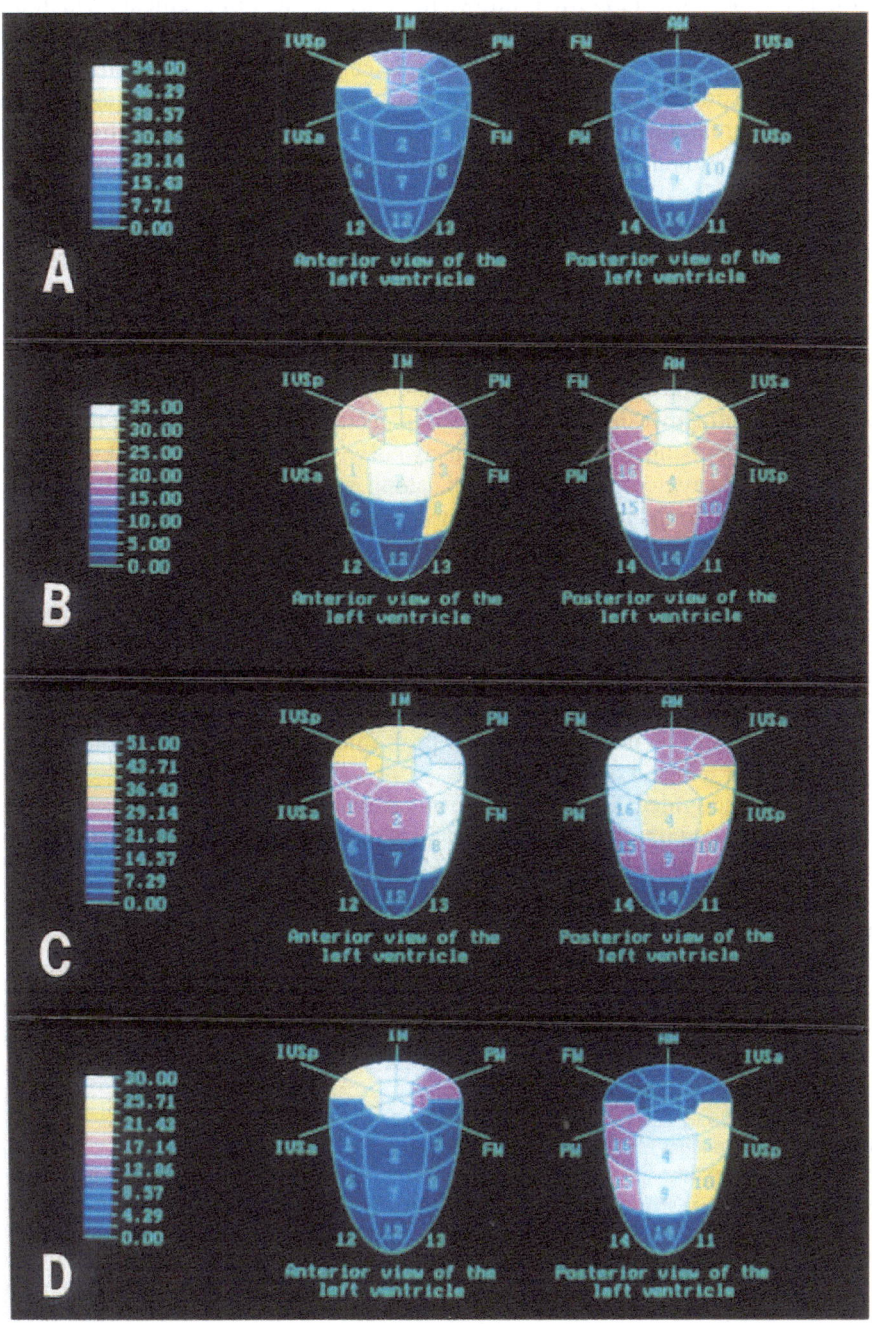

486   Aric A. Aiazian et al.

[10, 11]. Color-coded 3D MCE allows rapid assessment of segments which do not opacify following the injection into both left and right coronary arteries, indicating that these segments have no detectable microvascular flow. This technique is promising for the spatial estimation of the area at risk or size of myocardial infarction. Close correlation between two-dimensional myocardial contrast echocardiography and anatomical myocardial infarction has been demonstrated [12].

To shorten the acquisition time in the catheterization laboratory we utilized a minimal number of left ventricular views (three standard apical views) to reconstruct 16 standard left ventricular segments, similar to the method used in stress echo [9]. More views would allow reconstruction of 3D perfusion maps with more precise delineation of myocardial segments but the acquisition time would become longer. The dataset of 3D perfusion maps contains a complete information about the contrast wash-in and wash-out in each myocardial segment recorded. Complete processing of the time–intensity curves ensures that the peak intensity increase is always correctly identified, along with other relevant quantitative parameters (area under the time–intensity curve, contrast wash-out half-time, etc.). The results of this complex analysis are automatically displayed as a set of 3D images representing different aspects of myocardial perfusion.

The voxel or volume recording technique provides a different resultation of 3D perfusion maps and is based on a number of sequential left ventricular views acquired by a rotating ultrasound transducer. Because each of the views is obtained by the transducer at different times after the echocontrast injection this technique cannot be used to study the peak intensity increase and some other parameters of contrast opacification.

The data provided by quantitative measurements of myocardial intensity increase have certain limitations which have to be considered when interpreting the 3D dimensional perfusion images. Myocardial segments with a diminished increase in intensity on 3D myocardial perfusion maps were considered ischemic, relative to the segments perfused by a non-occluded artery. Three-dimensional myocardial contrast echocardiography is a feasible and attractive technique for perfusion imaging during cardiac catheterization since the other clinically used imaging modalities for 3D perfusion imaging have significant limitations. With

---

Figure 5. (Opposite) Three-dimensional perfusion maps in a patient with a right coronary artery occlusion. Abbreviations as in Figure 4. (A) The left coronary artery opacifies the whole left ventricle, a minimal intensity increase is observed in the inferior and posterior walls which are supplied by collaterals. Contrast opacification of the anterior wall and the anterior part of interventricular septum is bright and normal. (B) The right coronary artery injection shows only minimal intensity increase (up to 5 IU according to the color bar) which cannot be considered to be caused by the echocontrast injection. (C) Following successful opening of the occluded artery the left coronary artery injection shows significantly decreased collateral flow to the posterior and the inferior walls. (D) After the intervention right coronary artery supplies the inferior and the posterior walls. The established antegrade flow provides brighter opacification than the collateral flow from the left coronary artery.

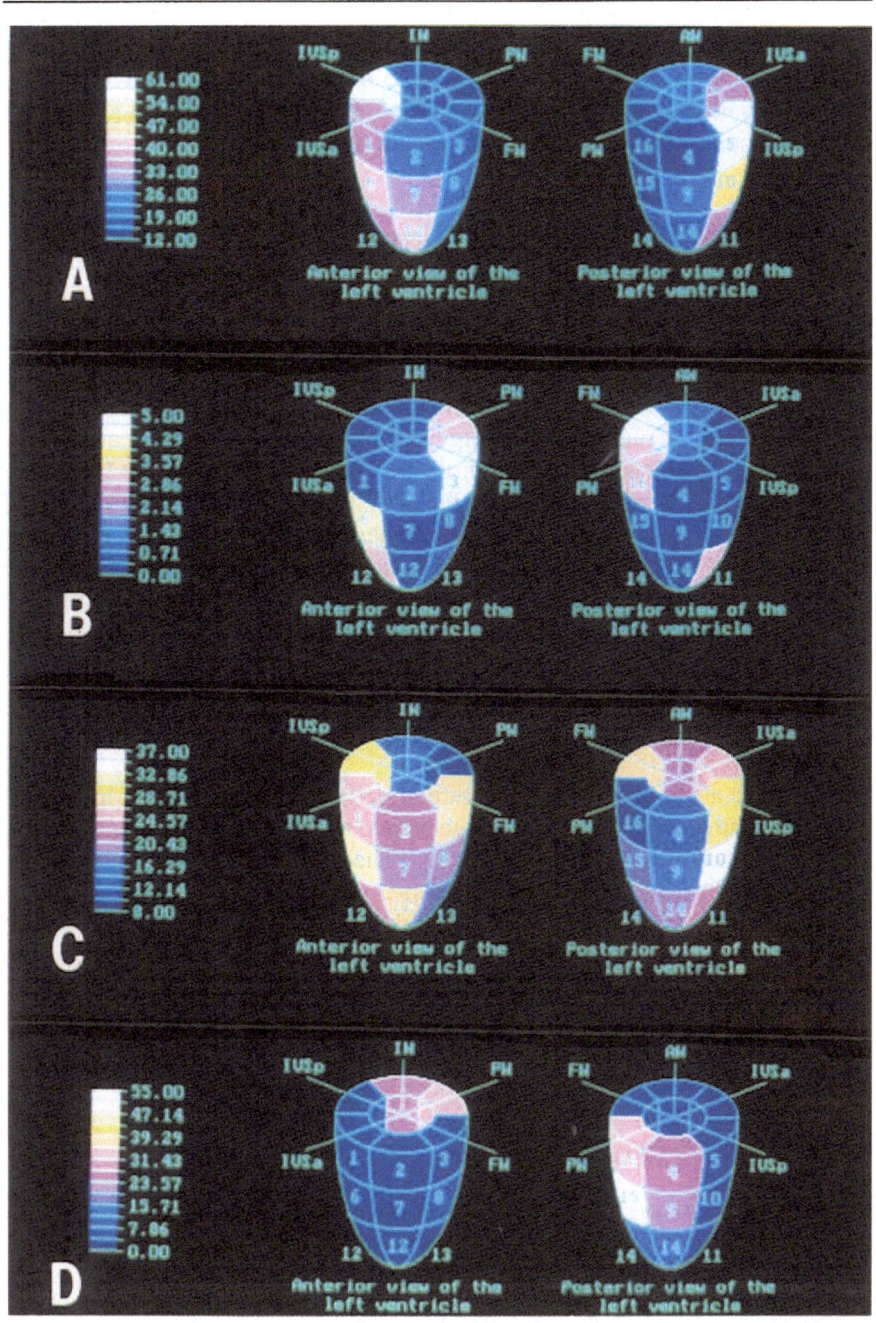

thallium-201 imaging (SPECT) positive scans are only applicable for 2–4 weeks after the intervention due to low specificity of the method [2, 13]. PET has limited resolution and a low signal-to-noise ratio. The equipment required for PET studies is also very expensive and cannot be used for the direct assessment of myocardial perfusion during an intervention [14].

CURRENT LIMITATIONS OF THE METHOD

The intensity of the ultrasound signals is influenced by electronic signal processing and attenuation [15]. The known limitations of two-dimensional myocardial contrast echocardiography therefore also influence image processing and data acquisition for the reconstruction of myocardial perfusion maps. Attenuation and poor lateral resolution in the original ultrasound images may cause false-positive and false-negative effects in the 3D perfusion map. Figure 5A shows false defects in the free wall due to the poor lateral resolution in the original four-chamber ultrasound view. Coronary hyperemia may itself increase contrast intensity [16]. These limitations are inherent in the current methodology and must be taken into account if 3D perfusion imaging is used. Another limitation is that current contrast agents do not allow quantitation of myocardial blood flow.

CONCLUSION

Comparative assessment of 3D myocardial perfusion maps obtained in patients after administration of echocontrast agents into the left and right coronary artery allows myocardial perfusion to be studied. The proposed method can be potentially used to evaluate the results of therapeutic intervention.

Further studies are required to validate the results of 3D myocardial contrast echocardiography using comparable techniques. New ultrasound transducers and signal processing techniques may facilitate 3D myocardial perfusion imaging; better image processing and new contrast agents may allow the quantitation of myocardial blood flow.

REFERENCES

1. Bloomfield ME, Gruntzig AR, Stertzer SH. Thallium-201 myocardial imaging and coronary angioplasty: Synergic diagnostics. DuPont-New England Medical Products Monograph no. 15C585A-2358. N. Billerica (MA): DuPont, May 1984:18–23.
2. Miller DD, Verani MS. Current status of myocardial perfusion imaging after percutaneous transluminal coronary angioplasty. J Am Coll Cardiol 1994;24:260–6.
3. Lang R, Feinstein S, Feldman T, Neumann A, Chua K, Boro D. Contrast echocardiography for evaluation of myocardial perfusion: Effects of coronary angioplasty. J Am Coll Cardiol 1986;8: 323–35.

4. Reisner S, Ong L, Lichtenberg G et al. Quantitative assessment of the immediate results of coronary angioplasty by myocardial contrast echocardiography. J Am Coll Cardiol 1989; 852–6.
5. Porter TR, D'Sa A, Pesko L et al. Usefulness of myocardial contrast echocardiography in detecting the immediate changes in anterograde blood flow reserve after coronary angioplasty. Am J Cardiol 1993;71:893–6.
6. Spotnitz WD, Kaul S. Intraoperative assessment of myocardial perfusion using contrast echocardiography. Echocardiography 1990;7:209–27.
7. Kasprzak JD, Ten Cate FJ, Sutton A, Vletter WB, van Meegen J. Dynamic three dimensional myocardial perfusion echocardiography using a deposit contrast agent. Circulation 1995; 92(Suppl):I-261.
8. Ten Cate FJ, Aiazian AA, Hamburger J, Ataoullakhanova D, Salustri A. Reconstruction of three-dimensional perfusion maps in humans by myocardial contrast echocardiography. J Am Coll Cardiol 1995;Febr(Suppl):247A.
9. Bourdillon PDV, Broderick TM, Sawada SG et al. Regional wall motion index for infarct and non-infarct regions after reperfusion in acute myocardial infarction: comparison with global wall motion index. J Am Soc Echocardiogr 1989;2:398–407.
10. Ragosta M, Camarano G, Kaul S, Powers ER, Sarembock IJ, Gimple LW. Microvascular integrity indicates myocellular viability in patients with recent myocardial infarction. New insights using myocardial contrast echocardiography. Circulation 1994;89:2562–9.
11. Vandenberg BF, Kerber RE, Skorton DJ. Detection of myocardial viability with ultrasound tissue characterization: myocardial contrast echocardiography and integrated backscatter imaging. Am J Cardiac Imag 1994;8:113–22.
12. Armstrong WF, West SR, Mueller TM, Dillon JC, Feigenbaum H. Assessment of location and size of myocardial infarction with contrast enhanced echocardiography. J Am Coll Cardiol 1983;2:63–9.
13. Garvin AA, Cullom SJ, Garcia EV. Myocardial perfusion imaging using single-photon emission computed tomography. Am J Cardiac Imag 1994;8:189–98.
14. Sinha S, Sinha U, Czernin J, Porenta G, Schelbert HR. Noninvasive assessment of myocardial perfusion and metabolism: feasibility of registering gated MR and PET images. Am J Roentgenol 1995;164:301–7.
15. Zwehl W, Areeda J, Schwartz G, Feinstein S, Ong K, Meerbaum S. Physical factors influencing quantitation of two-dimensional contrast-echo amplitudes. J Am Coll Cardiol 1984;4:157–64.
16. Kondo S, Tei C, Meerbaum S, Corday E, Shah PM. Hyperemic response of intracoronary contrast agents during two-dimensional echographic delineation of regional myocardium. J Am Coll Cardiol 1984;4:149–56.

Echo-enhancement agents in Doppler vascular sonography

PART THREE

Echo-enhancement agents in Doppler
vascular sonography

32. Usefulness of echo-enhancement in Doppler assessment of carotid arteries

GÜNTER FÜRST & MATTHIAS SITZER

INTRODUCTION

Extracranial carotid artery disease is an important risk factor for cerebral ischemia. Several investigators have shown that an asymptomatic internal carotid artery (ICA) stenosis carries a 1–2% annual risk of stroke [1–4]. The risk increases up to 12% per year if the patient develops a transient ischemic attack attributable to the ICA stenosis [1, 5]. The risk of cerebral ischemia also depends on the degree of ICA stenosis. For example, the risk of ipsilateral stroke in a patient with a symptomatic stenosis of 90–99% is more than double that for patients with 70–79% luminal narrowing [5]. Moreover, both large-scale clinical studies have shown benefit from operative carotid endarterectomy for patients with symptomatic high-grade (>70% luminal narrowing) ICA stenosis [1, 5], but not in those with low-grade stenosis (>50% luminal narrowing) [1]. Thus, for the purpose of individual risk stratification, it is important to measure accurately the severity of stenosis and to identify patients with high-grade lesions [6]. Until now, intra-arterial angiography has been the 'gold standard' for quantifying ICA stenosis, due to its high interobserver reliability, easy communicability between medical sub-specialities, and the simultaneous assessment of the extra- and intracranial circulation. However, several studies have shown that angiography carries a 1–2% risk of permanent neurological deficits [2, 7–9].

Since the late 1970s, a variety of non-invasive tests have been proposed for the detection and quantification of ICA disease. Among the most widely used are continuous-wave Doppler ultrasonography and color Doppler-assisted duplex imaging (CDDI). Both are capable of assessing such lesions with a high degree of accuracy [10, 11]. Nevertheless, the false-positive diagnosis of ICA occlusion due to minimal residual flow and impaired visualization of intrastenotic flow due to plaque calcification, very narrow residual flow lumina or increased absorption of ultrasonic energy by heterogeneous wall components still constitute important limitations.

Intravenous ultrasonic contrast media are known to improve the echogenicity

493

N. C. Nanda, R. Schlief and B. B. Goldberg (eds.), Advances in Echo Imaging Using Contrast Enhancement, Second Edition, 493–503.
© 1997 Kluwer Academic Publishers.

of flowing blood [12, 13]. However, until recently, short half-life times and their inability to survive pulmonary transit have limited the clinical use of such preparations. Two recently developed agents have overcome these shortcomings. One is based on microspheres of human albumin [14] and the other on a saccharide microparticle suspension [13]. The latter agent (SHU 508A) was evaluated with respect to its usefulness and limitations in the ultrasonographic assessment of the carotid system in two phase III trials [15–17].

CAROTID BLOOD ECHOGENICITY

The rationale underlying the use of i.v. contrast agents for enhancing blood echogenicity is that such agents amplify the reflection of ultrasonic energy by increasing the reflecting surface of flowing blood. To study the influence of the contrast agent on the echogenicity of carotid blood, continuous blood flow velocity spectra were obtained from the common cartotid artery in 12 subjects after i.v. injection of a standard 16 ml bolus into a cubital vein (concentration 200 mg/ml) [15, 16]. The time course of blood echogenicity is shown in Figure 1. The increase in reflected ultrasonic energy began a mean of 11 s (SD±2) after bolus application and reached a maximum of 21±2 dB compared with precontrast conditions 15±3 s. The subsequent decrease in echogenicity followed an exponential decay with a half-life of 75 s (Figure 2).

We also studied the influence of the echoenhancer on the reflected ultrasonic energy within different ranges of the Doppler frequency spectrum (Figure 3). Fast Fourier-transformed Doppler frequency spectrum analysis revealed a 1.56-fold greater increase in ultrasonic energy within the high frequency range (1500-2000 Hz) compared with the low frequency range (16–500 Hz) under enhanced conditions. There was also a significant increase in the measured systolic peak flow velocity by an average of 26±9% which paralleled the increase of ultrasonic energy.

In conclusion, the effect of the echoenhancer SHU 508A on the carotid blood echogenicity was three-fold. First, the pronounced enhancement of the entire Doppler spectrum by >20 dB, which lasted at least 3 min after a single injection. Second, the contrast agent led to a greater enhancement of high-frequency components of the reflected Doppler spectrum. This effect may be explained by the non-homogeneous distribution of the substance within the intravascular

Figure 1. (Opposite) A 128-point fast Fourier-transformed original Doppler velocity spectrum obtained from the common carotid artery (upper), three-dimensional diagram of the same measurement (lower). The X-axis shows blood flow velocity (cm/s); the Y-axis shows time (s); the Z-axis the reflected ultrasonic energy in arbitrary units. The inset gives velocity/energy relations (x/z planes) at two different times of the recording (arrows in the larger figure). The black area indicates non-enhanced, and the gray area enhanced conditions. Note that the increase in reflected ultrasonic energy is paralleled by an increase of the measurable and systolic peak velocity.

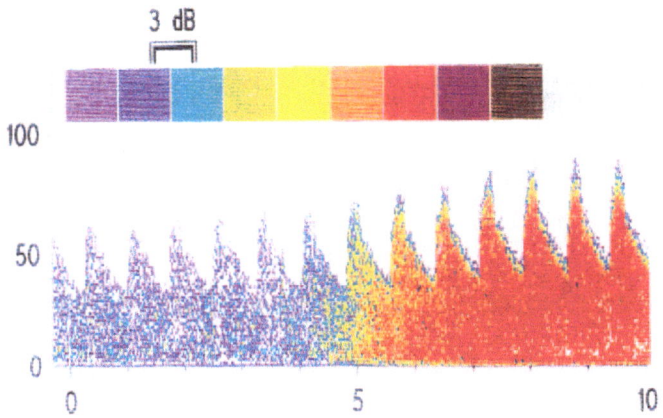

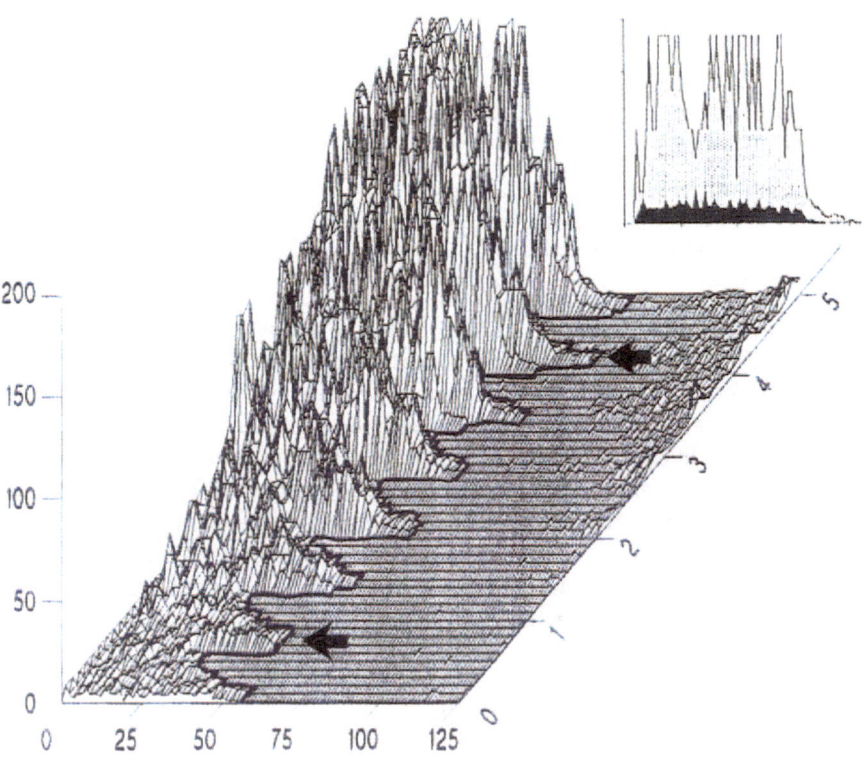

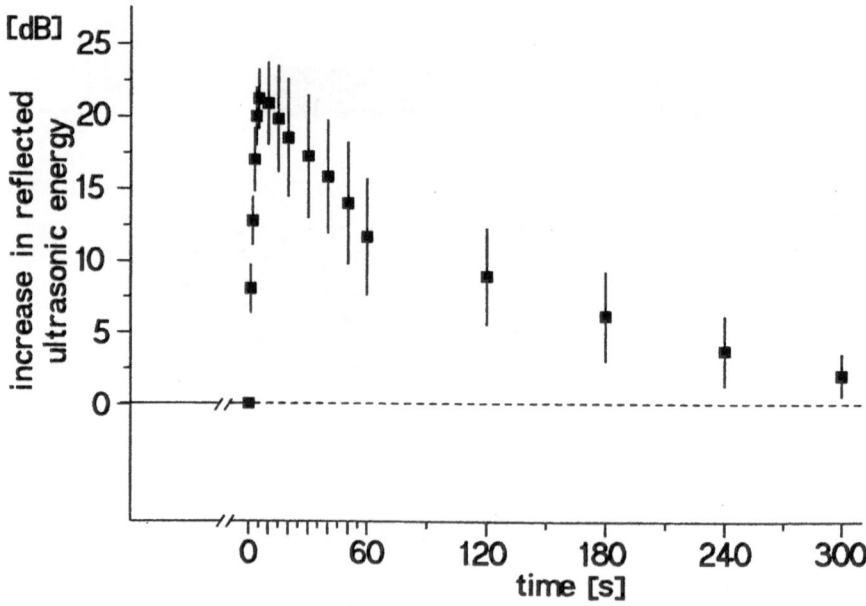

Figure 2. Time-course of SHU 508A-induced change in ultrasonic energy (in dB) recorded from the common carotid artery in 12 patients. Mean time interval between bolus injection of the agent (zero on the X-axis) and increase in ultrasonic energy was 11 s (SD±2); maximum increase in ultrasonic energy was 21±2 dB. For each point the mean value (solid squares) and standard deviation (vertical lines) are depicted.

quasi-parabolic velocity profile. Additionally, the finding of a significant increase of the measurable systolic peak flow velocity might be explained by the particular enhancement of the very fast components, which were below the detection threshold of the ultrasonic device under non-enhanced conditions [18]. This indicates that there was not only an increase of the reflected ultrasonic energy but also a broadening of the reflected frequency spectrum. However, it cannot be ruled out that non-linearities of the probe or amplifier equipment might lead to a similar effect. Nevertheless, from a practical point of view, stenosis quantification based on velocity measurements should be considered carefully when performed after echo-enhancement.

INTERNAL CAROTID ARTERY DISEASE

Carotid artery stenosis

Using color Doppler imaging, different ultrasonic applications and/or techniques can be used to detect and to quantify internal carotid stenosis. First, using B-mode

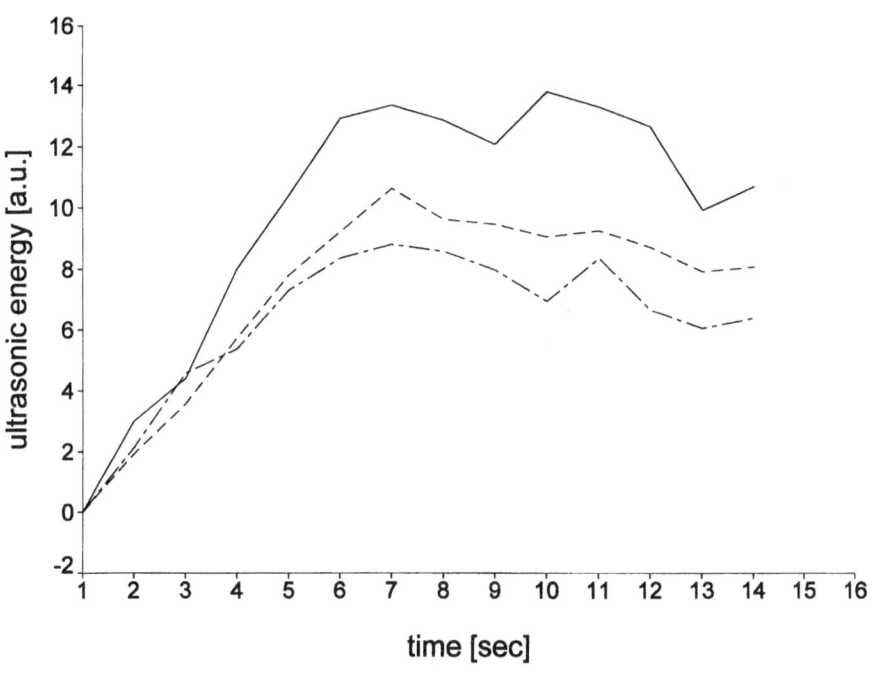

Figure 3. Time-course of the SHU 508A-induced change in reflected ultrasonic energy (Y-axis, in dB) within three ranges of the reflected Doppler spectrum:—— high frequency range, 1500–2000 Hz; —·—·— low frequency range, 16–500 Hz; ------ entire frequency range, 16–2000 Hz; high-pass filter was set at 16 Hz.

sonography, pulsed-wave Doppler can be anatomically placed to measure angle-corrected peak systolic and/or diastolic velocities after screening for flow abnormalities using the color-encoded velocity map. Second, the anatomical information (B-mode) and the superimposed color flow map can be used to measure cross-sectional luminal areas. From these measurements the percentage luminal narrowing can be calculated using the former, non-stenosed vessel lumen and the residual flow lumen at its tightest part. High reliabilities and good correlations with independent angiographic measurements have been reported for both techniques [10, 11, 19]. On the other hand, clinical practice often shows a deterioration of image quality caused by extended plaque calcification, very tight residual lumina or increased absorption of ultrasonic energy in case of fibrous plaques. Previous investigators reported unsatisfactory image quality in 7–21% of cases [10, 11, 19, 20], in which a final diagnosis was made from invasive angiography.

These clinically relevant shortcomings may be overcome by the use of intravenous echoenhancing contrast agents. In our study, we examined 32 patients with high-grade ICA stenoses using non-enhanced and echo-enhanced color

498   *Günter Fürst & Matthias Sitzer*

Table 1. Plaque characteristics of 32 high-grade (≥ 70% luminal narrowing) internal carotid artery stenosis under enhanced and unenhanced conditions.

	Unenhanced	Enhanced	
Irregular plaque surface	22/25 (88%)	29/30 (97%)	$p=0.21$
Ulcerative plaque surface	6/25 (24%)	12/30 (40%)	$p=0.21$
Insufficient image quality	7/32 (21%)	2/32 (6%)	$p=0.07$
Entire residual lumen visible	13/25 (52%)	25/30 (83%)	$p=0.01$

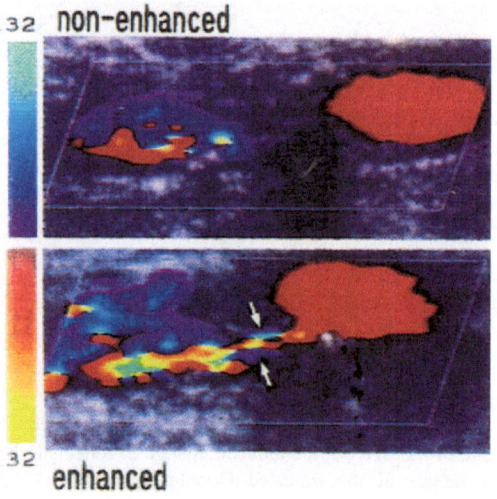

Figure 4. High-grade internal carotid artery stenosis (95% cross-sectional luminal area reduction). Non-enhanced (upper) and echo-enhanced (lower) CDDI (longitudinal views). Blood flow away from the transducer is coded in red, toward the transducer in blue, and aliasing phenomena in green; the ends of the color scale indicate mean blood flow velocities of 0.32 m/s. Characterization of the plaque surface and visualization of the intrastenotic residual lumen were impossible under non-enhanced conditions because of plaque calcification and very narrow residual flow lumen. After echo-enhancement plaque surface irregularities are indicated by areas of minor turbulences (arrows), and the residual flow lumen is now visualized in its entire length.

Doppler imaging. The percentage of investigations with insufficient image quality was reduced from 21% before to 6% after application of the contrast agent in our investigation ($p=0.07$) (Table 1). More importantly, the visualization of the residual flow lumen in its entire length was significantly improved from 52% in precontrast to 83% in postcontrast examinations ($p=0.01$) (Table 1, Figure 4). In patients with sufficient precontrast image quality, cross-sectional luminal area measurements were not influenced by the use of SHU 508A ($r=0.90$, linear regression analysis).

The results of this limited pilot study showed that the CDDI image quality was markedly improved by the use of the echoenhancer, which increased the proportion of quantifiable ICA stenoses. In addition, if the degree of ICA stenosis was determined from luminal area measurements no significant influence of the echoenhancer on these measurements was found.

Carotid artery occlusion

The false-positive diagnosis of ICA occlusion secondary to minimal residual flow constitutes an important limitation of all ultrasound techniques and is still the most relevant clinical pitfall of neurosonography. Compared with continuous wave Doppler sonography and conventional duplex B-mode imaging, CDDI has improved the sensitivity in detecting very slow residual flow in pre-occlusive conditions [11, 21, 22]. Nevertheless, misdiagnosis of ICA occlusion has a reported rate of up to 14% in clinical CDDI series [10, 19, 20]. The technical reason for this drawback is the insufficient sensitivity of conventional CDDI in detecting flowing blood at very low volume flow rates. Our own investigation using SHU 508A suggests that echoenhanced CDDI may be capable of detecting flow even within extremely narrow vascular portions and in the post-stenotic slow-flow segment where below-threshold flow signals may go undetected by non-enhanced CDDI. These suggestions are based on the finding of an improved visualization of the residual flow lumen in high-grade ICA stenosis and the influence of SHU 508A on the Doppler velocity spectrum (see above). Our own study included one patient whom we diagnosed as suffering from symptomatic ICA occlusion after a non-enhanced CDDI examination: echoenhanced CDDI showed a strand of residual blood flow. Intra-arterial angiography performed the same day showed the typical findings of an atheromatous pseudo-occlusion (Figure 5) [23]. This diagnosis was confirmed by surgical endarterectomy of the ICA segment in question. Further clinical investigations should be performed to clarify the value of enhanced CDDI in the diagnosis of almost occluded ICA.

Carotid plaque morphology

Both carotid B-mode sonography and CDDI have been proposed as suitable means of studying plaque morphology. In addition, the plaque ulceration, defined as abrupt disruption of plaque surface (plaque niche) filled with reversed flow, may be associated with an increased stroke risk [22]. A recently published validation study of the CDDI-based diagnosis of plaque ulceration in high-grade carotid stenosis revealed low sensitivities, specificities and predictive values compared with the pathoanatomical standard of reference [24]. This arises because CDDI cannot differentiate between plaque surface irregularities without intimal defects and plaque ulcerations. Thus, in its present state, non-enhanced CDDI cannot be regarded as a reliable or valid means for the diagnosis of plaque ulceration. Our results with echo-enhanced characterization of ICA plaque

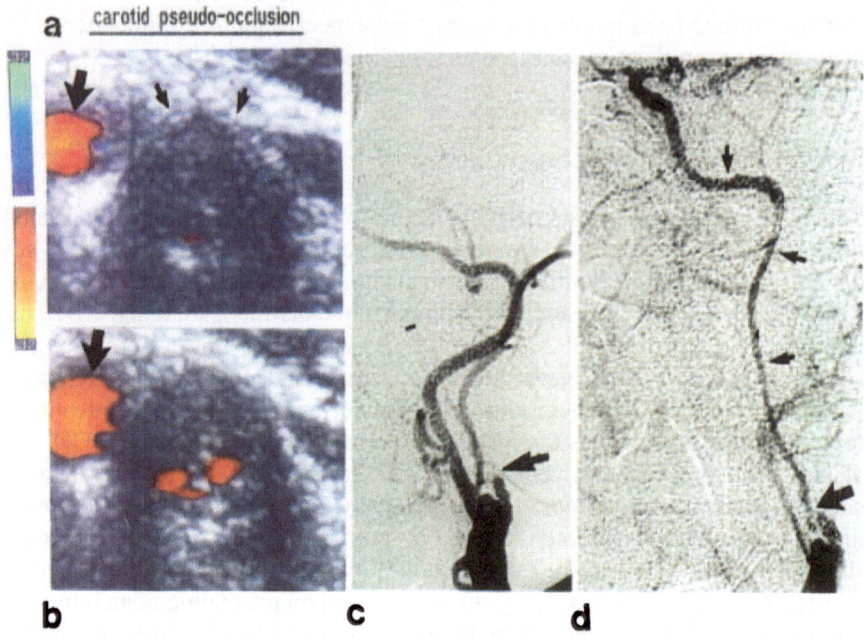

Figure 5. Atheromatous 'pseudoocclusion' of the internal carotid artery (ICA). Non-enhanced (a) and echo-enhanced (b) CDDI (transverse views). Selective intraarterial angiography (c, d). The early angiographic film (2 s after injection) (c) showed the ICA stump suggesting occlusion (arrow). In contrast, the later film (15 s after injection) (d) clearly showed residual flow through the ICA up to the skull base (small arrows). Non-enhanced CDDI (a) detected flow signals only in the external carotid artery (arrow), whereas no signals were obtained from the ICA (small arrows). Echo-enhanced CDDI, however, clearly depicted residual flow through the still patent ICA lumen (b).

morphology revealed a higher proportion of irregular and ulcerative plaques compared with non-enhanced conditions (Table 1, Figure 6) [15, 16]. This might be explained by the considerable enhancement of slow flow components. Plaque surface-related minor turbulences and reversed flow components were better visualized under enhanced conditions, which may increase the sensitivity of echo-enhanced CDDI in the detection of plaque surface abnormalities. The usefulness of echo-enhanced CDDI for the sonographic diagnosis of plaque ulceration should be evaluated in further studies.

CONCLUSIONS

The promising results of the above reported phase III study were threefold. First, the echo-enhancer SHU 508A led to a substantial increase of the reflected ultrasonic energy. This effect lasted more than 3 min, providing enough time for

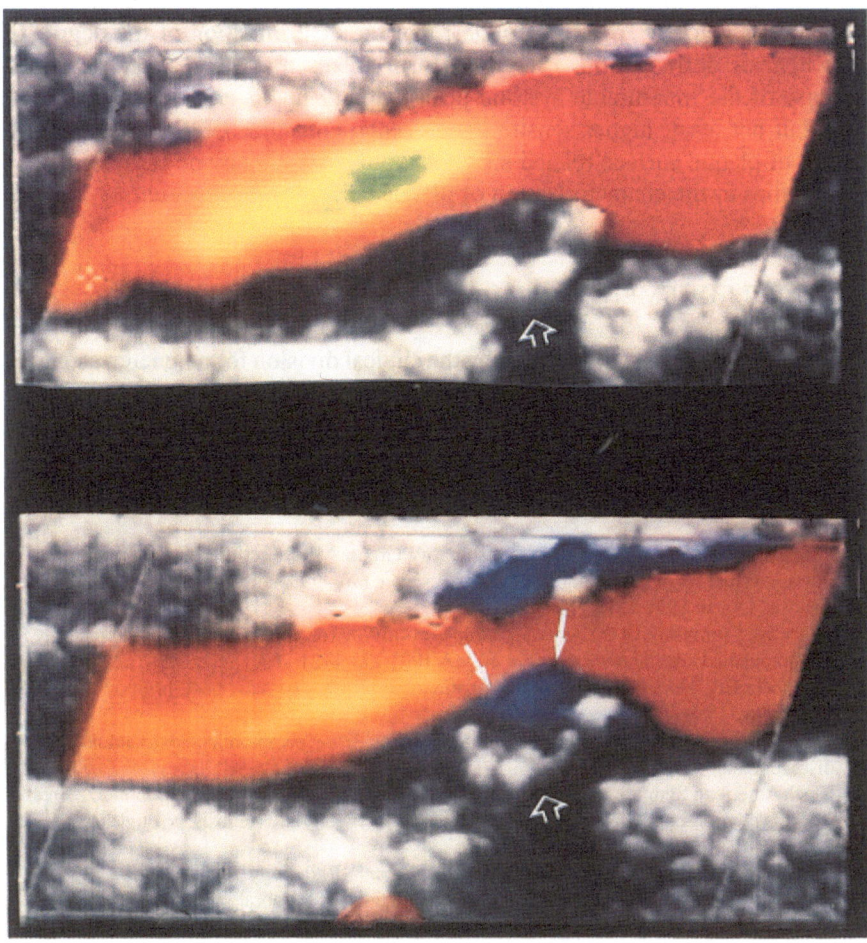

Figure 6. Non-enhanced (upper) and echo-enhanced (lower) CDDI (longitudinal views) in a patient with internal carotid artery stenosis. Blood flow away from the transducer is coded in red, toward the transducer in blue. An extended plaque formation with an area of calcification (arrowhead) is visible on both images, whereas a zone of slow reversed flow (arrows) within the plaque niche ('plaque ulceration') is demonstrated only under enhanced conditions (lower).

detailed sonographic examination. Second, CDDI image quality was significantly improved. This was mainly due to the better visualization of the residual flow lumen in its entire length, allowing cross-sectional area measurements in almost all patients. Thus, the use of an echoenhancer led to a significant improvement of CDDI imaging quality due to more detailed plaque characterization. Additionally, and equally important, the echo-enhancer did not influence the measurement of the residual flow lumen compared with non-enhanced conditions. On the other hand, our findings concerning the influence of the echo-enhancer on the Doppler velocity spectrum suggest that the determination of the degree of ICA stenosis derived from velocity measurements may be influenced by the increase of the measurable systolic peak velocity under enhanced conditions. Last, but not least, higher sensitivities of echo-enhanced CDDI in the visualization of plaque surface related slow-flow phenomena provide new diagnostic possibilities in the characterization of plaque surface abnormalities.

ACKNOWLEDGEMENTS

The pharmaceuticals were provided by the clinical division for magnetic resonance and ultrasonic contrast media of Schering AG, Berlin, Germany.

REFERENCES

1. European Carotid Surgery Trialists' Collaborative Group: MRC European Carotid Surgery Trial. Interim results for symptomatic patients with severe (70–99%) or with mild (0–29%) carotid stenosis. Lancet 1991;337:1235–43.
2. Executive Committee of the Asymptomatic Carotid Atherosclerosis Study. Endarterectomy for asymptomatic carotid artery stenosis. JAMA 1995;273:1421–8.
3. Chambers BR, Norris JW. Outcome of patients with asymptomatic neck bruits. New Engl J Med 1986;315:860–5.
4. Hennerici M, Rautenberg W, Mohr S. Stroke risk from symptomless extracranial arterial disease. Lancet 1982;ii:1180–3.
5. North American Symptomatic Carotid Endarterectomy Trial Collaborators. Beneficial effect of carotid endarterectomy in symptomatic patients with high grade stenosis. N Engl J Med 1991; 325:445–53.
6. Polak JF. Non-invasive carotid evaluation: carpe diem. Radiology 1993;186:329–31.
7. Earnest F, Forbes G, Sandok BA et al. Complications of cerebral angiography: prospective assessment of risk. Radiology 1984;142:247–53.
8. Dion JE, Gates PC, Fox AJ, Barnett HJM, Blom RJ. Clinical events following neuro-angiography: a prospective trial. Stroke 1987;18:997–1004.
9. Davies KN, Humphrey PR. Complications of cerebral angiography in patients with symptomatic carotid territory ischemia screened by carotid ultrasound. J Neurol Neurosurg Psychiat 1993; 56:967–72.
10. Polak JF, Dobkin GR, O'Leary DH, Wang AM, Cutler SS. Internal carotid artery stenosis: accuracy and reproducibility of color-Doppler-assisted duplex imaging. Radiology 1989;173: 793–8.
11. Sitzer M, Fürst G, Fischer H et al. Between-method correlation in quantifying internal carotid stenosis. Stroke 1993;24:1513–18.
12. Ophir J, Parker KJ. Contrast agents in diagnostic ultrasound. Ultrasound Med Biol 1989;15: 319–33.
13. Schlief R, Schürmann R, Niendorf HP. Blood-pool enhancement with SHU 508 A; results of phase II clinical trials. Invest Radiol 1991;26(Suppl 1):188–9.
14. Bleeker HJ, Shung KK, Barnhardt JL. Ultrasonic characterization of Albunex®, a new contrast agent. J Acoust Soc Am 1990;87:1792–7.
15. Sitzer M, Fürst G, Siebler M, Steinmetz H. Usefulness of an intravenous contrast medium in

the characterization of high-grade internal carotid strenosis with color Doppler-assisted duplex imaging. Stroke 1994;25:385−9.

16. Fürst G, Sitzer M, Hofer M, Steinmetz H, Hackländer T, Mödder U. Kontrastmittelverstärkte farbkodierte Duplexsonographie hochgradiger Karotisstenosen. Ultraschall in Med 1994;15: 140−4 (abstract in English).

17. Meents H, Burkard A. Contrast-enhanced colour duplex sonography of carotid arteries. Eur Radiol 1994;4:533−7.

18. Sitzer M, Rose G, Fürst G, Siebler M, Steinmetz H. Characteristics and clinical value of an intravenous contrast medium in the evaluation of high-grade internal carotid stenosis. J Neuroimag 1997 (in press).

19. Erickson SJ, Mewissen MW, Foley WD et al. Stenosis of internal carotid artery: assessment using color Doppler imaging compared with angiography. Am J Radiol 1989;152:1299−1305.

20. Steinke W, Kloetzsch C, Henerici M. Carotid artery disease assessed by color Doppler flow imaging: correlation with standard Doppler sonography and angiography. Am J Nucl Radiol 1990;11:259−66.

21. Kessler C, von Maravic C, von Maravic M, Koempf D. Colour Doppler flow imaging of the carotid arteries. Neuroradiology 1991;33:114−17.

22. Steinke W, Hennerici M, Rautenberg W, Mohr JP. Symptomatic and asymptomatic high-grade carotid stenoses in Doppler color-flow imaging. Neurology 1992;42:131−8.

23. O'Leary DH, Mattle H, Potter JE. Atheromatous pseudo-occlusion of the internal carotid artery. Stroke 1989;20:1168−73.

24. Sitzer M, Müller W, Rademacher J, Siebler M, Hort W, Steinmetz H. Color Doppler-assisted duplex imaging fails to detect ulceration in high-grade internal carotid stenosis. J Vasc Surg 1996;23:461−5.

33. Contrast-enhanced transcranial color Doppler imaging

ULRICH BOGDAHN, GEORG BECKER & ALBRECHT BAUER

INTRODUCTION

Transcranial color coded real time sonography (TCCS, synonym: transcranial duplex – TDS) has evolved from transcranial Doppler sonography (TCD) and color duplex imaging (CDI) as a completely new diagnostic entity for CNS disease [1–4]. Vascular pathology, such as ischemic and hemorrhagic stroke [5–7], arteriosclerotic vascular degeneration [8], arteriovenous malformations and aneurysms [9, 10], as well as parenchymal abnormalities such as tumors [11–14], degenerative disorders [15] and monopolar depression [16] can be detected. Disturbances of CSF circulation can also be characterized and monitored [17]. The main advantage of TCCS lies in its applicability for use at the bedside, its real time imaging function, and its combination of morphological and functional image representation. A major disadvantage of TCCS was the attenuation of the ultrasound signal by the skull, which produced an inadequate result in about 20% of patients due to the insufficient signal-to-noise ratio. Ultrasound contrast agents capable of crossing the pulmonary circulation, such as Levovist, have changed this situation completely, improving the signal intensity recovered from within the skull by up to 30 dB [18]. This chapter summarizes experience with Levovist in this new and promising field of application.

CNS SAFETY

Among the entire international study population of more than 1800 patients no specific CNS toxicity has been observed. In addition, transcranial sonography with Levovist did not aggravate symptoms or cause any further neurological complication [19–22]. Specifically, patients with brain edema due to brain neoplasms or infarcts experienced no signs and symptoms of an increase in edema. A single patient with epilepsy suffered one seizure within 24 h of administration of Levovist. Patients given up to six injections of Levovist during examination

505

N. C. Nanda, R. Schlief and B. B. Goldberg (eds.), Advances in Echo Imaging Using Contrast Enhancement, Second Edition, 505–524.
© 1997 *Kluwer Academic Publishers.*

showed no cognitive alterations. Levovist therefore appears to be extremely safe with respect to the CNS, considering all of the currently available information.

SIGNAL ENHANCEMENT

The crucial question when applying echo enhancing agents in CNS ultrasound diagnostics is signal-to-noise ratio. In order to define the potential of Levovist for signal enhancement, a cooperative study was conducted to measure signal enhancement by Levovist in different vascular territories. Suboptimal cardiac or vascular Doppler examinations were evaluated over 75 i.v. bolus injections performed in 30 patients. The contrast effects were quantified by spectral audio Doppler intensity measurements. The average organ/vessel increase in signal was approximately 16–20 dB and lasted more than 120 s; in the CNS (middle cerebral artery) the signal increased by up to 24.7 dB (range ±5.7 dB), with the longest duration of up to 270 s [18]. This promised a huge step forward in transcranial sonographic diagnosis, allowing 100% of patients to undergo at least a basic vascular investigation with success.

ARTERIAL SYSTEM

A principal advantage of transcranial Duplex techniques rests in their ability to determine angle-corrected Doppler flow velocities from intracranial vessels under anatomical control. Schöning and co-workers [23] studied age-dependent flow velocities in a large population of children and adolescents: the mean flow maximum velocities were 79 ± 17.7 cm/s ($\pm$ SD) in the anterior cerebral artery (ACA), 92.2 ± 13.0 cm/s in the middle cerebral artery (MCA) and 63.9 ± 13.6 cm/s in the posterior cerebral artery (PCA) in children under 10 years of age; children above 10 years of age and adolescents (up to 18 years of age) had respective mean flow maximum velocities of 69.4 ± 13.8 cm/s and 39.9 ± 10.5 cm/s in the ACA, 83.2 ± 11.9 cm/s and 50.8 ± 9.0 cm/s in the MCA, and 55.6 ± 10.1 cm/s and 33.1 ± 6.3 cm/s in the PCA. The time-averaged maximum velocity decreased significantly with age while resistance and pulsatility indices remained stable

---

Figure 1. (Opposite and following pages) Arterial system and echo-contrast enhancing agents. (A) Temporal acoustic bone window in transcranial color coded real time sonography. (B) Suboptimal TCCS and Doppler spectrum taken from M2-segment of right MCA (non-enhanced). (C) Levovist-enhanced TCCS and Doppler-spectrum taken from identical M2-segment of right MCA, showing right MCS, PCA and contralateral ACA (A1) in red, ipsilateral ACA (A1) and contralateral MCA and PCA in blue. (D) High-grade distal stenosis of the basilar artery detected by contrast-enhanced TCCS: color Doppler image (semicoronal scan from a right transtemporal approach) indicating turbulent high flow velocities ('aliasing effect') in the distal basilar artery (arrow). (E) Doppler sample from within the stenotic portion of the basilar artery (arrow) demonstrating systolic Doppler flow velocities of >256 cm/s. (F) Lateral MR – angiograms of the same patient with loss of the distal portion of the basilar artery (arrow), indicating a minimal net flow, which although high in velocity, cannot be detected.

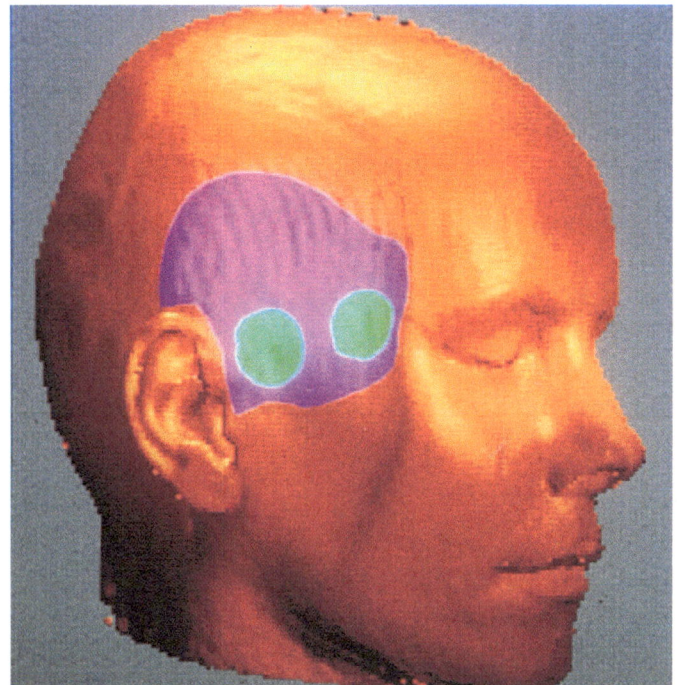

A

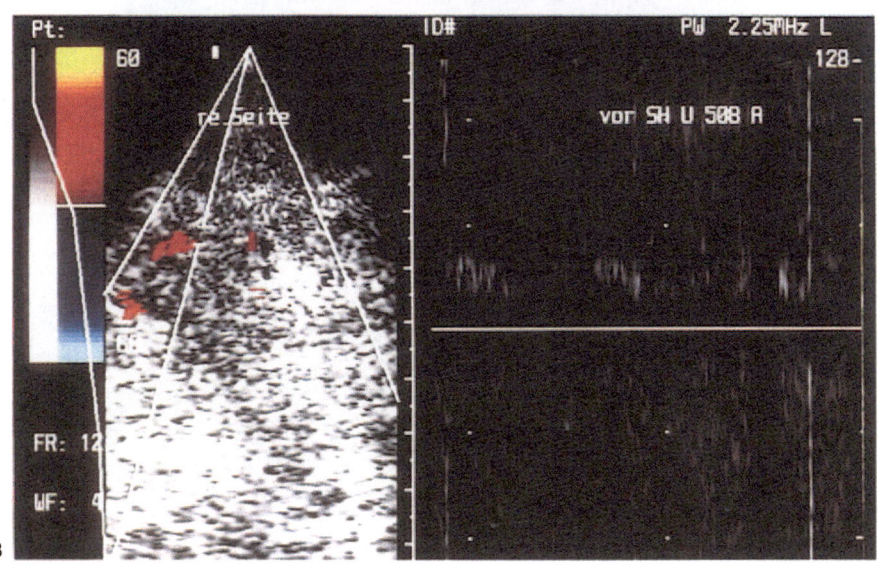

B

Figure 1.

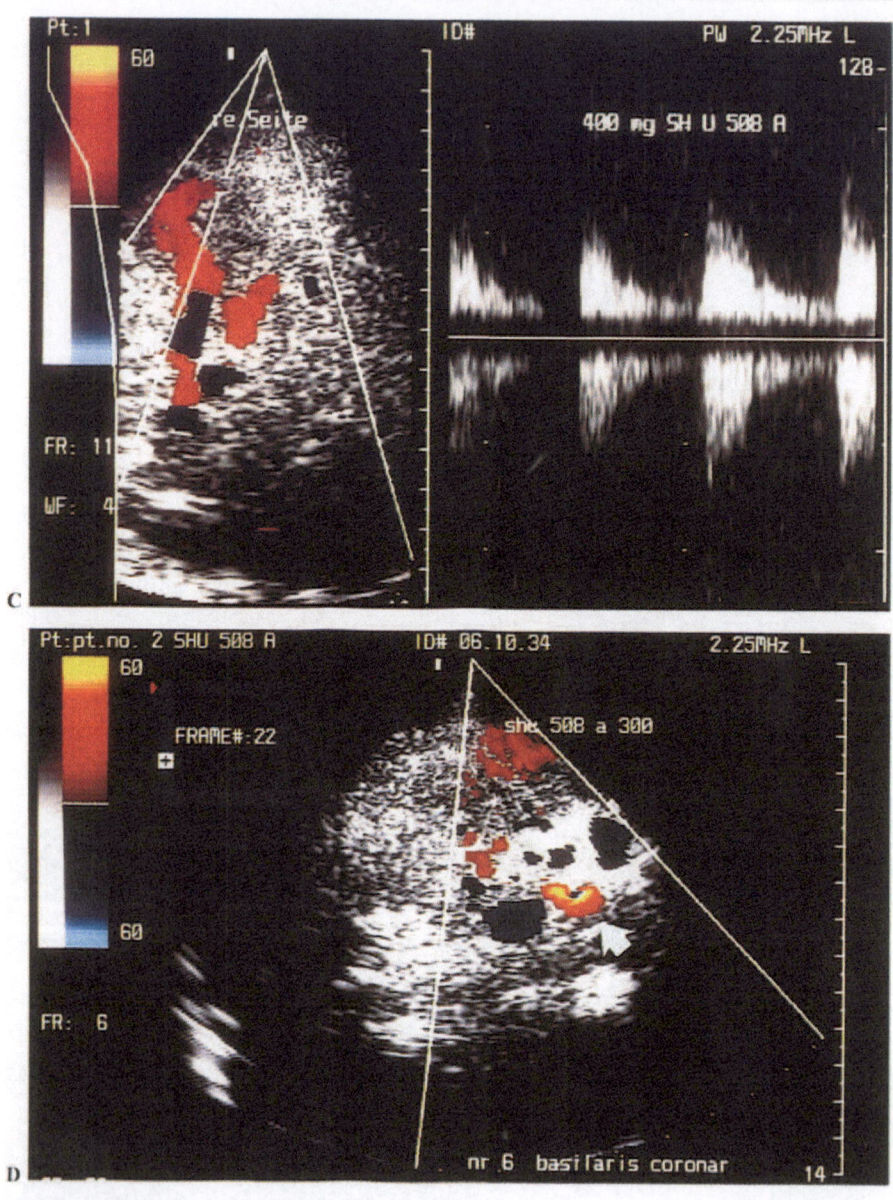

Figure 1. *Continued.*

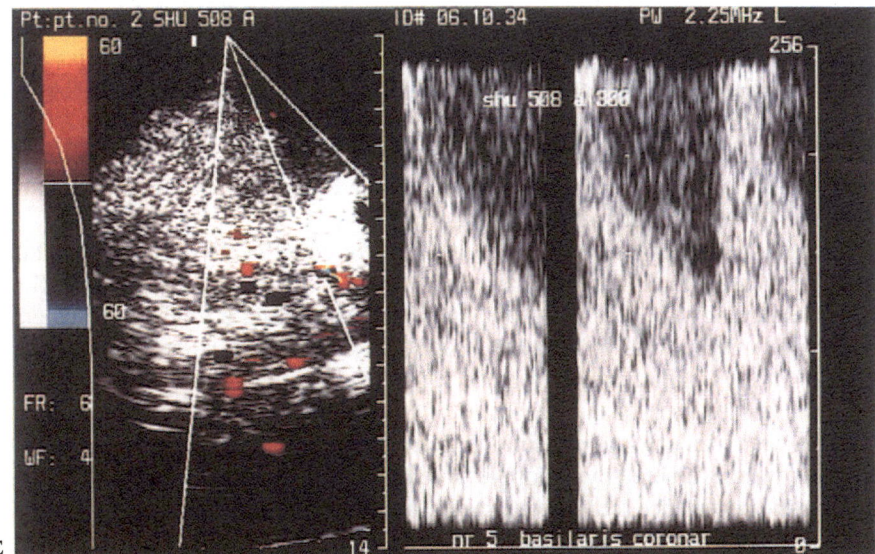

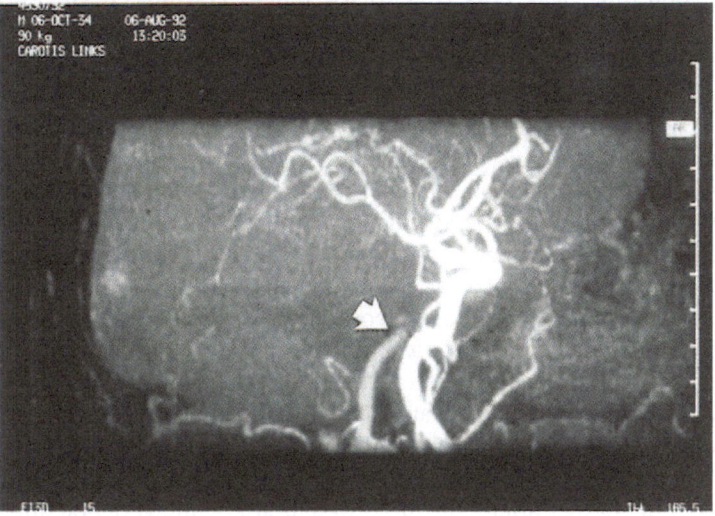

Figure 1. *Continued*

through childhood to adulthood. Schöning et al. also compared conventional transcranial Doppler sonography TCD and TCCS [24], and reported a good correlation of comparable parameters: methodological improvement was found for recording quality, between-examiner reproducibility and detection of anatomically more difficult locations (ACA). In addition, angle-corrected Doppler flow velocities were found to be 10–15% higher than those recorded with conventional TCD. Improved detection of the vertebrobasilar system by TCCS has been reported by several groups [25–27]; again, angle correction was an additional advantage. Angle-corrected flow velocities (peak systolic/end-diastolic) were calculated as 50/24 cm/s in the vertebral artery, 59/28 cm/s in the basilar artery and 56/30 cm/s in the posterior inferior cerebellar artery (PICA). Using TCCS the origin of the basilar artery (BA) could be focused at a depth of 70–75 mm, although previous TCD studies had suggested the BA origin at a depth of 80–90 mm [25–27].

A great deal of effort has been devoted to studying the effects of Levovist on investigation of the intracranial arterial system, using echo-enhanced TCCS or TCD technology [19–21, 27–29]. One study comprised a total of 86 patients with intracerebral vascular disease who had insufficient signals in previous TCD/TCCS examinations. Within the study protocol they received up to six i.v. injections of Levovist, usually at a dose of 10 ml (300 mg /ml) or 8 ml (400 mg/ml). Patients were evaluated at baseline, as well as before and after receiving Levovist. Results were evaluated by a Doppler intensity score as well as by the degree of diagnostic confidence (continuous scale). In the majority of cases 2–3 injections were enough to achieve sufficient diagnostic confidence. The mean viewing time (examination time) was 273 s (SD: 78.4 s, range: 117–468 s). The examiner's diagnostic confidence was 6.5±9.6 at baseline; after Levovist this rose to 87.5±13.9. Results for TCCS are very encouraging, and exhibit the advantages of visualizing peripheral branches of the basal arteries, the cerebellar arteries and the basal venous system (see below). In patients, for whom transcranial ultrasound examination was not feasible because of insufficient acoustic bone windows, at least the proximal branches of the large intracranial arteries could be evaluated. In patients with average or good acoustic bone windows, within the arterial system a number of abnormalities could be examined under highly improved conditions: this so far includes regions with anatomically difficult access, such as lesions in the distal portion of the basilar artery, the vertebrobasilar system, or intracanalicular lesions of the carotids. In conclusion, Levovist clearly improves the diagnostic potential of TCD/TCCS in patients with cerebrovascular disorders. Ultrasound examination times are increased to about 4.5 min, with sufficient signal-to-noise improvement.

The main improvements produced by Levovist are seen in examinations of the head of the basilar artery, the distal basilar artery, the vertebral arteries, the intracranial part (skull base) of the internal carotid artery, the small intracranial vessels (communicantes and peripheral branches, ophthalmic artery) and, of course, patients with insufficient acoustic windows (Figure 1).

VENOUS SYSTEM

Until recently, conventional transcranial vascular Doppler examinations were restricted to examinations of the large basal arterial system. Aaslid and co-workers [29] detected venous blood flow in the straight sinus by TCD. A preliminary report on the delineation of the basal cerebral venous system by TCCS [19] using a transtemporal approach and the injection of transpulmonary stable contrast agent SHU 508A disclosed a new perspective. In all seven patients examined, the inferior sagittal sinus, the straight sinus, the internal cerebral vein and the vein of Galen could be identified by color Doppler and further assessed by Doppler sampling. Due to its anatomical location, the confluence sinuum (torcular herophili) and the transverse and sigmoid sinus were only inconsistently identified by a transtemporal approach. Doppler recordings from the straight sinus (without contrast application) revealed an angle-corrected average peak-systolic flow of 19.8 cm/s and end-diastolic flow of 18.5 cm/s [30]; the Doppler frequency spectrum showed slight pulsatility. In a second study 10 individuals were evaluated after the application of the ultrasound contrast agent SHU 508A. An average angle-corrected peak-systolic flow of 20.2 cm/s and end-diastolic flow of 17.9 cm/s were found for the straight sinus; in a few cases flow in the confluence sinuum (mean systolic 12.0 cm/s, diastolic 10.3 cm/s, $n=3$; Figure 2), the inferior sagittal sinus (systolic 20.2 cm/s, diastolic 15.0 cm/s, angle-corrected, $n=1$) and the transverse sinus (systolic 12.7 cm/s, diastolic 9.9 cm/s, angle-corrected, $n=1$) was quantified. The superior sagittal sinus or bridging veins were only seen inconsistently through a foraminal approach; here the penetration of the transducers was the limiting factor.

So far only limited experience is available on transcranial ultrasound examination of cerebral venous thrombosis. Within a large phase III clinical protocol employing SHU 508A two patients with established superior sagittal sinus thrombosis were studied. In both patients, the Doppler flow velocities in the straight sinus (88 cm/s and 92 cm/s) and the great vein of Galen (83.7 cm/s and 92 cm/s) were elevated to approximately 4 times the normal values, indicating the presence of striking collateral venous flow. TCCS is, therefore, likely to allow diagnosis and monitoring of cerebral venous thrombosis [30].

In conclusion, echo-enhanced TCCS allows diagnostic imaging of the cerebral venous system, although there are still a number of different problems to be solved. First, slow flow detection of current sonographic equipment is not entirely satisfactory; second, there is too much high pass filtering in color Doppler or power Doppler imaging, allowing only insufficient detection of venous/slow flow; finally, the penetration of the current transducers is insufficient to allow imaging of the cortical sinuses or bridging veins through a foraminal approach.

ANEURYSMS AND ARTERIOVENOUS MALFORMATIONS

Arteriovenous malformations

The role of transcranial Duplex in screening and follow-up of patients with AVM is as yet undefined. Several reports [10, 31–32] indicate the potential of

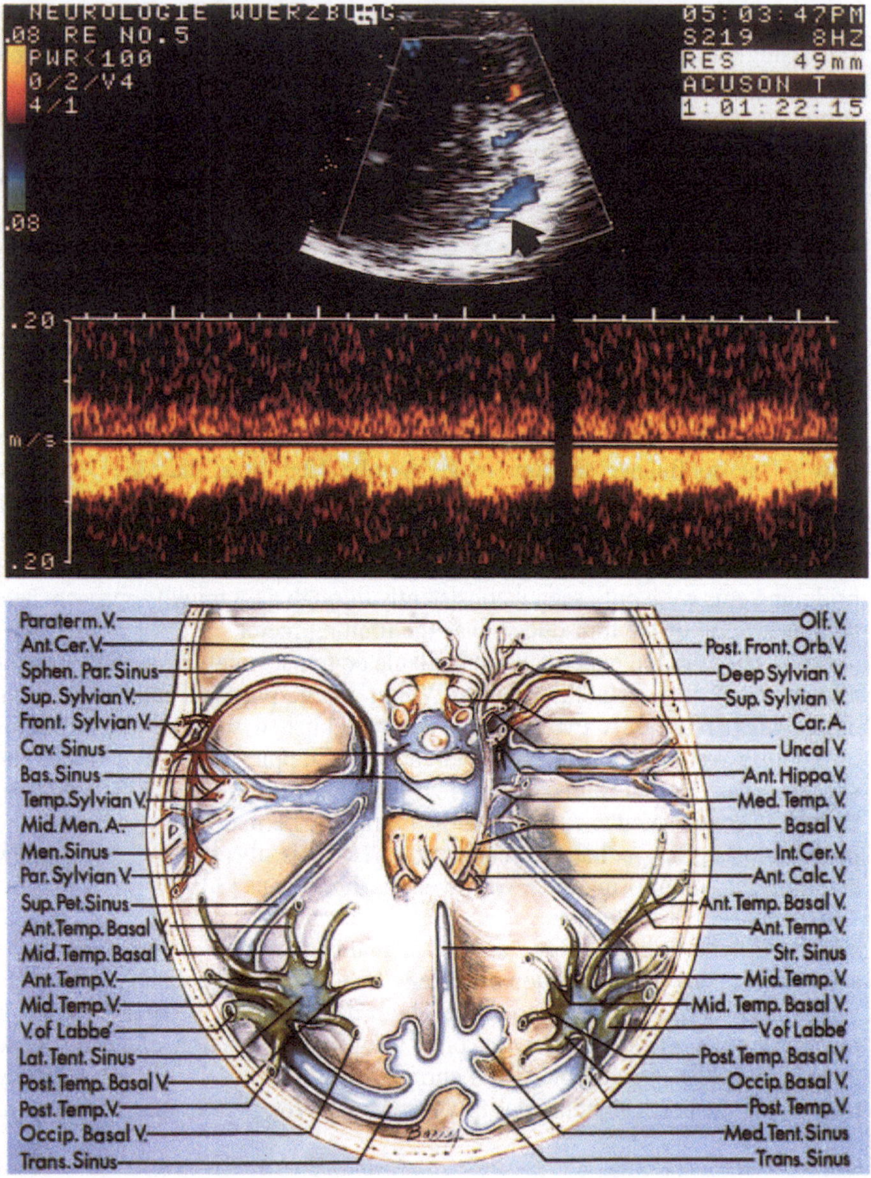

Figure 2. Imaging of the deep venous system by contrast-enhanced TCCS. (A) TCCS-image and Doppler spectrum of the contralateral transverse sinus (arrow) through a transtemporal approach; 400 mg/ml Levovist. (B) Scheme explaining venous anatomy.

unenhanced TCCS to detect the entire vascularization of intracranial AVM, improving the anatomical orientation and the steric configuration. The sensitivity of this method is unclear, but seems rather high. As long as the individual patient has an acoustic temporal bone window, an AVM can probably be detected. In B-mode, the lesions appear as hyperechogenic structures interspersed with zones of low density. Color Doppler displays flow within the AVM; feeding and draining vessels, as well as collateral supply can be identified anatomically. Doppler sampling frequently displays turbulent and bi/multi-directional flow; mean Doppler flow velocities within the main feeders of the AVM often are extremely high (up to 180 cm/s and more). Depending upon the location of the Doppler sample, the Doppler spectrum is more or less modulated by the pulse amplitude. The administration of ultrasound contrast agents may enhance the ability to delineate the entire AVM and to depict AVM not identified by non-enhanced TCCS. In addition, the main role of Levovist in diagnosis of AVM seems to rest in emergency medicine, by allowing identification of AVM within a CNS hemorrhage in the ICU setting, with a short time of diagnosis and allowing patients to be prepared correctly for angiography and putative neurosurgical approaches. Echo-enhanced sonographic diagnosis may also be used in patient follow-up after embolization of AVM with interventional neuroradiological approaches to detect or exclude early or late recurrences.

Aneurysms, which may already be seen with unenhanced TCCS [9], may be diagnosed with higher accuracy and possibly even to a smaller size of (<3 mm in diameter; Figure 3) with contrast-enhanced TCCS; in addition, the vascular anatomy may be analyzed with improved resolution. Contrast application may also assist in the differential diagnosis of intracranial hemorrhages and detection of possible sources of hemorrhage for emergency clot evacuation, although no data exist regarding this application as yet [9]. TCCS may have a role in the post-operative/post-interventional follow-up of AVM patients, since it allows unrestricted repeat examinations and real-time evaluation.

In conclusion, specific applications for emergency medicine, perioperative planning, and maybe for screening of patients exist or will soon develop. Many patients have not been able to be examined with these diagnostic entities, as they were hospitalized in ICUs in a critical condition: these were, however, exclusion criteria for the prelaunch studies performed with Levovist. Data will soon be accrued for these indications as Levovist becomes freely available. Another topic in this regard is clearly the missing capability to image in a 3-dimensional manner: this type of disorder definitively needs 3D datasets; only then can sonography compete with currently available techniques such as angiography or MRA/MRI.

CAVERNOUS HEMANGIOMAS

Cavernomas are diagnosed either electively by MRI or during preparation for the emergency surgery of an intracranial hematoma. There are no data in the

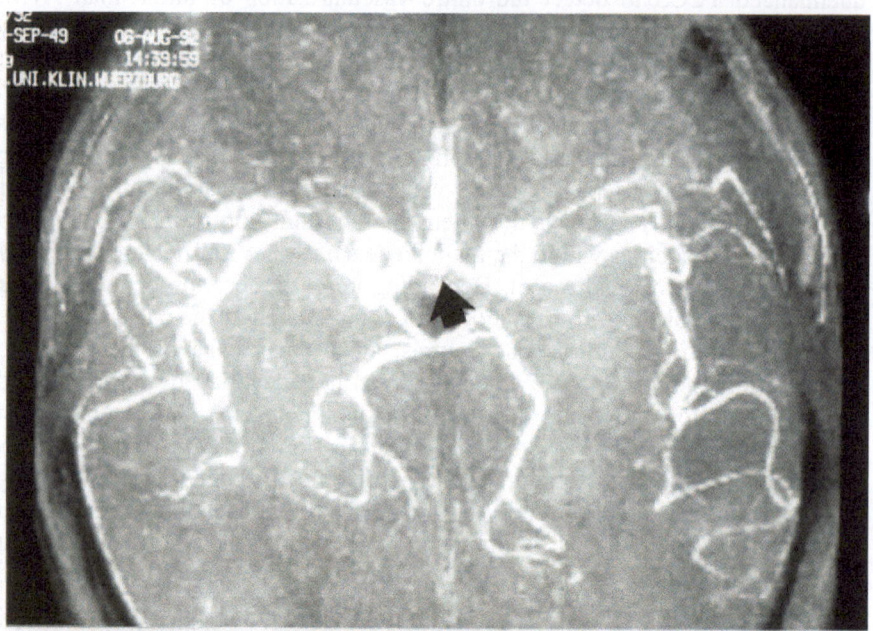

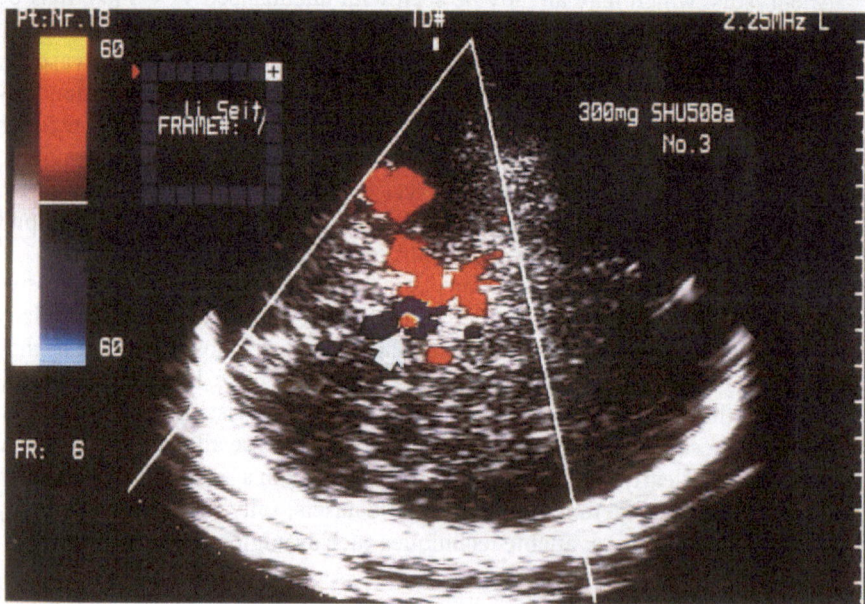

Figure 3. Imaging of intracerebral aneurysm by contrast-enhanced TCCS. (A) MRI-angiography of a small (2.3 mm diameter) anterior communicating artery aneurysm (arrow). (B) Contrast-enhanced color Doppler image (axial scan) showing the aneurysm as a color adnex to the regular vascular anatomy (arrow) of the anterior cerebral artery. All studies were performed after i.v. administration of the ultrasound contrast agent Levovist (SHU 508A; 400 mg/ml, 8 ml).

ultrasound literature relating to transcranial preoperative diagnosis of cavernomas. The potential of contrast-enhanced TCCS in delineating this vascular abnormality was, therefore, evaluated. In 10 patients with proven cavernomas (history of intracranial hematomas, typical MRI findings, surgery), nine lesions could be identified as hyperechogenic areas with dimensions and location correlating well with MRI findings. With unenhanced TCCS flow within cavernomas was undetectable. After application of ultrasound contrast agents (8 ml of 400 mg/ml SHU 508A), color Doppler imaging displayed slow flow within these lesions: Doppler sampling revealed mean peak systolic flow velocities of approximately 10–15 cm/s; diastolic flow had a venous flow pattern. Angle correction was not performed because flow was multidirectional in these lesions [21]. These velocities could be confirmed intraoperatively without contrast application. We suspect that when contrast agents become available for routine clinical use, contrast-enhanced TCCS may be especially helpful intraoperatively during the extirpation of the cavernoma, as well as during postoperative follow-up. Contrast-enhanced TCCS may also become a valuable tool for delineating the putative source of atypical parenchymal hemorrhage in the emergency room.

BRAIN TUMORS/TUMOR NEOANGIOGENESIS

Since the early 1960s, transcranial sonography has been devoted to the identification of brain tumors. In the pre-CT era several reports presented the sonographic findings of intracerebral space-occupying lesions and found a high frequency of cerebral tumors [11]. Although CT and MRI replaced sonography in the diagnosis of brain disorders in adults, ultrasound has become an established diagnostic method in pediatric and intraoperative neuroimaging due to the fine resolution, portability, ease of performance and relative cost-effectiveness of modern ultrasound systems [33–38]. In pediatric and intraoperative sonographic neuroimaging, detailed information has accumulated concerning the ultrasound pattern of brain tumors. With the introduction of modern transcranial Duplex ultrasound techniques, the experience gained from pediatric and intraoperative neurosonology has been applied to TCCS in adult brain tumors [32].

Despite a great range of histology, the echo-pattern of brain tumors in adults is similar: brain tumors normally exhibit increased echogenicity. The hyperechogenic matrix is often interspersed with more or less extensive hypoechogenic areas. According to the amount and distribution of hypoechogenic areas, the spectrum of sonographic appearance of brain tumors can extend from homogeneously hyperechogenic lesions to hypodense lesions surrounded by a thin, hyperechogenic margin. Comparing ultrasound findings with histopathology revealed that echogenicity of brain tumors may be related to the cell density and the extracellular constituents [12]: hyperechogenic areas reliably correspond to solid tumor tissue and hypodense lesions within the hyperdense matrix represent necrotic or cystic tumor areas. Cysts are anechogenic and distinctly circumscribed. Meningiomas, lymphomas, ependymomas and solid astrocytomas predominantly appear as hyperechogenic lesions; glioblastomas, high grade gliomas and metastatic lesions frequently exhibit non-homogeneous echotexture; cystic astrocytomas are almost always hypoechogenic. Regardless of their histological grading and biological behavior, preliminary experiences have revealed that about 5% of cerebral brain tumors are isoechogenic relative to normal brain tissue and therefore not detectable by TCCS, even when the bone window is adequate [32]. A study comparing CT, TCCS and histological findings indicated the superior sensitivity and fine resolution of TCCS over CCT in the differentiation of tumor components (necrotic/cystic areas, solid tumor, peritumoral tissue) [14].

Brain tumors are sometimes surrounded by a thin, hypoechogenic 'halo', a result of adjacent brain tissue compression and consequent changes in tissue impedance [32]. While some authors have reported that perifocal brain edema exhibits slightly increased [33, 39] or decreased echogenicity [40], most agree that brain edema is isoechogenic relative to normal brain parenchyma and therefore not detectable by TCCS [12]. According to our preliminary findings, increased echogenicity always implies solid tumor or gliosis [12–14, 35].

Brain tumor extension can be detected precisely by TCCS. Similar findings have been reported from intraoperative studies using ultrasound [12, 34, 35]. Comparative studies of CT and TCCS revealed some advantages of TCCS over CT in the estimation of tumor size, particularly in tumors with low attenuation on CT [14, 35]. The 'isoechogenicity' of brain edema and superior ability of TCCS to differentiate tissue components may serve as one explanation for the accuracy of TCCS in the delineation of solid tumor extent. However, it should be emphasized that TCCS, as with all other neuroimaging methods, also fails to depict the far peripheral tumor zone, where it diffusely infiltrates normal tissue.

Sonography after brain tumor resection, either through the intact skull or through craniotomy defects, allows unequivocal identification of residual tumor tissue and tumor regrowth [41–45]. The resection defect is displayed as an anechogenic area. The resection cavity is separated by a thin echogenic margin from adjacent brain parenchyma. The tumor-free resection line appears as a thin (<0.5 cm), slightly hyperechogenic margin of the resection defect. Residual tumor tissue appears as a hyperechogenic lesion adjacent to the resection cavity. Postoperative TCCS examinations are performed 1–2 weeks after surgery, since postoperative artifacts may interfere with the ability to detect residual tumor during the first days after surgery: air trapped in the resection cavity causes a distinct increase in echogenicity and prevents differentiation of tumor within the resection site. These artifacts usually resolve after approximately 5 days. Bleeding along the resection line or within the resection bed may also confound the interpretation of early postoperative TCCS studies. In contrast to the remaining tumor tissue, the increased echogenicity of blood gradually decreases within the first 2 postoperative weeks [5, 6].

As known from pediatric and intraoperative sonography, ultrasound is very sensitive not only in the identification of residual tumor but also in the detection of tumor recurrence [13, 34, 41–45]. Comparing CT, MRI and TCCS data in a series of 69 children, Dittrich and co-workers demonstrated that ultrasound and MRI are almost equally sensitive, and better than CT, in identifying residual tumor and tumor regrowth [41]. In addition, De Slegte presented data from post-operative sonographic follow-up examinations through a surgical defect and suggested that ultrasound provides valuable information about tumor response to chemotherapy and radiation: increasing irregularity of the matrix of residual tumor tissue indicated necrosis and reflected the effects of adjuvant therapy [45]. Our preliminary experience with CT, TCCS and histological findings suggested that TCCS was more sensitive than postoperative CT for the identification of residual tumor tissue. Tumor recurrence, on average, was detected earlier by TCCS than by CT (range 0–6 months).

The potential of TCCS to detect primary intracranial tumor neoangiogenesis was recently studied in a series of 28 patients, examined before and after application of Levovist (SHU 508A) [46]. Patients had conventional neuroradiological CCT and MRI examinations. All lesions were visualized as hyperechogenic lesions in B-mode and these correlated well with respect to extension and location with CCT/MRI-data. Color flow signals associated with these hyperechogenic lesions were detected in a single high grade and low grade tumor, each prior to contrast application: color-flow signals were observed after contrast in the majority of low grade and in all high grade lesions. High grade malignant tumors always displayed atypical arterial and venous Doppler spectra with irregular distribution of Doppler shift and Doppler signal intensities (Figure 4). These atypical flow patterns were also inconsistently detected in some low grade tumors. Contrast-enhanced TCCS was superior to intra-arterial digital subtraction angiography (DSA) in displaying tumor-related pathological circulation, indicating a high sensitivity of contrast-enhanced TCCS in the detection of tumor vascularization. Increased vascular proliferation as seen on histological examination always correlated with atypical flow phenomena in high grade lesions. These results suggest that data relating to tumor vascularization may supply new prognostic information about the biology of individual tumors. In addition, TCCS, as a real time non-invasive monitoring system, may find application in perioperative and treatment monitoring of CNS tumors, detection of tumor progression or recurrence following treatment, and in differentiation of radiotherapy related morbidity (radionecrosis vs. recurrence).

PERSPECTIVES OF TCCS AS A PARENCHYMAL NEUROIMAGING MODALITY

CT and MRI are based on X-ray tissue absorption respective excitation of magnetic moments; ultrasound, however, is based upon a different physical parameter, namely velocity and reflection of ultrasound waves in tissue, quantified by echo-

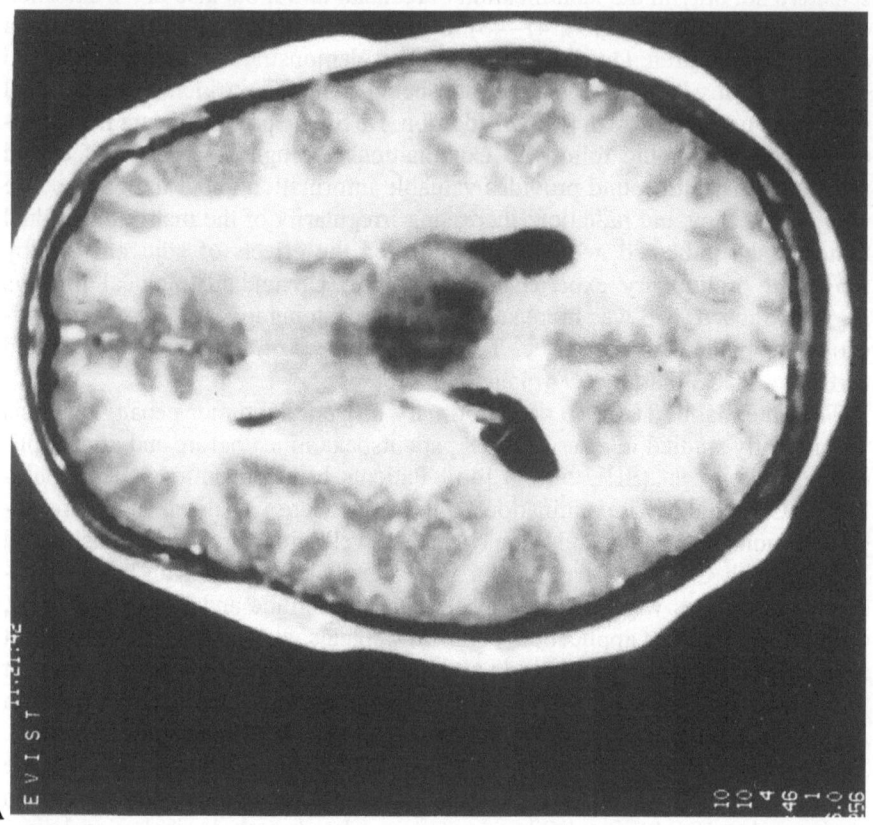

A

Figure 4. (Above and following pages) Imaging of intracranial neoplasms by contrast-enhanced TCCS. Tumor vascularization in a grade III astrocytoma of the splenium. (A) Axial MRI proton density (SE 460/20) depiction of astrocytic tumor within the splenium. (B) Transcranial B-mode image of almost identical scanning plane, taken from a right temporal acoustic bone window, showing the hyperechoic tumor of the splenium (arrow), as well as the lateral ventricles. The open arrow indicates the displaced pericallosal artery. (C) Detection of Doppler spectrum within the pericallosal artery (arrow) during injection of 8 ml of 400 mg/ml SHU 508A. Note the intensity of the Doppler spectrum and, observing the position of the sample volume, the displacement of the vessel by the tumor. (D) Transcranial duplex illustrating atypical arterial flow spectrum within the tumor parenchyma (Doppler sample volume indicated by arrowhead). Atypical arterial Doppler spectra displayed an irregular distribution of Doppler shift and an irregular distribution of signal intensities.

genicity. Therefore it may add new pathophysiological insights and TCCS may, therefore, unveil new etiopathogenetic considerations. This may be illustrated in affective disorders: a reduced echogenicity of the brain stem raphe was found in patients with unipolar depression [16]. No differences in raphe echogenicity were seen in healthy adults or in patients with a bipolar affective disorder. The

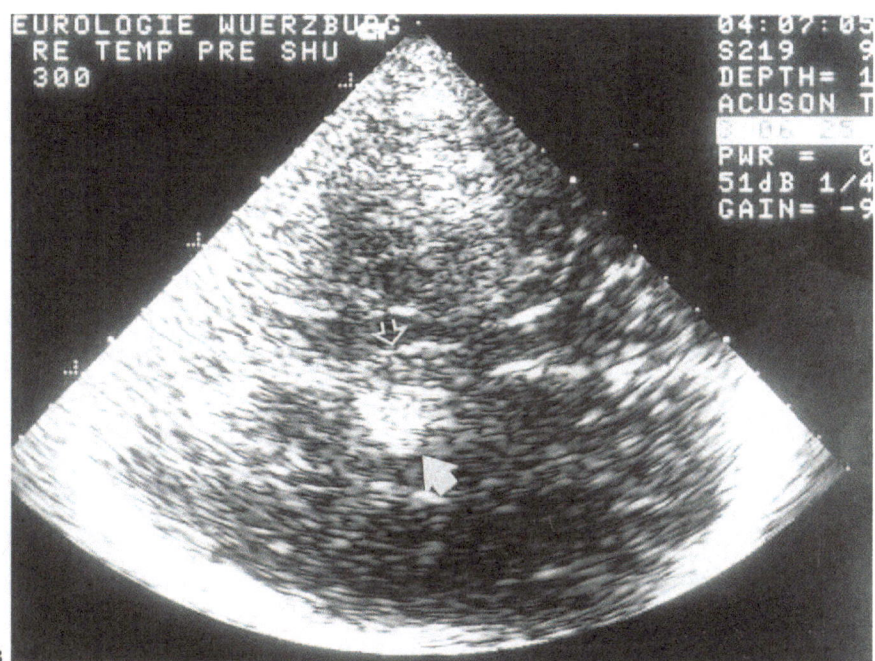

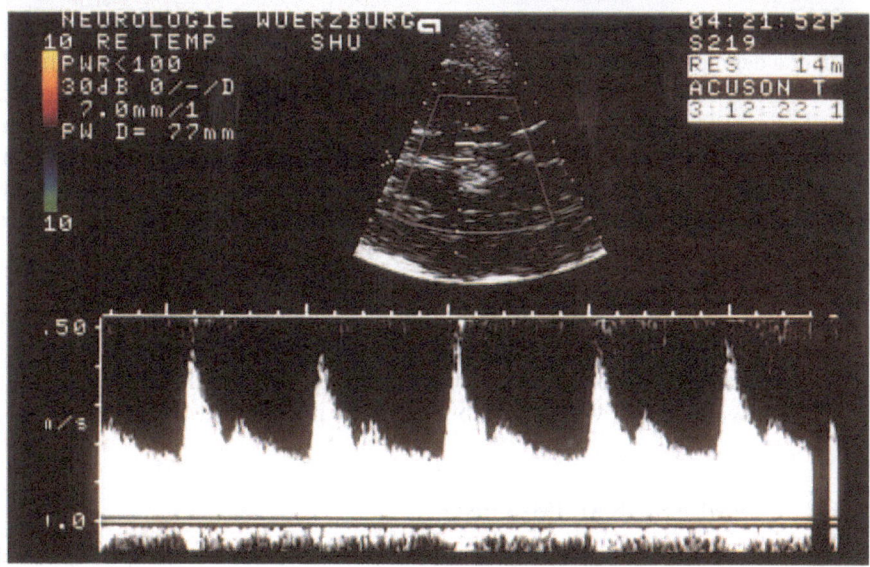

Figure 4. *Continued.*

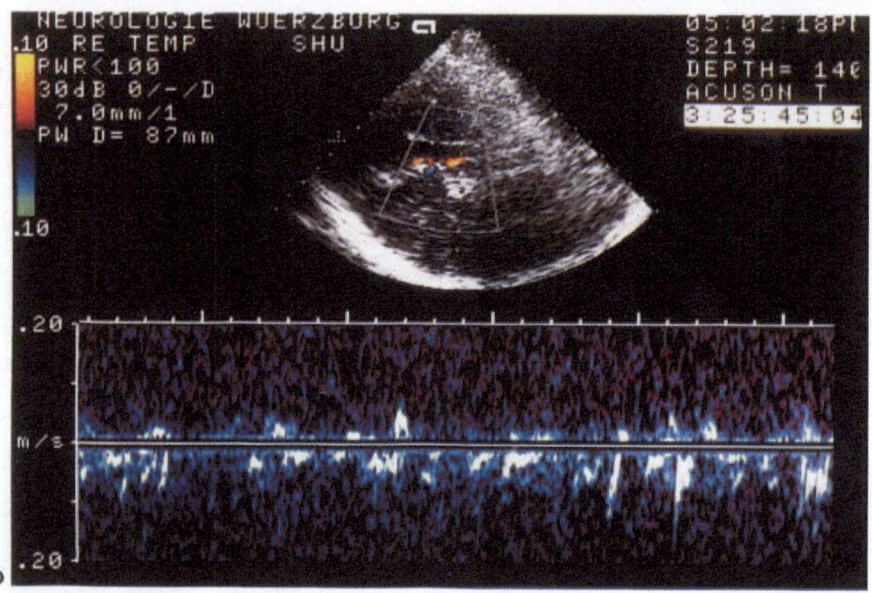

Figure 4. *Continued.*

distinct reduction in raphe echogenicity identified in unipolar depression may reflect a degree of structural disruption of the brain stem raphe, which is normally a sonographically homogenous territory. However, the raphe contains groups of serotoninergic raphe nuclei interwoven with the crossing tegmental fibers and the medial forebrain bundle, an accumulation of noradrenergic, cholinergic and serotonergic pathways. Interestingly, there is considerable evidence suggesting that the brain stem raphe may be involved in the pathogenesis of affective disorders [47–52]. In the clinical setting, TCCS was the first method to demonstrate this observation.

Neurodegenerative disorders are another application for TCCS as a new imaging modality. So far, no data have accrued from looking at CNS tissue perfusion or vascularization of these disorders employing echo-enhancing agents such as Levovist. This will clearly be necessary in the immediate future.

THREE-DIMENSIONAL IMAGING

The strong increase in signal achieved by echo enhancing agents allows imaging even of the peripheral small vessels through the intact skull. Three-dimensional imaging of the intracranial contents was therefore undertaken for continuous vessel imaging, projection plane imaging and to also improve parenchyma orientation. In a preliminary study, 11 patients with different neurological disorders were examined following administration of Levovist (2–6 injections; 10 ml of 300 mg/ml or 8 ml of 400 mg/ml) by TCCS (2.5 MHz, phased array, color duplex system) using velocity-based color Doppler (CD), amplitude-based power Doppler (PD) and spectral Doppler (SD). Scanning was performed through the intact temporal acoustic bone window in rotational geometry. The video output of the ultrasound scanner was digitized and the spatial position of the recorded frames was simultaneously registered using a mechanical position sensor. After automatic segmentation of the color information, the 3D datasets were reconstructed offline on a Unix workstation. Visualization was achieved by maximum intensity projection or surface visualization techniques. In all 11 patients Levovist strongly enhanced the vascular Doppler signal intensity. Echo-enhancement was sufficient for acquisition of a 3D dataset in all 11 patients with an imaging window of 3–5 min. For 3D reconstruction, contrast-enhanced power Doppler and intensity-weighted segmentation was identified as the best method. Three-dimensional visualization of the complete circle of Willis was achieved in all patients examined, and pathological vascularization of the parenchyma was encountered. Contrast-enhanced color Doppler imaging of cerebral vessels through the intact skull may be used to obtain 3D reconstructions of the basal circle of Willis and to recognize pathological parenchyma structures with atypical vascularization. Contrast-enhanced power Doppler and 3D analysis are therefore a suitable basis for quantification [53] (Figure 5).

REFERENCES

1. Berland LL, Bryan CR, Sekar BC, Moss CN. Sonographic examination of the adult brain. J Clin Ultrasound 1988;16: 337–45.
2. Schöning M, Grunert D, Stier B. Transkranielle real-time Sonography bei Kindern und Jungendlichen. Ultraschallanatomie des Gehirns. Ultraschall 1988;9:286–92.
3. Bogdahn U, Becker G, Winkler J, Greiner K, Perez J, Meurers B. Transcranial color-coded real-time sonography of adults. Stroke 1990;21:1680–8.
4. Becker G, Bogdahn U. Transcranial color-coded real-time sonography in adults. In Babikian V, Wechsler L, editors. Transcranial Doppler Ultrasonography. Mosby Year Book, Neurosonology 1993.
5. Seidel G, Kaps M, Dorndrof W. Transcranial color-coded Duplex sonography in intracerebral hematomas in adults. Stroke 1993;24:1519–27.
6. Becker G, Lindner A, Winkler J, Hoffmann E, Bogdahn U. Differentiation between ischemic and hemorrhagic stroke by transcranial color-coded real-time sonography. J Neuroimaging 1993;3:41–7.
7. Enzmann DR, Britt RH, Loyons BE et al. Natural history of experimental intracerebral hemorrhage: sonography, computed tomography and neuropathology. Am J Nuclear Radiol 1981;2:517–26.
8. Becker G, Lindner A, Hofmann E, Bogdahn U. Contribution of transcranial color-coded real-time sonography to the etiopathogenetic classification of middle cerebral artery stenosis. J Clin Ultrasound 1994;22:471–7.
9. Becker G, Greiner K, Kaune B et al. Diagnosis and monitoring of subarachnoid hemorrhage by transcranial color-coded real-time sonography. Neurosurgery 1991;28:814–20.
10. Becker G, Winkler J, Hoffmann E, Bogdahn U. Imaging of cerebral arterio-venous malforma-

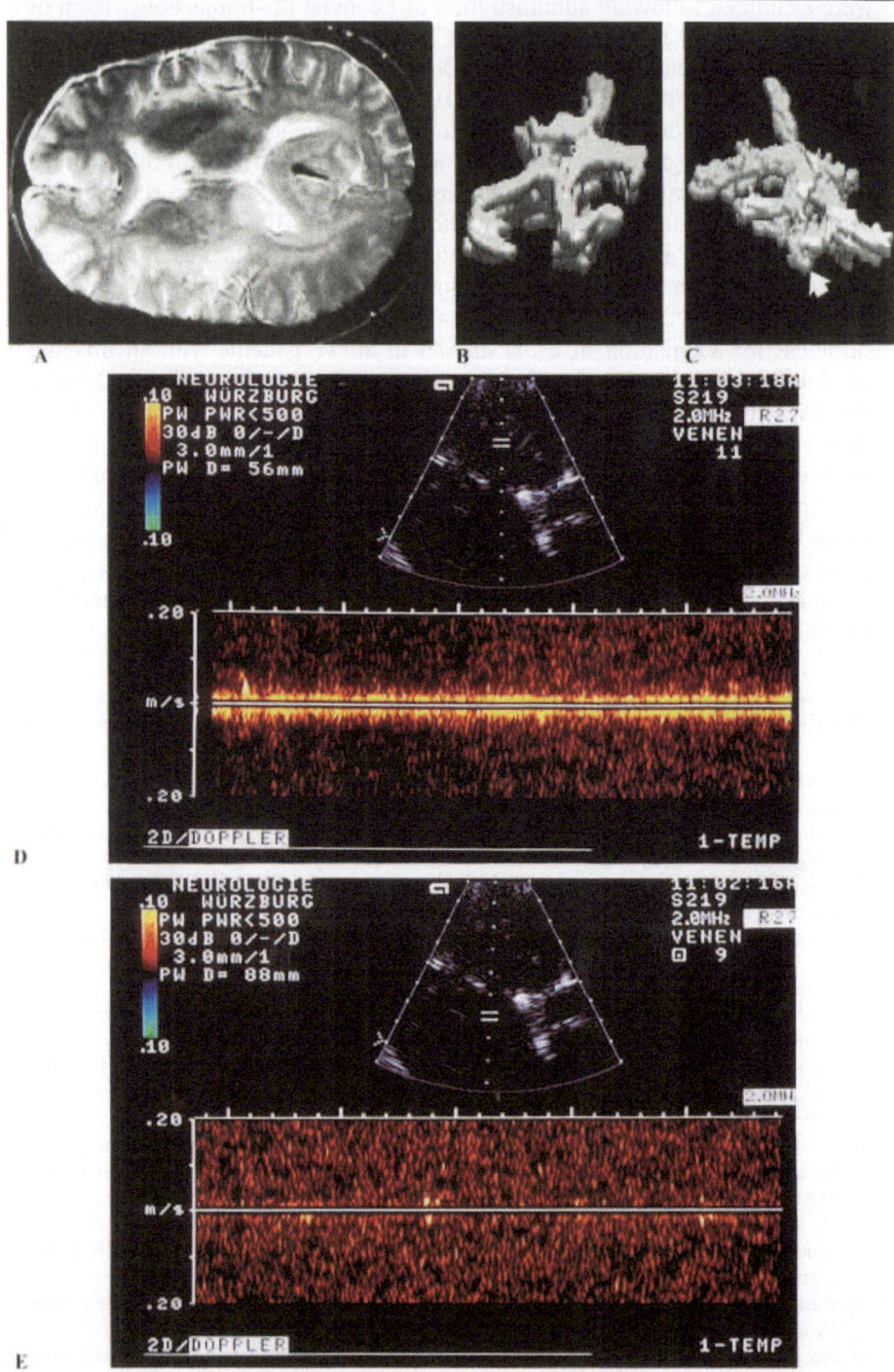

tions by transcranial colour-coded real-time sonography. Neuroradiology 1990;3:280–8.

11. French LA, Wild JJ, Neal D. The experimental application of ultrasonic to the localisation of brain tumors. J Neurosurg 1951;8:198–203.

12. Enzmann DR, Wheat R, Marshall WH et al. Tumors of the central nervous system studied by computed tomography and ultrasound. Radiology 1985;154:393–9.

13. LeRoux PD, Berger MS, Ojemann GA, Wang K, Mack LA. Correlation of intraoperative ultrasound tumor volumes and margins with preoperative computerized tomography scans. J Neurosurg 1989;71:691–8.

14. Becker G, Krone A, Koulis D, Hofmann E, Bogdahn U. Reliability of transcranial color-coded real-time sonography in the assessment of brain tumors: correlation between ultrasound, computerized tomography and biopsy findings. Neuroradiology 1995;365:90.

15. Becker G, Seufert J, Bogdahn U, Reichmann H, Reiners K. Degeneration of substantia nigra in chronic Parkinson's disease visualized by transcranial color-coded real-time sonography. Neurology 1995;45:182–4.

16. Becker G, Struck U, Bogdahn U, Becker T. Echogenicity of the brainstem raphe in patients with major depression. Psychiatry Res Neuroimaging 1994;55:75–8.

17. Becker G, Bogdahn U, Strassburg H-M et al. Identification of ventricular enlargement and estimation of ventricular pressure by transcranial color-coded real-time sonography. J Neuroimaging 1994;4:17–22.

18. Schwarz KQ, Bechar H, Schimpfky C, Vorwerk D, Bogdahn U, Schlief R. A study of the magnitude of Doppler enhancement with SH U 508 A in multiple vascular regions. Radiology 1994;193:195–201.

19. Bogdahn U, Becker G, Schlief R, Reddig J, Hassel W. Contrast-enhanced transcranial color-coded real-time sonography. Stroke 1993;24:676–84.

20. Otis S, Rush M, Boyajian R. Contrast-enhanced transcranial imaging. Results of an American phase-two study. Stroke 1995;26:203–9.

21. Bogdahn U, Becker G, Fröhlich T. Contrast enhanced transcranial color coded real time sonography of cerebrovascular disease. Echocardiography 1993;10:678.

22. Bogdahn U, Becker G, Fröhlich T et al. Contrast-enhanced transcranial color-coded real-time sonography allows detection of primary central nervous system tumor vascularisation. Radiology 1994;192(1):141–8.

23. Schöning M, Staab M, Walter J, Niemann G. Transcranial color duplex sonography in childhood and adolescence. Stroke 1993;24:1305–9.

24. Schöning M, Buchholz R, Walter J. Comparative study of transcranial color duplex sonography and transcranial Doppler sonography in adults. J Neurosurg 1993;78:776–84.

25. Schöning M, Walter J. Evaluation of the verterbro-basilar-posterior system by transcranial color Duplex sonography in adults. Stroke 1992;23:1280–6.

26. Kaps M, Seidel G, Bauer T, Behrmann B. Imaging of the intracranial vertebrobasilar system using color-coded ultrasound. Stroke 1993;23:1577–82.

27. Becker G, Bogdahn U, Lindner A. Imaging of the vertebro-basilar system by transcranial color-coded real-time sonography. J Ultrasound Med 1993;12:395–401.

28. Bauer A, Becker G, Krone A, Fröhlich T, Bogdahn U. Transcranial duplex sonography using ultrasound contrast enhancers. Clin Radiol 1996;51(Suppl 1):19–23.

29. Aaslid R, Newell DW, Stoss R, Sorteberg W, Lindegaard KF. Assessment of cerebral autoregulation dynamics from simultaneous arterial and venous transcranial Doppler recordings in humans. Stroke 1991;22:1148–54.

Figure 5. (Opposite) Contrast-enhanced 3D tumor vascularization. (A) T2-weighted MR-image of thalamic glioma (WHO Gr. II/III) (arrow). (B) 3D-visualization of circle of Willis (normal subject) with Levovist-enhanced 3D power Doppler. (C) 3D-visualization of circle of Willis with tumor vascularization (arrow) of thalamic glioma. (D) TCCS and Doppler spectrum within thalamic glioma (atypical vascularization). (E) TCCS and Doppler spectrum within normal thalamus.

30. Becker G, Bogdahn U, Gehlberg C, Fröhlich T, Hofmann E, Schlief R. Transcranial color-coded real-time sonography of intracranial veins. Normal values of blood flow velocities and findings in superior sagittal sinus thrombosis. J Neuroimaging 1995;5:87–94.
31. Bartels E, Rodiek SO, Flügel KA. Quantitative measurement of blood flow velocity in basal cerebral arteries with transcranial duplex flow-imaging. J Neuroimaging 1994;4:77–81.
32. Becker G, Perez J, Krone A et al. Transcranial color-coded real-time sonography in the evaluation of intracranial neoplasms and arteriovenous malformations. Neurosurgery 1992;31:420–8.
33. McGahan JP, Ellis WG, Budenz RW, Walter JP, Boggan J. Brain gliomas: Sonographic characterisation. Radiology 1986;159:485–92.
34. Gooding GA, Boggan JE, Weinstein PR. Characterisation of intracranial neoplasms by CT and intraoperative sonography. Am J Nuclear Radiol 1984;5:517–20.
35. Knake JE, Chandler WF, Gabrielsen TO, Latack JT, Gebarski SS. Intraoperative sonographic delineation of low-grade brain neoplasms defined poorly by computed tomography. Radiology 1984;151:735–9.
36. Strassburg HM, Sauer M, Weber S, Gilsbach J. Ultrasonographic diagnosis of brain tumor in infancy. Pediatr Radiol 1984;14:284–7.
37. Chuang S, Harwood-Nash D. Tumors and cysts. Neuroradiology 1986;28:463–75.
38. Han BK, Babock DS, Oestreich AE. Sonography of brain tumors in infants. AJNR 1984;143: 31–6.
39. Quencer RM, Montalvo BM. Intraoperative cranial sonography. Neuroradiology 1986;28: 528–50.
40. Chandler WF, Knake JE, McGillicuddy JE et al. Intraoperative use of ultrasound in neurosurgery. J Neurosurg 1982;57:157–63.
41. Dittrich M, Gutjahr P, Dinkel E, Higer HP, Müller W. Postoperative sonographische Verlaufsuntersuchungen im Kindesalter. Klin Pädiat 1987;199:403–10.
42. Hoffman RB, Laudau B. Ultrasound B-scan imaging of intracranial lesion through a bone flap defect. J Clin Ultrasound 1975;4:125–7.
43. Rubinstein JB, Pasto ME, Rifkin MD, Goldberg BB. Real-time neurosonography of brain through a calvarial defects with computed tomography correlation. J Ultrasound Med 1980; 137:831–2.
44. Winkler P, Helmke K. Ultrasonic diagnosis and follow-up of malignant brain tumors in childhood. Pediatr Radiol 1985;15:215–9.
45. De Slegte RGM, Valk J, Kaiser MC. Sonography of the postoperative brain: a report of 2 years of experiences. Neuroradiology 1986;28:591–8.
46. Bogdahn U, Becker G, Fröhlich T et al. Contrast-enhanced transcranial color-coded real-time sonography allows detection of primary central nervous system tumor vascularisation. Radiology 1994;192:141–8.
47. Ball WA, Whybrow PC. Biology of depression and mania. Curr Opin Psychiatry 1993;6: 27–34.
48. Hillegaart V. Functional topography of brain seritonergic pathways in rat. Acta Physiol Scand 1991;(Suppl);598:1–54.
49. Rizvi TA, Ennis M, Shipley MT. Reciprocal connections between medial preoptic area and the midbrain periaqueductal gray in rat. A WGA-HRP and PHA-L study. J Comp Neurol 1992; 315:1–15.
50. Simpson PE, Weiss JM. Altered activity of the locus coeruleus in animal model of depression. Neuropsychopharmacology 1988;1:287–95.
51. Chan-Palay V, Asan E. Quantification of catecholamine neurons in the locus coeruleus in human brains of normal young and older adults and in depression. J Comp Neurol 1989;287:357–72.
52. Paulus W, Jellinger K. The neuropathologic basis of different clinical subgroups of Parkinson's disease. J Neuropathol Exp Neurol 1991;50:743–55.
53. Bauer A, Bogdahn U, Hutzlmeier A et al. Three dimensional vascular imaging by echo-enhanced transcranial power Doppler sonography. Radiology, submitted 1996.

34. The promise of sonographic contrast agents in the kidney

MICHELLE L. ROBBIN

INTRODUCTION

The ability of current ultrasound scanners to detect tissue perfusion is limited. An attempt to overcome this restricted color and spectral Doppler sensitivity has given rise to recent 'power Doppler' techniques which demonstrate smaller, more peripheral vessels [1]. However, detection of parenchymal blood flow in organs that are not extremely superficial and relatively motionless remains difficult. Therefore, enhancement of the detection of blood flow by intravenous ultrasound contrast agents may allow a significant expansion of sonographic diagnostic capabilities. Such an agent may improve the ability of existing equipment to detect blood flow, perhaps avoiding costly upgrades.

Few currently available intravenous ultrasound contrast agents have a half-life long enough to provide substantial enhancement of renal parenchyma. This is because most of these agents have a significant first pass removal via pulmonary capillaries [2, 3]. The current drugs, which are either recently approved by the FDA or in clinical trials, have an intravascular half-life which can be measured in minutes [4–8]. Therefore, investigation regarding the utility of intravenous ultrasound contrast agents in renal applications is in its infancy. Nevertheless, preliminary work at this institution and elsewhere has begun to demonstrate some of the many exciting potential applications of intravenous contrast agents in the kidney.

In this chapter, applications of ultrasound contrast agents in the evaluation of solid and complex cystic renal masses will be illustrated. These will include enhancement of the solid renal mass as well as the differentiation of solid masses from pseudotumors such as columns of Bertin. The potential for the contrast-enhanced sonographic evaluation of the complex renal mass to replace CT or MRI will be explored. Opportunities to improve the sensitivity and specificity of the sonographic detection of renal artery stenosis will be discussed. Potential applications in pyelonephritis and renal transplants will be reviewed.

525

N. C. Nanda, R. Schlief and B. B. Goldberg (eds.), Advances in Echo Imaging Using Contrast Enhancement, Second Edition, 525–541.

RENAL MASSES

Solid masses

The evaluation of solid or complex renal masses is one of the most frequently
encountered problems encountered by the sonologist [9, 10]. These may be
suspected because of hematuria or prior studies, but are common incidental
findings on routine abdominal sonograms. At present, a solid isoechoic or hypoechoic lesion is assumed to be a renal cell carcinoma until proven otherwise,
although benign lesions may have a similar appearance [11]. Lack of demonstrable flow within the lesion does not exclude a carcinoma. A pre- and post-
contrast thin-section CT scan to search for fat is performed in those lesions that
contain hyperechoic elements to exclude an angiomyolipoma [12]. The cost of
the radiological investigation of a renal mass could be reduced if the initial
sonogram was definitive.

The vascularity of these renal masses is difficult to characterize with current
ultrasound technology [13–15]. Visualization of the main renal vein when looking
for tumor thrombus is frequently incomplete. Unfortunately, color Doppler is
not currently sensitive enough to assess the vasculature of the normal renal
parenchyma and to look for alterations in blood flow suggestive of a very small
tumor. At present, there is no objective way to characterize the degree of enhancement of complex cysts with contrast (except perhaps videodensitometry), such
as with the Hounsfield number on CT.

Small hyperechoic masses (<3.0 cm) are especially problematic [16]. Some
reports [17, 18] have shown that small renal cell carcinomas often mimic angiomyolipomas sonographically. Thus a CT must be performed to determine
whether fat is present [19]. Future study may show a difference in the vascularity
between angiomyolipomas and renal cell carcinomas, allowing hyperechoic
small renal masses to be evaluated solely with ultrasound [13]. This may be
possible with the greater sensitivity to flow provided by the ultrasound contrast.

When a definite renal tumor is identified, the presence of a second tumor must
be excluded. In the search for a second renal lesion, a non-contour deforming
lesion may not be detected if there is no gray-scale difference between the tumor
and normal renal parenchyma. CT or MRI has the ability to detect small tumors
based on the different contrast enhancement between tumor tissue and the
surrounding normal tissue. Thus a CT or MRI is almost always required to complete the investigation of a solid or complex cystic renal mass, not only to assess
the mass itself but to look for vascular invasion, metastases and second lesions.

When a renal lesion is detected by ultrasound, the injection of an intravenous
ultrasound contrast agent should allow improved detection of vascularity within
the solid mass or solid portion of the cystic mass. With greater sensitivity to
flow, absence of vascularity may have better negative predictive value. Other
renal masses may become more conspicuous secondary to increased vascularity
or alteration of normal vessel courses. Contrast enhancement can be expected

to increase the ability to see the main renal vein and detect tumor thrombus. Enhancement of small vessels may increase the ability to detect flow in renal tumor thrombus, thus differentiating it from bland thrombus.

Figure 1 shows vascular enhancement of a solid hypoechoic renal adeno-carcinoma before and after administration of sonographic contrast agent. Very little flow was identified within the mass prior to the administration of the ultra-sound contrast agent. Definite peripheral flow within the tumor was noted after injection. CT images of an enhancing renal mass before and after administration of contrast are shown in Figure 2A, B. Corresponding ultrasound images depict a marked increase in vascularity in the mass after contrast (Figure 2C, D). A significant increase in the ability to detect color flow within the renal vein after contrast enhancement was demonstrated (Figure 2E, F), thus virtually eliminating the possibility of a renal vein thrombus. On pathological examination, this neo-plasm was an oncocytoma.

Demonstration of intratumoral vascularity where none was detected pre-viously may not change the management of a patient with a solid renal lesion, although it does clearly indicate that the mass is a neoplasm. However, if the renal vein can be imaged confidently and clot excluded, a more complete staging could be performed with ultrasound. If hepatic metastases can be excluded sonographically, and both kidneys, hilar region and renal vein can be confi-dently seen, the need for an abdominal CT scan for staging is arguable. While the CT may still be needed as a baseline for postoperative follow-up, the diagnosis and management may be clear after the first imaging study in a higher pro-portion of cases.

Complex cysts

Complex renal cysts (Bosniak type 2–4) are typically evaluated with pre- and post-contrast CT to assess the degree of contrast enhancement present [16, 20]. Not infrequently, a lesion classified as Bosniak type 2 on ultrasound will appear to be a simple cyst on CT because of the lack of vascularity of the septations. This is less common with type 3 lesions. However, the inability of the CT scanner to detect the fine detail imaged on sonography as a result of volume averaging, even on thin CT sections, can lead to uncertainty, and follow-up studies are usually performed.

A complex multi-cystic renal cell carcinoma (Bosniak class 4) imaged after sonographic contrast agent injection is shown in Figure 3. The marked vascularity of the renal septations increases diagnostic confidence that the mass is malignant. The renal vein in this case was not well seen prior to ultrasound contrast, but was obvious after contrast administration. A renal vein thrombus could thus be confidently excluded in this case after imaging with a sonographic contrast agent.

After more experience with sonographic contrast analysis of these complex

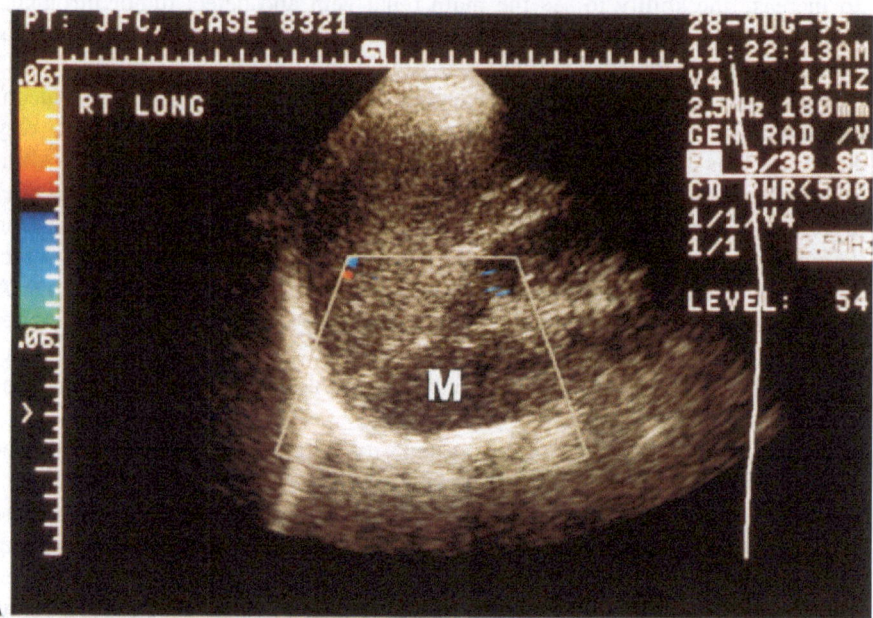

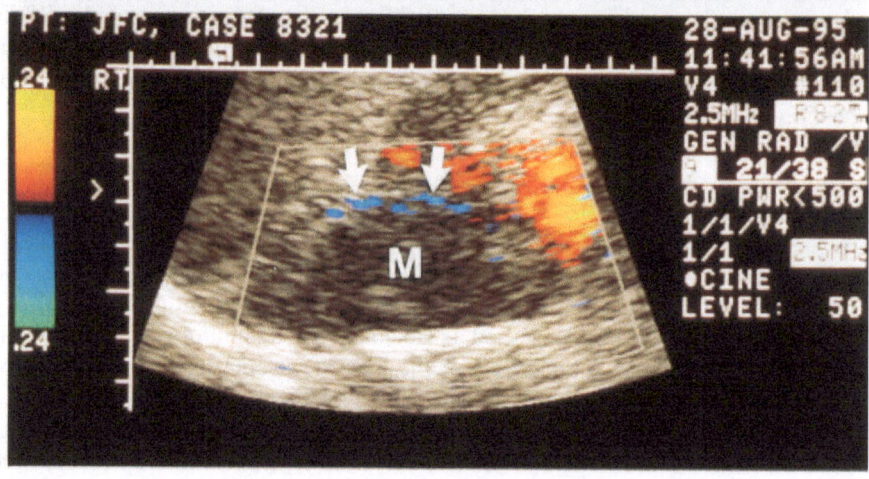

Figure 1. Renal adenocarcinoma. (A) Virtually no intra-tumoral color flow can be demonstrated in a 3.9 cm solid hypoechoic renal cell cancer using conventional color sonography (longitudinal, right kidney). M=Mass. (B) After ultrasound contrast (EchoGen) administration, a peripheral vessel in the tumor can be clearly identified (arrows).

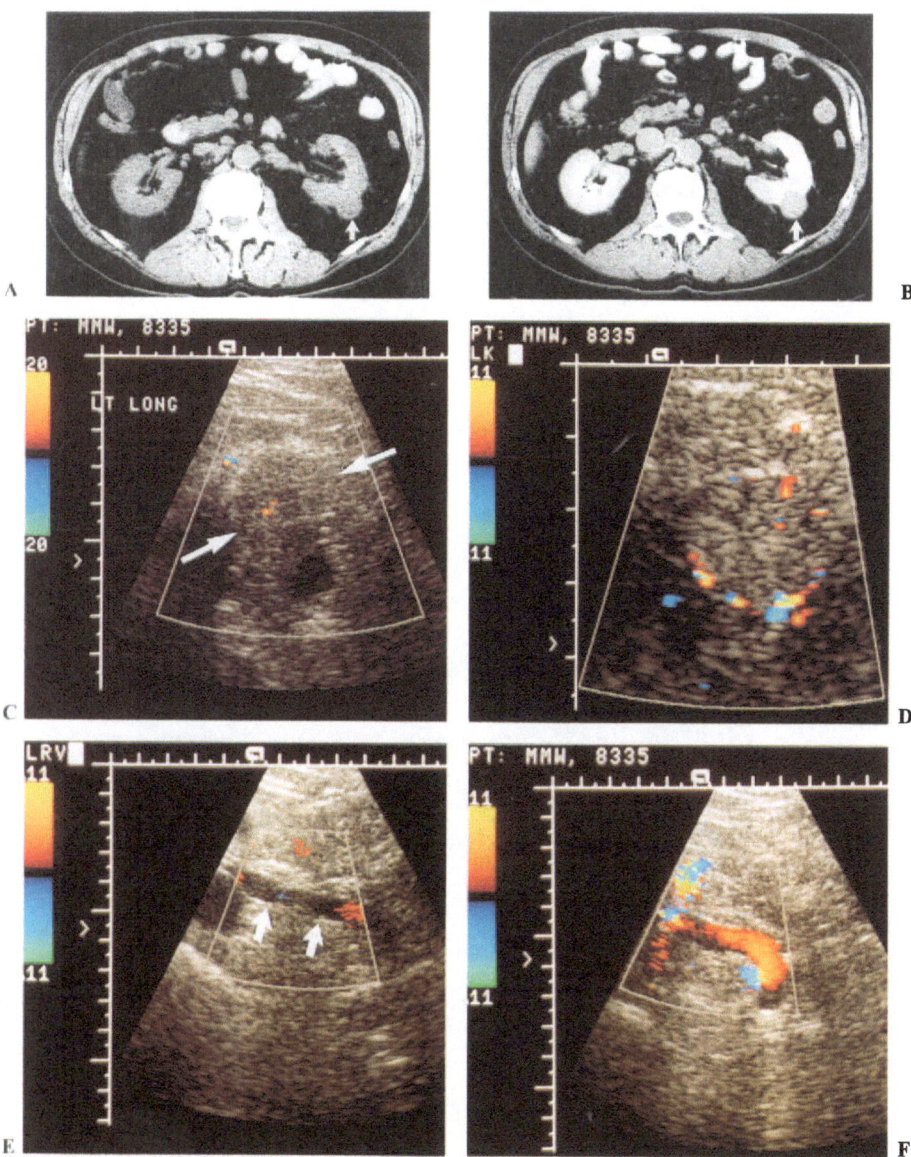

Figure 2. Oncocytoma. (A) Pre-contrast CT demonstrates 2.8 cm solid left renal mass (arrow), measuring 6 Hounsfield units. (B) Post-contrast CT shows enhancement of tumor to 62 Hounsfield units. (C) Pre-contrast transverse sonogram of the mass (arrows) shows very little color flow. (D) Post-ultrasound contrast (EchoGen) transverse sonogram shows enhanced peripheral and central color flow within the mass. (E) Pre-contrast, little color flow was seen in the renal vein (arrows). (F) Transverse sonogram after contrast demonstrates filling of the renal vein with color.

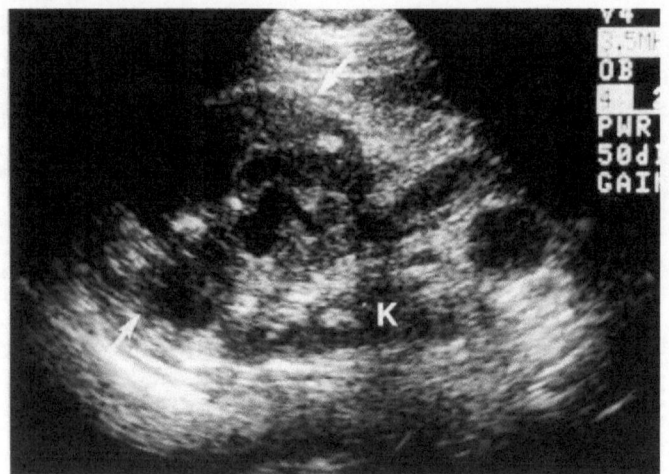

A

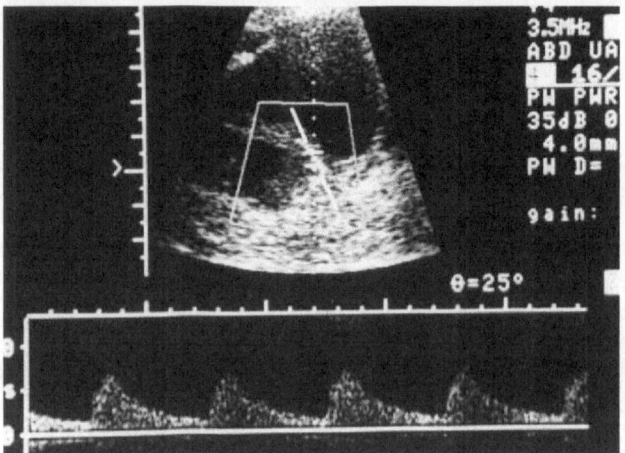

B

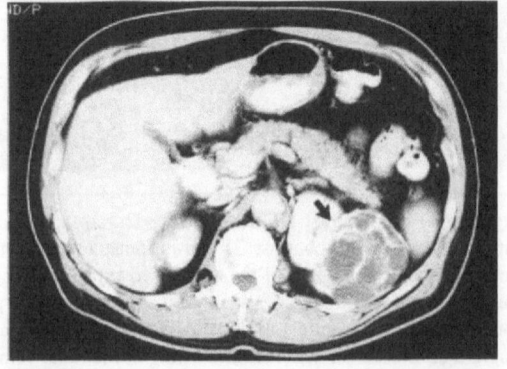

C

cysts has been gained, investigation of a non-simple renal cyst (Bosniak type 2 or 3) detected by ultrasound may be as follows: (1) Evaluation of the vascularity of the septations using an ultrasound contrast agent. (2) If no enhancement is seen on color Doppler or power Doppler in combination with an ultrasound contrast agent, follow-up with ultrasound in 3–4 months. If enhancement of septations is seen, consider surgical removal.

Questionable renal masses

Pseudotumors can be due to persistent fetal lobulations or prominent columns of Bertin. These can be difficult to distinguish from a focal renal mass. Slightly different echotexture may be seen because of disarray of the renal tubules due to anomalous embryology. Figure 4A shows a patient in whom a solid renal mass indistinguishable from a renal cell carcinoma was seen with ultrasound. The CT which followed showed the mass to be merely a prominent column of Bertin (Figure 4B). A subsequent ultrasound contrast study showed smooth branching of the vessels into and around the column with no increased vascularity in the region and no mass effect (Figure 4C, D). Note the significant increase in vascularity that was detected post-contrast. Ultrasound contrast agents may be a cost-effective way to exclude or diagnose a renal tumor without the need for additional verifying imaging studies in those cases of prominent renal fetal lobularity or columns of Bertin. This may significantly reduce the costs now incurred when incidental lesions or potential masses are discovered on sonography.

RENAL VASCULAR ABNORMALITIES

Renal artery stenosis

Approximately 1–2% of hypertensive individuals have elevated blood pressure secondary to renal artery stenosis. Such patients are potentially curable with revascularization techniques such as angioplasty, stenting or endarterectomy of the arterial lesion. However, definitive diagnosis requires a renal artery arteriogram, an invasive examination that may impair renal function in those with pre-existing renal disease. Many less invasive tests have been investigated as a potential screening study for renal artery stenosis, such as excretory

---

Figure 3. (Opposite) Cystic renal cell carcinoma, clear cell type. (A) Pre-contrast longitudinal sonogram shows a 7.7 cm complex multicystic upper pole renal mass (arrows). K, normal kidney. (B) Increased color and spectral flow was seen in the thick septations of the mass after ultrasound contrast (EchoGen) administration. (C) Contrast CT shows marked enhancement of the septations in the complex renal mass (arrow).

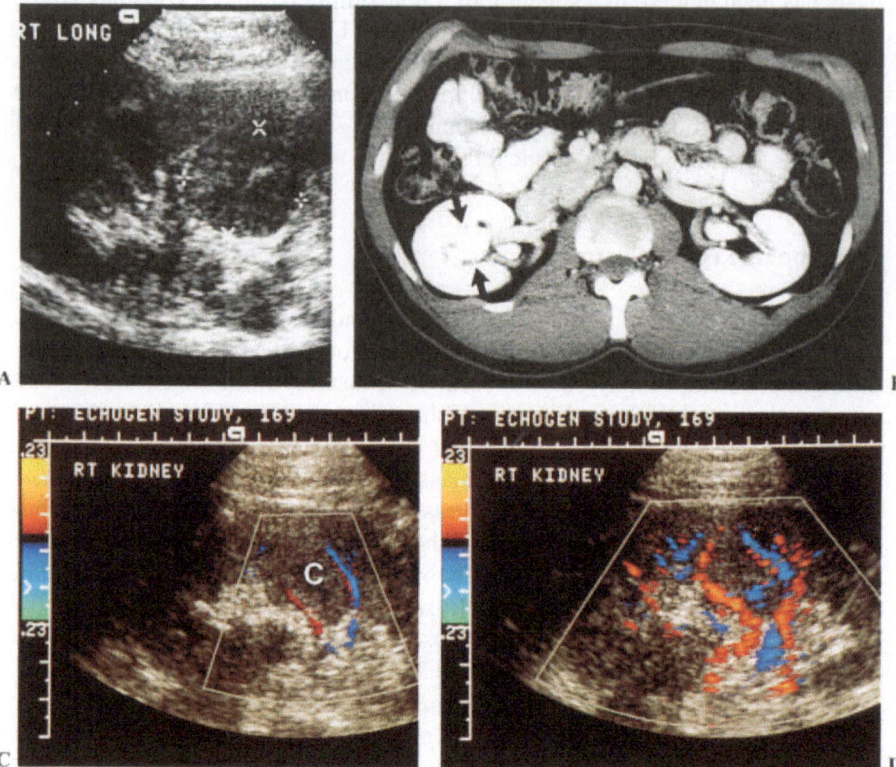

Figure 4. Prominent column of Bertin. (A) Longitudinal sonogram shows a renal mass indistinguishable from renal cell carcinoma (cursors). (B) Post-contrast CT shows a prominent column of Bertin (arrows). (C) Pre-contrast sonogram shows color flow in some normal vessels; however the vascularity shown is not sufficient to confidently exclude a neoplasm. C, Column of Bertin. (D) Post-contrast (EchoGen) color ultrasound shows smoothly branching vessel throughout the 'mass' without mass effect or increased vascularity in the region.

urography and sonography. One of the most often mentioned is captopril nuclear renography [21, 22]. Unfortunately, it may miss bilateral disease and is not sensitive in those patients with limited renal function [23].

A plethora of articles has recently been published concerning the ability to diagnose renal artery stenosis accurately by ultrasound [24–26]. The sonographic diagnosis of renal artery stenosis involves either direct interrogation of the main renal arteries or indirect assessment of the effects of the renal artery stenosis on the intrarenal segmental, interlobar or arcuate arteries.

The parameters most often used for detection of ≥60% stenosis in the main renal artery are an elevated peak systolic velocity (>180 cm/s) associated with post-stenotic turbulence, or a ratio of the renal artery peak systolic velocity to

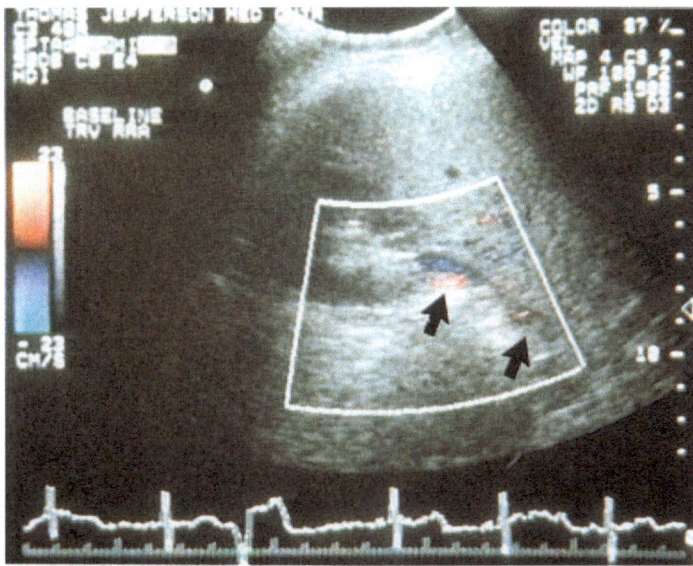

A

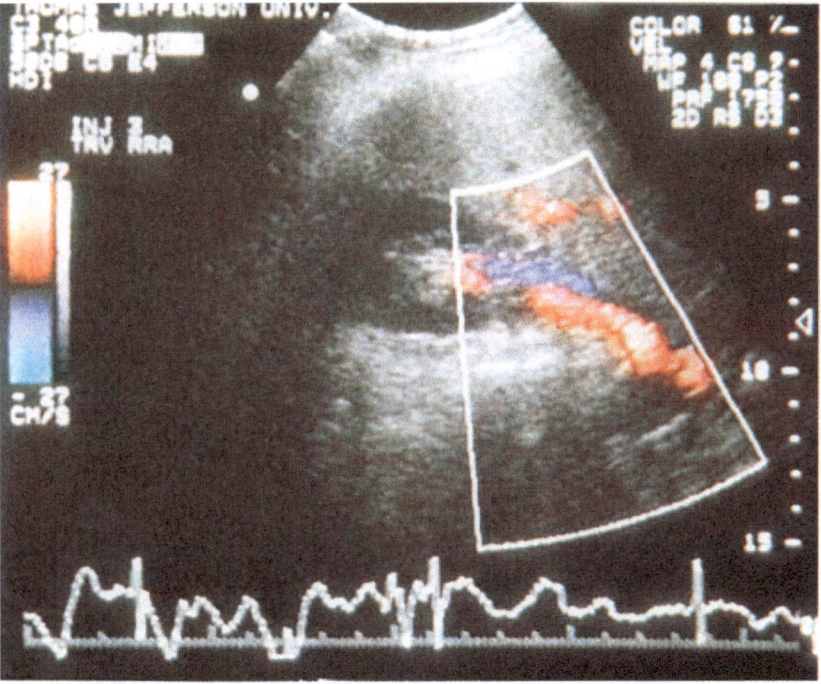

B

Figure 5. Renal artery. (A) Transverse sonogram of right kidney shows color flow in the hilar renal artery, with little color seen in the deeper renal artery (arrows) off the aorta. (B) Entire right renal artery is seen easily after contrast (Levovist) injection.

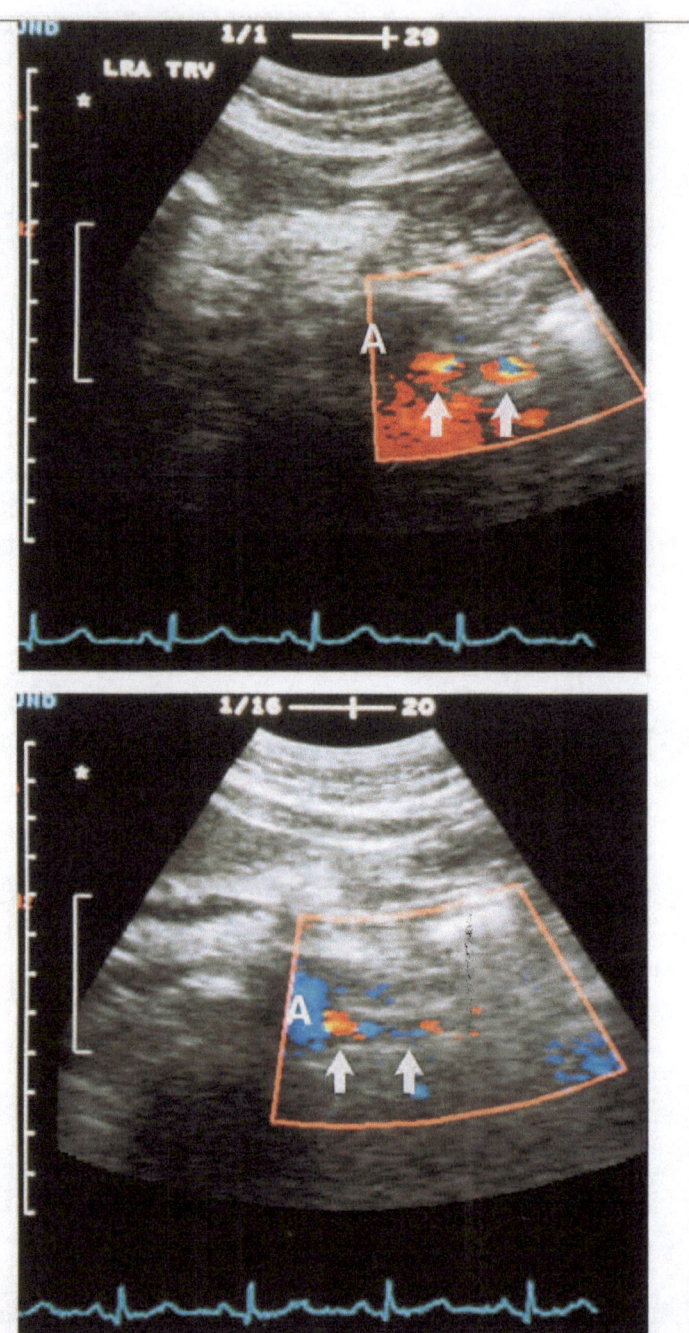

Figure 6. (Above and opposite) Renal artery stenosis (RAS), main renal artery technique. (A) Transverse sonogram of aorta shows some color flow in the left renal artery (arrows). A, Aorta. The image is noisy because of the high gain settings necessary for visualization. RAS was suspected but could not be proven sonographically. (B) Color flow in the main left renal artery (arrows) is seen better after sonographic contrast (Levovist) injection. (Continued on opposite page.)

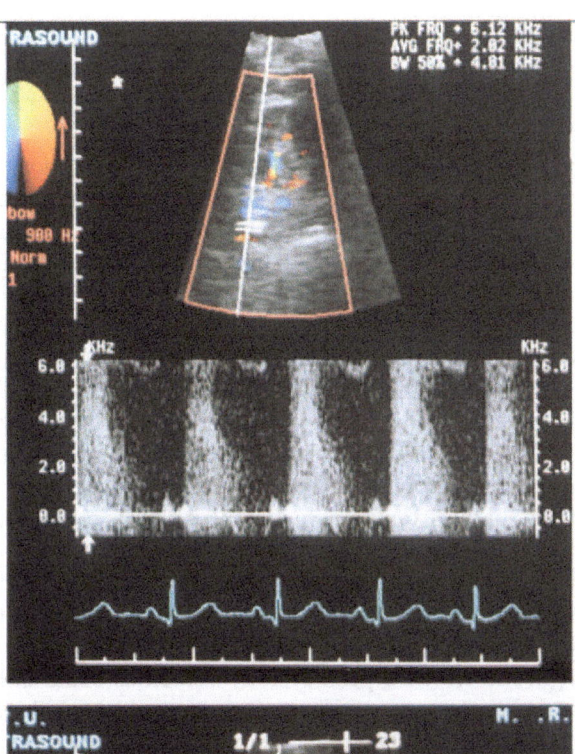

C

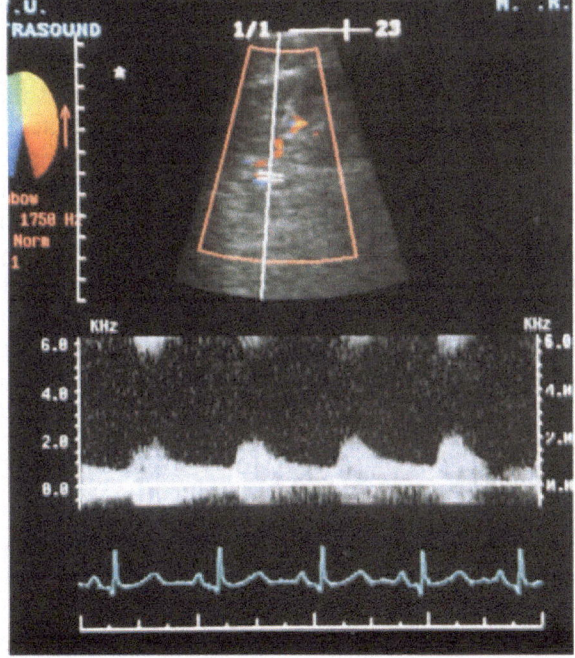

D

Figure 6. (*Continued*) (C) 6 KHz shift in main renal artery is diagnostic for RAS. (D) Distal to stenosis, lower resistance flow is seen with a 2 KHz shift. Only this region of the renal artery could be imaged pre-contrast.

aortic peak systolic velocity ≥3.5. Criteria in intrarenal arteries include analysis of the morphology of the waveform (loss of the early systolic peak), acceleration time <0.7 s, acceleration <3 m/s, and a change in the average resistive index of >0.5 between the kidneys.

No one sonographic criterion has been proven to be satisfactory when repeated by multiple laboratories. Visualization of the main renal arteries can be difficult secondary to overlying bowel gas, or in obese patients [27]. Accessory renal arteries, found in up to 20% of the population, are difficult to visualize. Intrarenal measurements can be difficult to obtain, particularly in patients who cannot hold their breath. Some of the intrarenal parameters, such as the change in resistive index, depend on the presence of unilateral stenosis. This is a problem as up to 30% of cases are bilateral [28]. Two of the seemingly most critical factors in the great variability of the study results are local expertise and available equipment. It is possible that an intravenous ultrasound contrast agent may make it easier to obtain both intrarenal and main renal artery waveforms and lessen the site-specific nature of this examination, making it a more reproducible technique, and thus more widely available.

Figure 5 shows a main renal artery before and after administration of ultrasound contrast agent. Note the significantly increased conspicuity of the main renal artery post-injection. Figure 6 shows a case of suspected renal artery stenosis where the diagnostic elevated peak systolic velocities could only be obtained with an ultrasound contrast agent. Figure 7 shows an abnormal intrarenal waveform in a patient with a 90% right renal artery stenosis. The difference in resistive index between the two sides was 0.06, also consistent with renal artery stenosis.

Preliminary work in this laboratory and others found that the use of ultrasound contrast agents enhanced the vascular signals detected in the kidney, making it quicker and easier to obtain both intrarenal and main renal artery and vein color and corresponding spectral Doppler signals. Contrast enhancement may decrease the difficulty encountered with overlying bowel gas obscuring the main renal arteries. Ease of detection of vascular signals may improve the proportion of successful examinations in patients who are unable to hold their breath. Improved detection of accessory renal arteries may allow a more complete examination of the renal arteries. Thus there may be some equalization of success between laboratories as the examination becomes technically easier to perform.

Renal vein thrombosis

Renal vein thrombosis is an entity difficult to recognize clinically [29]. Patients with renal vein thrombosis may have proteinuria, hematuria or acutely worsening renal function. Imaging is crucial for diagnosis, but intravenous urography is non-specific. Classic sonographic findings of renal vein thrombosis include an enlarged kidney and an enlarged renal vein filled with low level echoes. Failure to visualize flow within the renal vein is not diagnostic of renal vein thrombus,

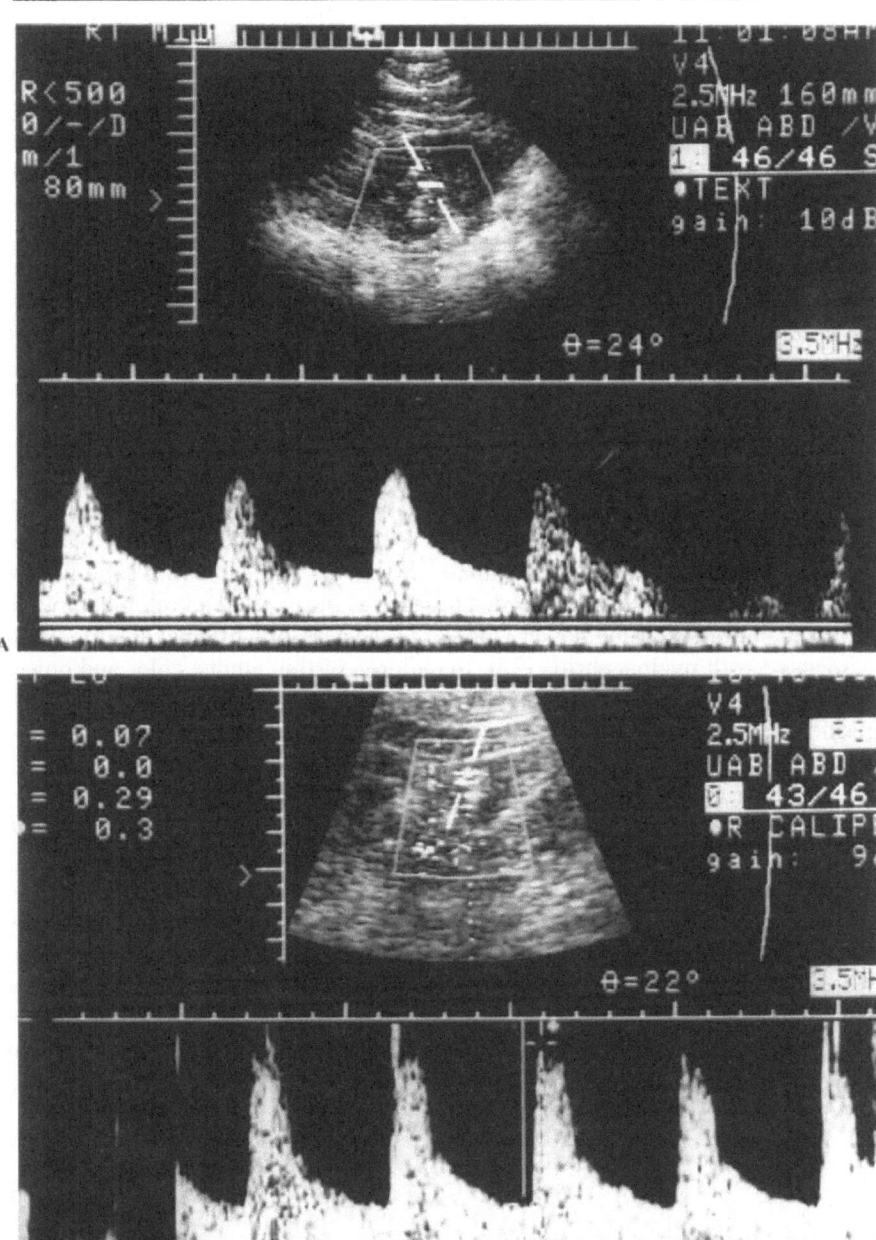

Figure 7. Renal artery stenosis, intrarenal waveform analysis. (A) Longitudinal sonogram of the right mid-kidney with an abnormal intrarenal arterial waveform post ultrasound contrast agent (EchoGen). (B) The left renal waveform appeared normal. Angiography showed a 90% right renal artery stenosis.

as the apparent lack of flow is sometimes secondary to poor arterial inflow or very slow venous outflow.

Visualization of the renal vein at the renal hilum is inadequate for the exclusion of renal vein thrombosis as extensive collaterals may develop relatively quickly in the presence of main renal vein thrombosis. Recent reports suggest renal arterial flow analysis is not helpful in the detection of native renal vein thrombosis, in contradistinction to the renal transplant [30].

In FDA phase 2 and 3 trials, markedly improved sonographic visualization of the course of the main renal vein after contrast injection was seen in several patients. More clinical experience is needed but ultrasound contrast agents may be useful in the detection and exclusion of renal vein thrombosis, without the need for administration of potentially nephrotoxic contrast at CT or venography.

RENAL INFECTIONS

To date, sonography has shown limited usefulness in the evaluation of acute renal infections [31]. CT is considered to be the more definitive procedure (although it is only used when needed because of the expense). Preliminary work has been performed using power Doppler in the detection of renal abnormalities associated with pyelonephritis [32]. In children or thin adults, color and power Doppler may show focal perfusion defects due to alterations in blood flow resulting from pyelonephritis. Abscess detection may also be improved. In the obese and even the normal-sized adult, adequate detection of very small intrarenal vessels such as the intralobular artery is difficult, even with power Doppler. Thus, its utility is now limited secondary to relatively poor color sensitivity in the usual clinical setting.

Intravenous ultrasound contrast agents may change the use of ultrasound in pyelonephritis as enhancement of even small vessels makes their detection relatively easier. Higher sensitivity and specificity of ultrasound for renal infections could then be possible. Thus a CT may not be necessary to look for complications of pyelonephritis or abscess. This is another example of a potential cost saving if the use of ultrasound contrast agents results in the routine use of ultrasound instead of CT in the diagnosis of the complications of pyelonephritis. Further prospective clinical studies will be needed to prove this hypothesis.

RENAL TRANSPLANTS

Beautiful images of the vascularity of some renal transplants can now be obtained with power Doppler, and somewhat less often, with color Doppler. Even small vessels reaching to within millimeters of the cortex surface can be detected. However, these fine images are today obtained only with the latest generation software and scanners, and only in patients with a relatively small amount of soft tissue overlying the transplant.

The imaging and spectral analysis of the intrarenal vessels currently plays a somewhat limited role in patient care. The resistive index is not specific for rejection [33–35], but may also be abnormally elevated as a result of renal insults such as cyclosporine toxicity and pyelonephritis. Assessing the main renal artery and vein for stenosis or occlusion [36] is usually not difficult, but can be a problem secondary to the sometimes deep location of these structures, which may also be obscured by overlying bowel gas.

The increased vascular enhancement provided by an intravenous ultrasound contrast agent may improve the visualization of the main renal artery and vein in problem cases where a stenosis or occlusion is suspected. The number of branch vessels present may prove a more sensitive indicator of rejection than the resistive index [37].

SUMMARY

There is a large potential for the use of intravenous ultrasound contrast agents in renal imaging. In any area in which improved sensitivity to flow is important, these agents may markedly increase the diagnostic capabilities of sonography. Investigations are being performed into the utility of contrast agents and the sonographic imaging of solid and complex cystic masses, renal artery stenosis, pyelonephritis and renal transplants. Improved visualization of renal vasculature may decrease the need for subsequent CT or other corroborating radiographic studies, thus yielding potential cost savings in this era of managed care. However, much research must be performed before the application of ultrasound contrast agents to the kidney becomes routine.

ACKNOWLEDGMENTS

I would like to acknowledge Mr Michael Clements BS, RDMS, RVT for his invaluable contribution to the Phase 2 and 3 FDA ultrasound contrast agent trials we have completed here at UAB. Review of this manuscript by Dr Lincoln Berland and Dr Phil Kenney was much appreciated. Special thanks to Dr Larry Needleman and Dr Aronson for the generous use of their cases. Acknowledgment also goes to Ms Angie French for help in manuscript preparation, and Mr Tony Zagar for photographic assistance.

REFERENCES

1. Rubin JM, Bude RO, Carson PL, Bree RL, Adler RS. Power Doppler US: a potentially useful alternative to mean frequency-based color Doppler US. Radiology 1994;190:853–6.

2. Wheatley MA, Schrope B, Shen P. Contrast agents for diagnostic ultrasound: development and evaluation of polymer-coated microbubbles. Biomaterials 1990;11:713–7.
3. de Jong N, Ten Cate FJ, Lancée CT, Roelandt JRTC, Bom N. Principles and recent developments in ultrasound contrast agents. Ultrasonics 1991;29:324–30.
4. Needleman L, Robbin ML, Beach KW, Platt JF, Merritt CB, Arger PH. Phase II clinical trial of a new perfluorochemical emulsion US contrast agent. Radiological Society of North America. Chicago, IL. Nov 25–Dec 1, 1995.
5. Lopez-Ben RR, Robbin ML, Needleman L, Berland LL. US contrast agent imaging of liver hemangiomas. Radiological Society of North America. Chicago, IL. Nov 25–Dec 1, 1995.
6. Aronson S, Wiencek JG, Feinstein SB. Assessment of renal blood flow with contrast ultrasonography. Anesth Analg 1993;76:964–70.
7. Schlief R. Ultrasound contrast agents. Curr Opin Radiol 1991;3:198–207.
8. Schwarz KQ, Becher H, Schimpfky C, Vorwerk D, Bogdahn U, Schlief R. Doppler enhancements with SHU 508 A in multiple vascular regions. Radiology 1994;193:195–201.
9. Rumack CM, Wilson SR, Charboneau JW. Diagnostic Ultrasound: volumes 1 and 2. New York, Mosby Year Book. 1991:220–6.
10. Hartman DS, Aronson S, Frazer H. Current status of imaging indeterminate renal masses. Radiol Clin N Am 1991;29:475–96.
11. Davidson AJ, Hayes WS, Hartman DS, McCarthy WF, Davis CJ Jr. Renal oncocytoma and carcinoma: failure of differentiation with CT. Radiology 1993;186:693–6.
12. McClennan BL, Rabin DN. Kidney. In: Lee JKT, Sagel SS, Stanley RJ, editors. Computed body tomography with MRI correlation, 2nd edn. New York: Raven Press, 1989:801–4.
13. Erden I, Bedük Y, Karalezli G, Aytaç S, Anafarta K, Safak M. Characterization of renal masses with colour flow Doppler untrasonography. Br J Urol 1993;71:661–3.
14. Takase K, Takajashi S, Tazawa S, Terasawa Y, Sakamoto K. Renal cell carcinoma associated with chronic renal failure: evaluation with sonographic angiography. Radiology 1994;192:787–92.
15. Kuijpers D, Jaspers R. Renal masses: differential diagnosis with pulsed Doppler US. Radiology 1989;170:59–60.
16. Bosniak MA. The small (≤3.0 cm) renal parenchymal tumor: detection, diagnosis, and controversies. Radiology 1991;179:307–17.
17. Forman HP, Middleton WD, Melson GL, McClennan BL. Hyperechoic renal cell carcinomas: increase in detection at US. Radiology 1993;188:431–4.
18. Strotzer M, Lehner KB, Becker K. Detection of fat in a renal cell carcinoma mimicking angiomyolipoma. Radiology 1993;188:427–8.
19. Hélénon O, Chrétien Y, Paraf F, Melki P, Denys A, Moreau JF. Renal cell carcinoma containing fat: demonstration with CT. Radiology 1993;188:429–30.
20. Yamashita Y, Watanabe T, Miyazaki T, Yamamoto H, Harada M, Takahashi M. Cystic renal cell carcinoma. Acta Radiol 1994;35:19–24.
21. Fommei E, Volterrani D. Renal nuclear medicine. Sem Nucl Med 1995;25:183–94.
22. Nally JV Jr. Provocative Captopril testing in the diagnosis of renovascular hypertension. Urol Clin N Am 1994;21:227–34.
23. Bourgoignie JJ, Rubbert K, Sfakianakis GN. Angiotensin-converting enzyme-inhibited renography for the diagnosis of ischemic kidneys. Am J Kidney Dis 1994;24:665–73.
24. Stavros T, Harshfield D. Renal Doppler, renal artery stenosis, and renovascular hypertension: direct and indirect duplex sonographic abnormalities in patients with renal artery stenosis. Ultrasound Q 1994;4:217–63.
25. Kliewer MA, Tupler RH, Carroll BA et al. Renal artery stenosis: analysis of Doppler waveform parameters and tardus-parvus pattern. Radiology 1993;189:779–87.
26. Stavros AT, Parker SH, Yakes WF et al. Segmental stenosis of the renal artery: pattern recognition of tardus and parvus abnormalities with duplex sonography. Radiology 1992;184:487–92.
27. Berland LL, Koslin DB, Routh WD, Keller FD. Renal artery stenosis: prospective evaluation of diagnosis with color duplex US compared with angiography. Radiology 1990;174:421–3.

28. Mann SJ, Pickering TG, Sos TA et al. Captopril renography in the diagnosis of renal stenosis: accuracy and limitations. Am J Med 1991;90:30–9.
29. Badr KF, Brenner BM. Vascular injury to the kidney. In: Wilson JD, Braunwald E, Isselbacher KJ et al., editors. Harrison's Principles of Internal Medicine. 12th edn. New York: McGraw-Hill, Inc. 1991;1192–6.
30. Platt JF, Ellis JH, Rubin JM. Intrarenal arterial Doppler sonography in the detection of renal vein thrombosis of the native kidney. Am J Radiol 1994;162:1367–70.
31. Lowe LH, Zagoria RJ, Baumgartner BR, Dyer RB. Role of imaging and intervention in complex infections of the urinary tract. Am J Radiol 1994;163:363–7.
32. Clautice-Engle TL, Jeffrey RB Jr. Power Doppler imaging in acute pyelonephritis. Radiological Society of North America. Chicago, IL. Nov 25–Dec 1. 1995.
33. Grant EG, Perrella RR. Wishing won't make it so: duplex Doppler sonography in the evaluation of renal transplant dysfunction. Am J Radiol 1990;155:538–9.
34. Perrella RR, Duerinckx AJ, Tessler FN et al. Evaluation of renal transplant dysfunction by duplex Doppler sonography: a prospective study and review of the literature. Am J Kidney Dis 1990;15:544–50.
35. Perchik JE, Baumgartner BR, Bernardino ME. Renal transplant rejection: limited value of duplex Doppler sonography. Invest Radiol 1991;26:422–6.
36. Kaveggial LP, Perrella RR, Grant EG et al. Duplex Doppler sonography in renal allografts: the significance of reversed flow in diastole. Am J Radiol 1990;155:295–8.
37. Aronson S, Thistlethwaite RJ, Walker R, Feinstein SB, Roizen MF. Safety and feasibility of renal blood flow determination during kidney transplant surgery with perfusion ultrasonography. Anesth Analg 1995;80:353–9.

35. Ultrasound contrast agents in peripheral vascular disease*

JÖRG P. LANGHOLZ

INTRODUCTION

Driven by rapid technical advances, color-coded Doppler ultrasonography is progressing fast. Early skepticism about the value of the technique has been replaced by growing enthusiasm, and Doppler color flow mapping is now accepted as a valuable aid to the investigation of peripheral arterial and venous disease. In the vast majority of routine investigations the technique has the potential to replace invasive arteriography and phlebography. Despite the ever-increasing sensitivity of ultrasound hardware, visualizing slow-flowing blood vessels in the deeper tissues and detecting flow signals that have been attenuated by overlying calcifications or pockets of bowel gas still poses severe problems. Indeed, the iliac region, the peripheral segments of the deep femoral arteries, the area around the adductor canal and the proximal calf, especially in obese patients, are frequently regarded as inaccessible to ultrasonography. Circulatory problems in these areas affect a certain number of the patients referred to a peripheral vascular disease clinic, so the intense interest in the development of an agent that enhances the Doppler signal from the arteries and veins of the leg is hardly surprising. If that enhancement proves clinically useful, and if it can be produced by a peripheral intravenous injection, it will offer several important advantages over X-ray arteriography. Gray-scale images from the peripheral circulation are too poor, even after enhancement, to give any useful information, and Doppler enhancement is the only effect that has any diagnostic value.

The only echo-enhancer that has been extensively investigated in clinical trials is Levovist (Schering AG, Berlin): this agent has been approved as an aid to the diagnosis of peripheral arterial disease in most European countries. Its toxicology and mode of action have been described in great detail elsewhere [1]. Other potential echo-enhancers, including the soya-based Spherosomes (Bracco-BYK GmbH, Constance) the albumin-stabilized Albunex (Nycomed

---

*This chapter has been published in Advances in Echo-Enhancement 1997;1:30–42

N. C. Nanda, R. Schlief and B. B. Goldberg (eds.), Advances in Echo Imaging Using Contrast Enhancement, Second Edition, 543–560.
© 1997 *Kluwer Academic Publishers.*

Imaging AG, Oslo) microbubbles and the perfluorocarbon emulsion Echogen (Sonos) are in development [2] but none has yet been shown to produce useful enhancement in the peripheral circulation. For reasons that remain unclear, the clinical development of Albunex in Europe was halted early in 1996 [3].

LEVOVIST ENHANCEMENT IN THE PELVIC AND LEG ARTERIES

Diagnostically useful Doppler enhancement appears within 12–15 s of an intravenous injection of Levovist. The degree of enhancement seen in the pelvic arteries and in the arteries of the lower leg and foot does not differ significantly [4]. It persists, on average, for more than 2 min (2.3 ± 1.3) and extends to the arteries of the foot. In some individuals, useful enhancement persists for more than 7 min [5]. Table 1 shows the duration of enhancement according to the concentration and volume of the Levovist injection.

A single injection sometimes fails to enhance the Doppler signal from the arteries of the calf, especially if the proximal segments are severely occluded, but repeated injections almost invariably produce a diagnostic result. In a large multicenter clinical trial of Levovist [6] a single injection at the recommended dose (200–400 mg/ml) produced marked enhancement in 74% of cases and weak enhancement in 22%. The investigators had no previous experience with the agent and sometimes chose too low a dose for the clinical problem with which they were faced. In the vast majority of cases in which enhancement was weak the enhancer had been used at the lower dose of 200 mg/ml. If the dose chosen is appropriate to the condition investigated, Levovist produces marked Doppler enhancement in 95% of cases.

In a few cases the higher dose of 400 mg/ml produced excessive enhancement that created imaging artefacts and color blooming that led to inaccurate diagnoses. Because of the possibility of over-enhancement, the higher dose should not be used to enhance the Doppler signal from the proximal segments of the iliac artery or from the aorta. In the arteries of the lower leg, however, although the lower dose of 200 mg/ml will sometimes give adequate enhancement, the higher dose is often essential. The concentration used depends on the clinical situation; broadly speaking the higher dose is needed when the iliac segments are severely occluded or when the vessels under investigation are at the distal end of the leg. In general, and especially when investigating the arteries of the foot and lower leg, the best results are achieved with a concentration of 300 mg/ml. Table 2 shows the relationship between dose and enhancement in the same group of patients as in Table 1.

Since the injection rate of Levovist is not critical and even bolus injections are well tolerated, physicians need not hesitate to inject the enhancer quite rapidly. In diagnostically complicated cases a slow injection rate can produce unsatisfactory enhancement. An injection rate of 1–2 ml/min usually produces satisfactory results.

One of the doubts sometimes expressed about the value of radiographic

Table 1. Duration of diagnostically useful enhancement of arteries of the lower leg.

Arteries	Concentration (mg/ml)	Volume (ml)	Duration of diagnostically usable enhancement (min)		No. injections
			Mean±SD	Range	
Iliac[a]	200	10	1.5±0.7	0.8–2.6	8
	200	16	2.0±1.1	0.4–4.0	23
	300	10	2.1±1.4	1.0–7.6	29
Upper leg[b]	200	10	2.1±0.8	1.2–2.7	3
	200	16	1.9±0.9	0.5–4.0	43
	300	10	2.3±1.7	0.7–5.9	21
	400	8	2.6±1.2	0.8–3.5	4
Lower leg[c]	200	16	1.4±0.8	0.5–3.5	18
	300	10	2.6±1.1	0.9–5.7	27
	400	8	2.2±1.3	0.9–4.4	5

[a]60 injections in 15 patients.
[b]71 injections in 25 patients. One additional injection produced no change in Doppler signal intensity.
[c]50 injections in 17 patients. Four additional injections produced no change in Doppler signal intensity.

Table 2. Doppler enhancement by concentration and volume.

Arteries	Concentration (mg/ml)	Volume (ml)	Increase in Doppler signal intensity				No. injections
			Unchanged	Slight	Considerable to optimal	Excessive	
Iliac[a]	200	10	0	0	8	0	8
	200	16	0	13	10	0	23
	300	10	0	2	27	0	29
Upper leg[b]	200	10	0	0	3	0	3
	200	16	1	9	32	2	44
	300	10	0	1	20	0	21
	400	8	0	0	4	0	4
Lower leg[c]	200	16	3	7	11	0	21
	300	10	1	7	18	2	28
	400	8	0	1	4	0	5

[a]60 injections in 15 patients.
[b]72 injections in 25 patients.
[c]54 injections in 17 patients.

arteriography is the possibility that the vascular anatomy and flow pattern seen during the investigation could have changed significantly during the months that pass before the patient is referred back to the hospital. Most physicians are understandably reluctant to subject their patients to a second arterial puncture and repeat dose of iodinated contrast medium after such a short period. Many patients with peripheral vascular disease have renal problems, and others have thyroid dysfunction, both of which limit the frequency with which contrast angiography can be repeated. The only other available information, from pulse waveforms and ankle systolic pressures, does not provide any morphological information. If the patient's vascular anatomy allows, duplex ultrasonography can help to close this worrying diagnostic gap. In some instances the conventional color Doppler scan is inadequate and in these instances an echo-enhancer such as Levovist can be extremely useful. As the following case reports demonstrate, Levovist enhancement can provide useful information on peripheral blood flow even in cases that appeared, diagnostically, to be quite hopeless.

Case 1

Bowel gas often overshadows the iliac arteries and attenuates the Doppler signal. This patient had undergone successful dilatation of the external iliac artery close to its origin but because the dilated segment was overshadowed by bowel gas it was impossible to determine whether or not it was still patent in conventional color Doppler scans. One day after the dilatation the patient complained of severe pain in the ipsilateral inguinal region, caused by a hematoma that prevented palpation of the inguinal pulse (Figures 1A–C). To avoid repeated arteriography, and in an attempt to clarify the issue, the patient was investigated by color Doppler before and after a peripheral i.v. injection of 8 ml Levovist (400 mg/ml). The resulting images (Figures 1D–G) showed quite clearly that the dilated arterial segment was still patent, and eliminated the need for repeat arteriography. The echo-enhancer also boosted the Doppler signal from the external iliac vein very substantially.

Case 2

Because they are so close to the surface, bifurcation of the femoral artery and the proximal course of the superficial femoral artery are normally quite easy to visualize sonographically. However, the distal segment of the profound femoral artery, deep within the muscles of the thigh, is usually difficult to demonstrate. In the present case, even the proximal segment was hidden behind mural calcifications (Figure 2C) and flow within it was very slow because the segments proximal to it (and a former cross-over bypass from the contralateral iliac artery) were occluded (Figure 2A,B). Consequently the vessel could not be

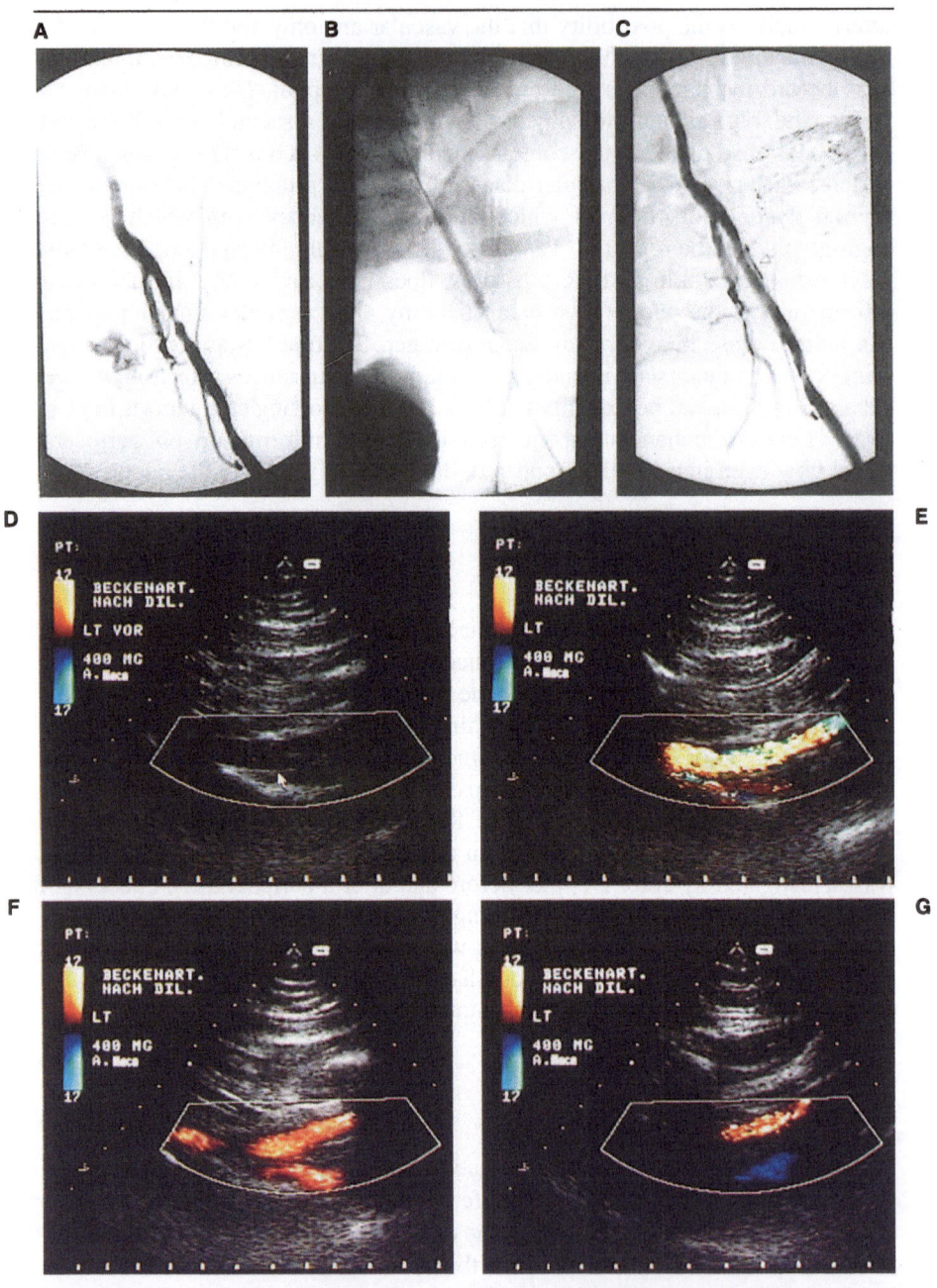

Figure 1. Case 1. (A) Stenosis of the external iliac artery before dilatation (arteriogram). (B) Balloon angioplasty of the stenosis shown in (A). (C) Left external iliac artery immediately after dilatation (arteriogram). (D) Absence of color pixels in the external iliac segment (unenhanced CCDS image). (E) Left iliac artery: blooming during the first enhancement phase after i.v. injection of Levovist (8 ml, 400 mg/ml). (F) As (E), about 60 s later. (G) Same segments as in (A–F): concomitant venous enhancement after Levovist injection.

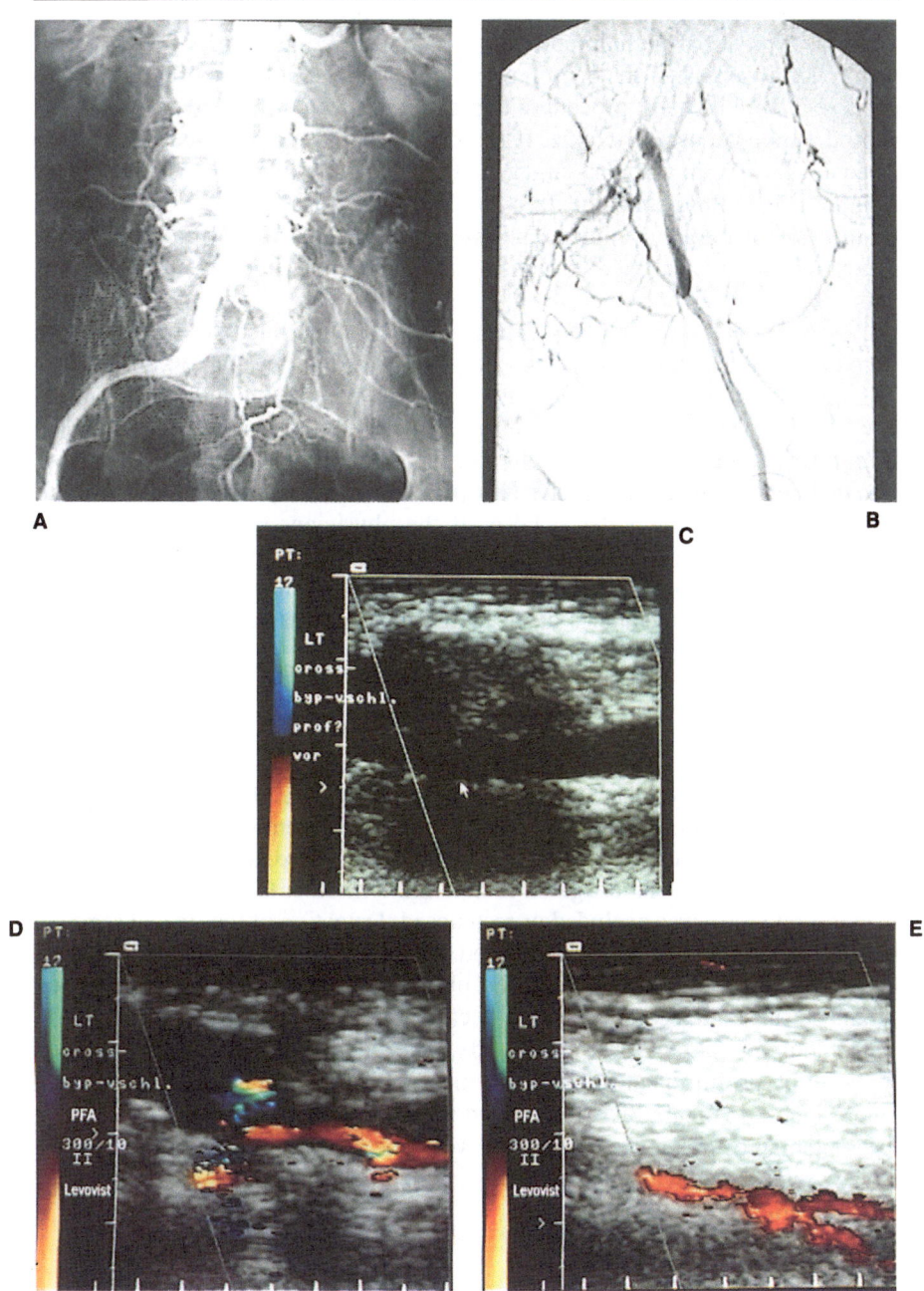

Figure 2. Case 2. (A) Arteriogram of the iliac arteries showing occlusion of all left iliac arteries and the left common femoral artery. (B) Arteriogram showing collateral filling of the left profound femoral artery. (C) Sonogram of the proximal segment of the left profound femoral artery before Levovist enhancement. (D) Longitudinal view of the proximal segment of the left profound femoral artery filled by collaterals (blue signal) after administration of Levovist (300 mg/ml). (E) Distal segment of the left profound femoral artery after Levovist enhancement.

visualized adequately in conventional color Doppler scans.

In this type of case it is important to reach a correct diagnosis when investigating these vessels. If the deep femoral artery is patent, it may be possible to use a relatively low risk procedure to reconstruct a cross-over bypass from the contralateral, patent pelvic axis. If the deep femoral artery is occluded, more extensive prosthetic vascular surgery will be necessary. These procedures carry a higher risk. In this patient the deep femoral artery was filled only via collaterals (Figure 2B) via color Doppler. The artery was clearly visualized after an i.v. injection of 10 ml Levovist (300 mg/ml) and the patient was successfully managed with a new cross-over bypass.

The value of color Doppler enhancement is underlined by the problems encountered in this case in which, after initial clinical evaluation, an attempt was made to establish the diagnosis by intravenous digital subtraction angiography (DSA). Unfortunately, because the deep femoral artery was separated from the rest of the vascular system and supplied only via collaterals, i.v. DSA was unable to visualize it clearly. This patient's vascular anatomy was eventually established by intraarterial DSA. If the physicians responsible had been experienced in the use of echo-enhanced, color-coded Duplex sonography this patient could have been spared the risk of repeated contrast angiography. This case illustrates the advantages of the virtually non-invasive technique of echo-enhanced color Doppler extremely clearly.

Case 3

Five years previously, this patient had undergone successful balloon angioplasty to treat a filiform stenosis in a segment of superficial femoral artery within the adductor canal (Figures 3A–B). The patient now presented with intermittent claudication, and was investigated to determine whether the dilated segment of the femoral artery was occluded or restenosed. Detection of the treated segment before enhanced color Doppler proved difficult. The scan showed very few color pixels, none of which could be placed anatomically. The first injection produced substantial color blooming because of over-enhancement (Figure 3D). This blooming lasted for roughly 30 s, after which it diminished and was replaced by useful diagnostic enhancement which confirmed that the segment has restenosed (Figure 3E). This raised the possibility of a repeat angioplasty. The outcome of this investigation was very encouraging for the patient. Again it shows how Levovist-enhanced color Doppler can avoid the necessity for invasive arteriography.

Case 4

In this diabetic patient, an unenhanced color Doppler scan of a longitudinal section near the point of origin of the anterior tibial artery showed

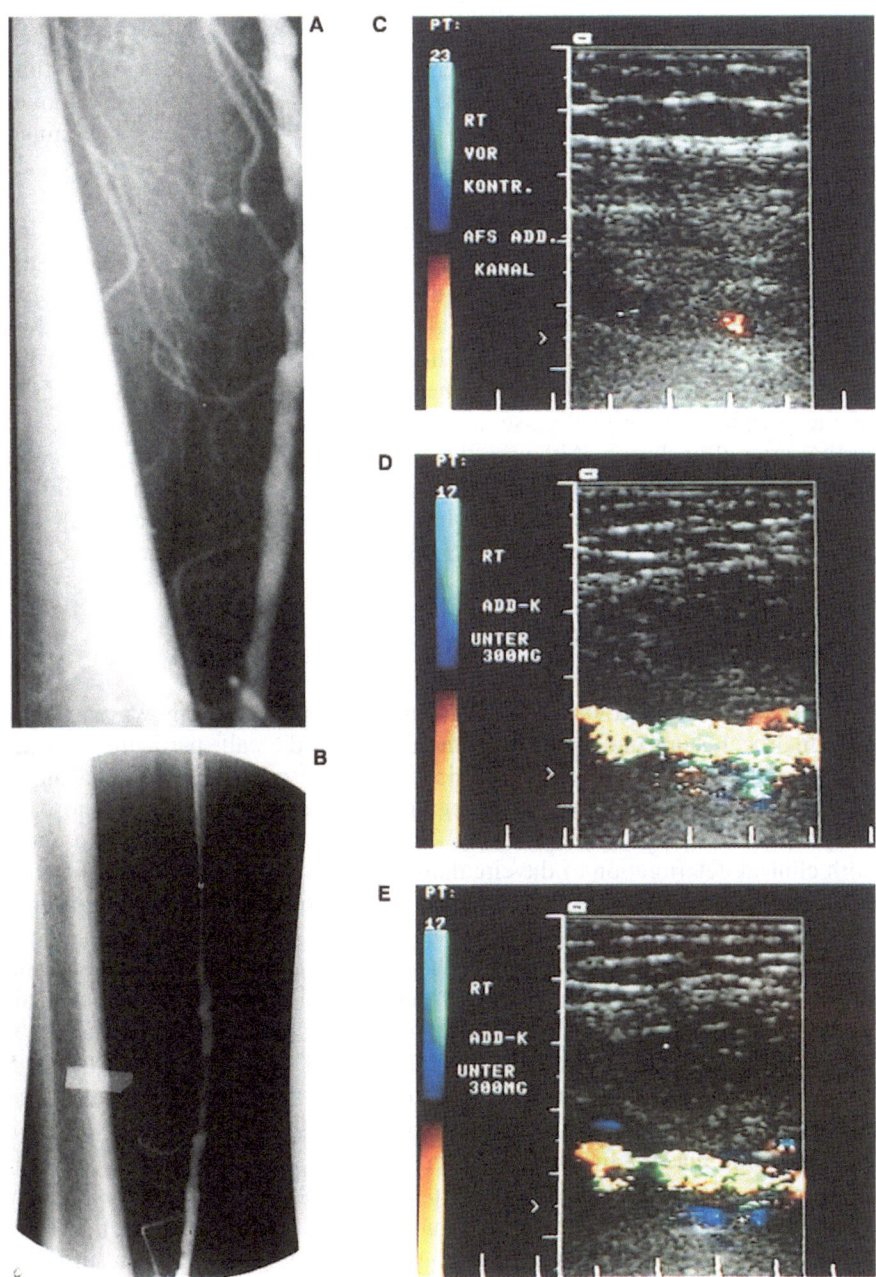

Figure 3.   Case 3. (A) Arteriogram of a stenosed right superficial femoral artery (5 years before the patient presented). (B) Dilatation of the stenosis shown in (A). (C) Five years later – unenhanced CCD image of the segment of the superficial femoral artery in the adductor canal. Weak color coding only. (D) Colour blooming in the superficial femoral artery 12 s after an i.v. injection of Levovist (10 ml, 300 mg/ml) into the arm. (E) Detection of restenosis of the superficial femoral artery in the adductor canal. Image obtained 60 s later than (D).

atherosclerotic calcifications of the arterial wall (Figure 4A). No color signals were detectable from the tissue behind the calcification. One year earlier, arteriography had shown the tibial artery to be patent (Figure 4B). At the time of the repeat examination, the patient suffered from gangrene of the forefoot of the leg under investigation and faced the possibility of a below-knee amputation. To plan the treatment it was essential to know whether the previously patent lower leg arteries remained patent.

After injecting 16 ml Levovist (200 mg/ml) (Figure 4C), followed by a second injection of 10 ml (300 mg/ml) (Figure 4D), good filling was visible in the proximal region of the tibial artery and flow was also detectable in the collateral arterioles. Good flow signals were also obtained from the proximal two branches of the anterior tibial veins. This demonstration of the patency of the arteries of the lower leg avoided unnecessary bypass surgery. The severity of the peripheral arterial disease was a consequence of local infection, not further arterial occlusion. Although the success of the contrast-enhanced color Doppler meant that repeat arteriography was unnecessary from a strictly clinical viewpoint, it was essential to a complete record of the case (Figure 4E). The angiographic results confirmed the validity of the color Doppler findings.

Case 5

In this patient arteriography several months previously had shown extensive proximal occlusions and stenoses in the thigh and the calf, particularly in the region of the P3 segment. The posterior tibial artery was also occluded in its proximal segment. Furthermore arteriography shows two high-grade stenoses at the proximal end of the refilled vessel (Figure 5A). The patient now presented with clinical deterioration of the circulation in the lower leg and foot. The clinical question to be answered was whether or not it might be possible to insert a crural bypass at the distally patent arterial segment. Unfortunately the patient's renal function was impaired and, after the first contrast arteriography, his serum levels of creatinine and urea increased markedly. Consequently the danger of severe kidney damage, possibly leading to hemodialysis, ruled out the further use of X-ray contrast agents and forced us to use Doppler ultrasound instead. Because of the slow flow produced by the extensive stenoses, unenhanced color Doppler scans (Figure 5B) did not provide enough information to enable a decision to be made. After echo-enhancement with 8 ml Levovist (400 mg/ml) flow in the distal crural segment of the posterior tibial artery was clearly visualized (Figure 5C). The enhanced color Doppler scan showed that the artery was refilled in spite of proximal occlusions and two stenoses. It kept a sufficient diameter to allow the construction of a crural bypass. This result confirms the place of echo-enhancement as an alternative to contrast arteriography when assessing arterial patency in the lower leg.

Echo-enhancement is as effective in the less accessible peroneal artery as it is

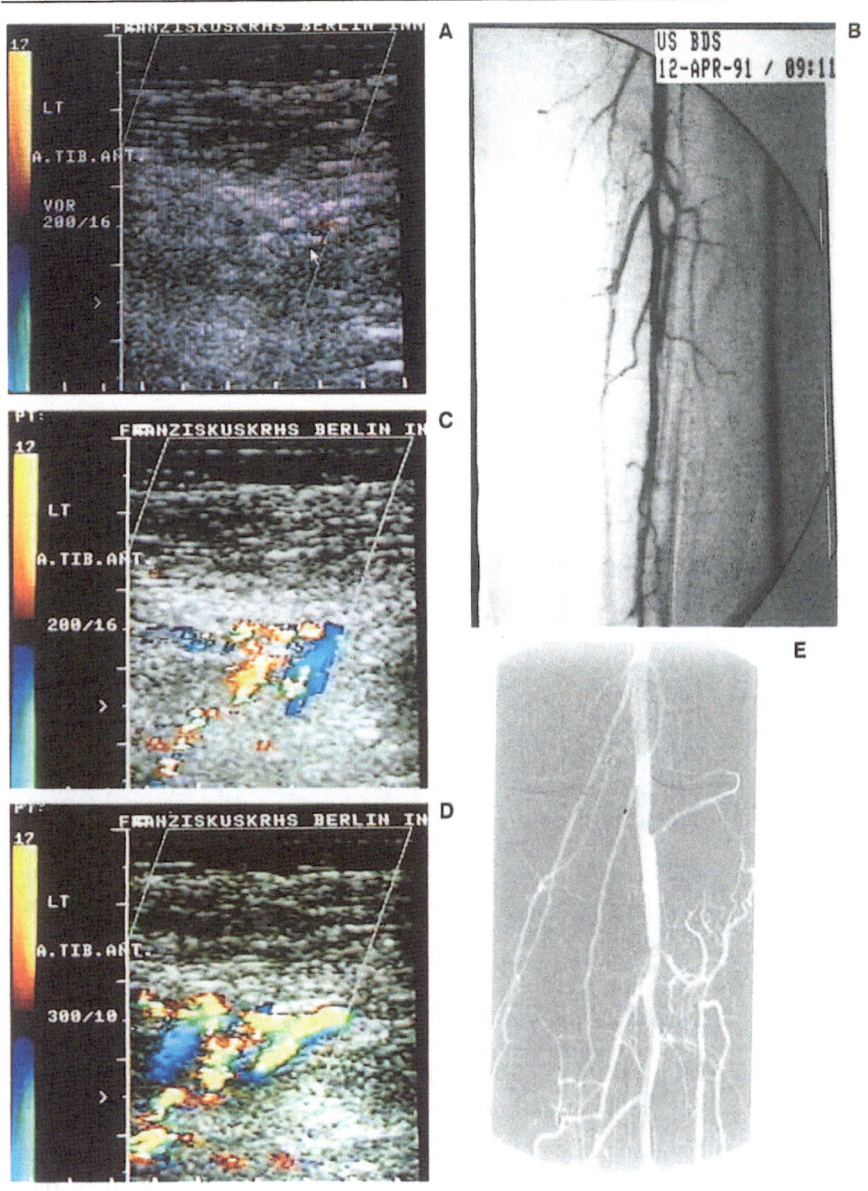

Figure 4. Case 4. (A) Longitudinal view of the proximal section of the anterior tibial artery before Levovist enhancement. Calcified walls of the artery (mediasclerosis) prevent the detection of color Doppler signals. (B) Arteriogram of the lower leg one year previously, showing a patent tibial artery. (C) Proximal segment of the left anterior tibial artery after the first i.v. injection of Levovist (200 mg/ml). Color flow visualized in the original artery, in collaterals and in both venous branches of the anterior tibial group. (D) Proximal segment of the left anterior tibial artery after the second injection of Levovist (300 mg/ml). Stronger flow signals than obtained in (C). (E) Repeat angiography.

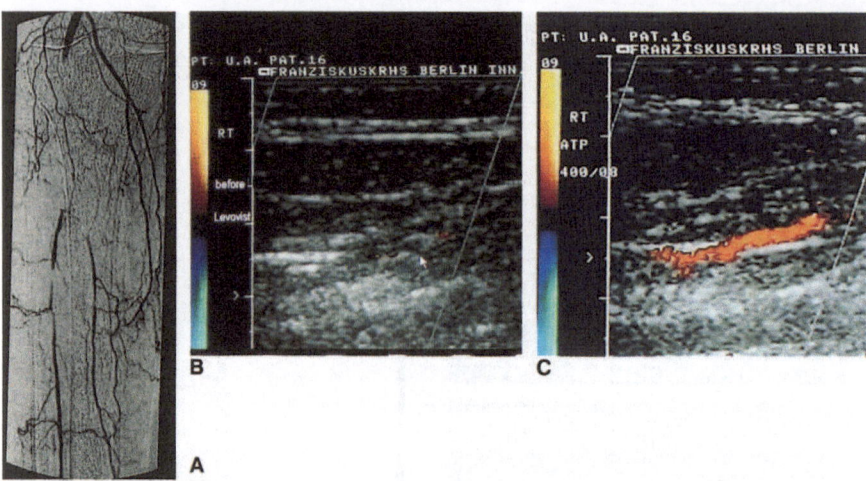

Figure 5. (A) Arteriogram of the lower leg demonstrating a patent posterior tibial artery distal to an occlusion and a high grade stenosis. (B) Unenhanced CDDS of the posterior tibial segment shown in (A). (C) Segment of the same segment as (A) after Levovist enhancement (400 mg/ml).

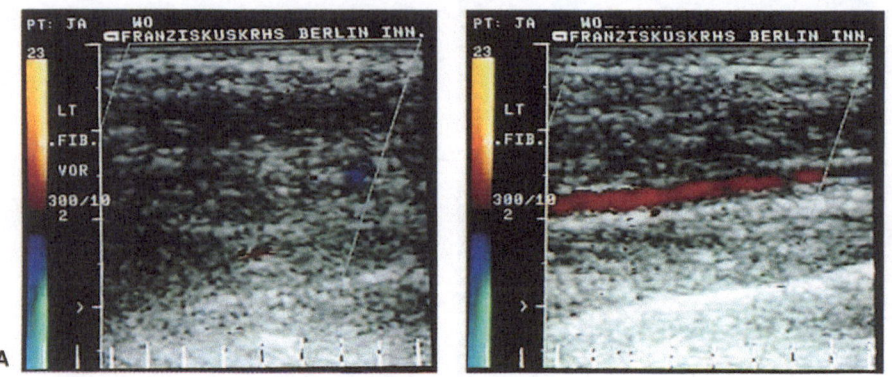

Figure 6. Peroneal artery before (A) and after (B) two i.v. injections of Levovist (300 mg/ml).

in the more readily visualized posterior tibial artery. Figure 6 shows the peroneal artery before and after enhancement. Flow which is almost invisible in the plain image is clearly visualized in the enhanced scan.

Case 6

The Doppler enhancement produced by Levovist extends right down to the most distal arteries of the foot. In this patient, the arterial circulation within the foot

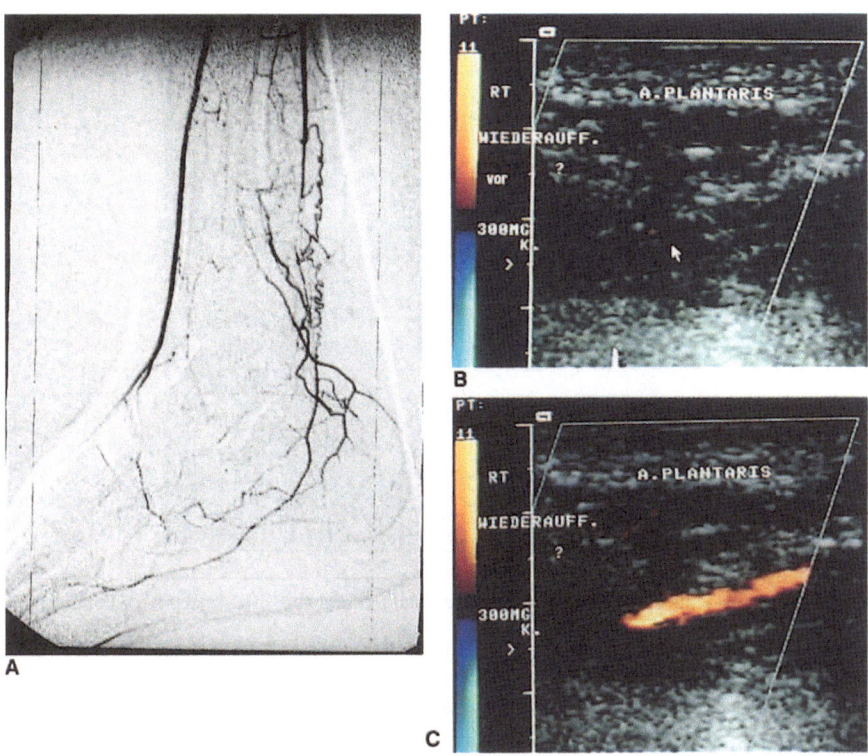

Figure 7. Patent plantar artery distal to multiple occlusions of crural arteries. (A) Arteriogram; (B) unenhanced CCDS of a segment of (A); (C) as (B), after administration of Levovist (16 ml, 200 mg/ml).

was severely impaired. Arteriography showed multiple occlusions of the crural arteries and there appeared to be a low blood supply to the plantar artery (Figure 7A). An unenhanced color Doppler scan of the foot similarly showed no evidence of flow within the plantar artery (Figure 7B). After an i.v. injection of 10 ml Levovist (300 mg/ml) the Doppler color coding showed some flow within the plantar artery, indicating the artery, or at least some of its regional branches, was still patent (Figure 7C). In this patient with critical ischemia, the result suggested that conservative therapy with a prostaglandin infusion might be effective and that the relatively high cost involved would be fully justified.

These examples show that Levovist enhancement of the Doppler signal from the peripheral circulation improves the accuracy of diagnosis and allows a decision regarding the most appropriate treatment to be reached. Occlusive disease of the peripheral arteries, which could previously be diagnosed only with radiographic arteriography, can now be visualized, even in anatomically or functionally difficult cases, by the far less invasive technique of echo-enhanced

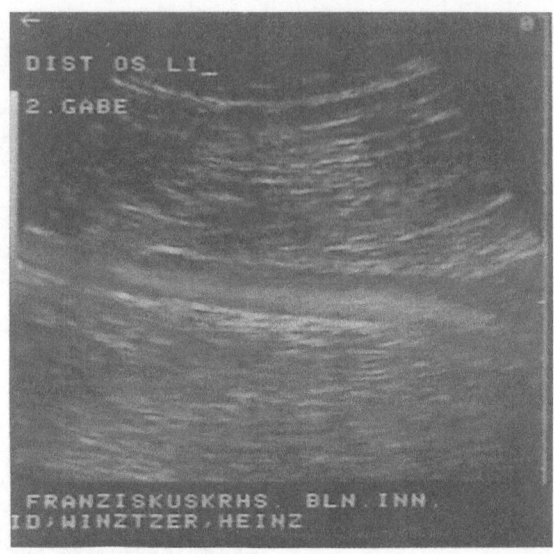

Figure 8.   Gray-scale enhancement of the superficial femoral vein after injection of Echovist into a dorsal foot vein.

color Doppler ultrasonography. However, echo-enhancement is only appropriate in a small minority of cases and it would be unfortunate if ultra-sonographers were to use an echo enhancer to compensate for the deficiencies of their ultrasound equipment. The color sensitivity of the best available ultrasound machines allows the satisfactory diagnosis of impaired peripheral vascular flow in 85–90% of cases. Most of the remaining 10–15% involve examinations of the pelvic region (where echo enhancement can help to over-come over-shadowing by bowel gas) or in the lower leg where flow is sometimes extremely slow.

LEVOVIST ENHANCEMENT IN THE LEG VEINS

In two of the cases described above (Cases 1 and 4) Levovist produced useful enhancement of the Doppler signal from the venous circulation. The earlier saccharide-based agent Echovist, injected into a vein on the back of the foot, enhance both the gray-scale and Doppler color-coded image (Figure 8). In 1989 we reported on results obtained with Echovist in the investigation of deep venous thrombosis [7]. This raises the possibility that there is some potential to venous enhancement that could lead to better diagnoses of the venous vascular bed.

Nevertheless, echo-enhancement has relatively limited application in the

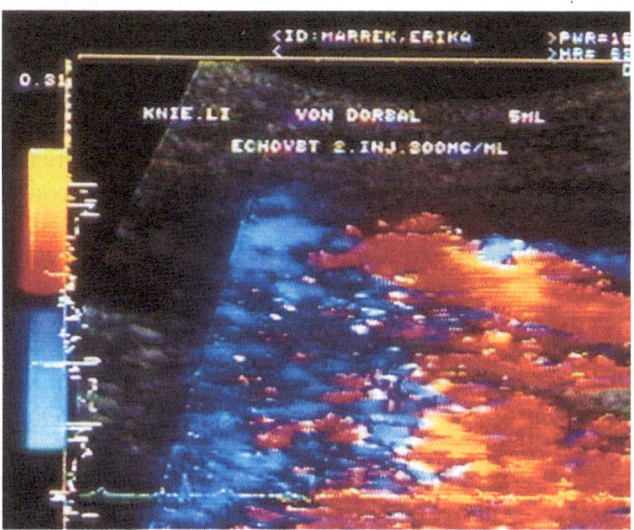

Figure 9.   Color blooming in the popliteal vein after injection of Echovist into a dorsal foot vein.

venous circulation because compression sonography or unenhanced color-coded Doppler are usually quite adequate. It also proved difficult to control the echo-enhancement produced by Echovist in the leg veins and the over-enhanced signal produced color blooming (Figure 9). On the other hand, an agent such as Levovist, that enhances the signal from the veins of the lower leg after a periph-eral intravenous injection, could help in the investigation of the venous circula-tion when radiographic phlebography is impractical because of conditions such as edema or lymphedema affecting the back of the foot. The increase seems to be diagnostically useful and, as in the arterial bed, it could enable a diagnosis to be reached in otherwise hopeless cases. Studies of venous echo-enhancement are continuing and the following cases illustrate some of the possible indications.

Case 8

This patient presented with recurrent swellings of the left lower leg but was unaware of any previous venous disease. Phlebography showed the great saph-enous vein but none of the deep venous system in the area investigated (Figure 10A). Evidence of older post-thrombotic changes at the crural and popliteal levels led to a diagnosis of venous stasis after earlier deep vein thrombosis. The femoral veins were considered to be still occluded at the time of examination. Unenhanced color Doppler scans sometimes showed extremely weak signals from a segment of the superficial femoral vein (Figure 10B) but provided no useful information. An initial injection of 10 ml Levovist (300 mg/ml) failed to improve the situation. After a second dose, however, it was apparent that the superficial femoral vein was patent, although there was evidence of post-thrombotic damage to the venous valves (Figures 10C,D). Figure 10C shows a clear color signal from the superficial femoral vein. Because compression was applied distally to the scanning point, proximal flow was orthograde and coded blue in comparison to the red signal from the superficial femoral artery. The scan shown in Figure 10D was made after release of compression distally to the scanning point. Flow is displayed as red, as in the signal from the accompanying artery, indicating the retrograde distal venous flow associated with valve damage that is often encountered in post-thrombotic states. In this case Levovist-enhanced color Doppler was superior to conventional phlebography in terms of the diagnostic information that it provided.

Levovist enhancement also improves the color coding of the Doppler signal from the crural veins, but whether or not this improvement is diagnostically useful remains to be established. In general, Levovist enhancement in the leg veins is at its most pronounced in the more proximal segments. Enhancement in the iliac veins is usually pronounced, even without venous compression, but in the veins of the crural region useful enhancement without compression is rather unusual. Compression distal to the scanning point is essential for adequate signal enhancement from the crural veins and their branches. A minor disadvantage of peripheral venous echo-enhancement with Levovist is that two or more injections are often needed to achieve a diagnostically useful result.

Levovist enhancement was necessary in only 31 (5.0%) patients of the 617 that we have investigated during the screening period. Our patients were referred to a specialist angiology clinic and therefore represent more serious cases. The proportion of an unselected population of patients presenting with suspected deep venous thrombosis who would benefit is probably between 3 and 5%. That proportion will, of course, be influenced by the cost of the echo-enhancer.

CONCLUSION

There are many situations in which Levovist could usefully enhance the Doppler signal from the vasculature of the legs. The indications for the use of echo-enhancement in the peripheral arterial circulation are well established and the indications for its use in the venous circulation are already becoming clearer. The diagnostically useful enhancement produced by an i.v. injection of 10–16 ml Levovist at its recommended concentration is around 2 min and this period can be prolonged by repeating the injection. Two or three injections are usually sufficient and five (the highest dose) allows a firm diagnosis in almost every case. Future research should concentrate on the development of agents that have a longer duration of action.

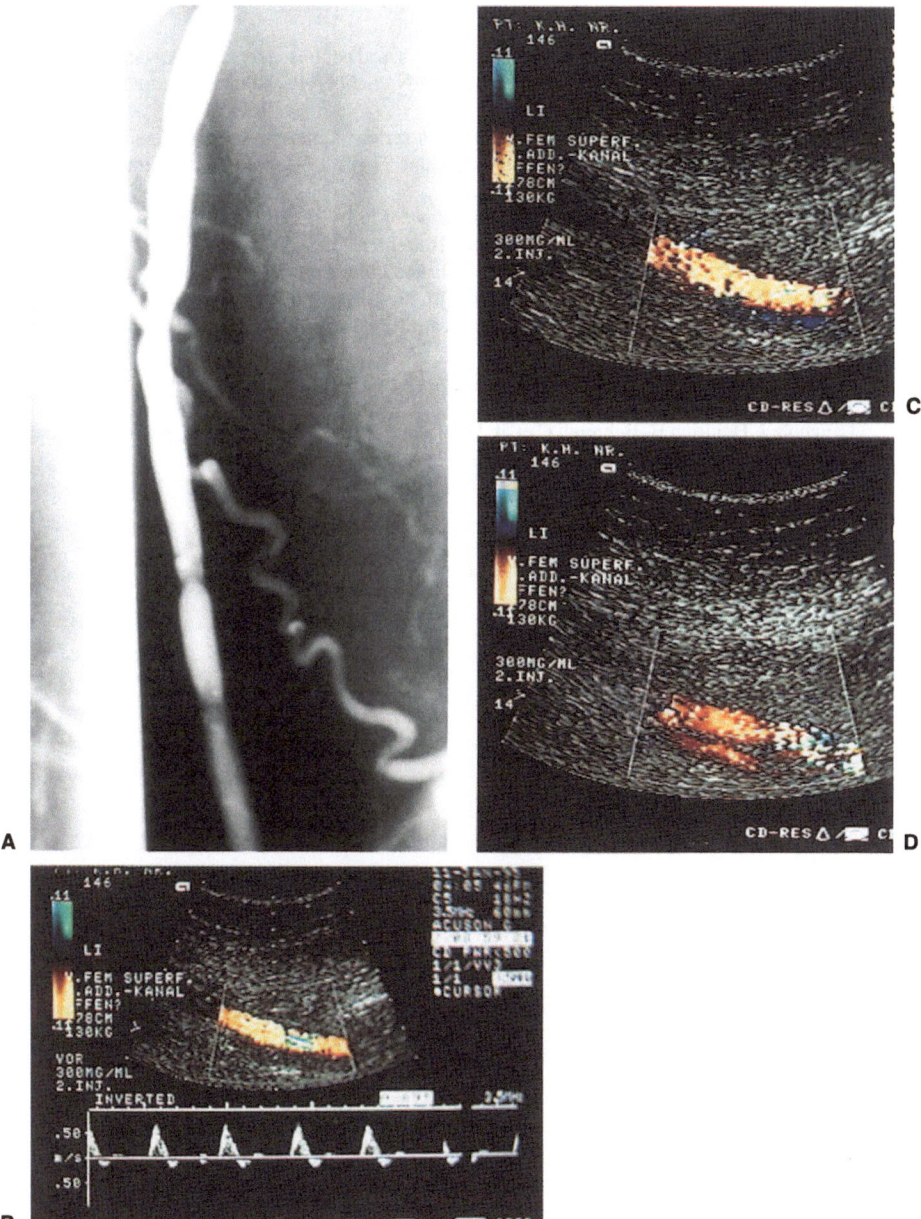

Figure 10. (A) Phlebography of the left upper leg of a patient (height 178 cm, weight 130 kg) presenting with recurrent swelling of the left lower knee. The great saphenous vein is shown; the deep veins are not visible. (B) Region of the superficial femoral vein in the adductor canal: unenhanced CCDS shows no color flow in the vein. The spectra represent the arterial flow. (C) After two i.v. injections of Levovist (300 mg/ml) into the arm, blue signals under the red color flow in the arterial segment indicate orthograde flow in the superficial femoral vein. (D) As (C), after decompression of the lower leg, red signals under the red color flow in the arterial segment represent retrograde venous flow, as it is seen in the post-thrombotic state.

REFERENCES

1. Schlief R, Schürman R, Balzer T, et al. Saccharide based contrast agents. In: Advances in Echo Imaging Using Contrast Enhancement. Nanda N, Schlief R, editors. Dordrecht, Kluwer Academic;1993:58–71.
2. Progress in contrast echocardiography I and II. A series of presentations delivered at the 11th symposium on echo-cardiography, Rotterdam, June 21–22 1995. Adv Cardiac Echo-Contrast. 1995;4:82–94.
3. Muan B. A survey of the results from clinical trials of Infoson. Presented at the First European Symposium on Ultrasound Contrast Imaging, Rotterdam, Jan 25–26 1996.
4. Langholz J, Wanke M, Petry J et al. Echo-enhanced ultrasound imaging of leg arteries with Levovist (SH U 508 A) – multicenter clinical trial results. Angiology 1996;47(7), Part 2.
5. Langholz J, Schlief R, Schürmann R et al. Contrast enhancement in leg vessels. Clin Radiol 1996;51(Suppl. 1):31–4.
6. Schlief R. Echo-enhanced Doppler imaging: the Levovist trial. Adv Echo-Enhancement 1996; 1:2–3.
7. Langholz J, Heidrich H, Behrendt C et al. Ultrasonic contrast medium in the diagnosis of deep venous thrombosis? Phlebologie 1989;89:366–8.

36. Clinical usefulness of color Doppler enhancement in the assessment of the liver and the portal system

RODOLFO CAMPANI, FABRIZIO CALLIADA, OLIVIA BOTTINELLI, GIULIA MARESCA, ENZO ANGELI, RIZZARDO ANGUISSOLA, ANNA BOZZINI & ALFREDO LA FIANZA

INTRODUCTION

Both the sensitivity and the specificity of imaging techniques in the study of focal hepatic lesions are relatively low (33–66%) [1]; ultrasonography is no exception. Defining the region(s) of interest (ROI) is sometimes difficult with both conventional and color Doppler sonography. Thus, approaching the study of liver vessels and of the portal system may be difficult because venous flow is often slow, poor, deep, or within an organ with high US beam absorption because of associated diffuse hepatopathy. The Doppler system must be sensitive enough to detect even very slow flows: this has been partly achieved in the latest equipment which has just been, or will soon be, marketed. However, the necessary increase in gain may worsen image quality dramatically, making the image useless for diagnostic purposes, because of the thickness of the organ under examination. Moreover, respiratory artifacts and, in the segments near the great vessels of the heart, pulsatile artifacts are common [2, 3].

In theory, the introduction of agents which improve the visibility of vessels, especially of the most difficult vessels to study, should solve most of the above problems [4]. The stronger enhancement and fewer artifacts obtained with the second harmonic imaging yield best results [2].

Our study, which was carried out partly in 1993 and partly in 1996, concentrated on focal and diffuse liver conditions, on the portal system and on the follow-up of patients undergoing chemoembolization or chemotherapy.

MATERIALS AND METHODS

One hundred and forty-four patients were examined using Levovist (Schering). Up to three doses were used, according to the result obtained after the first administration, and concentrations were 200 or 300 mg. A rapid injection was always given, as recommended by the manufacturer, and the manufacturer's guidelines

N. C. Nanda, R. Schlief and B. B. Goldberg (eds.), Advances in Echo Imaging Using Contrast Enhancement, Second Edition, 561–583.

were always followed when preparing the contrast agent. The images were acquired using a Toshiba SSA 270, and ATL Ultramark 9 and an Acuson XP10 in 1993 and a Toshiba SSA 270 Super HG and an ATL HDI 3000 in 1996; 2.5–5.0 MHz probes were used, depending on liver parenchyma thickness and depth of the lesion.

In the 1993 study, the US parameters were set to the color scale and this was maintained throughout the examinations. The PRF was set for slower flows and low velocity filters were used. The sample volume was as broad as possible, (about 50% of the vessel caliber), to detect even very slow and poor flows. In contrast, the increased sensitivity of the equipment available in 1996 made us reduce all parameters, some of them markedly, after contrast agent administration, to avoid possible blooming artifacts which would have nullified the diagnostic yield.

In both the 1993 experimental protocol and the 1996 study, gray-scale real-time ultrasonography was performed first to locate the known or suspected lesion(s), and then color Doppler was integrated with spectral Doppler. We kept the probe as still as possible, and patients were asked to hold their breath. Progressive enhancement of the vessels was monitored, alternating color and spectral Doppler to assess the presence/absence of vessels unquestionably. When any vessels were found, their shape and function were studied. Color Doppler signal enhancement was studied with scans as identical as possible. Each examination was video recorded and color images were also obtained by freezing the image. In addition to the operator using the probe and setting the unit, another operator is needed to prepare and inject the contrast agent and to finely tune parameter setting. When looking at video recorded material, the operators distinguished internal from peripheral tumor enhancement and scored it according to vessel conspicuity.

RESULTS

Spectral and color Doppler imaging have revolutionized sonography capability to detect and study liver vascular abnormalities. Color Doppler ultrasonography displays blood flow in real time, passively and automatically, in vessels which are inapparent or inconspicuous on gray-scale images, as well as in some structures whose vascular nature was not suspected at B-mode imaging.

Color Doppler sonography accurately shows transient alterations in flow dynamics and pulsatility, noninvasively and without affecting or altering flow dynamics, which no other imaging method can do. Diagnoses that were once made only with angiography, in addition to some which were previously impossible, can now be made noninvasively using color Doppler imaging [5]. The results and therapeutic adequacy of vascular surgery and interventional angiographic procedures can also be assessed.

The region of interest (ROI) is easy to find with color Doppler ultrasonography, which helps position the sample volume quickly and easily when further spectral Doppler studies are needed. Color Doppler imaging is used alone when a global view of vascular anatomy is needed, or to assess the presence/absence of flow to make the final diagnosis. The vascular features in two different areas can also be compared. Spectral Doppler sonography is used when flow velocity quantitation is required, or to detail the type of hemodynamic alteration. In spite of their unquestionable advantages, both spectral and color Doppler imaging exhibit some limitations. First, the access for the ultrasound beam may be limited, obscuring vessels and collateral branches, especially in the retro-peritoneum. Second, studying intraparenchymal liver vessels may be difficult because of marked acoustic absorption by fibrotic livers, even with low frequency probes which, in addition, worsen image quality. Finally, low rate and slow/very slow flows, as well as static flow, which are sometimes detected in the portal vein, are difficult to depict with conventional units and may be misinterpreted as anechoic clots (Figure 1). Contrast agents such as Levovist help to overcome these limitations, markedly increasing the unit sensitivity (+15 dB) and thus per-mitting better assessment of liver hemodynamics.

Portal hypertension

Portal hypertension results from portal flow obstruction, within or outside the liver. Liver cirrhosis and portal thrombosis are its most common causes. Chronic portal hypertension may be caused by the Budd–Chiari syndrome, veno-occlusive disease, schistosomiasis, chronic active hepatitis, nodular regenerative hyperplasia and nonfibrotic idiopathic liver fibrosis. In contrast, acute portal hypertension may be related to portal vein clots, severe acute or postalcoholic hepatitis, acute fatty liver of pregnancy and acute veno-occlusive disease; this condition is often reversible, unlike the chronic form [6]. Depicting portal collateral branches may be enough to diagnose even unsuspected portal hypertension, with normal liver findings at ultrasonography, or else it may substantiate diagnostic suspicion or the diagnosis itself. Moreover, depicting collateral branches is clinically useful in patients with unexplained jaundice, ascites or upper digestive tract bleeding [7].

Some spontaneous portosystemic shunts, such as splenorenal shunts [8] and para-umbilical vein recanalization [9], help reduce the risk of bleeding from esophageal varices by decreasing blood flow toward the portal vein. Color Doppler sonography is an irreplaceable tool for detecting portal collaterals, some of which are missed at B-mode ultrasonography. The most common collateral branches are the left gastric vein (i.e. the stomach coronary vein, by far the most frequent site of spontaneous portosystemic collaterals) and the para-umbilical vein. The former is best detected at the origin, usually at the splenoportal

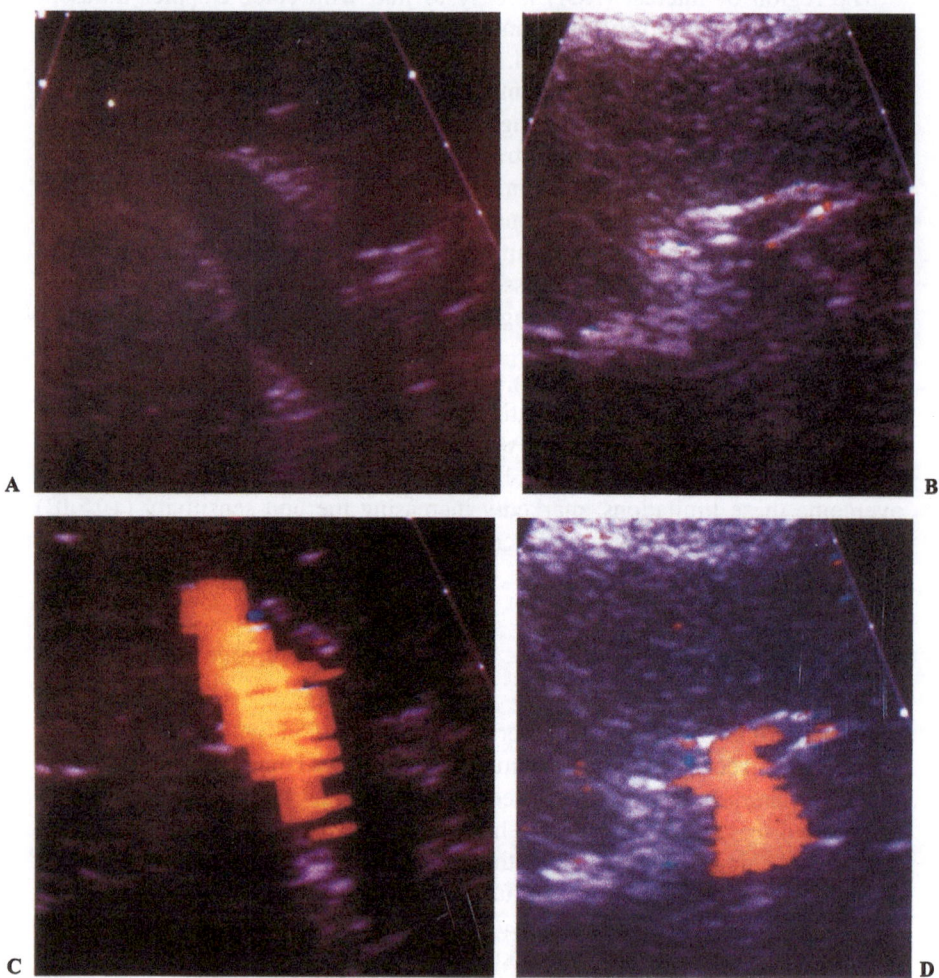

Figure 1. Severe portal hypertension with static flow. (A, B) Baseline color Doppler examination shows the main trunk of the portal vein and the umbilical tract (B). No color signals appear in the vascular lumen. (C, D) Enhanced color Doppler shows static flow in the vascular lumen.

confluence or in the most distal splenic vein, following its course cephalad within the lesser omentum. This vein is difficult to study without a contrast agent because of its deep site: in our series, the left gastric vein appeared on baseline images only in 33% of cases, vs. 100% after Levovist administration.

The para-umbilical vein is easier to depict due to its superficial course [9]; it originates from the umbilical end of the portal vein and usually runs caudad through the ligamentum teres, where it communicates with superficial peritoneal collaterals (caput Medusae). We observed the following types of portosystemic shunts in the patients given Levovist:

- Retroperitoneal: their site is deep and they are difficult to detect on baseline color Doppler images (11%); the detection rate was markedly higher after contrast agent injection (88%).
- Splenorenal: these are usually easy to depict even at baseline, due to their relatively superficial course and reasonable size (88%); nevertheless, our post-contrast detection rate reached 100%.
- Splenoretroperitoneal: these may be found in different, often deep, sites, which makes them difficult to display, with the exception of some small branches at the lower spleen pole which are relatively superficial. They appeared on 22% of baseline vs. 77% of postcontrast images.
- Short gastric: these often small branches are found at the gastric wall or between the latter and the retroperitoneum; they are best depicted with the patient in left lateral recumbency, after distending the stomach with water or using the spleen as an acoustic window. They were depicted on 33% of baseline, and 77% of enhanced images.

Normally, portal flow is hepatopetal in the main portal trunk and in all the segmentary branches, down to the smallest ones. Reversed (hepatofugal) flow even in one segmentary branch only, is pathognomonic of portal hypertension [5]. The main trunk flow was always easy to measure in our series, while small segmentary branches, especially in the deepest right lobe, were difficult to detect at baseline, which increased examination time considerably. Levovist enhancement quickly and accurately displayed flow direction in all the segmentary branches in all patients. Other portal flow abnormalities include bidirectional flow and, less frequently, nearly static, very slow flow: in such cases, blood flow is difficult to differentiate from a recent (and thus anechoic) portal thrombus, even with color Doppler ultrasonography. We had only one such case in our series: the condition was demonstrated at angiography, after baseline power Doppler had misdiagnosed it as a portal thrombus formed in the 2 days between angiography and ultrasonography. The pattern was perfectly depicted with enhanced power Doppler, which showed main portal trunk patency with partial thrombosis of some seg-mentary portal branches.

Portal thrombosis

Portal thrombosis is not uncommon in patients with liver cirrhosis and portal hypertension [10], although it more frequently results from infiltration from a HCC, with portacaval shunts, septic portal thrombosis (pylophlebitis), pancreatitis and traumas. Color Doppler ultrasonography easily detects portal lumen thrombi from the blood flow around them (Figure 2). If there is no flow, thrombosis can be diagnosed only by comparison with flow in a vessel with

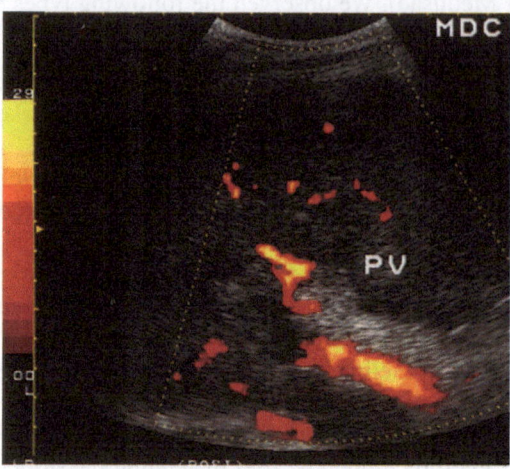

Figure 2. Portal hypertension. Power Doppler examination after contrast administration clearly shows thrombosis in the right branch of the portal vein.

similar features (i.e. depth, caliber, flow velocity) with the same scan angle and unit settings. A contrast agent makes the difference in these patients where the presence/absence of portal thrombosis means the (un)feasibility of liver transplant.

Therapeutic portosystemic shunts

Surgical shunts and those created with stents inserted at interventional angiography are created to prevent gastrointestinal bleeding from esophageal varices or to decrease blood pressure from varices elsewhere. The most common therapeutic portosystemic shunts are end-to-side portocaval, end-to-end portocaval and distal splenorenal. Color Doppler sonography demonstrates even the surgical shunts missed on B-mode images [11]. Their relatively deep location may make the results unreliable, but the contrast agent improves diagnostic accuracy and confidence.

Transjugular intrahepatic portosystemic shunt (TIPS) stents are inserted through the jugular vein. The stent, an expandable wire mesh cylinder, is positioned between a suprahepatic vein and the portal veins. TIPS stents have become increasingly popular because they interfere much less with possible subsequent liver transplantation; they are displayed at color Doppler sonography [11], which is useful before and during stent positioning as well as in the follow-up, to monitor shunt patency and, as far as possible, flow rate.

Stent patency can be studied reliably only when color Doppler ultrasonography shows flow at the anastomosis. However, the stent site sometimes permits only a narrow acoustic window, with a poor scan angle: contrast agent administration before color Doppler scanning could help increase the diagnostic confidence in such cases.

Increased arterial flow: arterialization

Hemodynamic changes in portal hypertension cause portal flow to decrease as a result of increased distal impedance and hepatic artery flow [9, 12] because of a homeostatic compensation mechanism which exists to maintain hepatic perfusion. Arterial flow may be very pronounced in a recent thrombosi and following surgical interruption of the main portal vein or portal flow reversal. In many cirrhotic patients, the hepatic artery has the typical corkscrew appearance: this phenomenon, known as arterialization of hepatic hemodynamics, is found only in advanced liver disease [9]. It is observed not only in cirrhosis, but also in chronic active or alcoholic hepatitis and intrahepatic arteriovenous shunts (from malignant tumors or iatrogenic causes). Arterial blood flow is increased to a lesser extent than in portal hypertension. Color Doppler easily depicts the increase in hepatic artery size and flow velocity; nevertheless, in our personal series, the contrast agent permitted better morphological detailing of arterial branches, detecting the tortuous course typical of advanced disease more easily and rapidly.

Focal and diffuse liver diseases

Primary tumors – benign

Angiomas
This type of lesion has been studied by several authors [13 – 17] since the introduction of color Doppler ultrasonography. The typical slow flow of angiomas was long believed not to yield any Doppler signal, and this diagnostic sign was declared reliable by Taylor et al. [15, 16] who first studied it. With the introduction of color units and the increased sensitivity to slow flows, some signals were detected in middle-sized to large angiomas, while no signal was still observed in small angiomas [13, 14, 18 – 20]. However, the very latest units do detect some vessels and measure their flow even in the smallest angiomas. Angioma vessels are tortuous and therefore difficult to see and depict along their whole course: the most typical pattern is internal color spots, with mainly arteriolar, sometimes even continuous, mid-slow flow (Figure 3). The contrast agent enhances the difference between the normal intraparenchymal vessels displaced by the angioma, which exhibit strong enhancement, and the absence of vessels or the presence of color spots inside the lesion. A small afferent vessel is sometimes depicted. Three-dimensional vessel reconstructions can depict long vessel tracts. Maximum flow velocity is lower than, or at most equal to that of

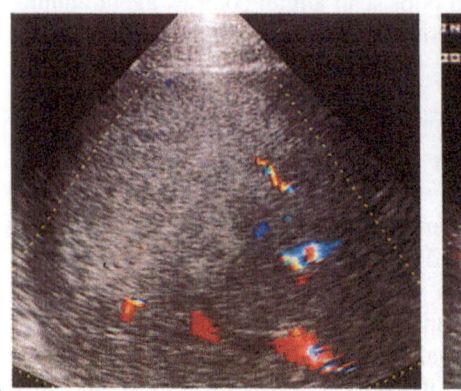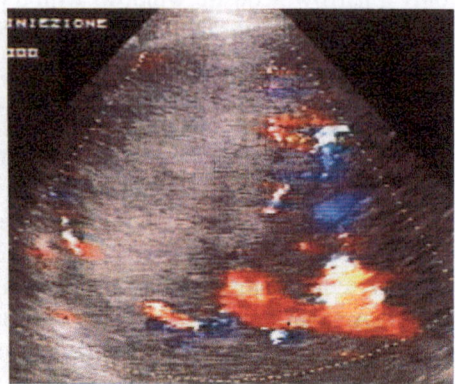

A   B

Figure 3. Hepatic angioma. (A) Baseline examination shows a gross hyperechoic lesion in the right lobe; no color spots are seen inside the lesion and some normal vessels appear in liver parenchyma. (B) After contrast administration the parenchymal vessels surrounding the lesion are markedly enhanced, but no vascular signals originate inside the angioma.

the hepatic artery (0.30–0.40 m/s), in agreement with the typical slow flow of this type of lesion, appearing as slow contrast agent uptake at CT and angiography [16–18]. Such features were confirmed in studies using carbon dioxide injections into the hepatic artery [21]. In our personal experience with i.v. administration of contrast agent (Levovist) [22], we observed color spots inside three of nine medium-large angiomas, all of which exhibited no signal at baseline examinations. Even after contrast agent administration, the color signal appeared as spots and no branches could be detected.

As reported in the literature [23], the patterns of cavernous angiomas are totally different; these lesions may grow rapidly, with major vascular proliferation, and then languish. In our study, we observed color Doppler hypervascularization, with similar high velocity flow to malignant lesions, in the former phase and fewer and slower vascular signals in the latter phase.

Other benign lesions
Focal nodular hyperplasia is the liver disease with the greatest number of recent literature reports [21, 24]. Its vascular pattern, as demonstrated at angiography, consists in an arterial afferent branch reaching the center of the lesion, where it branches centrifugally, giving a typical 'cart-wheel' appearance. Venous drainage is often found. The typical centrifugal vascularization is often demonstrated with color Doppler, when the afferent artery and its branches are depicted with the typical stellate pattern (Figure 4). This pattern is typical of focal nodular hyperplasia and of major diagnostic importance; unfortunately, the afferent arteriole is often only partly depicted on color Doppler images: the tortuous

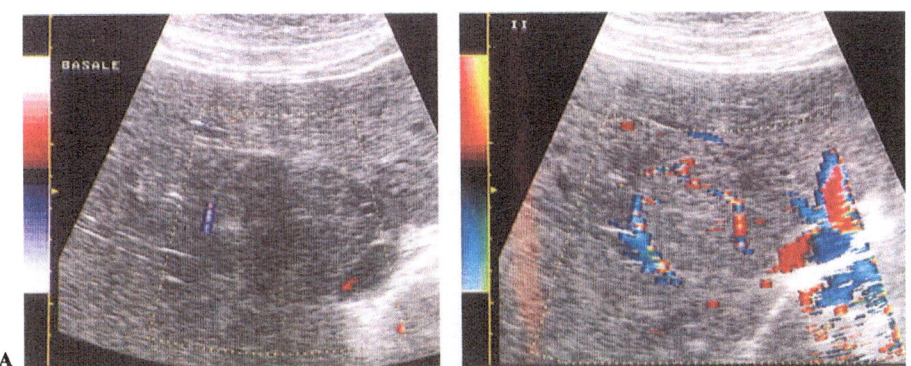

Figure 4. Focal nodular hyperplasia. (A) Baseline examination does not differentiate the lesion from normal liver parenchyma; no vessels are shown inside the lesion. (B) After contrast injection the normal vessels are enhanced and a vessel is demonstrated to reach lesion center.

course of the vessel makes its recognition as a single vessel difficult. This limitation is easily overcome with contrast agent administration.

The signals in the lesion center are mainly of the arterial type, with mid-high velocity [24] and similar values anywhere within the lesion; venous signals come only from the lesion periphery. This stellate pattern is absent only in a minority of cases, but even peripheral arterial signals are shown, sometimes following a basket pattern; in such lesions, the typical central scar is absent.

Hepatic adenomas

This type of lesion is characterized by rich vascularization, with frequent central bleeding areas. Angiography and CT show centripetal vascularization. Data on color Doppler patterns are poor: no unequivocal pathognomonic color Doppler pattern can be defined for adenomas because these lesions may present differently, with or without bleeding and/or fibrosis (Figure 5). Both vascular branches and color spots were depicted in adenomas. Once again, contrast agent administration makes it easier to depict the rich intratumoral vascularization of adenomas and shows the lack of vessels in possible bleeding areas within the lesion. Unfortunately, the hemorrhagic necrosis pattern is often not typical enough.

Hyperplastic nodules in cirrhosis

Defining the vascular pattern of this condition is important, but few literature reports deal with it [25]. We studied two cases [22]: one with no vessels at baseline and the other with peripheral color spots (Figure 6). No changes were observed in the former after contrast agent administration, while an arteriolar afferent branch was depicted in the latter, with no enhancement of intralesional vessels.

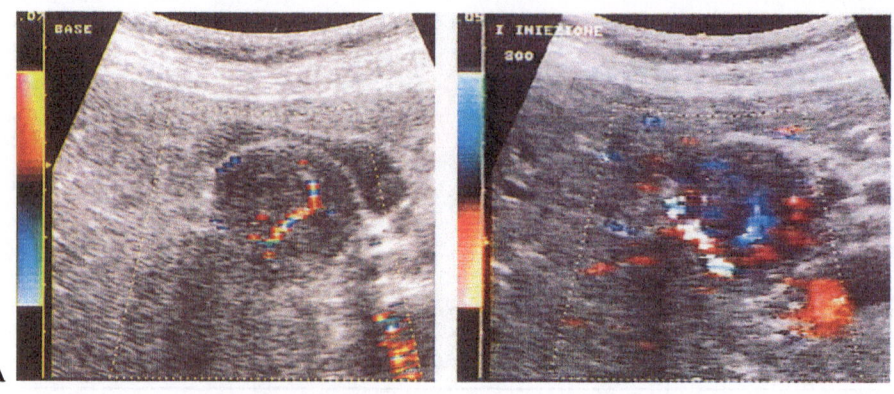

Figure 5. Hepatic adenoma. (A) Baseline examination shows a regular caliber vessel crossing through the lesion. (B) After contrast agent administration, the vessel is enhanced and its branches are depicted.

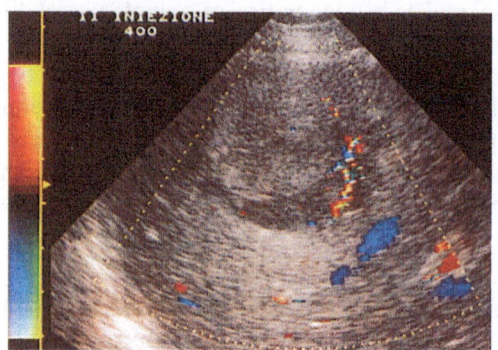

Figure 6. Regeneration nodule in cirrhosis. After two contrast agent injections, the whole perilesional vessel is visualized.

Primary tumors – malignant

Hepatocellular carcinomas

These are among the most richly vascularized lesions, which makes them the ideal subject for color Doppler investigations. The typical vascular patterns feature rich vascularization and fast flow [14, 15, 19]. The most characteristic appearance is the so-called basket pattern, featuring rich peripheral and peritumoral vascularization, and the hypoechoic ring often surrounding the lesion and from which intratumoral branches originate. The branches have irregular caliber and course: they are sometimes seen for long/short tracts, or they may appear as color spots within the tumor [14, 16, 17] (Figures 7–9). This typical

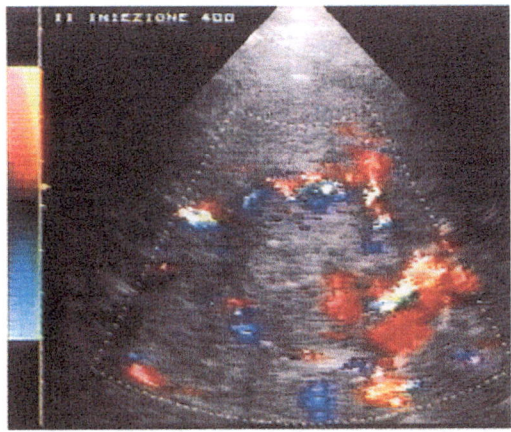

Figure 7. HCC. This nodule exhibited no intra-/perilesional vascular signals at baseline but, after contrast agent administration, the perilesional circle is enhanced and some intralesional color spots appear.

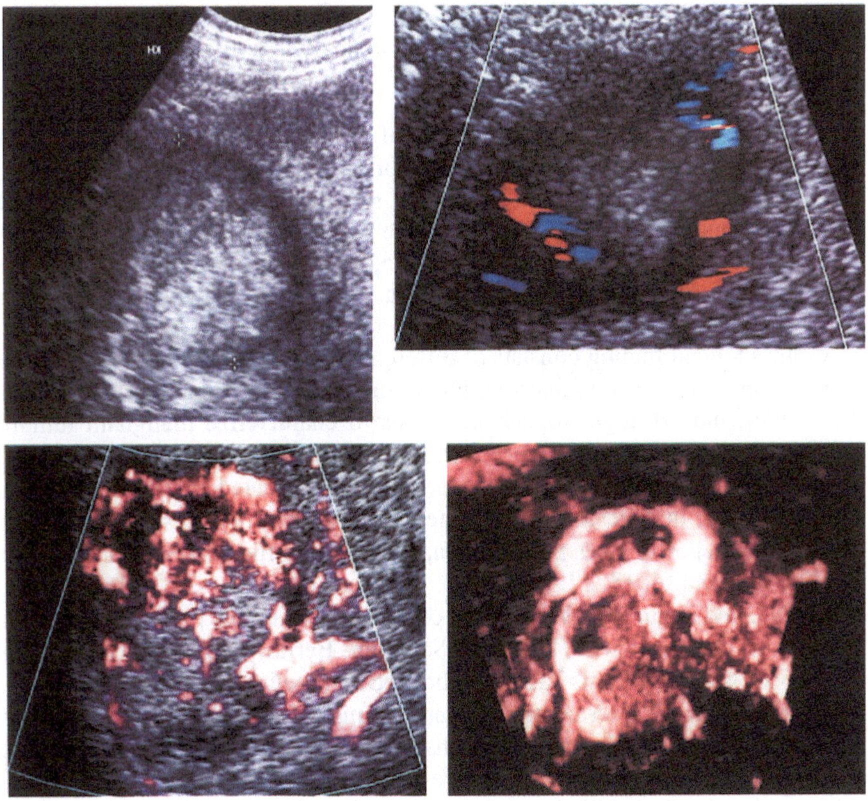

Figure 8. HCC. (A) Large isoechoic focal lesion with a hypoechoic halo. (B) Color Doppler shows poor, mainly peripheral, vascularization. (C) Power Doppler after contrast agent administration shows rich peripheral and internal vascularization (basket pattern), better depicted on 3D reconstruction (D).

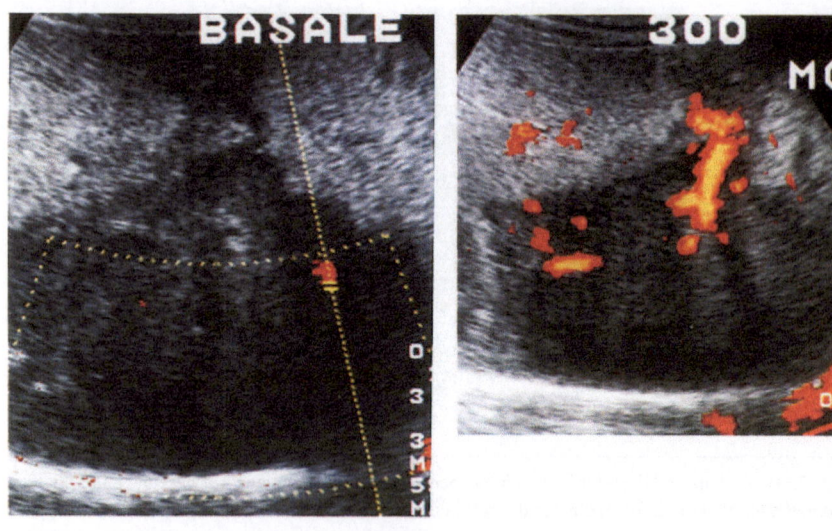

Figure 9. Large hypoechoic lesion in fatty liver. (A) Baseline power Doppler shows a small vascular spot inside the lesion. (B) Power Doppler after contrast agent administration shows multiple feeding vessels with 'tumoral' arterial flow on spectral Doppler.

pattern is not always observed. A minority of hepatocellular carcinomas (HCCs) is poorly vascularized because of necrosis or fatty degeneration.

HCC flow is mainly arterial, with high systolic velocity and an increased diastolic component. In the past, this feature was thought to be characteristic of HCC, as related to arteriovenous shunts [15, 16], but we now know that other malignant tumors exhibit similar vascular patterns too, while some HCCs exhibit medium or low velocity flow.

Another typical finding consists of different velocities in different sites of the same lesion, or reduced/no diastolic phase, thus providing high pulsatility index (PI) values; indeed, high impedance seems to characterize malignant tumors [18, 26, 27]. Contrast agent administration and 3D reconstructions (Figure 8) are major tools which better show the basket pattern represented by peripheral tumor vascularization; new vessels are depicted within the tumor mass, especially in the lesions deeper in liver parenchyma, which are difficult to demonstrate on baseline images [13, 18, 27, 28].

In 1993, and again in 1996, a group of Italian researchers [3, 4, 28] tested an intravenous contrast agent (Levovist, Schering, Berlin) in 96 HCC patients in whom poor vascular signals had been obtained on baseline color Doppler images. Signal enhancement after contrast agent administration was marked in 98% of patients, while no major change was observed in 2% because of tumor necrosis or portal thrombosis, which may reduce the number of fistulas and thus of high velocity signals [3, 4, 28]. Japanese trials with carbon dioxide injections [26]

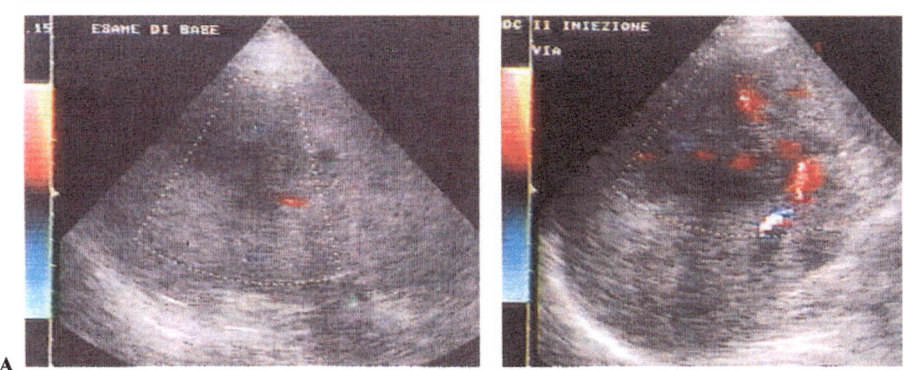

Figure 10. Cholangiocarcinoma. (A) Baseline examination shows a hypoechoic lesion with irregular outline and no signal inside. (B) After contrast agent administration, several horizontal and vermicular vascular spots are depicted inside the lesion; contrast agent wash-out was very rapid in this case.

also showed the intratumoral circulation to originate from an arterial peripheral vascular ring.

Color Doppler ultrasonography, optimized by the use of a contrast agent, depicts not only intra- and perilesional neoplastic circulation, but also the status of nearby, and especially portal vessels, as well as suprahepatic veins and inferior vena cava; it yields accurate data on the site and extent of possible venous thrombosis and the hypertrophy of the corresponding arterial branch [3, 4, 17, 18]. Arterial signals in the portal thrombus strongly suggest direct neoplastic infiltration of the vessel [13]. The contrast agents depicts such changes not only in the main branches, but also in segmentary and subsegmentary vessels.

Other malignant tumors

Data of the vascular patterns of malignant hepatic lesions other than HCC are scarce. We report on 4 cholangiograms, avascular on baseline images [22]: marked enhancement was observed in two of them after contrast agent injection (Figure 10), while the other two cholangiocarcinomas in the biliary confluence showed no enhancement. In the two enhancing lesions, at peak contrast agent effect, several horizontal intralesional vessels with rapid flow and rapid wash-out were demonstrated, giving the lesions a vermicular appearance. This pattern, which has never been observed without the contrast agent, was also seen in other primary malignant lesions and metastases [3, 4, 22, 28, 29].

Secondary tumors

Secondary hepatic tumors have been studied with color Doppler sonography [1, 3, 4, 14–16, 30–33] and often compared with primary lesions. Unfortunately,

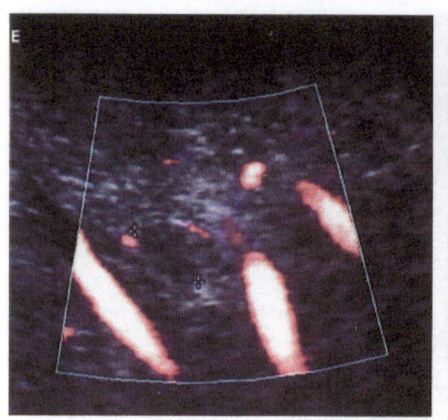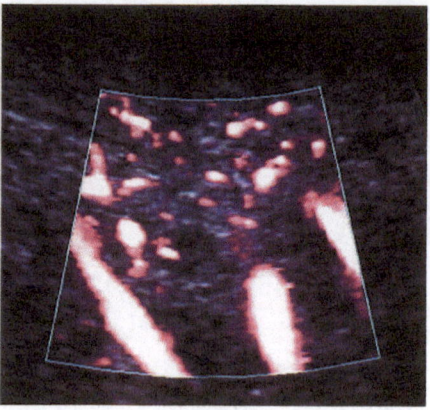

A   B

Figure 11. Gastric carcinoma metastasis. (A) Baseline power Doppler shows a small, fairly hypo-echoic, focal lesion (arrows) with a single peripheral vascular spot. (B) Power Doppler after contrast agent administration shows many peripheral and internal vascular spots.

their relatively poor vascularization prevents adequate morphological studies, and a large number of these lesions is usually classified as 'avascular' [31, 32]. Therefore, secondary hepatic tumors need to be studied following contrast agent administration.

Our personal experience with Levovist in 34 patients with secondary hepatic lesions from different primaries selected for poor Doppler signals on baseline images, yielded very interesting results and open new diagnostic prospects (Figures 11, 12). Thirteen of 34 lesions (38.2%) exhibited no central/peripheral vascularization on baseline images. Of the remaining 21 lesions, ten (29.4%) had poor vascular signals, seven (20.5%) moderate vascular signals and only four (11.7%) had rich and apparent vascularization. After contrast agent injection, only five of the 13 'avascular' lesions (14.7%) remained such, for reasons explained by other imaging methods: i.e., large necrotic areas with consequently low vascularized tissue rates in four cases and a small and very deep lesion. Our rate (14.7%) is significantly lower than that of Taylor et al. [15, 16] (20/37 cases, 54.1%) with no contrast agent, in a series with about the same number of patients. The rate of lesions remaining avascular even after contrast agent injection in our study is also lower than that recently reported by Nino Murcia [32] (26.5%) in 64 metastatic lesions: however, she had not selected those with a suboptimal signal. The latter rate lies about half-way between our rate at base-line (38.2%) and that obtained after contrast agent administration.

In our study, Levovist vascular enhancement appeared about 20 s after the injection and lasted about 240 s (Figures 11, 12), permitting accurate assessment of tumor vascularization in 29/34 lesions (85.3%). To our knowledge, no similar series have been reported to date in the literature. The vessels exhibited typically

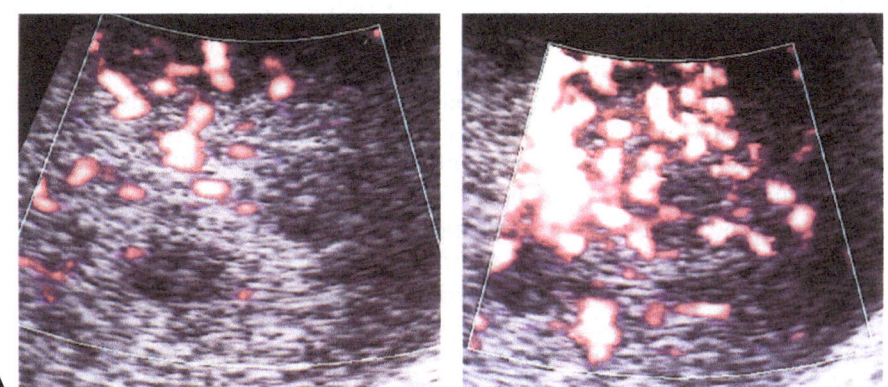

Figure 12. Lymphoma localization. (A) Baseline power Doppler shows two small avascular hypo-echoic lesions. (B) Power Doppler after contrast agent administration shows rich parenchymal vascularization with peripheral and internal vascular spots.

malignant features: i.e. irregular course, caliber and canalization, arteriovenous shunts, blood pools. Three main patterns of vessel distribution were fairly equally represented: nine cases (31.0%) showed vessels inside the lesion, nine (31%) showed vessels around the lesion and vessels inside and around the lesion were seen in 11 cases (37.9%). Pattern analysis according to lesion origin confirmed the second pattern as being most frequent in secondary lesions from breast cancers (5/8, 62.5%), while the last pattern was the most frequent in secondary lesions from colorectal cancers (8/14, 57.1%).

The flowmetric and morphological data from 12 lesions showed fair signal variability: thus, the signal was arterial in 10/12 cases, 'tumoral' (as defined by Taylor et al. [15, 16]) in four cases and of the portal venous type in only two cases. Our findings are in agreement with those of Dock et al. [31] in 35 meta-static lesions with flow velocity of $0-2.34$ m/s, and with Ralls et al. [1], who declared that secondary hepatic lesions cannot be distinguished from primary tumors on the basis of vascular signals alone, because the relative vascular patterns partially overlap.

Finally, on the basis of our results in a small series of patients, we believe a vascular pattern prevailing over another to lack statistical significance and that selecting the lesions among those with suboptimal baseline Doppler signal makes the sample not really representative of the general population.

Contrast agent administration increased diagnostic confidence from 27.4 ± 22.50 to 77.2 ± 22.50 in a subgroup of 513 of 1275 patients examined in Italy and all over Europe [4, 30]. Thus, the data so far encourage further in-depth studies to verify the current preliminary results, when Levovist and other possible contrast agents are available.

Echocontrast agents in post-chemotherapy follow-up

Several nonsurgical or interventional options such as transcatheter arterial chemoembolization (TAE) or percutaneous ethanol injection (PEI) are currently available for HCC patients who are considered inoperable because of advanced age, severe liver cirrhosis or lesion multifocality. TAE is generally performed in single/multifocal HCCs bigger than 2–3 cm which are hypervascular angiographically: the hepatic artery supplying the tumor is injected with 1–5 ml of iodized oil with a chemotherapy agent, followed by an absorbable embolic agent. The rationale for the use of hepatic arterial embolization is that HCCs are fed by arterial blood [34, 35]; occluding the feeding arteries results in tumor anoxia and leads to liquefactive tumor necrosis. TAE efficacy depends on tumor vascularity and it has been reported to be rather ineffective in some types of HCC, including very well differentiated, sclerosing, or poorly differentiated/undifferentiated HCC [36]. Histopathological studies of resected HCCs after TAE showed complete necrosis in only 22–50% of the tumors. It is particularly likely to fail when the lesions are large, non-capsulated, or supplied by collateral feeding arteries, or in the presence of vascular shunts between hepatic artery and portal vein [34, 37–39].

PEI is usually considered an appropriate treatment for patients with HCC lesions that are smaller than 3–5 cm, single (less than 3 for some authors) and easily detectable at ultrasonography. Ethanol exerts its local toxic action through cellular dehydration and protein denaturation followed by coagulative necrosis and vascular thrombosis. Incomplete necrosis due to poor ethanol spread is likely in sclerosing HCCs owing to the firm texture of the lesions [36] and in large HCCs with necrotic or septate areas [40–42]. Specific imaging techniques are required to assess whether the tumor has become completely necrotic after treatment. Lipiodol Computer Tomography (Lipiodol CT), contrast enhanced CT and Magnetic Resonance Imaging (MRI) are considered the most useful examin-ations in the short- and long-term follow-up of treated tumors [43]. Lipiodol CT can be performed no sooner than 2–3 weeks after TAE, to permit sufficient Lipiodol clearance by uninvolved liver parenchyma; persistent homogeneous iodized oil uptake is observed in well-treated tumors. Using contrast enhance-ment, the presence of unenhanced, low density (at CT) or low intensity (at MRI) areas suggests complete tumor necrosis; high density (CT) or high intensity (MR) areas within the lesion are usually a sign of viable tumor tissue and per-sistent tumor vascularization [42–44]. Sonography cannot be performed to confirm tumor regression, because necrotic and viable tumor tissue may have similar echogenicity on posttreatment US scans [42, 45].

Color Doppler ultrasonography has been recently advocated as a useful tool for the assessment of residual tumor after TAE: the persistence of flow signals 2 weeks after treatment correlated well with the presence of residual vascularization at CT and arteriography [46]. In this study, peritumoral and intratumoral pulsatile flows were depicted on baseline color Doppler examinations in 92%

and 55% of the lesions, respectively; one day after TAE, there was no residual color Doppler flow in any lesion [46]. Few false-negatives were observed in the 6–12 months' follow-up; these were probably due to small collateral arteries feeding the tumors or to low velocity of the persisting blood flow, below the level of detection by color Doppler sonography. Some of the above problems in the follow-up of treated HCCs can probably be solved with specific agents for color Doppler enhancement.

We administered Levovist to 43 HCC patients, selected for poor color Doppler signals on baseline images; four were studied after a single TAE [28]. At baseline examinations, vessels were depicted in 31 of 39 untreated tumors (79.5%); peritumoral and intratumoral flows were demonstrated in 24 (61.5%) and 16 (41%), respectively. All the untreated lesions exhibited color Doppler signs of peri-tumoral vascularization after i.v. Levovist administration, with intratumoral flows in 31 (79.5%) [4]. These results represent a major improvement over unenhanced color Doppler flow imaging findings, where up to 20% of HCCs may appear avascular [25, 46, 47]. The clear depiction of tumor vascularization is the basis for viable tissue detection in uncompletely treated HCCs. Moreover, the easier depiction of intratumoral flow reduces the risk of misdiagnosing uninvolved liver vascular structures as peritumoral vascularity.

We examined four patients with TAE-treated HCCs before and after Levovist injection. Three tumors (9 cm, 4.5 cm and 4 cm in greatest diameter, respectively) had been treated 2/3 days before, the other (2.5 cm) had been submitted to unsuccessful TAE 4 months earlier. All of these tumors appeared avascular on precontrast color Doppler images. The only tumor which remained avascular even after Levovist administration, was the 9 cm lesion submitted to TAE 3 days earlier, where Lipiodol CT and angiography subsequently demonstrated complete devascularization. The other three lesions exhibited moderate (two) or rich (one) vascularization after Levovist administration. Two lesions exhibited some intratumoral flow signals (Figure 13), while the other had a hyperechoic avascular center surrounded by a thick hypoechoic periphery with rich vascularization on color Doppler images (Figure 14A, B). Lipiodol CT and MR studies performed 3 weeks later demonstrated incomplete devascularization of the lesions; complete agreement was found between the early color Doppler study after Levovist administration and the MR study performed 3 weeks later using spin-echo and Gd-DTPA enhancement (Figure 14C). One of these patients underwent surgery, and histology of the resected specimen showed viable tissue persistence with capsular neoplastic invasion.

Our experience with early color Doppler follow-up of TAE-treated HCCs is in partial disagreement with a study [46] in which all the lesions studied one day after TAE appeared avascular, but confirms a recent observation by Taourel, indicating a rapid return of hepatic artery flow to baseline values 2 days after TAE [47]. Experimental studies based on enhanced Doppler sonography may help to develop more effective embolizing materials and procedures.

In conclusion, color Doppler signals enhance greatly after Levovist injection

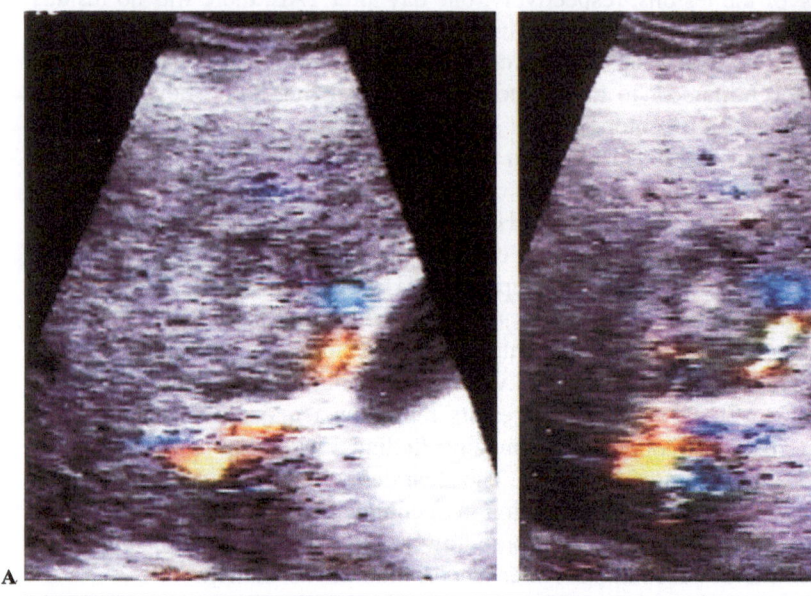

Figure 13. HCC (4.5 cm) of the right hepatic lobe, submitted to recent TAE. (A) Baseline color Doppler shows no flow signals inside the lesion, while some perilesional flow is depicted, possibly from uninvolved liver parenchyma. (B) After Levovist administration, an afferent collateral vessel is clearly demonstrated (arrows), originating from the hepatic hilum region and running into the tumor.

in patients with uncompletely treated HCCs. A color Doppler contrast agent is likely to improve the depiction of flows from smaller vessels of residual viable tumor tissue, and when they are deeply seated within liver parenchyma. Thus, a more confident diagnosis of persistent viable tissue can be made easily and noninvasively and in the immediate post-TAE period. These preliminary results need further confirmation in larger series of HCC patients, and in those treated with other therapeutic options, including PEI and physical energy techniques. A simple and early diagnosis can avoid patient rehospitalization and permit correct therapeutic decisions to be reached when previous treatment has failed.

CONCLUSIONS

Over the last 5 years, spectral and color Doppler imaging have revolutionized the ability of sonography to detect and study normal and abnormal liver vessels. Color Doppler ultrasonography displays in real time, passively and automatically, even vessels which are undetectable or inconspicuous on gray-scale images, and in structures whose vascular nature could not be suspected at B-mode imaging. Transient dynamic or pulsatility changes can be studied noninvasively

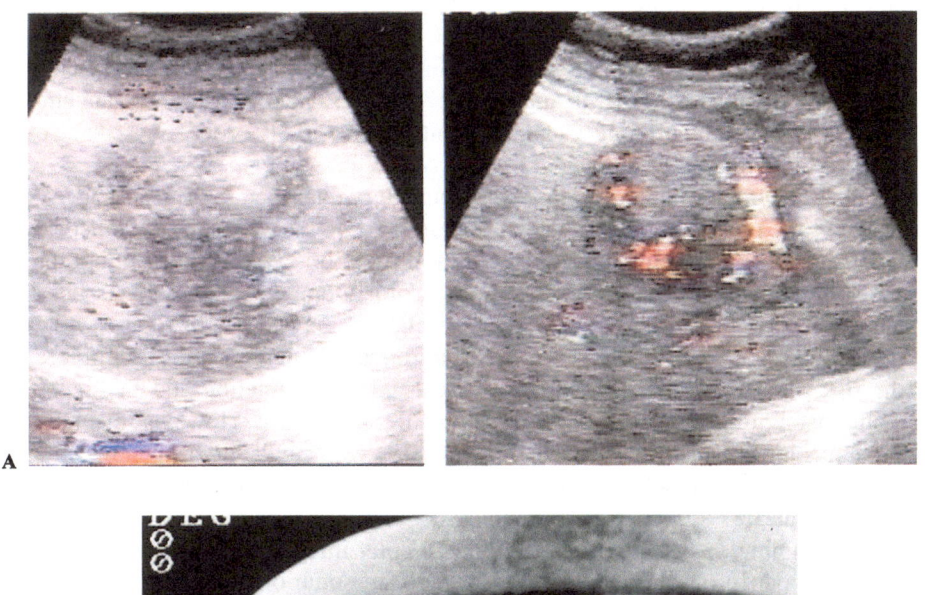

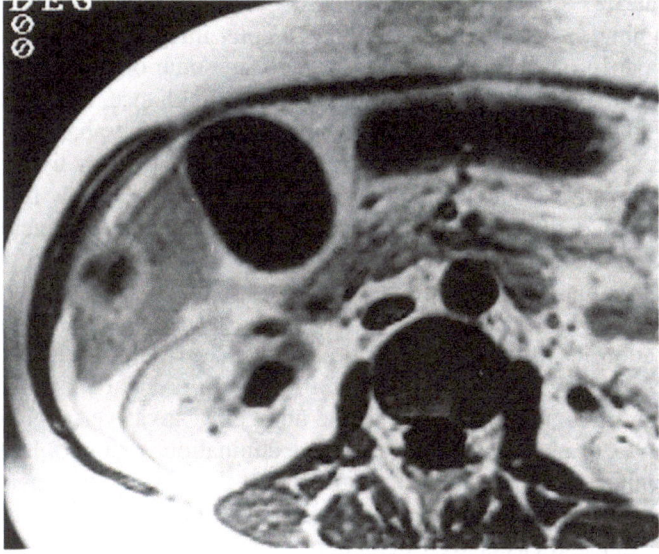

Figure 14. (Opposite) Subcapsular HCC (4 cm), submitted to recent TAE. (A) Baseline color Doppler, 3 days after TAE: the lesion center (arrows) and periphery are hyperechoic. No flow signals are depicted inside the lesion. (B) After contrast agent administration, the lesion center is still avascular, while its periphery is clearly vascularized. (C) Gd-DTPA MR study (SE T1-weighted sequence, 500/20), performed 3 weeks later: signal hyperintensity confirms persistent perilesional vascularization.

and with no effect on hemodynamics, which few or no other methods can do. Diagnoses that previously required invasive angiography can now be made non-invasively with color Doppler imaging. In some cases, color Doppler yields diagnostic findings where even angiography did not. The results and therapeutic adequacy of (vascular) surgery and interventional radiology procedures can also be assessed with color Doppler sonography.

Spectral Doppler imaging yields complementary quantitation of flow velocity or shows the type of hemodynamic alteration. The ROI is easy to find with color Doppler ultrasonography, which permits prompt sample volume positioning for possible further spectral Doppler studies. Color Doppler sonography is used alone when a global view of ROI vascular anatomy is needed, or to assess the presence/absence of flow to make the final diagnosis. This method minimizes the risk of missing blood flow in the area under examination and facilitates comparison of the vascular patterns of two different regions. However, even the latest machines are not sensitive enough to study very small and deep slow-flow and low-velocity vessels; technical limitations are likely to decrease the future progress rate.

Other limitations of color and spectral Doppler imaging are limited acoustic windows, partly obscuring vessels, especially in some organs and anatomical regions (e.g. the retroperitoneum), poor patient compliance, requiring examination time to be markedly reduced, difficult depiction of intraparenchymal liver vessels because of marked acoustic absorption by fibrotic/fatty liver, even with low frequency probes and difficult assessment of slow/very slow flows (e.g. in focal lesions). Contrast agents help overcome these limitations by increasing unit sensitivity (+15 dB), which is why we used an echocontrast agent (Levovist) whose clinical, diagnostic and possibly even therapeutic features make it a useful tool. It provides better assessment of liver hemodynamics and abnormal liver vascularization in diffuse chronic and acute hepatopathies, easier detection of diagnostic criteria and data collection for the differential diagnosis, shorter examination time and a reduction in the number of invasive (angiography) or more complex or expensive (CT and MR) examinations required. It also allows patient monitoring during and after any kind of therapy, monitoring the results of interventional procedures such as TAE, PEI, chemotherapy, stenting, TIPS, provides better and easier ROI depiction, increases diagnostic confidence and sensitivity and permits spectral Doppler display of antithetic flows and velocities within the same lesion.

These and many other considerations lead us to conclude that echocontrast agents play a major role in the qualitative and quantitative progress of ultrasonography, and may help spread its use.

If color Doppler sensitivity were increased, making the contrast agent less necessary for morphological purposes, (other) functional studies on load curve, wash-in and wash-out would yield further useful information for diagnosis and treatment, as occurs with CT and MRI.

REFERENCES

1. Ralls PW. Focal hepatic and pancreatic disease. In: Rifkin MD, editor. Syllabus special course: Ultrasound 1991. Oak Brook, IL: RSNA Publications. 1991:293–306.
2. Burns PN. Ultrasound contrast agents in radiological diagnosis. Radiol Med 1994;87(Suppl 1): 71–82.
3. Campani R, Bozzini A, Calliada F et al. Color Doppler imaging of liver metastases. The phase-II of a US contrast agent: SHU 508 A (Levovist) Schering. Radiol Med 1994;87(Suppl 1): 32–40.
4. Campani R, editor. Ultrasound contrast agents: state of the art. Radiol Med 1994;87(Suppl 1): 1–88.
5. Ralls PW. Color Doppler sonography of hepatic artery and portal venous system. Am J Roentgenol 1990;155:517–25.
6. Lebrec D, Benhamou JP. Noncirrhotic intrahepatic portal hypertension. Semin Liver Dis 1986; 6:101–9.
7. Lafortune M, Patriquin H, Pomier G et al. Hemodynamic changes in portal circulation after portosystemic shunts: use of duplex Doppler sonography in 43 patients. Am J Roentgenol 1987;149:701–6.
8. Calliada F, Bergonzi M, Passamonti C et al. Hepatic portal hypertension. Qualitative evaluation with real time and color Doppler US. Radiol Med 1990;79:339–45.
9. Morayasu F, Nishida O, Ban N et al. 'Congestion index' of portal vein. Am J Roentgenol 1986; 146:735–9.
10. Okuda K, Ohnishi K, Kimura K et al. Incidence of portal vein thrombosis in liver cirrhosis. Gastroenterology 1985;89:279–86.
11. Ferretti G, Rabbia C. IL sistema portale. In: Rabbia C, De Lucchi R, Cirillo R, editors. Eco Color Doppler Vascolare. Torino: Minerva Medica 1995;235–94.
12. Reuter SR, Berk RN, Orloff MJ. An angiographic study of pre and postoperative hemodynamics in patient with side to side portacaval shunts. Radiology 1975;116:33–9.
13. Maresca G, Danza FM, Vecchioli A. Tissue characterization by color Doppler. Rays 1991; 15:3–9.
14. Tanaka S, Kitamura T, Fujita M et al. Color Doppler flow imaging of liver tumors. Am J Roentgenol 1990;154:509–14.
15. Taylor KJW, Ramos I, Morse SS et al. Focal liver masses: differential diagnosis with pulsed Doppler US. Radiology 1987;164:643–7.
16. Taylor KJW, Ramos I, Carter D et al. Correlation of Doppler US tumor signals with neovascular morphology features. Radiology 1988;166:59–62.
17. Tomita S. Color Doppler imaging in hepatic disease. Med R 1991;43:28–32.
18. Maresca G, De Caetano AM, De Franco A et al. Lesioni focali epatiche. In: Corso monotematico di eco-color-Doppler. Appolonio, Brescia: APC-SIRM. 1994.
19. Nino Murcia M, Ralls PW, Jeffrey BR. Color flow characterization of focal hepatic lesions. Am J Roentgenol 1988;159:1196–8.
20. Tomita S. Color Doppler imaging of hepatic tumors. Med Radiol 1993;43:28–33.
21. Kudo M, Tomita S, Tochio H et al. Hepatic focal hyperplasia: specific findings of dynamic contrast enhanced US with carbon dioxide microbubbles. Radiology 1991;179:377–82.
22. Maresca G, Barbaro B, Summaria V et al. Color Doppler ultrasonography in the differential diagnosis of focal hepatic lesions. The SHU 508A (Levovist) experience. Radiol Med 1994;87 (Suppl 1):41–9.
23. Paltiel HJ, Patriquin HB, Keller MS et al. Infantile hepatic hemangioma: Doppler US. Radiology 1992;182:735–49.

24. Golli M, Mathieu D, Anglade MC et al. Focal nodular hyperplasia of the liver: value of color Doppler US in association with MR imaging. Radiology 1993;187:117–21.
25. Tanaka S, Kitamura T, Fujita M et al. Small hepatocellular carcinoma: differentiation from adenomatous hyperplastic nodule with color Doppler flow imaging. Radiology 1992;182: 161–5.
26. Kudo M, Tomita S, Tochio H et al. Small hepatocellular carcinoma: diagnosis with US angiography with intra-arterial CO_2 microbubbles. Radiology 1991;189:155–60.
27. Shimamoto K, Sakuma S, Ishigaki T et al. Hepatocellular carcinoma:evaluation with color Doppler US and MR imaging. Radiology 1992;182:149–53.
28. Angeli E, Campanelli R, Crespi G et al. Efficacy of SHU 508 A (Levovist) in color Doppler ultrasonography of hepatocellular carcinoma vascularization. Radiol Med 1994;87(Suppl 1): 24–31.
29. Goldberg BB, Hilpert PL, Burns PN et al. Hepatic tumors: signal enhancement at Doppler US after intravenous injection of contrast agent. Radiology 1990;177:713–7.
30. Campani R, Calliada F, Anguissola R et al. Improved signal-to-noise (S/N) ratio in the color-Doppler demonstration of normal and pathologic arterial and venous vessels in man by iv injection of SHU 508 A (Levovist) Atti Investigators Meeting Schering AG, Springer Verlag, Berlin. 1993.
31. Dock W, Grabenworger F, Metz V et al. Tumor vascularization: assessment with duplex sonography. Radiology 1991;181:241–4.
32. Nino-Murcia M, Ralls PW, Jeffrey BR. Color flow Doppler characterization of focal hepatic lesions. Am J Roentgenol 1992;159:1195–7.
33. Numata K, Tanaka K, Mitsui K et al. Flow characteristics of hepatic tumors at color Doppler sonography: correlation with arteriographic findings. Am J Roentgenol 1993;160:512–21.
34. Nakaura H, Hashimoto TM. Transcatheter oily chemoembolization of hepatocellular carcinoma. Radiology 1983;170:783–6.
35. Yamada R, Sato M, Kawabata M et al. Hepatic artery embolization in 120 patients with unresectable hepatoma. Radiology 1983;148:397–401.
36. Yamashita Y, Matsukawa T, Arakawa A et al. US-guided liver biopsy; predicting the effect of interventional treatment of hepatocellular carcinoma. Radiology 1995;196:799–804.
37. Takayasu K, Moriyama N, Muramatsu Y et al. Hepatic artery embolization for hepatocellular carcinoma: comparison of CT scans and resected specimens. Radiology 1984;150:661–5.
38. Nakamura H, Tanaka T, Hori S et al. Transcatheter embolization of hepatocellular carcinoma: assessment of efficacy in cases of resection following embolization. Radiology 1983;147: 401–5.
39. Sakurai M, Okamura J, Kuroda C. Transcatheter chemoembolization effective for treating hepatocellular carcinoma: a histopathologic study. Cancer 1984;54:387–92.
40. Tanaka K, Nakamura S, Numata K et al. Hepatocellular carcinoma: treatment with percutaneous ethanol injection and transcatheter arterial embolization. Radiology 1992;185:457–60.
41. Livraghi T, Festi D, Monti F et al. US-guided percutaneous alcohol injection of small hepatic and abdominal tumors. Radiology 1986;161:309–12.
42. Shiina S, Tagawa K, Unuma T et al. Percutaneous ethanol injection for the treatment of hepatocellular carcinoma: analysis of 77 patients. Am J Roentgenol 1990;155:1221–6.
43. Sironi S, Livraghi T, Angeli E et al. Small hepatocellular carcinoma: MR follow-up of treatment with percutaneous ethanol injection. Radiology 1993;187:119–23.
44. Livraghi T, Salmi A, Bolondi L et al. Small hepatocellular carcinoma: percutaneous alcohol injection – results in 23 patients. Radiology 1988;168:313–7.
45. Livraghi T, Vettori C. Percutaneous ethanol injection therapy of hepatoma. Cardiovasc Intervent Radiol 1990;13:146–52.

46. Tanaka K, Inoue S, Numata K et al. Color Doppler sonography of hepatocellular carcinoma before and after treatment by transcatheter arterial embolization. Am J Roentgenol 1992;158: 541–6.

47. Taourel P, Dauzat M, Lafortune M. Hemodynamic changes after transcatheter arterial embolization of hepatocellular carcinomas. Radiology 1994;191:189–92.

37. Echo-enhancing agents in tumors

BARRY B. GOLDBERG, JI-BIN LIU, DAVID O. COSGROVE & MARTIN J. K. BLOMLEY

INTRODUCTION

Interest in the vascular biology of tumors has focused on the demonstration of apparently unique properties of their morphology and make-up. Tumors require a neovasculature to continue growing beyond a diameter of only a few millimeters and this vascular supply is derived from adjacent normal vessels: 'angiogenesis factors' stimulate the ingrowth of arcades of new vessels that penetrate the tumor from its periphery. Distinguishing features are the relative abundance of the vessels (although this does not necessarily mean that tumors are well perfused) and their morphological disorder, characterized by the irregularity of their lumina and branching pattern, as well as direct intervascular shunts and often also blind-ending vessels. Although it is not yet clear whether all of these features are unique to malignant neovascularization (some chronic inflammatory processes may also have morphologically abnormal vessels) they are characteristically different from normal vasculature.

Functional abnormalities exist in addition to structural abnormalities: the normal smooth muscle layer which controls the lumen of all vessels, but most particularly the arterioles (and thus provides the vasomotor tone), is often incomplete so that control of perfusion is ineffective. In addition, the endothelium functions poorly and the vessels are leaky so that high molecular weight substances diffuse more readily into the interstitial space than from a normal vascular bed. This produces a high oncotic pressure within the tumor, which itself may be irregularly distributed. The overall result of these morphological and histological abnormalities is a disorganized vascular tree with lack of vasomotor control and the presence of shunts (which lead to increased blood flow) and, at the same time, regions with high interstitial pressure (which leads to reduction in flow). In addition, the neovascularization may be non-uniform, some parts of the tumor being more poorly vascularized than others. Overall, this leads to heterogeneity of blood flow and perfusion. Thus the heterogeneity of the vascular system mirrors the heterogeneity of the histology of the tumor tissue itself.

N. C. Nanda, R. Schlief and B. B. Goldberg (eds.), Advances in Echo Imaging Using Contrast Enhancement, Second Edition, 585–614.
© 1997 *Kluwer Academic Publishers.*

Many of these features can be demonstrated on ultrasound using color Doppler, but the classic means for demonstrating the morphology of tumor neovascularization is angiography. However, angiography cannot demonstrate the functional abnormalities that are revealed by spectral Doppler, particularly the lack of vasomotor control and the hemodynamic effects of arteriovenous shunting (producing rapid flow with a high diastolic component) that give high peak velocities and low resistance or pulsatility indices on spectral Doppler tracings. However, the patchiness of these changes means that many vessels have a low peak velocity and a high resistance index: this heterogeneity probably accounts for the confused reports in the literature on spectral indices in tumor vessels. The only feature that is not well demonstrated on ultrasound is the increased endothelial permeability: on angiography this shows as a tumor blush with persistent interstitial contrast retention but there is no corresponding ultrasonic abnormality because the microbubble echo-enhancing agents for ultrasound are $2-7\,\mu$m in diameter and so are confined to the vascular space. Unlike X-ray and MR contrast agents, microbubbles do not diffuse into the intercellular space but behave as blood-pool agents.

The potential of microbubble contrast agents for improving the delineation of these features of malignant neovascularization, and possibly extending them, is of great interest in oncology. It is important to note also that, apart from being a main controlling factor in tumor growth rate, neovascularity is also important in metastases: vascular tumors are probably more prone to metastasize, not only because the incomplete vasculature is more easily penetrated than a normal vessel structure would be but also because the high tissue fluid flow implies a high lymphatic drainage [1]. Many antitumor treatments actually damage the neovasculature, so that monitoring vascular changes can be expected to provide direct evidence of response to radio- and chemotherapy.

Crucial to the full exploitation of the enhancement of the display of tumor vasculature offered by blood-pool microbubble agents is the development of appropriate quantitative techniques. At present, while measurements are available for the Doppler spectrum (albeit rather simplistic ones that do not reflect the richness of the data they contain), only subjective evaluations can be made for color Doppler. Active research is, however, developing methods to measure the number of vessels and quantify their morphology on two-dimensional (and even on 3-D) Doppler images. For both of these the power Doppler ('energy', 'color power angio') mode is more appropriate than the conventional velocity mode because anatomical rather than dynamic information is paramount in this context. Analysis of the transit times of a bolus of a microbubble agent has the potential for offering entirely new features which can be extracted using techniques originally developed for nuclear medicine dynamic analysis (see below).

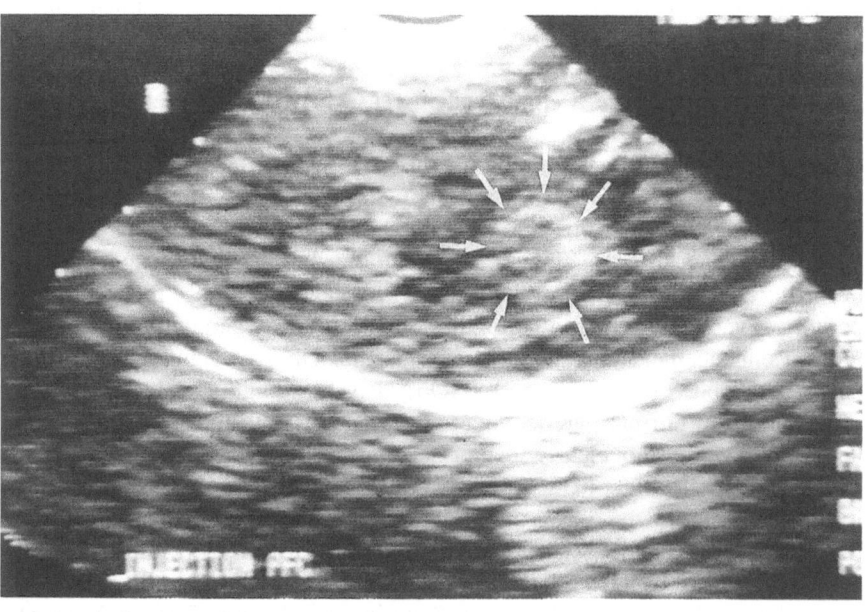

Figure 1. 24 h after administration of perfluorocarbon a rim (arrows) of increased reflectivity can be seen around a woodchuck hepatocellular tumor.

PRELIMINARY STUDIES IN ANIMAL TUMOR MODELS

The limitation of ultrasound without the use of contrast has been its inability, in many cases, to detect tumors within organs. However, with the use of an ultrasound contrast agent the ability of ultrasound to identify tumors and other abnormalities within various abdominal organs should improve.

The non-microbubble containing contrast agents, such as perfluoroctylbromide (PFOB), have been used for tumor evaluation in animal models. PFOB particles are much smaller than red blood cells and pass easily through the capillary system. They increase the reflectivity of gray-scale and Doppler signals due to their relative high density (1.9 g/ml) and low acoustic velocity (600 m/s) resulting in an acoustic impedance difference of 30% between them and the adjacent tissues [2]. Injection of a perfluorochemical agent into a peripheral vein produced gray-scale and Doppler enhancement in animal experiments. In addition, these perfluorochemicals were taken up by the reticuloendothelial system resulting in a progressive increase in liver echogenicity over time. This increase in the reflectivity of the normal liver tissue improved the visualization of tumors, which lack Kupffer cells. This agent can thus identify tumors, which appear hypoechogenic with a hyperechogenic rim (Figure 1) [3, 4].

Parker et al. used iodipamide ethyl particles, which act in a similar fashion to the perfluorochemicals [5]. The agent is taken up by the reticuloendothelial system, which thus increases in reflectivity, producing the potential for demonstrating tumors. More recently, this agent has been combined with gas microbubbles, which produce a significant increase in reflectivity, enhancing both gray-scale and Doppler images.

Several microbubble-containing agents have been evaluated with respect to their effectiveness in delineating tumors in animal models. Albumin-containing microbubble agents (Albunex and FSO69, Molecular Biosystems Inc., San Diego, CA) have been evaluated in VX2 tumor-bearing rabbits and naturally occurring hepatomas in woodchucks. These agents, when injected into a peripheral vein, showed circulation throughout the body and both produced significant enhancement of both spectral and color Doppler signals in both large and small blood vessels (Figure 2). FSO69 also showed gray-scale enhancement of signals from normal parenchyma and tumors. Within the woodchuck hepatoma, small vessels not clearly seen prior to intravenous injection of these agents were also demonstrated. In larger tumors the greatest increase in Doppler intensity was apparent at the tumor periphery since the central portion tended to be less vascular and, in some cases, was necrotic [6]. Early work in this area has shown the ability of this agent to improve visualization of tumor vessels.

A galactose-based agent, Levovist (Schering AG, Berlin, Germany), has also been evaluated in animal models. The gas microbubbles of this agent, which are coated with a thin layer of fatty acid, are $2-8\,\mu$m in size, 97% of the bubbles measuring $<6\,\mu$m. When injected into a peripheral vein, Levovist can circulate in excess of $3-5$ min [7, 8]. Recirculation allows complete evaluation of a variety of vessels in tumors and normal structures. In large animal models this agent enhanced Doppler signals in the vessels of the retina of the eye as well as the gallbladder, urinary bladder and bowel walls. Experiments with VX2 tumors in rabbits and hepatomas in woodchucks showed enhancement of both spectral and color Doppler signals, with the demonstration of small tumor vessels as well as displacement of normal vessels (Figure 3) [9]. However, there was no enhancement of parenchymal echogenicity of organs.

The effectiveness of ultrasound contrast agents in enhancing Doppler signals has been quantitatively evaluated by placing Doppler flow cuffs around both normal and abnormal vessels in animal models. In individual cases, the enhancement of the Doppler signals in a celiac vessel was as much as 24 dB and in a tumor vessel within a woodchuck hepatoma as much as 14 dB (Figures 4, 5). The possibility of measuring the uptake and the outflow of Levovist within masses, helping to differentiate benign from malignant tumors, has been reported for the evaluation of breast masses [10, 11]. Similar results have been reported using magnetic resonance imaging with gadolinium to differentiate benign from malignant breast masses [12]. More research is being conducted to confirm these initial reports and to evaluate the potential of ultrasound contrast agents in other areas of the body, including tumors within various abdominal organs.

Another contrast agent evaluated in animals has been QW3600, which is an

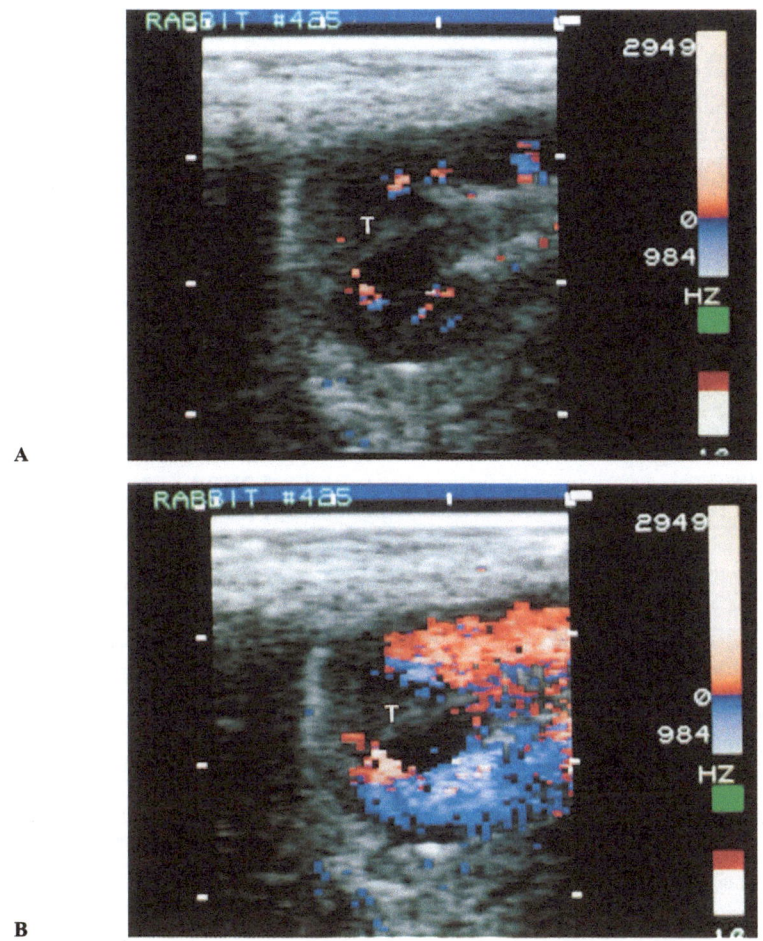

Figure 2. The differences in the color Doppler signal just before (A) and after (B) i.v. injection of air-filled albumin (Albunex). The color Doppler signal enhancement is seen around the periphery of the hypovascular rabbit VX2 tumor (T).

emulsion of a perfluorochemical (EchoGen, Sonus Pharmaceuticals, Bathell, WA). It has the ability to undergo a phase-shift from an emulsion containing a water-immiscible liquid to echogenic gas bubbles when injected into the blood. This temperature-dependent effect has the ability to enhance not only Doppler signals but also gray-scale images. VX2 tumors grown within the kidney could be clearly distinguished from the normal parenchyma due to the difference of the uptake of this agent. This agent recirculated in excess of 5 min [13]. It is anticipated from the animal experiments that tumor vessels in humans will also be enhanced, as has been demonstrated with other agents tested. Delineation of

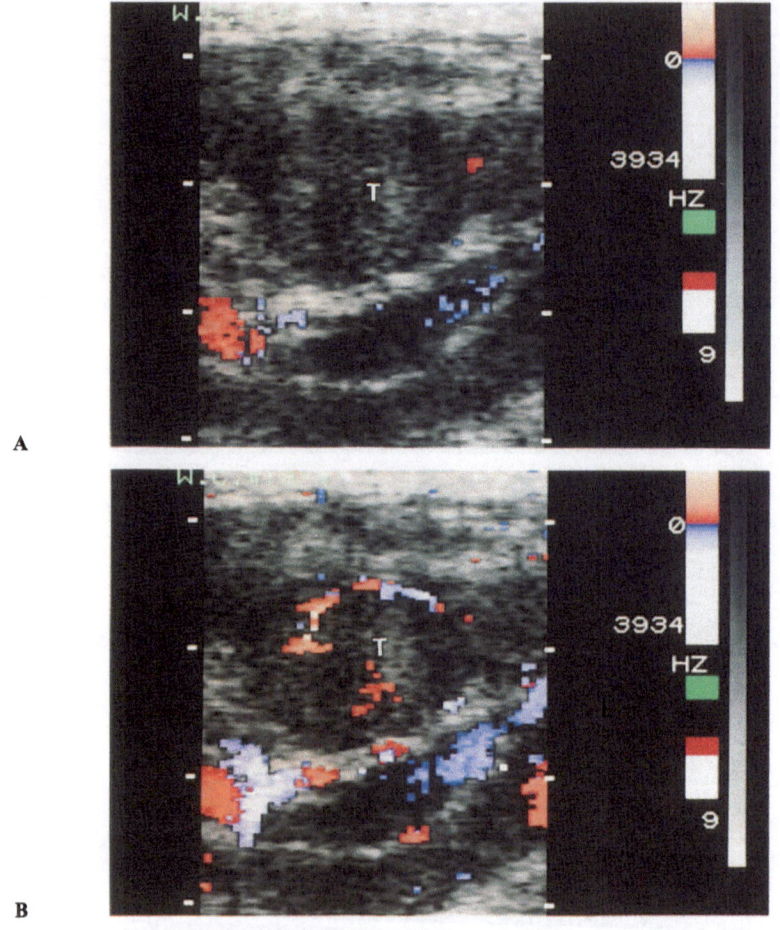

Figure 3. Color Doppler signal enhancement is seen within a woodchuck hepatocellular tumor (T) before (A) and after (B) an i.v. injection of 1.0 ml of SHU 508 (Levovist).

areas of focal ischemia produced within the kidney by tying off a segmental renal artery has been demonstrated using both gray-scale and Doppler ultrasound imaging (Figure 6).

An ultrasound contrast agent (Sonovist, Schering AG, Berlin, Germany) is taken up by the reticuloendothelial system of the liver and spleen in a variety of animal models. Using Doppler ultrasound these bubbles collapse, releasing energy – so-called 'induced acoustic emission'. Contrast-enhanced color imaging is displayed in areas of normal parenchyma. However, tumors which have destroyed the normal liver cells do not take up this agent (Figure 7). Initial work using VX2 tumors grown in the liver of rabbits and woodchuck hepatomas has

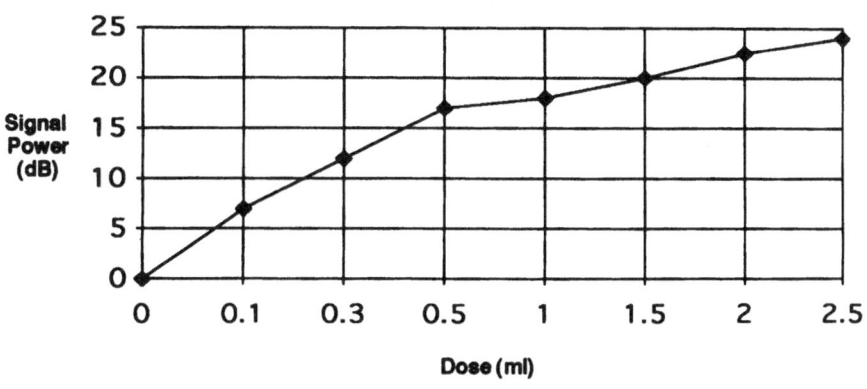

Figure 4. Doppler spectral measurements obtained from the celiac artery with increasing doses (0.1–2.5 ml i.v.) of contrast agent show progressive increase in the signal intensity (up to 24 dB).

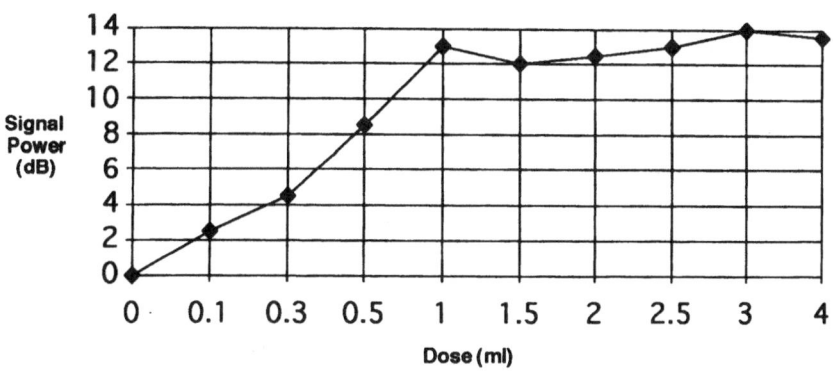

Figure 5. Doppler spectral measurements obtained from the tumor vessel in a woodchuck hepatocellular tumor with increasing doses (0.1–4 ml i.v.) of contrast agent show progressive increase in the signal intensity (up to 14 dB).

demonstrated the ability of this agent to detect tumors as small as 2–3 mm, which were not seen prior to the injection of this agent [14].

Ultrasound contrast agents that enhance gray-scale imaging within abdominal organs should have the same potential as iodinated contrast agents that are used

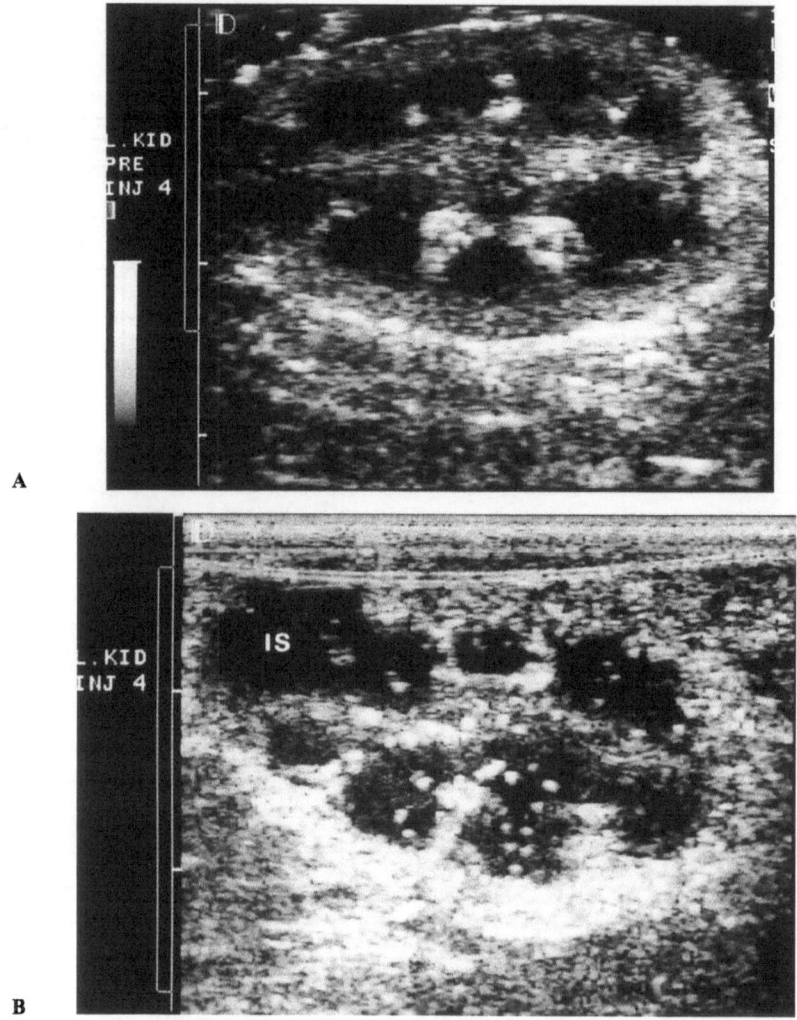

Figure 6. (A) Baseline gray-scale image of a rabbit kidney after ligation of a segmental renal artery. (B) After injection of 0.4 ml/kg i.v. of EchoGen the normal portion of the renal parenchymal enhancement is seen. Note the non-enhanced focal ischemic region (IS).

Figure 7. (Opposite) (A) Baseline color Doppler image of a woodchuck hepatoma (T). (B) Fifteen minutes following injection of 0.7 ml i.v. of Cavisomes, color Doppler imaging demonstrates color signals arising from the liver parenchyma. Because the hepatoma does not pick up the agent, no color is seen within the tumor (T).

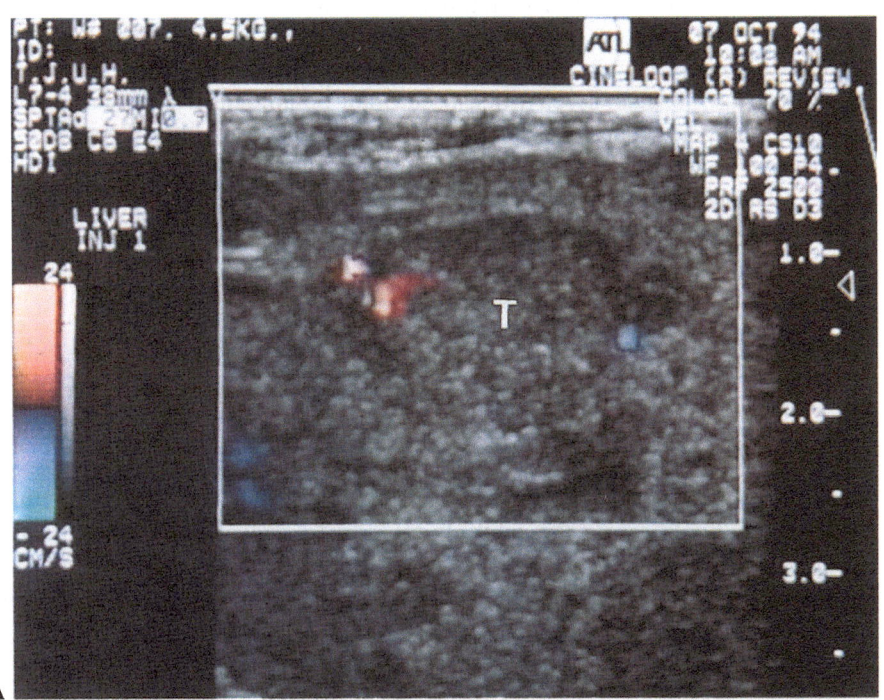

A

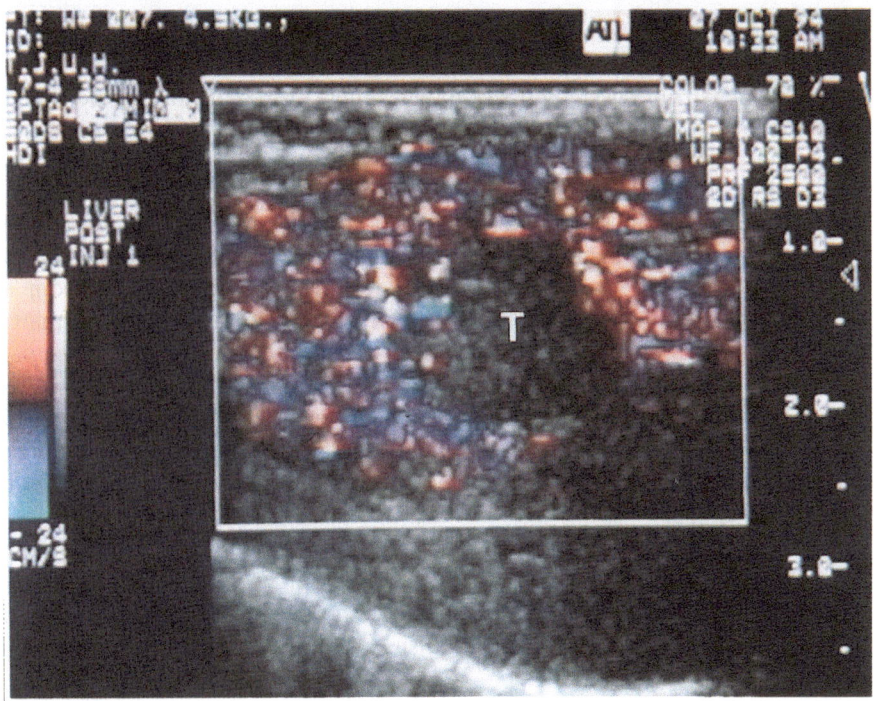

B

in conjunction with computerized tomography to improve delineation of tumors within these organs. As with non-enhanced contrast CT, many tumors have the same echogenicity on ultrasound as the normal parenchyma of the organs in which they are located. Even when tissues are pathologically different, their ultrasonic properties may be quite similar. It is not possible, therefore, to delineate all such tumors by conventional ultrasound techniques. Preliminary work in animals suggests that ultrasound contrast agents should improve the contrast between normal and tumor tissues.

CLINICAL APPLICATIONS OF MICROBUBBLE AGENTS BY TUMOR SITE

All of the studies reported to date have been conducted under clinical trial restrictions since they have been directed toward evaluating the safety and efficacy of the microbubble agents rather than toward demonstrating their diagnostic power. For this reason, only tentative comments can be made at this stage and, no doubt, larger clinical trials designed to evaluate the sensitivity and specificity of the microbubble methodologies will be mounted in the next few years. Intensive research activity can be predicted and many of these provisional conclusions are likely to be overturned as more experience is gained.

Brain

The acoustic barrier posed by the skull has almost prohibited ultrasound studies of parenchymal abnormalities in the adult brain, but the improved signal-to-noise ratio afforded by microbubble agents offers an opportunity to revisit this application. In a series of 28 patients with a variety of primary brain tumors studied with microbubble enhancement, color flow signals were observed in all high grade and in the majority of low grade lesions, while in the control examination only one case in each category gave detectable signals [15]. The Doppler spectra were described as 'atypical', with irregular distribution of the Doppler shifts and of the signal strengths in both arterial and venous traces; these patterns were seen consistently in the high grade lesions and also in some of the low grade tumors. Levovist-enhanced Doppler proved to be superior to intra-arterial digital subtraction angiography in displaying these pathological circulations and the vascular proliferation seen on histology correlated closely with the atypical signals.

The authors suggest that enhanced Doppler may supply new prognostic information on the biology of brain tumors and may form a real time non-invasive monitoring system for perioperative monitoring, for detection of progression or recurrence and to distinguish between radiotherapy damage and tumor recurrence.

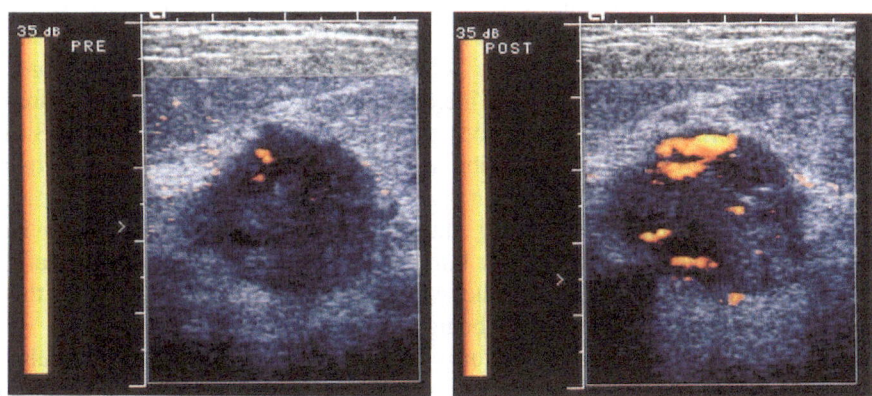

Figure 8. Breast carcinoma: enhancement with EchoGen. This large breast mass shows relatively little color signal on the baseline study (A) but following EchoGen (B) the rich vascularization is clearly shown. Carcinoma was found on biopsy.

Thyroid

No studies of enhancement in the thyroid have been reported but, in view of the current confusion in the ability of B-mode or Doppler ultrasound to discriminate between benign and malignant thyroid disease, this should be an important topic for future studies.

Breast

Thus far, five microbubble studies focused on breast masses have been reported, three using Levovist and one using InfoSon, and other agents are under investigation (Figure 8). Sixteen patients were entered into the InfoSon study and modest color Doppler enhancement was demonstrated [16]. The cancers tended to show more vessels than the benign solid lesions and the enhancement improved diagnostic accuracy from 50% to 69%, though it was not felt to be capable of demonstrating the small vessels associated with neovascularization. These results were confirmed in a study of 10 patients (four cancers) using Levovist: all seven with solid lesions showed enhancement on color Doppler [17]. In the Freiburg Phase III Levovist study of 14 carcinomas and eight masses that were considered suspicious on conventional ultrasound, a marked increase in vessel number was seen in the cancers (133%) compared with benign lesions (24%) [18]. An increase in the vessel number was also seen in the Italian multi-center study of 17 patients, though in this series there was considerable overlap between benign and malignant findings and no trend emerged [19].

The study reported by Kedar et al. included 36 cases, of which 18 were cancers, and was more ambitious because quantification was attempted [20]. Using a blinded double-observer review of the video tapes, the increased number and complexity of the vasculature was scored subjectively. The cancers scored higher on both counts, with a mean increase in vessel number of 3.2 compared with 0.9 in fibroadenomas, and a post-enhancement tortuosity scored as 2.3 in cancers compared with 0.4 in the fibroadenomas. An intriguing finding in this series was the description of apparent interconnections between adjacent vessels that the authors tentatively called shunts: these were seen after Levovist enhancement in every breast cancer but in none of the benign lesions. Clearly this preliminary result requires further evaluation but, if it is substantiated, this feature could provide an important benign/malignant differentiator.

This group also attempted to perform a dynamic analysis by timing the delay from the bolus injection to the moment of peak enhancement and to the point of return of the Doppler signal strength to baseline. An earlier peak enhancement was typical of the carcinomas (45 s compared with 1 min for fibroadenomas), though this difference was not statistically significant. The enhancement persisted for longer in the cancers (5 min 41 s) than in the fibroadenomas (3 min 3 s) and this difference was significant ($p < 0.0001$). These changes are exactly those predicted from a knowledge of tumor hemodynamics: the earlier peaking of enhancement could be due to shunts and vessels with a low resistance to flow while the prolongation of enhancement could be explained by tortuous vessels of wide caliber. This group has developed a computer-based quantitative system for estimating both the morphological and dynamic features and these are being applied retrospectively to the video-taped information from this study.

Liver

Free bubble approaches
One of the earliest uses of any bubble contrast agent was the direct intra-arterial injection of carbon dioxide to delineate metastatic liver disease in 184 Japanese patients [21]. The process depends on the different blood supply of metastases from the normal liver: the latter receives 75% of its blood from the portal venous system, whereas metastases are predominantly supplied by the hepatic artery. The microbubbles were prepared by vigorously mixing equal volumes of CO_2 and heparinized normal saline to which a small amount of the patient's own blood was added. The blood acts as a stabilizer, increasing the surface tension of the preparation and thereby producing small and relatively stable micro-bubbles; CO_2 is a safe gas because it is very soluble and is rapidly eliminated via the lungs.

Following injection of 5–20 ml of this preparation into a hepatic artery, the whole liver showed an increase in echogenicity; this rapidly faded in the normal tissue but persisted for many minutes in focal lesions, thus highlighting them. Hepatocellular carcinomas displayed a homogeneous or mosaic vascular pattern, with a peripheral arterial blood supply, while focal nodular hyperplastic nodules were characterized by an early central arterial supply, followed by uniform or lobulated enhancement. Adenomatous hyperplasia was hypovascular while hemangiomas and metastases displayed 'spotty pooling' and peripheral hypervascular patterns, respectively. Using the same technique, this group confirmed its ability to detect subcentimeter hepatocellular carcinomas which had been missed by conventional techniques [22].

In a more recent study, the injection of $10-20\,ml$ CO_2 directly into a hepatic artery during intraoperative scanning of 45 patients with hepatocellular carcinoma was compared with digital subtraction angiography, conventional ultrasound, CT and CT arterio-portography [23]. This technique, described as 'echocarbography', allowed the detection of more lesions than any of the other conventional modalities and no side-effect was reported.

The obvious limitation of the approach is its invasiveness and, though it can be used as an adjunct to conventional arteriography, it has most often been applied as an intraoperative technique, where it undoubtedly adds to the sensitivity of unenhanced scanning.

Microbubbles

Several studies using intravenous Levovist have been reported (Figure 9). Marked enhancement of color Doppler signals was seen in 25 of 28 patients with histologically proven liver tumors (two hepatocellular carcinomas, two carcinoid and 24 colorectal cancer metastases) [24]. Interestingly, there was no enhancement in three patients with colorectal cancer undergoing chemotherapy. Predominant enhancement of the tumor rim was characteristic of colorectal metastases while in carcinoid metastases homogeneous color Doppler signals were observed throughout the lesion (Figure 10). A 'basket-like' pattern of hypervascularity was observed in hepatocellular carcinomas and this was confirmed in other studies of 38, 18 and 10 patients with hepatocellular carcinomas [25–27]. These findings are similar to those reported by Tanaka et al. [28] using unenhanced Doppler imaging.

There is little information on the enhancement of benign lesions in the liver because of their relative rarity. In anecdotal reports, the characteristic hypervascular nature of both focal nodular hyperplasia and adenomas was clearly demonstrated following Levovist but there was little or no enhancement of Doppler signals in hemangiomas [27, 29]. In an intriguing but as yet unconfirmed study using Levovist [30] strong, transient signals on the spectral Doppler trace were noted from within hemangiomas but not from malignant tumors: the signals are reminiscent of so-called 'bubble noise' which is occasionally encountered in larger vessels and has been attributed to bubble destruction. It is plausible that the effect described in hemangiomas arises because of their slow flow such that the Levovist stays in the ultrasound beam for much longer than in malignancies or in the normal liver.

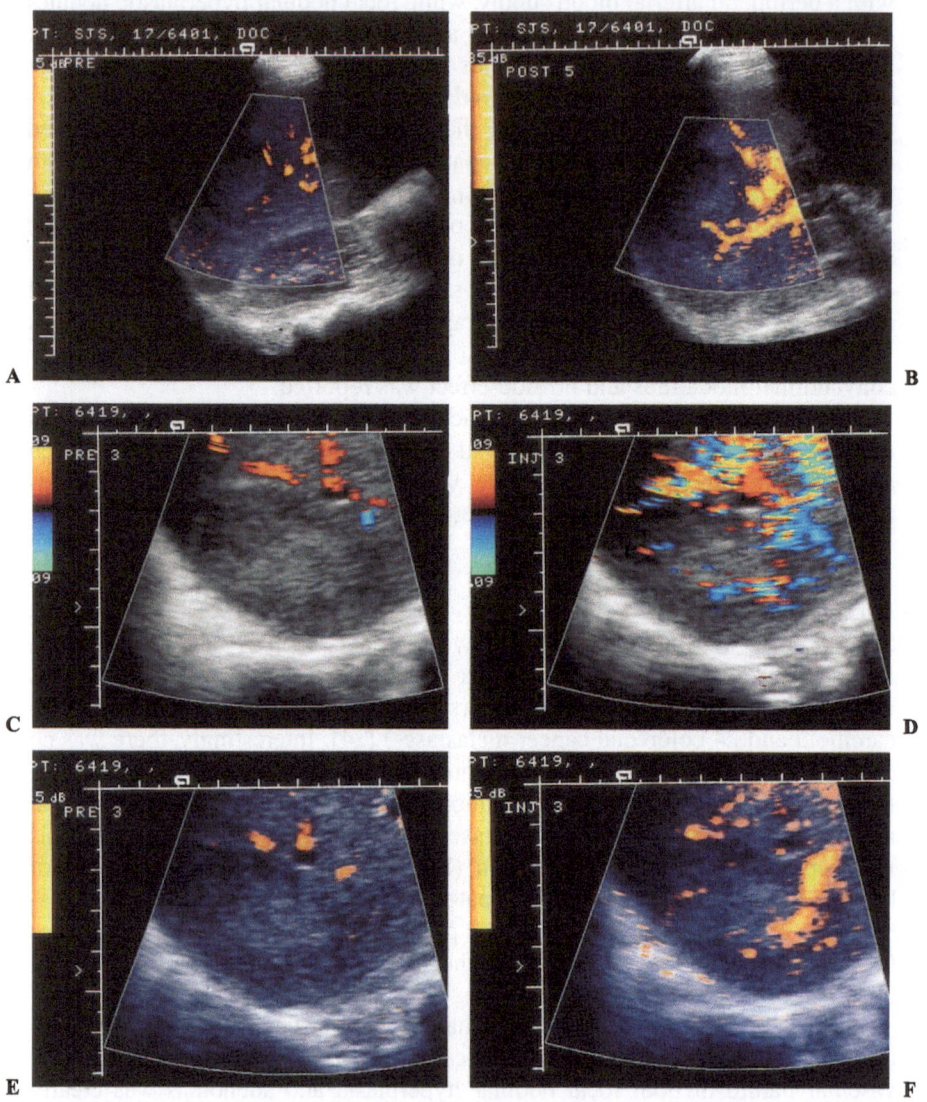

Figure 9. Enhancement in liver tumors. While many metastases, especially from colorectal cancers (A before, B after Levovist) show little enhancement, some are hypervascular and show the same 'basket pattern' as seen in hepatocellular carcinomas. (C, D) Velocity color Doppler. (E, F) Power color Doppler.

Non-bubble approaches

A non-microbubble approach to improving detection of liver metastases is the use of perfluroctylbromide (PFOB, Perfluobron) [31] which provides ultrasound contrast by virtue of its slow conduction speed. The agent is given as an emulsion

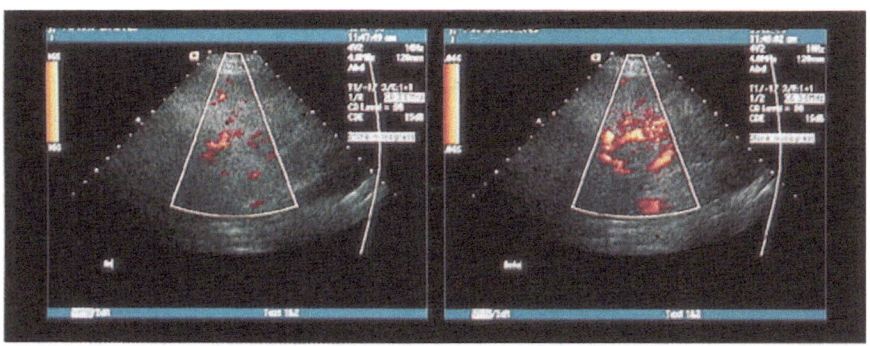

Figure 10. The neovascular pattern of these extensive carcinoid metastases (A) is much better seen after Levovist enhancement (B) in this scan taken with the Sequoia-512 system.

at a dose of 5 ml/kg and, following a phase of weak intravascular enhancement, is taken up by the reticuloendothelial system. In the liver a variety of appearances has been described, some metastases appearing as negative enhancement where the surrounding normal liver becomes increasingly echogenic because of Perfluobron within the Kupffer cells, other metastases showing rim enhancement, and yet others showing uniform enhancement.

The safety and dose–response of Perfluobron as an intravenous contrast agent for sonography were evaluated in a study of 14 patients with liver metastases [32]. The effects were dependent on the dose and on the time after infusion. Increased echogenicity of metastases compared with surrounding liver was demonstrated in 10 of 14 cases and a dose up to 2.5 g/kg was optimal for tumor detection. The need for such large doses and the 24–48 h delay between administration and development of contrast enhancement as well as its relative toxicity have meant that this approach has fallen out of favour. However, the selective uptake of agents by the reticuloendothelial system could be exploited further in the detection and characterization of liver tumors [33].

A somewhat similar approach underlies the use of iodipamide ethyl ester (IDE), a biodegradable substance which can be formulated as dense spherical particles with diameters in the range 0.8–1.2 μm. It increases ultrasound backscatter because of the impedance mismatch between the particles and soft tissue. Injected particles are taken up by the Kupffer cells in the liver within 10–20 min, and they increase the echogenicity of the hepatic parenchyma. In contrast, hepatic tumors (and other lesions without Kupffer cells) do not retain the particles and so are not enhanced [34]. Evaluation of IDE is still in the early stage, but it seems to be promising.

Sonovist
Another approach that exploits the affinity of the reticuloendothelial system for phagocytosing particulate material has recently been described. It uses one of

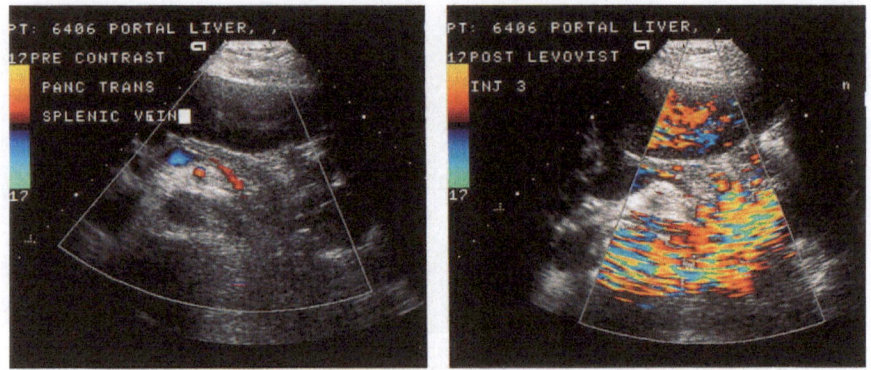

Figure 11. Islet cell tumor of the pancreas. This mass in the tail of the pancreas was unremarkable in the baseline Doppler study (A) but showed marked enhancement after Levovist injection (B). A metastasis in the left lobe of the liver also enhanced.

the third-generation microbubble agents developed by Schering (SHU 563A, known as Sonovist) which, by virtue of its rigid polymeric membrane, survives for long enough to be treated as foreign particles and so is phagocytosed by the reticuloendothelial system. Following a phase of intense vascular enhancement it is taken up by the bone marrow, spleen and liver, where it persists for long periods until it is degraded. The microbubbles of Sonovist have the intriguing property of emitting Doppler-shifted signals when insonated at relatively high power (i.e. at the upper range of conventional diagnostic Doppler systems). The oscillations thus produced result in a mosaic of color signals that highlights the normal liver, leaving lesions that are devoid of Kupffer cells as color voids. The effect is spectacular and, in animal studies, lesions as small as 2–3 mm are clearly demonstrated [35]. This approach is analogous to that of a sulfur colloid isotope scan, but with much finer spatial resolution and intrinsic tomography. It is likely, however, that it will not be completely specific: benign lesions that do not contain Kupffer cells will almost certainly show as areas of negative contrast: estrogen adenomas and inflammatory masses, as well as both primary and secondary malignancies and hemangiomas can be expected to show this effect. On the other hand, the fact that focal nodular hyperplasia and the regenerating liver nodules of cirrhosis do contain Kupffer cells suggests that they will not be highlighted by this technique, which thus is of great clinical interest. Sonovist has entered Phase I clinical trials and patient studies are eagerly awaited.

Pancreas

No pancreatic studies have been reported thus far but the potential for improving evaluation of the extent of pancreatic carcinomas as a preoperative staging procedure is of interest and prospective clinical trials are anticipated (Figure 11).

Levovist has been used to evaluate a variety of tumors in the abdomen. One of the more dramatic effects was seen in a retroperitoneal tumor in which it was initially difficult to delineate tumor vessels. After intravenous injection of this agent such vessels were clearly demonstrated by color Doppler ultrasound. In addition, pulsed Doppler was able to identify a high diastolic flow due to low peripheral resistance as a result of tumor arteriovenous communications. Within a poorly vascularized pancreatic mass no flow was demonstrated using Doppler imaging prior to injection of an ultrasound contrast agent. However, after a peripheral intravenous injection of Levovist vascularity was clearly demonstrated within the mass, and a pulsed Doppler pattern could then be recorded from these tumor vessels (Figure 12).

Kidney

Renal studies with microbubble enhancement have mainly been directed toward the clinically important and technically difficult problem of screening for renovascular hypertension and thus have focused on the main renal artery rather than on tumor vessels [36]. In one series of 34 renal masses, i.v. Levovist enhanced the color Doppler signals from the vessels within 22 of 28 renal cell carcinomas and from all four angiomyolipomas [37]. No patterns that offered diagnostic clues were discerned but a urothelial tumor and a leiomyosarcoma were considered 'hypovascular' after Levovist. Whether the use of enhancing agents will increase the current low sensitivity of ultrasound for detecting small renal tumors and for differentiating them from benign masses is an important question that remains to be addressed.

Gynecology

The effects of Levovist were studied in 13 patients with suspicious ovarian findings on transvaginal scanning (T. Bowne, personal communication). All three cancers in this series showed marked enhancement; this was also seen in five of the benign lesions, most markedly in two of three corpora lutea. An active follicle and a tubovarian abscess also showed enhancement as did one benign ovarian lesion but the enhancement lasted for 35 s, 40 s and 15 s respectively. Interestingly, this was much shorter than in the cancers (90–250 s, mean 187 s) and the two vascular corpora lutea (240 and 210 s), suggesting that transit time analysis might prove useful as a differential diagnostic tool. If the morphological and functional information potentially available from microbubble enhancement can reduce the false positive rate of transvaginal ultrasound and thus reduce the rate of unnecessary laparoscopy, ultrasonic screening for ovarian cancer could become a clinically useful method.

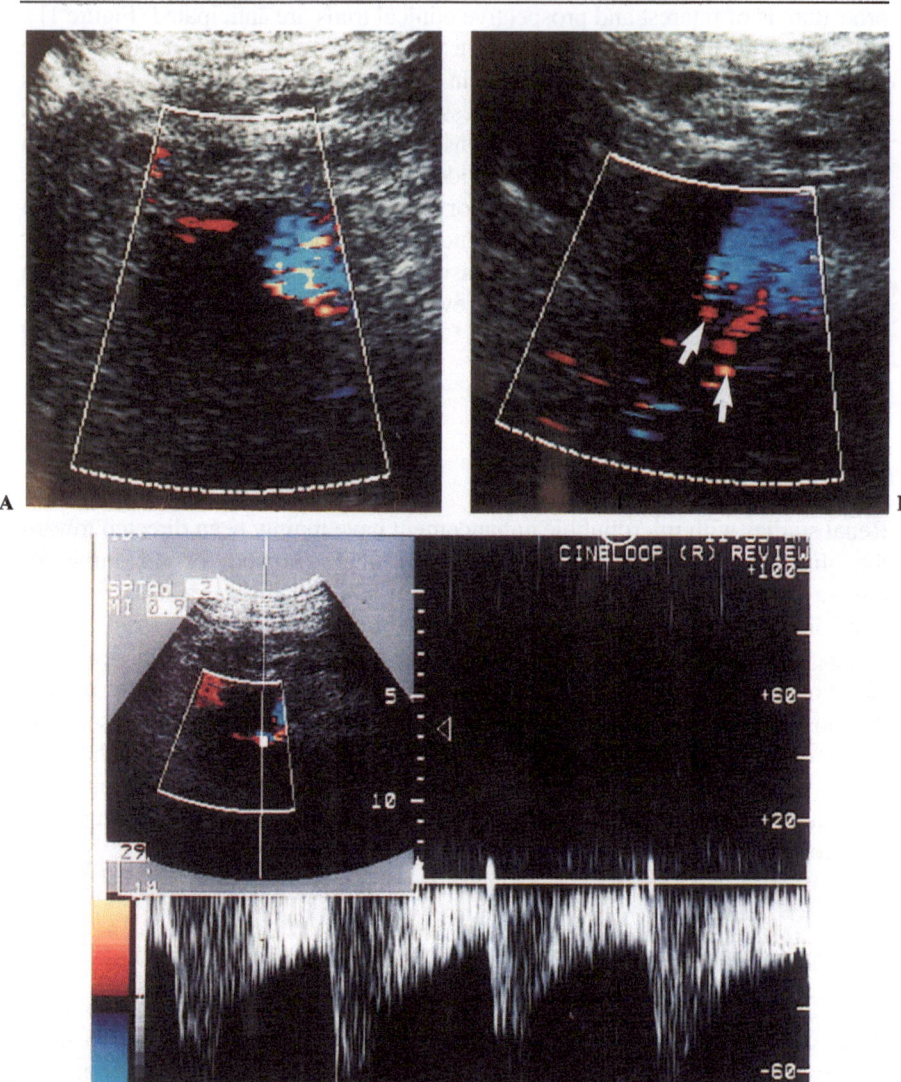

Figure 12. In a patient with pancreatic carcinoma, a 300 mg/ml i.v. injection demonstrated both color and spectral Doppler tumor vessel enhancement. (A) Color Doppler before injection (no spectral Doppler could be obtained from within the tumor); (B) color Doppler post-injection showing tumor vessels (arrows); (C) post-injection spectral Doppler signals could now be recorded from within the tumor.

In this same series eight uterine abnormalities were studied and with one exception (a lesion in a patient on hormone replacement therapy) all showed enhancement. In four fibroids the enhancement lasted 45–280 s (mean 161 s) compared with 220 and 300 s in two endometrial cancers. There was one false-positive result, in which the enhancement lasted only 30 s, again hinting at the potential for dynamic analysis.

Prostate

With the realization that prostatic biopsy targeted to regions with strong color Doppler signals has a higher yield of carcinoma than when the biopsy is directed by the B-mode appearances alone [38], interest has focused on the potential of microbubble enhancement to highlight abnormally vascularized areas. Thus far, only one study has been reported and this consisted of six patients with newly diagnosed prostate cancer, in whom the typical vascular patterns demonstrated elsewhere were also revealed [39]. These consisted of tangles of prominent vessels, some but not all of which also showed intervascular communications. A notable finding was the greater extent of the vascular abnormalities after Levovist enhancement than was suggested by the B-mode scan (Figure 13).

This series also included an additional seven patients who were receiving anti-androgenic hormones for tumor control. These showed strikingly less vascularity than the newly diagnosed cancers, with the exception of one patient whose prostate specific antigen was found to be rising rapidly and who, at the next clinical examination, had obviously escaped from hormone control. It thus seems likely that features of tumor vascularity in the prostate can be used to monitor response to medical treatment, though further studies will be needed to evaluate whether this is also true of radiotherapy, which also affects the vascularity of tumors, and whether this can be extended to other anatomical sites.

Testes

In a small study of testicular tumors, marked enhancement was noted with Levovist in three malignancies while a fourth showed only moderate enhancement [40]. The tumor types were not reported. The main clinical challenge is to differentiate focal inflammatory from malignant processes and this remains to be evaluated.

Malignant lymphadenopathy

Malignant lymph nodes may be more vascular than those affected by inflammatory lymphadenopathy [41] but there has been no report of the potential of microbubble enhancement in this area. A preliminary report of stimulated

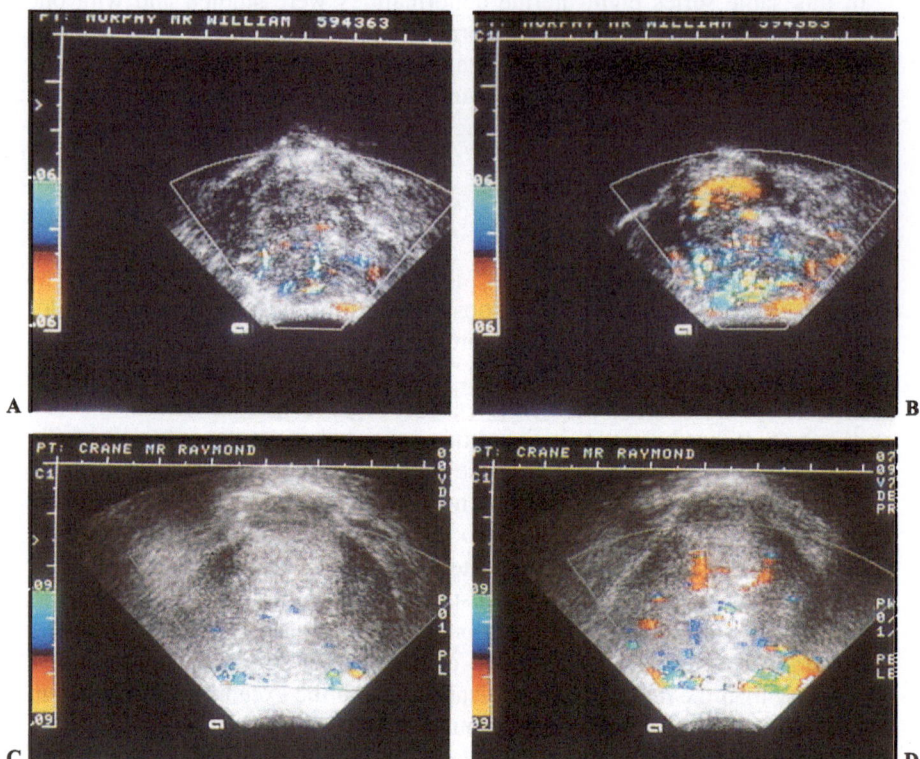

Figure 13. The vascularity of this extensive prostatic cancer (A) is greatly enhanced after Levovist injection (B) so that the complex architecture of the neovascularization is more clearly seen. The patient shown in (C) was under treatment with cyproterone, an anti-androgen: the tumor vascularity was suppressed and there was only minimal enhancement after Levovist (D).

emission of Doppler signals following injection of Sonovist into the interdigital web spaces in experimental animals suggests that this technique might be useful in selected areas of the body, particularly the inguinal regions and the axilla [35].

QUANTIFICATION OF MICROBUBBLE ENHANCEMENT IN TUMORS

Quantification of the effects of microbubbles has the potential to improve and augment ultrasound imaging of tumors in several ways. Objectivity clearly has advantages per se, both for research applications and in the eventual development of these techniques as clinical tools. Indices for measuring tumor neovascularity can be expected to be of value in tumor grading and in measuring the effects of therapy, as well as in primary diagnosis. The development of quantitative morphological indices that describe the branching pattern or the irregularity of vessels can be expected to improve the clinical value of the angiograms produced by Doppler ultrasound, with or without microbubble enhancement. Physiological quantities such as the fractional vascular volume and tissue perfusion can be compared with typical values or with those obtained using previously validated techniques such as microsphere embolization. A natural development is the mapping of such indices on a pixel-by-pixel basis to produce color functional images; this is particularly well suited to the correlation of anatomy and function in heterogeneous tissues such as tumors.

Such techniques are in their infancy in echo-enhanced ultrasound but have shown great promise in the other cross-sectional modalities, especially magnetic resonance imaging (MRI) using gadolinium chelates as contrast agents. For example, pharmacokinetic analyses with color mapping has recently shown potential in discriminating benign from malignant disease in suspected recurrent cervical carcinoma [42]. Similar results have been shown with carcinoma of the rectum [43], brain [44] and breast [45]. X-ray computed tomography (CT) has also shown promise: both Miles et al. [46] and Blomley et al. [47] have shown high hepatic arterial perfusion around liver tumors. Several algorithms and techniques have been developed for these MRI and CT applications but great caution should be used in extrapolating these to microbubble-enhanced ultrasound since, with very few exceptions, intravascular MRI and CT contrast agents are extracellular fluid (ECF) markers, which diffuse across the vascular endothelium within seconds, while microbubbles are blood-pool agents and never reach the ECF. Thus different mathematical approaches will need to be developed.

A number of potential strategies for using these agents to obtain quantitative indices can be suggested, but it should be emphasized that there is as yet relatively little clinical experience in this area and so many of these remarks are speculative. First, simply by improving the signal-to-noise ratio using microbubbles, measurements that previously were suboptimal can be expected to become reliable. For example, Leen et al. have suggested that the ratio of hepatic arterial to total liver blood flow (the Doppler perfusion index, DPI) is a sensitive test of liver metastases, including otherwise occult tumors [48]. Accurate measurements of these values would be helped by the use of echo-enhancing agents in many patients. Second, data (as discussed in this chapter) already suggest that echo-enhancing agents improve detection of the abnormal morphology of tumor neovascularity: it should be possible to calculate indices to quantify this. For example, the 'fractal dimension' of a branching structure is, in principle, easy to calculate, in both two and three dimensions, provided sufficiently detailed information is available [49, 50]. A large proportion of the tumor vascular tree would have to be displayed to make this type of computation feasible, however, and microbubble enhancement would be of great value, if not essential, even with state of the art ultrasound equipment. Power mode color Doppler, because of its sensitivity and relative independence of Doppler angle, would appear particularly well suited to this application.

Finally, and most ambitiously, completely new information should be revealed by the analysis of the time course of changes produced by echo-enhancers, if these changes can be quantified in a meaningful way. Two stages are necessary: a prerequisite is that the relationship between the microbubble concentrations and the ultrasound signals needs to be clearly defined so that an estimate of microbubble concentration (or, at least, relative concentration) can be made from the signal strength. Meaningful and practicable algorithms to enable this information to be used also need to be developed. These will be considered in turn.

Quantification considerations

Since on all clinical scanners the processing and display of ultrasound data is non-linear (being optimized for an esthetic display) the best way of tackling the first stage, quantifying relative microbubble numbers, is to use the unprocessed radiofrequency data. Such an approach, however, is complicated by the formidable problems of analyzing huge amounts of data and by the understandable reluctance of manufacturers to provide open access to the inner workings of their systems. A simpler and more pragmatic approach is to make the best use of the processed data provided by the scanners' audio or video output: several groups are pursuing this. Schwarz and colleagues have shown that the intensity (or power) of the spectral Doppler audio signal gives an accurate measurement of Levovist concentration in a phantom [51]. Rubin et al. recently showed that measurements derived from the color power Doppler display can estimate the regional concentration of scatterers [52]. Although performed with relatively large scattering particles (median size $44\,\mu$m), this suggests that power Doppler could be used to quantify microbubble concentration: we have confirmed this observation with Levovist in a phantom (unpublished data). Rubin and colleagues also suggest a method of correcting for attenuation with depth: if a purely vascular region can be identified, other areas at the same depth can be normalized with respect to this. In this way, the fractional vascular volume can be calculated.

Mathematical approaches

Given an estimate of microbubble concentration and following classical indicator dilution theory, three main types of index could be extracted, all of which are inter-related [53]. The first are purely temporal indices of transit or enhancement times, the second fractional vascular volume, and the third perfusion (i.e. flow per unit volume). Specifically, for an intravascular agent and when mean transit time is calculated as described in the next section, perfusion is precisely the ratio of fractional vascular volume to mean transit time (the 'central volume theorem') [54].

Temporal indices

Mean transit time, MTT, is a term used (and perhaps misused) in many different ways in the imaging literature. It approximates to the average persistence of a tracer within the circulation of a region as it travels from an inflow to an outflow vessel. Thus, MTT might be short if a tumor contained many arteriovenous shunts or long if it contained many extensive or tortuous channels. MTT is probably best defined and used in the context of the theory of time-invariant linear systems: that is an index with dimensions of time, calculated as the response of an outflow vessel after an instantaneous unit inflow of contrast into the supply vessel (the Dirac 'delta' or 'unit impulse' function). It may be calculated as the difference between the weighted first moments of the inflow and outflow vessels, where the weighted first moment of a function of time, $f(t)$, is the ratio:

$$\int_0^\infty tf(t)\mathrm{d}t / \int_0^\infty f(t)\mathrm{d}t \tag{1}$$

Although this may seem a somewhat abstract concept, the reader may find it helpful to think of it as the 'centre of gravity' (in the temporal sense) of a measurement. For example, suppose an artery (the input vessel) has a simple time–signal curve where a tracer appears at $t=0$, has a constant concentration of 1 unit (for simplicity we will simply consider the changes of concentration and not ultrasound signal), lasts 1 s and then disappears (Figure 14). The arterial time signal curve $a(t)$ will be a square wave (Figure 14). The area under the curve, formally expressed as $\int a(t)\mathrm{d}t$, is then 1 unit-second. If we now plot the value $ta(t)$ we will obtain a saw-tooth curve (Figure 14b). The area under the curve $\int_0^\infty ta(t)\mathrm{d}t$ is 0.5 unit-sec^2. The 'weighted first moment' of $a(t)$ (i.e. the mean transit time of the arterial bolus) is thus no more than the ratio of these two (see eqn. 1), or 0.5 s. If we examine the shape of $a(t)$, it is intuitively obvious that the average duration or 'centre of gravity' of $a(t)$ should be 0.5 s, but this formalism enables us to calculate this value for any time–signal curve. The example can be extended further: if the venous time–signal curve, $v(t)$, is equally simple, the venous mean transit time is 3 s and the mean transit time of the system is the difference between these two (2.5 s; Figure 14).

Mean transit time could thus be formally calculated if the input and output vessels could be studied simultaneously. Often, however, anatomical constraints mean that information on the parenchymal, input and output regions cannot be obtained in a single study. Mean transit times of tissue regions are sometimes estimated in other modalities by analyzing the 'residue function' or time–signal curves of tissue, for example by calculating the weighted first moment of the tissue time–signal curve. Unfortunately, this is prone to errors of up to 100%, as calculations of the true tissue MTT based on residue function analysis depend on assumptions about the distribution of tissue transit times [55]. Nevertheless,

608   *Barry B. Goldberg et al.*

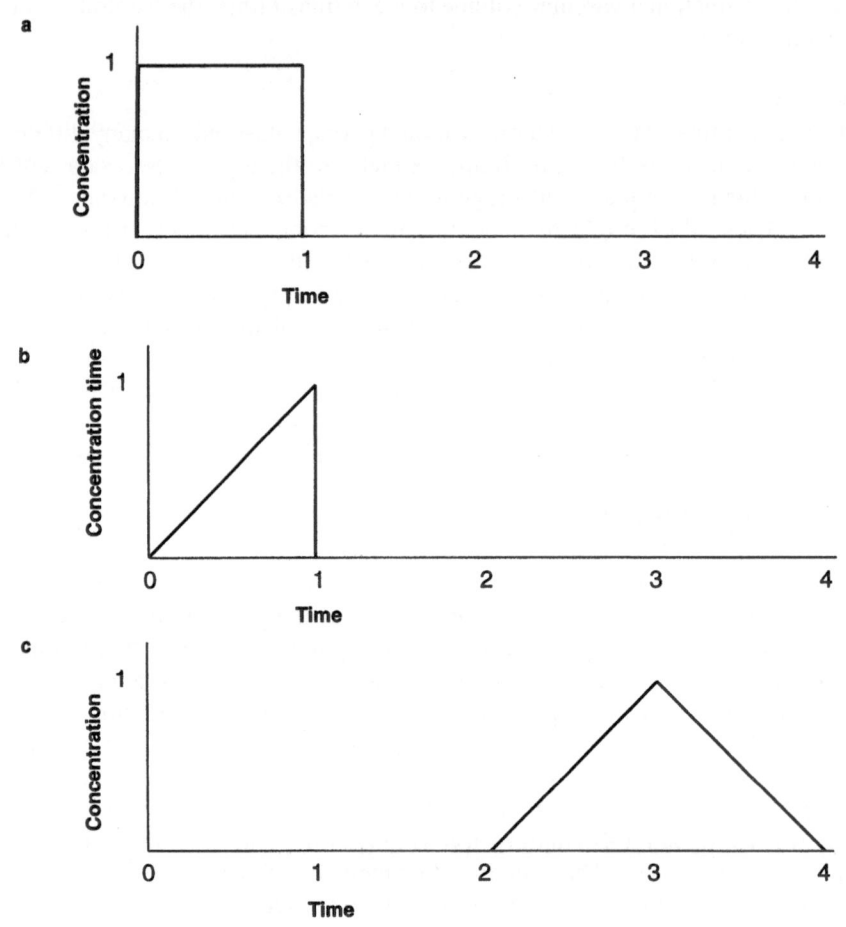

Figure 14. Calculating tissue transit: a simplified example. The arterial time–concentration curve for an idealized bolus of microbubbles that instantly reaches maximum concentration and washes out equally rapidly is a simple square wave (a). A plot of the product of time and concentration is then a saw tooth (b) which shows that the mean transit time for the bolus is 0.5 s. In more realistic flow profiles, calculation of the weighted first moment is required to extract what is intuitively obvious from this simplified case. If the venous time–concentration curve is equally simple (c) the mean transit time is 3 s and the mean transit time through the whole organ is the difference between the arterial and venous transit times, i.e. 2.5 s.

estimates of tissue transit times could still be made, for example, by fitting a gamma-variate curve to a tissue region of interest [56]. For a generalized gamma fit, where: $f(t) = K(t-AT)^{\alpha}e^{-(t-AT)/\beta}$, the weighted first moment of $f(t)$ is easily found as: $AT + \beta(a+1)$, where K, α and β are constants and AT is the arrival time of the contrast or the starting point of the curve.

In imaging, the term 'mean transit time' has also come to be used as an indicator of mean contrast duration or mean enhancement time. This is unfortu-nate as the duration of enhancement is very different, being substantially longer than the MTT and reflecting multiple recirculations of the contrast agent. It depends on several processes including the overall pharmacokinetics of the agent within the patient and 'true' tissue transit times when these are extremely long – for example with blind-ending vessels. In addition, color Doppler systems only display signals that are above their in-built thresholds: when power Doppler is used, the vascular volume of the tissue affects the display as a consequence of the thresholding inherent within ultrasound systems (signals from low concen-trations of microbubbles will be below the noise floor) and when velocity color Doppler is used, an important factor is the wall filter that removes low frequency shifts. Intriguingly, however, there is evidence that duration of enhancement provides useful discrimination between malignant and benign masses [20]: the reasons for this fascinating observation are not clear and need further clarifi-cation and research. One attractive aspect of measuring indices of enhancement time in this way is the ease of carrying them out: they can be performed indirectly by subjective assessment.

Another index which is relatively easy to calculate is the time to peak enhancement. Using Levovist, Kedar and Cosgrove showed a shorter time to peak in malignant breast masses than in benign disease [20] (albeit with small numbers of patients and without statistical significance) and in untreated car-cinoma of the prostate compared with treated cases [39]. Similar findings have also been reported using gadolinium-enhanced MRI mammography [45]. The time to peak is easily calculated from a gamma-variate fit; using the notation of equation 1 it is $AT + \alpha\beta$.

All these indices have the dimension of time only and so are, to a certain extent, independent of the means used to measure microbubble concentration. Thus, even if an indirect means of estimating microbubble concentration is used, such as quantification of color Doppler, they retain some validity. This is not true of other indices of fractional vascular volume or perfusion, which will be considered next.

Fractional vascular volume
Fractional vascular volume and perfusion are relatively easy concepts to understand: they are the proportion of tissue occupied by blood and the blood flow per unit volume (or mass) respectively. (Given that the specific gravity of most tissues is close to 1, there is little practical difference between normalizing blood flow with respect to mass or to volume.) The practical importance of these indices is that they are measures of the physiological 'vascularity' of a region. In general, most tumors are hypovascular but this partly reflects their hetero-geneity with small (presumably active) regions of high vascularity and larger ischemic regions. The vascularity of a tumor is often different from that of the host tissue. For example, the postmenopausal breast tends to be relatively poorly vascularized and this may explain why some carcinomas stand out by their

vascularity [1]. Since some anticancer therapies specifically target tumor vascularity, such indices may be of value in monitoring response to treatment as well as in the primary diagnosis, staging, and follow-up tumors.

In principle, these indices could be measured without the use of echo-enhancing agents. For example, power Doppler imaging, where the signal reflects the number of moving reflectors, should give a map of regional red cell concentration and thus of fractional vascular volume. A simple approach to this is to normalize a power Doppler image from a log-compressed to a linear scale and calculate the ratio of parenchymal to purely vascular Doppler intensity [52]. Difficulties include the problem of obtaining a sufficiently good signal-to-noise ratio and the effects of erythrocyte clumping.

Another problem often glossed over in the physiological imaging literature, is that of hematocrit differences: the hematocrit of capillary blood is 20–50% that of the systemic circulation [57] probably mainly due to the Fahreus effect, where red cells in small vessels travel faster than mean flow rate [58]. Estimates of fractional vascular volume based on erythrocyte differences will need to be corrected for this but, if an echo-enhancing agent is used with power Doppler, the estimates should be much better as the signal-to-noise ratio is improved. However, it is not yet clear whether the Fahreus effect also affects microbubble agents.

The fractional vascular volume can be calculated in other ways which are likely to be more cumbersome but less prone to such errors. For example, suppose the time–concentration curves for a parenchymal and a vascular region of interest are corrected for recirculation – for example with a gamma-variate fit – and integrated: the ratio of these two areas is the fractional vascular volume [54].

Perfusion
For convenience, perfusion will be defined as flow per unit volume, rather than mass. There are two different techniques for its calculation of which the first can be derived from the preceding discussion. For an intravascular tracer, such as a microbubble, perfusion can be found simply by dividing the fractional vascular volume by the mean transit time, if this can be calculated as above using the central volume theorem (perfusion = fractional vascular volume/mean transit time) [53]. There is, however, an alternative strategy: perfusion may be calculated directly if the time–concentration curves for parenchymal and vascular regions of interest can be defined. If the parenchymal (e.g. tumor) microbubble concentration at time t is $p(t)$ and that in a supply vessel is $a(t)$ (after correction for recirculation) then, provided the input bolus is fast compared to parenchymal transit and venous washout is reasonably slow, perfusion may be calculated as the peak value of $p(t)$ divided by the area under the curve of $a(t)$. Alternatively, perfusion may be calculated as the peak gradient of $p(t)$ divided by the peak value of $a(t)$, that is:

$$\frac{\frac{\mathrm{d}}{\mathrm{d}t}p(t)_{max}}{a(t)_{max}}$$

The second 'gradient' relationship is slightly less sensitive to errors from venous wash-out but, as it involves a differentiation process, is more noise sensitive. These two relationships do not seem to have been applied to contrast-enhanced ultrasound but have been usefully used in research studies using ultrafast CT of the myocardium and in some liver tumor applications [47, 59, 60] and the interested reader is referred to these sources for a fuller discussion including a theoretical derivation. Specifically, the gradient algorithm has given a good correlation with the results of PET and thermal clearance studies in tumor imaging [61].

Unlike the situation for blood-pool microbubble agents in large vessel applications, where their clinical value does not seem to be in doubt, in the more demanding problem of tumor evaluation, further clinical experience is obviously required. Nevertheless, the early indications are promising and the emerging ability to quantify the effects of the microbubbles and to extract unique functional information from the transit time curves makes this an intriguing area: fractional vascular volume, mean transit time and perfusion could, in principle, all be measured and displayed as color-coded functional images.

REFERENCES

 1. Vaupel PW. Blood Flow, Oxygenation, Tissue pH Distribution and Bioenergetic State of Tumors. Berlin: Ernst Schering Research Foundation No 23. 1994.
 2. Mattrey RF. Perfluoroctylbromide: a new contrast agent for CT, sonography, and MR imaging. Am J Radiol 1989;152:247–52.
 3. Mattrey RF, Scheible FW, Gosink BB, Leopold GR, Long DM, Higgins CB. Perfluoroctylbromide: a liver/spleen-specific and tumor-imaging ultrasound contrast material. Radiology 1982;145:759–62.
 4. Mattrey RF, Strich G, Shelton RE et al. Perfluorochemicals in US contrast agents for tumor imaging and hepatosplenography: preliminary clinical results. Radiology 1987;163:339–43.
 5. Parker KJ, Baggs RB, Lerner RM, Tuthill TA, Violante MR. Ultrasound contrast for hepatic tumors using IDE particals. Invest Radiol 1990;25:1135–9.
 6. Goldberg B, Hilpert P, Burns P et al. Hepatic tumors: signal enhancement of Doppler US after intravenous injection of a contrast agent. Radiology 1990;177:713–7.
 7. Schlief R, Staks T, Mahler M, Rufer M, Fritzsch T, Seifert W. Successful opacification of the left heart chambers on echocardiographic examination after intravenous injection of a new saccharide based contrast agent. Echocardiography 1990;7:61–4.
 8. Smith MD, Elion JL, McClure RR et al. Left heart opacification with peripheral venous injection of a new saccharide echo contrast agent in dogs. J Am Coll Cardiol 1989;13: 1622–8.
 9. Goldberg BB, Liu JB, Burns PN, Merton DA, Forsberg F. Galactose-based intravenous sonographic contrast agent: Experimental studies. J Ultrasound Med 1993;12:463–70.
10. Kedar RP, Cosgrove DO, McCready VR, Bamber JC. Microbubble Doppler angiography of breast masses: dynamic and morphologic features. Radiology 1993;189(P):154.
11. Duda VF, Rode G, Schlief R. Echocontrast agent enhanced color flow imaging of the breast. Ultrasound Obstet Gynecol 1993;3:191–4.
12. Stack JP, Redmond OM, Codd MB et al. Breast disease: tissue characterization with Gd-DTPA enhancement profiles. Radiology 1990;174:491–4.

13. Forsberg F, Liu JB, Merton DA, Rawool NM, Goldberg BB. Parenchymal enhancement and tumor visualization using a new sonographic contrast agent. J Ultrasound Med 1994;14: 949–57.

14. Goldberg BB, Forsberg F, Fritzsch T, Liu JB, Merton DA. Induced acoustic emission as a contrast mechanism for detection of hepatic abnormalities. J Ultrasound Med 1995;14:S7.

15. Bogdahn U, Becker G, Fröhlich T et al. Contrast-enhanced transcranial color-coded real-time sonography allows detection of primary central nervous system tumor vascularisation. Radiology 1994;192:141–8.

16. Pugh N, Holcombe C, Lyons K et al. Colour Doppler enhancement in breast tumours using a new transpulmonary ultrasound contrast agent, Infoson™: a preliminary report. In press 1996.

17. Duda VF, Rode G, Schlief R. Echocontrast agent enhanced color flow imaging of the breast. Ultrasound Obstet Gynecol 1993;3:191–4.

18. Madjar H, Prömpeler H, Schürmann R, Göppinger A, Breckwoldt M, Pfleiderer A. Verbesserung der Durchblutungsdiagnostic von brusttunoren durch echo-kontrastmittel. Geburtsh Frauenheilk 1993;53:866–9.

19. Campani R, Bozzini F, Calliada O et al. Color Doppler imaging of hepatic metastases: the value of a US contrast agent. Lam Radiol Med 1994;87(Suppl 1):32–40.

20. Kedar RP, Cosgrove DO, McCready VR, Bamber JC, Carter EC. Microbubble contrast agent for color Doppler US: effect on breast masses. Radiology 1996;198:679–86.

21. Kudo M, Tomita S, Tochio H et al. Sonography with intraarterial infusion of carbon dioxide microbubbles: Value in differential diagnosis of hepatic tumours. Am J Radiol 1992;158: 65–74.

22. Kudo M, Tomita S, Tochio H et al. Small hepatocellular carcinoma: Diagnosis with US angiography with intraarterial CO_2 microbubbles. Radiology 1992;182:155–60.

23. Garbagnati F, Milella M, Spreafico C et al. US contrast enhancement with intraarterial CO_2 injection in staging hepatocellular carcinomas. Radiol Med 1994;87(Suppl 1):65–70.

24. Leen E, Angerson WG, Warren H et al. Improved colour Doppler flow imaging of colorectal hepatic metastases using galactose microparticles: a preliminary report. Br J Surg 1994;81: 252–4.

25. Angelli E, Carpanelli R, Crespi G et al. Efficacy of SH U 508 A in colour Doppler ultrasonography of hepatocellular carcinoma vascularisation. Radiol Med 1994;87(Suppl 1):24–31.

26. Fujimoto M, Moriyasu F, Nishikawa K, Nada T, Okuma M. Color Doppler sonography of hepatic tumors with a galactose-based contrast agent. Am J Radiol 1994;163:1099–104.

27. Tano S, Ueno N, Tomiyama T, Kimura K. Possibility of differentiating small hyperechoic liver tumours using contrast-enhanced color Doppler ultrasonography. Clin Radiol (in press) 1996.

28. Tanaka S, Kitamura T, Fujita M et al. Colour Doppler flow imaging of liver tumours. Am J Radiol 1990;154:509–14.

29. Maresca G, Barbaro B, Summaria V et al. Colour Doppler ultrasonography in the differential diagnosis of focal hepatic lesions. The SHU 508A experience. Radiol Med 1994;87(Suppl 1):41–9.

30. Moriyasu F, Itoh Y, Nada T, Suginoshita Y, Matsumura T, Kobayashi K. Cavitation signals depicted from liver hemangioma in late phase of enhancement with galactose-based microbubble. Radiology 1995;197(P):230.

31. Mattrey RF, Scheible FW, Gosink BB et al. Perfluoroctylbromide: a liver/spleen-specific and tumour-imaging ultrasound contrast material. Radiology 1982;145:759–62.

32. Behan M, O'Connell D, Mattrey RF et al. Perfluoroctylbromide as a contrast agent for CT and sonography: Preliminary clinical results. Am J Radiol 1993;160:399–405.

33. Miller DL, Vermess M, Doppman JL et al. CT of the liver and spleen with EOE-13: review of 225 examinations. Am J Radiol 1984;143:235–43.

34. Parker KJ, Baggs RB, Lerner RM et al. Ultrasound contrast for hepatic tumours using IDE particles. Invest Radiol 1990;25:1135–9.

35. Fritzsch T, Schlief R. Future prospects for echo-enhancing agents. Clin Radiol 1996;51(Suppl 1):56–8.
36. Jackobsen J. Echo-enhancing agents in the renal tract. Clin Radiol 1996;51(Suppl 1):40–3.
37. Filippone A, Muzi M, Basilico V, Di Giandomenico V, Trapani AR, Bonomo L. Color Doppler flow imaging of renal disease. Value of a new intravenous contrast agent: SH U 508 A (Levovist). Radiol Med 1994;87:50–8.
38. Rickards D, Kelley IMG, Lees WR. Prostate cancer and the value of colour Doppler. Radiology 1993;189:153–6.
39. Cosgrove DO. TRUS with contrast agents in carcinoma of the prostate. In: Murphy G, Khoury S, Chatelain C, Denis L, editors. Proc 4th Int Symposium on Recent Advances in Urological cancer, 1994. Scientific Communication International, 1995:131–41.
40. Spreafico C, Lanocita R, Damascelli B. Increase in color Doppler signals in neoplastic lesions. Levovist Investigators Meeting, Berlin Jan 1993 (Schering AG).
41. Walsh JS, Dixon JM, Chetty U, Paterson D. Colour Doppler studies of axillary node metastases in breast carcinoma. Clin Radiol 1994;49:89–9.
42. Hawighorst H, Knapstein PG, Schaeffer U et al. Pelvic lesions in patients with recurrent cervical carcinoma: Efficacy of pharmacokinetic analyses in distinguishing recurrent tumors from benign conditions. Am J Radiol 1996;166:401–8.
43. Müller-Schlimpfe M, Brix G, Layer G et al. Recurrent rectal carcinoma: diagnosis with dynamic MR imaging. Radiology 1993;189:881– 9.
44. Gückel F, Brix G, Rempp K, Rother J, Georgi M. Assessment of cerebral haemodynamics in patients with vascular disorders and tumours. Adv MRI Contrast 1994;94:100.
45. Chevenert TL, Helvie MA, Aisen AM et al. Dynamic three-dimensional imaging with partial k-space sampling: initial applications for gadolinium-enhanced rate characterization of breast lesions. Radiology 1995;196:135–42.
46. Miles KA, Hayball MP, Dixon AK. Functional images of hepatic perfusion obtained with dynamic CT. Radiology 1993;188:405–11.
47. Blomley MJK, Coulden R, Dawson P et al. Liver perfusion studied with ultrafast CT. J Comput Axial Tomogr 1995;193:424–33.
48. Leen E, Angerson WJ, Wotherspoon H, Moule E, Cook TG, McArdle CS. Detection of colorectal liver metastases: comparison of laparotomy, CT, US and Doppler perfusion index and evaluation of post-operative follow-up results. Radiology 1995;195:113–6.
49. Lasuwerier H. Fractals. Princeton: Princeton University Press, 1991.
50. Smith TG Jr, Marks WB, Lange GD, Sherif WH Jr, Neale EA. A fractal analysis of cell images. J Neurosci Methods 1989;27:173–80.
51. Schwarz K, Bexante GP, Chen X, Schlief R. Quantitative echo contrast measurement by Doppler sonography. Ultrasound Med Biol 1993;4:289–97.
52. Rubin JM, Adler RS, Fowlkes JB et al. Fractional moving blood volume: estimation with power Doppler US. Radiology 1995;197:183–90.
53. Meier P, Zierler KL. On the theory of the indicator dilution method for measurement of blood flow and volume. J Appl Physiol 1954;6:731–44.
54. Axel L. Cerebral blood flow determination by rapid sequence CT. Radiology 1980;137:679–86.
55. Gobbel GT, Cann CE, Filke JR. Measurement of regional cerebral blood flow using ultrafast computed tomography: theoretical aspects. Stroke 1991;22:768–71.
56. Thompson HK, Starmer CF, Whalen RE, McIntosh HD. Indicator transit time considered as a gamma variate. Circ Res 1964;14:502–15.
57. Desjardines C, Duling BR. Microvessel hematocrit: measurement and implications for capillary oxygen transport. Am J Physiol 1987;252:H494–503.
58. Secomb TW, Gross JF. Theory of microvascular hematocrit fluctuations. Prog Microcirculatory Res 1984;2:155–60.
59. Miles KA, Hayball MP, Dixon AK. Functional images of hepatic perfusion obtained with dynamic computed tomography. Radiology 1993;188:405–11.

614   *Barry B. Goldberg et al.*

60. Blomley MJK, Coulden R, Dawson P et al. Liver perfusion studied with ultrafast CT. J Comput Asst Tomogr 1995;19:424–33.
61. Hattori H, Miyoski, Okada J et al. Tumor blood flow measured using dynamic computed tomography. Invest Radiol 1995;29:873–6.

38. Examination of efficacy of SHU 508A for color Doppler sonography of peripheral vascular diseases and soft tissue tumors in the pelvis and extremity

TOSHIKO HIRAI, KIMIHIKO KICHIKAWA, HAJIME OHISHI,
HIDEO UCHIDA, KYOICHI HIRAMATSU, KIYOSHI OHKUMA,
HITOSHI KATAYAMA, MASAHIRO SENMOTO,
RYOHEI KUWATSURU, TSUTOMU TAKASHIMA,
TOSHIFUMI GABATA, KATUHIDE ITOH, KYOKYO HAYAMIZU,
KUMIKO NAITOH, SYOJI YOSHIDA & NAOFUMI HISA

INTRODUCTION

Color Doppler sonography has been used clinically in the diagnosis of arterial occlusive diseases in the pelvis and extremity and of deep venous thrombosis in the lower extremities as well as for follow-up examination after bypass operation [1]. However, this method has certain limitations: it is difficult to delineate color display for intrapelvic vessels or vessels running in parallel with the probe, and only a poor color display is obtained for peripheral, small and low velocity flow vessels. Those limitations might be improved with use of contrast agents.

SHU 508A is a transpulmonary contrast agent for color Doppler sonography which contains stabilized bubbles as the active ingredient, with addition of a small amount of palmitic acid (0.1%) to galactose microgranules. Its safety has been confirmed in various animal species and in a clinical Phase I study, as a result of which its safety and positive signal enhancement effect in the left heart were established in healthy persons. In subsequent clinical Phase II and III studies, SHU 508A was found to be capable of enhancing the Doppler signal in vessels of the brain, liver, kidney and lower limbs, as well as in tumor vessels (liver, kidney, mammary gland) [2].

In a clinical Phase I study conducted in Japan, the safety and good efficacy of SHU 508A in the left ventricle were established in healthy volunteers. In a clinical Phase II study to determine the optimal dose, a concentration of 300 mg/ml administered at a dose of 5 ml was found to be capable of producing sufficient contrast effect to establish a diagnosis in patients with liver tumor, kidney disease and abnormalities of cerebral blood flow [3–5].

In the expectation that SHU 508A would produce a Doppler signal enhancement effect in peripheral vessels of the body, we conducted a clinical Phase III study using SHU 508A in the pelvis and extremity in order to determine its clinical usefulness.

N. C. Nanda, R. Schlief and B. B. Goldberg (eds.), Advances in Echo Imaging Using Contrast Enhancement,
Second Edition, 615–626.
© 1997 *Kluwer Academic Publishers.*

Table 1. Subjects.

Diagnosis	No. of cases (%)
Vascular disease	36 (87.8)
After operation for upper-arm aneurysm	1
Atherosclerosis obliteration (including the one after bypass operation)	23
Subclavian arterial stenosis	1
Common iliac stenosis	1
Burger disease	1
Pseudoaneurysm of popliteal artery	1
Malformation of femoral artery	1
Varices of lower extremity	1
Venous thrombosis of inferior vena cava (suspected)	1
After bypass operation for coronary artery using internal thoracic artery as graft	4
After replacement of femoral graft	1
Tumor lesions	5 (12.2)
Malignant tumor proximal to upper arm	1
Tibial tumor	1
Osteosarcoma	2
Cervical neurogenic tumor	1

SUBJECTS AND METHODS

Subjects

The study was conducted between February 1994 and January 1995 at six institutions and involved 41 patients undergoing color Doppler sonography for the diagnosis of arterial occlusive disease, venous thrombosis or some other vascular disease, or soft tissue tumor in the pelvis or extremities. In all cases it was impossible or difficult to detect a sufficient Doppler signal when a contrast agent was not used. There were 34 men and 7 women 15–74 years old (mean age, 58 years). The disease which was observed most commonly among the patients was atherosclerotic arterial occlusive diseases; other peripheral vascular diseases included pseudoaneurysm, arteriovenous malformation, varices, thrombosis of inferior vena cava, and follow up examination after coronary bypass using the internal thoracic artery as the graft or after replacement of the femoral artery graft. Soft tissue tumors included malignant tumor at the proximal site of the upper arm, bone tumors in the lower limbs and cervical soft tissue tumor (Table 1). Patients with galactosemia, advanced or unstable disease, suspected or confirmed pregnancy and lactating mothers were excluded from the study, as were patients who had previously received SHU 508A, certain other therapeutic agents, radiation therapy within 24h of administration, those with major changes in laboratory values, those who had undergone cardiac catheterization 48h before or 24h after administration of the contrast agent, patients given X-ray, MRI or ultrasound contrast agents as well as those judged unsuitable by the supervising physician.

Table 2. Recipe for SHU 508A.

Concentration of contrast agent (mg/ml)	Vial (g)	Amount of water for injection (ml)	Total volume of contrast agent after preparation) (ml)
300	2.5	7	8.6
	4.0	11	13.5
400	2.5	5	6.6
	4.0	8	10.5

Table 3. Dose of SHU 508A.

Round of injection	Concentration × volume
First	300 mg/ml × 5 ml
Second	400 mg/ml × 4 ml
Third*	300 mg/ml × 10 ml or 400 mg/ml × 8 ml

*Third injection was given at 300 mg/ml × 10 ml if it was considered to be of no use to raise the concentration based on overall considerations of the results with the first and second injections, and given at 400 mg/ml × 8 ml if it was considered to be useful to raise the concentration.

The contrast agent

The study was conducted with SHU 508A, a granular product for injection having palmitic acid added to galactose microparticle granules at a rate of 0.1% and a solution for dissolution. The product formulation used in this study contained either 2.5 g or 4.0 g in each vial, distilled water for injection being used for dissolution.

Method of injection

Immediately before SHU 508A was injected, the distilled water was added into the vial according to the recipe table (Table 2), and the mixture was shaken for 7–10 s. The mixture was left standing for 2 min to make it homogeneous, and it was then taken up into the injection syringe. Prior to injection, the absence of visible bubbles was confirmed, and the specified dose (Table 3) was injected into the vein at a rate of about 2 ml/s. Injection was immediately followed by flushing with normal saline (about 2 ml/s).

The number of injections per patient was limited to a maximum of three. Once a positive enhancement effect with a duration of the enhancement effect evaluated as '2+' was obtained, no more injection was given. If a second or a third injection had to be given, it was given only after the enhancement effect of the previous injection disappeared.

Items of evaluation and measurement/observation

Doppler signal enhancement effect
Using color Doppler sonography, the section which was the most useful for the diagnosis was selected. After contrast agent administration, the same section was evaluated under the same settings (PRF, color gain, others) for the enhancement effect and its duration. The enhancement effect was evaluated as 3+ (despite an adequate color gain, the blood flow signal was enhanced so much that it gave an unfavorable effect to the diagnosis), 2+ (adequate enhancement effect), + (minimum enhancement effect), – (blood flow signal remained almost unchanged) or x (not evaluated due to poor condition of the apparatus or differences in the section). The duration of enhancement effect was evaluated as 1, duration satisfactory; 2, duration unsatisfactory; 3, duration unsatisfactory or zero.

Contribution of the contrast agent to improvement of diagnosis was evaluated depending on whether any images useful for the diagnosis were obtained or not, according to the following criteria: 1, diagnosis definitely improved; 2, diagnosis slightly improved; 3, diagnosis remained unchanged; diagnosis deteriorated.

Laboratory tests
During the examination, laboratory tests were conducted on blood pressure, heart rate, general hematology, blood biochemistry and urinalysis.

Side effects
With regard to sign and symptoms which occurred during and after injection of the contrast agent, the time of onset, duration, progress and treatment were recorded in full; their severity, extent and causality were also evaluated. Except for those which fell into the category of not related – unlikely, all other signs and symptoms were classified as side effects.

Overall safety rating
Overall safety of SHU 508A was evaluated based on side effects and abnormal laboratory data as 1, safe (no side effects); 2, almost safe (mild side effects which caused no clinical problems); 3, certain problems with safety (moderate or severe side effects causing certain problems); 4, serious problems with safety (side effects serious enough to need discontinuation of the trial); 5, evaluation impossible.

Usefulness
Based on the blood flow signal enhancement effect, contribution to better diagnosis and overall safety rating, overall evaluation of usefulness was classified as 1, very useful; 2, useful; 3, fairly useful; 4, undesirable; 5, prohibitive; 6, evaluation impossible.

Principles for data computation and analysis
The incidence or frequency distribution of analyzed parameters was assessed, depending on their nature. In addition, descriptive statistical data (mean, standard

deviation, minimum, maximum, median, quartiles) were calculated to allow comparison between before and after injection (variance analysis, Wilcoxon 1-sample test, mark test) and to prepare diagrams of change. The level of significance was set at $p<0.05$. In the analysis of the enhancement effect and its duration, the efficacy was obtained from the cumulative number of cases up to the round of actual injection by designating 2+ or more or satisfactory as cases of positive efficacy, and the 95% confidence interval for the rate was calculated (Kaplan–Meier test, exact test).

RESULTS

Doppler signal enhancement effect

Doppler signal enhancement effect and its duration
With the first injection (300 mg/ml × 5 ml), a signal enhancement effect of designated 2+ or more was obtained in 38 of the 41 patients (92.7%). The duration of enhancement effect was considered satisfactory if it lasted for about 2 min from 30 s after injection; this was achieved in 35 of the 41 patients (85.4%). In the majority of the patients, sufficient enhancement effect and duration of effect were obtained with the first injection (Tables 4, 5).

Contribution to improvement of diagnosis
Forty of the 41 patients (97.6%) were classified as diagnosis definitely improved compared with baseline (Table 6). In those patients, no satisfactory color display had been obtained before the injection. Following injection of SHU 508A, the Doppler signal became enhanced, as a result of which the stenosis site could be confirmed along with the extent of stenosis and occlusion (Figures 1, 2). Follow-up after coronary bypass using the internal thoracic artery as the graft or after replacement of the femoral artery graft and diagnosis of pseudoaneurysm, venous thrombosis and intratumoral blood flow in osteosarcoma were made easier.

Safety

Signs, symptoms and side effects
Signs and symptoms included vascular pain and pain at the injection site, which occurred in one patient each. Both were mild and disappeared without treatment. The local pain at the injection site was probably due to the needle puncture and was thought to have no relationship with SHU 508A. There was, therefore, only one case of vascular pain (2.4%) as a side effect of SHU 508A.

Blood pressure and heart rate
Compared with baseline, heart rate was found significantly decreased 20–28 h after injection, but the range of variation was minimal. No other significant

Table 4. Doppler signal enhancement effect (cumulative efficacy rate).

Round of injection	Drug concentration ×volume (mg/ml×ml)	No. of cases	Blood flow signal enhancement effect					No. of cases analyzed	No. of cumulative cases for ≥2+	Cumulative efficacy rate (%)	95% confidence interval for cumulative efficacy rate (Kaplan–Meier's method)
			3+	2+	+	−	x				
First	300×5	41	1	37	3	0	0	41	38	92.7	80.1–98.5*
Second	400×4	3	0	2	1	0	0	3	39	95.1	88.5–100.0

*Calculated by the exact test

Table 5. Duration (cumulative efficacy rate).

Round of injection	Drug concentration ×volume (mg/ml×ml)	No. of cases	Duration				No. of cases analyzed	No. of cases satisfactory	Cumulative efficacy rate (%)	95% confidence interval for cumulative efficacy rate (Kaplan–Meier's method)
			Satisfactory	Rather unsatisfactory	Unsatisfactory none	x				
First	300×5	41	35	3	3	0	41	35	85.4	70.8–94.4*
Second	400×4	6	4	0	2	0	3	39	95.1	88.5–100.0

*Calculated by the exact test

Table 6. Contribution to better diagnosis.

Contribution to better diagnosis			Percentage of better diagnosis*	95% confidence interval (exact test)
No. of cases	Definitely improved	Slightly improved		
41	40	1	97.6	87.1 – 99.9

*Percentage of better diagnosis: percentage of definitely improved

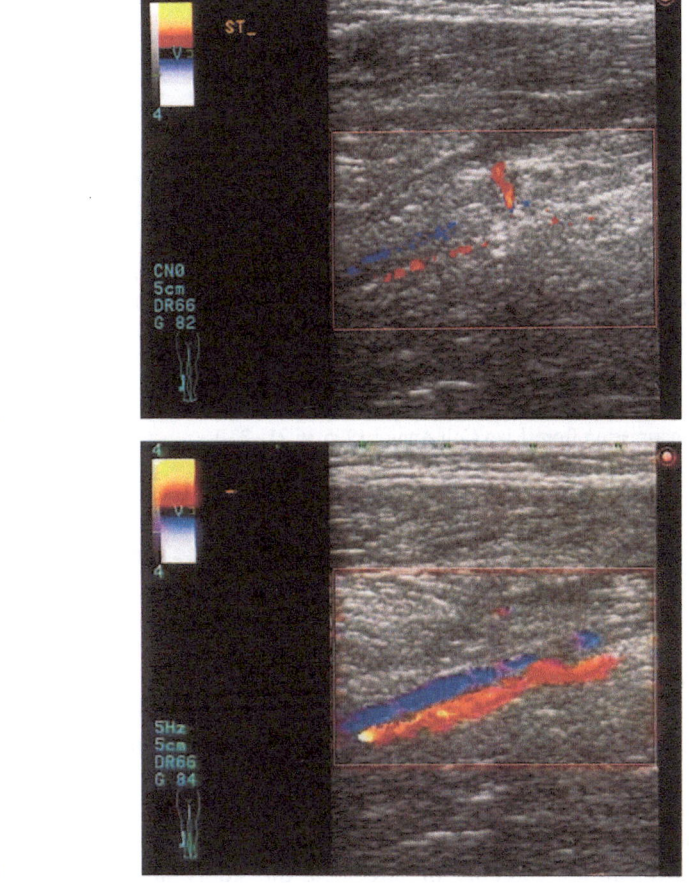

Figure 1. Color Doppler sonography of the left posterior tibial artery in a patient with occlusion of the left femoral artery. (A) Baseline. (B) After injection of SHU 508A at 300 mg/ml × 5 ml. Because of the presence of an occlusive lesion in the left femoral artery, the left posterior tibial artery with a slow blood flow could not be delineated before injection of SHU 508A, but a good color display was obtained after injection.

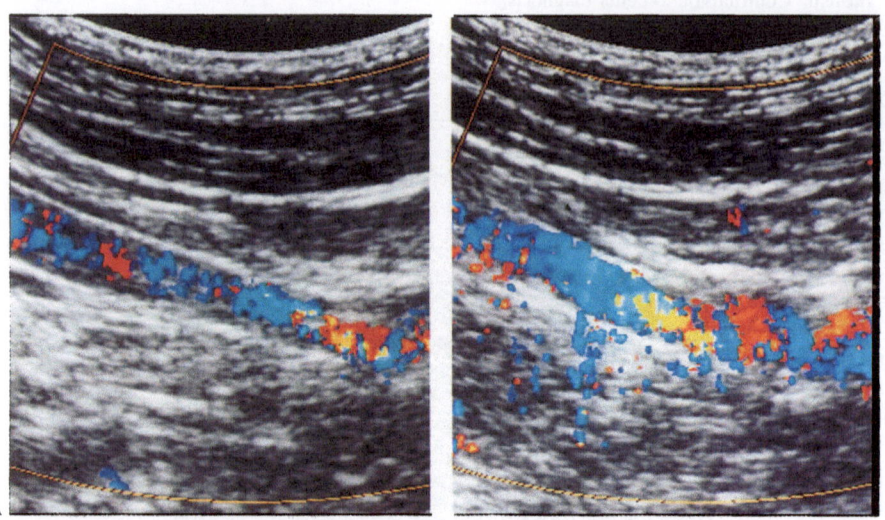

Figure 2. Color Doppler sonography in a patient with retention of PTA stent in the left common iliac artery. (A) Baseline. (B) After injection of SHU 508A at 300 mg/ml × 5 ml. Because the vessel was running almost in parallel with the probe, it was difficult to delineate the blood flow within the stent and the blood flow at the site of stenosis at the periphery of the stent simultaneously. After injection of the contrast agent, a good color display was obtained for the blood flow at the stenting site; at the same time, good delineation was also obtained for the blood flow at the site of stenosis. (Reproduced with permission from J Jap Coll Angiol 1996;36:21–33.)

changes were found in blood pressure and heart rate (variance analysis and LSD test), and thus it was concluded that SHU 508A had no clinically relevant effect on blood pressure and heart rate.

Laboratory tests
No significant changes were found in any tests between before and after injection, and thus it was concluded that SHU 508A had no effect on laboratory data.

Overall safety rating
One patient who developed vascular pain as a side effect was classified as 'almost safe', all others fell in the category of safe, with the safety rate amounting to 97.6%.

Usefulness
Usefulness was based on the enhancement effect and overall safety rating: 39 of the 41 patients were classified as very useful. Two patients were placed in the category of useful, one who developed vascular pain, although the enhancement effect was assessed as 2+, and one with arterial occlusive disease in which only

a minimum enhancement effect was obtained even after injection of SHU 508A; as a result, the usefulness rate amounted to 100%.

DISCUSSION

Color Doppler sonography is useful for the diagnosis of arterial occlusive diseases in carotid artery and the arteries in the extremities. This method enables the non-invasive observation of arterial stenosis or occlusion simultaneously with B mode findings at the lesion. The primary role of color Doppler sonography in the pelvis and extremity is to identify the presence or absence of vascular occlusion or the degree of stenosis [6, 7], and it is expected that color Doppler sonography may be useful in the identification of the presence or extent of stenosis or vascular occlusion before percutaneous intraluminal angioplasty (PTA) or bypass, as well as in postoperative follow-up [8]. Color Doppler sonography is also clinically applied to the diagnosis of pseudoaneurysm, arteriovenous fistula and venous thrombosis.

However, the currently available color Doppler sonography cannot definitely define the presence or absence of mild stenosis or vascular occlusion in intrapelvic tortuous arteries, small arteries and veins in the extremity with a slow blood flow. For that purpose, it seems necessary to enhance the blood flow signal with a contrast agent.

Special attention has been directed to microbubbles as a potential contrast agent for Doppler sonography, since their specific sound characteristics may be expected to provide a very effective ultrasound dispersion objective within the vascular lumen [9, 10]. Research has been focused on stabilization of bubbles. In the mid-1980s, a contrast agent using ultrasonically stirred human albumin was developed. After intravenous injection, a contrast agent passes through the pulmonary artery, and if the left heart is to be visualized, the contrast agent is required to have a bubble size of $< 5\,\mu$m. The newly developed albumin-based contrast agent (Albunex, Molecular Biosystems, USA) comprises micronizing the bubbles, obtained through ultrasound processing, with a mean diameter of $4\,\mu$m, which is close to the optimal target size. Albunex was marketed in Japan from November 1993 as a product capable of enhancing the contrast of the right and left ventricles in the ultrasound images [11].

Microgranules of sugar have also been tried as a means of stabilizing bubbles. This method is based on the principle that when microgranules with an irregular surface are suspended in water, part of their surface is kept from being drenched but microbubbles are formed on the microgranular surface. Galactose microgranules (Echovist, Schering AG, Germany) were developed as such a product, and was introduced into the German market in October 1991 as a contrast agent for the right heart, promising good retention of a sufficient amount of microbubbles at the time of suspension. It is used for echocardiography of right heart abnormalities (B mode and color Doppler) as well as for ultrasound contrast hysterosalpingography [12, 13]. Following intravenous injection, bubbles of this product disappear as it passes through the lung, so that the product is inappropriate for visualization of the left heart. However, SHU 508A, containing palmitic acid 0.1% to stabilize the galactose bubbles, can pass through capillary vessels of the lung [14]. In a clinical study conducted in Germany, SHU 508A was capable of not only visualizing the left heart but also enhancing the Doppler signal in the peripheral blood vessels by 15–20 dB 1–5 min after intravenous injection [15].

In this study, we administered SHU 508A at 300 mg/ml × 5 ml i.v. and obtained good results: a contrast effect in color Doppler sonography was obtained in 92.7% of patients and sufficient duration of the enhancement effect in 85.4%. One patient (2.4%) complained of vascular pain as a side effect, but this was generalized, temporary and mild and disappeared without treatment.

In the diagnosis of arterial occlusive diseases, standard color Doppler sonography promises efficient diagnosis in the region of the superficial femoral artery [8], but it is inadequate for the periphery distal to the popliteal artery, which was the target of this study. However, injection of SHU 508A successfully enhanced the Doppler signal within the artery, facilitating confirmation of the site of stenosis and its extent. Within the pelvis and in the region of the superficial femoral artery, for which a relatively good color display can be usually obtained, injection of SHU 508A yielded a good and diagnostically useful color display even in patients in whom a satisfactory color display could not be obtained because of a slower blood flow due to the presence of occlusive lesions on the proximal side. In patients in whom a stent was inserted into the common iliac artery, stenosis being present at the end of the stenting site, the Doppler signal was temporarily enhanced excessively immediately after injection of SHU 508A; this phenomenon is probably because the target artery was a central broad vessel in which the Doppler signal being enhanced more strongly and because the artifact due to the highly echogenic stent was enhanced. In the subsequent phase, a good contrast effect was obtained (Figure 2). In the diagnosis of arterial occlusive diseases in the pelvis and extremity, injection of SHU 508A was helpful in defining the blood flow of peripheral and fine vessels with a slower blood flow and facilitating the diagnosis of stenosis.

This ability of SHU 508A to improve visualization of arterial stenosis and occlusive sites was applied to observation of patients following coronary artery bypass in which the internal thoracic artery was grafted. The grafted site of the internal thoracic artery was seen only as the smallest color dot before the injection, but both the color signal and Doppler spectrum were enhanced after the injection, the entire length of the grafted internal thoracic artery being distinctly identified and the blood flow in the graft being clearly confirmed. In addition, in cases of arteriovenous malformation of the femoral region, a good contrast effect began to be obtained about 1 min after the entire contrast was prominently enhanced, and hence the structure of the abnormal blood vessels could be confirmed.

Usefulness of color Doppler sonography for the diagnosis of venous thrombosis in the lower extremity has been reported [16]. In a case of suspected venous thrombosis in the lower extremity, the color display of the target vein improved after injection of SHU 508A, as a result of which the presence of thrombosis could be excluded.

In patients with tumors, evaluation of intratumoral blood flow is useful in diagnosis and in evaluation of therapy [17]. In a study of a malignant tumor in the proximal site of the upper arm, a bone tumor (osteosarcoma) in the lower extremity and cervical neurogenic tumor, the rich blood flow within tumors could be confirmed through the enhancement produced following intravenous injection of SHU 508A [17]. This contrast agent allowed better evaluation of vascularity than at baseline, and it was also useful in evaluating the effect of chemotherapy.

In this study, we administered SHU 508A at 300 mg/ml × 5 ml for the first injection and at 400 mg/ml × 4 ml for the second injection. A sufficient contrast effect could be obtained with the first injection in 38 of the 41 patients (92.7%). The duration of the contrast effect was assessed as satisfactory in 35 of the 41 patients (85.4%) with the first injection. It was thus concluded that the effective dose for SHU 508A would be 300 mg/ml × 5 ml, as shown by the results of the clinical Phase II study.

In summary, SHU 508A (300 mg/ml × 5 ml, i.v.) produced a good contrast effect for color Doppler sonography of the pelvis and extremity. This contrast agent was useful for visualization of lesions in the peripheral vessels and also for observation of the postoperative progress. In addition, SHU 508A proved to be a useful and safe contrast agent for evaluation of therapy in tumors.

REFERENCES

1. Hirai T, Ohishi H. Doppler sonography for peripheral vascular lesions. In: Tomita S, editor. Textbook for Abdominal Color Doppler Diagnosis. Tokyo: Bunkodo, 1992:257–67.
2. Schlief R, Schürmann R, Niendorf HP. Multicenter, open-label clinical trial of SHU508A in patients for blood pool echo-enhancement in cardiac and vascular Doppler sonography. Schering AG Report No A737, 1994.
3. Oofuji M, Matsutani S, Mizumoto H et al. A study to determine the Doppler signal enhancement effect of SH/TA-508, an ultrasonic contrast agent, for liver tumor-Clinical phase II study. Med J Ultrasonogr 1995;22:461–71.
4. Matsuo H, Takahashi N, Yura T et al. Phase II study of SH/TA-508, an ultrasonic contrast agent – A study to determine the optimal dose in the renal region. Med J Ultrasonogr 1995;22: 637–48.
5. Yamaguchi T, Yasaka M, Moriyasu H et al. Clinical phase II study of SH/TA-508: Results of its clinical trial in the cerebrovascular region. Neurosonology 1995;121–7.
6. Polak JF, Karmel MI, Mannick JA et al. Determination of the extent of lower extremity peripheral arterial disease with color-assisted duplex sonography. Am J Radiol 1990;155: 1085–9.
7. Polak JF, Dobkin GR, O'Leary DH et al. Internal carotid artery stenosis: accuracy and reproducibility of color Doppler assisted duplex imaging. Radiology 1989;173:793–8.
8. Hirai T, Ohishi H, Kichikawa K et al. Evaluation of pretreatment and follow-up examination with color Doppler flow imaging in arterial occlusive diseases in the pelvis and lower extremity. Nippon Acta Radiol 1993;53:916–30.

9. Melzer RS, Tickner G, Sahires TP et al. The source of ultrasound contrast effect. J Clin Ultrasound 1988;8:121-7.
10. Gramiak R, Shah PM. Echocardiography of the aortic root. Invest Radiol 1968;3:356-66.
11. Feinstein SB, Lang RM, Cheirif J et al. Safety and efficacy of a new transpulmonary ultrasound contrast agent; initial multicenter clinical results. J Am Coll Cardiol 1990;16:316-24.
12. Schlief R. Ultrasound contrast agents. Curr Opinion Radiol 1991;3:198-207.
13. Schlief R, Schurmann R, Balzer T et al. Advances in echo imaging using contrast enhancement saccharide-based contrast agents. In: Navin CN, Schlief R, editors. Dordrecht: Kluwer Academic Publishers. 1993;71-96.
14. Ultrasound contrast agents – discussion. Invest Radiol 1991;26:S198-200.
15. Schlief R, Schurmann R, Niendorf HP. Blood pool enhancement with SHU 508A – results of phase II clinical trials. Invest Radiol 1991;26:S188-9.
16. Foley WD, Middleton WD, Lawson TL et al. Color Doppler ultrasound imaging of lower-extremity venous disease. Am J Radiol 1989;152:371-6.
17. Wonde HJ, Bloem JL, Schipper J et al. Changes in tumor perfusion induced by chemotherapy in bone sarcomas: color Doppler flow imaging compared with contrast enhanced MR imaging and three-phase bone scintigraphy. Radiology 1994;191:421-31.

39. Contrast-enhanced color Doppler in the diagnosis of liver tumors

SHOICHI MATSUTANI, HITOSHI MARUYAMA, MASAAKI EBARA,
MASAHARU YOSHIKAWA, HIROMITSU SAISHO & MASAO OHTO

INTRODUCTION

Recent advances in color Doppler have made it possible to demonstrate the blood flow signals within liver tumors [1, 2]. Some characteristic patterns in color Doppler have been demonstrated in malignant liver tumors, and characteristic findings in tumor invasion into the portal vein have been reported [3, 4]. However, the poor detectability of color Doppler signals prevents the wide use of this technique in the clinical investigation of liver tumors. Contrast enhancement is expected to improve the diagnostic ability of color Doppler in tumors. Carbon dioxide microbubbles have been already used for the B-mode enhancement of liver tumors, and allows precise evaluation of tumor vascularity [5]. However, it is available only by intra-arterial injection during hepatic angiography. Recently, new contrast agents which enhance the liver after intravenous injection have been developed. The enhancement of color Doppler signals in liver tumors after intravenous injection of sonicated albumin was observed in an experimental in vivo study [6]. In this article, recent clinical experience with a galactose-based contrast agent in the diagnosis of hepatocellular carcinoma (HCC) is presented and its significance for clinical practice is discussed.

CONVENTIONAL COLOR DOPPLER

Color blood flow signals, spotty or linear, were detected in 41 of 74 HCC (55.4%) by conventional color Doppler (Toshiba SSA-270A, 3.75 MHz ultrasonic probe, and 3.0–4.0 KHz pulse repetition frequency). Blood flow signals were detected less frequently in metastatic liver cancer and hepatic hemangioma than in HCC (metastatic liver cancer: 1/8, 12.5%; hepatic hemangioma: 4/26, 15.4%). As HCC is usually hypervascular, detection of blood flow was frequently possible in large tumors, but detection rate decreased in smaller tumors (Table 1). Demonstration of blood flow signals within the tumor also depended on the distance of

N. C. Nanda, R. Schlief and B. B. Goldberg (eds.), Advances in Echo Imaging Using Contrast Enhancement,
Second Edition, 627–634.
© 1997 *Kluwer Academic Publishers.*

628   *Shoichi Matsutani et al.*

Table 1. Detection of flow signals in HCC by color Doppler.

Tumor size (mm)	Detection of flow signals	
	Conventional	Enhanced
≤20	5/21 (24%)	4/5 (80%)
20–40	15/32 (47%)	9/9 (100%)

the tumor from the liver surface (≤30 mm: 30/41, 73.2%; >30 mm: 11/33, 33.3%; $p<0.05$). This low sensitivity of color Doppler in detecting blood flow signals in the tumor prevents adequate evaluation of tumor vascularity, which is important for the diagnosis of the malignant or benign nature, and evaluation of therapeutic effects. Color blood flow signals showing reversed flow were detected in tumor thrombus of the portal vein in nine of 14 patients (64.3%). Arterial blood supply in the tumor thrombus is thought to be the origin of these signals. As these findings were not observed in benign portal thrombus, it is a characteristic finding indicating intraportal invasion of HCC.

SACCHARIDE-BASED CONTRAST AGENT

A galactose-based contrast agent, SHU 508A (Schering AG, Germany) contains numerous and stable microbubbles smaller than red blood cells which can pass through the lung and produce enhancement of peripheral vascular color Doppler signals after peripheral venous injection [7]. Enhancement of color Doppler signals was confirmed in the liver during clinical trials [8]. Degree and duration of enhancement depended on the concentration of the contrast agent, not on the dose. In Phase II clinical trials in Japan, moderate enhancement of color Doppler signals, which allowed adequate evaluation of tumor vascularity, was obtained in 53.7–76.9% of patients. Duration of enhancement was 26.0–59.4 s (mean). The dose producing satisfactory enhancement of color Doppler images of the liver was determined to be 5 ml at a concentration of 300–400 mg/ml.

ENHANCED COLOR DOPPLER IN SMALL HCC

After i.v. injection of the galactose-based contrast agent SHU 508A at a dose of 1.5–1.6 g of a concentration of 300–400 mg/ml, color flow signals in HCCs were detected more frequently than with conventional color Doppler (Table 1, Figure 1). Blood flow signals were frequently detected even in small HCC (≤20 mm) by enhanced color Doppler (4/5 nodules, 80%). Degree of enhancement was moderate in three nodules and mild in one. Enhanced color flow signals were detected more abundantly in HCC than in metastatic liver cancer, in which enhanced signals were demonstrated only at the periphery of the tumor. In a regenerating nodule in liver cirrhosis and two hepatic hemangiomas <20 mm in

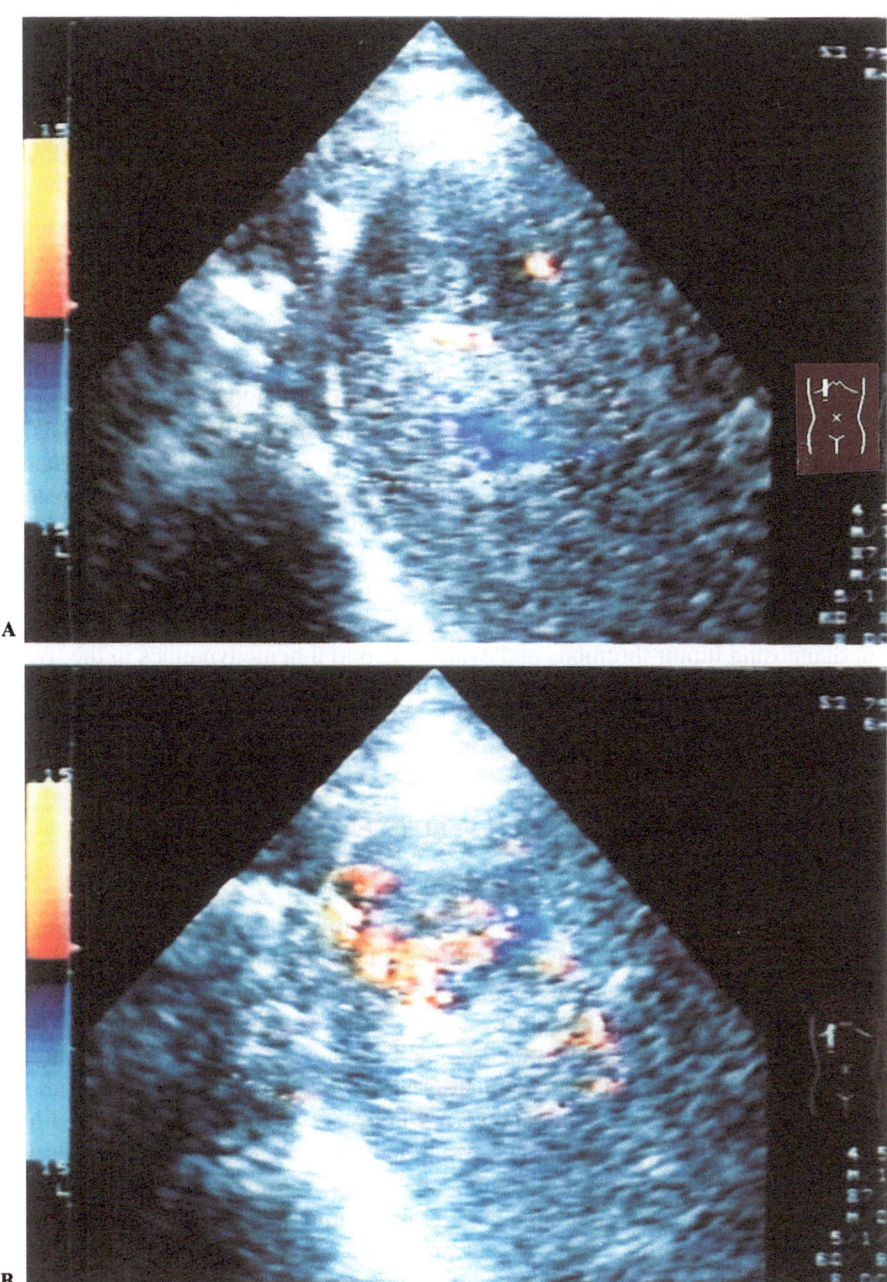

Figure 1. Contrast enhancement of color Doppler in a small HCC. (A) Before contrast enhancement. No blood flow signals were detected in the tumor. (B) After contrast enhancement. Abundant blood flow signals were demonstrated in the tumor after intravenous injection of SHU 508A.

size, no color flow signals could be demonstrated with either unenhanced or contrast-enhanced color Doppler. Demonstration of enhanced color flow signals in HCC correlated with the presence of arterial neovascular findings on angiography. These results suggest that enhanced color Doppler is useful in the evaluation of arterial vascularity of tumors and that the findings may improve the differential diagnosis of small liver tumors.

ENHANCED COLOR DOPPLER IN THE DIAGNOSIS OF INTRAVASCULAR TUMOR INVASION

Tumor thrombus in the portal vein is frequently found in patients with HCC and usually indicates a poor prognosis. Detection of reversed blood flow in the intraportal tumor thrombus by color Doppler is useful for the differential diagnosis of benign or malignant thrombus of the portal vein. However, half of the malignant portal thrombi did not show these color Doppler findings. Enhanced color Doppler increases the detection rate of these blood flow signals in tumor thrombus of the portal vein. In our clinical experience in one patient with HCC and portal thrombus, malignant findings were observed in the portal vein thrombus only with enhanced color Doppler (Figure 2). Enhanced color Doppler will be useful for diagnosing tumor thrombus and evaluating portal hemodynamics in the liver after intravascular tumor invasion by demonstrating the blood flow in and around the portal vein thrombus.

DIAGNOSIS OF THERAPEUTIC EFFECTS BY ENHANCED COLOR DOPPLER

In HCC, various non-surgical treatments are employed and effective therapeutic results are reported. Precise evaluation of the effects of non-surgical treatment by imaging is important in the management of patients with HCC. Contrast-enhanced CT is used for this purpose, allowing evaluation of hemodynamic changes in tumors due to treatment. Color Doppler is expected to allow non-invasive evaluation of the changes in arterial blood supply to liver tumors after treatment [9, 10]. In the patient shown in Figure 3, whose HCC was treated by transarterial embolization (TAE), conventional color Doppler showed no blood flow signals in the tumor after treatment. However, enhanced color Doppler demonstrated a hypervascular area in part of the tumor which was supplied from the cystic artery. In this case, residual tumor blood supply after treatment was correctly diagnosed by enhanced color Doppler. Diagnosis of the therapeutic changes in arterial tumor blood supply will be a good application of enhanced color Doppler by improving the sensitivity with which arterial tumor blood flow can be detected.

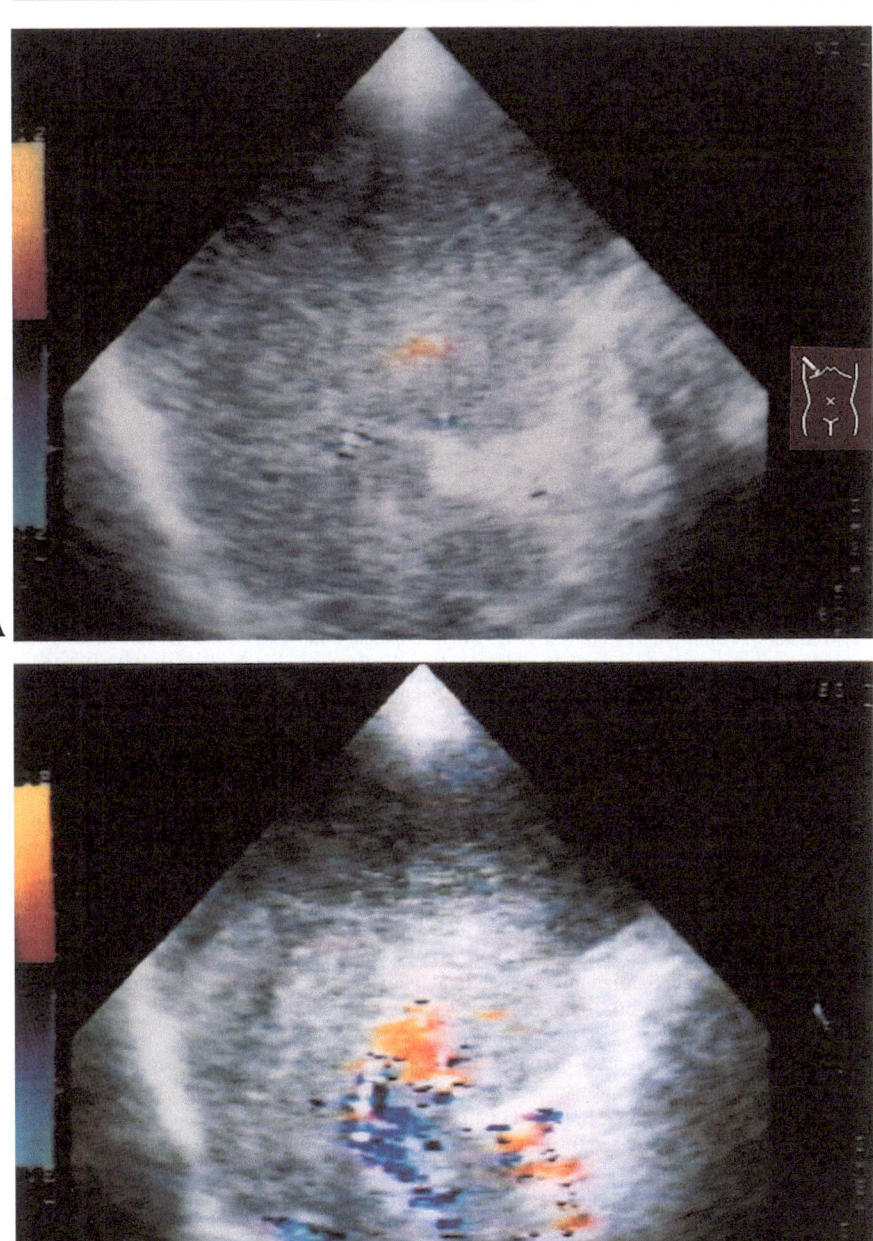

Figure 2. Enhanced color Doppler of a tumor thrombus in the portal vein in a patient with HCC. (A) Before contrast enhancement. The right portal vein is occluded by the tumor thrombus. No blood flow signals were detected in or around the right portal vein. (B) After contrast enhancement. Blood flow signals in the hepatic artery around the tumor thrombus were markedly enhanced. Reversed flow was demonstrated in the tumor thrombus after contrast enhancement.

632   *Shoichi Matsutani et al.*

CONCLUSION

Color Doppler is useful for the non-invasive diagnosis of various vascular abnormalities and tumor hemodynamics in liver diseases. However, limitations in the demonstration of blood flow in the peripheral vascular system or in deep vessel is a significant problem in clinical application. Ultrasound contrast agents will improve the sensitivity of color Doppler and detect blood flow in the liver and its diseases [11, 12]. Recently developed power Doppler, which displays the signal intensity of Doppler-shifted ultrasound echoes, may be more suitable for the use of contrast agents. Furthermore, contrast agents open up the possibility of evaluating perfusion of the liver using time-intensity measurements or harmonic ultrasound [13, 14]. Such techniques will provide new knowledge on hemodynamics in liver diseases. Contrast agents will be essential to the clinical practice of ultrasound and Doppler examination in the future.

REFERENCES

1. Tanaka S, Kitamura T, Fujita M et al. Color Doppler flow imaging of liver tumors. AJR 1990;154:509–14.
2. Tanaka S, Kitamura T, Fujita M et al. Small hepatocellular carcinoma: Differentiation from adenomatous hyperplastic nodule with color Doppler flow imaging. Radiology 1992;182:161–5.
3. Furuse J, Matsutani S, Yoshikawa M et al. Diagnosis of portal vein tumor thrombus by pulsed Doppler ultrasonography. J Clin Ultrasound 1992;20:439–46.
4. Dodd GD, Memel DS, Baron RL et al. Portal vein thrombosis in patients with cirrhosis: Does sonographic detection of intrathrombus flow allow differentiation of benign and malignant thrombus? AJR 1995;165:573–7.
5. Matsuda Y, Yabuuchi I. Hepatic tumors: US contrast enhancement with CO_2 microbubbles. Radiology 1986;161:701–5.
6. Goldberg BB, Hilpert PL, Burns PN et al. Hepatic tumors: Signal enhancement at Doppler US after intravenous injection of a contrast agent. Radiology 1990;177:713–7.
7. Schwarz KQ, Becher H, Schimpfky C et al. Doppler enhancement with SHU 508A in multiple vascular regions. Radiology 1994;193:195–201.
8. Ohto M, Matsutani S, Mizumoto H et al. Evaluation of color Doppler using a saccharide-based contrast agent, SH/TA-508, on liver tumor: A phase 2 clinical study. Jpn J Med Ultrasonics 1995; 22:461–71.
9. Tanaka K, Inoue S, Numata K et al. Color Doppler sonography of hepatocellular carcinoma before and after treatment by transcatheter arterial embolization. AJR 1992;158:541–6.
10. Lencioni R, Caramella D, Bartolozzi C. Hepatocellular carcinoma: Use of color Doppler US to evaluate response to treatment with percutaneous ethanol injection. Radiology 1995;194: 113–8.
11. Fujimoto M, Moriyasu F, Nishikawa K et al. Color Doppler sonography of hepatic tumors with

Figure 3. (Opposite) Assessment by enhanced color Doppler of the blood flow supply of an HCC after TAE. (A) Before contrast enhancement. No blood flow signals were detected by conventional color Doppler. (B) After contrast enhancement. Tumor blood supply from the cystic artery was demonstrated after contrast enhancement. (C) After contrast enhancement. Hypervascular area indicating residual tumor after TAE was demonstrated.

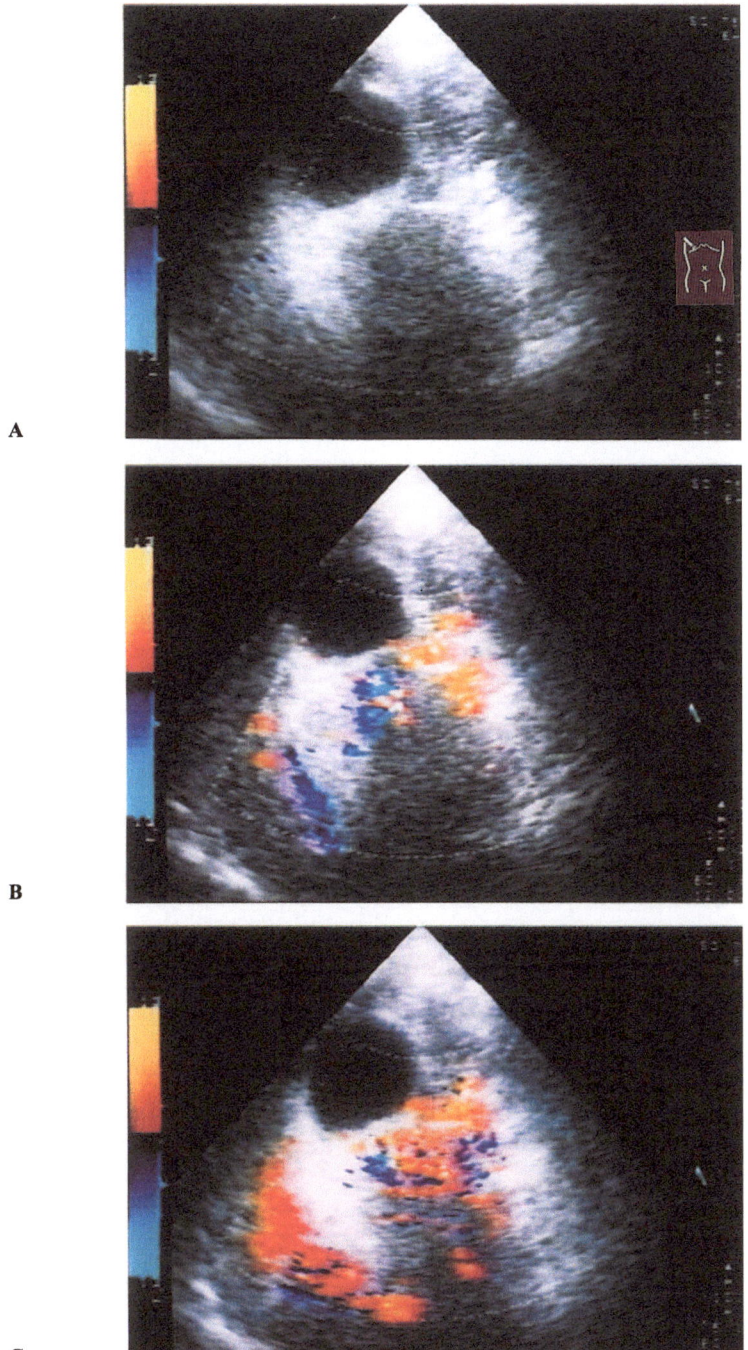

a galactose-based contrast agent: Correlation with angiographic findings. AJR 1994;163: 1099–104.

12. Tanaka S, Kitamura T, Yoshioka F et al. Effectiveness of galactose-based intravenous contrast medium on color Doppler sonography of deeply located hepatocellular carcinoma. Ultrasound Med Biol 1995;21:157–60.

13. Williams AR, Kubowicz G, Schlief R. Gas bubble dynamics in acoustic fields and their biological consequences. In: Nanda NC, Schlief R, editors. Advances in Echo Imaging Using Contrast Enhancement. Dordrecht: Kluwer Academic Publishers, 1993:111–31.

14. Burns PN. Contrast agents for Doppler ultrasound. In: Taylor KJW, Burns PN, Wells PNT, editors. Clinical Applications of Doppler Ultrasound, 2nd edn. New York: Raven Press, 1995: 367–79.

PART FOUR

Work in progress

40. Visualization of intramyocardial vasculature by contrast echocardiography

BRUNO COTTER, KOJI OHMORI & ANTHONY N. DeMARIA

INTRODUCTION

Contrast echocardiography has been applied since 1968, when the injection of agitated saline was observed to produce dense ultrasonic reflectances within cavities of the heart and great vessels [1, 2]. However, although intravenous contrast injections opacified the right-sided chambers, left-sided cardiac chambers were not affected due to filtering of the contrast by the lungs. The recognition that the disappearance of the contrast in the lungs was related to the propensity of very small microbubbles to dissolve in blood has led to the development of superior ultrasonic contrast agents with prolonged persistence times. These agents have proven capable of transiting the microcirculation of lungs and myocardium. In addition, instrumentation advances such as harmonic imaging have resulted in amplified signals from contrast compared with those from tissue [4, 5]. These advances have enabled myocardial opacification and vascular structure visualization.

The ability of contrast agents to enhance the backscatter intensity from the myocardium in a fashion similar to that of the blood pool was first demonstrated in 1980 [6]. The direct intracoronary injection of a variety of ultrasonic contrast agents was observed to produce dense myocardial opacification, which reverted to baseline as the agent cleared from the circulation. The enhancement was generally uniform in all regions of the myocardium, and was assumed to represent reflections from the microbubbles within the microcirculation of the myocardium. Although regional variability of intensity could be observed in individual images, such variability was generally related to attenuation and imaging characteristics, and intramyocardial contrast generally produced a relatively uniform myocardial blush. With the subsequent development of newer ultrasonic agents with greater persistence, intravenous injections proved to be capable of opacifying the myocardium [7–11]. Again, the resultant images revealed a uniform blush within heart muscle.

In the process of attempting to achieve superior myocardial enhancement by

637

N. C. Nanda, R. Schlief and B. B. Goldberg (eds.), Advances in Echo Imaging Using Contrast Enhancement, Second Edition, 637–647.
© 1997 Kluwer Academic Publishers.

intravenous contrast agent injection, a number of studies have evaluated newer ultrasonic agents in conjunction with harmonic imaging. During these studies, we and others have observed discreet linear or punctuate structures within the myocardium, rather than a diffuse blush [12, 13]. The findings were primarily obtained with certain ultrasonic contrast agents and during second harmonic imaging. Subsequent studies have yielded data suggesting that these linear structures represent intramyocardial coronary vessels. Although the precise mechanism of production and clinical significance of intramyocardial vessels visualized by contrast echocardiography is uncertain, a full understanding of these facts should lead to enhanced utilization of contrast ultrasound techniques in the assessment of the structure, physiology and pathophysiology of the coronary arteries.

NEW ULTRASONIC CONTRAST AGENTS

It is now well known that the mechanism of ultrasonic contrast is based upon the efficiency of backscatter produced by gas microbubbles: a gas–tissue or gas –blood interface has an efficiency which is an order of magnitude of 10^8–10^{12} greater than either a fluid–fluid or fluid–tissue interface. Previous studies have also demonstrated that injections of fluid devoid of microscopic gas particles do not produce ultrasonic contrast. The earliest ultrasonic contrast agents utilized room air as the gas microbubble, typically produced by agitation or sonication. Room air gas bubbles small enough to transit the microcirculation will have surface tension characteristics such that they will persist for less than 1 s prior to dissolving in blood. Development of advanced ultrasonic contrast agents has aimed to prolong the persistence of very small diameter microbubbles to achieve transmission to the left ventricle and myocardium.

The initial approach to increasing the persistence of room air gas microbubbles was to alter surface tension. Two approaches to enhanced persistence of room air microbubbles have been applied: protection of the bubble by human serum albumin or by a solution of polysaccharide [14, 15]. Human serum albumin was sonicated to yield a solution of microbubbles $<6\,\mu m$ in diameter capable of passing through the lungs: this is now commercially available as Albunex® (Molecular Biosystems, Inc., San Diego, CA). The other agent, Levovist® Injection (Berlex Laboratories Inc., Wayne, NJ), which has comparable characteristics to those of Albunex, comprises a powder of galactose particles and a small amount of palmitic acid which, when mixed with sterile water for injection, yields a fluid with a particle size $<8\,\mu m$. Levovist is currently being considered for market approval in the USA. Although both agents have been successful in producing left ventricular cavity opacification, they have not yet been able to produce myocardial opacification after peripheral intravenous injection.

The development of second-generation ultrasonic contrast agents has been based upon manipulation of the characteristics which determine the persistence of a gas bubble in a fluid. The interval during which a bubble remains intact in

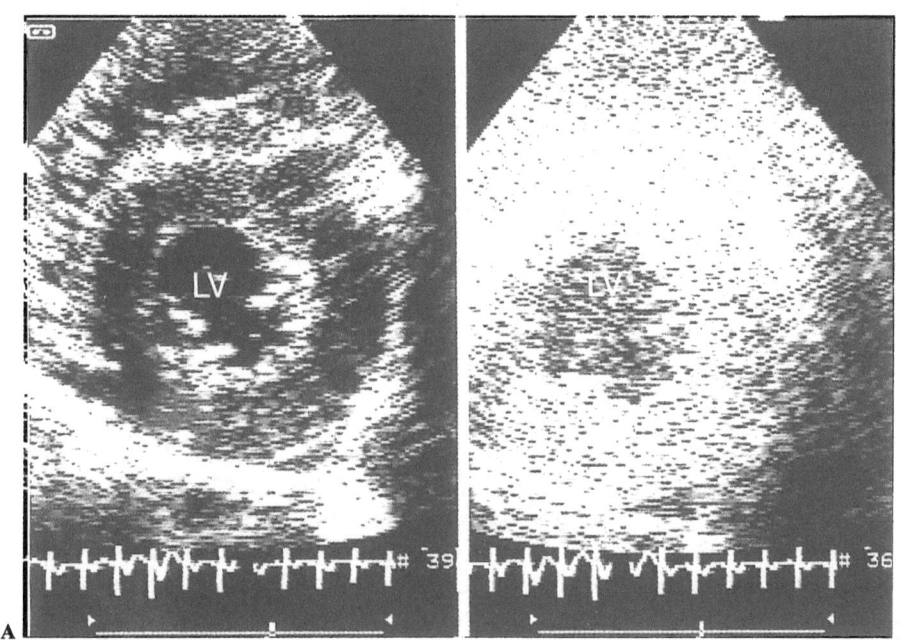

Figure 1. Opacification of the left ventricular myocardium following intravenous injection of EchoGen. (A) The left ventricular short-axis view obtained in a dog prior to contrast injection. (B) The same view 5 min following injection of the contrast agent. Dense and uniform opacification of the left ventricular myocardium is observed after the disappearance of most signals from the left ventricle (LV).

a solution is directly increased by the density of the gas and the radius of the gas bubble and is directly decreased by the ease of diffusivity of the gas and the magnitude of the saturation constant. Therefore, second-generation ultrasonic contrast agents have employed high-density gases with low diffusivity and low saturation constant to maximize the persistence of the bubble regardless of its size [16].

A number of new second-generation ultrasonic contrast agents have recently been developed which are capable of delivering microbubbles to the coronary arteries in sufficient concentration after intravenous injection so as to achieve myocardial opacification [7, 10, 17–19]. These agents have often exhibited striking differences in the opacification which they produce. EchoGen® or QW3600 (SONUS Pharmaceuticals, Bothell, WA) is a new agent consisting of dodeca-fluoropentane as the active agent. Peripheral intravenous injections of EchoGen have produced opacification of the myocardium which persisted for much longer than that exhibited by the blood pool of the left ventricular cavity (Figure 1). In contrast, FS069 (Molecular Biosystems Inc., San Diego, CA), which consists of perfluoropentane-filled albumin spheres produced by sonication, and MRX115

(ImaRx, Tucson, AZ), which combines fluorocarbons with microsomes, manifest myocardial opacification for a shorter duration, as would be produced by the transit of microbubbles through the circulation. These various agents thus resemble deposit and flow tracer types respectively. AF0150 or Imagent® US (Alliance Pharmaceutical Corp., San Diego, CA) is a new ultrasonic contrast agent consisting of perfluorohexane/nitrogen-filled, dry microspheres. When reconstituted this agent yields surfactant-coated microbubbles with a median size of 5 μm. AF0150 is relatively unique in the nature of the images it produces. Injections of Imagent US have yielded linear echoes indicative of intramyocardial vessels.

SECOND HARMONIC IMAGING

Second harmonic imaging is based on the non-linear emission of harmonics by resonant microbubbles pulsating in an ultrasonic field [20–22]. The resonance, or alternating expansion and contraction, of the ultrasonic bubble is dependent on the bubble radius, the composition of microbubble shell, the resonant frequency, the acoustic pressure, the velocity, the composition and the pressure of the surrounding medium. A gas microbubble exposed to an ultrasonic field will resonate at a given frequency. Microbubbles undergoing resonant oscillations have the ability to reflect significant ultrasonic energy at both the fundamental frequency of the ultrasonic beam transmitted and the frequency which results from the interaction between the transmitted signal and the resonating bubble. For most new ultrasonic agents, the resonating characteristics during normal physiological conditions within the circulation result in a significant reflection of the ultrasonic energy at the second harmonic of the transmitted frequency, but not all agents display a similar non-linear response. Key factors in the harmonic response of an agent are likely to be the size distribution of the bubbles and the mechanical properties of the bubble capsule (a stiff capsule, for example, will dampen the oscillations and attenuate the non-linear response) [23]. In this manner, imaging at a second harmonic frequency to the transmitted signal can amplify the reflections from the microbubbles relative to that from tissue and by programming the ultrasound to receive only the signals from the resonating bubbles, the contrast agent can be visualized and the surrounding tissue suppressed. Ideally, only the contrast agent would be visible since tissue does not emit significant harmonics. However, all current implementations allow some overlap between the transmitted fundamental frequency and the received harmonic frequency, so some tissue is visible even when no contrast is present. It is also clear that bubbles which exhibit vigorous resonance at a second harmonic of the transmitted frequency (smaller bubbles respond more strongly to higher frequencies) are also susceptible to a greater degree of destruction.

Harmonic ultrasound has been proposed as a method to enhance further the detection of ultrasound contrast agents within blood-containing cavities and vascularized structures. In this way, it has been hypothesized that real-time visualization of blood flow might be possible with ultrasound [24]. The non-linearity of many of the newer echocardiographic contrast agents has been examined and demonstrated to manifest such enhancement [25]. The use of second harmonic imaging with AF0150 has allowed identification of vascular structures.

VISUALIZATION OF INTRAMYOCARDIAL VASCULAR STRUCTURES

In the process of evaluating the efficacy and hemodynamic effects of AF0150 in our laboratories, various doses of the ultrasonic contrast agent were injected into dogs. As in other investigations of ultrasonic contrast agents, the animals were anesthetized, intubated closed chest dogs in whom the electrocardiogram was monitored, along with pulmonary artery pressure by a Swan–Ganz catheter and arterial pressure by a catheter placed in the femoral artery. Injections of the ultrasonic contrast agent were made into the femoral vein. Ultrasonic imaging was achieved in the parasternal short-axis projection by placing the animal in either the right or the left lateral decubitus position, whichever yielded superior imaging. Typical studies have used a variety of commercial and prototype instruments operating at frequencies ranging from 1.8 to 5.0 MHz. In addition to this standard fundamental imaging, several prototype instruments capable of providing images which represented a display of the ultrasonic energy returning at a second harmonic frequency of the signal transmitted were used. Thus, signals ranging from 1.8 to 2.5 MHz were transmitted and images were constructed from the signals reflected at 3.6–5.0 MHz. In addition, pulsed Doppler spectral recordings were obtained with a 2 mm sample volume positioned within specific areas of the myocardium. In all cases, the resultant ultrasonic images and data were recorded on 0.5 inch VHS or SVHS videotape for subsequent analysis.

The recognition of linear intramyocardial signals rather than a diffuse myocardial blush in response to the intravenous injection of ultrasonic contrast agents was initially reported by Mulvagh and co-workers [26], in animal studies using techniques similar to those described above. They reported that, following the injection of the new ultrasonic contrast agent AF0145 (Alliance Pharmaceutical Corp., San Diego, CA) during second harmonic imaging, long linear structures could be observed within the myocardium. Such structures were seen to branch and were accompanied by punctate signals consistent with linear structures imaged in cross-section. Since these echocardiographic structures resembled blood vessels both anatomically and in size, they were believed to represent actual intramyocardial coronary arteries. To verify this fact, Mulvagh and associates performed pulsed Doppler recordings from within these linear structures and obtained velocity recordings which were primarily diastolic in timing, conforming to the known flow pattern of the coronary arteries. Administration of vasodilator agents increased the velocity recorded from these

642   *Bruno Cotter et al.*

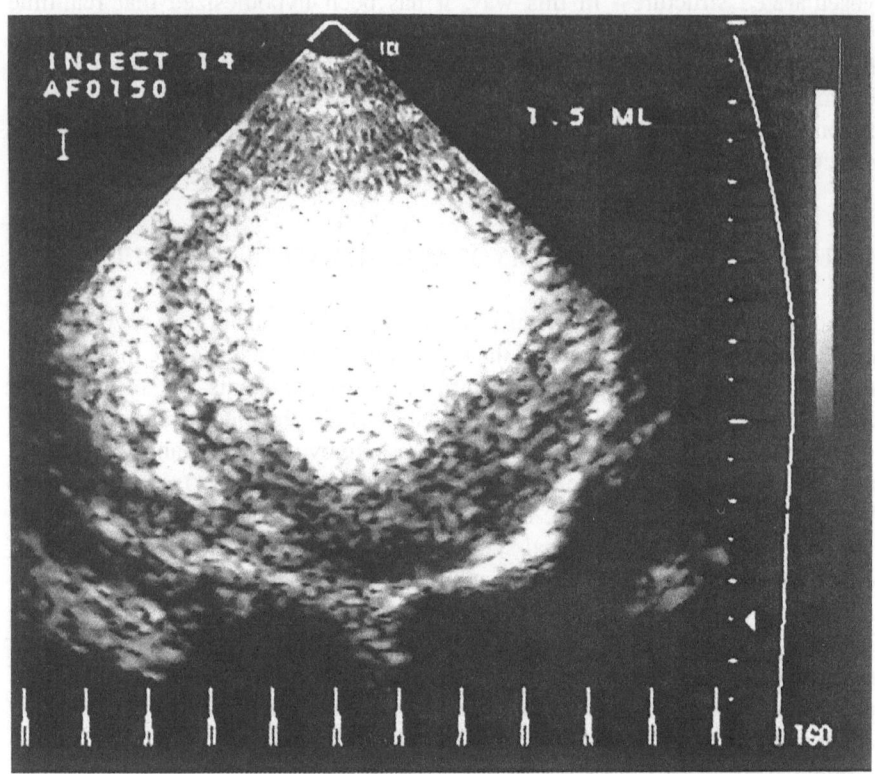

Figure 2. Left ventricular short-axis view obtained in a dog. A diffuse and intense myocardial blush can be seen following intravenous injection of the ultrasonic contrast agent Imagent US.

intramyocardial structures. The velocity recordings obtained from the contrast signals definitely behaved in a parallel fashion to those acquired from direct intra-vascular Doppler flow wire recordings. Accordingly, it was concluded that these structures clearly represent intramyocardial coronary vessels.

Studies in our laboratory using AF0150, which contains the same perfluoro-carbon gas as AF0145 and only differs in the composition of salts and buffers, yielded similar results [13]. Intravenous injections of AF0150 yielded a diffuse myocardial blush during fundamental imaging, comparable to that obtained by direct intracoronary injections or by intravenous injections of other ultrasonic agents (Figure 2). Although second harmonic imaging always tended to produce a mild diffuse blush opacification of the myocardium, discrete linear and punctuate signals consistent with vascular structures were often visualized. As in the prior study, structures recorded in our laboratory were of comparable size and branching to intramyocardial vessels (Figure 3). Doppler recordings from within these structures yielded flow signals which were predominantly diastolic in

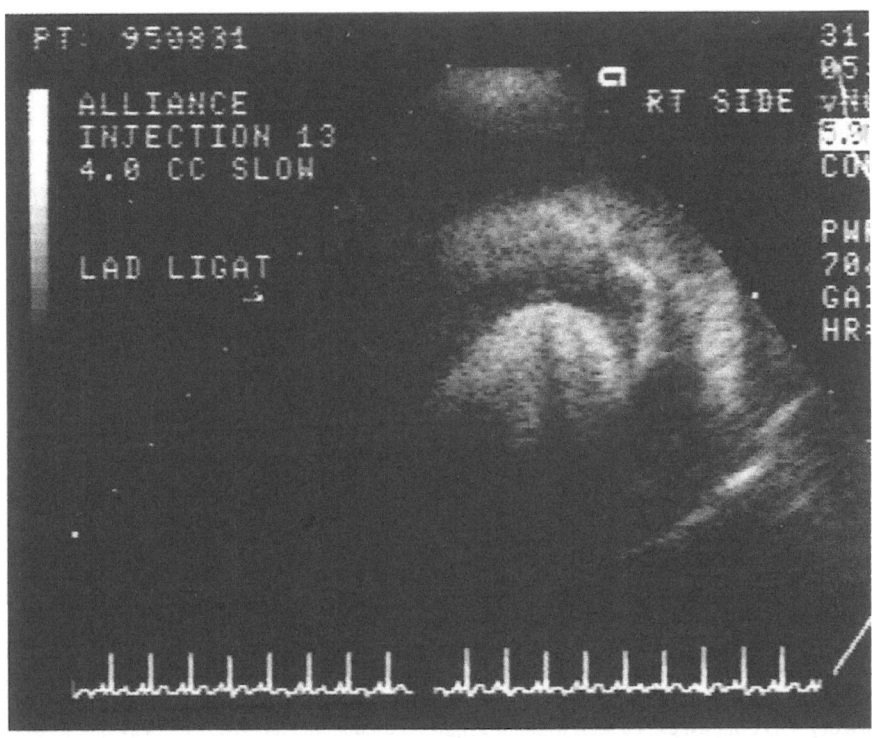

Figure 3. Left ventricular short-axis view obtained in a dog during second harmonic imaging following intravenous injection of Imagent® US. Contrast can be seen within the right ventricular and left ventricular cavities, as well as a prominent long curving signal within the interventricular septum extending from 3 to 12 o'clock consistent with an intramyocardial coronary artery.

velocity (Figure 4) and which increased nearly 2- to 3-fold after intravenous administration of dipyridamole. We concluded, therefore, that these structures clearly represent coronary vessels.

The mechanisms by which intramyocardial vascular structures are visualized following intravenous injection of AF0150 remain uncertain. Several possibilities exist, however, and preliminary observations have begun to shed some light upon this topic. It has long been recognized that ultrasonic energy is capable of destroying gas microbubbles, and the influence of ultrasonic energy upon AF0150 has been suggested as being responsible for the emergence of vascular structures from the contrast images. Such a concept would, of course, require that the ultrasonic energy had a differential affect upon the agent located in the microcirculation as compared to larger arterioles, a condition for which data do not exist at the current time.

That the effect of ultrasonic energy upon microbubbles plays a role in generating the vascular images from AF0150 is supported to some degree by the

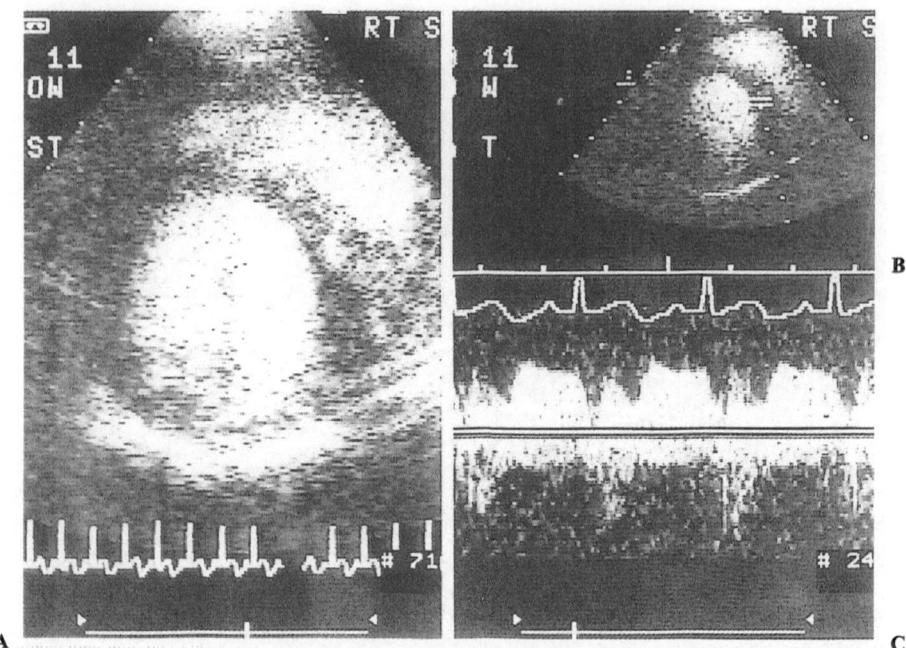

Figure 4. The echocardiographic images obtained following intravenous injection of Imagent US in a dog. (A) Contrast can be seen within the right ventricular and left ventricular cavities, as well as discrete linear and punctuate signals within the interventricular septum consistent with vascular structures. (B) Positioning of a pulsed Doppler sample volume within one vessel. (C) Doppler velocity recording obtained from this structure and the predominant diastolic flow consistent with coronary blood flow.

fact that these results are obtained only with second harmonic imaging. It could well be hypothesized that microbubbles of AF0150 resonate within the frequency of the transmitted signal and are, therefore, more easily destroyed. Destruction by ultrasonic energy would require bubbles within the microcirculation to be destroyed to a greater extent than those within the vessels. In this regard, preliminary studies in our laboratory have demonstrated that rapid switching from continuous to intermittent imaging during AF0150 injections with second harmonic yielded a diffuse myocardial blush which can envelop and veil the linear and punctuate structures. Since intermittent imaging (in this case one frame per cardiac cycle) subjects microbubbles to substantially less ultrasonic energy than continuous imaging at 20–30 frames per second, this observation is consistent with the role of destruction of microbubbles by ultrasonic energy in the generation of the appearance of vascular structures.

Although the mechanism by which intramyocardial coronary arteries are visualized by contrast echocardiography remains uncertain, the phenomenon provides a significant opportunity by which to understand the nature of contrast agents and their ability to produce myocardial opacification. It is now evident that the interaction between ultrasonic energy and contrast microbubbles is a major determinant of the resultant imaging. Thus, ultrasonic energy may destroy microbubbles, eliminating contrast enhancement. This effect may be particularly marked in the myocardium and has been demonstrated to be reversible by transient response imaging [27]. This relationship may be strongly affected by transmit power, the frequency of the signal, the velocity of blood flow and, most importantly, the nature of specific ultrasonic contrast agents. Specific studies investigating these variables are currently underway, and should yield important insights in regard to contrast echocardiography in general and myocardial blood flow in particular.

CONCLUSIONS

Myocardial contrast echocardiography promises to be a non-invasive method for the assessment of myocardial perfusion and coronary blood flow by cardiac ultrasound. Although initially requiring direct intracoronary injection, new ultrasonic contrast agents have been developed which enable myocardial opacification to be achieved by peripheral intravenous administration of contrast. These new agents provide prolonged persistence of bubbles and enable both short and long duration visualization of myocardial perfusion. Second harmonic imaging, by virtue of amplifying the signals returning from contrast bubbles in relation to tissue, offers the potential to further expand the role of contrast echocardiography in the evaluation of coronary flow. Recent studies have demonstrated that, in addition to the generalized myocardial blush typically produced by contrast enhancement, intramyocardial vascular structures can be imaged in the course of such procedures. Understanding the precise mechanisms of production of these phenomena should provide additional avenues by which to characterize coronary blood flow and myocardial perfusion by contrast echocardiography.

REFERENCES

1. Gramiak R, Shah PM. Echocardiography of the aortic root. Invest Radiol 1968;3:356–66.
2. Gramiak R, Shah PM, Kramer DH. Ultrasound cardiography: contrast studies in anatomy and function. Radiology 1969;92:939–48.
3. Crouse LJ. Sonicated serum albumin in contrast echocardiography: improved segmental wall motion depiction and implications for stress echocardiography. Am J Cardiol 1992;69:42H–5H.
4. Burns PN, Powers JE, Fritzsch T et al. Harmonic imaging: a new imaging and Doppler method for contrast enhanced ultrasound. Radiology 1992;185(P):142.
5. Mahmud E, Cotter B, Kimura B et al. Second harmonic imaging enhances contrast echocardiography in patients. Demonstration of feasibility. Am Coll Cardiol 1995;25:39A.
6. DeMaria AN, Bommer WJ, Riggs K et al. Echocardiography visualization of myocardial

perfusion by left heart and intracoronary injections of echo contrast agents. Circulation 1980; 62(Suppl III):143.

7. DeMaria AN, Dittrich H, Kwan OL, Kimura B. Myocardial opacification produced by peripheral venous injection of a new ultrasonic contrast agent. Circulation 1993;88:I-401.

8. Cotter B, Kwan OL, Cha YM, Dittrich H, Bhargava V, DeMaria AN. Dose-response characteristics, time-course, and hemodynamic response to QW3600, an ultrasonic contrast agent capable of myocardial opacification by intravenous injection. J Am Coll Cardiol 1994;23: 393A.

9. Cotter B, Keramati S, Kwan OL, Calisi C, Donaghey L, DeMaria AN. Myocardial opacification produced by intravenous injection of low dose activated EchoGen® in patients. Circulation 1995;92(Suppl):I-192.

10. Dittrich HC, Bales GL, Kuvelas T, Hunt RM, McFerran BA, Greener Y. Myocardial contrast echocardiography in experimental coronary artery occlusion with a new intravenously administered contrast agent. J Am Soc Echocardiogr 1995;8:465–74.

11. Grayburn PA, Erickson J, Escobar J, Womack L, Velasco CE. Peripheral intravenous myocardial contrast echocardiography using a 2% dodecafluoropentane emulsion: identification of myocardial risk area and infarct size in e canine model of ischemia. J Am Coll Cardiol 1995;26: 1340–7.

12. Mulvagh SL, Klarich KK, Aeschbacher BC, Tei C, Seward JB. Visualization of coronary arteries and measurement of coronary blood flow with transthoracic echocardiography after intravenous administration of a new echocardiographic contrast agent. J Am Coll Cardiol 1995;25:228A.

13. Cotter B, Duong A, Kwan OL, Wheeler K, Nozaki S, DeMaria AN. Visualization of intramyocardial coronary vessels by contrast echocardiography, observations using AF0150 (Imagent® US) during second harmonic imaging. J Am Coll Cardiol 1996;27:298A.

14. von Bibra H, Becher H, Firschke C, Schlief R, Emslander HP, Schomig A. Enhancement of mitral regurgitation and normal left atrial color Doppler flow signals with peripheral venous injection of a saccharide-based contrast agent. J Am Coll Cardiol 1993;22:521–8.

15. von Bibra H, Sutherland G, Becher H, Neudert J, Nihoyannopoulos P. Clinical evaluation of left heart Doppler contrast enhancement by a saccharide-based transpulmonary contrast agent. J Am Coll Cardiol 1995;25:500–8.

16. Cotter B, Kwan OL, DeMaria AN. Non invasive detection of myocardial perfusion by new echocontrast agents. Cardiovasc Imaging 1996;8:275–8.

17. Grauer SE, Xu J, Gong Z, Pantely Ga. Aerosomes™ MRX115 produces myocardial opacification without hemodynamic effects after intravenous injection in primates. Circulation 1995; 92(Suppl I):463.

18. Porter TR, Xie F, Kricsfeld A, Kilzer B. Noninvasive identification of acute myocardial ischemia and reperfusion with contrast ultrasound using intravenous perfluoropropane-exposed sonicated dextrose albumin. J Am Coll Cardiol 1995;26:33–40.

19. Rovai D, Lubrano V, Vassalle C et al. Detection of myocardial perfusion deficits by intravenous administration of the ultrasonic contrast agent BR1. Circulation 1995;92(Suppl I):659.

20. Miller DL. Ultrasonic detection of resonant cavitation bubbles in a flow tube by their second-harmonic emissions. Ultrasonics 1981;19:217–24.

21. Schrope B, Newhouse VL, Uhlendorf A. Simulated capillary blood flow measurement using a non linear ultrasonic contrast agent. Ultrasonic Imaging 1992;14:134–58.

22. Schrope B, Newhouse VL. Second harmonic ultrasonic blood perfusion measurement. Ultrasound Med Biol 1993;19:567–79.

23. Burns PN. Harmonic imaging with ultrasound contrast agents. Clin Radiol 1996;51(Suppl I): 50–5.

24. Burns PN, Powers JE, Simpson D et al. Harmonic power mode Doppler using microbubble contrast agents: an improved method for small vessel flow imaging. In: Proceedings 1994 IEEE Ultrasonics Symposium, Cannes, France, 1–4. IEEE 1994;3:1547–50.

25. Neppiras EA, Nyborg WL, Miller PL. Nonlinear behavior and stability of trapped micron-sized cylindrical gas bubbles in an ultrasound field. Ultrasonics 1983;21:109–15.
26. Mulvagh SL, Foley DA, Aeschbacher BC, Klarich KK, Seward JB. Second harmonic imaging of an intravenously administered echocardiographic contrast agent. Visualization of coronary arteries and measurement of coronary blood flow. J Am Coll Cardiol 1996;27:1519–25.
27. Porter TR, Xie F. Transient myocardial contrast after initial exposure to diagnostic ultrasound pressures with minute doses of intravenously injected microbubbles. Demonstration and potential mechanisms. Circulation 1995;92:2391–5.

41. Visualization of intramyocardial coronary vessels by color Doppler

NAVIN C. NANDA, CARLOS GARCIA DEL RIO,
GREGG W. TAYLOR, DIPAK I. AGRAWAL, GOPAL G. AGRAWAL,
CLAUDIA CARVALHO, MIGUEL ESPINAL, SANJAY MALHOTRA,
LANCE LaMOTTE & REINHARD SCHLIEF

Over the past few years echocardiographers have become interested in using echocontrast agents in the assessment of myocardial perfusion. In the clinical setting, however, this can currently be accomplished only by invasive means, requiring injections of these agents directly into the aortic root or the coronary arteries. Several contrast agents under development appear to have the potential of allowing myocardial perfusion to be assessed following their intravenous administration, a technique which is only minimally invasive when compared with left heart catheterization. Unfortunately, none of these agents have received approval for clinical use. Another interesting aspect of the work undertaken in this field has been the demonstration of intramyocardial coronary vessels in a closed chest dog model by intravenous B-mode contrast echocardiography using harmonic imaging, which has been found to be superior to fundamental or regular B-mode imaging in assessing contrast microbubbles [1]. We have found color Doppler imaging superior to B-mode imaging in the delineation of left ventricular cavity endocardium following intravenous injection of a contrast agent (Levovist, SHU 508A) [2].

Recently, a new echocardiographic system (Acuson Sequoia C256) has become commercially available. This utilizes multiple beam formers and a coherent image former for B-mode and color Doppler imaging and is expected to provide high-resolution echocardiographic images. In our limited clinical experience, we have found the system capable of demonstrating high quality flow signals in very small vessels, such as the epicardial coronary arteries in human patients, using color Doppler (Figure 1). In addition, in some patients intramyocardial vessels were demonstrated by color Doppler using both transesophageal and transthoracic approaches. In one patient, transthoracic color Doppler-guided pulsed Doppler interrogation of an intramyocardial vessel in the left ventricular free wall demonstrated a high diastolic velocity of 1.2 m/s, enabling us to make a diagnosis of myocardial compression by a large extracardiac mass lesion (Figure 2).

We have been successful in routinely imaging intramyocardial coronary arteries

N. C. Nanda, R. Schlief and B. B. Goldberg (eds.), Advances in Echo Imaging Using Contrast Enhancement, Second Edition, 649–654.

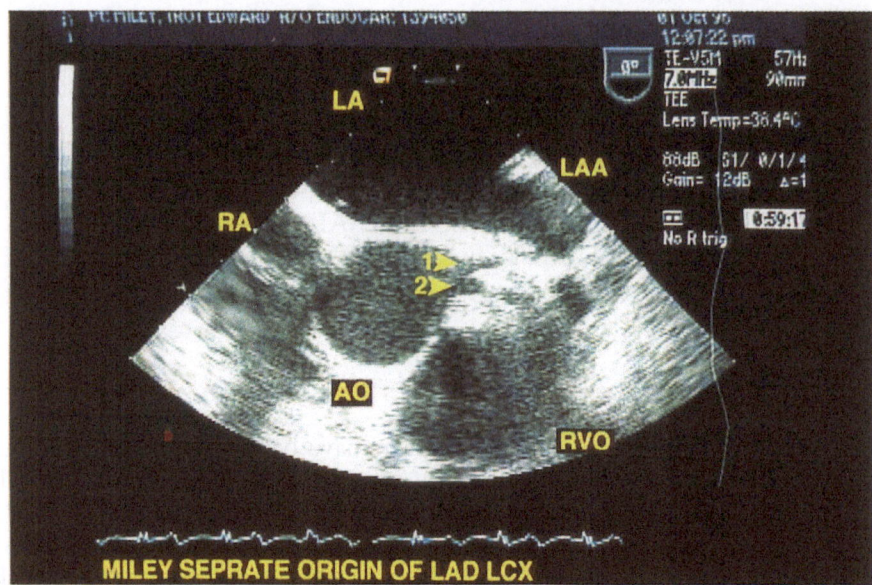

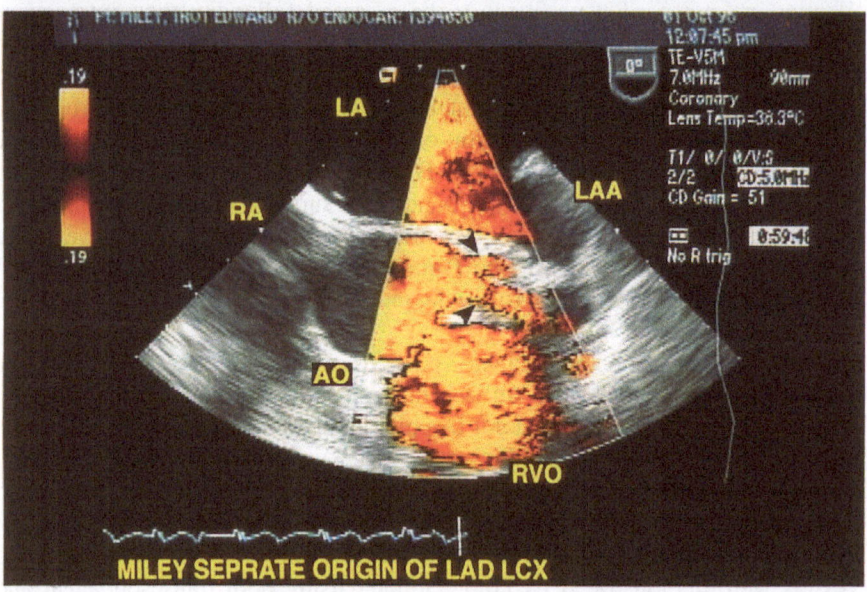

Figure 1. (A,B) Transesophageal echocardiographic imaging of epicardial coronary arteries. Both the circumflex (1) and left anterior descending (2) coronary arteries (arrowheads) have separate but adjacent origins from the aortic root in this adult patient. There is no left main coronary artery. AO = aorta; LA = left atrium; LAA = left atrial appendage; RA = right atrium; RVO = right ventricular outflow tract.

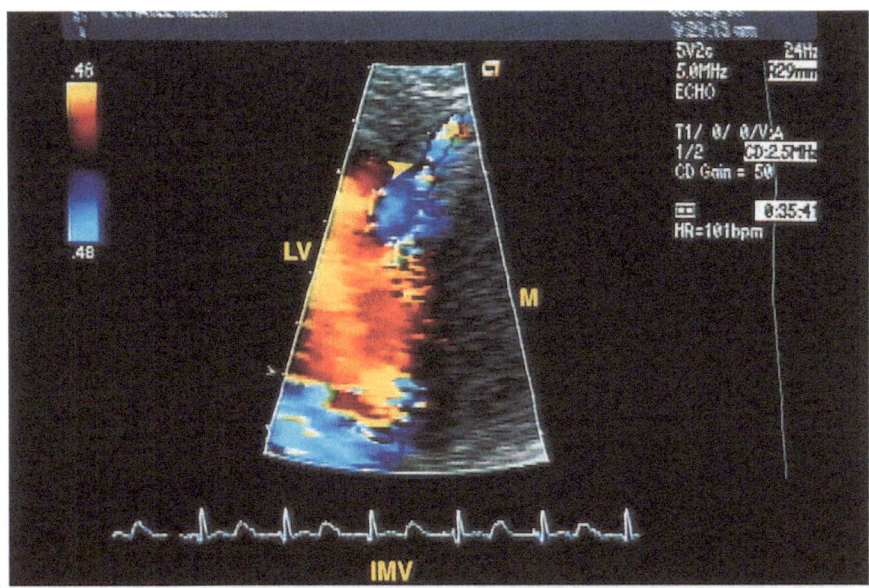

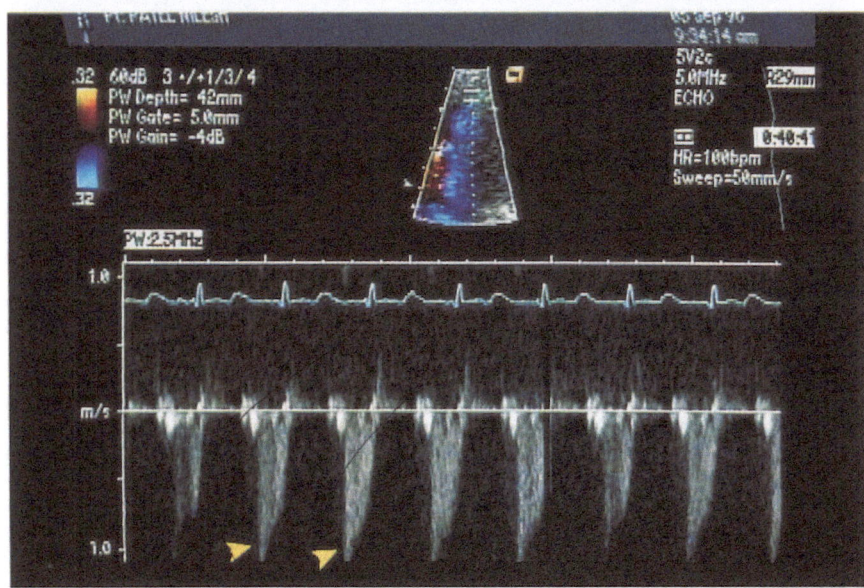

Figure 2. Transthoracic color Doppler examination of intramyocardial coronary artery. (A) Oblique view of the left ventricle (LV) demonstrates linear flow signals (arrow) within the left ventricular free wall consistent with an intramyocardial vessel. (B) Color Doppler-guided pulsed Doppler interrogation of this vessel reveals a high velocity of 1.2 m/s, indicative of some obstruction to flow produced by the adjacent extracardiac mass lesion (M) in this young adult.

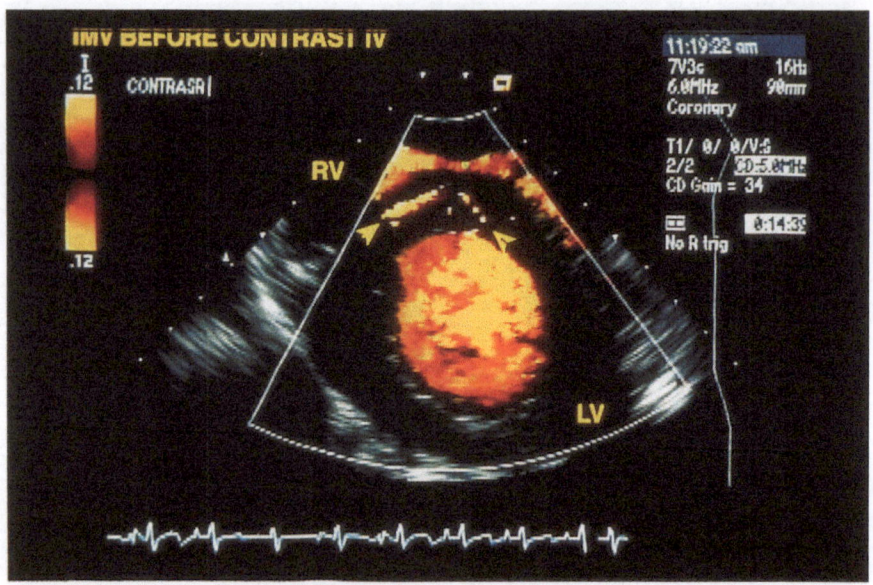

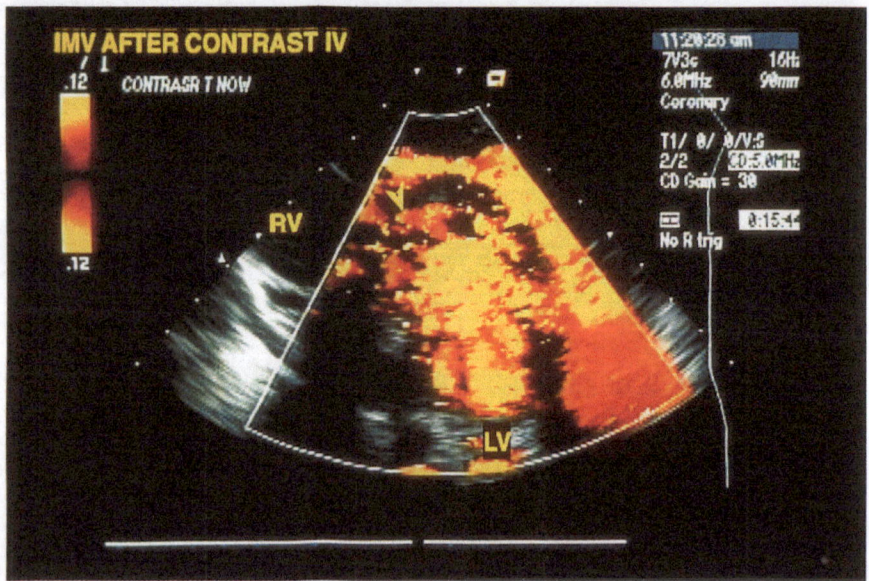

Figure 3. Transthoracic color Doppler examination of intramyocardial coronary vessels. (A) Baseline examination in an open-chest dog shows linear flow signals within the ventricular septum consistent with the septal perforator coronary artery (arrows). (B) Following intravenous injection of Levovist, significant contrast enhancement is noted and a larger extent of the perforator artery (arrow) together with some of its branches are now visualized. LV=left ventricle; RV=right ventricle.

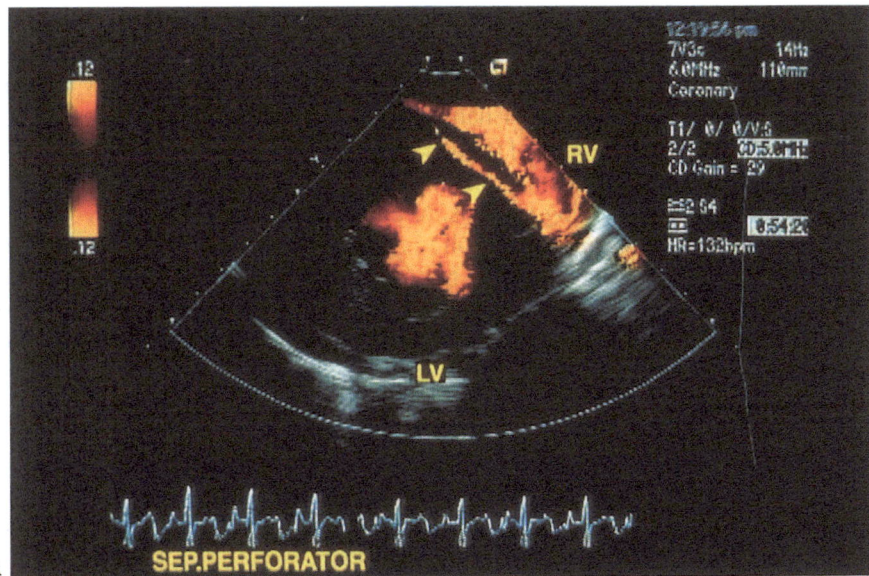

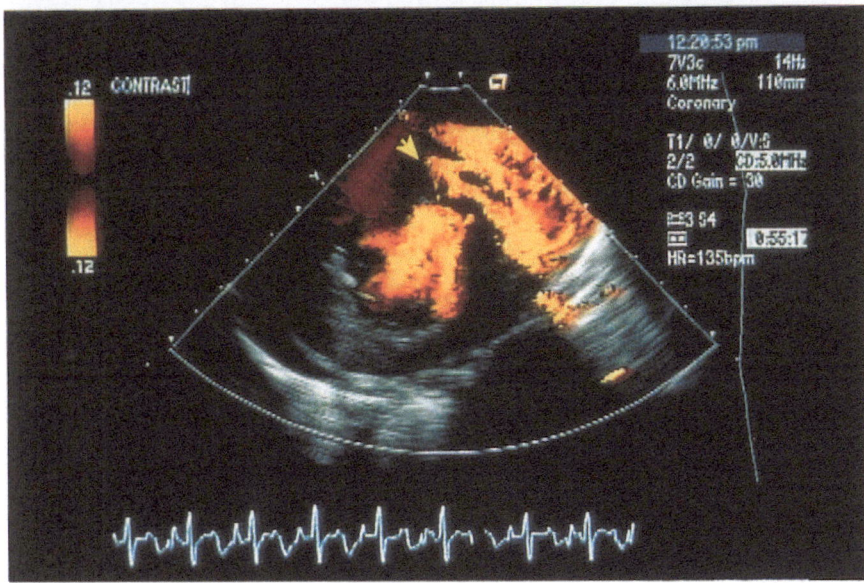

Figure 4. Transthoracic color Doppler examination of intramyocardial coronary vessels. (A) Baseline examination in an open-chest dog shows linear flow signals (arrows) within the anterior ventricular septum consistent with the septal perforator coronary artery. (B) Following injection of Levovist through a catheter placed in the left main coronary artery, the perforator artery not only becomes more prominent but also displays numerous branches (arrow). LV=left ventricle; RV=right ventricle. [Reproduced with permission from Garcia Del Rio et al. Echocardiography 1996;13.]

within the left ventricular walls of animals, especially the perforator vessels in the ventricular septum, using color Doppler and fundamental imaging in open-chest dogs and pigs. Both intravenous and intracoronary injections of a contrast agent such as Levovist produced significant enhancement of color flow signals in these vessels, resulting in visualization of a greater extent of the intra-myocardial coronary network [3] (Figures 3, 4). The contrast effect persisted for several minutes after a bolus injection of Levovist (30 mg/kg). A more prolonged effect was obtained by giving Levovist as an intravenous infusion.

It thus appears that echocardiography is now at the threshold of a new era in which not only epicardial but also intramyocardial coronary vessels can be visualized and evaluated using contrast-enhanced color Doppler imaging. The resulting benefit in assessing myocardial perfusion and coronary flow reserve should be very significant.

REFERENCES

1. Mulvagh SL, Foley DA, Aeschbacher BC et al. Second harmonic imaging of an intravenously administered echocardiographic contrast agent. J Am Coll Cardiol 1996;27:1519–25.
2. Agrawal G, Cape EG, Tirtaman C, Lee ET, Fan PH, Nanda NC. Color Doppler/contrast combination provides complete delineation of left ventricular cavity. J Am Coll Cardiol 1996; 27:77A.
3. Garcia del Rio C, Taylor GW, Nanda NC, Agrawal DI, Agrawal GG, Carvalho C, Espinal M, Becher H. Color Doppler visualization of intramyocardial coronary arteries using a new echo system: Effect of contrast enhancement and vasodilation. Echocardiography 1996;13:645–50.

42. Development of a novel site-targeted ultrasonic contrast agent

GREGORY M. LANZA, KIRK D. WALLACE, JAMES G. MILLER & SAMUEL A. WICKLINE

INTRODUCTION

Acoustic contrast agents were first introduced by Gramiak and Shah [1] and have since become an established tool for enhancing ultrasound image and Doppler sensitivity. The preponderance of systemic contrast agents described to date function to enhance the appearance of the blood pool and to define its distribution and integrity. A variety of perfusional contrast agents have been developed and one, Albunex, is available commercially [2–11].

The concept of site-directed ultrasonic contrast agents has been discussed by many authors but, until recently, not successfully demonstrated. Unlike a blood pool agent, a site-directed ultrasonic contrast agent is designed to specifically and sensitively enhance the acoustic reflectivity of a pathological tissue which would otherwise be difficult to distinguish from surrounding normal tissue. The rapid expansion of monoclonal antibodies has fuelled the recent emergence of site-targeted contrast agents by providing a plethora of specific and sensitive ligands directed against a wide spectrum of pathological molecular epitopes. As a result, the development of site-targeted ultrasonic contrast agents would be expected to expand the diagnostic capability and utility of all clinical ultrasound modalities.

We have recently described the development and in vivo application of the first targeted ultrasonic contrast agent for use across a broad range of ultrasonic frequencies [12–14]. The agent is a lipid-encapsulated, biotinylated perfluorocarbon emulsion which is non-gaseous and has an average particle diameter of approximately 250–300 nm, which are features that confer upon it a prolonged circulating half-life. Individual particles have inherently modest acoustic reflectivity in suspension but become markedly more echogenic at clinically relevant frequencies when cross-linked or bound to a surface. The acoustic agent is administered using a triphasic approach which takes advantage of the well-described avidin–biotin interactions [15–18]. In this chapter we will review the key contrast formulation strategies and research developments surrounding this novel acoustic agent.

655

N. C. Nanda, R. Schlief and B. B. Goldberg (eds.), Advances in Echo Imaging Using Contrast Enhancement, Second Edition, 655–667.
© 1997 *Kluwer Academic Publishers.*

FORMULATION OF SITE-TARGETED ULTRASONIC CONTRAST AGENTS: STRATEGY AND RESULTS

Conjugating ligands to acoustic particles

Site-targeted ultrasonic contrast agents are directed to their target by highly specific ligands, such as monoclonal antibodies, viruses, lectins, hormones, drugs and nucleic acids. When and how the acoustic particle is conjugated to the targeting ligand is an important design issue which can be addressed by one of two general approaches. The first group of methods employs direct chemical conjugation of the antibody to the acoustic particle prior to its i.v. administration. The alternative strategy entails complexing the ultrasonic contrast agent with the targeting ligand in vivo after i.v. administration.

Several chemical reactions have been developed to conjugate antibodies to particles, such as liposomes [19]. One of the more common methods utilizing thiolation has been adopted by McPherson and colleagues to target an acoustically active liposome to fibrinogen [20]. Although no significant tissue enhancement has been reported to date, in vitro retention of the liposome acoustic reflectivity and binding to a targeted epitope, as demonstrated by electron microscopy, have been successful [20].

We elected to target small perfluorocarbon particles to pathological tissue using an in vivo, three-step, modified pretargeting method [15–18], employing natural high affinity, high specificity avidin–biotin interactions to conjugate the targeting ligand to the acoustic particle. In the first step a biotinylated ligand, such as a monoclonal antibody or $F_{(ab)}$ fragment is administered and equilibrates with the target tissue (Figure 1). In step two, avidin is administered, which binds and cross-links the biotinylated antibodies, thus increasing their 'avidity' and stabilizing the antibody–avidin complex. In step three, the biotinylated perfluorocarbon emulsion particle is delivered and binds to avidin through unoccupied biotin receptor sites. The specific interaction of the acoustic particles with the tissue surface imparts increased acoustic contrast (i.e. reflectivity) that provides the unique ability to detect molecular ligands of interest with non-invasive ultrasound.

Regardless of the method of site targeting, the extent of ligand binding to tissue of interest is fundamental to the overall efficacy of the contrast system. The concentration of monoclonal antibody adhering to a specific tissue depends upon the accessibility of the tissue to the ligand, the affinity and avidity of the ligand for the epitope(s), the density of potentially accessible epitopes, the clearance rate of the ligand from the circulation, the dosage and route of ligand administered, and the time available for the ligand to equilibrate with the target tissue.

The approach of independently administering the ligand prior to the contrast agent, as in the triphasic contrast system, permits adequate time for the ligand to localize and bind to the epitopes. However, administration of the ligand after particle conjugation is expected to limit antibody binding to targeted epitopes secondary to the higher clearance rate of particles via the reticuloendothelial

Step 1   Step 2   Step 3

Biotinylated monoclonal antibodies "pretarget" tissue

Avidin binds to biotinylated monoclonal antibodies

Biotinylated perfluorocarbon emulsion binds to avidin

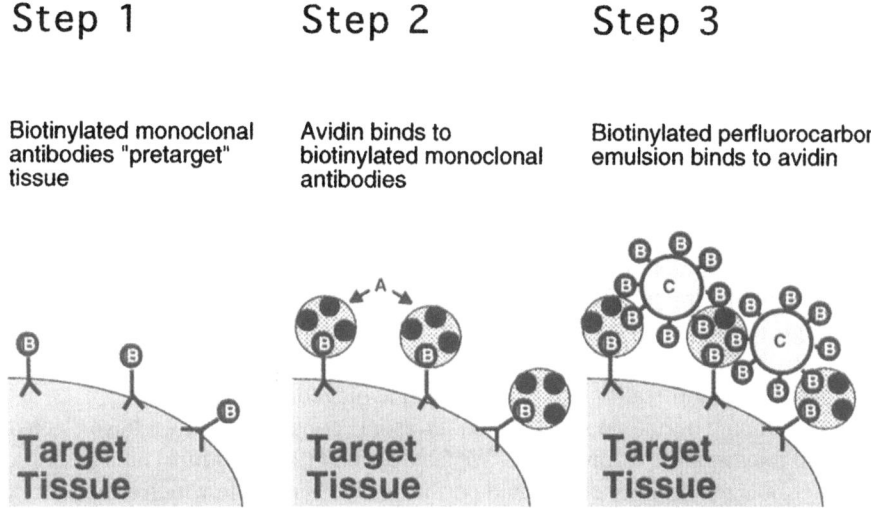

Figure 1. The use of avidin–biotin interaction to target specific molecular epitopes using a triphasic (three-step), pretargeting approach. A = avidin, B = biotin, C = contrast agent.

system, particularly for larger particles and for those expressing antibodies on their outer surface. Although 'stealthy' vesicles have been created by the addition of polyethylene glycol [21], the affinity of these particles for tissue markers is typically diminished, and this may diminish the overall contrast effect despite potential improvements in circulatory half-life.

Moreover, the strategic decision to adopt a modular targeting system rather than a direct chemical conjugation approach ultimately influences the final cost of the agent. In both ligand conjugation methods, the monoclonal antibody and the contrast particle must be derivatized or activated to allow cross-linking to occur. Covalent chemical conjugation prior to intravenous injection requires the antibody and the acoustic particles to be comixed for coupling, followed by subsequent purification by column chromatography or a centrifugation step. These processing steps engender reactant and product losses, as well as additional time and expense. These resource losses and the added processing steps are eliminated with the three-step targeting process system we have described [12–14].

The acoustic contrast: microbubbles versus non-gaseous particles

Conventional ultrasonic contrast agents have been prepared from a wide variety of substances including lipids, saccharides, proteins and co-polymers [2–11]. The most successful agents have incorporated an encapsulated gaseous phase and have utilized the acoustic resonance properties of microbubbles for contrast enhancement. Infusion of these particles into the circulation augments the acoustic signal of blood and allows the distribution of the blood pool to be assessed.

For most microbubble contrast agents, efficient transpulmonary passage after intravenous injection has been a difficult hurdle to overcome [22, 23]. Meltzer and colleagues [24] have suggested that microbubbles >18 µm in diameter are filtered out of the circulation by the sieve action of the pulmonary capillary bed. Submicron-sized microbubble particles appear to be quickly dissolved into the blood due to surface tension effects. Transpulmonary losses of up to 61% have been reported for sonicated albumin after intravenous injection [25].

Further diminution of the microbubble contrast effects occurs in the left heart as a result of structural instability under left ventricular pressures [26, 27]. Although left heart opacification after intravenous injection has been reported [28, 29], efficient, unambiguous and reproducible myocardial opacification amenable to quantification has not yet been achieved.

In developing our novel site-targeted ultrasonic contrast agent, we have elected to avoid microbubble formulations. We used a small (250–300 nm) non-gaseous, submicron-sized, lipid-encapsulated perfluorocarbon particle which experiences minimal losses due to transpulmonary entrapment and is dramatically tolerant of left ventricular pressures. Small particle size contributes to the extended circulatory half-life and allows the contrast to penetrate better into the interstices of targeted tissues.

Two almost identical perfluorocarbon formulations have been created, a biotinylated emulsion and a control (non-biotinylated) emulsion. The effect of increasing concentrations of free avidin on apparent particle size for the control and biotinylated emulsions were determined (Figure 2). As expected, the control emulsion particles were 0.234 ± 0.028 µm in diameter at baseline and remained unchanged in the presence of up to 2.0 µg avidin/ml (0.243 ± 0.012 µm), indicating a lack of agglutination due to the absence of the biotinylated group. Addition of avidin to the biotinylated emulsion resulted in a particle size increase from 0.263 ± 0.009 µm to $>2.142 \pm 0.175$ µm. Marked flocculation and sedimentation of the biotinylated perfluorocarbon emulsion was visible at avidin concentrations >2.0 µg/ml. These results document the small intrinsic particle size of the perfluorocarbon emulsions, typically one-tenth that of other ultrasonic contrast agents, and demonstrate their appropriate coalescence in the presence of titrated concentrations of avidin.

Maximizing ultrasonic signal-to-noise ratio

In contrast to our site-targeted ultrasonic technology, classical blood pool contrast agents are intended to have high, inherent acoustic reflectivity. In the case of microbubble agents, this effect is a function of the acoustic impedance mismatch between air and fluid/tissue and more so perhaps the resonant effects of microbubbles [9, 30]. When administered into the circulation, microbubble formulations

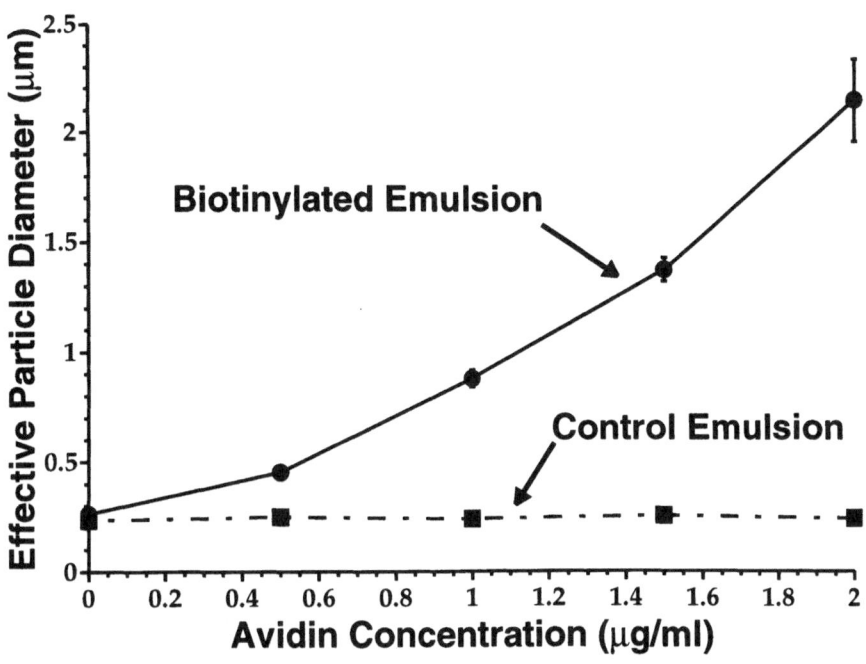

Figure 2. The effect of increasing concentrations of avidin on the apparent particle size of biotinylated and control perfluorocarbon emulsions.

tend to opacify lumens. Binding of microbubble agents to a targeted tissue would be expected to enhance the acoustic reflectivity of the tissue but these improvements in tissue signal might be suboptimal due to high background noise created by the unbound, circulating contrast. Thus, signal-to-noise issues could be problematic, depending on relative clearance rates from the blood pool and the tissue targeted. Non-gaseous, perfluorocarbon emulsions have low acoustic reflectivity and are perhaps poorly suited to function as ultrasonic blood pool contrast agents [31]. However, we have discovered that either cross-linking of these particles to themselves or binding of thin, perfluorocarbon films to a surface, markedly enhances the ultrasonic backscatter from surfaces.

The acoustic contrast effects of the biotinylated and control perfluorocarbon emulsions before and after exposure to avidin were delineated by suspending each in buffer within dialysis tubing and imaging the changes in reflected ultrasound due to cross-linking with a commercially available 7.5 MHz linear phased array, focused transducer (Figure 3). Although both suspensions were visually opaque, neither perfluorocarbon emulsion provided significant echo enhancement above the ultrasonic image noise floor for plain water (Figure 3a,b). Addition of avidin clearly increased the acoustic reflectivity of the biotinylated (Figure 3d) but not of the control emulsion (Figure 3c). The average gray-scale levels (0 = black,

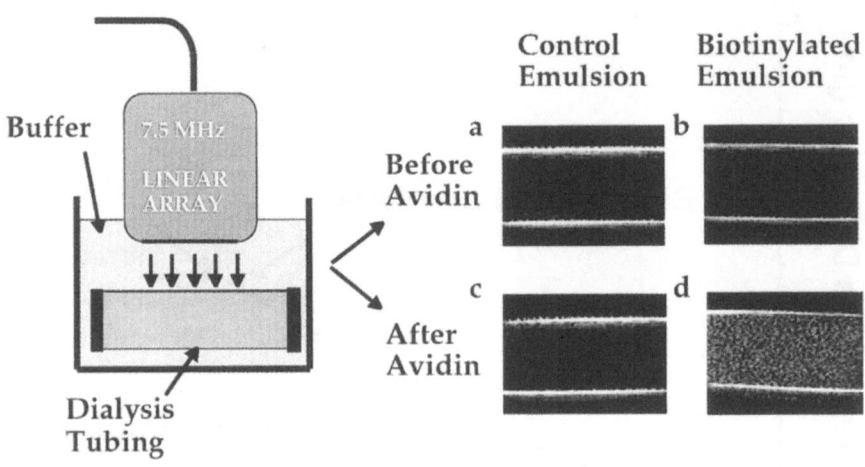

Figure 3. Ultrasonic imaging (7.5 MHz) of control and biotinylated perfluorocarbon emulsion before and after the addition of avidin.

255=white) of the control particles before (4.0±0.01) and after (4.0±0.01) avidin were similar. Mean gray-scale level of biotinylated particles increased ($p<0.05$) from 4.0±0.02 at baseline to 61.3±0.88 30 min after addition of avidin. These findings confirm the low inherent acoustic reflectivity of the perfluorocarbon emulsion particles at clinically relevant ultrasonic frequencies and illustrate the marked enhancement of ultrasonic scattering from the biotinylated contrast agent when cross-linked in suspension by avidin.

The enhanced acoustic backscatter from the cross-linked biotinylated emulsion indicated promise but could have resulted solely from the formation of large aggregates of perfluorocarbon particles. To determine whether the enhancement of reflectivity from a surface bound with a thin perfluorocarbon emulsion film was possible, the biotinylated emulsion was targeted to nitrocellulose membranes onto which avidin was covalently bonded. The nitrocellulose membrane discs were incubated with control or biotinylated emulsion particles and were examined with high frequency, high resolution 50 MHz broadband acoustic microscopy, according to methods previously used in our laboratory [32, 33]. Bandwidth-delimited backscattered acoustic power, a measure of the average efficiency of ultrasonic scattering from a surface expressed in decibels relative to the scattering from a near perfect reflector (a polished, steel plate), was computed over the useful bandwidth of the transducer (30–60 MHz) from the targeted regions of the coated discs. Ultrasonic backscatter power from control discs was −24.1±0.2 dB, compared with −17.8±0.2 dB from the biotinylated contrast agent discs, an increase in scattering of +6.3±0.1 dB, or approximately 4-fold ($p<0.05$), due to the specific attachment of biotinylated emulsion (Figure 4a,b). The power spectrum of ultrasound backscattered from the biotinylated perfluorocarbon

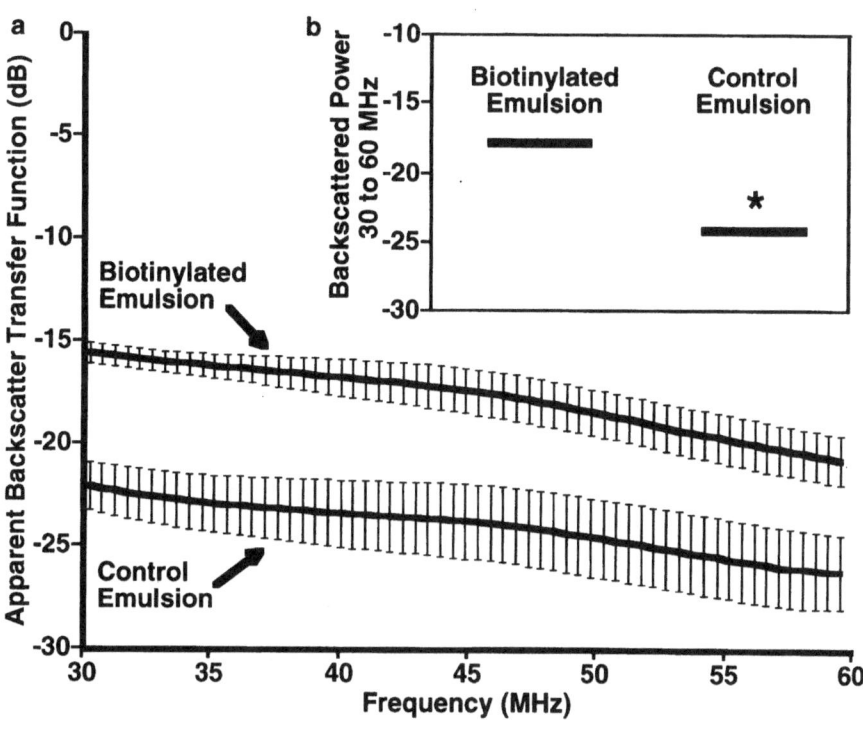

Figure 4. The apparent backscatter transfer function (a) and bandwidth delimited power (b) from avidin conjugated nitrocellulose membranes exposed to biotinylated or control perfluorocarbon emulsion.

contrast bound to nitrocellulose exceeded that for the control sample across the entire frequency spectrum. These results provided the first demonstration of an ultrasonic contrast agent specifically targeted to a membrane that dramatically augmented the reflectivity of the surface.

TARGETED ACOUSTIC ENHANCEMENT OF THROMBI

The biotinylated perfluorocarbon emulsion was targeted in vitro to porcine plasma clots with the use of biotinylated anti-fibrin monoclonal antibodies (generous donation of Dr Patrick Gaffney, National Institute for Biological Standards and Control, United Kingdom) [34, 35]. Imaging was performed with a 7.5 MHz vascular, focused linear array transducer. Plasma clots were created in vitro and imaged before and after exposure to control or biotinylated emulsion particles. Before exposure to emulsion, clots reflected little ultrasound. Five clots were

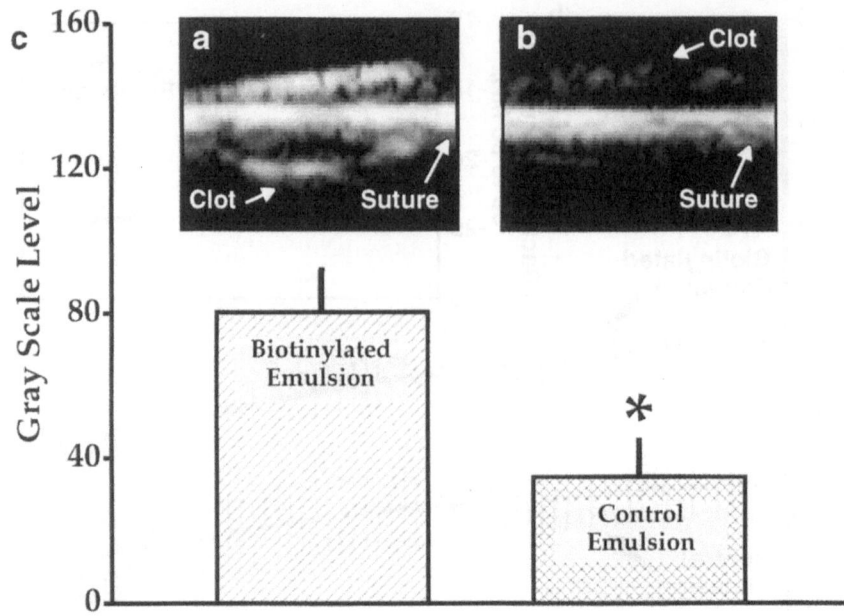

Figure 5. Ultrasonic images (7.5 MHz) of plasma clots pretargeted with antifibrin monoclonal antibody and exposed to biotinylated (a) or control (b) perfluorocarbon emulsion in vitro. The average pixel gray-scale level of plasma clot pretargeted with antifibrin monoclonal antibody and exposed to control or biotinylated perfluorocarbon emulsion (c).

incubated with biotinylated perfluorocarbon emulsion using the triphasic, targeting approach while four clots were exposed to the control emulsion alone. Anti-fibrin monoclonal antibody targeting of the biotinylated perfluorocarbon emulsion clearly enhanced the acoustic reflections along the clot perimeters. The clots exposed to the control perfluorocarbon emulsion remained unchanged (Figure 5a,b). The average pixel gray-scale level of plasma clot pretargeted with anti-fibrin monoclonal antibody and exposed to biotinylated or control perfluorocarbon emulsion is shown in Figure 5c.

Having demonstrated the triphasic acoustic contrast system against clots in vitro, we expanded this research to an in situ canine model. The effectiveness of the targeted biotinylated contrast was demonstrated for femoral artery thrombi, induced by electrical thermal injury [36, 37] and then exposed to control or antifibrin monoclonal antibody-targeted perfluorocarbon emulsion (Figure 6a). Thrombi were formed in six exposed femoral arteries and imaged before and after administration of the targeted ultrasonic contrast with a 7.5 MHz vascular, focused linear array transducer. Biotinylated anti-fibrin monoclonal antibodies (NIB 5F3, NIB 1H10) were pretargeted to thrombus and avidin was administered, followed by biotinylated perfluorocarbon emulsion. The control perfluorocarbon

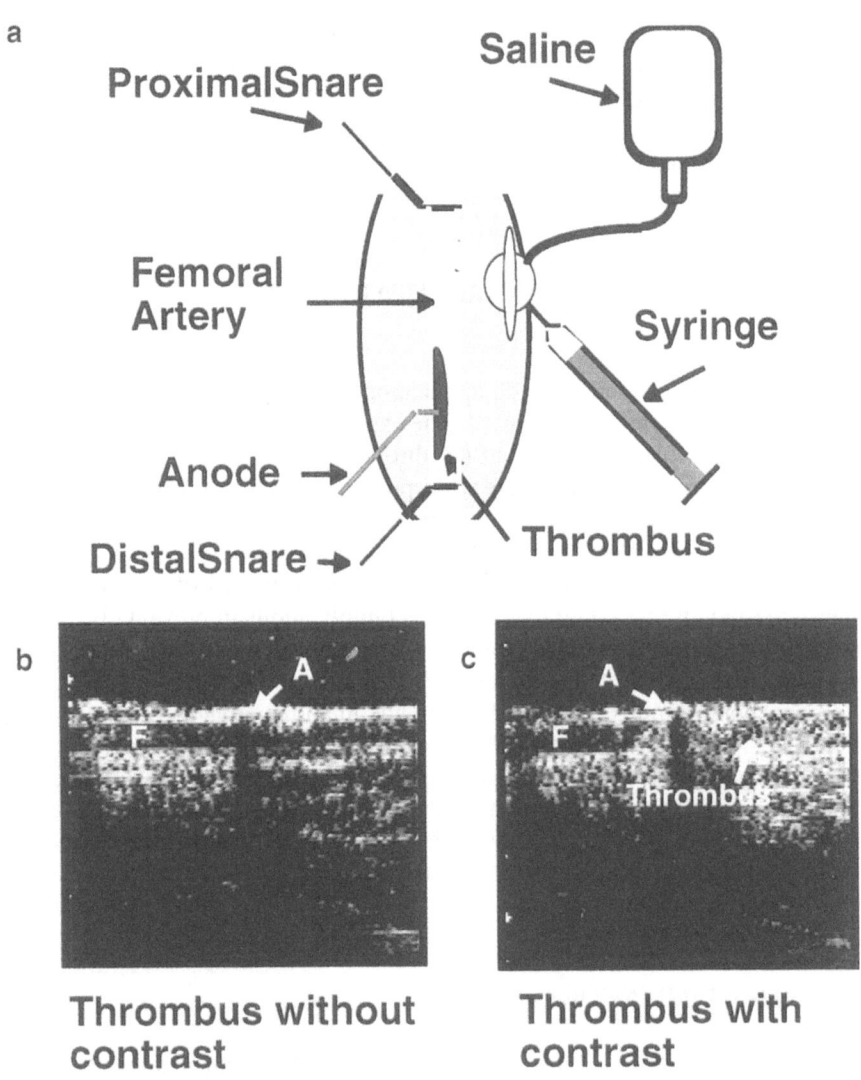

Figure 6. The acoustic reflectivity of canine femoral artery thrombus prepared in situ (a) and ultrasonically imaged at 7.5 MHz before (b) and after (c) exposure to biotinylated perfluorocarbon emulsion. A = transmural electrode, F = femoral artery.

emulsion was applied similarly to determine the extent of non-specific binding or entrapment by the non-biotinylated particles with thrombus.

Prior to contrast administration, acutely formed thrombi were detected in 3% of freeze-frame images obtained with the 7.5 MHz linear array transducer in both the control and biotinylated contrast group. Thrombi treated with two applications of the control perfluorocarbon emulsion were detected in only 5% of images.

However, the sensitivity for detecting thrombi treated with the targeted biotinyated contrast was markedly increased to 82% after only one application and further to 96% after the second emulsion application. A representative freeze-frame image of femoral artery thrombus before and after targeted contrast application is shown in Figure 6b,c.

CLINICAL POTENTIAL FOR TARGETED ULTRASONIC CONTRAST AGENTS

The potential clinical and research applications for site targeted ultrasonic contrast technology are broad and varied. The expected increased sensitivity for the detection of vascular and intracardiac thrombi, particularly residual thrombi following embolization, could have widespread ramifications in several medical and surgical specialties. In cardiology, the ability to diagnose cardiac transplant rejection, myocarditis, or small valvular vegetations would be of significant clinical benefit. The ability to screen for and monitor therapy of early or mature atherosclerotic disease could have profound implications in preventative health maintenance. Furthermore, the potential for conjugation of drugs, genes, paramagnetic materials or even radionuclides to the surface of these particles offers additional opportunities for medical treatment, including imaging with magnetic resonance. A role in the diagnosis and management of many cancers for site-targeted contrast can also be envisioned. We have already used high frequency acoustic microscopy to detect prostate, ovarian and colonic tumors in thin histological sections by linking the contrast agent to selected monoclonal antibodies and have confirmed its specificity with matched immunohistochemical analysis (Lanza et al, unpublished data).

POTENTIAL LIMITATIONS

Although perfluorocarbon emulsions are well suited for biomedical applications because of their stability, minimal toxicity and metabolism, the experimental use of avidin–biotin targeting in the present studies could present a drawback in terms of amnestic responses in the clinical situation. However, avidin and biotin have been used previously in humans for immunoscintigraphy without untoward effect [15–18]. Alternative modular targeting systems seem possible if the clinical utility of this technology evolves to even a fraction of its suggested potential.

Because site-targeted contrast agents are particles, access beyond the vascular system may be limited. The use of particulate contrast agents to target tissues beyond the circulatory system will depend on particle size and endothelial permeability. Although the particles used in the present system are quite small, further reductions may be possible down to 50–100 nm, depending on the particular perfluorocarbon utilized and/or process conditions. However, many clinically important sites can be accessed simply by employing other routes of administration, including intralymphatic, intraperitoneal, intrabiliary, transurethral, transbronchial, intrathoracic and intrathecal.

Finally, the amount of monoclonal antibody required to produce adequate signal at the intended target site may be prohibitively expensive or elicit toxic effects. The need for significant amounts of monoclonal antibody has previously been demonstrated by immunoscintigraphy studies but the toxicity of this antibody load is markedly diminished by 'humanizing' the antibody or by using $F_{(ab)}$ fragments. The expense of monoclonal antibody would be expected to diminish dramatically as a function of process scale for the most popular applications.

SUMMARY

Although numerous questions about their physical and biochemical properties must be addressed, site-targeted ultrasonic contrast agents appear to provide a new and powerful method for detecting molecular epitopes or receptors for which a biotinylated monoclonal antibody or ligand is available. Acoustic imaging of pathology enhanced with targeted acoustic contrast agents can be accomplished today with the full spectrum of commercially available ultrasonic technology.

ACKNOWLEDGMENTS

We thank Drs Patrick J. Gaffney, Tracey A. Edgell and Sanj Raut at the National Institute for Biological Standards and Control, UK for assistance in purification and preparation of the biotinylated antifibrin monoclonal antibodies. This work was supported by grant HL-42950. Dr Wickline is an Established Investigator of the American Heart Association, Dallas, TX. Dr Lanza is an American Heart Association, Missouri Affiliate Fellow.

REFERENCES

1. Gramiak R, Shah PM. Echocardiography of the aortic root. Invest Radiol 1968;3:356–66.
2. Bleeker HJ, Shung KK. Ultrasonic characterization of Albunex®, a new contrast agent. J Acoust Soc Am 1990;87:1792–7.
3. De Jong N, Hoff L, Skotland T, Bom N. Absorption and scatter of encapsulated gas filled microspheres: theoretical considerations and some measurements. Ultrasonics 1992;30:95–103.
4. De Jong N, Ten Cate FJ, Lancée CT, Roelandt JRCT, Bom N. Principles and recent developments in ultrasound contrast agents. Ultrasonics 1991;29:324–30.
5. Kaul S. Quantitation of myocardial perfusion with contrast echocardiography. Am J Card Imaging 1991;5:200–16.

6. Feinstein SB. New developments in ultrasonic contrast techniques: Transpulmonary passage of contrast agent and diagnostic implications. Echocardiography 1989;6:27–33.

7. Mattrey RF, Wrigley R, Steinbach GC, Schutt EG, Evitts DP. Gas emulsions as ultrasound contrast agents. Preliminary results in rabbits and dogs. Invest Radiol 1994;29(Suppl 2):S139–41.

8. Goldberg BB, Liu J-B, Burns PN, Merton DA, Forsberg F. Galactose-based intravenous sonographic contrast agent: experimental studies. J Ultrasound Med 1993;12:463–70.

9. Ophir J, Parker KJ. Contrast agents in diagnostic ultrasound. Ultrasound Med Biol 1989;15: 319–33.

10. Unger E, Shen D, Fritz T et al. Gas-filled lipid bilayers as ultrasound contrast agents. Invest Radiol 1994;29(Suppl 2):S134–6.

11. Kerber RE, Kioschos JM, Lauer RM. Use of an ultrasonic contrast method in the diagnosis of valvular regurgitation and intracardiac shunts. Am J Cardiol 1974;34:722–7.

12. Lanza GM, Wallace KD, Scott MJ et al. Initial description and validation of a novel site targeted ultrasonic contrast agent. Circulation 1995;92(Suppl 1):I-260.

13. Lanza GM, Wallace KD, Abendschein D et al. Specific acoustic enhancement of vascular thrombi in vivo with a novel site targeted ultrasonic contrast agent. Circulation 1995;92(Suppl 1):I-260.

14. Wallace KD, Lanza GM, Scott MJ et al. Initial description and validation of a novel site targeted ultrasonic contrast agent. Circulation 1995;92(Suppl 1):I-585.

15. Paganelli G, Magnani P, Fazio F. Pretargeting of carcinomas with the avidin-biotin system. Int J Biol Markers 1993;8:155–9.

16. Paganelli G, Magnani P, Zito F et al. Three-step monoclonal antibody tumor targeting in carcinoembryonic antigen-positive patients. Cancer Res 1991;51:5960–6.

17. Dosio F, Magnani P, Paganelli G, Samuel A, Chiesa G, Fazio F. Three-step tumor pre-targeting in lung cancer immunoscintigraphy. J Nucl Biol Med 1993;37:228–32.

18. Modorati G, Brancato R, Paganelli G, Magnani P, Pavoni R, Fazio F. Immunoscintigraphy with three step monoclonal pretargeting technique in diagnosis of uveal melanoma: preliminary results. Br J Ophthalmol 1994;78:19–23.

19. Martin FJ, Heath TD, New RRC. Covalent attachment of proteins to liposomes. In: New RRC, editor. Liposomes: A Practical Approach. New York: IRL Press, 1990:163–82.

20. Murer SE, Alkan-Onyuksel H, Kane BJ et al. Targeting of acoustically reflective liposomes for tissue specific ultrasonic enhancement. Circulation 1995;92(Suppl 1):I-659.

21. Allen TM, Hansen C. Pharmacokinetics of stealth versus conventional liposomes: effect of dose. Biochim Biophys Acta 1991;1068:133–41.

22. Meltzer RS, Sartorius OE, Lancee CT et al. Transmission of ultrasonic contrast through the lungs. Ultrasound Med Biol 1981;7:377–84.

23. Meltzer RS, Serruys PW, McGhie J, Verbaan N, Roelandt J. Pulmonary wedge injections yielding left-sided echocardiographic contrast. Br Heart J 1980;44:390–4.

24. Meltzer R, Tickner EG, Popp RL. Why do the lungs clear ultrasonic contrast? Ultrasound Med Biol 1980;6:263–9.

25. Keller MW, Feinstein SB, Watson DD. Successful left ventricular opacification following peripheral venous injection of sonicated agent: an experimental evaluation. Am Heart J 1987; 114:570–5.

26. Mottley J, Everbach EC, Schwarz KQ, Schlief R, Meltzer RS. Decay of ultrasound integrated backscatter from a saccharide contrast agent is accelerated by increased pressure. Circulation 1990;82(Suppl III):III-28.

27. Shandas R, Sahn DJ, Bales G, Elkadi T, Yau K-K, Gharib M. Persistence of Albunex (ALB) ultrasound contrast agent: in vitro study of the effects of pressure and acoustic power on particle size and the duration of contrast and Doppler enhancement. Circulation 1990;82(Suppl III): III-95.

28. Keller MW, Feinstein SB, Watson DD. Successful left ventricular opacification following peripheral venous injection of sonicated contrast agent: an experimental evaluation. Am Heart J 1987;114:570–5.

29. Feinstein SB, Cheirif J, Ten Cate FJ et al. Safety and efficacy of a new transpulmonary ultrasound contrast agent: initial multicenter clinical results. J Am Coll Cardiol 1990;16:316–24.
30. deJong N, Ten Cate F, Lancee CT, Roelandt JRCT, Bom N. Principles and recent developments in ultrasound contrast agents. Ultrasonics 1991;29:324–30.
31. Satterfield R, Tarter VM, Schumacher DJ, Tran P, Mattrey RF. Comparison of different per-fluorocarbons as ultrasound contrast agents. Invest Radiol 1993;28:325–31.
32. Wickline SA, Verdonk ED, Sobel BE, Miller JG. Identification of human myocardial infarction in vitro based on the frequency dependence of ultrasonic backscatter. J Acoust Soc Am 1992; 91:3018–25.
33. Wickline SA, Verdonk ED, Miller JG. Quantification of the transmural shift of myofiber orientation in normal human hearts with ultrasonic integrated backscatter. J Clin Invest 1991;88: 438–46.
34. Tymkewycz PM, Creighton Kempsford LJ, Hockley D, Gaffney PJ. Screen for fibrin specific monoclonal antibodies: the development of a new procedure. Thromb Haemostasis 1992;68: 48–53.
35. Tymkewycz PM, Creighton Kempsford LJ, Gaffney PJ. Generation and partial characterization of five monoclonal antibodies with high affinities for fibrin. Blood Coag Fibrinolysis 1993;4: 211–21.
36. Haskel EJ, Prager NA, Sobel BE, Abendschein DR. Relative efficacy of antithrombin compared with antiplatelet agents in accelerating coronary thrombolysis and preventing early reocclusion. Circulation 1991;83:1048–56.
37. Romson JL, Haack DW, Lucchesi BR. Electrical induction of coronary artery thrombosis in the ambulatory canine: a model for in vivo evaluation of antithrombotic agents. Thromb Res 1980;17:841–53.

43. Acoustically stimulated microbubbles in diagnostic ultrasound: properties and implications for diagnostic use

ALBRECHT BAUER, REINHARD SCHLIEF, MICHAEL ZOMACK,
ALBRECHT URBANK & HANS-PETER NIENDORF

INTRODUCTION

Ultrasound echo enhancement depends upon the enhancement of backscatter so that the intensity of the signal reflected back to the transducer is increased. However, the interaction between the incident ultrasound wave and the reflecting microbubbles is complex and dynamic. The incident ultrasound can drive the microbubbles into oscillation and, depending on the size of the microbubbles and the properties of their shells, that oscillation can become resonant. When some microbubble echo enhancers such as SHU 563A are exposed to high amplitude ultrasound this non-linear oscillation gives rise to a phenomenon called stimulated acoustic emission [1]. As the microbubbles oscillate they emit signals containing the second harmonic of the fundamental frequency to which they were exposed. With further increased transmit amplitude the microbubbles respond with high intensity broad-band signals and subsequent collapse. The signals originating from microbubbles due to stimulated acoustic emission can be used for specific detection of the ultrasound contrast agent with different techniques. This review will cover some of the basic principles involved and outline the diagnostic potential of stimulated acoustic emission and its use for microvascular imaging.

THE PRINCIPLES AND DIAGNOSTIC VALUE OF ECHO ENHANCEMENT

Ultrasound echo enhancers enable echogenic labeling of blood vessels and body cavities in B-mode, and increase the intensity of the Doppler signal throughout the blood pool. The individual microbubbles scatter the incident ultrasound beam, and the intensity of the backscattered ultrasound that returns to the transducer is the most important indicator of the efficacy of the enhancer. The increase in signal intensity produced by the backscatter is proportional to the number of scatterers

N. C. Nanda, R. Schlief and B. B. Goldberg (eds.), Advances in Echo Imaging Using Contrast Enhancement,
Second Edition, 669–684.
© 1997 Kluwer Academic Publishers.

in the region of interest. The effectiveness of the backscatter from a single micro-bubble is determined by its scattering cross-section σ. For a free gaseous bubble surrounded by water, σ is described as follows (Equation 1):

$$\sigma=\left[\frac{4}{9}R^2\pi\left(\frac{2\pi}{\lambda}R\right)^4\right]\times\left\{\left[\frac{k_s-k_w}{k_w}\right]^2+\frac{1}{3}\left[\frac{3(\rho_s-\rho_w)}{2\rho_s-\rho_w}\right]^2\right\} \tag{1}$$

in which:
σ = scattering cross section (m²)
R = microbubble radius (m)
λ = wavelength (M)
k_s = compressibility of bubble gas (m²/N)
k_w = compressibility of water (m²/N)
ρ_s = density of bubble gas (kg/m³)
ρ_w = density of water (kg/m³).

The scattering cross-section is proportional to the sixth power of the bubble radius, so the larger the bubbles, the more effective is the backscatter. The limit however is set by the necessity for transpulmonary passage; bubbles with a diameter $>7\,\mu$m cannot pass through the pulmonary capillaries.

An effective echo-enhancer will increase the intensity of the Doppler signal when the unenhanced signal is too weak to provide useful diagnostic information (Figures 1, 2). The enhanced signal usually contains enough flow information to establish the diagnosis. There are three common situations in which Doppler signal intensity is very weak (Figure 1).

The first of these is low flow volume, when the residual blood flow through a tight stenosis is difficult to detect. Erythrocytes scatter ultrasound poorly and when the number of moving erythrocytes falls below a certain level, flow is no longer registered by the Doppler system, even though the flow velocity is very high as for example in high grade stenosis. The second is high signal attenuation. When the vessel under investigation is obscured by overlying organs or tissues or layers of adipose tissue, the Doppler signal is often so attenuated that it can no longer be distinguished from the background noise. The third situation is slow flow velocity. Flow velocities in the venous system are often so low that, although the number of flowing erythrocytes is high enough to generate a high amplitude Doppler signal, that signal is eliminated by the wall filter (a high pass filter designed to eliminate the relatively high amplitude signals produced by normal tissue movement). This is due to the relatively low flow velocities resulting in a small Doppler shift, coming close to tissue movement.

The intravenous injection of an echo-enhancer such as Levovist overcomes these acoustic limitations by increasing the intensity of the Doppler signal. This enhancement applies equally to spectral Doppler velocity measurements and to color Doppler flow mapping, and the enhanced signals can be interpreted by applying exactly the same diagnostic criteria used in the examination of unenhanced images, because the velocities are not altered, only the signal strength is enhanced [13].

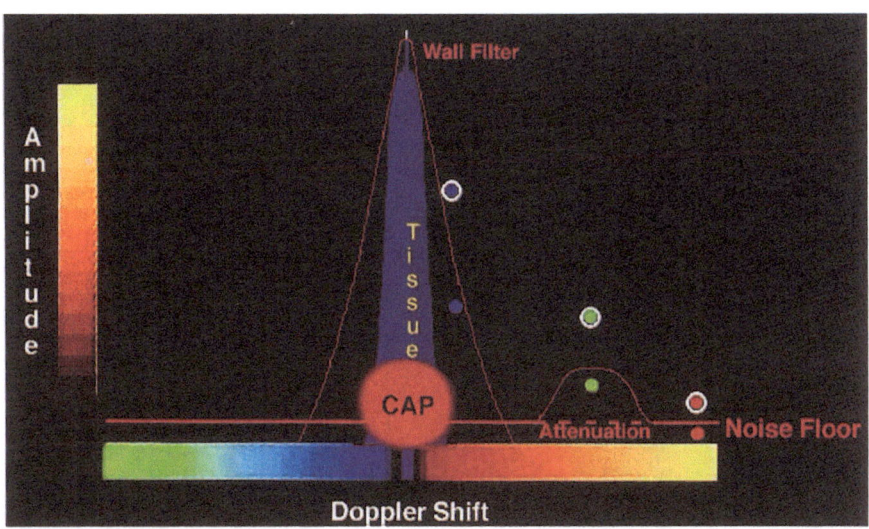

Figure 1. Doppler signal enhancement for different signal conditions. Red dots: high velocity/low volume arterial flow (i.e. a high grade stenosis). Low amplitude Doppler signal. Contrast enhancement (white rimmed dots) increases signal intensity and flow becomes detectable. Green dots: adequate flow volume and velocity but high attenuation makes signal undetectable. Contrast enhancement (white rimmed dots) increases signal intensity and compensates for the attenuation. Blue dots: high volume/low velocity venous flow. High amplitude Doppler signal eliminated by the wall filter. Contrast enhancement (white rimmed dots) increases signal intensity and overcomes effect of wall filter. CAP: The flow during the capillary passage is not detectable by conventional ultrasound. New agents like SHU 563A with acoustic emission properties enable detection of the agent even during the capillary passage, and open new diagnostic opportunities for ultrasound.

BUBBLE RESONANCE AND THE GENERATION OF HARMONICS

When the transmitted ultrasound power is low, the radius of the microbubbles remains constant over time. Mathematically, the bubbles are treated as static objects, and scattering is described as a linear phenomenon. As the transmitted ultrasound power rises the cyclic pressure variations become greater and the bubbles respond by oscillating, so that their radii change with time. Microbubble oscillation depends on the properties of the shell, the nature of the gas content and the amplitude and frequency of the ultrasound field. The oscillation of single gas bubbles within an incompressible fluid is described by the Raleigh–Plesset equation (Equation 2):

$$R\ddot{R}+\frac{3}{2}\dot{R}^2=\frac{1}{\rho_l}[p(R,t)-P(t)-P_0]-4v\frac{\dot{R}}{R}-\frac{2\sigma}{\rho_w R} \tag{2}$$

in which:
R = bubble radius over time, $\dot{R}$, $\ddot{R}$ = time derivatives of R

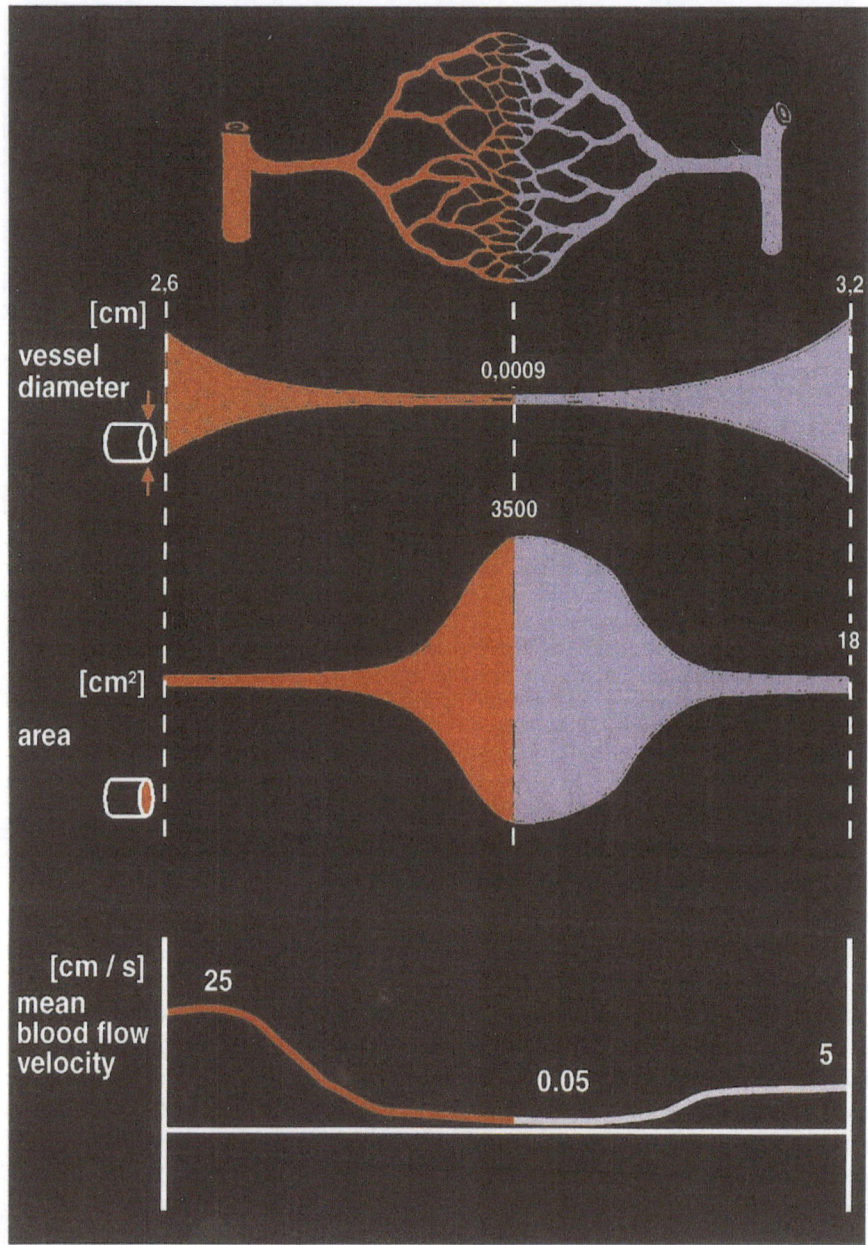

Figure 2. The blood pool compartment. Capillary flow, characterized by low flow velocities and low flow volume, is undetectable with conventional Doppler equipment without the use of ultrasound contrast agents. LOC imaging with SHU 563A shows the distribution of the microparticles in the circulation regardless of flow velocity and enables imaging of microvascular perfusion.

p   =pressure within bubble
$P(t)$=external driving pressure (ultrasound field)
p_w =ambient water pressure
v   =viscosity of water
σ =surface tension air–water interface
ρ_w =density of water.

These oscillations show a marked tendency to become resonant. If we neglect surface tension and assume that the pressure variations are small, the resonance frequency (f_0) of a free microbubble is described as follows (Equation 3) [2]:

$$f_0 = \frac{1}{2\pi R} \sqrt{\frac{3kp_w}{\rho_w}}$$ (3)

in which:
R =bubble radius
k =polytropic gas constant (adiabatic=1.4, isothermal=1.0)
p_w=ambient water pressure
ρ_w=density of water.

By a fortunate coincidence, the resonance frequency for bubbles of $1-5\,\mu m$ diameter (about 1–5 MHz) is within the range used for diagnostic ultrasound. If the amplitude of the signal at the resonant frequency is adequate, the microbubbles will begin non-linear oscillation and start to emit the harmonic components of the incident frequency [3]. Microbubble resonance also increases the intensity of the backscattered signal because the non-linear oscillation increases the average radius of the bubbles over time [4]. The overall effects of microbubble resonance are, therefore, an increase in the amplitude of the backscattered signal at its fundamental frequency (the frequency of the transmitted signal) and the generation of second harmonics (with frequencies two times that of the fundamental) or multiples of that frequency. The second harmonic response can also be produced by mechanisms other than microbubble resonance. The ultrasound transducer always generates some harmonic frequencies, and the non-linear propagation of the acoustic wave in tissue produces others. The harmonics generated by the transducer however contain mainly odd multiples (1, 3, 5 ...) of the transmit frequency. The resonant behavior of microbubbles such as those of Levovist or SHU 563A makes a far more important contribution to the production of harmonics than any of these other, secondary effects. The intensity of the higher order harmonics, i.e. the third or fourth harmonics, is usually below the second harmonic. In addition, the limited bandwidth of the transducer is less sensitive in detecting higher harmonics generated by the microbubbles. Therefore the use of the second harmonic is advantageous for diagnostic imaging with microbubbles.

Harmonic (or more correctly second harmonic) imaging is set up to transmit a signal at one frequency (the fundamental, or first harmonic) and to receive it at twice that frequency (the second harmonic). The technique offers the possibility

694 Albrecht Bauer et al.

of imaging based on the signal returned by the echo enhancer alone. The advantage in harmonic Doppler applications is that fewer artefacts from clutter and interference produced by the signals returned by tissue and blood appear. In B-mode, harmonic imaging may have the potential to give a picture of the entire blood compartment within an organ or tissue. In general, harmonic imaging is comparable to a real-time subtraction technique like digital subtraction angiography [5].

The combination of new digital imaging equipment and broadband transducers may give harmonic imaging a range of clinical applications exceeding that of conventional echo enhancement. The technique is still in its infancy and it is subject to some limitations, the most important of which is the necessity to exceed a threshold of signal intensity to induce non-linear resonance.

STIMULATED ACOUSTIC EMISSION

Like all other ultrasound contrast agents, the new polymeric agents such as SHU 563A increase the amount of backscatter and produce useful signal enhancement in both B-mode and Doppler applications. This effect is already observed when the acoustic output of the ultrasound machine is low. As the power rises, the rigid shells become leaky and release a free microbubble. These microbubbles oscillate far more readily than the rigid shells from which they have escaped. When the output exceeds a certain threshold the amplitude of the harmonic component begins to increase as the response of the contrast enhancer becomes non-linear [1], and becomes sufficiently intense to be detected by a harmonic imaging system. The generation of harmonics is mostly due to the active acoustic response of the microbubbles, which is part of the acoustic emission effect. The range of acoustic emissions starts at medium to high ultrasound amplitudes: the microbubbles undergo oscillations and generate a non-linear response to the incoming signal. The harmonic components are generated by the oscillations of the microbubbles by emitting acoustic signals in response to ultrasound. As the microbubbles act as sound sources in that process, the signals are termed 'stimulated acoustic emissions' from microbubbles [1].

The generation of harmonics is, however, only part of the effects observed due to stimulated acoustic emission. The non-linear response increases with rising amplitude of the incident ultrasound. At low amplitude (mechanical index <0.1) the bubbles act mostly as passive scatterers and enhancement is due to backscatter. At medium amplitudes the microbubbles begin to emit harmonics and they increase backscatter more effectively as their average cross-sectional area increases. For SHU 563A at medium to high amplitudes free microbubbles are liberated from the polymeric shell. The resulting free microbubbles are ideally suited to generate acoustic emission response. Stimulated by high ultrasound emission amplitude the bubbles respond with a high amplitude signal, containing a broad spectrum of frequencies. This signal is no longer limited to harmonics,

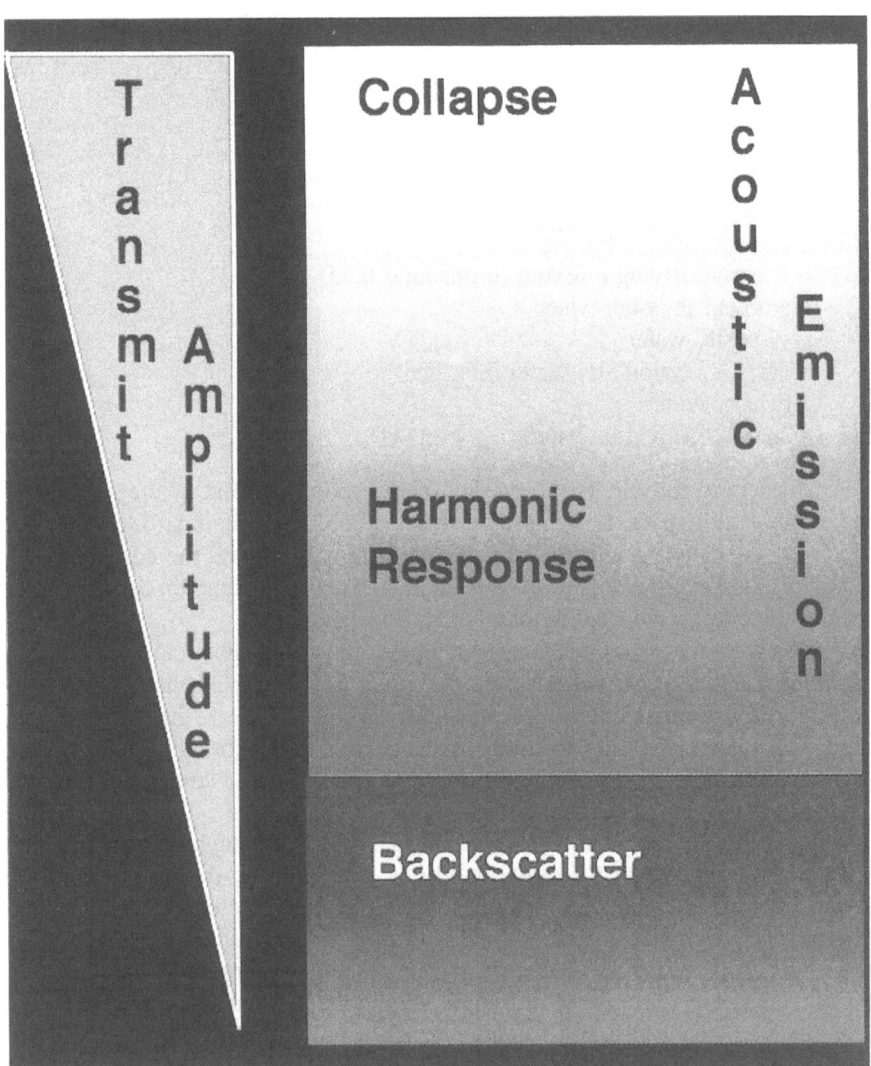

Figure 3. Transmit power dependence of acoustic response. At low insonation amplitudes, the main response is passive backscatter. As the amplitude rises, the microbubbles begin to oscillate and emit the harmonics of the transmitted fundamental. As the amplitude continues to increase the bubbles start emitting a characteristic broad-band signal and undergo subsequent collapse.

and contains all frequency components. The appearance of this signal is, however, limited to the time point of the ultrasound wave passing by. It is this property that makes the detection of this acoustic emission signal possible. Stimulated acoustic emission may be described by a modification of the Rayleigh–Plesset equation that defines the process in terms of energy loss (Equation 4):

$$R\ddot{R}+\frac{3}{2}\dot{R}^2=\frac{1}{\rho_l}[p(R,t)-P(t)-p_w]-4v\frac{\dot{R}}{R}-\frac{2\sigma}{\rho_w R}+\frac{R}{\rho_w c_w}\frac{d}{dt}[p(R,t)-P(t)] \quad (4)$$

in which:

R = bubble radius over time
$\dot{R}, \ddot{R}$ = time derivatives of R
$p(R,t)$ = pressure within bubble
$P(t)$ = external driving pressure (ultrasound field)
p_w = ambient pressure water
v = viscosity water
σ = surface tension air−water interface
ρ_w = density water
c_w = speed of sound in water.

The added term represents the energy of the sound emitted as the resonating microbubble collapses. If we assume that the bubbles contain an ideal van der Waals gas at constant temperature, this term may dominate the equation. The intensity of the energy radiated during stimulated acoustic emission can be greater than of the energy reflected by passive backscatter.

These equations all assume that the bubbles retain their spherical configuration as they expand and contract. The complexity of microbubble behavior in a non-uniform ultrasound field involves changes in bubble shape and the mathematical models that have been used to simulate this behavior are extremely sophisticated [8]. The experimental results obtained to date have detected stimulated acoustic emission as a high energy broad band signal.

Many kinds of microbubbles, including Levovist, are capable of stimulated acoustic emission under the right conditions. The intensity (or loudness) of the signal emitted and the ease with which it can be detected depends on the properties of the bubble shell and the nature of the gas content. In SHU 563A the composition of the shells and the properties and amount of the gas that they contain are designed to produce a loud and highly reproducible acoustic signal (stimulated acoustic emission) as they resonate and collapse.

SPECIFIC DETECTION BY LOC IMAGING

With an ultrasound agent such as SHU 563A, which shows stimulated acoustic emission properties, the exposure to high power color Doppler generates a specific response. During a sequence of color Doppler pulses the signal intensity varies substantially as the free gas bubbles are released from the microparticles, undergo acoustic stimulation and subsequently collapse. The color Doppler system interprets this varying intensity from pulse to pulse as random Doppler shifts [6].

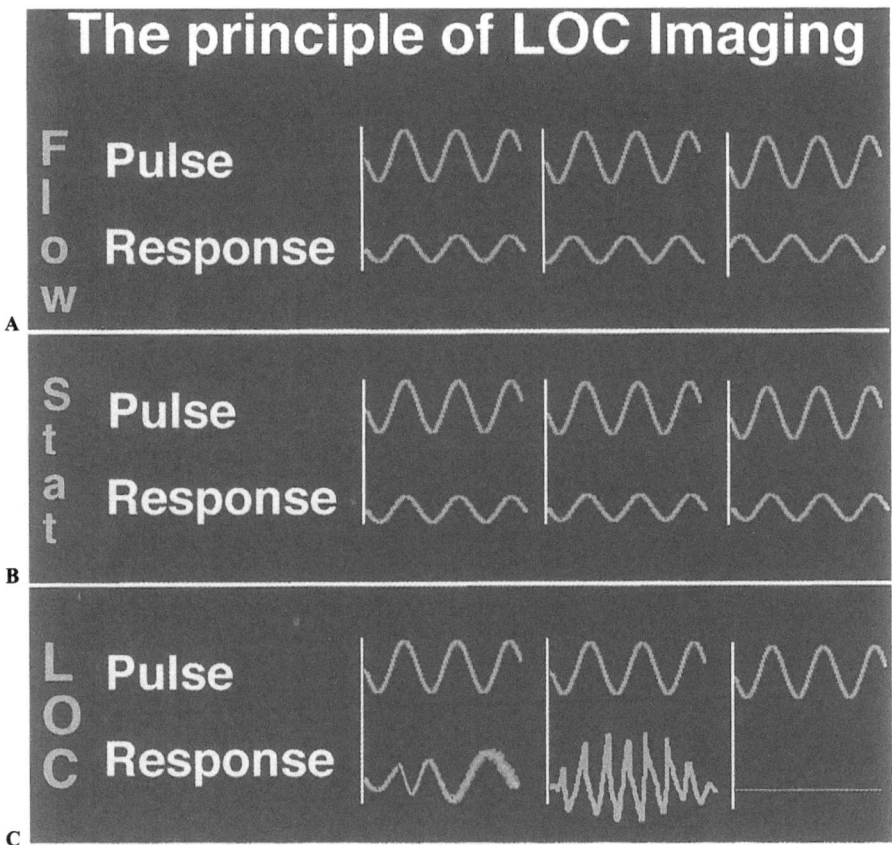

Figure 4. Generation of color Doppler signals. (A) Moving fluid containing an echo enhancer returns a signal with a frequency shift symbolized as a phase-change of the response to subsequent pulses. (B) Fluid at rest returns an unshifted signal. (C) When a fluid suspension of SHU 563A is exposed to high amplitude color/power Doppler the microparticles begin to leak and release free gas bubbles. The active response of the enhancer (acoustic emission) and the final collapse of the microbubbles results in a varying response to subsequent pulses. This loss of correlation is visualized as a random mosaic of pseudo-Doppler shifts, depicting the localization of the agent.

Because of the stimulated acoustic emission, the signal returned by one pulse no longer bears any correlation to the pulse sent immediately before it. The resulting loss of correlation pattern gives a map of the distribution of the contrast agent, the localization of the responding microspheres, and we describe this novel technique as LOC imaging (Figure 4). In short, this can be seen as the contrast agent inducing color Doppler signals by its acoustic response to high intensity ultrasound. Because the microparticles emit this distinctive signal in direct response to the high amplitude incident ultrasound the time taken by the sound

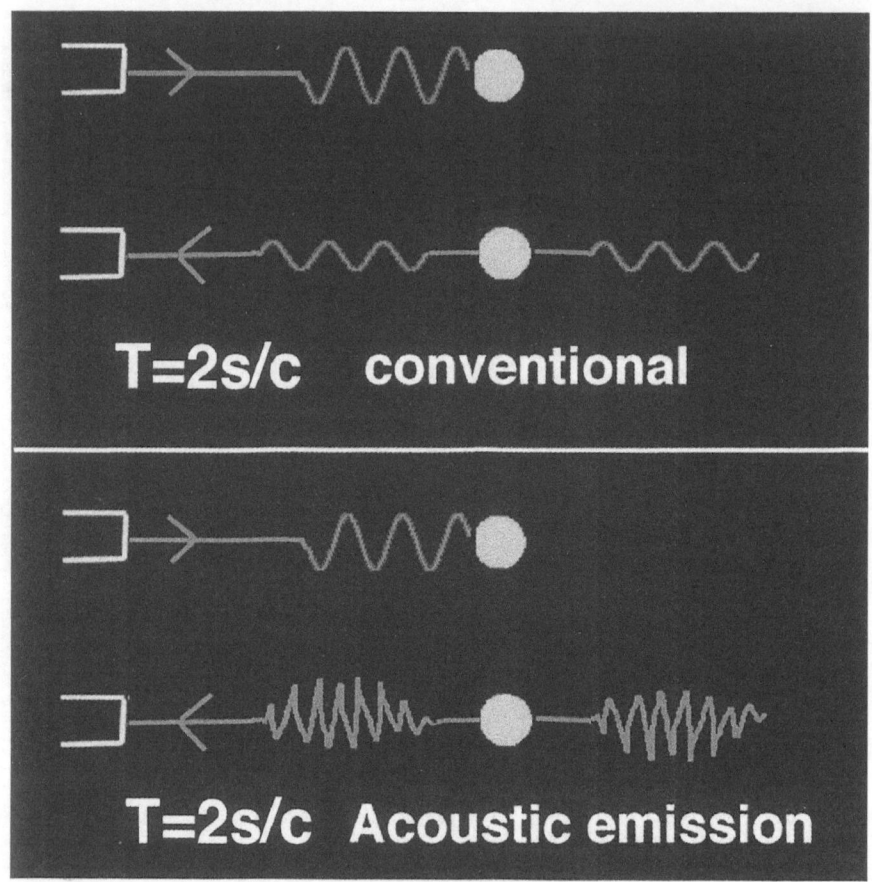

Figure 5. Presence mapping by acoustic emission response. Accurate mapping of microparticle distribution by mapping the acoustic emission response of the ultrasound contrast agent. The traveling time of the acoustic emission signal generated during ultrasonic stimulation is the same as that for reflected wave in conventional ultrasound.

to travel to and from the transducer is exactly the same as in conventional imaging. Because the traveling time is encoded as distance, LOC imaging gives an accurate map of microparticle distribution (Figure 5). Factors other than the amplitude of the incident ultrasound also influence the non-linear response of SHU 563; the combination of low transducer frequency and high pulse repetition frequency (PRF), for example, increases it substantially [7]. With the appropriate equipment settings, any of today's color Doppler ultrasound scanners is suitable for LOC imaging, provided a sufficiently high ultrasound amplitude is transmitted.

IMAGING TECHNIQUES FOR THE ACTIVE RESPONSE OF MICROBUBBLES

The two principal ultrasound imaging techniques are B-mode and Doppler. In B-mode imaging the intensity of the reflected signal is displayed as a gray-scale or 'brightness' map. Highly ecogenic regions are displayed as areas of brightness and less echogenic regions, like cavities containing blood or other fluid, appear dark. Doppler imaging exploits the shift in frequency (Doppler shift) between transmitted and received signals when the ultrasound is reflected by moving scatterers. The size of this frequency shift, which is proportional to the velocity of the scatters, is described in Equation 5:

$$f_D = \frac{2 f_t v \cos \Theta}{c} \qquad (5)$$

in which:
f_D = Doppler frequency shift
f_t = transmit frequency
c = speed of sound
cos = angle between flow vector and insonation.

Spectral Doppler imaging depends upon the measurement of the Doppler shift at a single point during a specific time period. The demodulation and analysis of the Doppler shift is based on Fourier transformation of the received signal. In practice we detect the Doppler shift from a sample volume within a region of interest and use it to calculate the velocity of the moving scatterers within the chosen sample.

  Color Doppler imaging uses signal processing based on the autocorrelation technique [9]. The autocorrelation function Φ is assumed to give an estimate of the mean flow velocity (i.e. the mean Doppler shift) during the period between successive ultrasound pulses. The mean Doppler shift estimated by the auto-correlation function is described in Equation 6:

$$\varpi = \frac{\int_{-\infty}^{+\infty} \omega P(\omega) d\omega}{\int_{-\infty}^{+\infty} P(\omega) d\omega} \cong \frac{\Phi(T)}{T}. \qquad (6)$$

Because the estimation of mean flow velocity depends upon averaging flow distribution in a very large area, in real time, it is a very complex mathematical procedure. The autocorrelation function can also be analyzed for the variance around the mean, used clinically to identify turbulent flow and any sudden changes in flow velocity or spectral broadening. The recent development of power Doppler (color Doppler energy or color Doppler amplitude) is based on the same principles as conventional Doppler imaging, but it displays the intensity of the Doppler signal rather than its frequency. This improves the dynamic range of the signal and lessens its angular dependence [10]. Resonance and stimulated acoustic emission, the active responses of microbubbles to ultrasound, produce characteristic changes in B-mode and Doppler images.

Active responses in B-mode imaging

In B-mode applications, the high intensity signals due to acoustic emissions are visualized as echogenic bright signals. When the scanner is switched from low to high amplitude the increase in brightness is greatest during the first few frames. This effect can be found by freezing and restarting the signal during high-amplitude B-mode imaging. This technique of intermittent or transient response imaging [11] offers several advantages. First, the ultrasound pulse destroys all the microbubbles, so the pulse triggering sequence can be arranged to allow areas of slow flow, such as capillary beds, to fill with intact microbubbles during the imaging pause. Consequently each pulse will produce the best possible imaging response. Second, stimulated acoustic emission is maximized because the high intensity ultrasound induces resonance and collapse in all the microbubbles in the imaging field simultaneously. This transient response, a signal with high amplitude, is exploited highly effectively by harmonic B-mode imaging [3]. Porter et al compared transient response harmonic imaging (2–2.5 MHz trans-mitted, 4.0–5.0 MHz received) with harmonic imaging in 15 patients whose resting left ventricular wall motion was normal [11]. They found that myocardial contrast was significantly greater in the transient response images than in images obtained with conventional pulse repetition. He suggests that the technique may enhance the ability of echo-enhancers to produce diagnostically useful myocardial contrast.

B-mode scans based on the harmonic returns from the microbubbles (harmonic B-mode) are advantageous for contrast agent visualization. The gray-scale difference, i.e. the observed enhancement, is much clearer in harmonic B-mode applications. The images also have a different speckle pattern because the harmonic returns are produced by active scattering by the resonating microbubbles.

Active responses in Doppler imaging

Color Doppler

Color Doppler provides a sensitive tool with which to detect stimulated acoustic emission. As the microbubbles collapse in response to successive Doppler pulses they produce random signal patterns. Signal processing produces an autocorrelation function with an arbitrary phase and a value for $R(T)$ that is very different from that for $R(0)$. The system interprets the response as a random Doppler shift which, because it originates from a static source, has been termed a pseudo-Doppler shift [6]. The colored spots that appear on the display map the spatial distribution of the microbubbles. This map is based on the loss of correlation between the response to Doppler pulses, so the mapping technique is referred to as LOC imaging, and the resulting mosaic as a LOCalization map of the distribution of the echo enhancer. The mapping process gives precedence to high value shifts because of the influence of the wall filter, so the display shows those colors that, in conventional color Doppler flow mapping, represent high velocities. The characteristic mosaic pattern shifts between frames as the random pattern changes in response to successive Doppler pulses. The pattern appears only at high power settings; reducing the power abolishes it completely because the microbubbles are no longer stimulated. Stimulated acoustic emission also influences the size of the variance. Because the events are singular, when they occur they produce very high estimates of apparent flow variance. Consequently, variance mapping provides an extremely sensitive tool with which to record and display stimulated acoustic emission. LOC mapping based on variance may distinguish the pseudo-Doppler shifts produced by stimulated acoustic emission from the real Doppler shifts produced by flow more clearly than mapping based on the pseudo-Doppler shift alone.

Power Doppler
The loss of correlation produced by stimulated acoustic emission influences the calculation of the Doppler phase angle and produces a non-zero value. Stimulated acoustic emission produces a signal with a relatively high amplitude. Since this signal is recognized as a color Doppler signal due to the loss of correlation, it is displayed at its high intensity in the power Doppler map. Power Doppler is, therefore, a very sensitive tool to delineate the presence of SHU 563A by LOC imaging with a better sensitivity than color Doppler. The conventional power Doppler map cannot, however, distinguish between the effects of real blood flow and those of stimulated acoustic emission in display. The variance is not imaged in power Doppler, and the characteristic fluctuating random color Doppler signals are replaced by a relatively homogeneous map, the color value being characteristic for the particle density in a certain area. To exploit a contrast specific presence map, Power Doppler based on the second harmonic is therefore preferred.

Spectral Doppler
Stimulated acoustic emission also influences the results of spectral Doppler analysis. During high power spectral Doppler it produces 'bubble noise', a series of high-intensity peaks that cover the entire spectrum. The effect is due to the high intensity of the signal emitted by the microbubbles, the limited dynamic range of spectral Doppler and receive gain saturation.

CLINICAL APPLICATIONS

The acoustic response of ultrasound contrast agents such as SHU 563A during high intensity ultrasound enables the presence of the agent to be detected by its

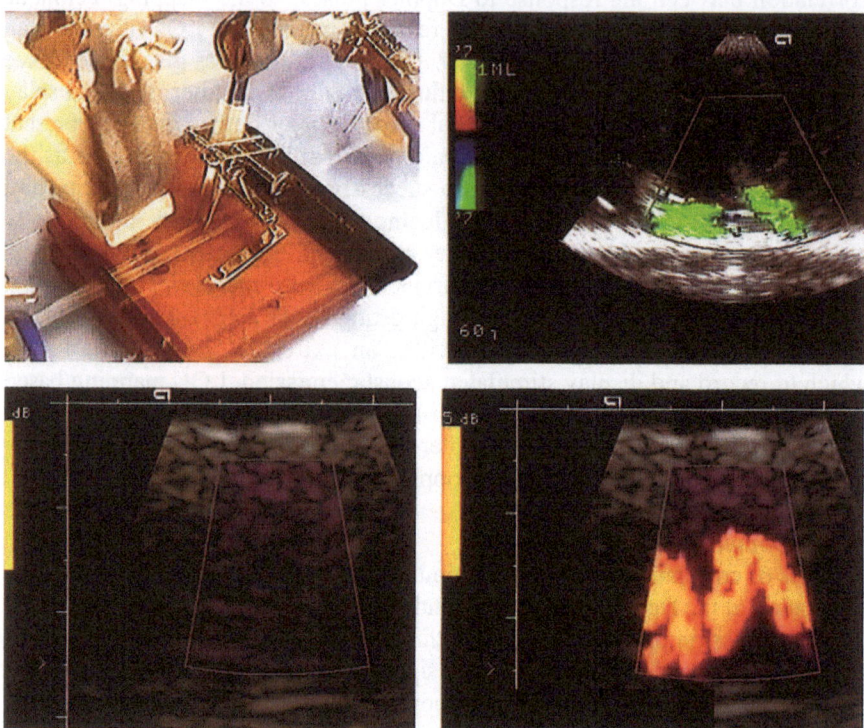

Figure 6. A model of capillary flow imaged with SHU 563A. (A) The flow model showing two bundles of approx. 200 capillaries each. Each capillary has a diameter of 200 μm. (B) LOC imaging shows the presence of SHU 563A in both bundles. In LOC imaging the color Doppler signals are generated by the acoustic emissions of the ultrasound contrast agent during high amplitude ultrasound scanning. The color Doppler is set to map the variance of the signal (encoded in green) delineating a homogeneous distribution of SHU 563A in the bundles. (C) With low amplitude power Doppler the capillaries (filled with SHU 563A) are not visualized and SHU 563A does not show acoustic emission response. The concentration of SHU 563A is too low to produce clearly discernible B-mode effects. (D) High amplitude power Doppler shows the presence of SHU 563A in the bundles. The same concentration of SHU 563A and instrument settings as in (C) were used, with a 50-fold increase in transmit power. Both bundles are visualized as homogenous separate structures, the spacing is less than 2 mm. As with color Doppler, the apparent Doppler shifts are due to the loss of correlation produced by active acoustic emission of SHU 563A, enabling microvascular imaging in this model.

specific acoustic signature, the acoustic emission. This aspect of microbubble behavior has important implications for diagnostic ultrasound. Stimulated acoustic emission is a broad band signal that would conventionally be classed as 'noise', but normal ultrasound noise is continuous, whereas stimulated acoustic emission is a short burst of sound, the source of which has a very specific location in time and space. This precise response gives a map of the spatial distribution of the microbubbles which can be used to outline the blood pool or, with an agent such as SHU 563A that is taken up by the Kupffer cells in a later phase, to delineate the RES compartment of the liver.

The acoustic emissions may be mapped in B mode, preferentially with harmonic intermittent scanning. An even more sensitive method is the use of color Doppler and power Doppler including harmonic imaging. The acoustic response of the agent generates color Doppler signals by means of LOC imaging. The color Doppler signals enable detection of the agent regardless of flow, producing a localization map of the agent. Imaging the microvascular compartment is therefore possible with this new type of ultrasound contrast agent such as SHU 563A. The image contrast in different regions is then based on the amount of agent present, representing an image of the perfusion distribution within an organ.

Criteria for the analysis of organ perfusion need to be established to prove the diagnostic usefulness of these new applications in clinical settings. The diagnostic potential is similar to that of nuclear imaging but localization mapping has the advantages of simplicity, relatively low cost and absence of ionizing radiation. Ultrasound also allows the analysis of 2D organ slices in any chosen plane, which suggests that the new imaging technique could provide more detailed anatomical information than conventional scintigraphy. It is possible that stimulated acoustic emission could provide diagnostic information similar to that available from positron emission tomography (PET) or single photon emission computerized tomography (SPECT), even with improved spatial resolution, although it may be unlikely to give the same degree of functional detail and will not allow access to metabolic information.

In defining LOC imaging as a technique for generation of a contrast agent presence map, we are mapping the blood volume distribution. The presence map can be utilized to image the relative amount of contrast agent in the organ microvasculature and therefore to determine organ perfusion. When using this new imaging technique we cannot use criteria from unenhanced images to guide us, as we can with conventional echo enhancement. SHU 563A and stimulated acoustic emission raise entirely new possibilities that demand the development of new diagnostic procedures and criteria.

Signal attenuation is an important factor influencing the presence mapping with SHU 563A. The active response of the microbubbles depends upon their exposure to high amplitude ultrasound. In anatomical areas that prove difficult to visualize with conventional ultrasound because of signal attenuation it will be difficult to achieve a signal strong enough to produce an active response. Modifications to existing equipment and new developments in transducer design may help to some extent but high attenuation conditions are likely to be a limiting factor for LOC imaging under certain conditions.

This new technique of localization mapping is the first ultrasound imaging technique to exploit the active response of an echo enhancer. Active detection may further expand the limits of resolution of ultrasound images, since imaging the response of single microbubbles may be principally possible.

SUMMARY AND FUTURE POSSIBILITIES

New types of ultrasound contrast agents such as SHU 563 A, with specific acoustic emission properties, will open up innovative diagnostic opportunities. The possibility of functional imaging (the cellular mechanism of RES uptake for example) and mapping of the RES compartment for detection of tumors will expand the horizons of diagnostic ultrasound considerably. The analysis of tissue perfusion in parenchymal organs represents a milestone in ultrasonic imaging, potentially providing information similar to blood pool scintigraphy. With the widely available ultrasound equipment this may lead to earlier diagnosis and enable therapeutic decisions at the earliest possible timepoint. These new possibilities are expected to extend the clinical applications of diagnostic ultrasound substantially.

REFERENCES

1. Uhlendorf V, Hoffmann C. Nonlinear acoustical response of coated microbubbles in diagnostic ultrasound. IEEE Ultrasonics Proc. Vol. 3 1994 (abstract).
2. Minaert M. On musical air bubbles and the sound of running water. Phil Mag 1933;16:235.
3. Burns PN, Powers JE, Fritzsch T. Harmonic imaging, a new imaging and Doppler method for contrast enhanced ultrasound. Radiology 1992;185:142.
4. de Jong N, ten Cate F, Lance C. Principles and recent developments in ultrasound contrast agents. Ultrasonics 1991;29:324–30.
5. Burns PN. Harmonic imaging with ultrasound contrast agents. Clin Radiol 1996;51:50–5.
6. Burns PN, Fritzsch T, Weitschies W, Uhlendorf V, Hope-Simpson D, Powers JE. Pseudo Doppler shifts from stationary tissue due to the stimulated emission of ultrasound from a new microsphere contrast agent. Radiology 1995;197:402.
7. Hoffmann C. Mechanische Charakterisierung von Ultraschallkontrastmitteln. Ultraschall Klin Prax 1995;9:208–10.
8. Chahine GL. Experimental and asymptotic study of nonspherical bubble collapse. In: Wijngaarden V, editor. Mechanics and Physics of Bubbles in Liquids. The Hague, Martinus Nijhoff, 1982.
9. Kasei C, Namekawa K. Real time two dimensional blood flow imaging using an autocorrelation technique. IEEE Ultrasonics Symp 1985:953–98.
10. Rubin JM, Bude RO, Carson PL, Bree RL, Adler RS. Power Doppler US: a potentially useful alternative to mean frequency-based color Doppler US. Radiology 1994;190:853–6.
11. Porter T, Xie F, Armbruster R, Kriesfeld D. Improved myocardial echocardiographic contrast with second harmonic transient response imaging in humans using intravenous perfluorcarbon-exposed sonicated dextrose albumin. (Rotterdam, January, 1996).
12. Williams RA, Kubowitz G, Schlief R. Gas bubble dynamics in acoustic fields and their biological consequences. In: Nanda NC, Schlief R, editors. Advances in Echo Imaging Using Contrast Enhancement. Dordrecht: Kluwer Academic Press, 1993:131.
13. Petrick J. Spektralanalyse von Ultraschall-Doppler-Signalen mit und ohne Ultraschallkontrastmittel an Modellen von Blutgefäßen. Berlin: Koester Verlag, 1996.

44. Microvascular imaging – results from a phase I study of the novel polymeric ultrasound contrast agent SHU 563A

ALBRECHT BAUER, MARIANNE MAHLER, ALBRECHT URBANK,
MICHAEL ZOMACK, REINHARD SCHLIEF &
HANS-PETER NIENDORF

INTRODUCTION

Ultrasound diagnosis uses B-mode images to provide anatomical information and Doppler analysis to measure the velocity of blood flow. In the capillaries, however, the velocity of blood flow ($0.03-0.3$ cm/s) is comparable to the speed of motion of the surrounding tissue, so capillary flow cannot be imaged by conventional Doppler techniques. Another limiting factor is the extreme spatial complexity of the capillary network. A new ultrasound contrast enhancer, SHU 563A, with non-linear acoustic properties, could overcome these limitations [1]. When the new agent is exposed to high power ultrasound it produces a characteristic sound which, on a color Doppler display, appears as pseudo-Doppler shifts. Color Doppler flow mapping depends upon the correlation between successive pulses which is infringed by the non-linear response of SHU 563A. The image based on pseudo-Doppler shifts shows the location of the contrast enhancer and gives a map of microparticle distribution. This new procedure may be described as '*LOC*' imaging (*loss* of *c*orrelation representing contrast agent *loc*alization). The image provided by a conventional color Doppler ultrasound scanner after injection of SHU 563A contains no information on fluid flow velocities. Instead it provides a color map of the distribution of the contrast agent. It gives a direct image of the microvascular compartment that contains anatomical and functional information on microvascular perfusion.

SHU 563A – PHYSICAL CHARACTERISTICS

All the ultrasound contrast agents introduced so far are formulations of microbubbles containing air or other gases that reflect ultrasound. It is essential that the microbubbles should cross the pulmonary vasculature after their peripheral intravenous injection. Both Levovist and SHU 563A have this ability and, in consequence, their effects are apparent throughout the entire blood pool [2].

N. C. Nanda, R. Schlief and B. B. Goldberg (eds.), *Advances in Echo Imaging Using Contrast Enhancement, Second Edition*, 685–690.

SHU 563A consists of air-filled microspheres with a mean diameter of $1-2\,\mu$m. The shells of the microspheres are formed by a thin layer of a biodegradeable cyanacrylate polymer. The microspheres are pre-formed and presented as a powder which is suspended by shaking in physiological saline for a few seconds before injection. This suspension is isotonic and remains stable for several hours. Unlike free gas bubbles and most other microbubble enhancers, the microspheres of SHU 563A circulate in the blood pool intact for up to 10 min after i.v. injection. The particles are eventually taken up by the reticuloendothelial system (RES), principally the Kupffer cells of the liver (60%), and the phagocytosing cells of the spleen [3]. The microparticles remain unaltered and retain their gaseous content after uptake by the RES, which gives them a diagnostic potential similar to that of microparticles of iron oxide in magnetic resonance imaging (Resovist®). In both cases, uptake shows the function and density of the macrophages within a specific organ and may help to differentiate tumors and other lesions from normal cells. A remarkably small number of SHU 563 microparticles is needed to produce clinically useful hepatic enhancement. No other ultrasound echo enhancer reported to date offers this potential.

INTERPRETATION OF LOC IMAGING – THE BLOOD POOL DISTRIBUTION

The color Doppler LOC image provides a localization map of the contrast agent that is independent of flow. The pseudo-Doppler color spots that constitute the map show the zones into which SHU 563A has perfused after injection. The majority of the blood volume (approximately two-thirds) is found in the microcirculation, consisting of arterioles, capillaries and venules [7]. Consequently, most color Doppler signals will originate from the microvascular blood pool shown in the LOC image. The LOC image shows all the blood vessels, including major arteries and veins. By its nature the technique is unable to distinguish between the capillary, arterial and venous blood pools but the LOC image may distinguish perfused from non-perfused tissue quite clearly. The identification of hypo- and hyperperfused areas is one of the most promising applications of LOC imaging with SHU 563A.

STUDY DESIGN AND METHODS

The Phase I trial of SHU 563A in microvascular blood pool imaging was designed as a single-blind intra-individual dose comparison in five healthy male volunteers. Color Doppler imaging of the whole liver as well as quantitative Doppler intensitometric measurements were performed throughout the study. An ATL Ultramark ultrasound scanner with a curved linear transducer array (C4-2) operating at 2–4 MHz was used and the transducer was maintained in the same position manually throughout the imaging sequence. The highest possible signal amplitude was used throughout this study, keeping the mechanical index (MI) above 1.0 throughout the imaging procedure. We used a high PRF of 3.5 KHz to clearly distinguish between color Doppler signals due to flow (no sudden change from frame to frame) and color Doppler signals generated by the acoustic response of SHU 563A (random variation in subsequent frames). Spectral Doppler scanning of the femoral artery was performed simultaneously. The gain was reduced to simulate the effect of insufficient signal intensity before the application of the contrast agent. A quantitative analysis (Doppler intensitometry) was performed during the application and the blood pool phase of SHU 563A.

A suspension of SHU 563 microparticles was given as a bolus injection in ascending doses of 0.1, 1 and 10 μl/kg bodyweight. Higher doses were given when necessary, with the maximum dose of 30 μl/kg. Each volunteer was given three injections at varying doses, with an interval of at least 1 h between doses to avoid any cumulative effect. Total volumes below 250 μl were injected via a Hamilton syringe. To ensure complete administration of the trial substance, the cannula was flushed with 0.5 ml of 0.9% saline after each injection. Safety assessment included heart rate, blood pressure, ECG, and laboratory analysis (hematology, coagulation and clinical chemistry) and urinanalysis.

STUDY RESULTS

The microvascular blood pool in the liver was delineated in all the volunteers. Pseudo-Doppler effects were already apparent at the lowest dose used (0.1 μl/kg bw). After a dose of 1 μl/kg bw the microvasculature was delineated for 2–3 min, but the effect was not homogeneous, and regarded as not diagnostically useful. At 10 μl/kg bw the delineation of parenchymal perfusion was homogeneous and lasted for up to 4.5 min. After higher doses the color Doppler delineation of perfused parenchyma was more persistent. The liver margin in the image field was well delineated, and it was possible to clearly distinguish perfused liver parenchyma from the surrounding organs and tissues (Figure 1). At lower ultrasound amplitudes, SHU 563A produced spectral Doppler enhancement without the generation of artefacts. Spectral Doppler enhancement was apparent after a dose of 1 μl/kg and when the dose was increased to 10 μl/kg the enhancement was more pronounced and long-lasting (Figure 2). The results suggest that SHU 563A might provide usefully prolonged enhancement in conventional spectral Doppler imaging.

SHU 563A was well tolerated and there were no adverse events or untoward effects that might have been related to the enhancer. There were no hemodynamic changes and no clinically relevant changes in any of the safety parameters at any time during the trial.

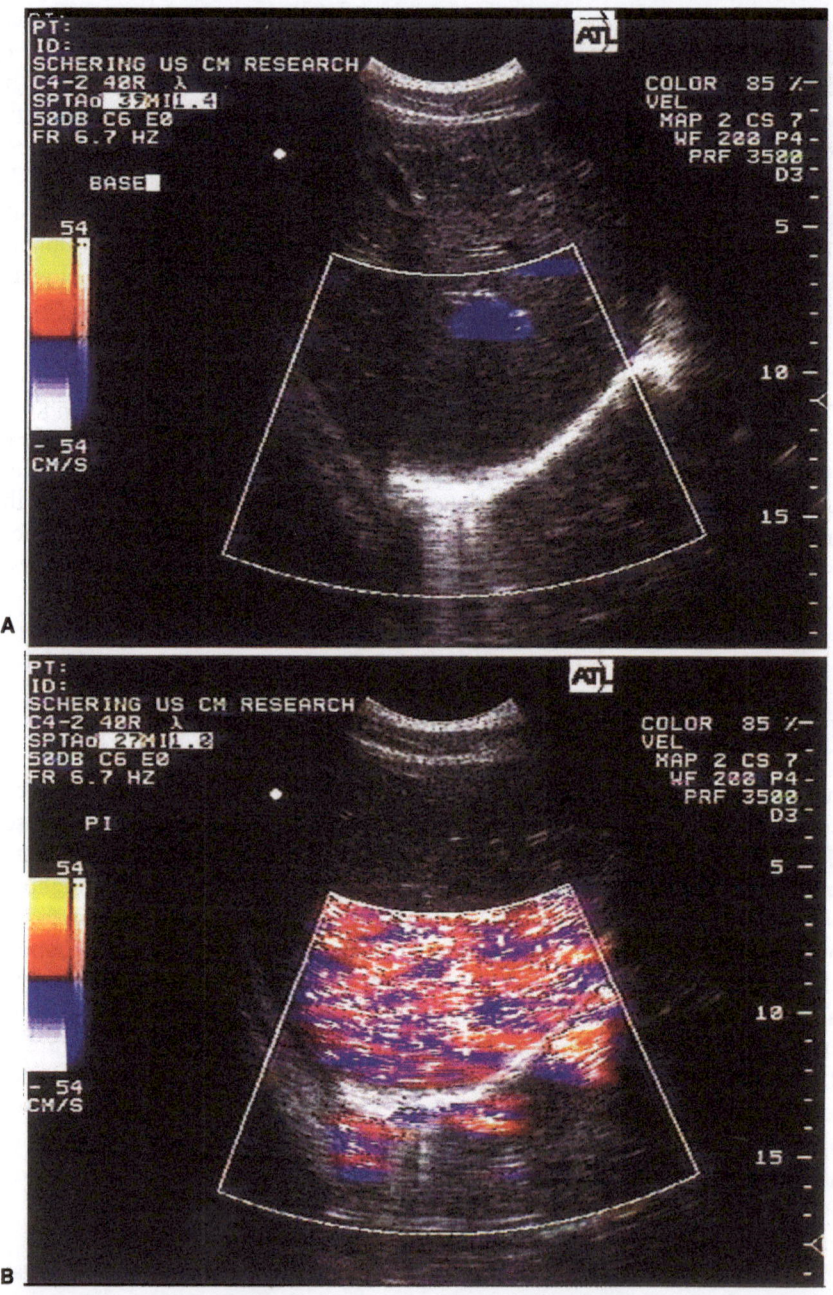

Figure 1. Liver imaging with SHU 563A. (A) Before injection of SHU 563A. A portal branch is visualized in color Doppler. (B) After injection of SHU 563A. A random color pattern delineates the *loc*alization map of the contrast agent due to the *los*s *o*f *c*orrelation of subsequent color Doppler pulses (*LOC* imaging) regardless of flow. The localization map represents mostly the microvascular compartment of the liver, and shows the pattern of perfusion.

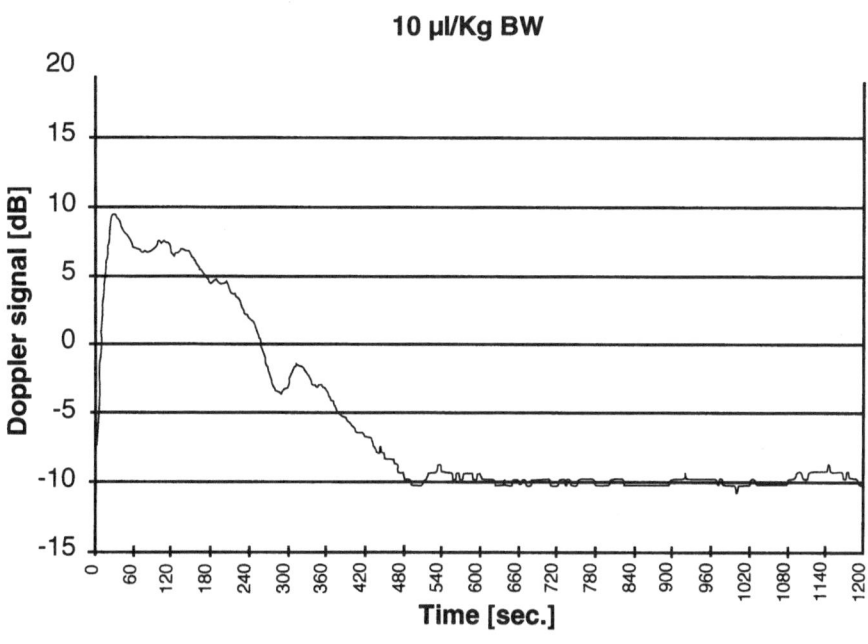

Figure 2. Doppler signal enhancement after SHU 563A (10 µl/kg) as shown by Doppler intensito-metric measurement of the femoral artery.

SUMMARY AND CONCLUSION

The polymer-based microspheres of SHU 563A represent a truly innovative step in the development of ultrasound contrast agents. The saccharide agents used for blood pool echo enhancement (Levovist®) have already proved their value. In addition to conventional echo enhancement, the microspheres of SHU 563A have shown in this study the ability to delineate the microvascular compartment of the blood pool, which was previously invisible to ultrasound. After peripheral intravenous injection the microspheres pass through the lungs and stay in the blood pool until they are taken up by the cells of the RES. This provides an imaging window of sufficient duration to allow the analysis of organ perfusion. The favorable acoustic properties of SHU 563A enable stimulated acoustic emission response during ultrasound exposure. This produces a loss of correlation between successive color Doppler pulses and therefore generates a color Doppler signal imaged by the ultrasound system as a presence map of the agent. Our results suggest that SHU 563A could have particular advantages in the assessment of organ and tissue perfusion.

REFERENCES

1. Uhlendorf V, Hoffmann C. Novel techniques of ultrasound imaging facilitated by contrast agents. First European symposium on ultrasound contrast imaging. Rotterdam. Jan 25–26 1996.
2. Schlief R, Schürmann R, Balzer T et al. Diagnostic value of contrast enhancement in vascular Doppler ultrasound. In: Nanda N, Schlief R, editors. Advances in Echo Imaging using Contrast Enhancement. Dordrecht: Kluwer Academic, 1993:309–23.
3. Schmidt R, Thews L. Physiologie des Menschen. Berlin: Springer, 1985:273.

Index

N. C. Nanda, R. Schlief and B. B. Goldberg (eds.), *Advances in Echo Imaging Using Contrast Enhancement,*
Second Edition, 691–698.
© 1997 *Kluwer Academic Publishers.*